Methods in Enzymology

Volume 187

ARACHIDONATE RELATED LIPID MEDIATORS

METHODS IN ENZYMOLOGY

Methods in Enzymology

Volume 187

Arachidonate Related Lipid Mediators

EDITED BY

Robert C. Murphy

DEPARTMENT OF PHARMACOLOGY
UNIVERSITY OF COLORADO HEALTH SCIENCES CENTER
DENVER, COLORADO

Frank A. Fitzpatrick

DEPARTMENT OF PHARMACOLOGY
UNIVERSITY OF COLORADO HEALTH SCIENCES CENTER
DENVER, COLORADO

ACADEMIC PRESS, INC.
Harcourt Brace Jovanovich, Publishers
San Diego New York Boston
London Sydney Tokyo Toronto

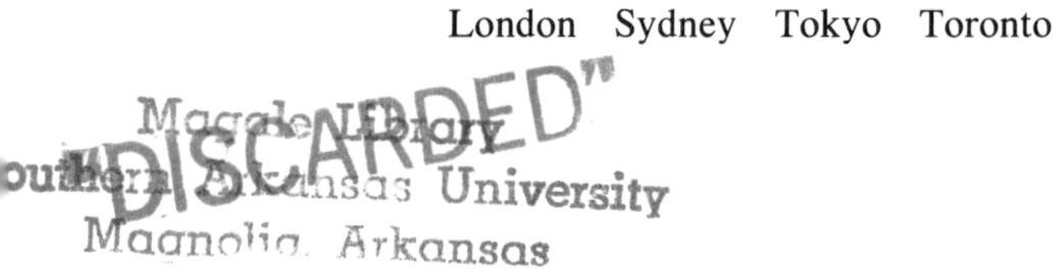

This book is printed on acid-free paper. 

ACADEMIC PRESS, INC.
San Diego, California 92101

United Kingdom Edition published by
ACADEMIC PRESS LIMITED
24-28 Oval Road, London NW1 7DX

LIBRARY OF CONGRESS CATALOG CARD NUMBER: 54-9110

ISBN 0-12-182088-2 (alk. paper)

PRINTED IN THE UNITED STATES OF AMERICA
90 91 92 93 9 8 7 6 5 4 3 2 1

Table of Contents

Contributors to Volume 187 xiii

Preface xxi

Volumes in Series xxiii

1. Nomenclature — William L. Smith, Pierre Borgeat, Mats Hamberg, L. Jackson Roberts II, Anthony L. Willis, Shozo Yamamoto, Peter W. Ramwell, Joshua Rokach, Bengt Samuelsson, E. J. Corey, and C. R. Pace-Asciak — 1

Section I. Assays

A. Prostaglandins

2. Electron-Capture Negative-Ion Chemical Ionization Mass Spectrometry of Lipid Mediators — Ian A. Blair — 13

3. Enzyme Immunoassays of Eicosanoids Using Acetylcholinesterase — Philippe Pradelles, Jacques Grassi, and Jacques Maclouf — 24

4. Radioimmunoassay of 11-Dehydrothromboxane B_2 — Giovanni Ciabattoni, Paola Patrignani, and Carlo Patrono — 34

5. Measurement of Thromboxane Metabolites by Gas Chromatography–Mass Spectrometry — Francesca Catella and Garret A. FitzGerald — 42

6. Quantification of $9\alpha,11\beta$-Prostaglandin F_2 by Stable Isotope Dilution Mass Spectrometric Assay — Daniel F. Wendelborn, Jason D. Morrow, and L. Jackson Roberts II — 51

7. Immunoaffinity Purification–Chromatographic Quantitative Analysis of Arachidonic Acid Metabolites — J. James Vrbanac, Jeffrey W. Cox, Thomas D. Eller, and Daniel R. Knapp — 62

B. Leukotrienes

8. Preparation of Tetradeuterated Leukotriene A_4 Methyl Ester: Methyl-[11,12,14,15-$^2H_4$]-(5*S*,6*S*)-Oxido-(7*E*,9*E*,11*Z*,14*Z*)-Eicosatetraenoate — J. P. Lellouche, J. P. Beaucourt, and A. Vanhove — 70

9. Quantitative Gas Chromatography–Mass Spectrometry Analysis of Leukotriene B_4 — W. Rodney Mathews — 76

10. Enzyme Immunoassays for Leukotrienes C_4 and E_4 Using Acetylcholinesterase — Philippe Pradelles, Catherine Antoine, Jean-Paul Lellouche, and Jacques Maclouf — 82

11. Quantitation of Sulfidopeptide Leukotrienes in Biological Fluids by Gas Chromatography–Mass Spectrometry — Robert C. Murphy and Angelo Sala — 90

12. Automated On-Line Extraction and Profiling of Lipoxygenase Products of Arachidonic Acid by High-Performance Liquid Chromatography — Pierre Borgeat, Serge Picard, Pierre Vallerand, Sylvain Bourgoin, Abdulrahman Odeimat, Pierre Sirois, and Patrice E. Poubelle — 98

13. Preparation of Antibodies Directed against Leukotrienes — Edward C. Hayes — 116

C. Platelet-Activating Factor

14. Bioassay of paf-Acether by Rabbit Platelet Aggregation — Marie-Jeanne Bossant, Ewa Ninio, Danièle Delautier, and Jacques Benveniste — 125

15. Bioassay of Platelet-Activating Factor by Release of [^{3}H]Serotonin — Peter M. Henson — 130

16. Quantitation of Platelet-Activating Factor by Gas Chromatography–Mass Spectrometry — Keith L. Clay — 134

17. Quantitative Analysis of Platelet-Activating Factor by Gas Chromatography–Negative-Ion Chemical Ionization Mass Spectrometry — Walter C. Pickett and Chakkodabylu S. Ramesha — 142

18. Isolation of Platelet-Activating Factor and Purification by Thin-Layer Chromatography — Donald J. Hanahan — 152

19. Separation and Characterization of Arachidonate-Containing Phosphoglycerides — Floyd H. Chilton — 157

D. Other Oxidative Products

20. High-Performance Liquid Chromatography Separation and Determination of Lipoxins — CHARLES N. SERHAN — 167

21. Quantitation of Epoxy- and Dihydroxyeicosatrienoic Acids by Stable Isotope-Dilution Mass Spectrometry — JOHN TURK, W. THOMAS STUMP, WENDY CONRAD-KESSEL, ROBERT R. SEABOLD, AND BRYAN A. WOLF — 175

22. High-Performance Liquid Chromatography for Chiral Analysis of Eicosanoids — ALAN R. BRASH AND DAN J. HAWKINS — 187

E. Phospholipase A_2 and Phospholipids

23. Extraction of Phospholipids and Analysis of Phospholipid Molecular Species — GEORGE M. PATTON AND SANDER J. ROBINS — 195

24. Macrophage Phospholipase A_2 Specific for *sn*-2-Arachidonic Acid — CHRISTINA C. LESLIE — 216

25. Measurement of Phosphoinositide-Specific Phospholipase C Activity — JOHN E. BLEASDALE, JAMES C. MCGUIRE, AND GREGORY A. BALA — 226

26. Solubilization of Arachidonate-CoA Ligase from Cell Membranes, Chromatographic Separation from Nonspecific Long-Chain Fatty Acid CoA Ligase, and Isolation of Mutant Cell Line Defective in Arachidonate-CoA Ligase — MICHAEL LAPOSATA — 237

Section II. Biosynthesis, Enzymology, and Chemical Synthesis

A. Prostaglandins

27. Preparation of Prostaglandin H_2: Extended Purification/Analysis Scheme — DUANE VENTON, GUY LE BRETON, AND ELIZABETH HALL — 245

28. Purification and Properties of Pregnancy-Inducible Rabbit Lung Cytochrome *P*-450 Prostaglandin ω-Hydroxylase — A. SCOTT MUERHOFF, DAVID E. WILLIAMS, AND BETTIE SUE SILER MASTERS — 253

B. Leukotrienes

29. Purification of Arachidonate 5-Lipoxygenase from Potato Tubers — PALLU REDDANNA, J. WHELAN, K. R. MADDIPATI, AND C. CHANNA REDDY — 268

30. Leukotriene Metabolism by Isolated Rat Hepatocytes — Michael A. Shirley and Danny O. Stene — 277

31. Purification and Characterization of Human Lung Leukotriene A_4 Hydrolase — Nobuya Ohishi, Takashi Izumi, Yousuke Seyama, and Takao Shimizu — 286

32. Potato Arachidonate 5-Lipoxygenase: Purification, Characterization, and Preparation of 5(*S*)-Hydroperoxyeicosatetraenoic Acid — Takao Shimizu, Zen-ichiro Honda, Ichiro Miki, Yousuke Seyama, Takashi Izumi, Olof Rådmark, and Bengt Samuelsson — 296

33. Leukotriene C_4 Synthase: Characterization in Mouse Mastocytoma Cells — Mats Söderström, Bengt Mannervik, and Sven Hammarström — 306

34. Leukocyte Arachidonate 5-Lipoxygenase: Isolation and Characterization — Carol A. Rouzer and Bengt Samuelsson — 312

35. Cytochrome $P\text{-}450_{LTB}$ and Inactivation of Leukotriene B_4 — Roy J. Soberman — 319

36. Cytosolic Liver Enzymes Catalyzing Hydrolysis of Leukotriene A_4 to Leukotriene B_4 and 5,6-Dihydroxyeicosatetraenoic Acid — Jesper Z. Haeggström — 324

37. Purification and Properties of Leukotriene C_4 Synthase from Guinea Pig Lung Microsomes — Roy J. Soberman — 335

38. Immunoaffinity Purification of Arachidonate 5-Lipoxygenase from Porcine Leukocytes — Natsuo Ueda and Shozo Yamamoto — 338

C. Platelet-Activating Factor

39. Platelet-Activating Factor Acetylhydrolase from Human Plasma — Diana M. Stafforini, Thomas M. McIntyre, and Stephen M. Prescott — 344

D. Other Oxidative Products

40. Synthesis of Epoxyeicosatrienoic Acids and Heteroatom Analogs — J. R. Falck, Pendri Yadagiri, and Jorge Capdevila — 357

41. Isolation of Rabbit Renomedullary Cells and Arachidonate Metabolism — Mairead A. Carroll, Elizabeth D. Drugge, Catherine E. Dunn, and John C. McGiff — 365

42. Ocular Cytochrome *P*-450 Metabolism of Arachidonate: Synthesis and Bioassay — Michal Laniado Schwartzman and Nader G. Abraham 372

43. Cytochrome *P*-450 Arachidonate Oxygenase — Jorge H. Capdevila, J. R. Falck, Elizabeth Dishman, and Armando Karara 385

Section III. Pharmacology: Antagonist and Synthesis Inhibitors

A. Prostaglandins

44. Radioligand Binding Assays for Thromboxane A_2/Prostaglandin H_2 Receptors — Perry V. Halushka, Thomas A. Morinelli, and Dale E. Mais 397

45. Thromboxane A_2/Prostaglandin H_2 Receptor Antagonists — Guy C. Le Breton, Chang T. Lim, Chitra M. Vaidya, and Duane L. Venton 406

B. Leukotrienes

46. Assessment of Leukotriene D_4 Receptor Antagonists — David Aharony 414

47. Sulfidopeptide Leukotriene Receptor Binding Assays — Seymour Mong 421

C. Platelet-Activating Factor

48. Ginkgolides and Platelet-Activating Factor Binding Sites — D. J. Hosford, M. T. Domingo, P. E. Chabrier, and P. Braquet 433

49. Kadsurenone and Other Related Lignans as Antagonists of Platelet-Activating Factor Receptor — T. Y. Shen and I. M. Hussaini 446

50. Use of WEB 2086 and WEB 2170 as Platelet-Activating Factor Antagonists — Jorge Casals-Stenzel and Hubert O. Heuer 455

Section IV. Molecular Biology

A. Prostaglandins

51. Cloning of Sheep and Mouse Prostaglandin Endoperoxide Synthases — David L. DeWitt and William L. Smith 469

52. Preparation and Proteolytic Cleavage of Apoprostaglandin Endoperoxide Synthase — Rebecca Odenwaller, Ying-Nan Pan Chen, and Lawrence J. Marnett 479

B. Leukotrienes

53. Cloning of Leukotriene A_4 Hydrolase cDNA — Olof Rådmark, Colin Funk, Ji Yi Fu, Takashi Matsumoto, Hans Jörnvall, Bengt Samuelsson, Michiko Minami, Shigeo Ohno, Hiroshi Kawasaki, Yousuke Seyama, Koichi Suzuki, and Takao Shimizu 486

54. Molecular Biology and Cloning of Arachidonate 5-Lipoxygenase — Colin D. Funk, Takashi Matsumoto, and Bengt Samuelsson 491

Section V. Cell Models of Lipid Mediator Production

A. Isolated Cell Preparations

55. Parietal Cell Preparation and Arachidonate Metabolism — N. Ann Payne and John G. Gerber 505

56. Culture of Bone Marrow-Derived Mast Cells: A Model for Studying Oxidative Metabolism of Arachidonic Acid and Synthesis of Other Molecules Derived from Membrane Phospholipids — Ehud Razin 514

57. Endothelial Cells for Studies of Platelet-Activating Factor and Arachidonate Metabolites — Guy A. Zimmerman, Ralph E. Whatley, Thomas M. McIntyre, Donelle M. Benson, and Stephen M. Prescott 520

58. Use of Cultured Cells to Study Arachidonic Acid Metabolism — Robert R. Gorman, Michael J. Bienkowski, and Christopher W. Benjamin 535

59. Eicosanoid Biochemistry in Cultured Glomerular Mesangial Cells — Michael S. Simonson and Michael J. Dunn 544

B. Cell–Cell Interactions

60. Leukotriene B_4 Biosynthesis by Erythrocyte–Neutrophil Interactions — David A. Jones and Frank A. Fitzpatrick 553

61. Leukotriene C_4 Biosynthesis during Polymorphonuclear Leukocyte–Vascular Cell Interactions — Steven J. Feinmark 559

62. Release and Metabolism of Leukotriene A_4 in Neutrophil–Mast Cell Interactions — Clemens A. Dahinden and Urs Wirthmueller 567

63. Interaction between Platelets and Lymphocytes in Biosynthesis of Prostacyclin — KENNETH K. WU AND AUDREY C. PAPP — 578

64. Eicosanoid Interactions between Platelets, Endothelial Cells, and Neutrophils — AARON J. MARCUS — 585

C. *In Vitro* Tissue Preparations

65. Isolated Perfused Rat Lung in Arachidonate Studies — SHIN-WEN CHANG AND NORBERT F. VOELKEL — 599

66. Isolated Coronary-Perfused Mammalian Heart: Assessment of Eicosanoid and Platelet-Activating Factor Release and Effects — ROBERTO LEVI AND KEVIN M. MULLANE — 610

67. Preparation of Human and Animal Lung Tissue for Eicosanoid Research — T. VIGANÒ, M. T. CRIVELLARI, M. MEZZETTI, AND G. C. FOLCO — 621

AUTHOR INDEX 629

SUBJECT INDEX 655

Contributors to Volume 187

Article numbers are in parentheses following the names of contributors.
Affiliations listed are current.

NADER G. ABRAHAM (42), *Department of Medicine, New York Medical College, Valhalla, New York 10595*

DAVID AHARONY (46), *Department of Pharmacology, ICI Pharmaceuticals Group, Wilmington, Delaware 19897*

CATHERINE ANTOINE (10), *Section de Pharmacologie et d'Immunologie, Commissariat a l'Energie Atomique, CEN/Saclay, 91191 Gif-sur-Yvette, France*

GREGORY A. BALA (25), *Department of Cell Biology, The Upjohn Company, Kalamazoo, Michigan 49001*

J. P. BEAUCOURT (8), *Service des Molécules Marquées, Département de Biologie, CEN/Saclay, 91191 Gif-sur-Yvette, France*

CHRISTOPHER W. BENJAMIN (58), *Department of Cell Biology, The Upjohn Company, Kalamazoo, Michigan 49001*

DONELLE M. BENSON (57), *Department of Medicine, Nora Eccles Harrison Cardiovascular Research and Training Institute, University of Utah, Salt Lake City, Utah 84112*

JACQUES BENVENISTE (14), *INSERUM U 200, 92140 Clamart, France*

MICHAEL J. BIENKOWSKI (58), *Department of Cell Biology, The Upjohn Company, Kalamazoo, Michigan 49001*

IAN A. BLAIR (2), *Department of Pharmacology, Vanderbilt University School of Medicine, Nashville, Tennessee 37232*

JOHN E. BLEASDALE (25), *Department of Cell Biology, The Upjohn Company, Kalamazoo, Michigan 49001*

PIERRE BORGEAT (1, 12), *Centre de Recherche du CHUL, Le Centre Hospitalier de l'Université Laval, Sainte-Foy, Quebec G1V 4G2, Canada*

MARIE-JEANNE BOSSANT (14), *INSERM U 200, 92140 Clamart, France*

SYLVAIN BOURGOIN (12), *Centre de Recherches du CHUL, Le Centre Hospitalier de l'Université Laval, Sainte-Foy, Quebec G1V 4G2, Canada*

P. BRAQUET (48), *Institut Henri Beaufour, F-92350 Le Plessis-Robinson, France*

ALAN R. BRASH (22), *Department of Pharmacology, Vanderbilt University School of Medicine, Nashville, Tennessee 37232*

JORGE CAPDEVILA (40, 43), *Division of Nephrology, Vanderbilt University School of Medicine, Nashville, Tennessee 37232*

MAIREAD A. CARROLL (41), *Department of Pharmacology, New York Medical College, Valhalla, New York 10595*

JORGE CASALS-STENZEL (50), *Department of Pharmacology, Boehringer Ingelheim KG, D-6507 Ingelheim am Rhein, Federal Republic of Germany*

FRANCESCA CATELLA (5), *Division of Clinical Pharmacology, Vanderbilt University School of Medicine, Nashville, Tennessee 37232*

P. E. CHABRIER (48), *Department of Biochemical Pharmacology, Institut Henri Beaufour, F-92350 Le Plessis-Robinson, France*

SHIN-WEN CHANG (65), *Pulmonary Section, Northwestern University Medical School, Chicago, Illinois 60611*

YING-NAN PAN CHEN (52), *Center for Biochemical and Biophysical Sciences and Medicine, Harvard Medical School, Boston, Massachusetts 02115*

FLOYD H. CHILTON (19), *Department of Medicine, The Johns Hopkins University, Baltimore, Maryland 21224*

GIOVANNI CIABATTONI (4), *Department of Pharmacology, Catholic University School of Medicine, 00168 Rome, Italy*

KEITH L. CLAY (16), *National Jewish Center for Immunology and Respiratory Medicine, Denver, Colorado 80206*

WENDY CONRAD-KESSEL (21), *Department of Pathology, Division of Laboratory Medicine, Washington University School of Medicine, St. Louis, Missouri 63110*

E. J. COREY (1), *Department of Chemistry, Harvard University, Cambridge, Massachusetts 02138*

JEFFREY W. COX (7), *Department of Pharmacological Sciences, Genentech Inc., South San Francisco, California 94080*

M. T. CRIVELLARI (67), *Institute of Pharmacological Sciences, University of Milan, 20133 Milan, Italy*

CLEMENS A. DAHINDEN (62), *Institute of Clinical Immunology, Inselspital University Hospital, CH-3010 Bern, Switzerland*

DANIÈLE DELAUTIER (14), *INSERM U 200, 92140 Clamart, France*

DAVID L. DEWITT (51), *Department of Biochemistry, Michigan State University, East Lansing, Michigan 48824*

ELIZABETH DISHMAN (43), *Department of Medicine, Vanderbilt University Medical Center, Nashville, Tennessee 37232*

M. T. DOMINGO (48), *Department of Biochemical Pharmacology, Institut Henri Beaufour, F-92350 Le Plessis-Robinson, France*

ELIZABETH D. DRUGGE (41), *Department of Pharmacology, New York Medical College, Valhalla, New York 10595*

CATHERINE E. DUNN (41), *Department of Pharmacology, New York Medical College, Valhalla, New York 10595*

MICHAEL J. DUNN (59), *Department of Medicine and Physiology and Biophysics, Case Western Reserve University, Cleveland, Ohio 44106*

THOMAS D. ELLER (7), *Department of Pharmacology, Medical University of South Carolina, Charleston, South Carolina 29425*

J. R. FALCK (40, 43), *Departments of Molecular Genetics and Pharmacology, University of Texas Southwestern Medical Center, Dallas, Texas 75235*

STEVEN J. FEINMARK (61), *Department of Pharmacology, Columbia University College of Physicians and Surgeons, New York, New York 10032*

GARRET A. FITZGERALD (5), *Division of Clinical Pharmacology, Vanderbilt University School of Medicine, Nashville, Tennessee 37232*

FRANK A. FITZPATRICK (60), *Department of Pharmacology, University of Colorado Health Sciences Center, Denver, Colorado 80262*

G. C. FOLCO (67), *Institute of Pharmacological Sciences, University of Milan, 20133 Milan, Italy*

JI YI FU (53), *Department of Physiology, Norman Betheine Medical University, Chingchun, Jilin, People's Republic of China*

COLIN FUNK (53, 54), *Department of Pharmacology, Vanderbilt University, Nashville, Tennessee 37232*

JOHN G. GERBER (55), *Division of Clinical Pharmacology, University of Colorado Health Sciences Center, Denver, Colorado 80262*

ROBERT R. GORMAN (58), *Department of Cell Biology, The Upjohn Company, Kalamazoo, Michigan 49001*

JACQUES GRASSI (3), *Section de Pharmacologie et d'Immunologie, Département de Biologie, Commissariat a l'Energie Atomique, CEN/Saclay, 91191 Gif-sur-Yvette, France*

JESPER Z. HAEGGSTRÖM (36), *Department of Physiological Chemistry, Karolinska Institutet, S-104 01 Stockholm, Sweden*

ELIZABETH HALL (27), *Hematology Division, University of Texas Medical School, Houston, Texas 77030*

PERRY V. HALUSHKA (44), *Department of Cell and Molecular Pharmacology and Experimental Therapeutics, Medical University of South Carolina, Charleston, South Carolina 29425*

MATS HAMBERG (1), *Department of Chemistry, Karolinska Institutet, S-104 01 Stockholm, Sweden*

SVEN HAMMARSTRÖM (33), *Department of Cell Biology, Faculty of Health Sciences, University of Linköping, S-581 85 Linköping, Sweden*

DONALD J. HANAHAN (18), *Department of Biochemistry, University of Texas Health Science Center, San Antonio, Texas 78284*

DAN J. HAWKINS (22), *Department of Pharmacology, Vanderbilt University School of Medicine, Nashville, Tennessee 37232*

EDWARD C. HAYES (13), *Department of Biochemical Parasitology, Merck Sharp & Dohme Research Laboratories, Merck & Co. Inc., Rahway, New Jersey 07065*

PETER M. HENSON (15), *National Jewish Center for Immunology and Respiratory Medicine, Denver, Colorado 80206*

HUBERT O. HEUER (50), *Department of Pharmacology, Boehringer Ingelheim KG, D-6507 Ingelheim am Rhein, Federal Republic of Germany*

ZEN-ICHIRO HONDA (32), *Department of Physiological Chemistry and Nutrition, Faculty of Medicine, University of Tokyo, Tokyo 113, Japan*

D. J. HOSFORD (48), *Institut Henri Beaufour, F-92350 Le Plessis-Robinson, France*

I. M. HUSSAINI (49), *Department of Chemistry, University of Virginia, Charlottesville, Virginia 22901*

TAKASHI IZUMI (31, 32), *Department of Physiological Chemistry and Nutrition, Faculty of Medicine, University of Tokyo, Tokyo 113, Japan*

DAVID A. JONES (60), *Department of Pharmacology, University of Colorado Health Sciences Center, Denver, Colorado 80262*

HANS JÖRNVALL (53), *Department of Physiological Chemistry, Karolinska Institutet, S-104 01 Stockholm, Sweden*

ARMANDO KARARA (43), *Department of Medicine, Vanderbilt University Medical Center, Nashville, Tennessee 37232*

HIROSHI KAWASAKI (53), *Department of Molecular Biology, Tokyo Metropolitan Institute of Medical Science, Tokyo 113, Japan*

DANIEL R. KNAPP (7), *Department of Pharmacology, Medical University of South Carolina, Charleston, South Carolina 29425*

MICHAEL LAPOSATA (26), *Department of Pathology, Massachusetts General Hospital, Harvard Medical School, Boston, Massachusetts 02114*

GUY C. LE BRETON (27, 45), *Department of Pharmacology, University of Illinois at Chicago, Chicago, Illinois 60680*

J. P. LELLOUCHE (8, 10), *Service des Molécules Marquees, Département de Biologie, CEN/Saclay, 91191 Gif-sur-Yvette, France*

CHRISTINA C. LESLIE (24), *Department of Pediatrics, National Jewish Center for Immunology and Respiratory Medicine, Denver, Colorado 80206*

ROBERTO LEVI (66), *Department of Pharmacology, Cornell University Medical College, New York, New York 10021*

CHANG T. LIM (45), *Department of Pharmacology, University of Illinois at Chicago, Chicago, Illinois 60612*

JACQUES MACLOUF (3, 10), *Unite 150 INSERM, UA 334 CNRS, Hôpital Lariboisière, 75475 Paris, France*

K. R. MADDIPATI (29), *Cayman Chemical Company, Ann Arbor, Michigan 48105*

DALE E. MAIS (44), *Cardiovascular Division, Eli Lilly Pharmaceutical Co., Indianapolis, Indiana 46285*

BENGT MANNERVIK (33), *Department of Biochemistry, Biomedical Center, University of Uppsala, S-751 23 Uppsala, Sweden*

AARON J. MARCUS (64), *Hematology/Oncology Section, New York VA Medical Center, New York, New York 10010*

LAWRENCE J. MARNETT (52), *Department of Biochemistry, Center in Molecular Toxicology, Vanderbilt University School of Medicine, Nashville, Tennessee 37232*

BETTIE SUE SILER MASTERS (28), *Department of Biochemistry, University of Texas Health Science Center at San Antonio, San Antonio, Texas 78284*

W. RODNEY MATHEWS (9), *Biopolymer Chemistry, The Upjohn Co., Kalamazoo, Michigan 49001*

TAKASHI MATSUMOTO (53, 54), *Central Research Institute, Japan Tobacco Inc., Midori-ku, Yokohama, Japan*

JOHN C. MCGIFF (41), *Department of Pharmacology, New York Medical College, Valhalla, New York 10595*

JAMES C. MCGUIRE (25), *Department of Cell Biology, The Upjohn Company, Kalamazoo, Michigan 49001*

THOMAS M. MCINTYRE (39, 57), *Departments of Medicine and Biochemistry, Nora Eccles Harrison Cardiovascular Research and Training Institute, University of Utah School of Medicine, Salt Lake City, Utah 84112*

M. MEZZETTI (67), *IV Clinic of Thoracic Surgery, Policlinico di Milano, 20100 Milan, Italy*

ICHIRO MIKI (32), *Department of Physiological Chemistry and Nutrition, Faculty of Medicine, University of Tokyo, Tokyo 113, Japan*

MICHIKO MINAMI (53), *Department of Physiological Chemistry and Nutrition, Faculty of Medicine, University of Tokyo, Tokyo 113, Japan*

SEYMOUR MONG (47), *Department of Immunology, Smith Kline & French Laboratories, King of Prussia, Pennsylvania 19406*

THOMAS A. MORINELLI (44), *Department of Cell and Molecular Pharmacology and Experimental Therapeutics, Medical University of South Carolina, Charleston, South Carolina 29425*

JASON D. MORROW (6), *Departments of Pharmacology and Medicine, Vanderbilt University School of Medicine, Nashville, Tennessee 37232*

A. SCOTT MUERHOFF (28), *Division of Biochemistry, Research Institute of Scripps Clinic, La Jolla, California 92037*

KEVIN M. MULLANE (66), *Gensia Pharmaceuticals Inc., San Diego, California 92121*

ROBERT C. MURPHY (11), *Department of Pharmacology, University of Colorado Health Sciences Center, Denver, Colorado 80262*

EWA NINIO (14), *INSERM U 200, 92140 Clamart, France*

ABDULRAHMAN ODEIMAT (12), *Centre de Recherches du CHUL, Le Centre Hospitalier de l'Université Laval, Sainte-Foy, Quebec G1V 4G2, Canada*

REBECCA ODENWALLER (52), *Department of Biochemistry, Center in Molecular Toxicology, Vanderbilt University School of Medicine, Nashville, Tennessee 37232*

NOBUYA OHISHI (31), *Department of Physiological Chemistry and Nutrition, Faculty of Medicine, University of Tokyo, Tokyo 113, Japan*

SHIGEO OHNO (53), *Department of Molecular Biology, Tokyo Metropolitan Institute of Medical Science, Tokyo 113, Japan*

C. R. PACE-ASCIAK (1), *Departments of Pediatrics and Pharmacology, The Hospital for Sick Children, Toronto, Ontario M5G 1X8, Canada*

AUDREY C. PAPP (63), *Department of Internal Medicine, Hematology and Vascular Disease Research Center, University of Texas Medical School at Houston, Houston, Texas 77030*

PAOLA PATRIGNANI (4), *Department of Pharmacology, Catholic University School of Medicine, 00168 Rome, Italy*

CARLO PATRONO (4), *Department of Pharmacology, Catholic University School of Medicine, 00168 Rome, Italy*

GEORGE M. PATTON (23), *Department of Medicine, Boston Veterans Administration Medical Center, Boston, Massachusetts 02130*

N. ANN PAYNE (55), *Department of Clinical Pharmacology, University of Colorado Health Sciences Center, Denver, Colorado 80262*

SERGE PICARD (12), *Centre de Recherches du CHUL, Le Centre Hospitalier de l'Université Laval, Sainte-Foy, Quebec G1V 4G2, Canada*

WALTER C. PICKETT (17), *Department of Oncology and Immunology, Lederle Laboratories, Pearl River, New York 10965*

PATRICE E. POUBELLE (12), *Centre de Recherches du CHUL, Le Centre Hospitalier de l'Université Laval, Sainte-Foy, Quebec G1V 4G2, Canada*

PHILIPPE PRADELLES (3, 10), *Section de Pharmacologie et d'Immunologie, Département de Biologie, Commissariat a l'Energie Atomique, CEN/Saclay, 91191 Gif-sur-Yvette, France*

STEPHEN M. PRESCOTT (39, 57), *Departments of Medicine and Biochemistry, Nora Eccles Harrison Cardiovascular Research and Training Institute, University of Utah, Salt Lake City, Utah 84112*

OLOF RÅDMARK (32, 53), *Department of Physiological Chemistry, Karolinska Institutet, S-104 01 Stockholm, Sweden*

CHAKKODABYLU S. RAMESHA (17), *Department of Inflammation Biology, Syntex Corporation, Palo Alto, California 94303*

PETER W. RAMWELL (1), *Department of Physiology and Biophysics, Georgetown University Medical Center, Washington, D.C. 20007*

EHUD RAZIN (56), *Institute of Biochemistry, The Hebrew University of Hadassah Medical School, Jerusalem 91010, Israel*

PALLU REDDANNA (29), *Department of Veterinary Science, The Pennsylvania State University, University Park, Pennsylvania 16802*

C. CHANNA REDDY (29), *Department of Veterinary Science and Environmental Resources Research Institute, The Pennsylvania State University, University Park, Pennsylvania 16802*

L. JACKSON ROBERTS II (1, 6), *Department of Pharmacology, Vanderbilt University School of Medicine, Nashville, Tennessee 37232*

SANDER J. ROBINS (23), *Department of Medicine, Boston Veterans Administration Medical Center, Boston, Massachusetts 02130*

JOSHUA ROKACH (1), *Merck Frosst Canada Inc., Dorval, Quebec H9R 4P8, Canada*

CAROL A. ROUZER (34), *Department of Pharmacology, Merck Frosst Centre for Therapeutic Research, Kirkland, Quebec H9R 4P8, Canada*

ANGELO SALA (11), *Department of Pharmacology, University of Colorado Health Sciences Center, Denver, Colorado 80262*

BENGT SAMUELSSON (1, 32, 34, 53, 54), *Department of Physiological Chemistry, Karolinska Institutet, S-104 01 Stockholm, Sweden*

MICHAL LANIADO SCHWARTZMAN (42), *Department of Pharmacology, New York Medical College, Valhalla, New York 10595*

ROBERT R. SEABOLD (21), *Department of Pathology, Division of Laboratory Medicine, Washington University School of Medicine, St. Louis, Missouri 63110*

CHARLES N. SERHAN (20), *Department of Medicine, Hematology Division, Brigham and Women's Hospital and, Harvard Medical School, Boston, Massachusetts 02115*

YOUSUKE SEYAMA (31, 32, 53), *Department of Physiological Chemistry and Nutrition, Faculty of Medicine, University of Tokyo, Tokyo 113, Japan*

T. Y. SHEN (49), *Department of Chemistry, University of Virginia, Charlottesville, Virginia 22901*

TAKAO SHIMIZU (31, 32, 53), *Department of Physiological Chemistry and Nutrition, Faculty of Medicine, University of Tokyo, Tokyo 113, Japan*

MICHAEL A. SHIRLEY (30), *Department of Pharmacology, University of Colorado Health Sciences Center, Denver, Colorado 80262*

MICHAEL S. SIMONSON (59), *Department of Medicine, Case Western Reserve University School of Medicine, Cleveland, Ohio 44106*

PIERRE SIROIS (12), *Département de Pharmacologie, Faculté de Médecine, Université de Sherbrooke, Sherbrooke, Quebec J1H 5N4, Canada*

WILLIAM L. SMITH (1, 51), *Department of Biochemistry, Michigan State University, East Lansing, Michigan 48824*

ROY J. SOBERMAN (35, 37), *Department of Rheumatology, Immunology and Medicine, Harvard Medical School, Boston, Massachusetts 02115*

MATS SÖDERSTRÖM (33), *Department of Biochemistry, Wallenberg Laboratory, University of Stockholm, S-106 91 Stockholm, Sweden*

DIANA M. STAFFORINI (39), *Nora Eccles Harrison Cardiovascular Research and Training Institute, University of Utah, Salt Lake City, Utah 84112*

DANNY O. STENE (30), *Department of Pharmacology, University of Colorado Health Sciences Center, Denver, Colorado 80262*

W. THOMAS STUMP (21), *Department of Pathology, Division of Laboratory Medicine, Washington University School of Medicine, St. Louis, Missouri 63110*

KOICHI SUZUKI (53), *Department of Molecular Biology, Tokyo Metropolitan Institute of Medical Science, Tokyo 113, Japan*

JOHN TURK (21), *Department of Pathology, Division of Laboratory Medicine, Washington University School of Medicine, St. Louis, Missouri 63110*

NATSUO UEDA (38), *Department of Biochemistry, Tokushima University School of Medicine, Kuramoto-cho, Tokushima 770, Japan*

CHITRA M. VAIDYA (45), *Department of Pharmacology, University of Illinois at Chicago, Chicago, Illinois 60612*

PIERRE VALLERAND (12), *Centre de Recherches du CHUL, Le Centre Hospitalier de l'Université Laval, Sainte-Foy, Quebec G1V 4G2, Canada*

A. VANHOVE (8), *Service des Molécules Marquées, Département de Biologie, CEN/Saclay, 91191 Gif-sur-Yvette, France*

DUANE VENTON (27, 45), *Department of Pharmacology, University of Illinois at Chicago, Chicago, Illinois 60612*

T. VIGANÒ (67), *Institute of Pharmacological Sciences, University of Milan, 20133 Milan, Italy*

NORBERT F. VOELKEL (65), *Cardiovascular Pulmonary Research Laboratory, University of Colorado Health Sciences Center, Denver, Colorado 80262*

J. JAMES VRBANAC (7), *Drug Metabolism Research, The Upjohn Co., Kalamazoo, Michigan 49001*

DANIEL F. WENDELBORN (6), *Departments of Pharmacology and Medicine, Vanderbilt University School of Medicine, Nashville, Tennessee 37232*

RALPH E. WHATLEY (57), *Department of Medicine, Nora Eccles Harrison Cardiovascular Research and Training Institute, University of Utah School of Medicine, Salt Lake City, Utah 84112*

J. WHELAN (29), *Department of Food Science, Cornell University, Ithaca, New York 14850*

DAVID E. WILLIAMS (28), *Department of Food Science and Technology, Oregon State University, Corvallis, Oregon 97331*

ANTHONY L. WILLIS (1), *Institute of Experimental Pharmacology, Syntex Research, Palo Alto, California 94303*

URS WIRTHMUELLER (62), *Institute of Clinical Immunology, Inselspital University Hospital, CH-3010 Bern, Switzerland*

BRYAN A. WOLF (21), *Department of Pathology and Laboratory Medicine, University of Pennsylvania, Philadelphia, Pennsylvania 19104*

KENNETH K. WU (63), *Department of Internal Medicine, Hematology and Vascular Disease Research Center, University of Texas Medical School at Houston, Houston, Texas 77030*

PENDRI YADAGIRI (40), *Basic Research and Development Center, International Flavors and Fragrances, Union Beach, New Jersey 07735*

SHOZO YAMAMOTO (1, 38), *Department of Biochemistry, Tokushima University, Kuramoto-cho, Tokushima 770, Japan*

GUY A. ZIMMERMAN (57), *Department of Medicine, Nora Eccles Harrison Cardiovascular Research and Training Institute, University of Utah School of Medicine, Salt Lake City, Utah 84112*

Preface

Volume 86 of the *Methods in Enzymology,* Prostaglandins and Arachidonate Metabolites, has become an important reference for methods used by individuals interested in the study of arachidonic acid metabolism. The success of that volume was due largely to the efforts of the editors, Dr. William E. M. Lands and Dr. William L. Smith, and to the contributions made by the various authors.

In the intervening years, considerable advances have been made in the field of eicosanoids and lipid mediators. Developments in analytical, biochemical, and molecular biological techniques have had a significant impact on current research. In addition, we know a great deal more about the biological roles that eicosanoids play in health and disease and the complexity with which they are synthesized *in vivo*. Notable advances have also been made in the appreciation of cellular cooperation during eicosanoid biosynthesis, and more specific receptor antagonists have been synthesized and characterized. For these reasons, this volume was designed to supplement the earlier one with the addition of new techniques in the measurement of eicosanoids, of new molecular biological techniques applied to eicosanoid research, and of descriptions of methods used to study the action and formation of eicosanoids in isolated cell and tissue systems.

We have expanded the scope of Volume 86 by including methods for the analysis of platelet-activating factor (PAF) and descriptions of the purification of enzymes involved in PAF biosynthesis and metabolism. In addition, the analysis of phospholipid precursors for both platelet-activating factor and eicosanoids is described and an isolation procedure for phospholipase A_2 specific for the release of arachidonate is presented. It now appears that PAF and eicosanoids are closely linked biochemically and it therefore seemed suitable to include platelet-activating factor methodology in this volume.

We thank Drs. Lands and Smith for providing important suggestions and advice used in the preparation of this volume. The organization and correspondence required in developing this work would not have been possible without the skillful and cheerful assistance of Ms. Deborah Beckworth. We thank her for her exceptional help.

Robert C. Murphy
Frank A. Fitzpatrick

METHODS IN ENZYMOLOGY

VOLUME I. Preparation and Assay of Enzymes
Edited by SIDNEY P. COLOWICK AND NATHAN O. KAPLAN

VOLUME II. Preparation and Assay of Enzymes
Edited by SIDNEY P. COLOWICK AND NATHAN O. KAPLAN

VOLUME III. Preparation and Assay of Substrates
Edited by SIDNEY P. COLOWICK AND NATHAN O. KAPLAN

VOLUME IV. Special Techniques for the Enzymologist
Edited by SIDNEY P. COLOWICK AND NATHAN O. KAPLAN

VOLUME V. Preparation and Assay of Enzymes
Edited by SIDNEY P. COLOWICK AND NATHAN O. KAPLAN

VOLUME VI. Preparation and Assay of Enzymes (*Continued*)
Preparation and Assay of Substrates
Special Techniques
Edited by SIDNEY P. COLOWICK AND NATHAN O. KAPLAN

VOLUME VII. Cumulative Subject Index
Edited by SIDNEY P. COLOWICK AND NATHAN O. KAPLAN

VOLUME VIII. Complex Carbohydrates
Edited by ELIZABETH F. NEUFELD AND VICTOR GINSBURG

VOLUME IX. Carbohydrate Metabolism
Edited by WILLIS A. WOOD

VOLUME X. Oxidation and Phosphorylation
Edited by RONALD W. ESTABROOK AND MAYNARD E. PULLMAN

VOLUME XI. Enzyme Structure
Edited by C. H. W. HIRS

VOLUME XII. Nucleic Acids (Parts A and B)
Edited by LAWRENCE GROSSMAN AND KIVIE MOLDAVE

VOLUME XIII. Citric Acid Cycle
Edited by J. M. LOWENSTEIN

VOLUME XIV. Lipids
Edited by J. M. LOWENSTEIN

VOLUME XV. Steroids and Terpenoids
Edited by RAYMOND B. CLAYTON

VOLUME XVI. Fast Reactions
Edited by KENNETH KUSTIN

VOLUME XVII. Metabolism of Amino Acids and Amines (Parts A and B)
Edited by HERBERT TABOR AND CELIA WHITE TABOR

VOLUME XVIII. Vitamins and Coenzymes (Parts A, B, and C)
Edited by DONALD B. MCCORMICK AND LEMUEL D. WRIGHT

VOLUME XIX. Proteolytic Enzymes
Edited by GERTRUDE E. PERLMANN AND LASZLO LORAND

VOLUME XX. Nucleic Acids and Protein Synthesis (Part C)
Edited by KIVIE MOLDAVE AND LAWRENCE GROSSMAN

VOLUME XXI. Nucleic Acids (Part D)
Edited by LAWRENCE GROSSMAN AND KIVIE MOLDAVE

VOLUME XXII. Enzyme Purification and Related Techniques
Edited by WILLIAM B. JAKOBY

VOLUME XXIII. Photosynthesis (Part A)
Edited by ANTHONY SAN PIETRO

VOLUME XXIV. Photosynthesis and Nitrogen Fixation (Part B)
Edited by ANTHONY SAN PIETRO

VOLUME XXV. Enzyme Structure (Part B)
Edited by C. H. W. HIRS AND SERGE N. TIMASHEFF

VOLUME XXVI. Enzyme Structure (Part C)
Edited by C. H. W. HIRS AND SERGE N. TIMASHEFF

VOLUME XXVII. Enzyme Structure (Part D)
Edited by C. H. W. HIRS AND SERGE N. TIMASHEFF

VOLUME XXVIII. Complex Carbohydrates (Part B)
Edited by VICTOR GINSBURG

VOLUME XXIX. Nucleic Acids and Protein Synthesis (Part E)
Edited by LAWRENCE GROSSMAN AND KIVIE MOLDAVE

VOLUME XXX. Nucleic Acids and Protein Synthesis (Part F)
Edited by KIVIE MOLDAVE AND LAWRENCE GROSSMAN

VOLUME XXXI. Biomembranes (Part A)
Edited by SIDNEY FLEISCHER AND LESTER PACKER

VOLUME XXXII. Biomembranes (Part B)
Edited by SIDNEY FLEISCHER AND LESTER PACKER

VOLUME XXXIII. Cumulative Subject Index Volumes I–XXX
Edited by MARTHA G. DENNIS AND EDWARD A. DENNIS

VOLUME XXXIV. Affinity Techniques (Enzyme Purification: Part B)
Edited by WILLIAM B. JAKOBY AND MEIR WILCHEK

VOLUME XXXV. Lipids (Part B)
Edited by JOHN M. LOWENSTEIN

VOLUME XXXVI. Hormone Action (Part A: Steroid Hormones)
Edited by BERT W. O'MALLEY AND JOEL G. HARDMAN

VOLUME XXXVII. Hormone Action (Part B: Peptide Hormones)
Edited by BERT W. O'MALLEY AND JOEL G. HARDMAN

VOLUME XXXVIII. Hormone Action (Part C: Cyclic Nucleotides)
Edited by JOEL G. HARDMAN AND BERT W. O'MALLEY

VOLUME XXXIX. Hormone Action (Part D: Isolated Cells, Tissues, and Organ Systems)
Edited by JOEL G. HARDMAN AND BERT W. O'MALLEY

VOLUME XL. Hormone Action (Part E: Nuclear Structure and Function)
Edited by BERT W. O'MALLEY AND JOEL G. HARDMAN

VOLUME XLI. Carbohydrate Metabolism (Part B)
Edited by W. A. WOOD

VOLUME XLII. Carbohydrate Metabolism (Part C)
Edited by W. A. WOOD

VOLUME XLIII. Antibiotics
Edited by JOHN H. HASH

VOLUME XLIV. Immobilized Enzymes
Edited by KLAUS MOSBACH

VOLUME XLV. Proteolytic Enzymes (Part B)
Edited by LASZLO LORAND

VOLUME XLVI. Affinity Labeling
Edited by WILLIAM B. JAKOBY AND MEIR WILCHEK

VOLUME XLVII. Enzyme Structure (Part E)
Edited by C. H. W. HIRS AND SERGE N. TIMASHEFF

VOLUME XLVIII. Enzyme Structure (Part F)
Edited by C. H. W. HIRS AND SERGE N. TIMASHEFF

VOLUME XLIX. Enzyme Structure (Part G)
Edited by C. H. W. HIRS AND SERGE N. TIMASHEFF

VOLUME L. Complex Carbohydrates (Part C)
Edited by VICTOR GINSBURG

VOLUME LI. Purine and Pyrimidine Nucleotide Metabolism
Edited by PATRICIA A. HOFFEE AND MARY ELLEN JONES

VOLUME LII. Biomembranes (Part C: Biological Oxidations)
Edited by SIDNEY FLEISCHER AND LESTER PACKER

VOLUME LIII. Biomembranes (Part D: Biological Oxidations)
Edited by SIDNEY FLEISCHER AND LESTER PACKER

VOLUME LIV. Biomembranes (Part E: Biological Oxidations)
Edited by SIDNEY FLEISCHER AND LESTER PACKER

VOLUME LV. Biomembranes (Part F: Bioenergetics)
Edited by SIDNEY FLEISCHER AND LESTER PACKER

VOLUME LVI. Biomembranes (Part G: Bioenergetics)
Edited by SIDNEY FLEISCHER AND LESTER PACKER

VOLUME LVII. Bioluminescence and Chemiluminescence
Edited by MARLENE A. DELUCA

VOLUME LVIII. Cell Culture
Edited by WILLIAM B. JAKOBY AND IRA PASTAN

VOLUME LIX. Nucleic Acids and Protein Synthesis (Part G)
Edited by KIVIE MOLDAVE AND LAWRENCE GROSSMAN

VOLUME LX. Nucleic Acids and Protein Synthesis (Part H)
Edited by KIVIE MOLDAVE AND LAWRENCE GROSSMAN

VOLUME 61. Enzyme Structure (Part H)
Edited by C. H. W. HIRS AND SERGE N. TIMASHEFF

VOLUME 62. Vitamins and Coenzymes (Part D)
Edited by DONALD B. MCCORMICK AND LEMUEL D. WRIGHT

VOLUME 63. Enzyme Kinetics and Mechanism (Part A: Initial Rate and Inhibitor Methods)
Edited by DANIEL L. PURICH

VOLUME 64. Enzyme Kinetics and Mechanism (Part B: Isotopic Probes and Complex Enzyme Systems)
Edited by DANIEL L. PURICH

VOLUME 65. Nucleic Acids (Part I)
Edited by LAWRENCE GROSSMAN AND KIVIE MOLDAVE

VOLUME 66. Vitamins and Coenzymes (Part E)
Edited by DONALD B. MCCORMICK AND LEMUEL D. WRIGHT

VOLUME 67. Vitamins and Coenzymes (Part F)
Edited by DONALD B. MCCORMICK AND LEMUEL D. WRIGHT

VOLUME 68. Recombinant DNA
Edited by RAY WU

VOLUME 69. Photosynthesis and Nitrogen Fixation (Part C)
Edited by ANTHONY SAN PIETRO

VOLUME 70. Immunochemical Techniques (Part A)
Edited by HELEN VAN VUNAKIS AND JOHN J. LANGONE

VOLUME 71. Lipids (Part C)
Edited by JOHN M. LOWENSTEIN

VOLUME 72. Lipids (Part D)
Edited by JOHN M. LOWENSTEIN

VOLUME 73. Immunochemical Techniques (Part B)
Edited by JOHN J. LANGONE AND HELEN VAN VUNAKIS

VOLUME 74. Immunochemical Techniques (Part C)
Edited by JOHN J. LANGONE AND HELEN VAN VUNAKIS

VOLUME 75. Cumulative Subject Index Volumes XXXI, XXXII, XXXIV–LX
Edited by EDWARD A. DENNIS AND MARTHA G. DENNIS

VOLUME 76. Hemoglobins
Edited by ERALDO ANTONINI, LUIGI ROSSI-BERNARDI, AND EMILIA CHIANCONE

VOLUME 77. Detoxication and Drug Metabolism
Edited by WILLIAM B. JAKOBY

VOLUME 78. Interferons (Part A)
Edited by SIDNEY PESTKA

VOLUME 79. Interferons (Part B)
Edited by SIDNEY PESTKA

VOLUME 80. Proteolytic Enzymes (Part C)
Edited by LASZLO LORAND

VOLUME 81. Biomembranes (Part H: Visual Pigments and Purple Membranes, I)
Edited by LESTER PACKER

VOLUME 82. Structural and Contractile Proteins (Part A: Extracellular Matrix)
Edited by LEON W. CUNNINGHAM AND DIXIE W. FREDERIKSEN

VOLUME 83. Complex Carbohydrates (Part D)
Edited by VICTOR GINSBURG

VOLUME 84. Immunochemical Techniques (Part D: Selected Immunoassays)
Edited by JOHN J. LANGONE AND HELEN VAN VUNAKIS

VOLUME 85. Structural and Contractile Proteins (Part B: The Contractile Apparatus and the Cytoskeleton)
Edited by DIXIE W. FREDERIKSEN AND LEON W. CUNNINGHAM

VOLUME 86. Prostaglandins and Arachidonate Metabolites
Edited by WILLIAM E. M. LANDS AND WILLIAM L. SMITH

VOLUME 87. Enzyme Kinetics and Mechanism (Part C: Intermediates, Stereochemistry, and Rate Studies)
Edited by DANIEL L. PURICH

VOLUME 88. Biomembranes (Part I: Visual Pigments and Purple Membranes, II)
Edited by LESTER PACKER

VOLUME 89. Carbohydrate Metabolism (Part D)
Edited by WILLIS A. WOOD

VOLUME 90. Carbohydrate Metabolism (Part E)
Edited by WILLIS A. WOOD

VOLUME 91. Enzyme Structure (Part I)
Edited by C. H. W. HIRS AND SERGE N. TIMASHEFF

VOLUME 92. Immunochemical Techniques (Part E: Monoclonal Antibodies and General Immunoassay Methods)
Edited by JOHN J. LANGONE AND HELEN VAN VUNAKIS

VOLUME 93. Immunochemical Techniques (Part F: Conventional Antibodies, Fc Receptors, and Cytotoxicity)
Edited by JOHN J. LANGONE AND HELEN VAN VUNAKIS

VOLUME 94. Polyamines
Edited by HERBERT TABOR AND CELIA WHITE TABOR

VOLUME 95. Cumulative Subject Index Volumes 61–74, 76–80
Edited by EDWARD A. DENNIS AND MARTHA G. DENNIS

VOLUME 96. Biomembranes [Part J: Membrane Biogenesis: Assembly and Targeting (General Methods; Eukaryotes)]
Edited by SIDNEY FLEISCHER AND BECCA FLEISCHER

VOLUME 97. Biomembranes [Part K: Membrane Biogenesis: Assembly and Targeting (Prokaryotes, Mitochondria, and Chloroplasts)]
Edited by SIDNEY FLEISCHER AND BECCA FLEISCHER

VOLUME 98. Biomembranes (Part L: Membrane Biogenesis: Processing and Recycling)
Edited by SIDNEY FLEISCHER AND BECCA FLEISCHER

VOLUME 99. Hormone Action (Part F: Protein Kinases)
Edited by JACKIE D. CORBIN AND JOEL G. HARDMAN

VOLUME 100. Recombinant DNA (Part B)
Edited by RAY WU, LAWRENCE GROSSMAN, AND KIVIE MOLDAVE

VOLUME 101. Recombinant DNA (Part C)
Edited by RAY WU, LAWRENCE GROSSMAN, AND KIVIE MOLDAVE

VOLUME 102. Hormone Action (Part G: Calmodulin and Calcium-Binding Proteins)
Edited by ANTHONY R. MEANS AND BERT W. O'MALLEY

VOLUME 103. Hormone Action (Part H: Neuroendocrine Peptides)
Edited by P. MICHAEL CONN

VOLUME 104. Enzyme Purification and Related Techniques (Part C)
Edited by WILLIAM B. JAKOBY

VOLUME 105. Oxygen Radicals in Biological Systems
Edited by LESTER PACKER

VOLUME 106. Posttranslational Modifications (Part A)
Edited by FINN WOLD AND KIVIE MOLDAVE

VOLUME 107. Posttranslational Modifications (Part B)
Edited by FINN WOLD AND KIVIE MOLDAVE

VOLUME 108. Immunochemical Techniques (Part G: Separation and Characterization of Lymphoid Cells)
Edited by GIOVANNI DI SABATO, JOHN J. LANGONE, AND HELEN VAN VUNAKIS

VOLUME 109. Hormone Action (Part I: Peptide Hormones)
Edited by LUTZ BIRNBAUMER AND BERT W. O'MALLEY

VOLUME 110. Steroids and Isoprenoids (Part A)
Edited by JOHN H. LAW AND HANS C. RILLING

VOLUME 111. Steroids and Isoprenoids (Part B)
Edited by JOHN H. LAW AND HANS C. RILLING

VOLUME 112. Drug and Enzyme Targeting (Part A)
Edited by KENNETH J. WIDDER AND RALPH GREEN

VOLUME 113. Glutamate, Glutamine, Glutathione, and Related Compounds
Edited by ALTON MEISTER

VOLUME 114. Diffraction Methods for Biological Macromolecules (Part A)
Edited by HAROLD W. WYCKOFF, C. H. W. HIRS, AND SERGE N. TIMASHEFF

VOLUME 115. Diffraction Methods for Biological Macromolecules (Part B)
Edited by HAROLD W. WYCKOFF, C. H. W. HIRS, AND SERGE N. TIMASHEFF

VOLUME 116. Immunochemical Techniques (Part H: Effectors and Mediators of Lymphoid Cell Functions)
Edited by GIOVANNI DI SABATO, JOHN J. LANGONE, AND HELEN VAN VUNAKIS

VOLUME 117. Enzyme Structure (Part J)
Edited by C. H. W. HIRS AND SERGE N. TIMASHEFF

VOLUME 118. Plant Molecular Biology
Edited by ARTHUR WEISSBACH AND HERBERT WEISSBACH

VOLUME 119. Interferons (Part C)
Edited by SIDNEY PESTKA

VOLUME 120. Cumulative Subject Index Volumes 81–94, 96–101

VOLUME 121. Immunochemical Techniques (Part I: Hybridoma Technology and Monoclonal Antibodies)
Edited by JOHN J. LANGONE AND HELEN VAN VUNAKIS

VOLUME 122. Vitamins and Coenzymes (Part G)
Edited by FRANK CHYTIL AND DONALD B. MCCORMICK

VOLUME 123. Vitamins and Coenzymes (Part H)
Edited by FRANK CHYTIL AND DONALD B. MCCORMICK

VOLUME 124. Hormone Action (Part J: Neuroendocrine Peptides)
Edited by P. MICHAEL CONN

VOLUME 125. Biomembranes (Part M: Transport in Bacteria, Mitochondria, and Chloroplasts: General Approaches and Transport Systems)
Edited by SIDNEY FLEISCHER AND BECCA FLEISCHER

VOLUME 126. Biomembranes (Part N: Transport in Bacteria, Mitochondria, and Chloroplasts: Protonmotive Force)
Edited by SIDNEY FLEISCHER AND BECCA FLEISCHER

VOLUME 127. Biomembranes (Part O: Protons and Water: Structure and Translocation)
Edited by LESTER PACKER

VOLUME 128. Plasma Lipoproteins (Part A: Preparation, Structure, and Molecular Biology)
Edited by JERE P. SEGREST AND JOHN J. ALBERS

VOLUME 129. Plasma Lipoproteins (Part B: Characterization, Cell Biology, and Metabolism)
Edited by JOHN J. ALBERS AND JERE P. SEGREST

VOLUME 130. Enzyme Structure (Part K)
Edited by C. H. W. HIRS AND SERGE N. TIMASHEFF

VOLUME 131. Enzyme Structure (Part L)
Edited by C. H. W. HIRS AND SERGE N. TIMASHEFF

VOLUME 132. Immunochemical Techniques (Part J: Phagocytosis and Cell-Mediated Cytotoxicity)
Edited by GIOVANNI DI SABATO AND JOHANNES EVERSE

VOLUME 133. Bioluminescence and Chemiluminescence (Part B)
Edited by MARLENE DELUCA AND WILLIAM D. MCELROY

VOLUME 134. Structural and Contractile Proteins (Part C: The Contractile Apparatus and the Cytoskeleton)
Edited by RICHARD B. VALLEE

VOLUME 135. Immobilized Enzymes and Cells (Part B)
Edited by KLAUS MOSBACH

VOLUME 136. Immobilized Enzymes and Cells (Part C)
Edited by KLAUS MOSBACH

VOLUME 137. Immobilized Enzymes and Cells (Part D)
Edited by KLAUS MOSBACH

VOLUME 138. Complex Carbohydrates (Part E)
Edited by VICTOR GINSBURG

VOLUME 139. Cellular Regulators (Part A: Calcium- and Calmodulin-Binding Proteins)
Edited by ANTHONY R. MEANS AND P. MICHAEL CONN

VOLUME 140. Cumulative Subject Index Volumes 102–119, 121–134

VOLUME 141. Cellular Regulators (Part B: Calcium and Lipids)
Edited by P. MICHAEL CONN AND ANTHONY R. MEANS

VOLUME 142. Metabolism of Aromatic Amino Acids and Amines
Edited by SEYMOUR KAUFMAN

VOLUME 143. Sulfur and Sulfur Amino Acids
Edited by WILLIAM B. JAKOBY AND OWEN GRIFFITH

VOLUME 144. Structural and Contractile Proteins (Part D: Extracellular Matrix)
Edited by LEON W. CUNNINGHAM

VOLUME 145. Structural and Contractile Proteins (Part E: Extracellular Matrix)
Edited by LEON W. CUNNINGHAM

VOLUME 146. Peptide Growth Factors (Part A)
Edited by DAVID BARNES AND DAVID A. SIRBASKU

VOLUME 147. Peptide Growth Factors (Part B)
Edited by DAVID BARNES AND DAVID A. SIRBASKU

VOLUME 148. Plant Cell Membranes
Edited by LESTER PACKER AND ROLAND DOUCE

VOLUME 149. Drug and Enzyme Targeting (Part B)
Edited by RALPH GREEN AND KENNETH J. WIDDER

VOLUME 150. Immunochemical Techniques (Part K: *In Vitro* Models of B and T Cell Functions and Lymphoid Cell Receptors)
Edited by GIOVANNI DI SABATO

VOLUME 151. Molecular Genetics of Mammalian Cells
Edited by MICHAEL M. GOTTESMAN

VOLUME 152. Guide to Molecular Cloning Techniques
Edited by SHELBY L. BERGER AND ALAN R. KIMMEL

Volume 153. Recombinant DNA (Part D)
Edited by Ray Wu and Lawrence Grossman

Volume 154. Recombinant DNA (Part E)
Edited by Ray Wu and Lawrence Grossman

Volume 155. Recombinant DNA (Part F)
Edited by Ray Wu

Volume 156. Biomembranes (Part P: ATP-Driven Pumps and Related Transport: The Na,K-Pump)
Edited by Sidney Fleischer and Becca Fleischer

Volume 157. Biomembranes (Part Q: ATP-Driven Pumps and Related Transport: Calcium, Proton, and Potassium Pumps)
Edited by Sidney Fleischer and Becca Fleischer

Volume 158. Metalloproteins (Part A)
Edited by James F. Riordan and Bert L. Vallee

Volume 159. Initiation and Termination of Cyclic Nucleotide Action
Edited by Jackie D. Corbin and Roger A. Johnson

Volume 160. Biomass (Part A: Cellulose and Hemicellulose)
Edited by Willis A. Wood and Scott T. Kellogg

Volume 161. Biomass (Part B: Lignin, Pectin, and Chitin)
Edited by Willis A. Wood and Scott T. Kellogg

Volume 162. Immunochemical Techniques (Part L: Chemotaxis and Inflammation)
Edited by Giovanni Di Sabato

Volume 163. Immunochemical Techniques (Part M: Chemotaxis and Inflammation)
Edited by Giovanni Di Sabato

Volume 164. Ribosomes
Edited by Harry F. Noller, Jr., and Kivie Moldave

Volume 165. Microbial Toxins: Tools for Enzymology
Edited by Sidney Harshman

VOLUME 166. Branched-Chain Amino Acids
Edited by ROBERT HARRIS AND JOHN R. SOKATCH

VOLUME 167. Cyanobacteria
Edited by LESTER PACKER AND ALEXANDER N. GLAZER

VOLUME 168. Hormone Action (Part K: Neuroendocrine Peptides)
Edited by P. MICHAEL CONN

VOLUME 169. Platelets: Receptors, Adhesion, Secretion (Part A)
Edited by JACEK HAWIGER

VOLUME 170. Nucleosomes
Edited by PAUL M. WASSARMAN AND ROGER D. KORNBERG

VOLUME 171. Biomembranes (Part R: Transport Theory: Cells and Model Membranes)
Edited by SIDNEY FLEISCHER AND BECCA FLEISCHER

VOLUME 172. Biomembranes (Part S: Transport: Membrane Isolation and Characterization)
Edited by SIDNEY FLEISCHER AND BECCA FLEISCHER

VOLUME 173. Biomembranes [Part T: Cellular and Subcellular Transport: Eukaryotic (Nonepithelial) Cells]
Edited by SIDNEY FLEISCHER AND BECCA FLEISCHER

VOLUME 174. Biomembranes [Part U: Cellular and Subcellular Transport: Eukaryotic (Nonepithelial) Cells]
Edited by SIDNEY FLEISCHER AND BECCA FLEISCHER

VOLUME 175. Cumulative Subject Index Volumes 135–139, 141–167 (in preparation)

VOLUME 176. Nuclear Magnetic Resonance (Part A: Spectral Techniques and Dynamics)
Edited by NORMAN J. OPPENHEIMER AND THOMAS L. JAMES

VOLUME 177. Nuclear Magnetic Resonance (Part B: Structure and Mechanism)
Edited by NORMAN J. OPPENHEIMER AND THOMAS L. JAMES

VOLUME 178. Antibodies, Antigens, and Molecular Mimicry
Edited by JOHN J. LANGONE

VOLUME 179. Complex Carbohydrates (Part F)
Edited by VICTOR GINSBURG

VOLUME 180. RNA Processing (Part A: General Methods)
Edited by JAMES E. DAHLBERG AND JOHN N. ABELSON

VOLUME 181. RNA Processing (Part B: Specific Methods)
Edited by JAMES E. DAHLBERG AND JOHN N. ABELSON

VOLUME 182. Guide to Protein Purification
Edited by MURRAY P. DEUTSCHER

VOLUME 183. Molecular Evolution: Computer Analysis of Protein and Nucleic Acid Sequences
Edited by RUSSELL F. DOOLITTLE

VOLUME 184. Avidin–Biotin Technology
Edited by MEIR WILCHEK AND EDWARD A. BAYER

VOLUME 185. Gene Expression Technology
Edited by DAVID V. GOEDDEL

VOLUME 186. Oxygen Radicals in Biological Systems (Part B: Oxygen Radicals and Antioxidants)
Edited by LESTER PACKER AND ALEXANDER N. GLAZER

VOLUME 187. Arachidonate Related Lipid Mediators
Edited by ROBERT C. MURPHY AND FRANK A. FITZPATRICK

VOLUME 188. Hydrocarbons and Methylotrophy (in preparation)
Edited by MARY E. LIDSTROM

VOLUME 189. Retinoids (Part A: Molecular and Metabolic Aspects) (in preparation)
Edited by LESTER PACKER

VOLUME 190. Retinods (Part B: Cell Differentiation and Clinical Applications) (in preparation)
Edited by LESTER PACKER

VOLUME 191. Biomembranes (Part V: Cellular and Subcellular Transport: Epithelial Cells) (in preparation)
Edited by SIDNEY FLEISCHER AND BECCA FLEISCHER

VOLUME 192. Biomembranes (Part W: Cellular and Subcellular Transport: Epithelial Cells) (in preparation)
Edited by SIDNEY FLEISCHER AND BECCA FLEISCHER

VOLUME 193. Mass Spectrometry (in preparation)
Edited by JAMES A. MCCLOSKEY

VOLUME 194. Guide to Yeast Genetics and Molecular Biology (in preparation)
Edited by CHRISTINE GUTHRIE AND GERALD R. FINK

[1] Nomenclature

By WILLIAM L. SMITH, PIERRE BORGEAT, MATS HAMBERG, L. JACKSON ROBERTS II, ANTHONY L. WILLIS, SHOZO YAMAMOTO, PETER W. RAMWELL, JOSHUA ROKACH, BENGT SAMUELSSON, E. J. COREY, and C. R. PACE-ASCIAK

In 1988, the ad hoc Committee on Eicosanoid Nomenclature, originally appointed by the Editorial Board of *Prostaglandins,* was constituted by the Joint Commission on Biochemical Nomenclature (JCBN) of the International Union of Pure and Applied Chemistry and the International Union of Biochemistry, as an advisory panel on the nomenclature of leukotrienes. In addition to making proposals regarding a systematic leukotriene nomenclature, the advisory panel was requested to provide suggestions on the nomenclature of enzymes involved in eicosanoid metabolism. The committee reviewed and revised three consecutive draft proposals each on the naming of leukotriene, hydroxy, epoxy, and oxo fatty acids and on the naming of enzymes. The committee also solicited advice from investigators attending the 1988 and 1989 Keystone Winter Prostaglandin Conferences. The following guidelines were adopted as a result of these deliberations. They have been presented to the JCBN, and are now under consideration.

Nomenclature of Leukotrienes, and Related Hydroxy, Hydroperoxy, Epoxy, and Oxo Fatty Acids

Guideline 1A. The nomenclature for the leukotrienes LTA, LTB, LTC, LTD, and LTE, including the numerical subscript designations, will be as proposed by Samuelsson and Hammarstrom[1] and by Samuelsson *et al.*[2] This nomenclature applies to products of the arachidonate 5-lipoxygenase pathway.

Guideline 1B. The use of 6-*trans*-LTB_3, 6-*trans*-LTB_4, 6-*trans*-LTB_5 as well as 6-*trans*-12-epi-LTB_3, 6-*trans*-12-epi-LTB_4, and 6-*trans*-12-epi-LTB_5 will be used to denote the 6-*trans*- and 6-*trans*-12-epi derivatives of LTB_3, LTB_4, and LTB_5, respectively. Other 5,6-, 5,12-, and 5,15-dihydroxyeicosatrienoic, dihydroxyeicosatetratenoic, and dihydroxyei-

[1] B. Samuelsson and S. Hammarstrom, *Prostaglandins* **19,** 645 (1980).

[2] B. Samuelsson, P. Borgeat, S. Hammarstrom, and R. C. Murphy, *Prostaglandins* **17,** 785 (1979).

cosapentaenoic acids will be named according to the rules described in Guideline 3 below.

Guideline 1C. The terms 11-*trans*-LTC_4, 11-*trans*-LTD_4, and 11-*trans*-LTE_4 will be applied to the 11-*trans* isomers of LTC_4, LTD_4, and LTE_4, respectively.[3]

Guideline 2. The nomenclature for the lipoxins LXA and LXB and their derivatives, including the numerical subscripts, will be as proposed by Serhan *et al.*[4] The term lipoxene, which was originally used for eicosapentaenoic acid derivatives,[5] will no longer be used.

Guideline 3. Hydroxy, hydroperoxy, epoxy, and oxo unsaturated fatty acids which do not qualify as leukotrienes or lipoxins according to Guidelines 1 or 2 are to be abbreviated according to the proposals of Smith and Willis[6] with minor modifications.

General Structures

Number of substituents	Substituent name	Number of carbons	Number of double bonds
1: no name	Hydroxy: H	12: D (dodeca)	1: ME (monoenoic)
2: Di	Hydroperoxy: Hp	14: T (tetradeca)	2: DE (dienoic)
3: Tri	Epoxy: Ep	15: P (pentadeca)	3: TrE (trienoic)
4: Tetra	Keto: Oxo	16: Hx (hexadeca)	4: TE (tetraenoic)
5: Penta		17: H (heptadeca)	5: PE (pentaenoic)
6: Hexa		18: O (octadeca)	6: HE (hexaenoic)
		19: N (nonadeca)	
		20: E (eicosa)	
		22: Do (docosa)	

Position of Substituent Groups

The position of the substituent is to be indicated by a number, counting, as usual, from the carboxyl group. For epoxides, the numbers should be of the form # (#) [e.g., 5(6)-EpETE]; for hydroxyl, hydroperoxyl, and oxo groups, the number should take the usual form (e.g., 5-HETE).

[3] D. A. Clark, G. Goto, A. Marfat, E. J., Corey, S. Hammarstrom, and B. Samuelsson, *Biochem. Biophys. Res. Commun.* **94,** 1133 (1980).

[4] C. N. Serhan, P. Y-K. Wong, and B. Samuelsson, *Prostaglandins* **34,** 201 (1987).

[5] P. Y-K. Wong, R. Hughes, and B. Lam, *Biochem. Biophys. Res. Commun.* **126,** 763 (1985).

[6] D. L. Smith and A. L. Willis, *Lipids* **22,** 983 (1987).

Double-Bond Positions

The positions of the double bonds should be indicated with a number to designate the location of carbon nearest the carboxyl end of the molecule and a letter, either *Z* or *E*, to indicate cis and trans geometry, respectively. The designation for the position of the double bond should be in parentheses at the beginning of the shorthand name. For example, 5-HETE should be (6*E*,8*Z*,11*Z*,14*Z*)5-HETE. The designation for the double-bond position need only be used in the abbreviations section of a manuscript, unless it is appropriate to use this designation throughout for the purpose of clarity.

Nomenclature of Enzymes Involved in Eicosanoid Metabolism

Tables I–III present relevant information regarding the enzymes involved in eicosanoid metabolism. The tables are formatted as in "Enzyme Nomenclature 1984."[48] Enzymes involved in prostaglandin and throm-

[7] S. Ohki, N. Ogino, S. Yamamoto, and O. Hayaishi, *J. Biol. Chem.* **254,** 829 (1979).
[8] D. L. DeWitt and W. L. Smith, *Proc. Natl. Acad. Sci. U.S.A.* **85,** 1412 (1988).
[9] T. Shimizu, S. Yamamoto, and O. Hayaishi, *J. Biol. Chem.* **254,** 5222 (1979).
[10] E. Christ-Hazelhof and D. H. Nugteren, *Biochim. Biophys. Acta* **572,** 43 (1979).
[11] N. Ogino, T. Miyamoto, S. Yamamoto, and O. Hayaishi, *J. Biol. Chem.* **252,** 890 (1977).
[12] Y. Tanaka, S. L. Ward, and W. L. Smith, *J. Biol. Chem.* **262,** 1374 (1987).
[13] V. Ullrich, L. Castle, and P. Weber, *Biochem. Pharmacol.* **30,** 2033 (1981).
[14] D. L. DeWitt and W. L. Smith, *J. Biol. Chem.* **258,** 3285 (1983).
[15] V. Ullrich and M. Haurand, *Adv. Prostaglandin Thromboxane Leukotriene Res.* **11,** 105 (1983).
[16] R-F. Shen and H-H. Tai, *J. Biol. Chem.* **261,** 11592 (1986).
[17] K. Watanabe, R. Yoshida, T. Shimizu, and O. Hayaishi, *Adv. Prostaglandin Thromboxane Leukotriene Res.* **15,** 151 (1985).
[18] P. Y-K. Wong, this series, Vol. 86, p. 117.
[19] D. F. Reingold, A. Kawasaki, and P. Needleman, *Biochim. Biophys. Acta* **659,** 179 (1981).
[20] E. Anggard and B. Samuelsson, *Ark. Kemi* **25,** 293 (1966).
[21] S. S. Braithwaite and J. Jarabak, *J. Biol. Chem.* **250,** 2315 (1975).
[22] S-C. Lee and L. Levine, *J. Biol. Chem.* **250,** 548 (1975).
[23] S-C. Lee, S-S. Pong, D. Katzen, K-Y. Wu, and L. Levine, *Biochemistry* **14,** 142 (1975).
[24] D. B. B. Chang and H-H. Tai, *Biochem. Biophys. Res. Commun.* **99,** 745 (1981).
[25] Y. M. Lin and J. Jarabak, *Biochem. Biophys. Res. Commun.* **81,** 1227 (1978).
[26] K. Watanabe, T. Shimizu, S. Iguchi, H. Wakatsuka, M. Hayashi, and O. Hayaishi, *J. Biol. Chem.* **255,** 1779 (1980).
[27] J. M. Korff and J. Jarabak, this series, Vol. 86, p. 152.
[28] H. S. Hansen, this series, Vol. 86, p. 156.
[29] J. Jarabak, this series, Vol. 86, p. 163.
[30] S-C. Lee and L. Levine, *J. Biol. Chem.* **250,** 4549 (1975).
[31] H-H. Tai and B. Yuan, this series, Vol. 86, p. 113.
[32] J. D. Wadkins and J. Jarabak, *Prostaglandins* **30,** 335 (1985).

boxane biosynthesis are presented in Table I, enzymes involved in prostaglandin catabolism in Table II, and enzymes involved in hydroxy acid and leukotriene formation in Table III. Changes and new listings are also indicated.

[33] R. W. Bryant, J. M. Bailey, T. Schewe, and S. M. Rapoport, *J. Biol. Chem.* **257,** 6050 (1982).
[34] T. Matsumoto, C. D. Funk, O. Radmark, J-O Hoog, H. Jornvall, and B. Samuelsson, *Proc. Natl. Acad. Sci. U.S.A.* **85,** 26 (1988).
[35] M. Hamberg and B. Samuelsson, *Proc. Natl. Acad. Sci. U.S.A.* **71,** 3400 (1974).
[36] D. H. Nugteren, *Biochim. Biophys. Acta* **380,** 299 (1975).
[37] D. P. Wallach and V. R. Brown, *Biochim. Biophys. Acta* **663,** 361 (1981).
[38] D. H. Nugteren, this series, Vol. 86, p. 49.
[39] D. Shabata, J. Steczko, F. E. Dixon, M. Hermodson, R. Yasdanparast, and B. Axelrod, *J. Biol. Chem.* **262,** 10080 (1987).
[40] S. Narumiya and J. A. Salmon, this series, Vol. 86, p. 45.
[41] M. Minami, S. Ohno, H. Kawasaki, O. Radmark, B. Samuelsson, H. Jornvall, T. Shimizu, Y. Seyama, and K. Suzuki, *J. Biol. Chem.* **263,** 13873 (1988).
[42] N. Ohishi, T. Izumi, M. Minami, S. Kitamura, Y. Seyama, S. Ohkawa, S. Terao, H. Yotsumoto, F. Takaku, and T. Shimizu, *J. Biol. Chem.* **262,** 10200 (1987).
[43] T. Shimizu, *Int. J. Biochem.* **20,** 661 (1988).
[44] M. Y. Goore and J. F. Thompson, *Biochim. Biophys. Acta* **132,** 15 (1967).
[45] F. H. Leibach and F. Binkley, *Arch. Biochem. Biophys.* **127,** 292 (1968).
[46] L. Orning, S. Hammarstrom, and B. Samuelsson, *Proc. Natl. Acad. Sci. U.S.A.* **77,** 2014 (1980).
[47] S. Hammarstrom, *Annu. Rev. Biochem.* **52,** 355 (1983).
[48] "Enzyme Nomenclature" Recommendations (1984) of the Nomenclature Committee of the International Union of Biochemistry Academic Press, San Diego, 1984.

TABLE I

ENZYMES INVOLVED IN PROSTAGLANDIN AND THROMBOXANE BIOSYNTHESIS

EC Number	Recommended name	Reaction	Other names	Systematic name	Comments	Refs.
1.14.99.1	Prostaglandin endoperoxide synthase[a]	5,8,11,14-Eicosatetraenoate + AH_2 + $2O_2 \rightarrow$ prostaglandin H_2 + H_2O + A	Prostaglandin G/H synthase; fatty acid cyclooxygenase; prostaglandin synthetase	5,8,11,14-Eicosatetraenoate, hydrogen donor: oxygen oxidoreductase	Exhibits both dioxygenase and hydroperoxidase activities	7,8
5.3.99.2	Prostaglandin D synthase[a] Type I: requires reduced glutathione Type II: no known cofactor requirement	$PGH_x \rightarrow PGD_x$ (x = 1, 2, or 3); (5*Z*, 13*E*)-(15*S*)-9α,11α-epidioxy-15-hydroxyprosta-5,13-dienoate = (5*Z*,13*E*)-(15*S*)-9α,15-dihydroxy-11-oxoprosta-5,13-dienoate	PGH-PGD isomerase	(5*Z*,13*E*)-(15*S*)-9α,11α-Epidioxy-15-hydroxyprosta-5,13-dienoate D-isomerase	Opens epidioxy bridge	9,10
5.3.99.3	Prostaglandin E synthase[a]	$PGH_x \rightarrow PGE_x$ (x = 1, 2, or 3); (5*Z*,13*E*)-(15*S*)-9α,11α-epidioxy-15-hydroxyprosta-5,13-dienoate = (5*Z*,13*E*)-(15*S*)-11α,15-dihydroxy-9-oxoprosta-5,13-dienoate	PGH-PGE isomerase, endoperoxide isomerase	(5*Z*,13*E*)-(15*S*)-9α,11α-Epidioxy-15-hydroxyprosta-5,13-dienoate E-isomerase	Opens epidioxy bridge; requires glutathione	11,12
5.3.99.4	Prostaglandin I synthase[a]	$PGH_x \rightarrow PGI_x$ (x = 2 or 3); (5*Z*,13*E*)-(15*S*)-9α,11α-epidioxy-15-hydroxyprosta-5,13-dienoate = (5*Z*,13*E*)-(15*S*)-6,9α-epoxy-11α,15-dihydroxyprosta-5,13-dienoate	Prostacyclin synthase	(5*Z*,13*E*)-(15*S*)-9α,11α-Epidioxy-15-hydroxyprosta-5,13-dienoate 6-isomerase	Converts prostaglandin H_2 to prostaglandin I_2; a heme-thiolate protein	13,14

(continued)

TABLE I (*Continued*)

EC Number	Recommended name	Reaction	Other names	Systematic name	Comments	Refs.
5.3.99.5	Thromboxane A synthase	$PGH_x \rightarrow TxA_x$ (x = 1, 2, or 3); (5*Z*,13*E*)-(15*S*)-9α,11α-epidioxy-15-hydroxyprosta-5,13-dienoate = (5*Z*,13*E*)-(15*S*)-9α,11α-epoxy-15-hydroxythromba-5,13-dienoate	Thromboxane synthetase, thromboxane A_2 isomerase	(5*Z*,13*E*)-(15*S*)-9α,11α-Epidioxy-15-hydroxyprosta-5,13-dienoate thromboxane-A_2-isomerase	Converts prostaglandin H_2 to thromboxane A_2; a heme-thiolate protein	15,16
1.1.1.188	Prostaglandin F synthase[a]	$PGF_\alpha + NADP^+ \rightarrow PGH_x + NADPH + H^+$ and $PGF_\alpha + NADP^+ \rightarrow PGD_x + NADPH$; (5*Z*, 13*E*)-(15*S*)-9α,11α,15-trihydroxyprosta-5,13-dienoate + $NADP^+$ = (5*Z*,13*E*)-(15*S*)-9α,15-dihydroxy-11-oxoprosta-5,13-dienoate + NADPH	PGF reductase; prostaglandin D 11-ketoreductase; prostaglandin F synthetase	(5*Z*,13*E*)-(15*S*)-9α, 11α, 15-Trihydroxyprosta-5,13-dienoate: $NADP^+$ 11-oxidoreductase and NADPH: (5*Z*,13*E*)-(15*S*)-9α, 11α-epidioxy-15-hydroxyprosta-5,13-dienoate oxidoreductase	Reduces prostaglandin D_2 and prostaglandin H_2 to prostaglandin $F_{2\alpha}$	17–19

[a] Represents change from 1984 IUB Nomenclature List (see ref. 48).

TABLE II
Enzymes Involved in Prostaglandin Catabolism

EC Number	Recommended name	Reaction	Other names	Systematic name	Comments	Refs.
1.1.1.141	15-Hydroxyprostaglandin dehydrogenase (NAD^+)	PGE/PGF/PGB + NAD^+ → 15-keto-PGE/PGF/PGB + NADH + H^+; (5*Z*,13*E*)-(15*S*)-11α,15-dihydroxy-9-oxoprost-5,13-dienoate + NAD^+ = (5*Z*,13*E*)-11α-hydroxy-9,15-dioxoprost-5,13-dienoate + NADH + H^+	Type I 15-hydroxyprostaglandin dehydrogenase	(5*Z*,13*E*)-(15*S*)-11α, 15-Dihydroxy-9-oxoprost-5,13-dienoate:NAD^+ 15-oxidoreductase	Acts on prostaglandins E_2, $F_{2\alpha}$, B_1; not on prostaglandin D_2	20–24
1.1.1.197	15-Hydroxyprostaglandin dehydrogenase ($NADP^+$)	PGE/PGF/PGB + $NADP^+$ → 15-keto-PGE/PGF/PGB + NADPH + H^+;(5*Z*,13*E*)-(15*S*)-11α,15-dihydroxy-9-oxoprost-5,13-dienoate + $NADP^+$ = (5*Z*,13*E*)-11α-hydroxy-9.15-dioxoprost-5,13-dienoate + NADPH + H^+	Type II 15-hydroxyprostaglandin dehydrogenase	(5*Z*,13*E*)-(15*S*)-11α,15-Dihydroxy-9-oxoprost-5,13-dienoate:$NADP^+$ 15-oxidoreductase	Acts on prostaglandins E_2, $F_{2\alpha}$, B_1; not on prostaglandin D_2	22–25
1.1.1.196	15-Hydroxyprostaglandin D dehydrogenase ($NADP^+$)[a]	PGD + $NADP^+$ → 15-keto-PGD + NADPH + H^+; (5*Z*,13*E*)-(15*S*)-9α,15-dihydroxy-11-oxoprosta-5,13-dienoate + $NADP^+$ = (5*Z*,13*E*)-9α-hydroxy-11,15-dioxoprost-13-enoate + NADPH	Prostaglandin D 15-dehydrogenase ($NADP^+$)	(5*Z*,13*E*)-(15*S*)-9α,15-Dihydroxy-11-oxoprost-13-enoate:$NADP^+$ 15-oxidoreductase	Specific for prostaglandins D; (cf. EC 1.1.1.141 and 1.1.1.197)	26

(continued)

TABLE II (*Continued*)

EC Number	Recommended name	Reaction	Other names	Systematic name	Comments	Refs.
1.1.1.231[b]	15-Hydroxyprostaglandin I dehydrogenase ($NADP^+$)	PGI_2 + $NADP^+$ → 15-keto-PGI_2 + NADPH + H^+; (5*Z*,13*E*)-(15*S*)-6,9α-epoxy-11α,15-dihydroxyprosta-5,13-dienoate + $NADP^+$ = (5*Z*,13*E*)-6,9α-epoxy-11α-hydroxy-15-oxoprosta-5,13-dienoate + NADPH	—	(5*Z*,13*E*)-(15*S*)-6,9α-Epoxy-11α,15-dihydroxyprosta-5,13-dienoate 15-oxidoreductase	Specific for prostaglandin I_2	27
1.3.1.48[b]	15-OxoprostaglandinΔ^{13}-reductase	15-Keto-13,14-dihydro-PGE/PGF/PGD + $NADP^+$ → 15-keto-PGE/PGF/PGD + NADPH + H^+; (5*Z*)-(15*S*-11α-hydroxy-9,15-dioxoprostanoate + $NAD(P)^+$ = (5*Z*)-(15*S*)-11α-hydroxy-9,15-dioxoprosta-13-enoate + NAD(P)H	—	(5*Z*)-(15*S*)-11α-Hydroxy-9,15-dioxoprostanoate: $NAD(P)^+$ Δ^{13}-oxidoreductase	Reduces 15-oxoprostaglandins to 13,14-dihydro derivatives	28,29
1.1.1.189	Prostaglandin E 9-ketoreductase[a]	PGF_α + $NADP^+$ → PGE + NADPH + H^+; (5*Z*,13*E*)-(15*S*)-9α,11α,15-trihydroxyprosta-5,13-dienoate + $NADP^+$ = (5*Z*,13*E*)-(15*S*)-11α,15-dihydroxy-9-oxoprosta-5,13-dienoate + NADPH	—	(5*Z*,13*E*)-(15*S*)-9α, 11α, 15-Trihydroxyprosta-5,13-dienoate:$NADP^+$ 9-oxidoreductase	May be same as 1.1.1.197	30–32

[a] Represents change from 1984 IUB Nomenclature List (see ref. 48).
[b] Represents new listing from 1984 IUB Nomenclature List (see ref. 48).

TABLE III

ENZYMES INVOLVED IN HYDROXY ACID AND LEUKOTRIENE FORMATION

EC Number	Recommended name	Reaction	Other names	Systematic name	Comments	Refs.
1.13.11.34	Arachidonate 5-lipoxygenase	a. Arachidonate + $O_2 \rightarrow$ (*S*)-5-hydroperoxyarachidonate b. (*S*)-5-Hydroperoxyarachidonate $\rightarrow$ leukotriene A_4 + H_2O	5-Lipoxygenase, LTA_4 synthase	Arachidonate:oxygen 5-oxidoreductase	Has two activities: (a) an oxygenase and (b) a dehydrase catalyzing LTA_4 formation from 5-HpETE	33,34
1.13.11.31	Arachidonate 12-lipoxygenase	Arachidonate + $O_2 \rightarrow$ (*S*)-12-hydroperoxyarachidonate	12-Lipoxygenase	Arachidonate:oxygen 12-oxidoreductase	Has two activities: (a) an oxygenase and (b) a dehydrase catalyzing formation of (5*Z*,8*Z*,10*E*,12*E*)-14(15)-EpETE from 15-HpETE	35–38
1.13.11.33	Arachidonate 15-lipoxygenase Type I: plant enzyme Type II: mammalian enzyme	Arachidonate + $O_2 \rightarrow$ (*S*)-15-hydroperoxyarachidonate	15-Lipoxygenase	Arachidonate:oxygen 15-oxidoreductase		39,40
3.3.2.6[a]	Leukotriene A_4 hydrolase	Leukotriene A_4 + $H_2O \rightarrow$ leukotriene B_4;(7*E*,9*E*,11*Z*,14*Z*)-(5*S*,6*S*)-5,6-epoxyeicosa-7,9,11,14-tetraenoate + H_2O = (6*Z*,8*E*,10*E*,14*Z*)-(5*S*,12*R*)-5,12-dihydroxy-6,8,10,14-tetraenoate	LTA_4 hydrolase	(7*E*,9*E*,11*Z*,14*Z*)-(5*S*,6*S*)-5,6-Epoxyeicosa-7,9,11,14-tetraenoate hydrolase	Converts 4,5-leukotriene A_4 into leukotriene hydrolase B_4	41,42
2.5.1.37[a]	Leukotriene C_4 synthase	Leukotriene A_4 + glutathione $\rightarrow$ leukotriene C_4 + H_2O	Glutathione *S*-transferase	(7*E*,9*E*,11*Z*,14*Z*)-(5*S*,6*S*)-5,6-Epoxyeicosa-7,9,11,14-tetraenoate: glutathione leukotriene-transferase (epoxide ring opening)	Different from soluble glutathione *S*-transferase	43
2.3.2.2	γ-Glutamyltransferase	Leukotriene C_4 + $H_2O \rightarrow$ leukotriene D_4 + glutamate	γ-Glutamyltranspeptidase	(5-Glutamyl)-peptide: amino acid 5-glutamyltransferase		44–46
3.4.13.6	Cysteinyl-glycine dipeptidase	Leukotriene D_4 + $H_2O \rightarrow$ leukotriene E_4 + glycine	LTD_4 dipeptidase	L-cysteinyl-glycine hydrolase		47

[a] Represents new listing from 1984 IUB Nomenclature List (see Ref. 48).

Section I

Assays

A. Prostaglandins
Articles 2 through 7

B. Leukotrienes
Articles 8 through 13

C. Platelet-Activating Factor
Articles 14 through 19

D. Other Oxidative Products
Articles 20 through 22

E. Phospholipase A_2 and Phospholipids
Articles 23 through 26

[2] Electron-Capture Negative-Ion Chemical Ionization Mass Spectrometry of Lipid Mediators

By IAN A. BLAIR

Introduction

Lipid mediators are often present in biological fluids in extremely low concentrations; thus, highly sensitive methods are required for their analysis. The structural similarity of many of the mediators, together with the presence in their biological milieu of compounds with similar activities, has stimulated the use of mass spectrometry for such analyses. Min *et al.*[1] recognized that the attachment of an electron-capturing group to a prostaglandin (PG) moiety, coupled with the use of electron-capture negative-ion chemical ionization-mass spectrometry (NCI–MS), increased sensitivity of analysis by almost two orders of magnitude compared with electron-impact ionization. The high sensitivity of NCI–MS makes it the technique of choice for quantitative analyses of PGs and other lipid mediators in the picomolar range when high specificity is required.[2,3] The pentafluorobenzyl (PFB) ester is the most widely used electron-capturing derivative for eicosanoids.[4,5] For glycerolipids the pentafluorobenzoyl (PFBO) ester is normally used.[6] Paradoxically, the PFB derivative that is introduced to facilitate ionization of the eicosanoid is lost during dissociative electron capture. The carboxylate anion that is formed during this process contains the intact eicosanoid molecule and is generally the base peak in the mass spectrum.[7] In contrast to eicosanoid PFB derivatives, molecular anions are the base peaks in the spectra of glycerolipid PFBO derivatives.[6]

Lipid mediators often contain polar hydroxyl groups that cause tailing during analysis by gas chromatography (GC). This problem can be readily overcome by conversion of the hydroxyl groups to trimethylsilyl (TMS)

[1] B. H. Min, J. Pao, W. A. Garland, J. A. F. de Silva, and M. Parsonnet, *J. Chromatogr.* **183,** 411 (1980).

[2] S. E. Barrow, K. A. Waddell, M. Ennis, C. T. Dollery, and I. A. Blair, *J. Chromatogr.* **239,** 71 (1982.)

[3] I. A. Blair, *Br. Med. Bull.* **39,** 223 (1983).

[4] K. A. Waddell, I. A. Blair, and J. Wellby, *Biomed. Mass Spectrum.* **10,** 83 (1983).

[5] R. J. Strife and R. C. Murphy, *J. Chromatogr.* **305,** 3 (1984).

[6] C. S. Ramesha and W. C. Pickett, *Biomed. Environ. Mass Spectrom.* **13,** 107 (1986).

[7] C. R. Pace-Asciak, *Adv. Prostaglandin Thromboxane Leukotriene Res.* **18,** 1 (1989).

ethers. Silyl ethers possess excellent GC–MS characteristics and can be formed in quantitative yield with suitable derivatizing reagents. Unfortunately, if a ketone is also present in the molecule it can form unstable enol-TMS derivatives. This can be prevented by conversion of the ketone to a methoxime (MO) derivative prior to silylation.[8] A mixture of *syn*- and *anti*-MO derivatives is formed. MO derivatives are also useful for stabilizing labile eicosanoids during purification by preventing the intramolecular dehydration reactions that may occur.[9] However, peak broadening during GC–MS analysis may be observed if resolution between the *syn* and *anti* isomers is poor.

A number of precise, accurate, and sensitive methods for the quantitative analysis of lipid mediators have been developed using NCI–MS. An important requirement for these assays has been the availability of appropriate stable isotope analogs.[10] These compounds are efficient internal standards and also act as carriers to prevent losses during isolation, purification, and GC analysis. The availability of an appropriate standard can be crucial for assay development when analyzing picomolar amounts of material. The high sensitivity of NCI–MS increases the risk of interfering substances from the biological matrix eluting at the same GC retention time as the analyte. Such interference can be minimized by the use of suitable purification techniques. Nonspecific interference can be minimized by heating the GC column to temperatures above 300° at the end of each chromatographic run. GC columns with bonded phases have become widely used in NCI–MS assays because of their stability at these high temperatures.

Arachidonic acid is the precursor for the formation of an enormous range of potent biologically active lipid mediators. Analyses of arachidonic acid itself have been carried out in order to quantify incorporation into phospholipids[11] and to investigate the potential role of arachidonic acid as a mediator in its own right.[12] The NCI mass spectrum of arachidonic acid shows a typical [M − PFB] anion at m/z 303 (Fig. 1a). Heavy isotope analogs have been synthesized for use as internal standards either through catalytic reduction of eicosatetraenoic acid with deuterium gas[13] or by

[8] K. Green, *Chem. Phys. Lipids* **4,** 254 (1969).

[9] I. A. Blair, S. E. Barrow, K. A. Waddell, P. J. Lewis, and C. T. Dollery, *Prostaglandins* **23,** 579 (1982).

[10] C. O. Meese, *J. Label. Compound Radiopharmacol.* **23,** 295 (1986).

[11] J. S. Hadley, A. Fradin, and R. C. Murphy, *Biomed. Environ. Mass Spectrum.* **15,** 175 (1988).

[12] B. A. Wolf, J. Turk, W. R. Sherman, and M. L. McDaniel, *J. Biol. Chem.* **261,** 3501 (1986).

[13] D. F. Taber, M. A. Phillips, and W. A. Hubbard, *Prostaglandins* **22,** 349 (1981).

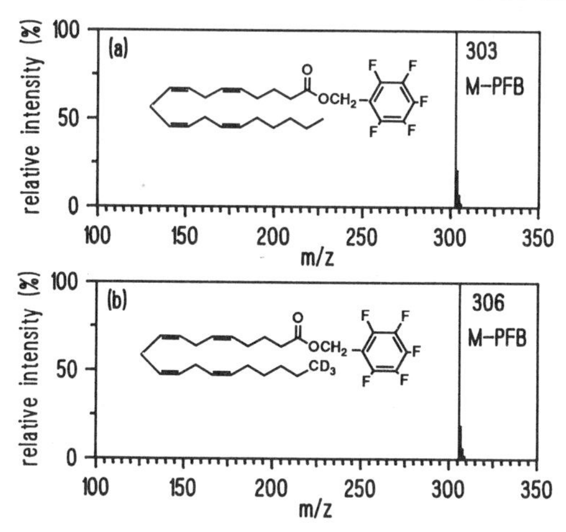

FIG. 1. NCI mass spectra of (a) arachidonic acid PFB ester, (b) C-20-$^2H_3$-labeled arachidonic acid PFB ester.

total synthesis in which deuterium was incorporated specifically at C-20.[14] C-20 trideuterated arachidonic acid showed <0.5% of protiem impurity at *m/z* 303 and the [M − PFB] anion at *m/z* 306 was the base peak in the NCI mass spectrum (Fig. 1b).

Metabolites of arachidonic acid, including leukotrienes (LTs), hydroxyeicosatetraenoic acids (HETEs), prostaglandins (PGs), thromboxanes (Txs), and epoxyeicosatrienoic acids (EETs) have all been quantified by capillary column GC–NCI–MS. LTB_4 derivatized as the bis-TMS ether, PFB ester gave an NCI mass spectrum with the expected [M − PFB] anion at *m/z* 479.[15] Fragment ions that corresponded to sequential losses of TMSOH were observed at *m/z* 389 and *m/z* 299 (Fig. 2). It was noted that these fragment ions eluted slightly after the [M − PFB] anion on GC–MS analysis and that their intensity was dependent on the length of the capillary column that was used.[16] This suggested that they were formed as a consequence of a thermal rearrangement that occurred during chromatography. In support of this idea, it was found that fragmentation could be minimized by the use of relatively short capillary columns. Interestingly, this type of fragmentation also occurred with the bis-*tert*-

[14] C. Prakash, S. Saleh, B. J. Sweetman, D. F. Taber, and I. A. Blair, *J. Label. Compound Radiopharmacol.*, **27,** 539 (1989).

[15] J. MacDermot, C. R. Kelsey, K. A. Waddell, R. Richmond, R. K. Knight, P. J. Cole, C. T. Dollery, D. N. Landon, and I. A. Blair, *Prostaglandins* **27,** 163 (1984).

[16] I. A. Blair, A. R. Brash, J. Daugherty, and G. A. FitzGerald, *Adv. Prostaglandin Thromboxane Leukotriene Res.* **15,** 61 (1985).

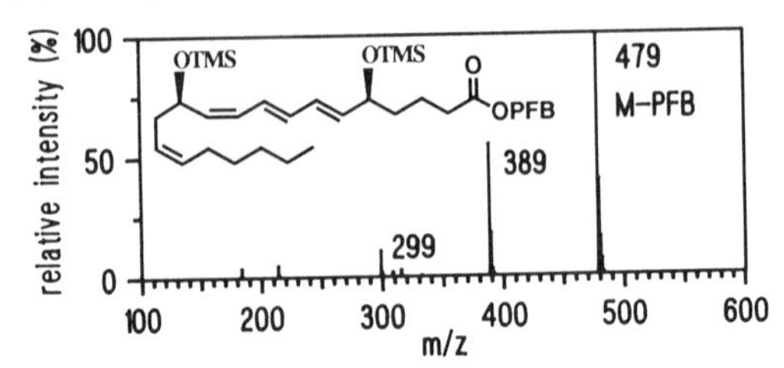

FIG. 2. NCI mass spectrum of the bis-TMS ether, PFB ester derivative of LTB_4.

butyldimethylsilyl ether and was employed in a selected reaction monitoring assay for LTB_4 using tandem mass spectrometry.[17]

There are several reports on the quantitative analysis of HETEs by NCI–MS.[18,19] Spectra were intermediate between those obtained for arachidonic acid and dihydroxy-LTs. For example, 12(*S*)-HETE-TMS-PFB had an [M − PFB] anion at *m/z* 391 and a fragment ion at *m/z* 301 corresponding to M-PFB-TMSOH (Fig. 3a). Unfortunately, the capillary GC properties of HETEs are not ideal as extensive tailing and decomposition are often observed. These problems can be overcome by hydrogenation of the HETEs prior to GC–MS analysis. For quantitative assays, a stable isotope analog internal standard must be used, since side reactions occur to varying amounts with different HETEs. The method of hydrogenation (Rh/Al_2O_3 catalyst) described by Balazy and Murphy[20] provides the highest yield of desired product. The mass spectrum of hydrogenated 12(*S*)-HETE-TMS-PFB showed major ions at *m/z* 399 (M − PFB) and *m/z* 309 (M-PFB-TMSOH) (Fig. 3b). ^{18}O-labeled HETE internal standards have found wide utility in the analysis of these compounds because of their stability during hydrogenation. Deuterium-labeled compounds are subject to exchange during this procedure even with mild catalysts such as Rh/Al_2O_3.[20]

Negative-ion methodology has found particular utility in the analysis of PGs[21] Txs and their metabolites.[22,23] For example, GC–NCI–MS pro-

[17] M. Dawson, C. M. McGee, P. M. Brooks, J. H. Vine, and T. R. Watson, *Biomed. Environ. Mass Spectrom.* **17,** 205 (1988).

[18] R. J. Strife and R. C. Murphy, *Prostaglandin Leukotriene Med.* **13,** 1 (1984).

[19] E. Malle, J. Nimpf, H. H. Leis, M. Wurm, H. Gleispach, and G. M. Kostner, *Prostaglandin Leukotriene Med.* **27,** 53 (1987).

[20] M. Balazy and R. C. Murphy, *Anal. Chem.* **58,** 1098 (1986).

[21] C. R. Pace-Asciak and Z. Domazet, *Biochim. Biophys. Acta* **796,** 129 (1984).

[22] D. F. Wendelborn, K. Seibert, and L. J. Roberts II, *Proc. Natl. Acad. Sci. U.S.A.* **85,** 304 (1988).

[23] D. J. FitzGerald, J. Fragetta, and G. A. FitzGerald, *J. Clin. Invest.* **82,** 1708 (1988).

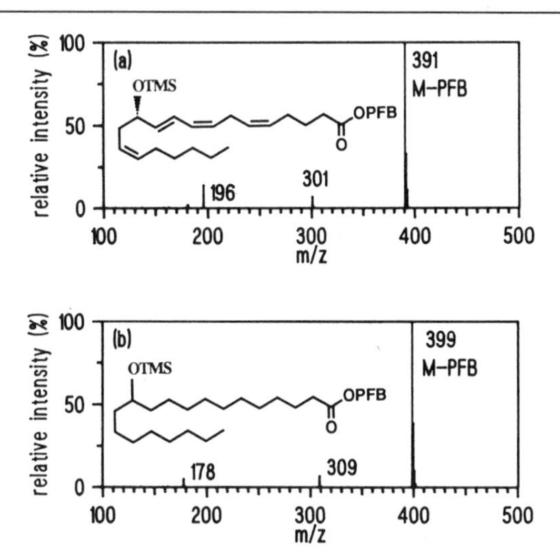

FIG. 3. NCI mass spectrum of (a) TMS ether, PFB ester derivative of 12(*S*)-HETE; (b) TMS ether, PFB ester derivative of H_8-12(*S*)-HETE.

vided the specificity and sensitivity to establish that the concentrations of 6-oxo-$PGF_{1\alpha}$ present in normal human plasma were in the range <0.5–2.4 pg/ml.[9] Prior to this study, literature values for normal concentrations of plasma 6-oxo-$PGF_{1\alpha}$ spanned a range of several orders of magnitude. The NCI mass spectrum of 6-oxo-$PGF_{1\alpha}$-MO-tris-TMS-PFB showed one major ion at *m/z* 614 [M − PFB] (Fig. 4a). This spectrum was almost identical with that obtained for TXB_2-MO-tris-TMS-PFB (Fig. 4b). However, the two compounds did not interfere with each other during analysis since they could be readily separated by capillary GC. GC–NCI–MS methodology was employed for the analysis of the dinor metabolites of 6-oxo-$PGF_{1\alpha}$ and TxB_2[23] and the 11-oxo metabolite of TxB_2.[24] Analysis of biological samples presents a major technical problem because of the extensive purification that is required prior to NCI–MS analysis. Urinary measurements are particularly challenging because of the potential for interfering substances in the urine. Relatively clean samples can be obtained if the eicosanoids are first isolated by reversed-phase chromatography on C_{18} minicolumns and then purified by TLC. Such methodology can be extremely tedious and several approaches have been adopted to speed up sample preparation time. These include prior chromatography

[24] F. Catella, D. Healy, J. A. Lawson, and G. A. FitzGerald, *Proc. Natl. Acad. Sci. U.S.A.* **83,** 5861 (1986).

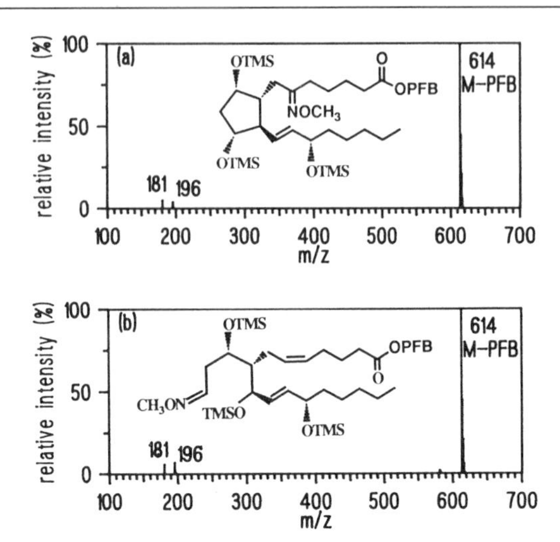

FIG. 4. HCI mass spectrum of (a) MO, Tris-TMS ether, PFB ester derivative of 6-oxo-$PGF_{1\alpha}$; (b) MO, Tris-TMS ether, PFB ester derivative of TxB_2.

on phenyl boronate minicolumns,[25] immunoaffinity chromatography,[26,27] and tandem mass spectrometry.[28]

The excellent resolution that can be obtained with capillary GC has encouraged investigators to develop assays that allow quantification of several mediators in a single chromatography run.[29,30] The method is less sensitive because extensive cleanup of samples cannot be carried out. One example where this technique proved to be particularly useful was in the quantification of the small pool of arachidonate that was released on stimulation of human platelets with epinephrine.[31] Eicosanoids were analyzed after stimulation of the platelets in the presence of a Tx receptor antagonist and a Tx synthase inhibitor. Under these conditions any re-

[25] J. A. Lawson, A. R. Brash, J. Doran, and G. A. FitzGerald, *Anal. Biochem.* **150,** 463 (1985).

[26] H. L. Hubbard, T. D. Eller, D. E. Mais, P. V. Halushka, R. H. Baker, I. A. Blair, J. J. Vrbanac, and D. R. Knapp, *Prostaglandins* **33,** 149 (1987).

[27] J. Nowak, J. J. Murray, J. A. Oates, and G. A. FitzGerald, *Circulation* **76,** 6 (1987).

[28] H. Schweer, C. O. Meese, O. Furst, P. G. Kuhl, and H. W. Seyberth, *Anal. Biochem.* **164,** 156 (1987).

[29] K. A. Waddell, S. E. Barrow, C. Robinson, M. A. Orchard, C. T. Dollery, and I. A. Blair, *Biomed. Mass Spectrom.* **11,** 68 (1984).

[30] C. R. Pace-Asciak and S. Micallef, *J. Chromatogr.* **310,** 233 (1984).

[31] J. D. Sweatt, I. A. Blair, E. J. Cragoe, and L. E. Limbird, *J. Biol. Chem.* **261,** 8660–8666 (1986).

leased arachidonate was converted to PGD_2, PGE_2, and $PGF_{2\alpha}$. Quantification of the PGs by capillary column GC–NCI–MS allowed the size of the initial pool of released arachidonate to be quantified in the presence of a relatively high background of TxB_2.

Interest in the analysis of trace quantities of the four regioisomeric EETs (cytochrome *P*-450 metabolites of arachidonic acid) stimulated the development of NCI–MS methods for these compounds.[32,33] The NCI characteristics of the EET-PFB esters were a little different from the other eicosanoid PFB derivatives. 14,15-EET-PFB gave an intense [M − PFB] anion at *m/z* 319, together with fragment ions at *m/z* 303 (M-PFB-O) and *m/z* 301 (M-PFB-H_2O) (Fig. 5a). Identical mass spectra were obtained for the other three EET-PFB regioisomers. The four regioisomers could not be separated by capillary column GC. However, they could be separated as PFB esters on an HP17 fused silica column after hydrogenation with Rh/Al_2O_3.[34] In addition, characteristic fragmentation reactions at the epoxide moiety were observed.[32] This provided further confirmatory evidence for the structures of the individual regioisomers. For example, H_6-14,15-EET-PFB gave fragment ions at *m/z* 225 and *m/z* 241, in addition to the [M − PFB] anion at *m/z* 325 (Fig. 5b).

Quantitative methods for the analysis of platelet-activating factor (PAF) in biological fluids using GC–NCI–MS have been reported.[6] The inherent polarity of the molecule prevented direct analysis, so the PAF was first treated with phospholipase C and the resulting diglyceride analyzed as the PFBO derivative. It was found that purification of the diglyceride on TLC resulted in partial rearrangement of the acetyl group from C-2 to C-3. The rearrangement went almost to completion if the derivative was allowed to stand overnight on the TLC plate.[35] The NCI mass spectrum of the rearranged PFBO derivative showed an intense ion at *m/z* 552 (Fig. 6), with only minor fragment ions present in the mass spectrum. Analysis of lyso-PAF by NCI–MS using the novel *p*-nitrophenylacetal electron-capturing group has also been reported.[36]

In summary, NCI–MS is the method of choice for trace analysis of PGs and other lipid mediators. In combination with stable isotope dilution capillary GC, highly specific and sensitive analyses can be carried out.

[32] R. Toto, A. Siddhanta, S. Manna, B. Pramanik, J. R. Falck, and J. H. Capdevila, *Biochim. Biophys. Acta* **919,** 132 (1987).

[33] J. H. Capdevila, V. Kishore, E. Dishman, I. A. Blair, and J. R. Falck, *Biochem. Biophys. Res. Commun.* **146,** 638 (1987).

[34] K. Yamashita, J. H. Capdevila, and I. A. Blair, unpublished observations.

[35] B. W. Christman and I. A. Blair, *Biomed. Environ. Mass Spectrom.* **18,** 258 (1989).

[36] A. I. Mallet, *Biomed. Environ. Mass Spectrom.* **16,** 207 (1988).

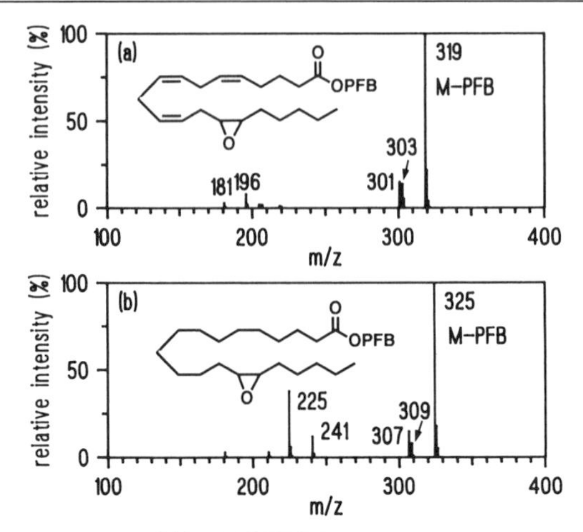

FIG. 5. NCI mass spectra of (a) 14,15-EET PFB ester; (b) H_6-14,15-EET PFB ester.

This methodology has facilitated a number of studies on the role that lipid mediators play in both normal and pathophysiological processes.

Method for Quantitative Analysis of 6-Oxo-PGF$_{1\alpha}$ by Capillary GC–NCI–MS

General Procedure

HPLC-grade solvents (B & J, American Scientific, McGaw Park, IL) were used throughout the study. Pyridine (HPLC grade) was obtained from Aldrich Chemical Co. (Milwaukee, WI). 6-Oxo-PGF$_{1\alpha}$ (Sigma, St. Louis, MO) and 6-oxo[$^2H_4$] PGF$_{1\alpha}$ (Merck Sharpe and Dohme, Quebec, Canada) were dissolved in ethanol and stored as 1 mg/ml stock solutions at $-70°$. Dilutions for standards were made from these solutions. 6-Oxo

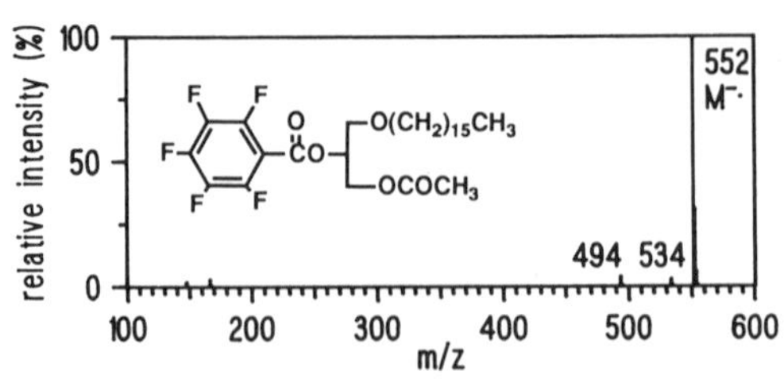

FIG. 6. NCI mass spectrum of 1-*O*-hexadecyl-3-acetylglycerol PFBO derivative from C_{16}-PAF.

[^{3}H]PGF$_{1\alpha}$ (141 Ci/mmol) was obtained from Amersham Corporation (Arlington Heights, IL). Methoxyamine hydrochloride (Aldrich) was recrystallized prior to use from ethanol containing 1% HCI. Bistrimethylsilyltrifluoroacetamide (Pierce Chemical Co., Rockford, IL), pentafluorobenzyl bromide, and diisopropylethylamine (Aldrich) were used without further purification. Evaporations were carried out under dry nitrogen using an N-evap apparatus (Organomation, South Berlin, MA). Reversed-phased octadecyl PrepSep cartridges were obtained from Thomas Scientific (Swedsburgh, NJ). TLC was carried out on Whatman (Clifton, NJ) LK6D channel plates with preconcentration zones. Radiochromatograms were obtained on a Bioscan (Washington, DC) System 200 Image Scanner.

Isolation and Derivatization

To the biological fluid (up to 20 ml) is added 6-oxo[^{2}H$_{4}$]PGF$_{1\alpha}$ (10 ng). The standards are prepared by adding 6-oxo[^{2}H$_{3}$]PGF$_{1\alpha}$ (10 ng) to each of six tubes containing water (the same volume as the biological fluid undergoing analysis). Increasing amounts of authentic 6-oxo-PGF$_{1\alpha}$ (0, 50, 200, 500 and 1000 pg) are then added to five of the individual tubes. 6-oxo-[^{3}H]PGF$_{1\alpha}$ (100,000 cpm) is added to the sixth tube to act as a marker during TLC purification. Samples and standards are followed for 20 min in order to allow equilibration to take place.

In a typical analysis, the pH of the sample is lowered to 4.5 with 0.5% formic acid. The sample is loaded on a PrepSep cartridge that has been preconditioned by washing with 10 ml of methanol followed by 10 ml of water. The PrepSep cartridge is washed with water (5 ml), methanol/water (5 ml; 2 : 3, v/v), and dried under vacuum from a water pump for 5 min. Eicosanoids are eluted with ethyl acetate (1.7 ml) and the ethyl acetate evaporated. A solution of methoxyamine hydrochloride in pyridine (50 μl; 1 : 99, w/v) is added to the residue and allowed to stand overnight at room temperature. The pyridine is evaporated and residual pyridine removed by azeotroping with a further 0.5 ml of hexane. The residue is dissolved in ethyl acetate (30 μl) and applied to the preconcentration zone of a silica TLC plate. Residual material is dissolved in methanol (20 μl) and also applied to the plate. The plate is developed with chloroform/methanol/water/acetic acid (45.0:5.0:0.2:0.25, v/v/v/v) and the band containing 6-oxo-PGF$_{1\alpha}$-MO localized by radiochromatogram scanning (R_f 0.33; the R_f values for PGE$_2$-MO and TxB$_2$-MO derivatives are 0.46 and 0.43, respectively). The band is scraped, suspended in 0.5% formic acid (0.5 ml), and extracted with ethyl acetate (1 ml followed by 0.5 ml). The combined organic extracts are evaporated to dryness. Methanol (1 ml) is added and evaporated to dryness again. To the residue is added acetonitrile (20 μl) followed by 20 μl of pentafluorobenzyl bromide in acetonitrile (1:9, v/v) and 20 μl of diisopropylethylamine in acetonitrile (1:19, v/v). The solution is stored under nitrogen for 30 min at room temperature and evaporated. The residue is suspended in water (0.5 ml) and extracted with 1 ml followed by 0.5 ml of ethyl acetate/hexane (1:4, v/v). The combined organic extracts are evaporated, redissolved in ethyl acetate (30 μl), and applied to the preconcentration zone of a silica TLC plate. Residual material is dissolved in methanol (20 μl) and also applied to the plate. The plate is developed with chloroform/methanol/water (46.0:4.0:0.2), v/v/v) and the band containing 6-oxo-$PGF_{1\alpha}$-MO-PFB localized by radiochromatogram scanning (R_f 0.28). The band is scraped, suspended in water (0.5 ml), and extracted with ethyl acetate (1 ml followed by 0.5 ml). The combined organic extracts are evaporated to dryness. Methanol (1 ml) is added and again evaporated to dryness. Recovery of 6-oxo[^{3}H]$PGF_{1\alpha}$ is in the range 60–65%. The residue is dissolved in pyridine (10 μl), bis-trimethylsilyltrifluoroacetamide (10 μl) is added, and derivatization to the Tris-TMS ether allowed to proceed for 2 hr at room temperature. Samples are stored at − 20° until the day of analysis. The silylating reagent is then carefully evaporated under nitrogen until there is a residue of 1 to 2 μl. The residue is dissolved in decane (10 μl) ready for GC–MS analysis.

Gas Chromatography–Mass Spectrometry

GC–MS is carried out on a Nermag R1010C quadrupole mass spectrometer (Delsi Instruments, TX) interfaced with a Varian Vista gas chromatograph (Varian, Palo Alto, CA). Data reduction and analysis are carried out using Nermag SIDAR software. A 15 m SPB-1 fused silica column (0.25 mm id; 0.25 μm coating thickness, Supelco, Bellefonte, PA) is passed through the transfer line from the gas chromatograph until the end is just inside the ion source. The glass inserts for the gas chromatograph injector are cleaned with chromic acid, silanized with dimethyldichlorosilane in toluene (1 : 9 v/v) for 30 min, deactivated in methanol for 10 min, washed with methanol, and dried at 100°. A clean glass liner is inserted in the injector immediately prior to analysis and the temperature maintained at 250°. Helium (Alphagaz, San Francisco, CA; high-purity grade), dried and oxygen-free (gas purifier, Supelco), is used as the carrier gas at a flow rate of 1 ml/min (head pressure 10 psi). The source and transfer lines of the mass spectrometer are maintained at 250° and 260°, respectively. The conversion dynode is set at +5 kV and the multiplier at 2.4 kV.

Methane (Alphagaz; research-grade) is introduced into the source of the mass spectrometer to a final pressure of approximately 6.0×10^{-6} torr. A small amount of calibration gas (FC 43) is introduced into the calibration gas line through the automatic solenoid valve. It is then allowed to pass into the mass spectrometer by carefully opening an in-line Nupro SS-4BRW (Nupro, Willoughby, OH) needle valve. Great care is taken to avoid contamination of the mass spectrometer source with excess FC 43. The current on the rhenium filament is set at 0.25 mA and tuning carried out on the ion at m/z 633. The electron energy is optimized for maximum sensitivity (60–90 eV, depending on the cleanliness of the source) and the source controls are then tuned for maximum sensitivity. The lenses in the quadrupole assembly are tuned for optimal peak shape and peak heights across the mass range (100–633 D). The resolution is adjusted to give a 20% valley between the following pairs of ions: m/z 633/634, m/z 595/596, m/z 562/563 m/z 524/525, m/z 452/453, m/z 414/415, and m/z 281/282. The methane pressure is adjusted for optimal sensitivity. Tuning of the source and lenses is then repeated in order to ensure that all parameters are optimal for sensitivity and resolution. The instrument is then calibrated. Finally, the calibration gas needle valve is closed until the peak for m/z 633 has almost disappeared and the tuning procedure carried out once again.

Injections are made in the splitless mode. Flow of helium through the sweep and split valve is stopped for 0.2 min prior to injection. It is opened 0.8 min later and the oven temperature program initiated from 150 to 310°at 15°/min. The column is held at 310° for 5 min at the end of the chromatographic run to remove potential interfering substances. The retention time for 6-oxo-$PGF_{1\alpha}$-MO-Tris-TMS-PFB is 7.5 min under these conditions. Selected-ion monitoring is carried out on m/z 614 (6-oxo-$PGF_{1\alpha}$) and m/z 618 (6-oxo-[$^2H_4$]-$PGF_{1\alpha}$). A calibration curve is prepared by analyzing the extracted standards and plotting the relative peak heights (or peak areas) of m/z 614 to m/z 618 against concentration. The biological samples are then analyzed. Great care is taken to avoid cross-contamination of samples with the injection syringe. "Memory effects" can be minimized by flushing the syringe first with methanol and then with ethyl acetate. The relative peak heights (or peak areas) of m/z 614 to m/z 618 are determined for the biological samples. Concentrations of 6-oxo-PGF 1_{α} are then calculated by interpolation from the standard curve.

Acknowledgment

This work was supported by NIH grants DK 38226, ES 00267, GM 15431, and HL 19153.

[3] Enzyme Immunoassays of Eicosanoids Using Acetylcholinesterase

By PHILIPPE PRADELLES, JACQUES GRASSI, and JACQUES MACLOUF

The quality of measurement in immunoassays depends primarily on the distinct qualities of two molecules: a specific and high-affinity antibody and the corresponding labeled molecule with high specific activity. In recent years, nonisotopic labeling techniques[1] have been developed to overcome certain problems associated with the use of radioactive tracers. Among these methods, enzyme immunoassays (EIA) appear to represent an important alternative to radioimmunoassay. This approach has been widely used for all compounds that can be measured by radioimmunoassays; more recently such assays have emerged for eicosanoids.[2] We have developed such systems for these compounds and will describe our procedure using acetylcholinesterase (AChE, EC 3.1.1.7) from *Electrophorus electricus* as a label with high specific activity.[3] The same approach has been used in developing these reagents for leukotrienes and the strategy that we have used for obtaining polyclonal antisera as well as for the preparation of corresponding tracers is described later in this volume (see [10]).

Reagents

Polyclonal Antisera. All antisera were developed using eicosanoids coupled by their carboxylic group to bovine serum albumin as an antigenic carrier. These low molecular weight substances (around 350) need this covalent attachment to a macromolecule to elicit antibody production. Since this chapter is concerned with the preparation and use of a nonisotopic label, the reader is referred to other procedures for antibody production.[4]

Preparation of Enzymatic Tracers. AChE is present in both the central and peripheral nervous systems of vertebrates and is found at even higher concentrations in the electric organ of electric fishes such as *Electrophorus*

[1] J. Grassi, J. Maclouf, and P. Pradelles, *in* "Radioimmunoassay in Basic and Clinical Pharmacology" (C. Patrono and B. A. Peskar, eds.), p. 91. Springer-Verlag, New York, 1987.

[2] Y. Hayashi, T. Yano, and S. Yamamoto, *Biochim. Biophys. Acta* **663,** 661 (1981).

[3] P. Pradelles, J. Grassi, and J. Maclouf, *Anal. Chem.* **57,** 1170 (1985).

[4] J. Maclouf, this series, Vol. 86, p. 273.

electricus. This enzyme exists naturally in multiple molecular forms categorized as "globular" and "asymmetric." Globular forms (G1, G2, G4) consist of one (molecular weight 80,000), two, or four catalytic subunits, while asymmetric forms (A4, A8, A12) contain one, two, or three tetramers (G4) attached to a rodlike tail segment (molecular weight approximately 100,000) partly collagenous in nature (for a complete review and purification, see Massoulié and Bon.[5,6] The G4 form of AChE obtained following trypsinization of the naturally abundant asymmetric forms A8 and A12 is covalently attached to the eicosanoid.[7] Briefly, the carboxylic function of a prostanoid or thromboxane is transformed into an activated ester using stoichiometric amounts of *N*-hydroxysuccinimide and *N,N'*-dicyclohexylcarbodiimide, in anhydrous dimethylformamide (3–6 μmol/ml).[3] Over 90% of the eicosanoid can be converted into the corresponding active ester. The success of the preparation of the esters can be verified by thin-layer chromatographic analysis on silica gel (solvent, ethyl acetate/hexane, 8 : 2, v/v) using ^{3}H-labeled eicosanoid as a tracer. When stored anhydrous at −20°, the active esters are stable for 4 to 5 months. The ester (100–200 nmol) is subsequently added to the globular G4 forms of acetylcholinesterase (0.4–0.6 nmol) in 500 μl of 0.1 *M* borate buffer, pH 9 (except for prostaglandin E_2 which is treated in a pH 7 buffer to avoid its dehydration into PGA_2 or PGB_2). After 1 hr at 22°, 1 ml of EIA buffer (see following section) is added to stop the reaction and to avoid adsorption of the enzyme during subsequent handling. Purification of the conjugates is performed with a BioGel A 15-m column (90 × 1.5 cm) eluted with 10^{-2} *M* Tris buffer, pH 7.4, containing 1 *M* NaCl, 10^{-2} *M* $MgCl_2$, and 0.01% sodium azide.[3]

Materials and Reagents for Enzyme Immunoassay

Materials

Microtitration plates with 96 flat-bottomed wells (Microwell Immuno NUNC 96 F with certificate Cat. No 4 394 54, Nunc, Roskilde, Denmark)
Microtiter plate plastic film (Flow Lab, Cat No 77 400 05, Helsinki, Finland)
Microplate reader able to read at 412–414 nm
Microplate shaker
Polypropylene, polystyrene or glass test tubes, 5 ml

[5] J. Massoulié and S. Bon, *Annu. Rev. Neuro Sci.* **5,** 57 (1982).
[6] J. Massoulié and S. Bon, *Eur. J. Biochem.* **68,** 531 (1976).
[7] L. Mc Laughlin, Y. Wei, P. T. Stockmann, K. M. Leahy, P. Needleman, J. Grassi, and P. Pradelles, *Biochem. Biophys. Res. Commun.* **144,** 469 (1987).

Reagents

Affinity-purified anti-rabbit IgG (H + L) polyclonal (e.g., Jackson Immuno Research, Baltimore, MD; Zymed, San Francisco, CA) or mouse monoclonal (Cayman Chemicals, Ann Arbor, MI).

Enzyme tracers and corresponding specific antisera. The tracers are commercially available (Cayman Chemicals) and the antisera can be purchased from the same or from others sources.

Stock potassium phosphate buffer 1 *M*, pH 7.4. This buffer can be kept for several weeks at 4°.

EIA buffer (0.1 *M* phosphate), pH 7.4, containing 0.1 g of sodium azide, 23.4 g sodium chloride, 1 g tetrasodium EDTA, and 1 g bovine serum albumin in 1 liter. After addition of all salts, the pH should be readjusted to 7.4. This buffer is fairly stable for more than 4–8 weeks.

Washing buffer (0.01 *M* phosphate), pH 7.4, containing Tween 20 (0.05%, v/v).

Enzyme substrate (Ellman's reagent). Concentrated Ellman's reagent (enough for five plates) is prepared by dissolving in 2 ml potassium phosphate (0.5 *M*) buffer, pH 7.4, 0.175 g NaCl, 21.5 mg 5,5′-dithiobis(2-nitrobenzoic acid) (Sigma, St. Louis, MO); when dissolved, 20 mg acetylthiocholine iodide (Sigma) are added. Immediately after this last addition, the mixture is rapidly frozen to avoid the hydrolysis of acetylthiocholine which may occur in the concentrated buffer. This concentrated reagent can be lyophilized and stored for several months at 4°, protected from light. Just before use, the contents of a vial is dissolved in 100 ml distilled water and eventually kept up to 1 week at 4° in the dark.

Standard solutions of precisely known concentrations of eicosanoids (e.g., Cayman Chemicals, Biomol Research Laboratories, Philadelphia, PA).

Preparation of Plates

Coating with Mouse Monoclonal Anti-Rabbit IgG Antibodies. Add 20 ml potassium phosphate buffer 5×10^{-2} *M* to 200 μg of monoclonal IgG, and distribute 200 μl of this solution into each well using the automatic dispenser. (*Note:* Considerable time can be saved by using an automatic dispensing system and a microplate washer. If not available, these can be substituted by an 8-tip microwell pipette.) Cover with plastic film. Allow this to incubate at least 18 hr at room temperature prior to the saturation step.

Saturation of Plates. Add 100 μl of EIA buffer containing 3g/liter of bovine serum albumin and 0.3 g/liter sodium azide. Cover with microtiter plate, plastic film, and leave for at least 18 hr at 4° before use. [*Note:* For some affinity-purified polyclonal anti-rabbit IgG antibodies, the use of glutaraldehyde (2%) added to the coating buffer solution usually increases the yield of the coating reaction.[3]]

It is critical that plates are saturated with BSA prior to use in order to minimize nonspecific adsorption. In this condition (i.e., tightly sealed to avoid erratic evaporation from well to well and containing EIA buffer), the plates may be stored for several weeks (4–6 months) at 4°.

We recommend that the following volumes be employed: 200 μl for coating, 300 μl for saturation, 150 μl for competitive binding, and 200 μl for the enzyme reaction.

Competitive Binding Step

Determination of Antiserum Titer/Assay Dilution

Doubling dilutions of antiserum (starting from 1/100, 1/200, 1/400, ... , 1/204,800) are prepared in EIA buffer. The protocol used on the plate is essentially similar to the one described below except that all wells are treated as if they were B_0 (see below for definition of these terms). Figure 1 typifies titration curves. These curves differ from classical antibody dilution curves in two ways: first, there is a reduction of bound enzyme activity at high concentrations of antisera; second, the solid-phase bound-enzyme activity (B_0) never exceeds a few percentage of the total enzyme activity (T) introduced [it may be as low as 3% of total enzyme activity (see Table I) without influencing the quality of the assay]. For a discussion of the likely explanation for this phenomenon, see Ref. 8. The relative value of B_0/T is unimportant in this procedure and the selection of the antiserum assay dilution is the one that gives between 0.2 and 1 absorbance unit (A) for given conditions of immunoreaction (time and temperature) and for a predetermined period of enzymatic reaction (usually between 30 min and 2 hr).

Preparation of Standards

Dissolve a precisely weighed eicosanoid standard of known purity in acetonitrile/water (50 : 50, v/v) for 6-keto-PGF$_{1\alpha}$ or in methanol/water (50 : 50, v/v) for other eicosanoids at a concentration of 100 μg/ml. Store this stock solution at −70°. From this stock solution, prepare further dilutions in EIA buffer down to the "working highest concentration"

[8] P. Pradelles, J. Grassi, D. Chabardes, and N. Guiso, *Anal. Chem.* **61,** 447 (1989).

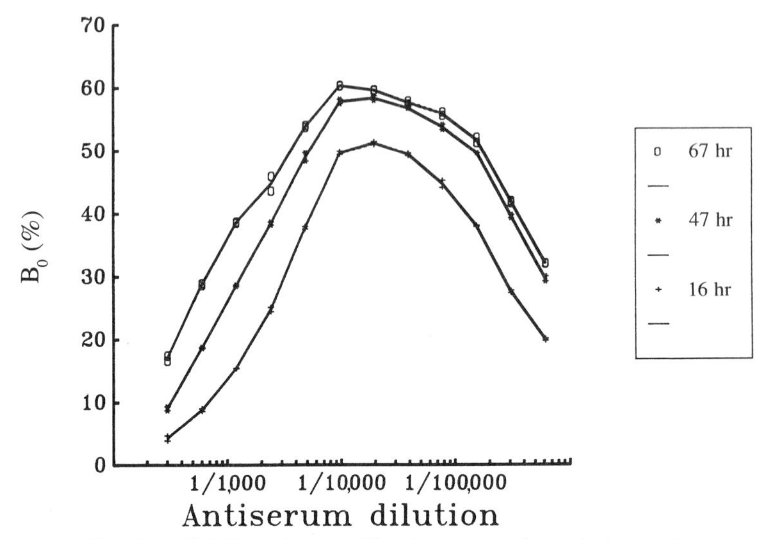

FIG. 1. Titration of TxB_2 antiserum. Titration was performed at room temperature as described in the text. The time for the immunological reaction are indicated in the box. Results are expressed in B_0 (percentage of bound tracer in the absence of competitor) as a function of the final dilution of antiserum.

(usually 0.5–2 ng/ml depending on the eicosanoid and on the antiserum). See considerations on stability in Ref. 9.

Prepare serial dilutions from 1 ml of the standard solution according the following scheme. Place 500 μl of EIA buffer in tubes 2–8. Transfer 500 μl of standard solution from tube 1 to tube 2 and vortex. Continue the same procedure from tube 2 to 3, then 3 to 4, etc.

In order to improve the reproducibility of the assay, we recommend preparation of several tubes of the "highest working concentration" in EIA buffer and storage at $-70°$ for 4–6 months. For each assay, the standard solution will be thawed and the serial dilutions performed from this tube as described above.

Enzyme Immunoassay Procedure

Definitions of abbreviations are as follows: BK, blank of instrument (empty wells) with Ellman's reagent at the last step; B, solid-phase bound enzymatic activity after competition binding; B_0, solid-phase bound enzymatic activity (no competition with eicosanoid); T, total activity of enzy-

[9] R. G. Stehle, this series, 86, p. 436.

TABLE I
BINDING PARAMETERS OF ENZYME IMMUNOASSAYS OF EICOSANOIDS USING ACETYLCHOLINESTERASE AS LABEL

Compound	IC_{50} [b]	Antiserum[a]	
		B_0(%)	Final dilution
TxB_2	4	40	1/60,000
6-Keto-$PGF_{1\alpha}$	1	3	1/300,000
PGE_2	1	5	1/30,000
PGD_2-MO	2	25	1/600,000
$PGF_{2\alpha}$	2	11	1/300,000
Bicyclo-PGE_2	13	6	1/60,000
11-dehydro-TxB_2	2	3	1/1,500,000
dinor-TxB_2	2.5	7	1/60,000

[a] Final working dilution to obtain 40% of maximal binding.
[b] Picogram of eicosanoids corresponding to 50% of initial tracer binding.

matic tracer added (a fraction of the tracer is added to designated wells after all washing steps); NSB, nonspecific binding, i.e., binding of tracer to wells without specific antibodies present; and S1–8, standard solutions.

Incubating the Standards and the Samples

Because this method can be automated, we suggest a constant arrangement of the samples on the 96-well plate (Fig. 2) as shown in the tabulation below. With the exception of the B_0, all points are run in duplicate.

Row/plate number	Preparation
1A to 1H	Used for the blank of the Ellman's reagent (see below)
2A and 2B	Usually kept to have an evaluation of the total enzymatic activity; 5 μl of tracer are added *after* washing the plate and before addition of the Ellman's reagent
2C to 2D	(NSB, nonspecific binding) 100 μl of EIA buffer
2E to 2H	(B_0) 50 μl of EIA buffer
4G, 4H	50 μl of the least concentrated standard (S 8)
4E, 4F	50 μl of the next concentration (S7)
3A to 4D	50 μl of the other 6 points of the standard curve (S1–6)
5A to 12H	50 μl of each sample (*) to be measured at the appropriate dilution

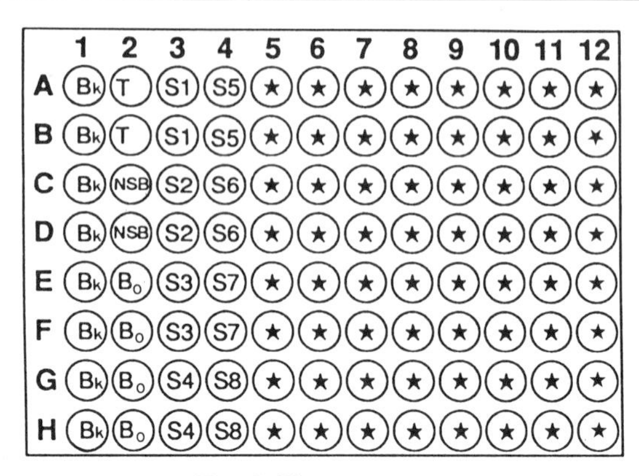

FIG. 2. Plate setup.

Add 50 μl of the enzyme tracer to each well from column 2 to 12 with the exception of row 2A,B. *Then,* add 50 μl of the diluted specific anti-eicosanoid antiserum from row 2E-H and in all subsequent rows.

For the addition of tracer and antiserum, we recommend the use of a repeating dispenser (Multipet, Eppendorf, Hamburg, FRG) that allows good precision in the pipetting.

Measurement of the Solid-Phase Bound AChE (B)

After completion of the competitive binding step, the plate is washed as described above in "preparation of the plates"; 5 μl of enzymatic tracer are added in wells 2A,2B (for total activity assessment) and 200 μl of the Ellman's reagent are added to all wells. The plate is agitated gently using a 96-well plate shaker. At this stage, the enzymatic reaction should be carried out in the absence of uv light, including fluorescent tubes or direct sunlight. In order to achieve sufficient accuracy with the optical density reading, the absorption of the B_0 should have reached at least 0.2 A at 412–414 nm (Fig. 3). This step usually takes approximately from 30 min to 2 hr depending on conditions. The blank of the reader (BK) is set to the absorbance of the wells in the first row (1A to 1H). Care should be taken to keep the optical surface (the bottom) of the microtiter plates clean. If the absorbance is too high, the plate can be washed and the enzymatic reaction restarted.

Calculation of the Results. This is done following the classical protocol of competitive binding studies using a linear log–logit transformation.[10] The mean of the nonspecific binding (NSB) is subtracted from all sample

[10] D. W. Rodbard, W. Bridson, and P. Rayford, *J. Lab. Clin. Med.* **74,** 770 (1969).

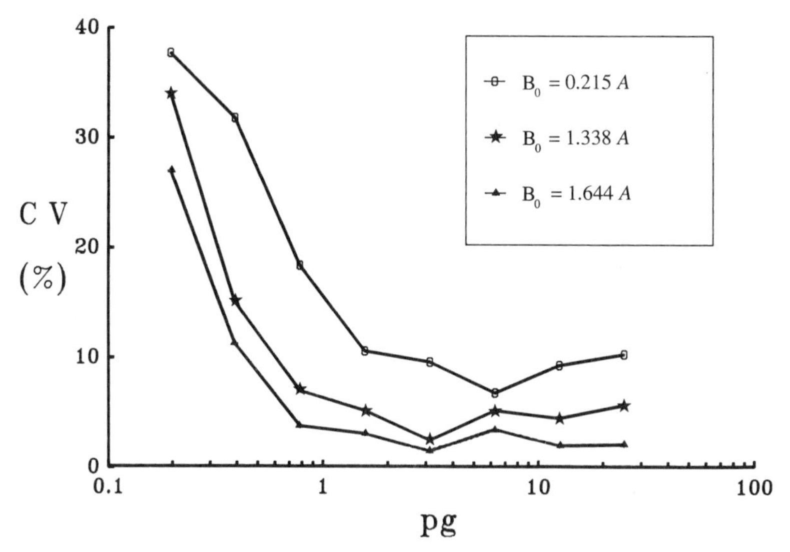

FIG. 3. Precision profile of the TxB_2 assay at different stages of enzymatic reaction. The values of absorbance for B_0 of colorimetric reaction are indicated in the box.

values, but should not exceed more than 0.1% of the total enzymatic activity. The mean of the enzymatic activity (A) of each replicate is calculated and the percentage of the bound fraction is calculated from the following equation:

$$B/B_0\ (\%) = \frac{(A \text{ of sample}) - (A \text{ NSB}) \times 100}{(AB_0) - (A \text{ NSB})}$$

Fitting of the curves is done following Ref. 10.

Assay Assessment

Titer and Sensitivity of Different Antisera Using Enzyme Tracer

As we can see from Table I, the use of acetylcholinesterase–eicosanoid as a label for different eicosanoids permits the use of a high dilution of all antisera that have been tested. In addition, the sensitivity obtained with the various eicosanoids is in the low picogram range. In all cases when comparisons are made, sensitivity obtained using the AChE–eicosanoid tracer is equal to or exceeds that obtained using the high specific radioactivity ^{125}I-labeled eicosanoids.[11] In addition, in most cases the sensitivity

[11] J. Maclouf, M. Pradel, P. Pradelles, and F. Dray, *Biochem. Biophys. Acta* **431,** 139 (1976).

TABLE II
EFFECT OF ANTISERUM CONCENTRATION ON SENSITIVITY OF TxB_2 ASSAY[a]

Antiserum dilution	B_0(%)	B/B_0, 50% (pg)
1/60,000	40.6	7.8
1/120,000	26.5	5.4
1/240,000	19.6	3.6

[a] Immunological reaction was performed for 16 hr at room temperature.

of these assays follow the fundamental principles of competitive immunoassay which may be affected either by diluting the specific antibody (Table II) or by changes in the temperature of the immunological reaction (Fig. 4).

Inter- and Intraassay Reproducibility

Table III shows the reproducibility of measurement of different eicosanoids at low, medium, and high concentration and assayed on different days. As can be seen, the coefficient of variation (*CV*) was less than 10% for most systems at most concentrations.

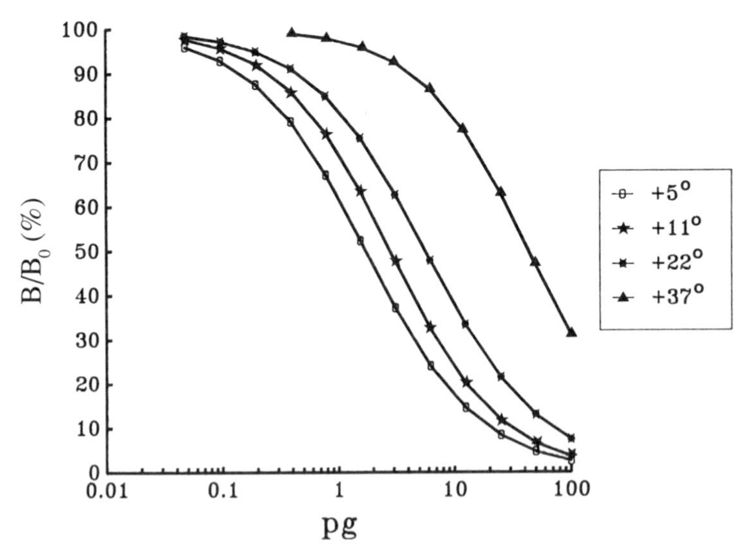

FIG. 4. Effect of the temperature on the sensitivity of the TxB_2 assay. The antiserum was used at 1/150,000 and the immunological reaction performed for 18 hr.

TABLE III
INTERASSAY VARIATION OF ENZYME IMMUNOASSAYS AT LOW, MEDIUM, AND HIGH CONCENTRATION EICOSANOIDS

Eicosanoid	Concentration (pg)	Coefficient of variation (%)
TxB_2 (n = 63)	12.5 ± 0.5	8
	5 ± 0.25	5
	1.2 ± 0.1	4
6-Keto-$PGF_{1\alpha}$ (n = 8)	4.8 ± 0.45	13
	1.1 ± 0.1	9
	0.45 ± 0.05	9
Bicyclo-PGE_2 (n = 8)	51 ± 2.5	5
	13 ± 0.75	6
	2.6 ± 0.15	5

Specificity of Assay: Validation with Gas Chromatography–Mass Spectrometry

The optimal way of establishing the specificity of an immunoassay should be performed by an independent assay method in the relevant biological fluid. Comparison with gas chromatography–mass spectrometry represents an excellent validation criterion. The results obtained after measuring the same rat lung extracts by these two methods is shown

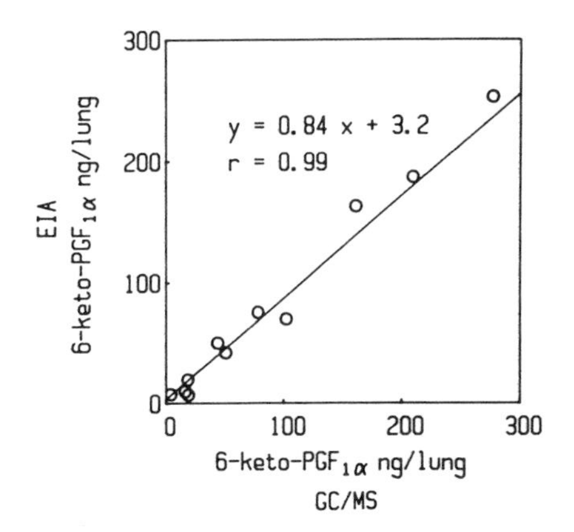

FIG. 5. Correlation between EIA and GC-MS for the analysis of 6-keto-$PGF_{1\alpha}$ from rat lung samples (from Ref. 12 with permission).

in Figure 5. As can be observed, there is a very good correlation between the measurements obtained by enzyme immunoassay and GC-MS.[12]

Concluding Remarks

The protocol presented here is a reliable nonisotopic alternative to radioimmunoassays. In situations where we have made a direct comparison with radioimmunoassays using ^{125}I-labeled eicosanoids,[3] enzymatic tracers have provided sensitivity equal to or better than with radioactivity. With growing concerns of the use of radioactivity in the laboratory and its inherent costs, an enzyme label is an interesting option. Furthermore, the techniques described are technically straightforward and have a potential for automation, representing a major advantage over radioimmunoassay. Although the acetylcholinesterase tracers provide very high sensitivity, all guidelines of validation for immunoassays in a particular biological fluid should be strictly applied before one can rely on the quantitative information provided by this technique.[13]

Acknowledgments

These studies were made possible by financial support from Commissariat à l'Energie Atomique and grants from CNAMTS/INSERM and CNRS.

[12] J. Y. Wescott, S. Chang, M. Balazy, D. O. Stene, P. Pradelles, J. Maclouf, N. F. Voelkel, and R. C. Murphy, *Prostaglandins* **32,** 857 (1986).

[13] C. Patrono, *in* "Radioimmunoassay in Basic and Clinical Pharmacology" (C. Patrono and B. A. Peskar, eds.), p. 213. Springer-Verlag, New York, 1987.

[4] Radioimmunoassay of 11-Dehydrothromboxane B_2

By GIOVANNI CIABATTONI, PAOLA PATRIGNANI, and CARLO PATRONO

Introduction

11-Dehydrothromboxane (TxB_2) is a major enzymatic derivative of TxB_2 identified in the circulating plasma and/or urine of several animal species, including the rat, rabbit, nonhuman primates, and humans. The rationale for measurement of 11-dehydro-TxB_2 as a reflection of changes in TxA_2 production in humans, as well as the experimental evidence supporting this contention, are reviewed elsewhere (Patrono *et al.*[1]). In this chap-

[1] C. Patrono, G. Ciabattoni, and P. Patrignani, *in* "Platelets and Vascular Occlusion" (G. A. FitzGerald and C. Patrono, eds.), p. 193. Raven Press, New York, 1989.

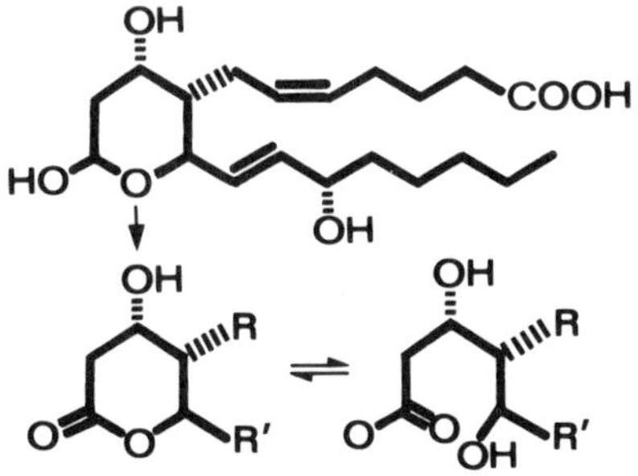

FIG. 1. Dehydrogenation of the hemiacetal alcohol group of TxB_2 (top structure) results in the formation of 11-dehydro-TxB_2. The equilibrium between the closed, δ-lactone form (bottom, left) and the open, acyclic form (bottom, right) is pH-dependent.

ter, we shall focus on the methodological aspects of 11-dehydro-TxB_2 measurements by radioimmunoassay (RIA) techniques, as developed and currently used in our laboratory (Ciabattoni *et al.*[2]; Ciabattoni *et al.*[3]; Patrignani *et al.*[4]).

Chemical Properties of 11-Dehydro-TxB_2

9α,15(*S*)-Dihydroxy-11-oxothromba-5*Z*,13*E*-dienoic acid or 11-dehydro-TxB_2 is formed as a result of dehydrogenation of the hemiacetal alcohol group of TxB_2 and occurs in two forms in a pH-dependent equilibrium (Roberts *et al.*[5]). At acidic pH it exists in the δ-lactone form, represented by an intact thromboxane ring (Fig. 1), whereas at basic pH it is turned into an open ring structure, characterized by a carboxyl group at C-11.

Kumlin and Granström[6] characterized the time course of hydrolysis of the δ-lactone ring at different pH values and found that hydrolysis at pH 8.6 was complete after 3 hr, as was the lactonization at pH 2.0. However, the same authors noted that hydrolysis as well as lactonization required progressively more time when incubation was carried out at pH values approaching neutrality. Thus, at pH 6.0, equilibrium between open and

[2] G. Ciabattoni, J. Maclouf, F. Catella, G. A. FitzGerald, and C. Patrono, *Biochim. Biophys. Acta* **918,** 293 (1987).

[3] G. Ciabattoni, F. Pugliese, G. Davì, A. Pierucci, B. M. Simonetti, and C. Patrono, *Biochim. Biophys. Acta* **992,** 66 (1989).

[4] P. Patrignani, H. Morton, M. Cirino, A. Lord, L. Charette, J. Gillard, J. Rokach, and C. Patrono, *Biochim. Biophys. Acta* **992,** 71 (1989).

[5] L. J. Roberts II, B. J. Sweetman, and J. A. Oates, *J. Biol. Chem.* **253,** 5305 (1978).

[6] M. Kumlin and E. Granström, *Prostaglandins* **32,** 741 (1986).

closed forms required 20 days. Furthermore, the hydrolysis process is temperature dependent, increasing progressively with higher temperature.

These peculiar features are essential in developing a RIA procedure for 11-dehydro-TxB_2, since a carboxyl group in position 11 represents a strong antigenic determinant influencing the immune response in animals as well as the affinity and specificity of antisera.

Materials

Unlabeled 11-dehydro-TxB_2 for immunization purposes and as a standard in the RIA can be purchased from Cayman Chemicals (Ann Arbor, MI). Radiolabeled 11-dehydro-TxB_2 can be synthesized by incubating commercially available [^{3}H]TxB_2 (New England Nuclear, Boston, MA; specific activity, 150 Ci/mmol) with the high-speed supernatant of guinea pig liver homogenate (Lawson *et al.*[7]). 11-Dehydro[^{3}H]TxB_2 (120 Ci/mmol) as well as a ^{125}I-labeled derivative of higher specific activity can now be purchased from Amersham International (Amersham, UK). Anti-11-dehydro-TxB_2 sera and complete RIA kits are commercially available (e.g., New England Nuclear). However, we have prepared and continue to use our own antisera as described below.

Antibody Production

Covalent Coupling of 11-Dehydro-TxB_2 to Protein Carriers

Haptens containing a reactive carboxyl group can form peptide bonds with the amino terminus or the side amino groups of protein carriers. Carbodiimide and *N,N'*-carbonyldiimidazole methods have been successfully employed by Kumlin and Granström.[6] Also, conjugation of the hapten with keyhole limpet hemocyanin (KLH) using the *N*-hydroxysuccinimide ester procedure has been described (Ciabattoni *et al.*[2]).

Theoretically, 11-dehydro-TxB_2 may be coupled to the carrier protein either at the C-1 carboxyl or at the ring (C-11) carboxyl group. However, since coupling conditions induce lactonization because of the relatively low pH of the solution (particularly if a water-soluble carbodiimide is employed), it is likely that the conjugation takes place exclusively at the C-1 carboxyl, regardless of the coupling reagent employed. This is an obvious advantage since, when the conjugate is injected into the animal, hydrolysis at the pH of the connective tissue (7.4) will turn approximately

[7] J. A. Lawson, C. Patrono, G. Ciabattoni, and G. A. FitzGerald, *Anal. Biochem.* **155,** 198 (1986).

90% of lactone form into the open form. The hydrolysis reaction probably occurs within the dermis of the immunized animal because (a) being emulsified in Freund's adjuvant, the conjugate is slowly adsorbed within several days and (b) body temperature strongly increases the conversion rate.

This particular behavior of 11-dehydro-TxB_2 has two main consequences: (a) hydrolysis unmasks a strong antigenic determinant in a strategic position and the molecule attached through the C-1 terminus protrudes out of the carrier protein, thus leaving the C-11 carboxyl exposed to macrophage recognition; (b) antibody specificity will be mainly directed against this portion of the molecule, which is unique among the metabolites produced by cyclooxygenation of arachidonate.

In our laboratory, conjugates are produced by coupling 11-dehydro-TxB_2 to human serum albumin (HSA) in a molar ratio of 10 : 1 by reaction with 1-ethyl-3-(3-dimethylaminopropyl)-carbodiimide (Calbiochem, San Diego CA). One mg of 11-dehydro-TxB_2, 2 mg of HSA, and 1 mg of carbodiimide are dissolved in 1 ml of distilled water. About 0.1 μCi of [^{3}H]TxB_2 is added in order to calculate the final yield of the reaction and to identify the presence of conjugate in column eluates. The mixture is stirred at 4° for 24 hr and coupled 11-dehydro-TxB_2 is separated from uncoupled material and carbodiimide by means of Sephadex G-50F (Pharmacia, Sweden) column chromatography. The final yield is in the range of 10 to 15%, thus allowing a molar ratio of 11-dehydro-TxB_2 : HSA in the conjugate of 9 : 1–14 : 1.

Immunization

Two New Zealand White rabbits receive three injections of 11-dehydro-TxB_2–HSA conjugate emulsified in Freund's complete adjuvant at 3-week intervals. About 10–15 μg of coupled 11-dehydro-TxB_2 is given to each animal by multiple intradermic injections into the back. The first immunization is performed when the animals are still very young (about 2 months of age). Fifteen days after the third injection, animals are bled and reimmunized with about one-half of the previous dose of immunogen. A detectable rise in antibody titer is observed in both animals. After extensive bleeding, animals are reimmunized with a low dose of immunogen.

Titer and Affinity of Antisera

It is reasonable to assume that the antibodies would recognize the two chemical forms of 11-dehydro-TxB_2 differently. Thus, their affinity might be expected to show pH dependence. To test this hypothesis, dilution tests are carried out in the ordinary RIA buffer (phosphate buffer 0.02 *M*, pH 7.5) and in a series of buffers with pH values ranging from 6.0 to 10.0. Approximately 3500 dpm of 11-dehydro[^{3}H]TxB_2 is mixed with appropriately diluted antiserum in a volume of 1.5 ml of assay buffer and incubated for 20–24 hr at 4°, in order to reach equilibrium of antigen–antibody reaction. Separation of antibody bound from free 11-dehydro[^{3}H]TxB$_2$ is achieved by rapidly adding 0.1 ml of blood bank plasma and 0.1 ml of a charcoal suspension (100 mg/ml) and centrifuging at 4°. Because of recent difficulties in handling large volumes of human plasma safely, plasma is replaced with a 5% BSA solution.

As could be anticipated, the binding of the labeled hapten to the antibodies is strictly dependent on the pH of the incubation medium. When incubation is carried out in phosphate buffer at pH 6.0, only a weak binding, not exceeding 20–30% of labeled antigen, can be detected at low dilutions of antisera (1:1,000).

At pH 7.0, the percentage binding is significantly increased, although the titer of tested antisera remains too low for RIA purposes. Increasing the pH above the neutral range results in a progressive increase in titer, i.e., the antiserum dilution at which 40–50% of the radioactive hapten is bound. Figure 2 shows the behavior for one antiserum routinely employed in our laboratory (L4). The maximum binding is found in pH range 9.0–9.5. A further increase in pH leads to a progressive decrease of antibody titer, which is consistent with the notion that immune complexes are dissociated at both extremes of the pH scale.

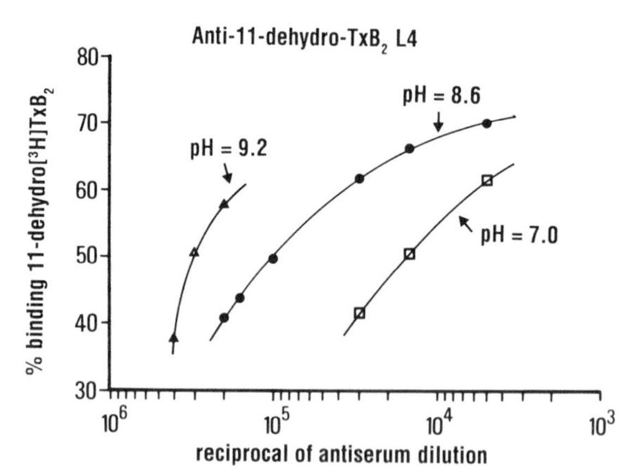

FIG. 2. pH-dependence of 11-dehydro-[^{3}H]-TxB$_2$ binding to antiserum L4. The three lines depict the relationship between the antiserum dilution (the reciprocal of which is reported on a logarithmic scale in the abscissa) and percentage binding of labeled hapten, when the pH value of the RIA buffer is varied between 7.0 and 9.2. Incubation conditions and separation procedures are detailed in the text.

At pH 9.2, antiserum L4 apparently has a single class of antibody-combining sites with an association constant of 0.1×10^{-12} liter/mole, as determined by Scatchard's plot. This is presumably due to the high dilution achieved in the RIA system (1:400,000), whereby the antigen-antibody reaction involves almost exclusively a single class of immunoglobulins with the highest energy of binding.

Antisera with higher titers have been produced in our laboratory. In particular, one antiserum (R1) shows a titer of 1:1,000,000 with ^{3}H-labeled 11-dehydro-TxB_2 and the titer rose to 1:4,000,000 when tested with ^{125}I-labeled 11-dehydro-TxB_2. However, both sensitivity and specificity of this antiserum are not quite as good as those of L4 antiserum.

Sensitivity and Specificity

Sensitivity of the RIA system is influenced by pH. When a tris–phosphate buffer (0.025 *M*, pH 8.6) is used, the best available antiserum (L4) binds 47% of the labeled antigen at a final dilution of 1:250,000. Unlabeled 11-dehydro-TxB_2 displaces this binding with an IC_{50} (the concentration required to displace 50% of the initially bound tracer) of 20 pg/ml. When the same buffer is adjusted to pH 9.2, the same antiserum binds 42% of the tracer at a final dilution of 1:400,000. Under these conditions, unlabeled 11-dehydro-TxB_2 displaces the homologous tracer in a linear fashion over the range of 1 to 50 pg/ml with an IC_{50} of 8.0 pg/ml. This sensitivity, obtained with an homemade tracer, significantly increases when 11-dehydro[^{3}H]TxB_2 of higher specific activity (120 Ci/mmol) becomes available (Fig. 3).

Immunological cross-reactivity of antiserum L4 is illustrated in Fig. 4, which shows a set of standard curves obtained with thromboxane metabolites whose structure is related to that of 11-dehydro-TxB_2. The cross-reactivities of 2,3-dinor-TxB_2, TxB_2, and 11,15-bisdehydro-TxB_2 are 0.1, 0.006, and <0.002%, respectively. Other eicosanoids tested show a cross-reactivity <0.001% and are not reported in Fig. 4.

Measurement of 11-Dehydro-TxB_2 in Plasma and Urine

RIA measurements of 11-dehydro-TxB_2 in plasma and urine have been performed in our laboratory in order to characterize the rate of entry of TxB_2 into the human circulation and fractional conversion to its major urinary metabolite (Ciabattoni *et al.*[2,3]). The detection limit (5 pg/ml) for assays performed in unextracted samples of human plasma (diluted 1:10 in the RIA buffer) is not adequate for measuring the very low

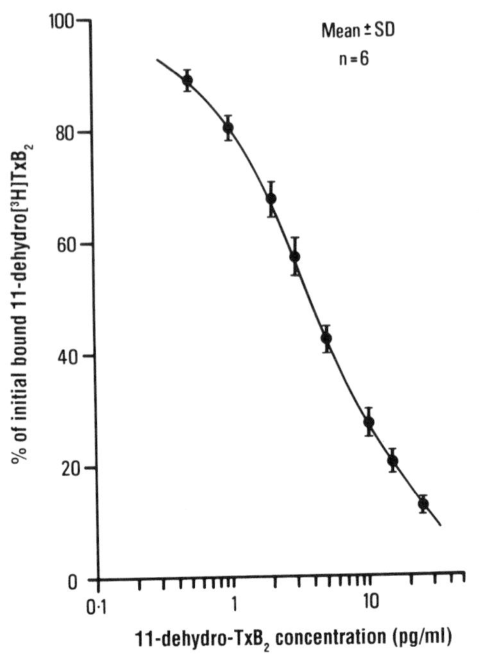

FIG. 3. Standard curve of the radioimmunoassay for 11-dehydro-TxB_2 with antiserum L4. The mean ± SD of six separate experiments is depcited.

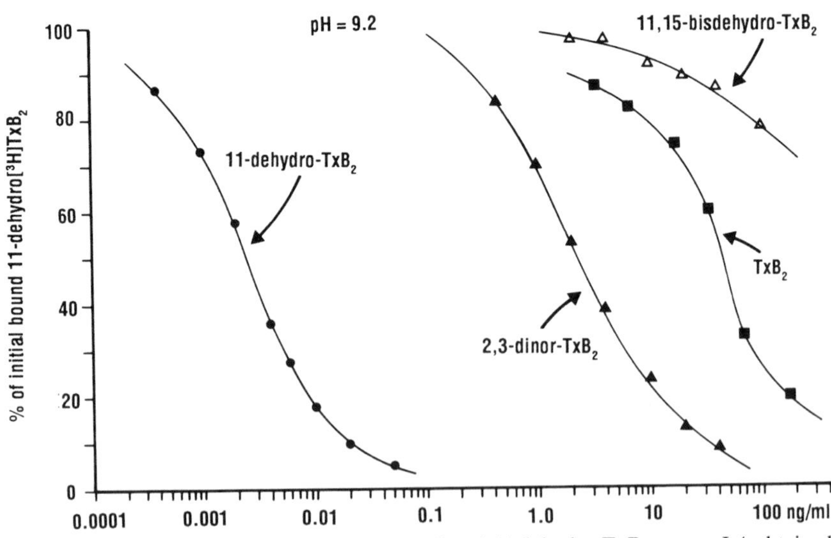

FIG. 4. Family of dose–response curves of anti-11-dehydro-TxB_2 serum L4 obtained with different thromboxane metabolites, at pH 9.2.

"basal" circulating levels of this metabolite, i.e., 1–2 pg/ml (Catella *et al.*[8]).

In contrast, urinary excretion (180–400 pg/mg creatinine) of 11-dehydro-TxB_2 can be easily measured in extracted and chromatographed samples, with adequate recovery (70–80%) and satisfactory reproducibility on repeated sampling (intrasubject coefficient of variation: 15–25% in healthy volunteers). Details of extraction and chromatographic procedures are described in Ref. 2. Reversed-phase high-performance liquid chromatography (RP-HPLC) separation of 11-dehydro-TxB_2 extracted from the urine of cynomolgus monkeys is also detailed elsewhere (Patrignani *et al.*[4]).

Validation of Assay

We have used several independent criteria in order to validate measurements of 11-dehydro-TxB_2 performed by RIA: (a) recovery of exogenously added authentic 11-dehydro-TxB_2 to plasma and urine samples; (b) identical immunochemical behavior of the endogenous immunoreactive material and the reference standard on 10- to 50-fold dilution (i.e., apparent concentration of the "unknown" independent of the dilution factor); (c) identical chromatographic behavior of the extracted urinary immunoreactivity and authentic 11-dehydro-TxB_2 on reversed-phase HPLC (i.e., immunoreactivity detected only in the fractions corresponding to the retention time of the standard); (d) statistically significant, linear correlation between paired measurements performed by RIA in our laboratory and by gas chromatography–mass spectrometry in the negative-ion chemical ionization mode (GC–NCI–MS) in Dr. G. A. FitzGerald's laboratory. Such a comparison, which represents the ultimate criterion for establishing the validity of RIA measurements, should be performed in a blind fashion and involve a reasonable number of samples (i.e., 10–15) encompassing the whole range of concentrations of potential pathophysiological and pharmacological interest (Ciabattoni *et al.*[2]).

Conclusions

The rationale underlying the development of RIA for 11-dehydro-TxB_2 is related to the need for a convenient biochemical index of the *in vivo* synthesis and release of TxA_2. The long half-life (approx. 45 min) of 11-dehydro-TxB_2 in the human circulation and its relative abundance in

[8] F. Catella, D. Healy, J. A. Lawson, and G. A. FitzGerald, *Proc. Natl. Acad. Sci. U.S.A.* **83,** 5861 (1986).

urine, together with a fractional elimination comparable to that of 2,3-dinor-TxB_2, make these measurements an extremely convenient approach to monitoring thromboxane production both in healthy and disease conditions. The unique chemical properties of 11-dehydro-TxB_2 are associated with greater than usual specificity of the immunological recognition, which combined with adequate chromatographic procedures, can allow reliable assessment of metabolite levels in complex biological fluids.

Acknowledgments

The expert editorial assistance of G. Protasoni and M. L. Bonanomi is gratefully acknowledged. Our studies were supported by grants from Consiglio Nazionale delle Ricerche (Progetto Finalizzato Medicina Preventiva e Riabilitativa).

[5] Measurement of Thromboxane Metabolites by Gas Chromatography–Mass Spectrometry

By FRANCESCA CATELLA and GARRET A. FITZGERALD

Thromboxane (Tx)A_2 is a potent vasoconstrictor and proaggregatory substance that, by virtue of these properties, may play a role in human syndromes of vascular occlusion. Consistent with this concept, aspirin (acetylsalicylic acid) which inhibits TxA_2, has proved to be beneficial in syndromes associated with platelet activation, such as unstable angina.

Biochemical assessment of thromboxane biosynthesis complements the use of pharmacological probes to elucidate the biological importance of this compound. Unfortunately, due to its chemical instability, TxA_2 cannot be directly measured in biological fluids. Its hydration product, TxB_2, is more stable, but its assessment in plasma is confounded by artifactual formation during blood withdrawal.[1–3] Although urinary excretion of TxB_2 may be used as an index of systemic TxA_2 formation in certain settings, such as platelet activation, it largely reflects renal production under physiological conditions.

[1] G. A. FitzGerald, A. K. Pedersen, and C. Patrono, *Circulation* **67,** 1174 (1983).

[2] E. Granstrom, P. Westlund, K. Kumlin, and A. Nordenstrom, *Adv. Prostaglandins Thromboxane Leukotriene Res.* **15,** 67 (1985).

[3] F. Catella, D. Healy, J. Lawson, and G. A. FitzGerald, *Proc. Natl. Acad. Sci. U.S.A.* **83,** 5861 (1986).

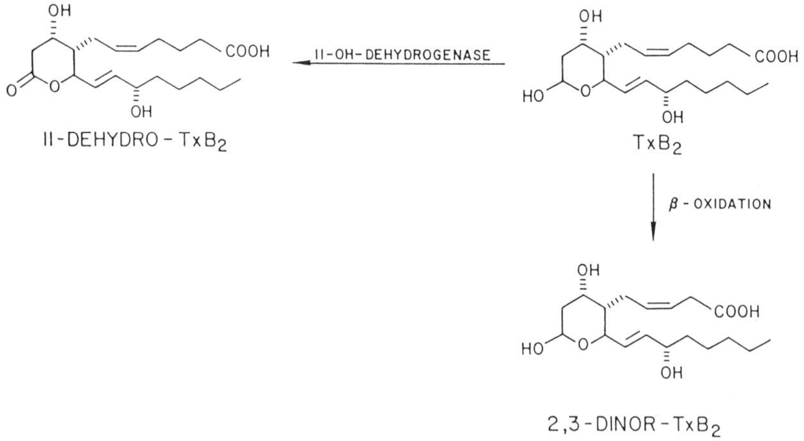

OH
COOH
II-OH-DEHYDROGENASE
O
OH
II-DEHYDRO-TxB2
HO
TxB2
β-OXIDATION
2,3-DINOR-TxB2

FIG. 1. 2,3-Dinor-TxB_2 and 11-dehydro-TxB_2 are indexes of the two major pathways of metabolism of TxB_2.

Roberts *et al.*[4] identified twenty compounds formed via two major and one minor pathway of metabolism, by infusing [^{3}H]TxB_2 in a healthy volunteer. Indexes of the two major pathways are 2,3-dinor-TxB_2, metabolized via β-oxidation, and 11-dehydro-TxB_2, resulting from dehydrogenation of the hemiacetal alcohol group at C-11 (Fig. 1). More recently, these metabolites have also been shown to be the major metabolites of TxA_2 itself in the monkey.[5]

The small concentrations of these metabolites in plasma render 11-dehydro-TxB_2 the only realistic target for analysis in plasma with available methodology.[3] However, measurement of either metabolite in urine has proved to offer a reliable index of *in vivo* TxA_2 formation by platelets, although the kidney is likely to possess the capacity to form both the dinor and dehydro derivatives.[6] Urinary excretion of both metabolites is almost, but not completely, inhibited following chronic administration of aspirin at a dose which completely inhibits platelet cyclooxygenase.[7] Furthermore, their excretion is significantly elevated in syndromes associated with platelet activation, such as peripheral vascular disease, unstable angina, thrombolysis, and systemic sclerosis.

[4] L. J. Roberts, B. J. Sweetman, and J. A. Oates, *J. Biol. Chem.* **256,** 8384 (1981).

[5] P. Patrignani, H. Morton, M. Cirino, A. Lord, L. Charette, J. Gillard, J. Rokach, and C. Patrono, *Clin. Res.* **37,** 341A (1989).

[6] A. Benigni, C. Chiabrando, N. Perico, R. Fanelli, C. Patrono, G. A. FitzGerald, and G. Remuzzi, *Am. J. Physiol.* **257,** F77 (1989).

[7] F. Catella and G. A. FitzGerald, *Thromb. Res.* **47,** 647 (1987).

Gas Chromatography–Mass Spectrometry versus Immunoassay

Measurement of 2,3-dinor-TxB_2 and 11-dehydro-TxB_2 can be accomplished by gas chromatographic (GC)–mass spectrometric (MS) techniques or by radioimmunoassays (RIA). Enzyme immunoassays, already available for 11-dehydro-TxB_2, are currently being developed for 2,3-dinor-TxB_2.

The immunoassays can usually be performed on a larger scale in a relatively short time and do not require such expensive equipment. On the other hand, these methods may generate misleading results. This is generally due to lack of specificity resulting from inappropriate purification or from the use of antisera with high cross-reactivity with other metabolites of related compounds. To avoid these problems, any new immunoassays should be validated by characterization of the thin-layer chromatography (TLC) pattern of distribution of the extracted eicosanoid-like immunoreactivity, by using multiple antisera and by comparison with GC–MS determinations. RIA results which are highly comparable with GC–MS determinations have indeed been attained when the metabolites have been properly purified from biological matrixes and a highly specific antiserum has been employed.[8,9]

On the other hand, even highly validated immunoassays have practical limitations. Recently, for example, fish oils, rich in the ω-3 fatty acid, eicosapentaenoic acid (EPA), have been extensively investigated for their potential antithrombotic properties. Dietary supplementation with fish oil results in decreased formation of TxA_2 (derived from arachidonic acid; AA), and increased TxA_3 (derived from EPA and reported to be less thrombogenic than TxA_2). Commonly available immunoassays cannot discriminate between eicosanoids derived from AA and those derived from EPA, which differ only by one double bond. By contrast, GC–MS techniques can easily discriminate between these two series of compounds.[10]

Another limitation of immunoassay methods overcome by GC–MS is their poor sensitivity. It was suggested that analysis of eicosanoids in plasma could facilitate the investigation of their temporal formation *in vivo:* this would be desirable when phasic eicosanoid formation might be of interest, such as during ischemic episodes in patients with unstable

[8] C. Patrono, G. Ciabattoni, F. Pugliese, I. A. Blair, and G. A. FitzGerald, *J. Clin. Invest.* **77,** 590 (1986).

[9] G. Ciabattoni, J. Maclouf, F. Catella, G. A. FitzGerald, and C. Patrono, *Biochim. Biophys. Acta* **918,** 293 (1987).

[10] H. R. Knapp, I. A. G. Reilly, P. Alessandrini, and G. A. FitzGerald, *N. Engl. J. Med.* **314,** 937 (1986).

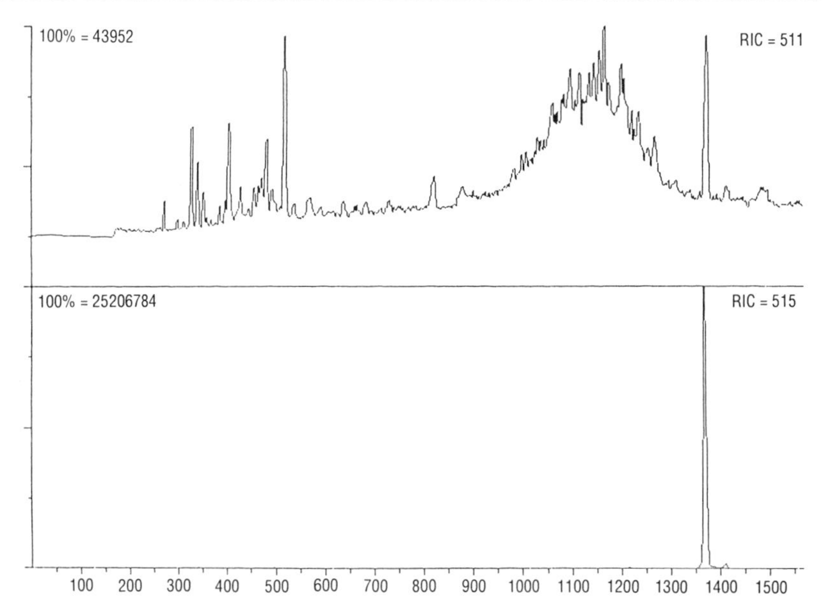

FIG. 2. Selected-ion monitoring of 11-dehydro-TxB_2 (m/z 511) and its tetradeuterated internal standard (m/z 515). The plasma concentration of this sample corresponds to 1.5 pg/ml. (By permission, Ref. 3.)

angina. As far as TxA_2 is concerned, plasma concentrations of its hydration product, TxB_2, are readily confounded by *ex vivo* platelet activation.[1-3] However, 11-dehydro-TxB_2, a long-lived enzymatic metabolite in plasma,[11] has been shown to minimize this problem.[3] Its measurement in plasma requires a detection limit of less than 1 pg/ml. This can be attained by capillary gas chromatography–negative-ion chemical ionization mass spectrometry (GC–NCI–MS) (Fig. 2), but it is far below the limit of sensitivity achieved by most commercially available immunoassays. Combined analysis of 11-dehydro-TxB_2 and 2,3-dinor-TxB_2, indexes of the two major pathways of thromboxane metabolism in humans, permits one to distinguish between altered metabolism and increased biosynthesis of TxA_2.[7] Moreover, combined analysis of metabolites in plasma and urine can be utilized to identify alterations in the volume of distribution and renal clearance of TxA_2, as might occur due to drug administration or disease.

On the other hand, although the methods have been refined, GC-MS remains a time-consuming approach. Samples have to undergo extraction, purification, and derivatization prior to injection onto the instrument.

[11] J. Lawson, C. Patrono, G. Ciabattoni, and G. A. FitzGerald, *Anal. Biochem.* **155,** 198 (1986).

Internal Standards

The first step in quantitative analysis by GC–MS is to spike the samples with an internal standard. This permits accurate quantitation, irrespective of losses during purification or due to incomplete derivatization. The most satisfactory standards are analogs incorporating stable isotopes, because they have physiochemical properties almost identical to their natural isomers.[12]

Some stable isotopes are commercially available (Biomol, Plymouth Meeting, PA; Cayman Chemicals, Ann Arbor, MI). The remainder must be chemically or biologically synthesized. Deuterated 11-dehydro-TxB_2 can be biologically synthesized by incubation of deuterated TxB_2 with the high-speed supernatant of guinea pig liver homogenate in the presence of NAD.[11] Deuterated 2,3-dinor-TxB_2 can be produced by prolonged incubation of the tetradeuterated TxB_2 in a culture of *Mycobacterium rhodochrous*.[13]

Another example of biological generation of standards is the formation of the dinor metabolite of tetradeuterated 6-keto-$PGF_{1\alpha}$ by cultured hepatocytes.[14]

Alternatively, 2,3-dinor-TxB_2 and 11-dehydro-TxB_2 can be labeled by exchange of the two carboxylic oxygen atoms with ^{18}O using labeled water as a donor.[15] These exchange reactions can be catalyzed by acid, base, or enzymatically via esterases as described by Murphy and Clay.[16] A limitation to the use of ^{18}O-labeled compounds is the back-exchange loss of the label that can occur rapidly in plasma and tissues containing esterases. This undesired enzymatic exchange does not represent a problem in urine, due to the lack of esterase. In other biological matrixes, it can be avoided easily by lowering the pH of the medium to approximately 3.5 before adding the ^{18}O-labeled standard.

Extraction

The selective extraction of 2,3-dinor-TxB_2 can be achieved by relying on the chemical properties of the thromboxane ring; in a pH-dependent equilibrium, it exists with an open ring form. At pH 5, the open form, stabilized by derivatization as the methoxime, results in a hydroxyl con-

[12] W. A. Garland and M. L. Powell, *J. Chromatogr. Sci.* **19,** 392 (1981).

[13] F. F. Sun, B. M. Taylor, F. H. Lincoln, and O. K. Sebek, *Prostaglandins* **20,** 729 (1980).

[14] M. Balazy, E. P. Brass, J. G. Gerber, and A. S. Nies, *Prostaglandins* **36,** 421 (1988).

[15] H. J. Leis, E. Malle, R. Moser, J. Nimpf, G. M. Kostner, H. Esterbauer, and H. Gleispach, *Biomed. Environ. Mass Spectrom.* **13,** 483 (1986).

[16] R. C. Murphy and K. L. Clay, this series, Vol. 86, pp. 547.

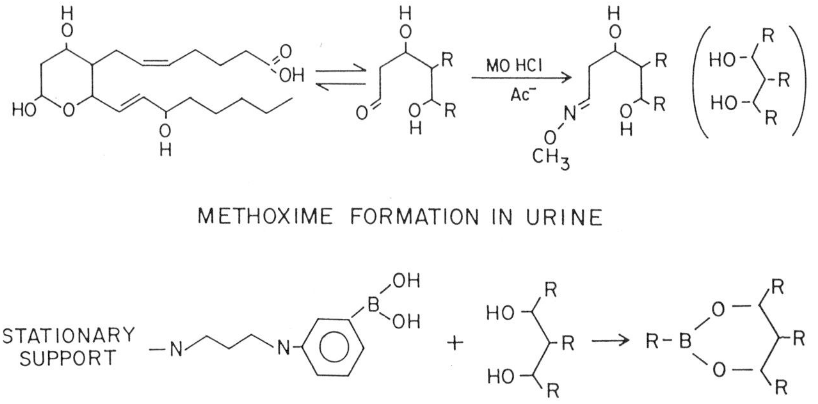

FIG. 3. 2,3-Dinor-TxB_2 exists in equilibrium with an open ring form which is derivatized as the methoxime. Following derivatization, the hydroxyl configuration permits condensation with bonded-phase phenylboronic acid to form a stable complex. (By permission, Ref. 17.)

figuration favorable to the formation of a stable complex with bonded-phase phenylboronic acid (Fig. 3).[17] This procedure has been employed for the purification of prostaglandin F metabolites as well as for carbohydrate and catecholamine analysis.

Reagents and Materials

Methoxyamine Hydrochloride (MOHCl) (Sigma, St. Louis, MO)
Phenylboronic acid bonded-phase (PBA) columns (Infolab, Fort Oglethorpe, GA)
Prep-Sep columns (Fischer Scientific, Fair Lawn, NJ)

Procedure. To each 5-ml urinary sample, 125 mg MOHCl dissolved in 1.5 ml sodium acetate 1.5 *M* (pH 5) is added. Following vigorous mixing, the samples are allowed to equilibrate at room temperature for 20 min and then applied to PBA columns, which have been previously washed with 3 ml methanol and 3 ml of 0.1 *N* hydrochloric acid (HCl). The PBA columns are then washed with 2 ml of 1 part of 1 *M* NaCl and 1 part of 0.1 *N* HCl, followed by another wash with 2.5 ml of methanol. The samples are then eluted with 4 ml of 1 part 0.1 *M* NaOH and 1 part methanol and subsequently diluted to 15 ml with water. The pH is then adjusted to approximately 3 with 10% formic acid and the sample applied to a C_{18} Prep-Sep that has been preconditioned with 3 ml of methanol and 3 ml of

[17] J. A. Lawson, A. R. Brash, J. Doran, and G. A. FitzGerald, *Anal. Biochem.* **150,** 463 (1985).

distilled water. The Prep-Sep is then washed with 3 ml of water and eluted with 3 ml of ethyl acetate.

This technique, although highly specific, requires extensive purification following extraction, prior to mass spectrometric analysis. This represents a major drawback to its widespread application. The cleanup time has been greatly reduced by the use of immunopurification techniques.[18,19] Biological samples are applied to a reversed-phase cartridge, interfering substances are partly removed by specific solvents, and the compound can then be eluted and applied to a stationary phase (hydroxysuccinimidyl-SP500 silica gel) where polyclonal antibodies are chemically immobilized.

Reagents and Materials

Polyvinylpyrrolidone (PVP-40) (Sigma, St. Louis, MO)
Antibody columns prepared as in Ref. 18

Procedure. Urine samples are extracted by C_{18} Prep-Sep, eluted with ethyl acetate, dried under nitrogen, and redissolved in 0.5 ml of buffer consisting of 1 g/liter PVP-40, 0.15 *M* NaCl, 0.01 *M* Trizma hydrochloride, 0.1 *M* $CaCl_2$, and 0.5 *M* Mg_2SO_4. This is applied to the antibody affinity columns and eluted with 4 ml of 4 parts acetone and 1 part 1% acetic acid. After drying the acetone under nitrogen, the samples are extracted into 4 ml ethyl acetate. The columns are washed with phosphate-buffered saline (PBS) and stored in PBS at 4°.

These techniques allow selective and simultaneous extraction of two or more metabolites, have an extraction efficiency from urine of about 50%, and are highly reproducible [coefficient of variation (*CV*) $<10\%$].[19]

Purification

Following extraction of the metabolites from biological matrixes, further purification is necessary to obtain an interference-free chromatogram. As far as 2,3-dinor-TxB_2 is concerned, purification usually relies on two TLC steps if a previous extraction has been performed with PBA columns. Only one TLC step is necessary following antibody column extraction. In the first case, samples are applied as the methoxime derivatives to the preadsorbent zone of silicic acid plates (Linear-K silica gel, Whatman) and developed in a mobile phase consisting of ethyl acetate : acetic acid : hex-

[18] H. L. Hubbard, T. D. Eller, D. E. Mais, P. V. Halushka, R. H. Baker, I. A. Blair, J. J. Vrbanac, and D. R. Knapp, *Prostaglandins* **33,** 149 (1987).

[19] C. Chiabrando, A. Benigni, C. Piccinelli, C. Carminati, E. Cozzi, G. Remuzzi, and R. Fanelli, *Anal. Biochem.* **163,** 255 (1987).

ane : water (54 : 12 : 25 : 60). Sample R_f is identified by 2,3-dinor-TxB_2 methoxime standard (5 μg) spotted on a separate plate and visualized by 10% phosphomolybdic acid in ethanol. The samples are then scraped from the appropriate area and extracted into ethyl acetate.

A second TLC step, the only one following antibody column extraction, is performed following further derivatization as pentafluorobenzyl (PFB) ester. The developing solvent is the organic layer of isooctane : ethyl acetate : water (65 : 85 : 100). The R_f if 2,3-dinor-TxB_2–methoxime–PFB derivative is visualized as described above.

In the case of 11-dehydro-TxB_2, urinary or acidified plasma samples are extracted by reversed-phase cartridge, allowed to stand at room temperature in 10% formic acid for 2 hr to ensure ring closure, and finally purified by two TLC steps before [Solvent A = ethyl acetate : acetic acid (48 : 1)] and after [Solvent B = ethyl acetate : heptane (75 : 25)] derivatization as PFB ester. Only the second TLC is necessary following antibody column extraction.

The purification procedure for both metabolites can also be considerably facilitated by the use of highly selective triple-stage quadrupole MS, instead of the single-quadrupole MS.[20]

Derivatization

The polar functions of 2,3-dinor-TxB_2 and 11-dehydro-TxB_2 need to be derivatized in order to improve both their gas chromatograhic characteristics and mass spectrometric fragmentation patterns. Usually the ketone group of 2,3-dinor-TxB_2 is derivatized to methoxime and, in both molecules, the hydroxyl groups are derivatized to trimethylsilyl (TMS) ethers and the terminal carboxyl group to methyl or pentafluorobenzyl (PFB) esters (Fig. 4).

One way to perform methoxime derivatization has been already described as a prerequisite for PBA column extraction. Alternatively, methoxime derivatives can be formed by adding μl of 0.5% methoxyamine hydrochloride in pyridine to the dried samples and allowing the reaction mixture to stand at room temperature overnight.

The pentafluorobenzyl (PFB) ester is formed by allowing the samples to stand at room temperature for 30 min in 10 μl diisopropylethylamine and 20 μl of 12.5% PFB Br in acetonitrile. The PFB group is electron capturing and has the advantage of allowing analysis in the negative-ion, chemical ionization (NCI) mode. It also directs fragmentation, essentially re-

[20] H. Schweer, C. Meese, O. Furst, P. G. Kuhl, and H. W. Seyberth, *Anal. Biochem.* **164,** 156 (1987).

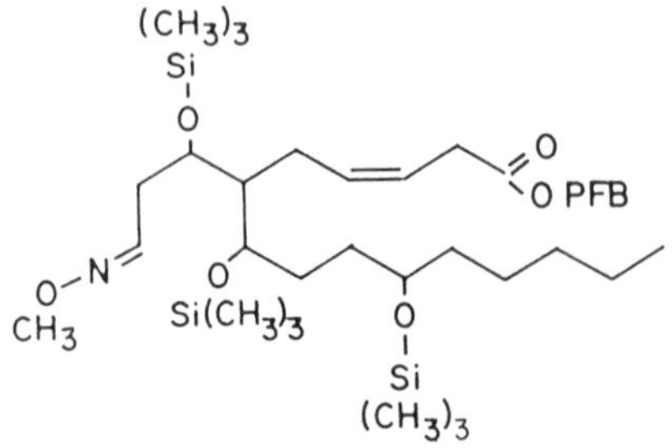

FIG. 4. Derivatization of 2,3-dinor-TxB_2 as the methoxime, TMS ether, PFB ester.

stricting it to the loss of the PFB group. Taken together, these factors enhance sensitivity by several orders of magnitude over electron-impact techniques.

The trimethylsilyl (TMS) ether is formed by leaving the sample at room temperature for 1 hr in 10 μl dry pyridine and 10 μl *N,O*-bistrimethylsilyltrifluoroacetamide (BSTFA).

Quantitation

Quantitation by GC–MS is generally performed in the selected-ion monitoring (SIM) mode by measuring the ratio of the integrated areas of the peaks corresponding to the ion for the endogenous material and the ion for the internal standard. This represents a highly specific technique, allowing discrimination between the target compound and other contaminant peaks with different GC retention times.

Specificity is also enhanced by the use of capillary rather than packed columns. Capillary columns enhance specificity by increasing efficiency, thereby resolving the metabolites from contaminants originating in biological fluids. They also improve sensitivity by concentrating the compound into a sharp peak. These features are apparent in Fig. 2.

In conclusion, measurement of thromboxane metabolites by GC–MS provides a reliable, specific, and sensitive index of thromboxane biosynthesis. It has permitted considerable insight into the pathophysiological role of thromboxane A_2 in human disease.

Acknowledgments

Supported by grants (HL 30400, GM15431) from the National Institutes of Health and Daiichi Seiyaku. Dr. Catella is the recipient of a Faculty Development Award from the Pharmaceutical Manufacturers Association Foundation. Dr. FitzGerald is an Established Investigator of the American Heart Association and the William Stokes Professor of Experimental Therapeutics.

[6] Quantification of $9\alpha,11\beta$-Prostaglandin F_2 by Stable Isotope Dilution Mass Spectrometric Assay

By DANIEL F. WENDELBORN, JASON D. MORROW, and L. JACKSON ROBERTS

Introduction

Prostaglandin D_2 (PGD_2) is produced by a number of tissues and cells and exerts a variety of biological actions. The pharmacology of PGD_2 has been extensively reviewed.[1] One area of considerable interest has been the finding that PGD_2 is the major cyclooxygenase product produced by mast cells.[2] It has been shown to be released in large quantities into the lower respiratory tract in humans following antigen challenge and it is markedly overproduced in patients with systemic mastocytosis, a disorder characterized by an abnormal proliferation of tissue mast cells.[3,4] Patients with systemic mastocytosis frequently experience severe episodes of vasodilation and PGD_2 has been found to be the major mediator of these episodes.[5]

PGD_2 is metabolized *in vivo* in humans predominantly via a 11-ketoreductase pathway to metabolites with a PGF ring. Subsequently, *in vitro* studies revealed that PGD_2 is transformed stereoselectively by 11-ketoreductase to $9\alpha,11\beta$-PGF_2 rather than to $PGF_{2\alpha}$.[6] $9\alpha,11\beta$-PGF_2 has been shown to be produced *in vivo* in humans and has been found to be formed in markedly increased quantities during episodes of systemic mast cell activation, which is associated with the release of large amounts of PGD_2.[4,6]

Of interest and importance is that $9\alpha,11\beta$-PGF_2 has been found to be biologically active. Transformation of an eicosanoid formed directly from PGH_2 to a biologically active metabolite is unique to PGD_2. $9\alpha,11\beta$-PGF_2

[1] H. Giles and P. Leff, *Prostaglandins* **35,** 277 (1988).

[2] L. J. Roberts II, R. A. Lewis, J. A. Oates, and K. F. Austin, *Biochim. Biophys. Acta* **575,** 183 (1979).

[3] K. Seibert, J. R. Sheller, and L. J. Roberts II, *Proc. Natl. Acad. Sci. U.S.A.* **84,** 256 (1987).

[4] D. F. Wendelborn, K. Seibert, and L. J. Roberts II, *Proc. Natl. Acad. Sci. U.S.A.* **85,** 304 (1988).

[5] L. J. Roberts II, B. J. Sweetman, R. A. Lewis, K. F. Austen, and J. A. Oates, *N. Engl. J. Med.* **303,** 1400 (1980).

[6] T. E. Liston and L. J. Roberts II, *Proc. Natl. Acad. Sci. U.S.A.* **82,** 6030 (1985).

has been found to contract human bronchial smooth muscle, contract coronary arteries, inhibit platelet aggregation, induce natruresis, and it is a pressor substance.[3,6–9] Because it is a metabolite of PGD_2, quantification of $9\alpha,11\beta$-PGF_2 can provide a means to assess endogenous production of PGD_2 in humans. Because $9\alpha,11\beta$-PGF_2 is biologically active, assessing its formation also is of value in understanding the biological pharmacology of PGD_2 release *in vivo*. Furthermore, quantification of the formation of $9\alpha,11\beta$-PGF_2, is an essential tool for *in vitro* studies involving identification and characterization of the 11-ketoreductase enzyme.

For these reasons we have developed a stable isotope dilution mass spectrometric assay for quantification of $9\alpha,11\beta$-PGF_2 in biological fluids. This chapter details the preparation of the deuterium-labeled $9\alpha,11\beta$-PGF_2 internal standard, the procedures employed for purification of $9\alpha,11\beta$-PGF_2 in biological fluids, and the analysis of $9\alpha,11\beta$-PGF_2 by GC–MS. Specific examples are given demonstrating both the utility and the limitations of the assay.

Preparation of $[^2H_7]9\alpha,11\beta$-PGF_2 Internal Standard

Preparation of $[^2H_7]PGD_2$

$[^2H_7]9\alpha,11\beta$-PGF_2 is obtained by conversion of $[^2H_8]$arachidonic acid to $[^2H_7]PGD_2$ followed by reduction of $[^2H_7]PGD_2$ to $[^2H_7]9\alpha,11\beta$-PGF_2 by 11-ketoreductase. $[5,6,8,9,11,12,14,15$-$^2H_8]$Arachidonic acid can be prepared by reduction of tetraacetylene eicosatetrayenoic acid (ETYA) according to the method of Taber *et al.*[10] More recently $[^2H_8]$arachidonic acid has become commercially available (Cayman Chemicals, Ann Arbor, MI). The deuterated arachidonate is then incubated with sheep seminal vesicle microsomes to obtain deuterated PGD_2. The microsomes are prepared according the method of Hamberg and Samuelsson[11] with the following modifications. Microsomes from 5 g of defatted sheep seminal vesicle glands are obtained and resuspended in 0.1 *M* EDTA buffer, pH 8.0, containing 2% fatty acid-free bovine serum albumin (BSA). The mi-

[7] L. J. Roberts, II, K. Seibert, T. E. Liston, M. V. Tantengco, and R. M. Robertson, *Adv. Prostaglandin Thromboxane Leukotriene Res.* **17,** 427 (1987).

[8] G. Bugliese, E. G. Spokas, E. Marcinkiewicz, and P. Y.-K. Wong, *J. Biol. Chem.* **260,** 14621 (1985).

[9] C. T. Stier, Jr., L. J. Roberts, II, and P. Y.-K. Wong, *J. Pharmacol. Exp. Therap.* **243,** 487 (1987).

[10] D. F. Taber, M. A. Phillips, and W. C. Hubbard, this series, Vol. 86, p. 366.

[11] M. Hamberg and B. Samuelsson, *Proc. Natl. Acad. Sci. U.S.A.* **70,** 889 (1973).

crosomes are incubated in this albumin-containing buffer for 1 hr at 4°. This step is incorporated to remove arachidonic acid loosely bound to the microsomes (albumin binds arachidonic acid). This is important since it greatly reduces the $[^2H_0]/[^2H_7]$ ratio (blank) of the deuterated PGD_2 formed. Following this, the microsomes are centrifuged again at 100,000 *g* for 60 min and resuspended in 2.5 ml 0.1 *M* EDTA buffer (pH 8.0) containing 2 m*M* phenol, 3 m*M* hemoglobin, 1 m*M* L-tryptophan and 1 m*M* *p*-hydroxymercuribenzoate at 4°.

The microsomes are then incubated with stirring in a 37° water bath. When the temperature of the microsome suspension has reached 37°, 1 mg of $[^2H_8]$arachidonic acid and several microcuries of $[^3H_8]$arachidonic acid are then added as the sodium salt. The sodium salt is prepared by dissolving the arachidonic acid in a volume of 100 m*M* Na_2CO_3 sufficient to yield a 50 m*M* solution of arachidonic acid. The mixture is incubated at 37° for 1 min and then added to 1 ml of buffer containing 2% (w/v) fatty acid-free bovine serum albumin and allowed to stir at 37° for 5 min so that the endoperoxide formed is isomerized to PGD_2 and PGE_2. The presence of bovine serum albumin increases the relative proportion of PGD_2 compared to PGE_2 obtained from isomerization of the endoperoxides.

The mixture is then acidified to pH 3 with 10% formic acid (w/v) and extracted twice with an equal volume of ethyl acetate. The ethyl acetate is dried over anhydrous Na_2SO_4 and evaporated under reduced pressure. The residue is then subjected to normal-phase HPLC using a 5-μm silica (25 cm $\times$ 4.6 mm) column (Alltech, Deerfield, IL). A linear gradient is employed of solvent A (chloroform: acetic acid, 100 : 0.01, v/v) to 100% solvent B (chloroform: methanol: acetic acid, 95 : 5 : 0.01, v/v/v) over 1 hr, 1 ml/min, 1-ml fractions. PGD_2 characteristically elutes at 30–35 ml and PGE_2 at 40–45 ml.

Conversion of $[^2H_7]PGD_2$ to $[^2H_7]9\alpha,11\beta$-PGF_2

Conversion of $[^2H_7]PGD_2$ to $[^2H_7]9\alpha,11\beta$-PGF_2 is then accomplished using 11-ketoreductase in the 100,000 *g* supernatant of either human or rabbit liver. Approximately 18 g of liver are homogenized (1 : 3, w/v) in 0.1 *M* potassium phosphate buffer (pH 7.4). The homogenate is centrifuged at 10,000 *g* at 4° for 30 min. The 10,000 *g* supernatant is then centrifuged for 90 min at 100,000 *g* and the supernatant is decanted and stored at −70° until needed. 11-Ketoreductase activity remains stable stored at −70° for several months but eventually decays.

$[^2H_7]9\alpha,11\beta$-PGF_2 is then obtained by incubating $[^2H_7]PGD_2$ with liver supernatant containing a NADPH-generating system. Approximately 50 μg of $[^2H_7]PGD_2$ is incubated with 0.5 ml of liver supernatant diluted to a final volume of 3 ml with 0.1 *M* phosphate buffer (pH 7.4) containing NADPH (0.5 m*M*), glucose-6-phosphate dehydrogenase (9 units), and glucose 6-phosphate (5 m*M*). The incubation is carried out at 37° for 60 min which results in near-quantitative conversion of $[^2H_7]PGD_2$ to $[^2H_7]9\alpha,11\beta$-PGF_2. The solution is then acidified to pH 3 with 1 *N* HCl and extracted using a C_{18} Sep-Pak (Waters Associates, Milford, MA) as described below (Analysis of $9\alpha,11\beta$-PGF_2 in Biological Fluids). The $[^2H_7]9\alpha,11\beta$-PGF_2 is then purified by reversed-phase HPLC using a 5 μm particle size, 250 × 4.6 mm Alltech C_{18} column with the solvent system acetonitrile/water/acetic acid, 29 : 71 : 0.01 (v/v) run at 1 ml/min, 1-ml fractions. The $9\alpha,11\beta$-PGF_2 characteristically elutes with a retention volume of approximately 35 ml. The $9\alpha,11\beta$-PGF_2 is extracted from the HPLC solvent two times with an equal volume of ethyl acetate and evaporated under nitrogen.

The $[^2H_0]/[^2H_7]$ ratio (blank) of the $[^2H_7]9\alpha,11\beta$-PGF_2 obtained following these procedures is less than 0.010 (less than 1%). Standardization of the $[^2H_7]9\alpha,11\beta$-PGF_2 is performed against unlabeled $9\alpha,11\beta$-PGF_2 by gas chromatography–negative-ion chemical ionization mass spectrometry (GC–NCI–MS). Unlabeled $9\alpha,11\beta$-PGF_2 (Cayman Chemicals, Ann Arbor, MI) is accurately weighed and dissolved in ethanol. The amount of $[^2H_7]9\alpha,11\beta$-PGF_2 obtained is estimated from the specific activity of the $[^3H_8]$ and $[^2H_8]$ arachidonic acid used at the beginning. It is important to recognize that this is only an approximation since there can be isotope discrimination between deuterium and tritium during enzymatic conversion of arachidonic acid to PGD_2 and then to $9\alpha,11\beta$-PGF_2 and some separation of tritium and deuterium can occur during HPLC purification of the products formed. A known amount of $[^2H_0]9\alpha,11\beta$-PGF_2 is aliquoted and mixed with what is estimated to be approximately an equivalent amount of the $[^2H_7]9\alpha,11\beta$-PGF_2 and coderivatized to a pentafluorobenzyl (PBF) ester, trimethylsilyl (Me_3Si) ether derivative and analyzed by GC–NCI–MS as described below. From the ratio of m/z 569 (unlabeled $9\alpha,11\beta$-PGF_2) to m/z 576, ($[^2H_7]9\alpha,11\beta$-PGF_2) the amount of $[^2H_7]9\alpha,11\beta$-PGF_2 can be calculated.

Analysis of $9\alpha,11\beta$-PGF_2 in Biological Fluids

Many types of biological samples can be analyzed for $9\alpha,11\beta$-PGF_2 although certain precautions must be taken. The most important of these precautions is that samples, once obtained, must be analyzed immediately as nonenzymatic generation of PGF_2 compounds *ex vivo* from arachidonic acid in biological fluids occurs very readily.[12] This problem is further

[12] J. D. Morrow, T. M. Harris, and L. J. Roberts II, *Anal. Biochem.* **184,** 1 (1990).

discussed later. The procedures outlined below have been successfully applied for analysis of 9α,11β-PGF_2 in urine, cerebrospinal fluid, plasma, and to assess conversion of PGF_2 to 9α,11β-PGF_2 in *in vitro* cell incubations. Basic techniques for each sample type are similar.

Purification of 9α,11β-PGF_2 from Biological Fluids

Using plasma as an example, to 3 ml of platelet-poor plasma is added 200–600 pg of [$^2H_7$]9α,11β-PGF_2. The mixture is acidified to pH 3 with 1 *N* HCl and applied to C_{18} Sep-Pak columns (Waters Associations, Milford, MA) preconditioned with 5 ml methanol and 5 ml of water (pH 3). The Sep Pak columns are then washed with 10 ml of water (pH 3), 10 ml of acetonitrile/water (15 : 85, v/v) (to remove polar impurities when present such as in urine), 10 ml heptane, and then eluted with 10 ml ethyl acetate/heptane (50 : 50, v/v).

The ethyl acetate/heptane eluate from the C_{18} Sep-Pak is dried over anhydrous Na_2SO_4 and applied to a silica Sep-Pak (Waters Associates, Milford, MA). The Sep-Pak is then washed with 5 ml ethyl acetate and subsequently with 5 ml of ethyl acetate/methanol (50 : 50 v/v). The ethyl acetate/methanol eluate from the silica Sep-Pak is evaporated under a stream of nitrogen and the residue subjected to TLC using Whatman (Clifton, NJ) LK6D silica gel TLC plates with the solvent chloroform/methanol/acetic acid/water (86 : 14 : 1 : 0.8, v/v/v/v). A few micrograms of $PGF_{2\alpha}$ are spotted and run on a separate TLC plate and visualized by spraying with a 10% solution of phosphomolybdic acid in ethanol followed by heating. Compounds migrating in the region of $PGF_2\alpha$ (R_f about .33) are scraped and extracted from the silica gel with ethanol/ethyl acetate (50:50, v/v). Routinely an area 1 cm on either side of the $PGF_2\alpha$ standard spot is scraped from the sample TLC plates. $PGF_{2\alpha}$ and 9α,11β-PGF_2 are incompletely resolved on silica TLC. $PGF_{2\alpha}$ is used as the standard for visualization instead of 9α,11β-PGF_2 to avoid contamination of the samples with 9α,11β-PGF_2 since microgram quantities of the standard are handled and samples being analyzed frequently contain only picogram quantities. If some contamination of samples with $PGF_{2\alpha}$ occurs, this does not interfere since $PGF_{2\alpha}$ and 9α,11β-PGF_2 are completely resolved on the capillary GC column when analyzed by GC-MS. This TLC separation of the free acid is usually only required when the biological fluid being analyzed contains a large amount of polar impurities, such as urine. Otherwise this step can be omitted and sufficient purification can be accomplished by using only the second TLC procedure outlined below in which the compounds are subjected to TLC after conversion to PFB ester.

After Sep-Pak extraction or free-acid TLC the sample is dried under nitrogen and converted to a PFB ester by treatment with a mixture of 40 μl of 10% pentafluorobenzyl bromide in acetonitrile and 20 μl of 10% *N,N*-diisopropylethylamine in acetonitrile at room temperature for 30 min. The reagents are then dried under nitrogen and the procedure is repeated a second time. We have observed that performing the esterification procedure once does not result in quantitative esterification even if the reaction is carried out longer than 30 min or if it is heated. However, repeating the procedure does result in near quantitative conversion to the PFB ester. After the second esterification, the material is dried under nitrogen and subjected to TLC using the solvent chloroform/ethanol (93:7, v/v). Compounds migrating in the region of the methyl ester of $PGF_{2\alpha}$ (R_f 0.15) and the adjacent areas 1 cm on either side are scraped and extracted from the silica gel with ethyl acetate. The methyl ester of $PGF_{2\alpha}$ is used as the TLC standard instead of a PFB ester as a further precaution to avoid contamination of the samples being analyzed since the methyl ester will not be detected when analyzed by GC–NCI–MS.

GC–MS Analysis of 9α,11β-PGF₂

The 9α,11β-PGF_2 is then converted to the trimethylsilyl ether derivative by adding 20 μl *N,O*-bis(trimethylsilyl)trifluoroacetamide (BSTFA) and 20 μl dimethylformamide and incubating at 40° for about 20 min. We have found that the use of solvents other than dimethylformamide such as pyridine with BSTFA results in incomplete silyation of 9α,11β-PGF_2 even though $PGF_{2\alpha}$is quantitatively silylated by a mixture of pyridine and BSTFA. Replacing pyridine with dimethylformamide, however, results in quantitative silyation of 9α,11β-PGF_2. The sample is then dried under nitrogen and redissolved for GC–MS analysis in 10–20 μl of undecane or dodecane which has been dried over calcium hydride.

For analysis of 9α,11β-PGF_2, we use a Nermag R10-10C GC–MS in strument interfaced with a DEC-PDP 11/23 plus computer system. GC is done with a 15 m DB1701 fused silica capillary column (J & W Scientific, Folsom, CA). We have found that this column gives a far superior separation of 9α,11β-PGF_2 from other PGF_2 compounds when compared to other columns such as DB-1. For example, $PGF_{2\alpha}$ elutes at a retention time of approximately 8 sec longer than 9α,11β-PGF_2. The column temperature is programmed from 190° to 300° at 20°/min. Methane is used as the carrier gas for negative-ion chemical ionization (NCI) at a flow rate of about 1 ml/min. Ion source temperature is 250°, electron energy is 70 eV, and filament current 0.25mA. Ions monitored for endogenous 9α,11β-PGF_2 and the [$^2H_7$]9α,11β-PGF_2 internal standard are the carboxylate anions *m/z* 569 and *m/z* 576, respectively (M − 181, loss of $\cdot CH_2C_6F_5$).

Application of Assay for Analysis of 9α,11β-PGF_2 in Various Biological Fluids

As mentioned previously, we have successfully applied this assay procedure for analysis of 9α,11β-PGF_2 in a number of diverse biological fluids, such as plasma, urine, cerebrospinal fluid, cell incubates, and other fluids.

Shown in Fig. 1 are selected ion-current chromatograms obtained from skin blister fluid of a child suffering from cutaneous bullous mastocytosis, a disorder characterized by an excess of skin mast cells associated with the formation of large blisters on the skin. The peak in the upper m/z 569 selected ion-current chromatogram represents the endogenous 9α,11β-PGF_2 present. 9α,11β-PGF_2 has a slightly longer retention time (1 sec) than the 9α,11β-PGF_2 internal standard shown in the lower m/z 576 chromatogram because of GC isotope discrimination between protium and deuterium. The level of 9α,11β-PGF_2 measured in the blister fluid was 290 ng/ml.

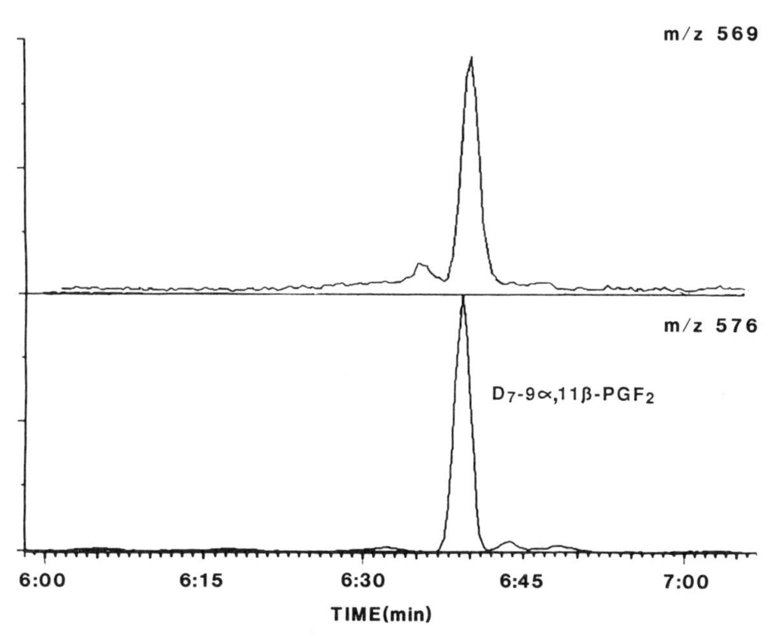

FIG. 1. Analysis of blister fluid from a child with cutaneous bullous mastocytosis by GC–NCI–MS. The m/z 569 selected-ion current chromatogram represents endogenous 9α,11β-PGF_2 and the m/z 576 chromatogram represents the [$^2H_7$]9α,11β-PGF_2 internal standard. The level of 9α,11β-PGF_2 is 290 ng/ml blister fluid.

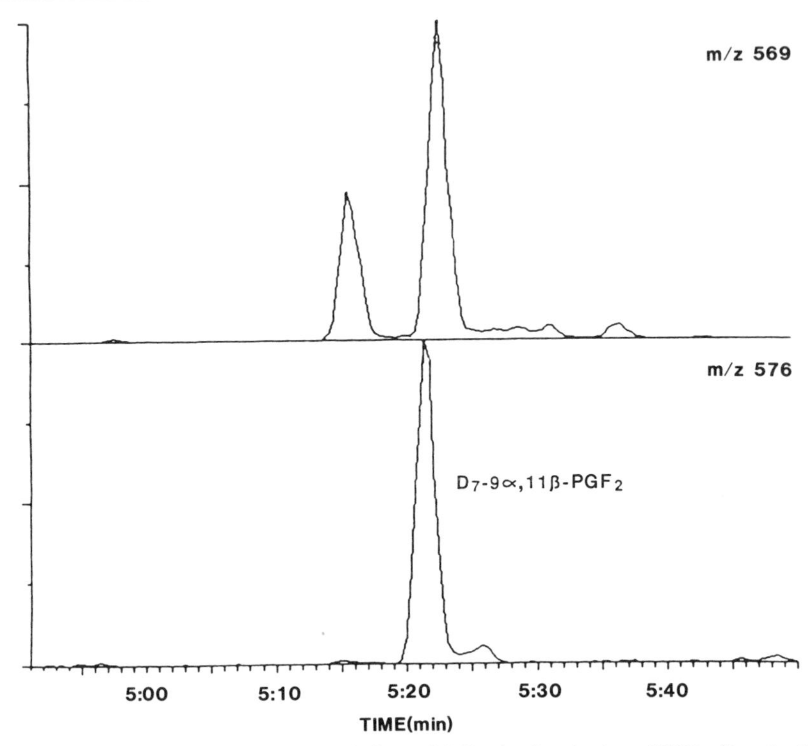

FIG. 2. Analysis of PGF_2 compounds formed following incubation of PGD_2 (2 μm) with 5×10^5 human eosinophils. The upper *m/z* 569 chromatogram shows two peaks. The one directly above the *m/z* 576 $[^2H_7]9\alpha,11\beta$-PGF_2 internal standard represents $9\alpha,11\beta$-PGF_2.

We recently found that the human eosinophil contains 11-ketoreductase activity.[13] In Fig. 2 the upper *m/z* 569 chromatogram reveals PGF_2 formed following incubation of PGD_2 with human eosinophils. The peak directly above the *m/z* 576 internal standard peak shown in the lower chromatogram again represents $9\alpha,11\beta$-PGF_2 formed. Also formed from PGD_2 in these incubations is an isomer of $9\alpha,11\beta$-PGF_2 which is represented by the prominent peak eluting at a slightly shorter retention time than the $9\alpha,11\beta$-PGF_2. This compound was tentatively identified as 12-epi-$9\alpha,11\beta$-PGF_2 by demonstrating cochromatography of this compound on capillary GC with known 12-epi-$9\alpha,11\beta$-PGF_2. Recently we have found that PGD_2 undergoes isomerization and that albumin (and probably other proteins) can catalyze the isomerization. Albumin has also

[13] W. G. Parsons III and L. J. Roberts II, *J. Immunol.* **141,** 2413 (1988).

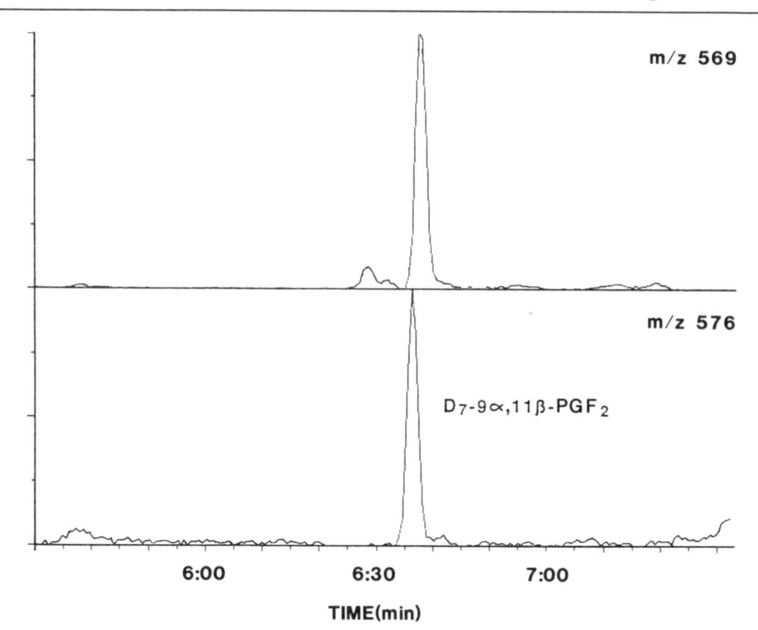

FIG. 3. Analysis of plasma obtained from an individual during pharmacological intervention thought to be associated with release of prostaglandins *in vivo*. The *m/z* 569 chromatogram represents endogenous $9\alpha,11\beta$-PGF_2 and the *m/z* 576 chromatogram represents the $[^2H_7]9\alpha,11\beta$-PGF_2 internal standard. The level of endogenous $9\alpha,11\beta$-PGF_2 was 560 pg/ml (normal = approximately 1 pg/ml or less).

been shown to catalyze the dehydration of PGD_2.[14] Following isomerization of PGD_2 in the presence of albumin, reduction by 11-ketoreductase results in the formation of isomeric PGF_2 compounds. One of the major isomers formed in the presence of albumin is 12-epi-$9\alpha,11\beta$-PGF_2.

Figure 3 shows an example of measurement of $9\alpha,11\beta$-PGF_2 in plasma obtained from an individual during a pharmacological intervention thought to possibly be associated with the release of prostaglandins *in vivo*. Again at the top is the *m/z* 569 chromatogram representing endogenous $9\alpha,11\beta$-PGF_2. As discussed below, normal circulating levels of $9\alpha,11\beta$-PGF_2 in plasma are probably in the range of 1 pg/ml or less. Levels present in this plasma sample obtained during the pharmacological intervention were markedly elevated at 560 pg/ml, indicating that large quantities of PGD_2 had been released.

In contrast to Fig. 3 where a single prominent $9\alpha,11\beta$-PGF_2 peak was present in plasma during a situation in which large quantities of PGD_2 are

[14] F. A. Fitzpatrick and M. A. Wynalda, *J. Biol. Chem.* **258,** 11713 (1983).

released, analysis of plasma from a normal individual reveals the presence of multiple *m/z* 569 peaks, including a very small peak with a retention time relative to the internal standard where $9\alpha,11\beta$-PGF_2 elutes (Fig. 4). However, we have recently found that essentially all or almost all of the compounds represented by these peaks are PGF_2 compounds formed by noncyclooxygenase oxidation of arachidonic acid and are not derived from reduction of PGD_2. The levels of these compounds in fresh plasma range from approximately 5–40 pg/ml and evidence has been obtained that the levels of these compounds formed by this noncyclooxygenase mechanism are actually circulating *in vivo*.[12] In addition, analysis of urine reveals a similar pattern of compounds ranging from about 500–1000 pg/ml and our recent evidence also suggests that almost all of these compounds also are formed via the noncyclooxygenase oxidation mechanism and do not derive from PGD_2. The formation of these compounds also occurs very readily *ex vivo* in plasma and, therefore, probably in any biological fluid containing lipids. For example, levels of these compounds in plasma that has been stored for a period of time at $-20°$ are characteristically in the range of 1–5 ng/ml.

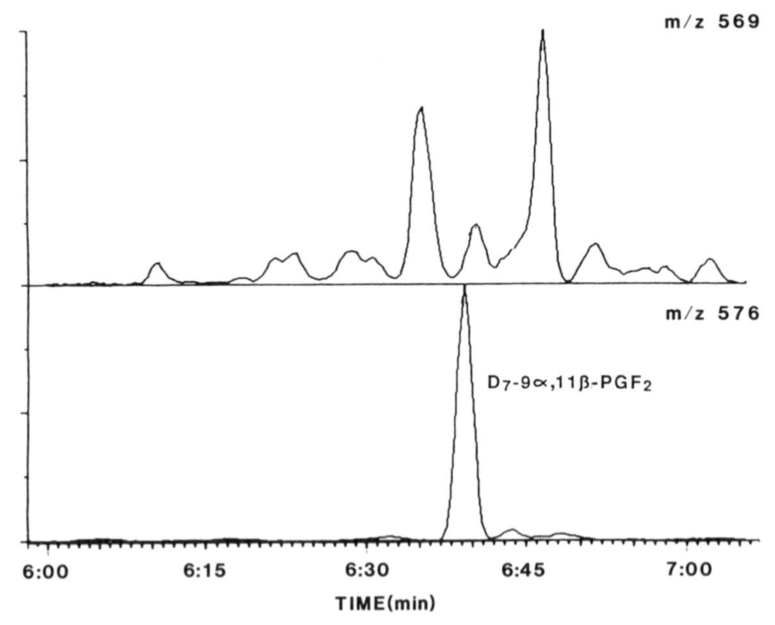

FIG. 4. Analysis of fresh plasma from a normal individual. The multiple *m/z* 569 chromatographic peaks almost entirely represent PGF_2 compounds formed by noncyclooxygenase oxidation of arachidonic acid rather than from 11-ketoreductase metabolism of PGD_2. The levels of these compounds range from approximately 5–40 pg/ml. The lower *m/z* 576 chromatogram again represents the $[^2H_7]9\alpha,11\beta$-PGF_2 internal standard.

It is important to recognize that PGF_2 compounds formed via noncyclooxygenase oxidation of arachidonic acid may be easily misinterpreted as isomeric PGF_2 compounds arising from PGD_2. The mechanism of formation of the compounds involves the formation of bicyclic endoperoxides, which are directly reduced to 4 regioisomers of PGF_2 compounds, each of which is comprised of a mixture of 8 racemic diastereomers.[12] Because these compounds arise from direct reduction of intermediate endoperoxides, the cyclopentane ring hydroxyls are oriented cis in all of these compounds, in contrast to $9\alpha,11\beta$-PGF_2 in which the cyclopentane ring hydroxyls are trans. Therefore, one way that these compounds can be differentiated from PGF_2 compounds arising from 11-ketoreductase metabolism of PGD_2 is to access their ability to form a cyclic boronate derivative. Treatment of PGF_2 compounds in which the cyclopentane hydroxyls are cis with butylboronic acid forms a cyclic butylbornate derivative bridging the oxygens at C-9 and C-11 whereas $9\alpha,11\beta$-PGF_2 will not form this derivative.[15] Thus, it is possible to treat samples with butylboronic acid and then quantify the material that did boronate. The PGF_2 compounds are initially converted to a PFB ester and then treated with 50 μl butylboronic acid in acetonitrile:dimethylformamide (4:1) (v/v) for 60 min at 20°. The reagents are then evaporated under nitrogen, silylated as previously described, and analyzed as before with selected-ion monitoring of *m/z* 569 and *m/z* 576. However, if this procedure is used, it is important to add a control to the sample to determine completion of the boronation reaction. One such control that we have employed is addition of $[^2H_4]PGF_{2\alpha}$ which is commercially available (Cayman Chemicals, Ann Arbor, MI). The PFB Me_3Si ether derivative of $[^2H_4]PGF_{2\alpha}$ generates an ion at *m/z* 573 when analyzed by GC–MS–NCI and thus does not interfere with PGF_2 compounds present at *m/z* 569. The PFB ester, butylboronate, Me_3Si ether derivative of $[^2H_4]PGF_{2\alpha}$ generates an ion at *m/z* 495 representing the M − 181 ion of this derivative. From the ratio of *m/z* 573 to *m/z* 495 it is, therefore, possible to determine whether the *cis*-hydroxyl PGF_2 compounds were quantitatively converted to a cyclic boronate derivative. It may also be of value to mention that we have observed that treatment of $9\alpha,11\beta$-PGF_2 with butylboronic acid results in what appears to be considerable destruction of the compound, although the compound does not form a cyclic boronate derivative. The mechanism of this is unclear. In any event, if the approach using a boronation reaction is employed for measurement of $9\alpha,11\beta$-PGF_2, substantial quantities of the compound need to be present in the sample to permit its detection after treatment with butylboronic acid.

[15] C. Pace-Asciak and L. S. Wolfe, *J. Chromatogr.* **56,** 129 (1971).

One might question whether an RIA for $9\alpha,11\beta$-PGF_2 would possibly be superior to the GC–MS assay to differentiate $9\alpha,11\beta$-PGF_2 from the *cis*-hydroxyl-PGF_2 compounds formed by the noncyclooxygenase oxidative mechanism. An RIA kit has recently become commercially available for quantification of $9\alpha,11\beta$-PGF_2 (Amersham, UK). The antibody has only 0.0006% cross-reactivity with $PGF_{2\alpha}$. However, we have found that there is substantial cross-reactivity of this antibody with the *cis*-hydroxyl-PGF_2 compounds formed by the noncyclooxygenase oxidative mechanism. Therefore, it would appear that the GC–MS assay employing the boronation procedure is superior to RIA in differentiating these compounds from $9\alpha,11\beta$-PGF_2.

In conclusion, the mass spectrometric assay we have developed provides a sensitive and accurate means for the analysis of $9\alpha,11\beta$-PGF_2 in a variety of biological fluids. However, special precautions must be taken to differentiate $9\alpha,11\beta$-PGF_2 from PGF_2 compounds present in biological fluids that arise from noncyclooxygenase oxidation of arachidonic acid rather than from 11-ketoreductase metabolism of PGD_2.

[7] Immunoaffinity Purification–Chromatographic Quantitative Analysis of Arachidonic Acid Metabolites

By J. JAMES VRBANAC, JEFFREY W. COX, THOMAS D. ELLER, and DANIEL R. KNAPP

Quantitative analysis of arachidonic acid metabolites, and structurally related compounds of biological and synthetic origin, can be divided into two basic steps. These are extraction–purification procedures and analytical measurement by such techniques as radioimmunoassy (RIA), gas chromatography (GC), GC–mass spectrometry (GC–MS), and high-performance liquid chromatography (HPLC) with ultraviolet (UV) or fluorescence (FL) detection.[1] General extraction techniques are simple, efficient, and rapid. Unfortunately purification procedures are time-consuming, can result in significant loss of analyte, and may involve more than one chromatographic step for chemically complex samples such as urine. For example, measurement of urinary eicosanoids by the highly specific analytical technique of gas chromatography–negative-ion chemical ionization mass spectrometry (GC–NCI–MS) requires an analytical chromatographic purification step prior to derivatization for GC.

[1] W. E. M. Lands and W. L. Smith, this series, Vol. 86 Sections II, IV, and V.

Immunoaffinity purification is an attractive alternative to the more traditional purification techniques such as TLC, LC, and HPLC.[2–7] Quantitative analytical procedures which utilize immunoaffinity purification offer the potential of greatly decreased analysis time. Immunoaffinity purification techniques are, in general, robust, fast, and easy to perform.

Immunoaffinity purification is a simple and powerful analytical technique when applied as the initial step in a three-stage analytical process which includes a high-performance chromatography step followed by a specific detection method, for example, GC–NCI–MS, or HPLC-FL detection. The degree of analytical specificity provided by this strategy can be exceptional due to the differences in the physicochemical principles involved in the extraction, chromatography, and detection phases of the process.[2–14] For example, immunoaffinity extraction followed by GC–NCI–MS discriminates analytes on the basis of three-dimensional structure, volatility, polarity, and atomic weight. This chapter will describe in detail the immunoaffinity purification step of this analytical technique. Methods of immunoaffinity purification using both solid-phase extraction and immunoaffinity precipitation will be described.

Procedure

Materials and Reagents. Immobilized protein A on Sepharose CL-4B (Pierce Chemical, Rockford, IL); Sep-Pak octadecasilyl cartridges (ODS) (Waters, Milford, MA); and Affi-Gel (10 hydroxysuccinimidyl coupling gel

[2] H. L. Hubbard, T. D. Eller, D. E. Mais, P. V. Halushka, R. H. Baker, I. A. Blair, J. J. Vrbanac, and D. R. Knapp, *Prostaglandins* **33,** 149 (1987).

[3] J. J. Vrbanac, T. D. Eller, and D. R. Knapp, *J. Chromatogr.* **425,** 1 (1988).

[4] J. W. Cox, R. H. Pullen, and M. E. Royer, *Anal. Chem.* **57,** 2365 (1985).

[5] J. W. Cox and R. H. Pullen, *J. Pharm. Biomed. Anal.* **4,** 653 (1986).

[6] J. W. Cox, W. M. Bothwell, R. H. Pullen, M. A. Wynalda, F. A. Fitzpatrick, and J. T. Vanderlugt, *J. Pharm. Sci.* **75,** 1107 (1986).

[7] S. J. Gaskell and B. G. Bronsey, *Clin. Chem.* **29,** 677 (1983).

[8] W. Krause, U. Jakobs, P. E. Schulze, B. Nieuweboer, and M. Humple, *Prostaglandins Leukotriene Med.* **17,** 167 (1985).

[9] A. Nakagawa, Y. Matsushita, S. Muramatsu, Y. Tanishima, T. Hirota, W. Takasaki, Y. Kawahara, and H. Takahagi, *Biomed. Chromatogr.* **2,** 203 (1987).

[10] U. Lemm, J. Tenczer, H. Baudisch, and W. Krause, *J. Chromatogr.* **342,** 393 (1985).

[11] C. Chiabrando, A. Benigni, A. Piccinelli, C. Carminati, E. Cozzi, G. Remazzi, and R. Fanelli, *Anal. Biochem.* **163,** 255 (1987).

[12] J. W. Cox and R. H. Pullen, *Anal. Chem.* **56,** 1866 (1984).

[13] J. W. Cox, N. A. Andriadis, R. C. Bone, R. J. Maunder, R. H. Pullen, J. J. Ursprung, and M. J. Vassar, *Am. Rev. Resp. Dis.* **137,** 5 (1988).

[14] R. H. Pullen and J. W. Cox, *J. Chromatogr.* **343,** 271 (1985).

(Bio-Rad Laboratories, Richmond, CA) can be purchased along with other materials from standard sources. Polyvinylpyrrolidone (PVP) buffer, pH 7.4, consists of 8.76 g sodium chloride, 6 g Trizma hydrochloride, 1 g sodium azide, and 1 g PVP per liter of water and the pH is adjusted with 1 *M* HCl. Phosphate-buffered saline (PBS) consists of 0.05 *M* potassium dihydrogen phosphate and 0.15 *M* sodium chloride adjusted to pH 7.4 with 1 *M* NaOH.

Antiserum

The techniques for preparation of antiserum for arachidonic acid metabolites have been described previously in this series and will not be discussed here.[1] There are two important issues related to antiserum production that need to be discussed in the present context. First, cross-reactivity with structurally similar compounds is not necessarily a problem as it is in RIA procedures and, in fact, could be desirable if more than one structurally related analyte is to be measured. For example, polyclonal antisera to 6-ketoprostaglandin $F_{1\alpha}$ (6K-$PGF_{1\alpha}$) and thromboxane B_2 (TxB_2) have significant cross-reactivity with their 2,3-dinor metabolites. Although this is an undesirable situation if the antisera is to be used in RIA it is, in fact, a highly desirable situation when using immunoaffinity purification coupled with GC–MS analysis since both metabolites, or all four metabolites if the immunoaffinity gels are combined, can be measured simultaneously. A similar situation exists for PGE_1 and PGE_2. Thus, samples of antisera with higher cross-reactivity could be more desirable. A second important consideration relates to the amount of antiserum needed for a series of research projects. As will be described later, if the antiserum is bound to a solid support matrix it can be used repeatedly in a series of binding, elution, and reconstitution steps. Thus, one rabbit with an average titer can supply enough immunoaffinity gel to construct several thousand columns if the total binding capacity of each column is between 100–200 ng. On the other hand, it is more difficult to reuse antibody from immunoprecipitation extraction procedures. Because of its simplicity, however, the latter approach may be used advantageously when only a few hundred samples are to be processed or the antibody supply is not limited. Finally, if a double-antibody precipitation method is to be used, then a second antiserum, such as goat anti-rabbit IgG, is needed.

Isolation of IgG Fraction

The polyclonal antibody contained in raw serum must be partially purified before immobilization on a solid-phase support. Dilute 1 ml of antiserum with 2 ml of PBS buffer and apply the mixture to a column containing 5 ml of immobilized protein A on Sepharose. Filter the serum solution through a plug of glass wool in a pipette during the addition to prevent insoluble aggregates from accumulating on the gel. Vortex the suspension and let stand for 15 min. Open the column and wash the unbound proteins from the gel with 50 ml of PBS. Rewash the gel with 15 ml of isotonic saline to remove most of the buffer salts. The bound IgG fraction is eluted with 15 ml of 1% acetic acid in isotonic saline at a rate of about 3 ml/min. The eluate is immediately adjusted to pH 7–8 with 2.5 *M* bicarbonate solution. The IgG may be used in this form, or it may be rapidly frozen with a −70° Dry Ice–acetone bath and stored frozen for later coupling. The protein A column need only be washed with 20 ml of PBS and it is ready for reuse. It is strongly recommended that the protein A columns be dedicated to one antiserum because some IgG is retained after each use and will appear in subsequent elutions, even if the gel is well washed with 1% acetic acid.

It is helpful to be able to estimate how much of a given antiserum is required to prepare a given quantity of immunoaffinity gel with a particular binding capacity. This information can be obtained from the standard binding assay data used to determine the titer of an antiserum. In the following example the binding data for an antiserum to 6K-$PGF_{1\alpha}$ is used to calculate the anticipated yield. The point on the assay curve where half the the total counts per minute (CPM) added was bound by the antiserum is selected to determine what dilution of serum binds a given mass of antigen. In this case, 0.1 ml of serum diluted 1:2700 was found to bind 11,108 CPM of tracer out of 22,579 present. The tracer has a molecular weight of 370 and specific activity of 150 Ci/m*M*. One microcurie is equivalent to about 1,100,000 CPM; therefore, 11,100 CPM corresponds to about 0.067 pmol or 25 pg. The IgG obtained from 1 ml of serum would therefore be expected to bind on the order of 27,000 × 25 pg = 680 ng. The actual gel capacity obtained would depend on the efficiency of both protein A extraction and coupling to the activated support. The general procedure for partial purification of the IgG fraction, immobilization of the antiserum, and the use of immunoaffinity gels is outlined in Fig. 1.

Immobilization of Antiserum

Purified IgG in about 20 ml of isotonic saline is treated with approximately 200 mg of sodium bicarbonate to bring the pH to 8.0 and the solution is added to a cake of about 12 ml of *N*-hydroxysuccinimidyl coupling gel which was washed just previously with ice-cold distilled water. The mixture is rotated gently at 4° overnight. (Rotation about the long axis of the container is recommended.) Any remaining chemically reactive sites on the gel are subsequently deactivated by addition of 1 ml of

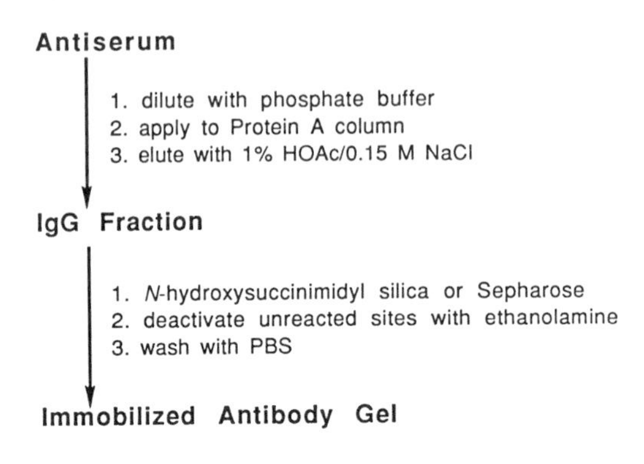

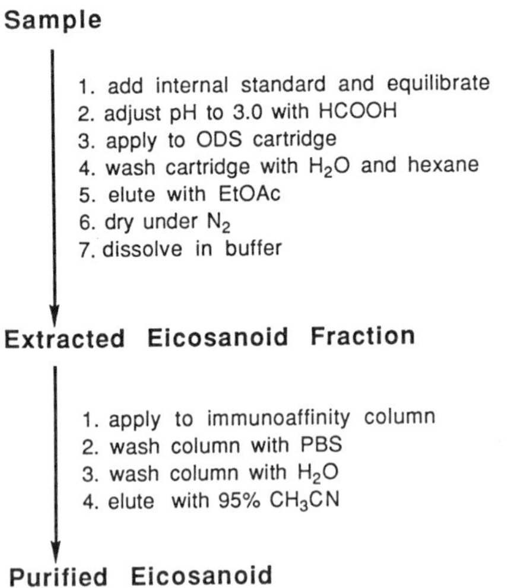

Immunoaffinity Extraction
Sample
1. add internal standard and equilibrate
2. adjust pH to 3.0 with HCOOH
3. apply to ODS cartridge
4. wash cartridge with H2O and hexane
5. elute with EtOAc
6. dry under N2
7. dissolve in buffer
Extracted Eicosanoid Fraction
1. apply to immunoaffinity column
2. wash column with PBS
3. wash column with H2O
4. elute with 95% CH3CN
Purified Eicosanoid

FIG. 1. Diagram outlining the general procedure for preparation and use of immunoaffinity purification columns.

10% aqueous ethanolamine which has been previously adjusted to pH 8 with concentrated HCl. The mixture is rotated at room temperature for 1 hr to complete deactivation and the gel is washed with 10 volumes of PBS, followed by 2 volumes of acetonitrile:water (19:1), and finally with 5 volumes of PBS again.

Binding Capacity of the Gel

The binding capacity of the immunoaffinity gels often far exceeds the amount required and this capacity can be adjusted by dilution with inactive blank support. Dilution is problematical with commercial silica-based functional supports, however, due to the particle sizes available. Sepharose supports are easier to use in this regard (we have also noted less background as a result of nonspecific adsorption for GC–MS analysis when using Sepharose gels compared to silica gels). To determine the approximate binding capacity of a batch of gel, prepare two columns containing 2 ml of immunoaffinity gel each. Prepare a solution consisting of about 200,000 CPM of ^{3}H-labeled antigen in 1.1 ml of PVP buffer. Pipette 0.5 ml of this solution into two containers and add 0.5 ml of PVP buffer to one tube (control) and 0.5 ml of PVP buffer containing 4 μg of unlabeled antigen to the second. Mix well and add each solution to one of the columns without disturbing the gel and incubate 15 min at room temperature. Wash the gels with 30 ml of PBS buffer and 15 ml of distilled water each, then elute the bound material with 15 ml of acetonitrile:water (19:1). *Never wash the gels with absolute organic solvents which will irreversibly denature the immobilized protein.* Evaporate the sample under a stream of nitrogen or on a rotary evaporator while maintaining the bath temperature at no more than 40°. Count the evaporated eluates to determine the total CPM bound in the control column and the column with excess antigen present. The ratio of the recovered counts reflects the amount of unlabeled antigen bound. The control column will bind about 97% of the tracer added. If it was determined, for example, that the spiked column CPM recovered was 25% of the control, then the column capacity is $0.25 \times 4 = 1.0$ μg or about 0.5 μg/ml of gel. This capacity is high for many applications and the gel could, therefore, be diluted.

Extraction-Purification of Samples

Solid Phase

The following method was developed for measurement of urinary eicosanoids using GC–MS but is generally applicable to other matrixes and methods.[2,3] Add a stable isotope-labeled internal standard to the sample and vortex mix. Adjust the pH to 3.0 with 10 *M* formic acid. Slowly pass the sample through preconditioned ODS cartridges (Waters Sep-Pak or equivalent, washed with methanol followed by distilled water) at the rate of approximately 1–3 drops per sec. Wash the cartridge with 40 ml of distilled water and 10 ml of hexane. Elute eicosanoids with 8–10 ml of ethyl acetate and evaporate to dryness under a stream of nitrogen. Redissolve the residue in 1.2 ml of PVP buffer. Adjust the pH of the solution to between 7.0 and 7.4 with 2.5 *M* sodium bicarbonate. The residue is acidic and will frequently exceed the buffer capacity. The immunoaffinity columns will not function optimally at low pH or in the presence of organic solvents. If the sample still contains insoluble material after pH adjustment, filter it before application to the affinity column to avoid contaminating the bed.

For purification of biological samples, use 2 ml of immunoaffinity gel contained in a silanized all-glass column with silanized frit measuring about 1–2 cm wide and 8–10 cm in height (we have found silanized all-glass columns to be important if GC–MS analysis is to be used). Apply the sample slowly to avoid disturbing the gel. Optimal results are obtained if the volume of sample added is equivalent to the void volume of the gel (this is determined by slow addition of a measured volume of a solution of food coloring to the column and observing when the color reaches the frit). The volume for a 2-ml Sepharose gel bed is typically between 1.2 and 1.4 ml. Allow the samples to equilibrate with the gel for 15 min, then wash the column with 30 ml of PBS followed by 10 ml of distilled water. Push out the water remaining in the column dead volume, elute the bound sample with 15 ml of acetonitrile : water (v:v,19 : 1), and evaporate the solvents under nitrogen or with a rotary evaporator. Wash the gel with an additional 10–15 ml of acetonitrile : water to remove any traces of the sample, then wash with 20 ml of PVP buffer to ready the column for another sample. If suction or pressure is used to accelerate the washing and elution process, do not dry out the gel by moving large amounts of air through the column as this can cause irreversible denaturation of the IgG. The gel may be vortexed to help release trapped bubbles. Store the columns under PBS saturated with nitrogen at 4°.

Immunoprecipitation

The following method was developed for quantitation of PGE_1[5] but has been used for other E-type prostaglandins[4,6] and, in principle, is applicable to a wide variety of analytes and matrixes. Rabbit anti-PGE_1 (0.06 ml of a 1 : 10 dilution of 0.1 *M* Tris-HCl buffer, pH 7.8, and containing 0.1 mg/ml Thimerosal) is added to thawed plasma samples (0.5 ml) at room temperature. Vortex for 30 sec. Shake on a horizontal shaker at 3 cycles/sec for 30 min. Add 0.03 ml of undiluted goat anti-rabbit serum and vortex for 30 sec. Incubate overnight at 4°. Vortex samples for 1 min and centrifuge for 10 min at 1500 *g* at room temperature. Decant the supernatant by inversion and touch the lip of the tube to a paper towel to remove the last drop. Resuspend the pellet in 1.0 ml of distilled water by vortexing for 1 min. Centrifuge as before. Decant the supernatant and repeat the wash cycle once more. Decant the wash and add 1.0 ml of acetonitrile, vortex, and centrifuge for 10 min at 1500 *g*. Transfer the acetonitrile extract (0.9 ml) to a tube and evaporate to dryness under the stream of nitrogen. The antiserum dilution factors used with this technique should be optimized for each analyte and batch of antiserum. Using a radioactive analyte concentration that is at least twice the anticipated level in test samples, independently vary the dilution factors for primary and secondary antisera and measure the recovery of radioactivity in the precipitate. The objective is to determine the highest antiserum dilution factor that is compatible with maximum analyte recovery. In our hands, the optimal dilution factor for the primary antiserum is typically at least 100 times lower than would be required for radioimmunoassay with the same antiserum. The precipitation efficiency should be verified to be constant over the analyte concentration range of interest, and it is also advisable to screen several solvents to determine which provides maximum recovery of the analyte from the precipitate. In the case of E-type prostaglandins, acetonitrile, tetrahydrofuran, 2-propanol, and ethyl acetate all gave greater than 95% recovery for radioactive analyte from the immunoprecipitate, while toluene and chlorinated hydrocarbons gave recoveries of less than 30%.[4]

Quantitative Analysis

Gas Chromatography–Mass Spectrometry

Following immunoaffinity purification of the biological sample, eicosanoids can be quantitated by GC–MS with a high degree of analytical specificity. The method of choice with respect to sensitivity and specificity is high-resolution GC–NCI–MS. Form methoxime-pentafluorobenzyl ester-trimethylsilyl ether (MO-PFB-TMS) derivatives for these analyses (form the methoxime derivative only if a carbonyl group is present) as previously described.[3] NCl mass spectra of these derivatives are very simple, with the majority of the ion current present in the carboxylate anion formed by dissociative electron capture. An alternative recommended approach if Cl is not available is to form MO-methyl ester-*tert*-butyldimethylsilyl ether derivatives and monitor $M^+ - 57$ using electron ionization.[15] These and other GC–MS methods have been extensively described in the literature.[16]

High-Performance Liquid Chromatography–Fluorescence Detection

As with GC–MS methods there exists a number of different HPLC methodologies that could be developed to quantitate eicosanoids following

[15] A. C. Bazan and D. R. Knapp, *J. Chromatogr.* **236,** 201 (1982).

[16] A. L. Burlingame, D. Maltby, P. T. Holland, and D. H. Russell, *Anal. Chem.* **60,** 319R (1988).

immunoaffinity purification. One method with proved sensitivity in the picogram per milliliter range is to derivatize with panacyl bromide and quantify the panacyl ester derivatives by normal-phase HPLC-FL.[12] This method utilizes column-switching technology to perform on-line separation of the excess derivatizing reagent and to impart additional specificity, and has been used successfully to quantify E-type prostaglandins in plasma at low picogram per milliliter levels following immunoextraction.[5,6,13] Whether such a sophisticated HPLC system will be required for a given problem will depend on the analyte, the matrix, and the concentration. It is clear, however, that the complexity of the overall procedure will be reduced by substituting immunoaffinity purification procedures for conventional extraction and cleanup procedures. For example, the conventional method for extracting and quantitating 15-methyl-PGE_2 from plasma prior to panacyl bromide derivatization and HPLC-FL analysis involved a two-stage solid-phase extraction followed by reversed-phase HPLC purification.[14] In contrast, the immunoaffinity purification method involved only a simple immunoprecipitation/extraction procedure and produced a cleaner simple for HPLC analysis in only a fraction of the time.[4]

[8] Preparation of Tetradeuterated Leukotriene A_4 Methyl Ester: Methyl-[11,12,14,15-$^2H_4$]-(5*S*,6*S*)-*Oxido*-(7*E*,9*E*,11*Z*,14*Z*)-Eicosatetraenoate

By J. P. LELLOUCHE, J. P. BEAUCOURT, and A. VANHOVE

PPh_3, Br^- —(n-BuLi)→ PPh_3

CO_2CH_3 —(− Ph_3PO)→ CO_2CH_3 (1) + CO_2CH_3 (2) (1)

H_2 or D_2 → (2)

(3) R = - H
(4) R = - D

Procedure

Wittig Coupling Step

A three-neck 100-ml round-bottom flask is oven-dried overnight and cooled in a glove box under nitrogen. The flask, equipped with a rubber septum, an alcohol thermometer, and a magnetic stir bar is loaded with 518.0 mg (1.0 mmol) of 2*E*-tridecene-(4,7)-diynyl triphenylphosphonium bromide.[1] Crystalline phosphonium bromide (mp 175°) is stable for over 1 year when stored in a stoppered flask at 0° under nitrogen. The flask is connected to a high-vacuum five-port line fitted with a dry nitrogen inlet. The phosphonium bromide is dried at 10^{-2} mm Hg for 1 hr, then a slight positive pressure of dry nitrogen is applied and maintained throughout the reaction. Anhydrous tetrahydrofuran (25 ml), distilled from sodium benzophenone ketyl immediately before use is added with a syringe. The tetrahydrofuran suspension is cooled to −70° in a dry ice–ethanol bath. Then, 0.62 ml (0.99 mmol) of a 1.6 *M* solution of *n*-butyllithium (assayed according to the Gilman and Cartledge procedure[2]) in hexane (Merck Schuchardt) is added dropwise over 10 min with stirring. The addition rate maintains the internal temperature at −70°. The solution which turns black is stirred for an additional 10 min at −70°. Methyl 7-oxo-(5*S*,6*R*)-oxidoheptanoate[3] (171.0 mg, 0.99 mmol) dissolved in anhydrous tetrahydrofuran (12 ml) is added dropwise with a syringe over 30 min. The addition rate maintains the internal temperature at −70°. The solution is stirred 20 min at −70°. After removal of the cooling bath, the black solution is allowed to warm up to 0° within 35–45 min. The reaction mixture is maintained at 0° by an ice–water bath for 20 min. It is then poured into an ice-cold mixture of water–triethylamine (50 ml–3 ml) containing ammonium acetate (14 g). Ammonium acetate is washed with ether before use.

[1] J. P. Lellouche, J. Deschamps, C. Boullais, and J. P. Beaucourt, *Tetrahedron Lett.* **29,** 3073 (1988).
[2] H. Gilman and F. K. Cartledge, *J. Organomet. Chem.* **2,** 447 (1964).
[3] J. Rokach, R. Zamboni, C. K. Lau, and Y. Guindon, *Tetrahedron Lett.* **22,** 2759 (1981).

The mixture is decanted in a separatory funnel containing 150 ml of ether. The aqueous layer is extracted again with 50 ml of ether. The combined organic layers are washed with water saturated with sodium chloride (3×25 ml) and dried over anhydrous sodium sulfate. The mixture is filtered through a millipore filter (LS type, 5 μm) and concentrated under reduced pressure. The crude mixture should be used as soon as all the volatile solvents have evaporated since concentration damages the coupling compounds at this stage. The residue oil is dissolved in 2 ml of the mixture hexane/ethyl acetate/triethylamine (70/30/2,v/v/v) and filtered using the same solvent on 16 g of deactivated silica gel Merck 60 type V activity [100 g of silica gel Merck 60 (70-230 mesh) stirred for 2 hr with 20 g of deionized water]. Before use the column of deactivated silica gel is eluted first with 50 ml of the mixed solvent hexane/triethylamine (80/20, v/v) and then 100 ml of the mixture hexane/ethyl acetate/triethylamine (70/30/2,v/v/v). The two isomeric 7*Z*- and 7*E*-diyne epoxides are eluted in the first 35 ml. The same procedure is performed again using 10 g of deactivated silica gel in order to ensure the total elimination of triphenylphosphine oxide. After concentration under reduced pressure, the crude oil (about 340 mg) is immediately dissolved in 3 ml of the mixed HPLC purification solvent hexane/ethyl acetate/triethylamine (98/2/0.5,v/v/v). The HPLC system is composed of a solvent delivery pump Waters model 590, a U6K injector, and a UV detector λ_{max} 481 Waters equipped with a semipreparative or an analytical cell, depending on need. HPLC purification performed on a preparative silica gel column S5W Prolabo (20 mm $\times$ 250 mm, 5 μm) using three successive injections (3×1 ml) affords the two chromatographically homogeneous diyne epoxides 7*Z*- (**1**) (78.0 mg) and 7*E* (**2**) (45.0 mg) with an overall yield of 37.5% (7*E*/7*Z* = 37/63). The chromatographic conditions of the purification are as follows: solvent, hexane/ethyl acetate/triethylamine (98/2/0.5,v/v/v); flow rate, 10.0 ml/mn; UV detection, λ 256.5 nm (sensitivity 2). The solvent fraction eluted between 424 and 490 ml contains the 7*Z* isomer, whereas the fraction eluted between 504 and 578 ml contains the 7*E* isomer. The chromatographic purity (better than 99.0%) can be checked by HPLC analysis using an analytical silica gel column S5W Prolabo (5 mm $\times$ 250 mm, 5 μm) eluted with hexane/ethyl acetate/triethylamine (98/2/0.5) at a flow rate of 1.0 ml/min. The UV detection is performed at 256.5 nm;rat;7*Z*-(**1**) (t_R = 17.8 mn), 7*E*-(**2**) (t_R = 22.2 mn). 7*Z*-(**1**) and 7*E*-(**2**) exhibit the spectral properties as indicated below.

Methyl (5*S*,6*S*)-oxido-(11,14)-diyne-(7*Z*,9*E*)-eicosadienoate (**1**): UV (C_2H_5OH): λ_{max1} = 271.5 nm (32800); λ_{max2} = 283.5 nm (39900). $[\alpha]_D^{20} = -8.5°$ ($CHCl_3$; 6.1 g/100 ml); MS;rat;m/z = 328.5 ($M^{+}\cdot$; 19%); ^1H NMR (250 MHz; $CDCl_3$; δ in ppm; TMS standard); 0.92 [t, 3H (H_{20}), $J_{19,20}$ = 7.0 Hz]; 1.32 to 1.83 [broad signal, 10H ($H_3 + H_4 + H_{17} +$

H_{18} + H_{19})]; 2.17 [tt, 2H (H_{16}), $J_{16,17}$ = 7.0 Hz, $J_{13,16}$ = 3.0 Hz]; 2.40 [t, 2H (H_2), $J_{2,3}$ = 7.2 Hz]; 2.88 [multiplet, 1H (H_5)]; 3.33 [multiplet, 2H (H_{13})]; 3.48 [dd, 1H (H_6), $J_{5,6}$ = 2.0 Hz, $J_{6,7}$ = 9.0 Hz]; 3.70 [s, 3H ($-OCH_3$)]; 5.10 [t, 1H (H_7), $J_{7,8}$ = 10.0 Hz]; 5.67 [d, 1H (H_{10}), $J_{9,10}$ = 16.0 Hz]; 6.23 [t, 1H (H_8), $J_{8,9}$ = 10.0 Hz]; 6.95 [dd, 1H (H_9)].

Methyl (5*S*,6*S*)-oxido-(11,14)-diyne-(7*E*,9*E*-eicosadienoate (**2**): UV (C_2H_5OH): λ_{max_1} = 271.0 nm (44500); λ_{max_2} = 283.0 nm (35800). $[\alpha]_D^{20}$ = $-43°$ ($CHCl_3$; 1.14 g/100 ml); MS: *m/z* = 328.5 ($M^{+}\cdot$; 14.4%); ^1H NMR (250 MHz; $CDCl_3$; δ in ppm; TMS standard); 0.90 [t, 3H (H_{20}), $J_{19,20}$ = 7.0 Hz]; 1.32 to 1.80 [broad signal, 10H (H_3 + H_4 + H_{17} + H_{18} + H_{19})]; 2.15 [t, 2H (H_{16}), $J_{16,17}$ = 7.0 Hz]; 2.38 [(t, 2H (H_2), $J_{2,3}$ = 7.2 Hz]; 2.87 [t, 1H (H_5), $J_{4,5}$ = 5.0 Hz]; 3.13 [dd, 1H (H_6), $J_{5,6}$ = 2.0 Hz, $J_{6,7}$ = 8.0 Hz]; 3.32 [s, 2H (H_{13})]; 3.68 [s, 3H ($-OCH_3$)]; 5.47 [dd, 1H (H_7), $J_{7,8}$ = 15.0 Hz]; 5.63 [d, 1H (H_{10}), $J_{9,10}$ = 15.0 Hz]; 6.40 [dd, 1H (H_8), $J_{8,9}$ = 11.0 Hz]; 6.53 [dd, 1H (H_9)]. Structures (**1**), (**2**), (**3**), and (**4**) are very stable compounds when stored in the HPLC purification solvent at −80° under nitrogen. Regular HPLC controls show no noticeable decomposition after more than 1 year for these compounds.

Semireduction Step

A dry, 15-ml round-bottom flask (see above) equipped with a magnetic stir bar (5 mm × 10 mm) is loaded with 5 ml of anhydrous hexane containing 1/10,000 anhydrous pyridine (an increased percentage of pyridine in hexane inhibits the semireduction) and 19.88 mg (60.5 μmol) of (**2**). Anhydrous hexane and pyridine are dried overnight on regenerated 4 Å molecular sieves. After addition of 2.0 mg of Lindlar catalyst (Fluka), the flask is connected to a Toepler pump and frozen in liquid nitrogen. This apparatus has been especially devised for catalytic tritiations at normal pressure since it allowed a precise monitoring of the absorbed gas (scale precision is ±0.01 ml) (see Ref. 4).

High-vacuum (10^{-2} mm Hg) is applied above the mixture and hydrogen or deuterium gas (99.98% isotopic enrichment) is flushed. The overall procedure is repeated twice in order to ensure the total purging of the apparatus. The flask is allowed to warm to 20° using a water bath and vigorously stirred (2000 rmp). The semireduction is stopped after 30 min when the absorbed volume of gas (2.53 ml) reaches 85% of the theoretical stoichiometric gas volume (2.90 ml). This percentage has been determined after careful optimization of the semireduction (see Ref. 5). The medium is filtered on a Millipore filter (LS type, 5 μm). After washing the filter with hexane (2 × 2 ml), the combined organic phases are concen-

[4] E. A. Evans, "Tritium and Its Compounds." Butterworths, Stoneham, MA, 1980.
[5] J. P. Lellouche, F. Aubert, and J. P. Beaucourt, *Tetrahedron Lett.* **29**, 3069 (1988).

trated under reduced pressure and immediately dissolved in 1 ml of the HPLC purification mixed solvent hexane/ethyl acetate/triethylamine (100/1/1,v/v/v). The HPLC purification (see above) of the crude product performed on a semipreparative silica gel column Lichrosorb SI 60 (10 mm × 250 mm, 7 μm) affords the chromatographically homogeneous (see below) methyl ester of LTA_4 (**3**) or (**4**) (5.50 mg) with a yield of 32%. The isotopic purity, as checked by ^1H NMR (300 MHz, $CDCl_3$) and mass spectrometry (CI), is better than 97.0%. Deuterium is detected only in positions 11,12,14, and 15.

The chromatographic conditions of the purification are as follows: solvent, hexane/ethyl acetate/triethylamine (100/1/1,v/v/v); flow rate, 5.0 ml/mm; UV detection; λ = 280.0 nm (sensitivity 2). The solvent fraction eluted between 60 and 69 ml contains (**3**) or (**4**). The chromatographic purity (better than 99.0%) can be checked by HPLC analysis using an analytical silica gel column S5W Prolabo (5 mm × 250 mm, 5 μm) eluted with hexane/ethyl acetate/triethylamine (100/1/1,v/v/v) at a flow rate of 1.0 ml/min. The UV detection is performed at 280.0 nm : (**3**) or (**4**) (t_R = 11.6 min). Structures (**3**) and (**4**) coelute in these conditions with an authentic standard of LTA_4 methyl ester prepared via another route.[6]

The spectral properties of (**3**) correlate well with those described previously.[7] Structure (**4**) shows spectral data given below.

Methyl [11,12,14,15-$^2H_4$]-(5*S*,6*S*)-oxido-(7*E*,9*E*,11*Z*,14*Z*)-eicosatetraenoate: ^1H NMR (300 MHz, $CDCl_3$, δ in ppm): 0.90 (t, 3H H_{20}, $J_{19,20}$ = 7.0 Hz); 1.20–1.40 (broad signal, 6H, $H_{17} + H_{18} + H_{19}$); 1.65 (multiplet, 2H, H_4); 1.80 (multiplet, 2H, H_3); 2.05 (t, 2H, H_{16}, $J_{16,17}$ = 6.9 Hz); 2.40 (t, 2H, H_2, $J_{2,3}$ = 7.5 Hz); 2.85 (t, 1H, H_5, $J_{4,5}$ = 5.5 Hz); 2.90 (s, 2H, H_{13}); 3.15 (dd, 1H, H_6, $J_{5,6}$ = 2.1 Hz, $J_{6,7}$ = 7.9 Hz), 3.65 (s, 3H, $-CO_2CH_3$); 5.40 (dd, 1H, H_7, $J_{7,8}$ = 15.1 Hz); 6.20 (dd, 1H, H_9, $J_{8,9}$ = 10.8 Hz, $J_{9,10}$ = 14.8 Hz); 6.45 (dd, 1H, H_8); 6.55 (d, 1H, H_{10}). MS : m/z = 336 ($M^{+}\cdot$); $[\alpha]_D^{20}$ = −28.4° (0.46 g in 100 ml hexane); UV hexane) : λ_{max_1} = 268.0 nm (41300); λ_{max_2} = 278.5 nm (51900); λ_{max_3} = 290.5 nm (38700). The yield has been determined with regard to the theoretical stoichiometric gas volume calculated at normal conditions of temperature and pressure (20°, 760 mm Hg).

Discussion

Regular use of GC–MS coupling for the elucidation of the complex arachidonic acid metabolism prompted us to prepare multideuterated leukotrienes as standard compounds. GC–MS coupling offers several advan-

[6] I. Ernest, A. J. Main, and R. Menasse, *Tetrahedron Lett.* **23,** 167 (1982).

[7] M. Rosenberger and C. Neukom, *J. Am. Chem. Soc.* **102,** 5426 (1980).

tages over other methods such as HPLC, RIA, or EIA including the use of gas chromatography for high resolving power and mass fragmentometry to monitor ions highly specific for a compound. Furthermore, amounts of 10 pg to 1 ng can be assayed by this method.

The two-step procedure described here allows the short and efficient preparation of tetradeuterated chiral LTA_4 methyl ester. The synthesis of the racemic form of this compound was previously described via sulfonium ylid chemistry,[7] but the critical reaction conditions reported suggested that the Wittig reaction presented here would be more suitable. The obtained mixture of alkenes 7*Z*-(**1**) and 7*E*-(**2**) and the ratio of 7*E*/7*Z* (37/63) are characteristic of semistabilized ylids.[8] This procedure has the advantage of giving reproducible yields and using stable coupling components. The stereochemistry of the $\Delta^{7,8}$ double bond has been determined by high-field ^1H NMR (see above): $J_{7,8}$ = 10.0 Hz for 7*Z*-(**1**) and 15.0 Hz for 7*E*-(**2**).

The semihydrogenation or deuteration of the diyne precursor 7*E*-(**2**) presented an interesting challenge since semireduction of acetylenic compounds is a difficult reaction.[9] Yields and reaction conditions can vary greatly according to the stability and chemical structure of the precursors. In our case, it was best performed in anhydrous hexane containing 1/10,000 of pyridine over a classical Lindlar catalyst. Pyridine deactivates the catalyst and stabilizes the epoxidic compounds. The yields of labeled or unlabeled LTA_4 methyl ester were reproducible and found to vary only from 32 to 40% provided that the hydrogen or deuterium absorption was carefully controlled. (No isotopic effect is observed during the semireduction step.) Overreduction and decomposition of (**3**) or (**4**) during the purification step lowered the yield of the semireduction step. During the HPLC purification step on deactivated silica gel, we noticed that the epoxides (**1**) and (**2**) are much more stable than (**3**) and (**4**). The stereochemistry of the triene unit (7*E*, 9*E*, 11*Z*) and of the $\Delta^{14,15}$ *Z* double bond in (**3**) has been confirmed using high-field ^1H NMR techniques: $J_{7,8}$ = 15.0 Hz, $J_{9,10}$ = 15.0 Hz, $J_{11,12}$ = 11.0 Hz, $J_{14,15}$ = 10.9 Hz. The $J_{14,15}$ value has been determined by two-dimensional NMR (300 MHz) chemical shifts and *J*-resolved homonuclear correlations. The UV spectra of (**3**) and (**4**) show the characteristic triplet of the triene unit (7*E*, 9*E*, 11*Z*)[10] λ_{max} (hexane) = 268.0, 278.5, 290.5 nm.

[8] "Organophosphorous Reagents in Organic Synthesis" (J. L. G. Cadogan, ed.), Chap. 2, p. 66. Academic Press, New York, 1979.

[9] See, for example, P. W. Collins, S. W. Kramer, A. Gasiecki, R. M. Meier, P. H. Jonces, G. W. Gullikson, R. G. Bianki, and R. F. Bauer, *J. Med. Chem.* **30,** 193 (1987); K. E. Harding and D. R. Hollingsworth, *Tetrahedron Lett.* **29,** 3789 (1988).

[10] E. J. Corey, D. A. Clark, A. Marfat, C. Mioskowski, B. Samuelsson, and S. Hammarström, *J. Am. Chem. Soc.* **102,** 1436 (1980).

Tetradeuterated (5*S*,6*S*)-LTA_4 methyl ester is accessible via this two-step procedure. The yield of each step is modest but reproducible according to the known chemical instability of these epoxidic compounds. Nevertheless, when stored under specific conditions, no decomposition of this compound has been found to occur over 1 year.

Acknowledgment

The authors wish to thank Mrs. Th. Beaucourt and F. Toupet for their assistance in the preparation of this manuscript.

[9] Quantitative Gas Chromatography–Mass Spectrometry Analysis of Leukotriene B_4

By W. RODNEY MATHEWS

Introduction

Leukotriene B_4, (5*S*,12*R*)-5,12-dihydroxy-(6*Z*,8*E*,10*E*,14*Z*) eicosatetraenoic acid, a metabolite of arachidonic acid, is a mediator in a variety of inflammatory disorders.[1] As leukotriene B_4 (LTB_4) is an extremely potent compound and there are many structurally similar metabolites of arachidonic acid, sensitive and selective assays are required in order to establish its role in inflammation. Assays utilizing GC–MS have been used successfully to analyze a variety of eicosanoids and the technique of negative-ion chemical ionization (NCI)-mass spectrometry provides an especially sensitive method for the analysis of eicosanoids.[2] This chapter describes a GC–NCI–MS assay for LTB_4 employing $[^2H_4]LTB_4$ as an internal standard which is both sensitive and selective for quantifying LTB_4 in biological matrixes.[3] In order to give LTB_4 suitable electron capture and chromatographic properties necessary for GC–NCI–MS, the pentafluorobenzyl (PFB) ester of the carboxylic acid and the trimethylsilyl (TMS) ethers of the alcoholic hydroxyl groups of LTB_4 are formed.

[1] M. K. Bach, *Annu. Rev. Microbiol.* **36,** 371 (1982).
[2] R. C. Murphy, *Progr. Biochem. Pharmacol.* **20,** 84 (1985).
[3] W. R. Mathews, G. L. Bundy, M. A. Wynalda, D. M. Guido, W. P. Schneider, and F. A. Fitzpatrick, *Anal. Chem.* **60,** 348 (1988).

Assay Method

Principle. The GC–MS assay for LTB_4 described here is based on GC–NCI-MS of the PFB,TMS derivative of LTB_4 with selected-ion monitoring using a stable isotope analog of LTB_4 as an internal standard. The general steps are as follows: (1) Addition of the $[^2H_4]LTB_4$ internal standard to the sample; (2) partial purification of the sample; (3) formation of the PFB ester of the carboxylic acid group of LTB_4; (4) purification by silicic acid chromatography; (5) silylation of the alcohol hydroxyls of LTB_4; (6) determination of the ratio of unlabeled to labeled LTB_4 internal standard by selected-ion monitoring GC–MS.

Reagents

$[^2H_4]LTB_4$ (10.0 μg/ml ethanol) or other stable isotope-labeled internal standard
LTB_4 (Cayman Chemical, Ann Arbor, MI)
Pentafluorobenzyl Bromide (PFBBr, Pierce, Rockford, IL)
N,N'-Diisopropylethylamine, sequanal-grade (Pierce)
Bis(trimethylsilyl)trifluoroacetamide (BSTFA, Pierce)
10% BSTFA in hexane
Silicic Acid (Silicar CC-4, Mallinckrodt)
Octadecyl columns (Baker-10 SPE, J. T. Baker Chemical Co, Phillipsburg, NJ), 1.0 ml
DB-1 fused silica capillary column, 12 m (J&W Scientific, Folsom, CA)

Internal Standard

At the present time, stable isotopically labeled leukotriene B_4 is not commercially available, although $[^2H_4]LTB_4$ should be available soon from Cayman Chemical. In the mean time, a suitable isotopically labeled standard can be prepared. Several options for preparation of an internal standard are available using either ^2H or ^{18}O. $[^2H_4]LTB_4$ can be chemically synthesized. A method starting with 2,5-undecadiynol has been described.[3] Alternatively, $[^2H_8]LTB_4$ can be generated biosynthetically from commercially available $[^2H_8]$arachidonic acid.[4] A biosynthetic method for converting LTA_4 to LTB_4 is available[5] and $[^2H_4]LTA_4$ can be purchased from Cayman Chemical or synthesized.[6] An alternative to deuterium-labeled LTB_4 is $[^{18}O_2]LTB_4$. This can be easily prepared from LTB_4 by ^{18}O

[4] M. Dawson, J. H. Vine, M. J. Forest, C. M. McGee, P. M. Brooks, and T. R. Watson, *J. Label. Comp. Radiopharmacol.* **3,** 291 (1987).

[5] A. L. Maycock, M. S. Anderson, D. M. DeSousa, and F. A. Kuehl, Jr., *J. Biol. Chem.* **257,** 13711 (1982).

[6] J. P. Lellouche, F. Aubert, and J. P. Beaucourt, *Tetrahedron Lett.* **29,** 3069 (1988).

exchange,[7] although care must be taken to avoid hydrolysis and loss of the label during the sample workup.

After a stable isotopic analog of LTB_4 is obtained, it should be examined for isotopic distribution by derivatization and mass spectrometry as described below. The internal standard should be completely free of nonlabeled LTB_4 in order to obtain a linear standard curve. Deuterated LTB_4 can be stored as an ethanolic solution under argon at $-70°$ for several months without noticeable degradation. We use a solution containing 10 $\mu g/ml$ $[^2H_4]LTB_4$ in ethanol which can be accurately pipetted.

Sample Preparation

Prior to the GC-MS analysis of LTB_4, a certain amount of sample preparation is required. The amount and type of sample preparation necessary depends on the nature of the sample matrix, and ranges from relatively minimal preparation when determining the amount of LTB_4 produced by isolated cell suspensions to a more difficult and extensive one when LTB_4 levels in more complex matrixes such as urine and blood are the goal. A method that has been successful for the analysis of LTB_4 production by isolated neutrophils is described here. This method has also been used to determine LTB_4 levels in lung lavageates from guinea pigs. When analyzing more complex samples, preparation methods employing solid-phase extraction techniques[8] and immunoaffinity columns[9] should be considered.

Neutrophils

Following stimulation of neutrophils, the cells are removed by centrifugation and 100 μl $[^2H_4]LTB_4$ (10 ng) is added to an aliquot (0.5 ml) of the supernatant. The supernatant is adjusted to pH 3 with HCl and extracted twice with ethyl acetate (1.0 ml). The ethyl acetate layers are pooled and transferred to a 13 $\times$ 100 mm culture tube. The ethyl acetate is removed under a stream of nitrogen. The residue is reconstituted in 1.0 ml of methanol/water (10:90, v/v) and applied to a conditioned 1.0 ml Baker-10 SPE column. These columns are conditioned by washing with 5 ml water followed by 5 ml methanol and finally 5 ml water. After the sample is loaded, the columns are washed with 2.0 ml of methanol/water (20:80, v/v) and then eluted with 2.0 ml of methanol/water (80:20, v/v). The 80% methanol eluant is transferred to a screw-capped

[7] R. C. Murphy and K. L. Clay, this series, Vol. 86, p. 547.

[8] H. Salari and S. Steffenrud, *J. Chromatogr.* **378,** 35 (1986).

[9] J. J. Vrbanac, J. W. Cox, T. D. Eller, and D. R. Knapp, this volume [7].

13 × 100 mm culture tube topped with a Teflon-lined cap for derivatization.

Derivatization

The PFB ester, TMS ether derivative of LTB_4 is formed as follows. The sample is first dried under a stream of nitrogen and then any residual moisture is removed by adding dichloromethane, 0.25 ml, to the tube and drying under a stream of nitrogen. The residue is dissolved in 50 μl acetonitrile and 20 μl *N,N'*-diisopropylethylamine and 10 μl PFBBr are added. The tube is capped and the reaction allowed to proceed for 10 min at room temperature. Excess reagent is removed with a stream of nitrogen and the residue dissolved in 100 μl ethyl acetate. A small Silicar CC-4 column (ca. 0.5 × 1.0 cm) is prepared in a Pasteur pipette by slurry packing the silicic acid in ethyl acetate. The sample is applied to the column which is washed with 2.0 ml ethyl acetate and then the PFB-LTB_4 is eluted with 2.0 ml dichloromethane/ethyl acetate (1 : 1, v/v). The dichloromethane/ethyl acetate is concentrated with a stream of nitrogen and transferred to a clean 1.0-ml Reacti-vial. The remainder of the solvent is removed and the TMS derivative formed. BSTFA, 50 μl, and acetonitrile, 50 μl, are added to the Reacti-vial which is closed with a Teflon-lined cap and heated for 30 min at 60° in a heating block. After cooling, the excess reagent is removed under nitrogen and the residue dissolved in 50 μl hexane/BSTFA (90 : 10, v/v) for GC–MS. At this point, the sample may be stored under argon at −70° and is stable for several weeks.

GC–MS Analysis

General

The NCl mass spectrum of PFB,TMS-LTB_4 is quite simple, consisting of the base peak at *m/z* 479 corresponding to the loss of the PFB moiety ($C_6F_6CH_2$) and ions at *m/z* 389 and *m/z* 299 corresponding to successive losses of trimethylsilyl alcohol [$(CH_3)_3SiOH$]. This can be seen in the spectrum presented elsewhere in this volume by Ian Blair.[10] LTB_4 is quantified by monitoring the (M-PFB) ion at *m/z* 479 for LTB_4 and the corresponding ion at *m/z* 483 for the [$^2H_4$]LTB_4 internal standard using the selected-ion-monitoring scan mode. A standard curve is constructed by adding a constant amount of [$^2H_5$]LTB_4 (10 ng;2.9 pmol) to 0–100 ng

[10] I. A. Blair, this volume [2].

(0–298 pmol) LTB_4 followed by derivatization as described above. The areas of the m/z 479 and m/z 483 peaks are determined and a plot of the amount of LTB_4 vs the peak area ratio (m/z 389)/(m/z 483) constructed. An example of a standard curve is shown in Fig. 1.

Standard Conditions

GC–NCI–MS is carried out using a VG 70SE mass spectrometer equipped with a HP 5890 GC and a VG 11-250J data system. The GC column is a 12 m DB-1 (0.25 mm id, 0.25 μm film thickness, J&W Scientific) fused silica capillary column. The column is equipped with a dedicated HP on-column injector and the column is introduced directly into the ion source of the mass spectrometer. Samples (1.0 μl) are injected with the oven temperature at 50°. After 1 min the temperature is increased to 265° at 20°/min and then programmed to 290° at 5°/min. Under these conditions LTB_4 elutes at approximately 9.3 min. The heated transfer line to the ion source is maintained at 270° and the ion source itself is heated to 180°. Methane is used as the CI reagent gas at a pressure of 1×10^{-5} torr measured at the source housing. The mass spectrometer operating conditions are typically: accelerating voltage, 8 kV; ionization energy, 80 eV;

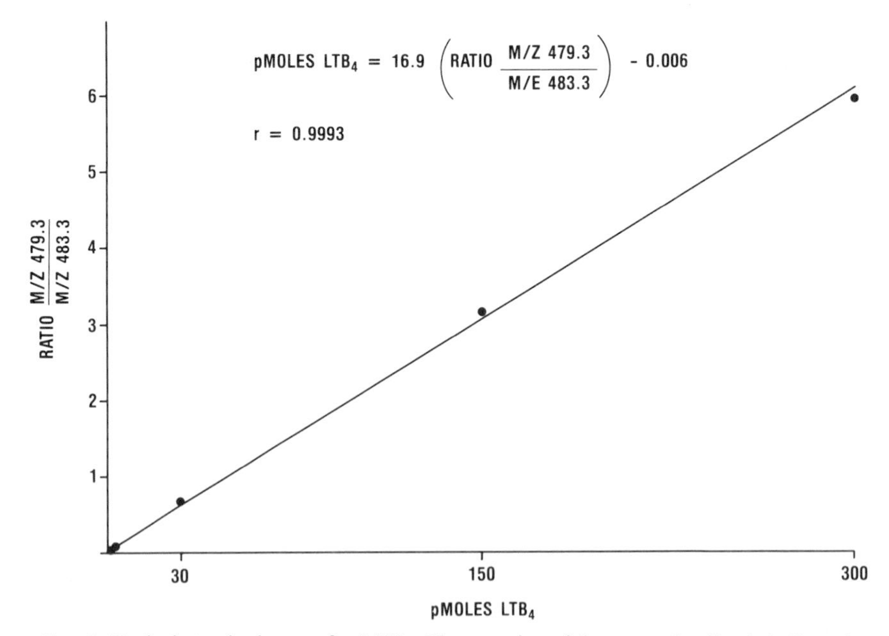

FIG. 1. Typical standard curve for LTB_4. The equation of the regression line is indicated.

emission current, 1 mA; resolution, 1000. The instrument is operated in the selected-ion-monitoring mode under the control of the data system and peak areas for *m/z* 479.3 and 483.3 are calculated by the data system.

Optimization of Sensitivity

Although the spectra of LTB_4 is not complex and the derivatization procedure is relatively straightforward, obtaining optimal sensitivity can be difficult. In order to obtain optimal sensitivity in the GC–MS analysis of LTB_4, care must be taken to ensure that both the chromatograph and mass spectrometer are being operated efficiently. A few general comments may help to obtain the necessary sensitivity. Relatively few problems have been encountered in the derivatization of LTB_4 with the exception of the hydrolysis of the TMS ethers. This can be minimized by using a solution of hexane containing BSTFA instead of hexane to dissolve the samples for injection.

Since derivatized LTB_4 is relatively stable, it is useful to generate a stock of derivatized LTB_4 (1 mg/ml) as well as serial dilutions of this standard. The overall GC–MS sensitivity should be checked with these standards.

The chromatography of PFB,TMS-LTB_4 can be a problem especially at low levels. The use of a new, short GC column (12 m) coupled with the use of an on-column injector and a direct interface to the MS ion source will usually be successful. The performance of a used GC column can often be recovered by removing the first meter. Since even 6-m columns can provide sufficient resolution for LTB_4 analysis,[3] this can be repeated several times. If poor chromatography is suspected it may be useful to prepare a PFB ester derivative of a saturated long-chain fatty acid (*e.g.*, stearic acid). The PFB fatty acids are less sensitive to active sites on the column and can help distinguish GC from MS problems.

A clean ion source is essential for optimal sensitivity in the NCI mode. The use of fluorinated hydrocarbons for tuning and calibration can rapidly contaminate the ion source and should be kept to a minimum. A clean source can be used for approximately 1 week before loss of sensitivity becomes a problem. Exact MS operating conditions will vary from instrument to instrument, however, the general discussion of negative-ion chemical ionization-mass spectrometry found in the review by Oehme[11] discusses several important factors and is quite useful.

[11] M. Oehme, *in* "Mass Spectrometry of Large Molecules" (S. Facchetti, ed.), p. 233. Elsevier, Amsterdam, 1985.

[10] Enzyme Immunoassays for Leukotrienes C_4 and E_4 Using Acetylcholinesterase

By PHILIPPE PRADELLES, CATHERINE ANTOINE, JEAN-PAUL LELLOUCHE, and JACQUES MACLOUF

Although quantitation of sulfidopeptide leukotrienes (LTs) can be achieved by the use of conventional bioassay, or gas chromatography-mass spectrometry after catalytic desulfurization,[1] radioimmunoassay analyses have turned out to be the most convenient method to measure LTs. However, the generation of antisera against LTC_4, LTD_4, and LTE_4 with sensitivities adapted to biological problems has remained difficult. This is due to several reasons including limitations in availability of synthetic material necessary to generate antibodies as well as the availability of ^3H tracers with specific radioactivity limited to approximately 2.22 TBq/mmol. This latter feature limits radioimmunoassay sensitivity (IC_{50}) to 0.3 pmol which should be able to be improved upon.[2–6]

We have successfully developed acetylcholinesterase from *Electrophorus electricus* as a label for various eicosanoids, thereby providing an enzyme immunoassay with sensitivities equal to or superior to those achieved with ^{125}I radioactive tracers.[7,8] We have undertaken a similar approach for LTC_4 and LTE_4. However, because of specific problems inherent with these molecules, we have combined a dual strategy to prepare the protein–LT conjugates necessary for the generation of antibodies and corresponding enzyme tracers. We will describe the development of such assays below.

[1] M. Balazy and R. C. Murphy, *Anal. Chem.* **58,** 1098 (1986).
[2] L. Levine, R. A. Morgan, R. A. Lewis, K. F. Austen, D. A. Clark, A. Marfat, and E. J. Corey, *Proc. Natl. Acad. Sci. U.S.A.* **78,** 7692 (1981).
[3] E. C. Hayes, D. L. Lombardo, Y. Girard, A. L. Maycock, J. Rokach, A. L. Rosenthal, R. N. Young, R. W. Egan, and H. J. Zweering, *J. Immunol. Methods* **131,** 429 (1983).
[4] M. A. Wynalda, J. R. Brashler, M. K. Bach, D. R. Morton, and F. A. Fitzpatrick, *Anal. Chem.* **56,** 1852 (1984).
[5] W. Aehringhaus, R. Wolbling, W. Konig, C. Patrono, B. M. Peskar, and B. A. Beskar, *FEBS Lett.* **146,** 114 (1982).
[6] J. Lindgren, S. Hammarstrom,and E. Goetzl, *FEBS Lett.* **152,** 83 (1983).
[7] P. Pradelles, J. Grassi, and J. Maclouf, *Anal. Chem.* **57,** 1170 (1985).
[8] P. Pradelles, J. Grassi, and J. Malouf, this volume, [3].

Assay Procedure

Reagents and Materials

Glutaraldehyde (Merck Darmstadt, FRG)

N,N'-Dicyclohexylcarbodiimide (DCC), *N*-hydroxysuccinimide ester (NHS), *S*-acetylmercaptosuccinic anhydride (Sigma Chemicals, St. Louis, MO), succinimidyl-4-(*N*-maleimidomethyl)cyclohexane 1-carboxylate (SMCC) (Pierce Chemicals, Rockford, IL)

N-Succinimidyl-*S*-acetylthioacetate (SATA) (Calbiochem, San Diego, CA)

Hydroxylamine-HCl (Sigma)

Dimethylformamide and methanol (Merck) kept anhydrous with molecular sieve

Bovine serum albumin fraction V (Sigma)

Buffers. Two buffers will be used for the couplings throughout the experiments and will be referred to as phosphate buffer (0.1 *M* potassium phosphate buffer, pH 7.4) and borate buffer (0.1 *M* borate buffer, pH 9).

Enzyme. Acetylcholinesterase is prepared by a one-step affinity chromatography as described by Massoulié and Bon.[9] The tetrameric form of the enzyme (G4 form) is obtained from the crude purified preparation by incubation with trypsin; its maleimidated form is prepared as described by McLaughlin *et al.*[10]

LTC_4 and LTE_4 are synthesized as described by Corey *et al.*[11] and Rokach *et al.*[12]

Gel filtration equipment. BioGel A15-m column (90 × 1.5 cm) (Bio-Rad, Richmond, CA); GF 0.5 column (20 × 1 cm) (IBF, France)

NOTE. All the reagents, equipment, and other technical details necessary for the enzyme immunoassay are described in [3] of this volume.

Coupling Procedures

Production of Antisera

Peptido-LTs haptens require coupling to an antigenic molecule in order to elicit antibodies. Because LTs possess both amino and carboxylic

[9] J. Massoulié and S. Bon, *Eur. J. Biochem.* **68,** 531 (1976).

[10] L. Mc Laughlin, Y. Wei, P. T. Stockman, K. M. Leahy, P. Needleman, J. Grassi, and P. Pradelles, *Biochem. Biophys. Res. Commun.* **144,** 469 (1987).

[11] E. J. Corey, D. A. Clark, G. Goto, A. Marfat, C. Miokowski, B. Samuelsson, and S. Hammarström, *J. Am. Chem. Soc.* **102,** 1436 (1980).

[12] J. Rokach, Y. Girard, Y. Guindon, J. G. Atkinson, M. Larue, R. N. Young, P. Masson, and G. Holme, *Tetrahedron Lett.*, p. 1485 (1980).

functions to which they can be attached to the carrier, the final selection of the method would depend on two important criteria: the uniqueness of the function carried on the LT (there is only 1 amino group/molecule of LT and 3 or 2 carboxylic groups for either LTC_4 or LTE_4) and the final yield of the coupling. In addition, we use bovine serum albumin as the carrier throughout these experiments. All tests are performed on LTE_4 and the general approach extrapolated to LTC_4.

Coupling of LTE_4 by Its Carboxylic Groups. LTE_4 (25 nmol) is reacted overnight at room temperature in the dark with 250 nmol of NHS and 25 nmol of DCC in 60 μl of dimethylformamide. The ester is then reacted with 2 nmol of bovine serum albumin in solution in 300 μl borate buffer.

Coupling of LTE_4 by Its Amino Group. LTE_4 (25 nmol) and bovine serum albumin (2 nmol) in solution in 100 μl phosphate buffer, are reacted overnight at room temperature in the dark with 100 μl of a 0.12% glutaraldehyde solution.

Coupling of LTE_4 to Thiolated Albumin. Bovine serum albumin (0.83 μmol) in 2 ml of borate buffer is reacted with 40 mmol of *S*-acetylmercaptosuccinic anhydride during 30 min at room temperature. The ester is then hydrolyzed using 3 ml of a 1 *M* hydroxylamine solution, pH 7.2, overnight at room temperature and the modified SH-albumin purified by gel filtration on a Sephadex G-25 (Pharmacia) column (1 $\times$ 30 cm). Colorimetric analysis at 414 nm reveals that approximately 7 thiols are present per molecule albumin. LTE_4 (25 nmol), in 100 μl of phosphate buffer is mixed with 25 nmol of SMCC (in 5 μl dimethylformamide). After 30 min at room temperature, thiolated albumin (3 nmol) is then added in 500 μl borate buffer for overnight reaction.

The three conjugates are purified by gel filtration (GF 0.5 column 20 $\times$ 1 cm) to allow separation of free LTE_4 from LT covalently linked to albumin. All fractions are measured for their immunoreactivity using an antiserum and enzyme tracer previously prepared (P. Pradelles and J. Maclouf, unpublished data[12a]). Calculation of the percentage of bound LTE_4 vs free showed that the yield of coupling is most efficient using glutaraldehyde ($>$86%) (Fig. 1), contrasting with less than 0.1% reaction by the carboxylic functions of the LT and undetectable reaction for the thiolated albumin. Because of the limited amount of our LTs supply, we did not investigate further these unsuccessful coupling approaches. Glutaraldehyde is thus selected for both LTC_4 and LTE_4 starting from 3.2 and

[12a] In preliminary experiments, a small amount of an immunogen of LTE_4 (kindly provided by Drs. Rokach and Young, Merck Frosst, Montreal, Canada) allowed us to generate enough antiserum and corresponding immunoreactive tracer to perform the initial tests of immunoreactivity described in Fig. 1.

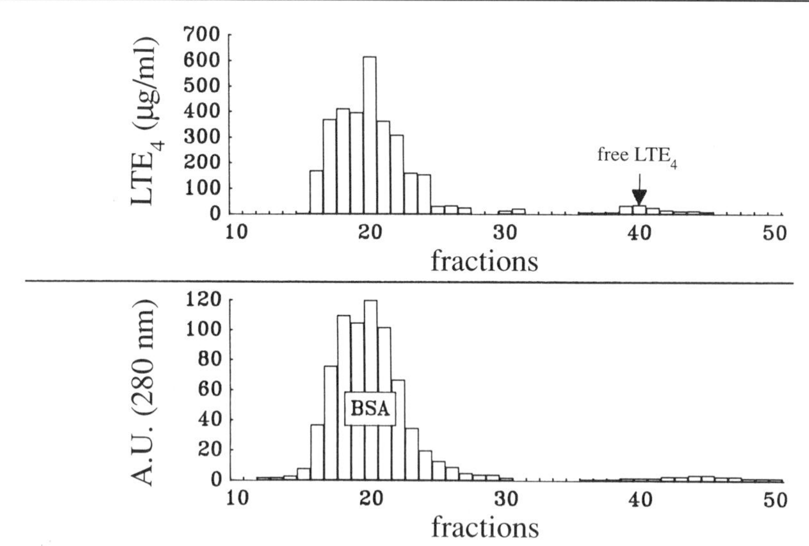

FIG. 1. Purification of the albumin conjugates of LTE_4 by gel filtration. Elution is performed using 0.1 *M* phosphate buffer, pH 7.4, containing 0.4 *M* NaCl, 0.01 *M* EDTA, bovine serum albumin 0.1% as described in the text. Top panel represents the immunoreactive LTE_4 as a function of the fractions collected during the purification. Bottom panel represents the absorbance at 280 nm of the same fractions. (*A.U.*, absorbance units.)

4.5 μmol of LT, respectively, using the ratio of the different reagents in the coupling process as previously found successful.

Immunization. Immunization of rabbits followed the procedure described by Vaïtukaitis.[13] The LT–conjugate (1 mg in 1 ml water) is emulsified with 1 ml of Freund's complete adjuvant (Difco, Detroit, MI) and injected subcutaneously to three rabbits. Six weeks later, the first booster is performed using the same dose of conjugate; rabbits are bled on a weekly basis. At each bleeding, the serum is analyzed for titer and sensitivity. When the titer drops, a new booster is performed and the same follow-up protocol used. Antisera are kept at 4° after addition of sodium azide (0.02% final).

Preparation of Tracers

LTC_4 and LTE_4 can be coupled to the AChE taking advantage of the same structural moiety as that used to generate the immunogen. However, it is critical to assess whether the enzyme activity is impaired during the

[13] J. Vaïtukaitis, J. B. Robbins, and T. Ross, *J. Clin. Endocrinol.* **33,** 988 (1971).

coupling procedure (e.g., homo-bifunctional reagents such as glutaraldehyde polymerize the enzyme and may lead to a complete loss of activity). In addition, it is our experience that the coupling reagent should be different for the immunogen and for the tracer in order to minimize any undesirable recognition of the coupling moiety (i.e., spacer molecule) on the label by the antibodies. Therefore distinct methods of coupling are used for the preparation of immunogen and tracer. Finally, the stoichiometry of the reaction of the enzyme with the LT is ideally 1/1 in order to achieve the highest specific activity of the tracer and hence maximize sensitivity.

Coupling of LTE_4 by Its Carboxylic Groups to Amino Groups of Enzyme. This reaction involves a first-stage formation of an activated ester (500 nmol of NHS and 50 nmol DCC in 20 μl are reacted with 50 nmol of LTE_4 in 20 μl dimethylformamide). After 30 min at room temperature, 0.08 nmol of the G4 form is added in 300 μl of borate buffer for 1 hr at room temperature.

Coupling of Thiolated LTE_4 to Maleimidated Enzyme. LTE_4 (8.5 nmol in 100 μl of borate buffer) is added to 17 nmol of SATA in 10 μl dimethylformamide and incubated at room temperature for 30 min. The resulting LT-thioester is hydrolyzed using 100 μl of 1 *M* hydroxylamine, pH 7, for 30 min and then added to the maleimidated G4 form (0.28 nmol in 800 μl 0.1 *M* phosphate buffer, pH 6, containing 5 m*M* EDTA).

Coupling of LTE_4 to Enzyme Using SMCC. LTE_4 (100 nmol in 100 μl phosphate buffer is added to 100 nmol of SMCC in 10 μl dimethylformamide during 30 min at room temperature in the dark. The G4 form (0.3 nmol in 500 μl borate buffer is subsequently reacted overnight at room temperature.

Each of these tracers is then purified by gel filtration on a BioGel A 15m column. For each of the conjugates, the fractions corresponding to the enzyme activity are pooled and their capacity to bind the antibodies is further tested as well as displacement of bound activity by unlabeled LTs. From these data, the coupling using SMCC is equivalent to the thiolated LTE_4 method. However, the first method is much simpler to perform and is subsequently retained to prepare tracers. We could not obtain any results with the coupling by the carboxylic moieties of LTE_4. However, as for the immunogen, insufficient material (i.e., LTE_4) prevented further investigation of this approach.

Generation of Enzyme Immunoassay for Leukotrienes C_4 and E_4

Since the protocol of enzyme immunoassay for peptido-LTs as well as its practical aspects (materials, reagents, incubation) are identical to other

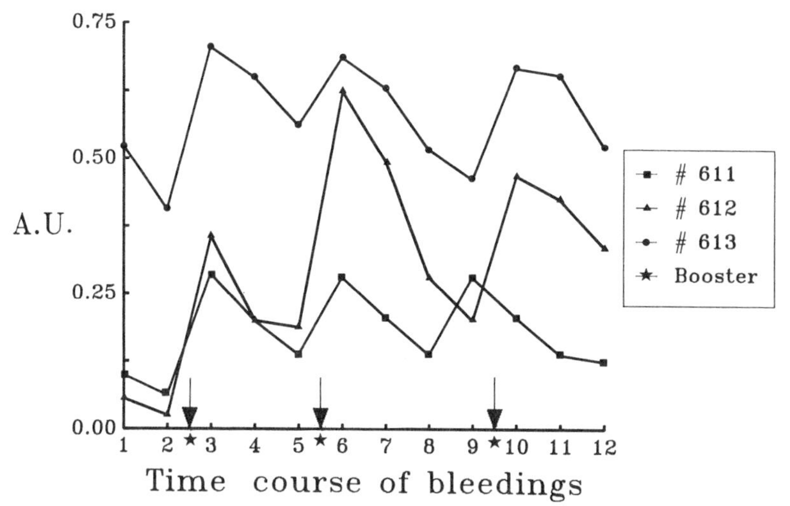

FIG. 2. Evolution of the titer of three different rabbits immunized against LTE_4–BSA as a function of the time course of immunization. The box indicates the rabbit number corresponding to each curve and the arrows the various boosters of immunizations. The numbers on the abscissa (1–12) represent the various bleedings performed after the first booster.

eicosanoids, the reader should consult chapter [3] of this volume for corresponding details.

Time-Course Evolution of Titer

Figure 2 shows the absorbance units of bound tracer (LTE_4–acetylcholinesterase) for three different antisera used at 1/15,000 (final dilution) during the course of immunization. For this experiment, incubation followed the protocol described for the titration of antiserum ([3] in this volume). The absorbance in Fig. 2 represents the solid-phase bound enzyme activity and, as expected, the titer increased after each booster.

Sensitivity and Optimization

Various procedures allow one to maximize sensitivity,[14] among these, *preincubation* (i.e., incubation of antibody in the presence of the antigen and delayed addition of labeled antigen) sometimes improves the sensitivity of various immunoassays. This procedure significantly improves the sensitivity of the LTE_4 system (i.e., IC_{50} = 10 pg for 1 hr preincubation as compared to 50 pg for the control curve) (Fig. 3). Similarly, the sensitivity

[14] G. Ciabattoni, *in* "Radioimmunoassay in Basic and Clinical Pharmacology" (C. Patrono and B. A. Peskar, eds.), p. 181. Springer-Verlag, New York, 1987.

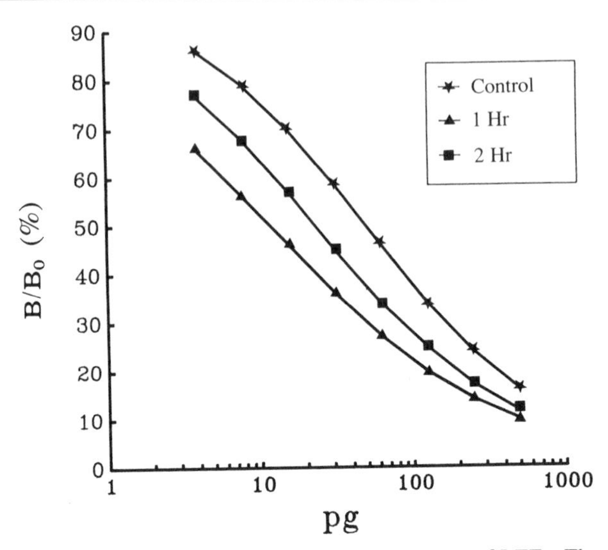

FIG. 3. Effect of preincubation on the dose–response curve of LTE_4. The anti-LTE_4 is incubated 18 hr at 4° with the standard; the tracer is then added and incubated as indicated in the box. Separation of bound from free and enzymatic reaction are carried out as previously described ([3] elsewhere in this volume). The control curve is generated following the standard protocol in [3].

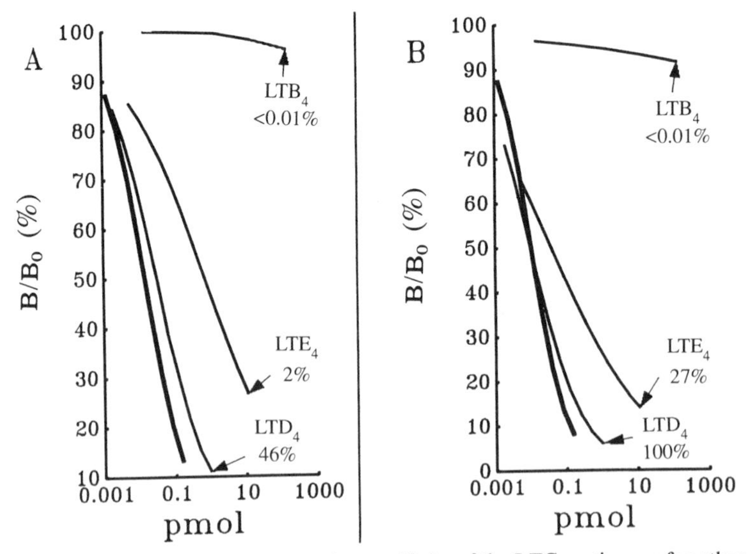

FIG. 4. Influence of temperature on the specificity of the LTC_4 antiserum for other LTs. (A) Room temperature; (B) 4°. The numbers represent the degree of cross-reactivity of each compound (expressed as percentage) at 50% displacement of bound tracer.

for the LTC_4 is increased from 5 pg (IC_{50}) to less than 2 pg with preincubation (not shown).

Specificity

Cross-reactivities of the anti-LTC_4 with other LTs are shown in Fig. 4. As can be seen, there is minimal cross-reactivity with respect to LTB_4. However, in the case of other sulfidopeptide LTs, these results are dramatically influenced by the temperature since the selectivity is better at room temperature (Fig. 4A) as compared to 4° (Fig. 4B).

Concluding Remarks

We have succeeded in developing enzyme immunoassays for peptido-LTs by combining separate approaches to attach these haptens to macromolecules. From previous experience, we can ascertain that the success of these assays is due to the use of distinct coupling reagents to generate antibodies and to prepare the enzyme label. Preparation of the immunogen was mainly concerned with optimizing the coupling reaction because of the limited amount of material. The final strategy employed in labeling the LTs with the enzyme was arrived at as a compromise between the efficiency of the coupling reaction and preservation of the enzymatic activity. Although LTC_4 and LTE_4 may not be the relevant metabolites to evaluate the *in vivo* production of peptido LTs, there are a number of *in vitro* situations when such assays are needed. In most systems, isolated cells only generate LTC_4 whereas in all other *in vitro* situations, the overall synthesis of peptido LTs can easily be estimated after enzymatic transformation of all peptido-LTs into LTE_4.[15]

Acknowledgements

These studies were made possible by financial support from Commissariat à l'Energie Atomique and grants from CNAMTS/INSERM and CNRS.

[15] D. J. Heavey, R. J. Soberman, R. A. Lewis, B. Spur, and K. F. Austen, *Prostaglandins* **33,** 693 (1987).

[11] Quantitation of Sulfidopeptide Leukotrienes in Biological Fluids by Gas Chromatography–Mass Spectrometry

By ROBERT C. MURPHY and ANGELO SALA

Sulfidopeptide leukotrienes can be quantitated in biological fluids by several means. First, the biological activity of sulfidopeptide leukotrienes can be used as a basis for quantitative analysis. These assays which have been described previously[1] typically involve contraction of a smooth muscle and the response pharmacologically quantitated. However, following the structure elucidation of slow-reacting substance of anaphylaxis (SRS-A), more specific assays are typically employed. The second general analytical protocol involves the use of specific antibody clones to recognize metabolites of arachidonic acid. Such immunoassay techniques can involve the use of radiolabeled tracers (radioimmunoassay)[2] or enzyme tracers (enzyme immunoassays).[3] The analytical protocols developed using these techniques are highly sensitive (for example, a few picograms can be detected with some immunoassays), however, cross-reactivity of known or unknown molecules can compromise interpretation of results.[4] Third, analytical techniques based on physicochemical properties of these molecules have been developed for quantitative analysis. The analysis of sulfidopeptide leukotrienes by HPLC separation with on-line UV detection can quantitate these substances when a few nanograms are present.[5] Mass spectrometry can also be adapted for quantitative analysis of these molecules. Such procedures can be sensitive (for example, detect a few picograms), quite precise, and accurate. A major drawback in the direct analysis by mass spectrometry is the fact that the sulfidopeptide leukotrienes are not volatile substances natively, nor do simple derivatives of the carboxyl, amino, and hydroxyl moieties of these molecules impart appreciable volativity.[6,7] Thus, in order to use combined gas chro-

[1] C. W. Parker, M. M. Huber, and S. F. Falkenhein, this series, Vol. 86, p. 655.

[2] F. A. Fitzpatrick, this series, Vol. 86, p. 286.

[3] P. Pradelles, J. Grassi, and J. Maclouf. *Anal. Chem.* **57,** 1170 (1985).

[4] J. Y. Westcott, S. Chang, M. Balazy, D. O. Stene, P. Pradelles, J. Maclouf, N. F. Voelkel, and R. C. Murphy, *Prostaglandins* **32,** 857 (1986).

[5] W. R. Mathews, J. Rokach, and R. C. Murphy, *Anal. Biochem.* **118,** 96 (1981).

[6] R. C. Murphy, S. Hammarstrom, and B. Samuelsson, *Proc. Natl. Acad. Sci. U.S.A.* **76,** 4275 (1979).

[7] R. C. Murphy, W. R. Mathews, J. Rokach, and C. Fenselau, *Prostaglandins* **23,** 201 (1982).

METHODS IN ENZYMOLOGY, VOL. 187

matography–mass spectrometry (GC-MS) for the analysis of sulfidopeptide leukotrienes it is necessary to change substantially the chemical structure of these molecules into derivatives that can be made volatile.[8] The overall procedure described here is one which relies on the facile cleavage of the carbon–sulfur bond by noble metal catalysts in the presence of hydrogen gas (hydrogenolysis).

Reagents

Rhodium on alumina, Rh/Al_2O_3 (Aldrich Chemical Co., Milwaukee, WI), rhodium content, 5%
Diisopropylethylamine (Aldrich, WI)
Bis(trimethylsilyl)trifluoroacetamide (BSTFA) (Supelco Chemical Co., Bellefonte, PA)
Pentafluorobenzyl bromide (PFB-Br) (Supelco)
Leukotriene C_4, D_4, E_4 (Biomol Research Labs., Plymouth Meeting, PA)
[^{3}H]LTC_4, [^{3}H]LTD_4, and [^{3}H]LTE_4 (>20 mCi/mmol) (New England Nuclear Research Products, Boston, MA
5-Hydroxyeicosatetraenoic acid (5-HETE) (Cayman Chemical Co., Ann Arbor, MI)
Sep-Pak, reversed-phase solid extraction cartridges (Waters Associates, Milford, MA)
$H_2{}^{18}O$ (Isotec Inc., Miamisburg, OH)
Methanol, dichloromethane, acetonitrile (HPLC-grade, Fisher Chemical Co.)

Principle of the Method. The assay for the sulfidopeptide leukotrienes described below is based on stable isotope dilution with quantitative analysis by selected-ion monitoring GC–MS. The steps (shown in Fig. 1) are as follows: 1. Addition of [^{3}H]leukotriene to the biological fluid as initial recovery internal standard; 2. solid-phase extraction of leukotrienes from biological fluid using reversed-phase Sep-Pak, or optional reversed-phase HPLC purification to separate LTC_4, LTD_4, and LTE_4; 3. recovery of [^{3}H]leukotriene determined in Sep-Pak methanol fraction; 4. addition of [$^{18}O_2$]5-HETE as mass spectrometric internal standard; 5. catalytic desulfuration and reduction with Rh/Al_2O_3; 6. extraction of 5-HEA; 7. conversion to the pentafluorobenzyl ester, trimethylsilyl ether derivative; and 8. quantitative analysis of the ratio of unlabeled to oxygen-18 internal standard by selected-ion monitoring GC–MS (m/z 399 and 403).

[8] M. Balazy and R. C. Murphy, *Anal. Chem.* **58,** 1098 (1986).

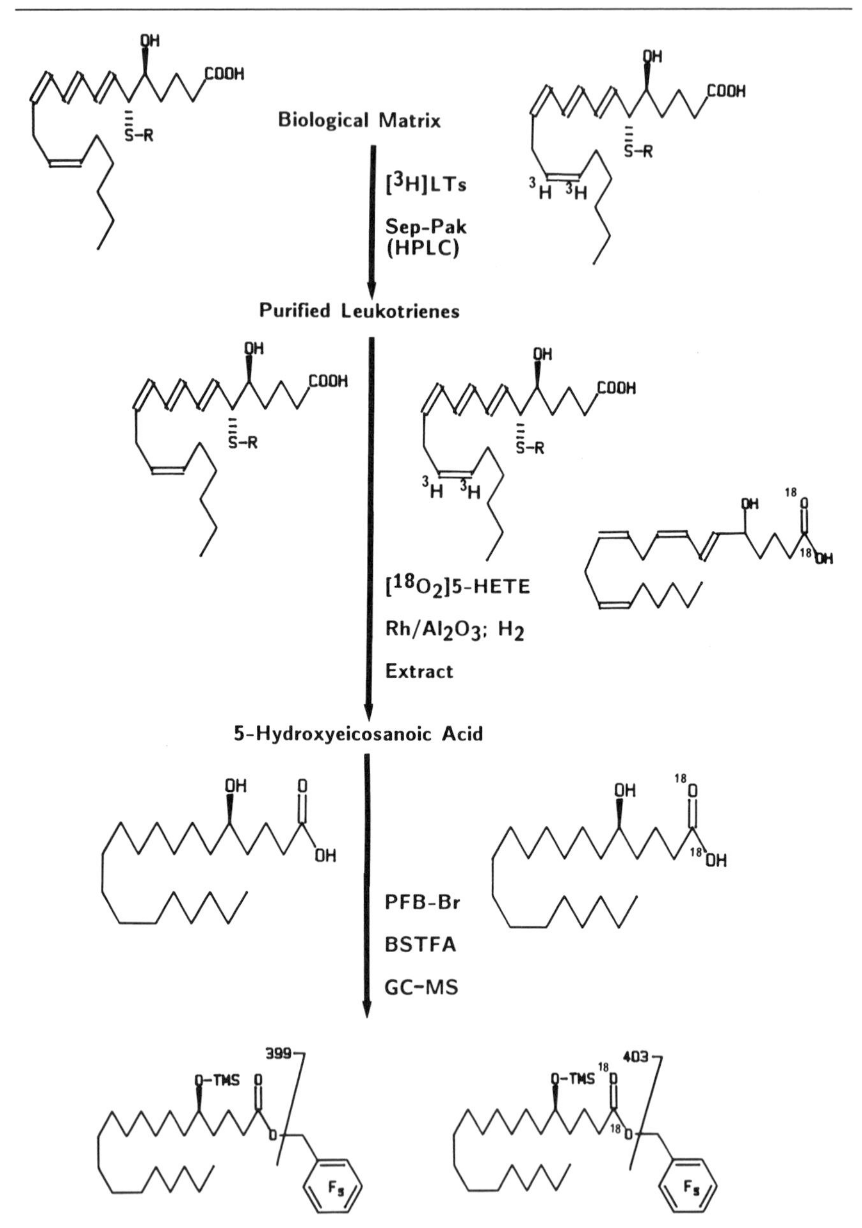

FIG. 1. Protocol for the quantitation of sulfidopeptide leukotrienes by negative-ion electron capture-mass spectrometry. The structures of the sulfidopeptide leukotrienes and the various internal standards (radiolabeled sulfidopeptide leukotrienes and ^{18}O-labeled 5-HETE) are indicated along with the chemical modifications during the process.

Internal Standards

The quantitative analysis of leukotrienes by stable isotope dilution techniques requires the availability of an appropriate stable isotopically labeled species of the analyte to be quantitated. The commercially available deuterium-labeled compounds, in which the deuterium atoms are present either on a vinylic or allylic position, are unsuitable for use in this assay. This is due to the fact that hydrogen atoms in such positions are completely lost during the catalytic reduction process because of exchange with hydrogen gas at the catalytic surface. The formation of π-allyl intermediate at a coordination site on the catalyst surface facilitates both hydrogenolysis and deuterium exchange.[9] For this reason, the ^{18}O-labeled 5-HETE is employed which has stable isotopes (^{18}O) that cannot be lost during the catalytic reduction step. Synthesis of ^{18}O-labeled intact leukotriene C_4, D_4, or E_4 has been difficult to achieve routinely. The synthesis of [$^{18}O_2$]5-HETE has been described in a previous volume of this series.[10]

The [$^{18}O_2$]5-HETE has two oxygen-18 atoms in the carboxylic acid moiety and after reduction yields the identical compound, 5-hydroxyeicosanoic acid, as that obtained from any sulfidopeptide leukotriene except that it has a molecular weight four units higher. Since this internal standard is structurally different from sulfidopeptide leukotrienes, it is added to the sample just prior to the reduction step and thus cannot account for any loss of the leukotrienes prior to this conversion step. In order to circumvent this problem, tritium-labeled sulfidopeptide leukotrienes must be added to the biological fluid to account for recovery prior to this reduction step. Of critical concern is the isotopic purity of the [$^{18}O_2$]5-HETE and the absence of any back-exchange during the assay protocol. The standard curve becomes a critical step in the assessment of these points.

Assay of Sulfidopeptide Leukotrienes in Biological Fluids

Step 1. To an aliquot of 1 to 5 ml of physiological fluid is added [^{3}H]LTC$_4$ (or [^{3}H]LTD$_4$, [^{3}H]LTE$_4$ depending on the exact species to quantitate) at a precisely known level (e.g., 20,000 dpm). The mixture is vigorously shaken for 2 min to effect mixing.

Steps 2 and 3. The biological fluid is passed through a Sep-Pak which had been previously washed with 10 ml of methanol, 5 ml 0.1 m*M* EDTA, and 10 ml water. The Sep-Pak is then eluted with 10 ml distilled water

[9] H. O. House, "Modern Synthetic Reactions," p. 23. W. A. Benjamin, Menlo Park, CA, 1972.

[10] R. C. Murphy and K. L. Clay, this series, Vol. 86, p. 547.

followed by 10 ml of hexane and 10 ml ethyl acetate containing 1% methanol (v/v) to remove prostaglandins and dihydroxyleukotrienes as well as other hydroxy acids including any 5-hydroxyeicosatetraenoic acid in the biological system.[11] Finally, the Sep-Pak is eluted with 5 ml of 80% methanol/water and the radioactivity recovery determined in an aliquot of this fraction.

A modification of the above protocol is to separate each sulfidopeptide leukotriene prior to reduction. This involves addition of all three radiolabeled internal standards [^{3}H]LTC$_4$, [^{3}H]LTD$_4$, and [^{3}H]LTE$_4$ followed by separation of the sulfidopeptide leukotrienes by reversed-phase HPLC using methanol/water (65/35/0.02 acetic acid), pH 5.7. The HPLC fractions corresponding to each sulfidopeptide leukotriene are analyzed separately by adding [$^{18}O_2$]5-HETE and carrying out the catalytic reduction. Even though the identical molecules are quantitated in this modified protocol, namely 5-HEA and [$^{18}O_2$]5-HEA, the prior separation and use of radiolabeled sulfidopeptide leukotriene to correct for recoveries of the separation step, allows quantitation of each sulfidopeptide leukotriene.

Steps 4 and 5. A precise amount of [$^{18}O_2$]5-HETE (typically 0.5–5.0 ng) is added to the methanol Sep-Pak eluate. The sample is adjusted to 50% methanol/water (v/v) either by drying the sample and dissolving it in this solvent system or by adding water to the eluate fraction. Reduction and desulfurization of leukotrienes is carried out at room temperature by adding 2–5 mg of Rh/Al$_2$O$_3$ and slowly bubbling hydrogen into the sample for 25 min. Special attention is paid to the contact between catalyst and hydrogen which is necessary for maximum desulfurization. The rate of hydrogen addition is maintained as that necessary for suspension of the catalyst.

Step 6. The solution is then made basic with 50 μl of 1 *N* KOH, in order to release the product completely from the catalyst surface. This step destroys a portion of the catalytic surface itself and any chemisorption of 5-HEA is eliminated. The catalyst is removed by centrifugation in a microcentrifuge (12,000 *g*, 5 min) and the supernatant removed. The supernatant is acidified with 10 μl formic acid and then extracted with 2 volumes of dichloromethane. The two phases are separated after centrifugation and the lower phase, which contains the 5-HEA, is dried under a stream of nitrogen.

Step 7. The extracted 5-HEA is derivatized first into the pentafluorobenzyl ester by the addition of 25 μl 10% diisopropylethylamine in acetonitrile followed by 25 μl of a 10% solution (v/v) of pentafluorobenzyl bromide (PFB-Br) in acetonitrile. Esterification proceeds for 25 min at room temperature. Excess reagents are then removed by a stream of

[11] W. S. Powell, this series, Vol. 86, p. 467.

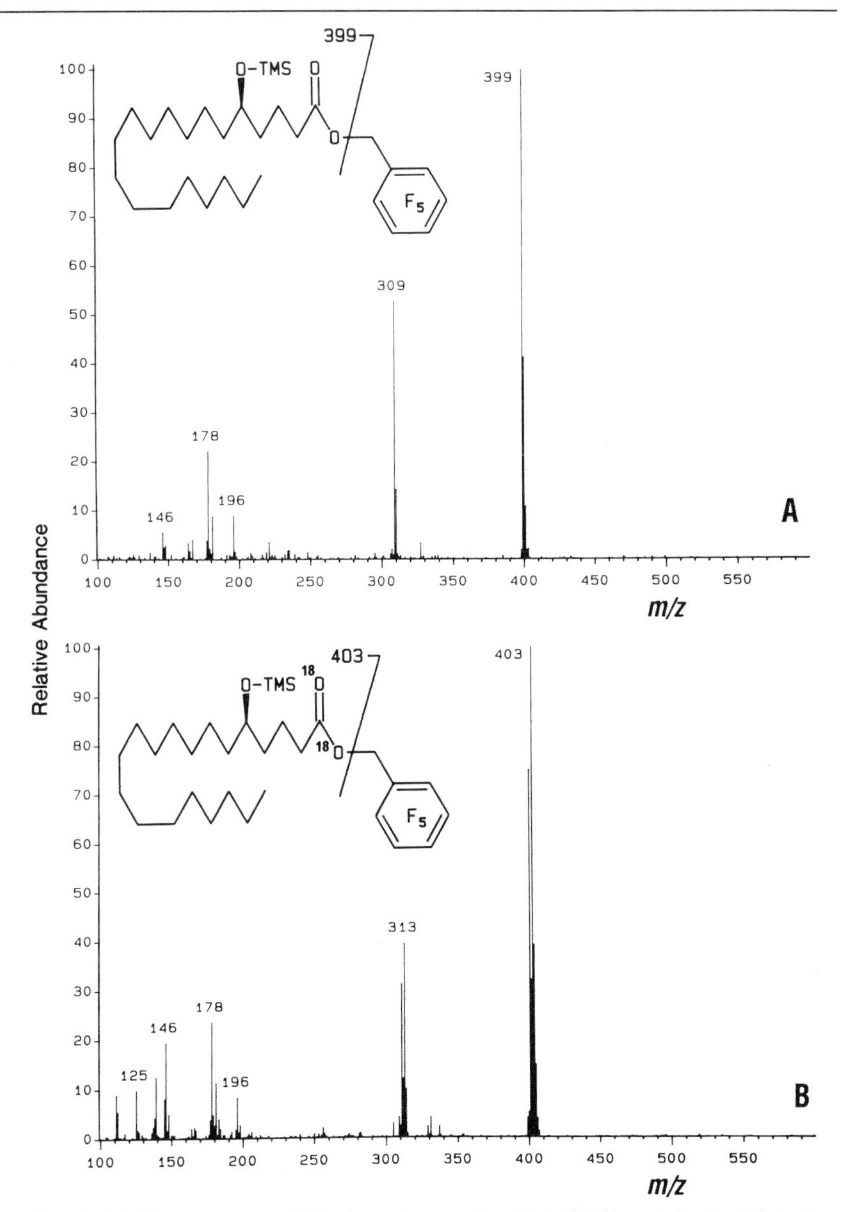

FIG. 2. (A) Mass spectrum of 5-hydroxyeicosanoic acid (5-HEA) as a trimethylsilyl ether, pentafluorobenzyl ester using electron capture (CH_4) and negative-ion detection. (B) The mass spectrum of the oxygen-18 internal standard of 5-HEA. The two ions at m/z 401 and 403 indicate one and two oxygen-18 atoms incorporated in the carboxy moiety. These derivatives were analyzed by capillary GC–MS.

nitrogen. The hydroxy group in 5-HEA is then converted to the trimethylsilyl ether derivative by the addition of 50 μl of BSTFA and 50 μl of acetonitrile. The reaction is heated to 60° for 1 hr. This final solution can be directly injected into the gas chromatograph or alternatively to effect concentration, the sample can be dried under a stream of nitrogen and dissolved in a smaller volume of hexane containing 5% BSTFA.

Step 8. The gas chromatographic separation of the pentafluorobenzyl ester, trimethylsilyl ether of 5-hydroxyeicosanoic acid is carried out on a gas chromatograph equipped with a 10 m × 0.25 mm id capillary column having a 25 μm film thickness of DB1 (J & W Scientific, Rancho Cordova, CA). The gas chromatographic column is programmed from 150° to 300° at a linear rate of 15°/min using helium as carrier gas (50 cm/sec). This column is directly interfaced to a quadrupole mass spectrometer (Nermag 1010C, Paris, France) operated in the negative-ion electron-capture mode using methane as moderating gas at a source pressure 0.1 Torr. Source temperature is maintained at 180° and electron energy typically at 80 eV. Selected-ion monitoring of negative ions is performed at *m/z* 399 corresponding to the carboxylate anion of endogenous 5-HEA from the sulfidopeptide leukotriene and *m/z* 403 corresponding to the added [$^{18}O_2$]5-HETE. The resultant mass spectra for these derivatives is shown in Fig. 2.

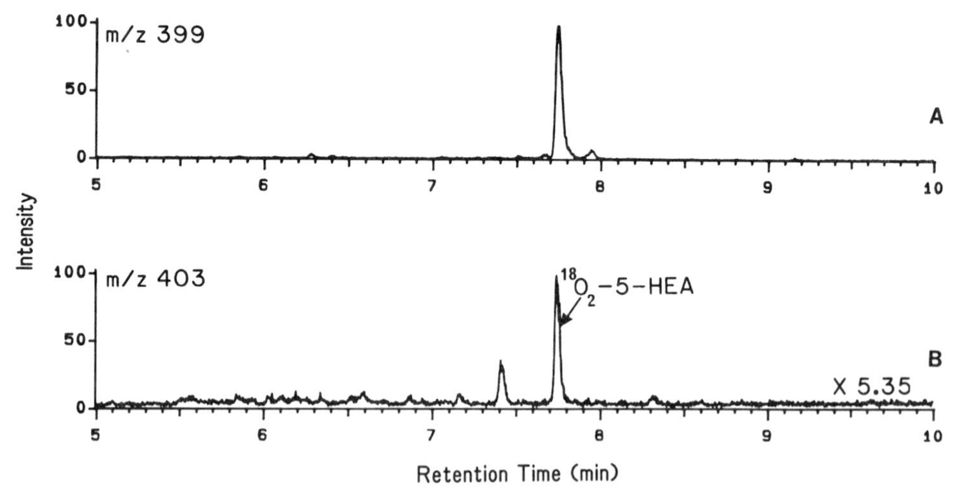

FIG. 3. (A) Selected-ion recording of *m/z* 399 indicating the presence of a sulfidopeptide leukotriene as analyzed as the 5-HEA-TMS, pentafluorobenzyl ester. The sulfidopeptide leukotriene was isolated from a murine mast cell preparation passively sensitized with IgE and challenged with DNP–albumin. (B) Selected-ion recording trace of *m/z* 403 derived from 0.5 ng of [$^{18}O_2$]5-HETE added as internal standard to the leukotriene extract before reduction and derivatization.

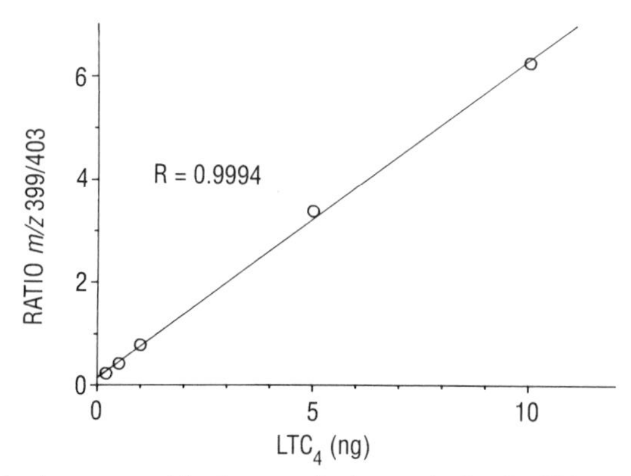

FIG. 4. Standard curve used in the quantitative analysis of sulfidopeptide leukotrienes with $[^{18}O_2]$5-HETE (0.5 ng) as internal standard. The protocol for preparation of the sample for analysis is illustrated in Fig. 1.

Retention time of the 5-HEA derivative in the conditions listed above is approximately 8 min as seen in Fig. 3.

Preparation of Standard Curve

For each batch of samples, a standard curve is generated where a fixed amount of $[^{18}O_2]$5-HETE (1 ng) is added to known concentrations of the sulfidopeptide leukotriene (typically 0.2, 0.5, 1, 5, 10 ng). The volume of internal standard solution ($[^{18}O_2]$5-HETE) used for this curve is precisely that used for the biological samples. These standard solutions, made in 50% methanol/water (v/v), are treated with Rh/Al_2O_3 starting at step 6 described above. The resulting ratio of m/z 399 to 403 is then plotted against the sulfidopeptide leukotriene concentration as seen in Fig. 4. Five different quantities of sulfidopeptide leukotrienes are typically used for such standard curves. The Y-intercept corresponds to the isotopic purity of the $[^{18}O_2]$5-HETE and such curves should be quite linear.

Quantitation of Sulfidopeptide Leukotrienes from Mast Cells

The protocol described above was used to quantitate LTC_4 from murine-transformed mast cells passively sensitized with monoclonal mouse IgE anti dinitrophenol (DNP) conjugate, then challenged with DNP–albumin. Following challenge, the cell supernatant was taken for analysis as in Fig. 1. As seen in Fig. 3 the selected-ion recording trace for the ions corresponding to the sulfidopeptide leukotriene (5-HEA derivative) with 1 ng of [$^{18}O_2$]5-HETE internal standard indicated production of 2.8 ng/ml LTC_4 eq./10^6 mast cells following immunological challenge. This assay can be used to separately quantitate sulfidopeptide leukotrienes if one institutes a prior separation of each sulfidopeptide leukotriene typically by reversed-phase HPLC.

Acknowledgement

This work was supported by a grant from the National Institutes of Health (HL25785).

[12] Automated On-Line Extraction and Profiling of Lipoxygenase Products of Arachidonic Acid by High-Performance Liquid Chromatography

By PIERRE BORGEAT, SERGE PICARD, PIERRE VALLERAND, SYLVAIN BOURGOIN, ABDULRAHMAN ODEIMAT, PIERRE SIROIS, and PATRICE E. POUBELLE

Because of the potential importance of the lipoxygenase products in several human diseases, substantial effort has been put into the development of methods for their assay. Immunoassays and mass spectrometric (MS) assays enabling detection of picogram amounts of material are now available for some compounds. Reversed-phase high-performance liquid chromatography (RP-HPLC) is also widely used for analysis of lipoxygenase products; the present popularity of reversed-phase HPLC for the analysis of leukotrienes (LTs) and related compounds arises from the fact that most lipoxygenase products carry a UV chromophore (Fig. 1) which enables their measurement by photometry.

Materials and Methods

Standards of the lipoxygenase products (5*S*,12*R*)-dihydroxy-(6*Z*,8*E*,10*E*,14*Z*)-eicosatetraenoic acid (LTB_4), 20-hydroxy-LTB_4 (20-OH-LTB_4), 20-carboxy-LTB_4 (20-COOH-LTB_4), (5*S*)-hydroxy-(6*R*)-*S*-glutathionyl-(7*E*,9*E*,11*Z*,14*Z*)-eicosatetraenoic acid (LTC_4), (5*S*)-hydroxy-(6*R*)-*S*-cysteinylglycine-(7*E*,9*E*,11*Z*,14*Z*)-eicosatetraenoic acid (LTD_4), (5*S*)-hydroxy-(6*R*)-*S*-cysteinyl-(7*E*,9*E*,11*Z*,14*Z*)-eicosatetraenoic acid (LTE_4), the *N*-acetyl derivatives of LTC_4, D_4, E_4, the sulfoxide forms of LTC_4, D_4, E_4 (LTC_4SO,LTD_4SO,LTE_4SO), (5*S*)-hydroxy-(6*E*,8*Z*,11*Z*,14*Z*)-eicosatetraenoic acid (5-HETE), (12*S*)-hydroxy-(5*Z*,8*Z*,10*E*,14*Z*)-eicosatetra-

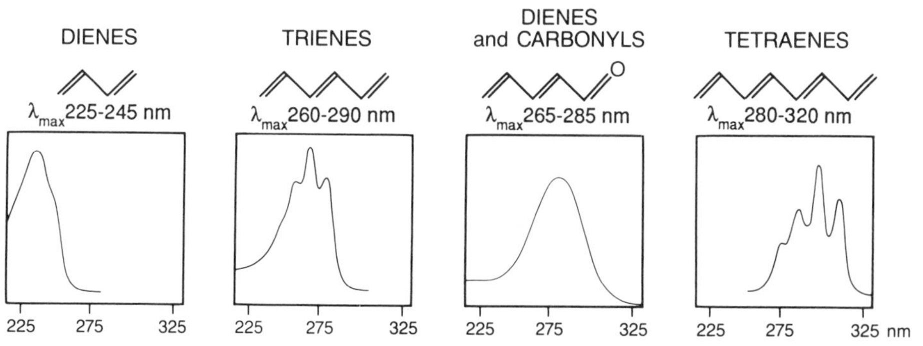

FIG. 1. Ultraviolet chromophores of lipoxygenase (and cyclooxygenase) products. Several lipoxygenase products carry conjugated dienes (monohydroxy or monohydroperoxy polyunsaturated fatty acids), conjugated trienes (leukotrienes and other dihydroxy polyunsaturated fatty acids) or conjugated tetraenes (lipoxins). (12*S*)-Hydroxy (5*Z*,8*E*,10*E*)-heptadecatrienoic acid (HHT), a cyclooxygenase product, also carries a conjugated diene chromophore [W. S. Powell, *J. Biol. Chem.* **257,** 9465 (1982)]. The conjugated diene and carbonyl chromophore is present in various carbonyl compounds derived from hydroperoxy polyunsaturated fatty acids [B. Fruteau de Laclos, J. Maclouf, P. Poubelle, and P. Borgeat, *Prostaglandins* **33,** 315 (1987)], in the cyclooxygenase product 12-keto-(5*Z*,8*E*,10*E*)-heptadecatrienoic acid (12-keto-HT) or its dienoic analog [H. Salari, P. Borgeat, P. Braquet, and P. Sirois, *J. Allergy Clin. Immunol.* **77,** 720 (1986) and W. S. Powell, *J. Biol. Chem.* **257,** 9465 (1982)], and PGB-type compounds widely used as internal standards in reversed-phase HPLC of lipoxygenase products [P. Borgeat and S. Picard, *Anal. Biochem.* **171,** 283 (1988)].

enoic acid (12-HETE), as well as dihydroxyeicosatetraenoic acids (DiHETEs) and lipoxins (LXs) were kindly provided by Drs. J. Rokach and R. Young of Merck Frosst Laboratories, Pointe-Claire, Quebec, Canada. The ionophore A23187, zymosan, and prostaglandin B_2 (PGB_2) were obtained from Sigma Co. (St. Louis, MO). 19-Hydroxy-PGB_2 (19-OH-PGB_2) was prepared from seminal fluid as described previously.[1] ^{3}H-Labeled compounds were obtained from Amersham Canada (Oakville, Ontario).

Equipment

Unless otherwise indicated, the HPLC apparatus used consists of two Beckman Model 100A HPLC pumps (Beckman Instruments, San Ramos, CA); two Beckman Model 160 fixed-wavelength UV photometers; a WISP Model 110B autosampler (Waters Millipore, Bedford, MA) equipped with a 1-ml injection syringe; two pumps Model 300 (Scientific System Inc., State College, PA); a double 3-way air-actuated Rheodyne Model 7030

[1] P. Borgeat and S. Picard, *Anal. Biochem.* **171,** 283 (1988).

high-pressure valve (Rheodyne Inc., Cotati, CA), and two Waters Millipore SSV solvent selection valves. The instrument is controlled by a Beckman Model 420 HPLC controller containing a Beckman pneumatic board interface with the capacity to control eight external events by means of 4 contact closures and 4 pneumatic flags (which can also be used as contact closures). The Model 420 controls the solvent flow of the two main HPLC pumps, the 2 solvent selection valves (2 contact closures are required for each valve distributing 4 solvents) and the accessory pumps 3 and 4 by means of 2 contact closures (on/off switching). Two pneumatic flags are used to control the air-actuated double 3-way valve. The in-line filters (2 μm stainless steel frit) are from Rheodyne. The static mixer is a 250-μl dead volume Visco mixer from The Lee Company (Westbrook, CT).

HPLC Cartridges and Columns

The silica guard cartridge and the octadecylsilyl silica (C_{18}) extraction cartridge used are Waters Millipore Guard Pak inserts; the C_{18} extraction cartridge is the Resolve C_{18} Guard Park. HPLC is performed using the following columns or cartridges: Waters Millipore radial compression cartridges, Resolve C_{18}, 10 μm particles, 5 × 100 mm, and the Waters Millipore steel column, Resolve C_{18}, 5 μm particles, 3.9 × 150 mm. The Resolve C_{18} is a non-end-capped packing and is always used for analysis of samples containing peptido-LTs with free amino group (LTC_4, D_4, and E_4). However, for analysis of samples containing oxo, hydroxy, hydroperoxy, or carboxy derivatives of fatty acids, as well as *N*-acetylated derivatives of the peptido-LTs, columns or cartridges packed with either non-end-capped or end-capped supports, such as the Waters Millipore Nova Pak C_{18} or the Beckman Ultrasphere C_{18}, are use interchangeably. The requirement for using non-end-capped C_{18} support in the analysis of LTC_4, D_4, and E_4 is essentially linked to the use of a specific solvent system and elution program (see HPLC procedures).

Preparation of Mobile Phases

Water is obtained from a Milli-Q water purification system (Millipore). Organic solvents and phosphoric acid are HPLC grade from Anachemia (Montreal) and Fisher Scientific (Fair Lawn, NJ), respectively. Acetonitrile and dimethyl sulfoxide (DMSO) are further purified by filtration through a bed of alumina. Ammonium hydroxide (reagent grade, Anachemia) is used without further treatment. The various mixtures of organic solvents and water required are made up as indicated in Table I; the pH of

TABLE I
COMPOSITION OF SOLVENT MIXTURES

Solvent	Volume (%)			
	A	B	C[a]	D
Water	54	5	20	100
Methanol	23	32	30	
Acetonitrile	23	63	50	
DMSO	0.003[d]			
H_3PO_4[b]	0.01	0.01	0.05	0.05
NH_4OH[c]			to pH 4.6[e]	

[a] See under "Elution program" for possible modification of solvent mixture C.
[b] 85% aqueous solution.
[c] 25% solution of aqueous NH_3.
[d] The composition of the solvent mixtures (% composition in water, methanol, acetonitrile) does not take into account the small volumes of DMSO, H_3PO_4, and NH_4OH added.
[e] Apparent pH using a meter and a glass electrode. The pH of mixtures A and B was not adjusted (pH ≈3.1 and ≈3.2, respectively).

the solvent mixture C is adjusted to pH 4.6 using ammonium hydroxide and the solvent mixtures A and C, prepared in 4-liter batches, are then further purified by pumping through a 1 × 10 cm steel column packed with octylsilyl silica particles (37–45 μm). This column is previously washed by passing 100 ml of HPLC-grade acetonitrile filtered through alumina (1 × 5 cm bed); the first 100 ml of the solvent mixture are discarded and the column washed between each 4-liter batch.

Determination of Recoveries

^{3}H-labeled LTC_4, LTB_4, and 5-HETE (specific activities >100 Ci/mmol) are purified separately using the reversed-phase HPLC system described in this chapter. Aliquots (≈25,000 DPM) of freshly purified compounds are transferred to autosampler vials and the volume and composition are adjusted to 2 ml and to methanol/acetonitrile/water (12.5/12.5/75,v/v/v). The samples are injected and the appropriate fractions collected for radioactivity counting. Recovery measurements are performed in triplicate for each compound.

Determination of UV Signal versus Mass Correlation

Aliquots, 25 μg, of standard LTs are purified separately using the reversed-phase HPLC system described in this chapter. Peaks of purified compounds are collected and the solutions calibrated by UV photometry. Solutions of concentrations ranging from 0.5 to 25 ng/ml are prepared in methanol/acetonitrile/water (12.5/12.5/75,v/v/v) and PGB_2 and 19-OH-PGB_2 are added to the final concentration of 5 ng/ml. Injection volume is 1.7 ml and analyses are performed in duplicate.

HPLC Procedures

The HPLC methodology was developed with the following goals including (1) minimization of sample preparation procedures, and (2) the ability to provide a profile of lipoxygenase products present in biological samples. Figure 2 shows a scheme of the apparatus used for on-line extraction and reversed-phase HPLC of lipoxygenase products. In principle, the procedure used is similar to the precolumn extraction techniques described by other investigators.[2,3] An extraction cartridge is connected or disconnected from the analytical column by means of a double 3-way valve. The silica and C_{18} cartridges connected in series (GC and EC) both act as guard cartridges, while the C_{18} cartridge is also used as the extraction support. The two accessory pumps (pumps 3 and 4) provide increased efficiency and flexibility. Indeed, the use of the system configuration shown is less time consuming since the two cartridges are washed and loaded with the next sample while the analytical column is washed and equilibrated. In addition, pump 4 provides the possibility to dilute samples with water or to execute more elaborate sample clean-up procedures on the extraction cartridge. It was previously reported that in order to achieve quantitative on-line extraction of polar arachidonic acid metabolites, such as 20-OH-LTB_4 or 19-OH-PGB_2 (used as internal standard), the mobile phase content in methanol, ethanol, or acetonitrile should be in the range of 15 to 25%.[1,3] Therefore, the capacity of this apparatus to perform on-line dilution of samples with water prior to introduction into the extraction cartridge allows for the possibility of directly analyzing samples with a high (>25%) methanol, ethanol or acetonitrile content. This was found to be very useful for biological samples subjected to solid-phase extraction prior to reversed-phase HPLC analysis or samples denatured with several volumes of organic solvents (see Figs. 6 and 8), containing 80% or more of

[2] W. C. Pickett and M. B. Douglas, *Prostaglandins* **29,** 83 (1985).

[3] W. S. Powell, *Anal. Biochem.* **164,** 117 (1987).

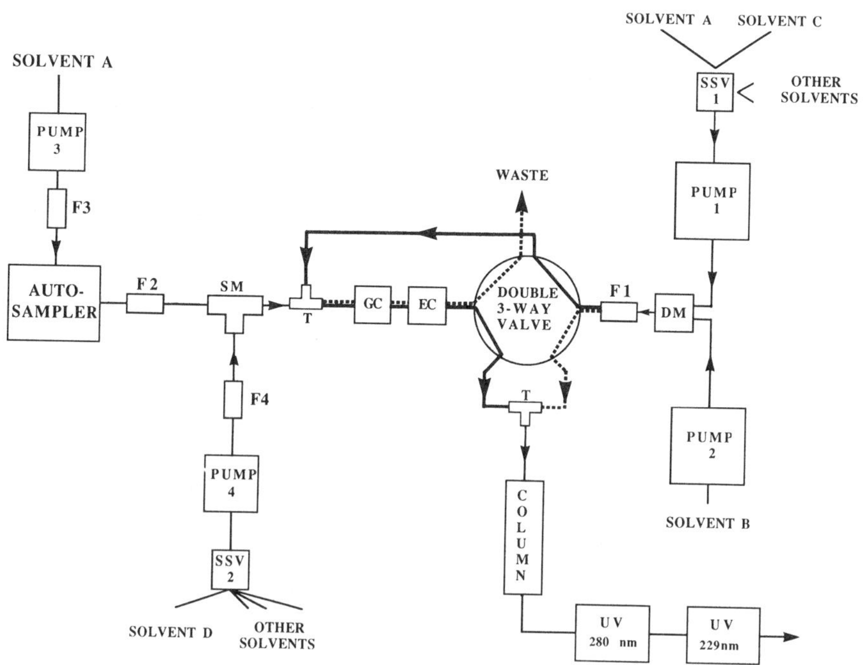

FIG. 2. Scheme of the apparatus for automated on-line extraction and profiling of lipoxygenase products. Pumps 1 and 2 are the main pumps that generate the gradients for elution of the various products. The pump 1 inlet is connected to a solvent selection valve (SSV) which enables selection between the solvent mixtures A or C (see Table I for composition of solvent mixtures). The pump 2 inlet is connected to solvent mixture B. The effluents of pump A and B are mixed in the dynamic mixer (DM) and then pass through an in-line filter (F1) prior to entering into one of the two inlets of the double 3-way valve, which can either be in "extraction" position (dashed line), or in "elution" position (solid line). With the double 3-way valve in "extraction" position, pumps 1 and 2 effluents are directed to the column and the UV detectors. With the double 3-way valve in "elution" position, pumps 1 and 2 effluents are directed to the silica guard cartridge (GC) and the C_{18} extraction (EC) through a 3-way connection (T) and then to the column and the UV detectors through the second inlet of the double 3-way valve. Pump 3 is an accessory pump used to move samples out of the autosampler injection loop through an in-line filter (F2) and a static mixer (SM) into GC and EC. Pump 4 is also an accessory pump and is used to reequilibriate EC and to dilute the sample with water (solvent D) or to perform some automated clean-up procedures of EC prior to or after loading of the samples; for this purpose another SSV valve connected to the pump 4 inlet provides the ability to use different solvents. Pumps 3 and 4 are used with the double 3-way valve in "extraction" position and effluents are directed to waste through the double 3-way valve. It is noteworthy that some types of autosampler will not withstand the backpressure of the system with the double 3-way valve in elution position (analytical column in circuit); in these cases a high-pressure valve should be installed between F2 and SM to protect the autosampler during elution of samples (a Rheodyne Model 7030 3-way valve is suitable).

methanol, ethanol or acetonitrile; indeed, such samples can be analyzed directly or following minimal preparation procedures (such as partial evaporation to reduce volume). In addition, it is noteworthy that in the analysis of samples containing hydrophobic solutes of interest such as arachidonic acid, it is critical that the percentage content of organic solvent in the samples be high enough (>50%) to prevent adsorption of the solutes to autosampler vials (data not shown).

Composition of Mobile Phase

The composition of the mobile phase is designed to enable separation and quantitation of the biologically active lipoxygenase products. In this regard, LTB_4 poses a particular problem, because it is often present in biological samples together with several stereoisomers having similar UV absorption properties, but little or no biological activity, i.e., 6-*trans*-LTB_4, 12-epi-6-*trans*-LTB_4, and the (5*S*,12*S*)-dihydroxy-(6*E*,8*Z*,10*E*,14*Z*)-eicosatetraenoic acid [(5*S*,12*S*)-DiHETE]. The use of both methanol and acetonitrile in solvent mixtures A and B (Table I) is necessary to enable the separation of the four isomers. Indeed, we found that LTB_4 comigrates with 6-*trans*-12-epi-LTB_4 and (5*S*,12*S*)-DiHETE, respectively (data not shown) when solvent mixtures containing either acetonitrile only or methanol only (as the organic solvents) are used. In reversed-phase HPLC analysis of the peptido-LTs, the pH and ion strength of the mobile phase have dramatic effects on capacity factors. Indeed, because of the presence of the amino groups, which undergo ionic interactions with free silanols in C_{18} silica supports, the peptido-LTs have a more complex chromatographic behavior than other lipoxygenase products (see below). These interactions must be controlled to obtain the desired separation and measurements at maximum sensitivity. We observe that solvent mixture acidification with H_3PO_4 allows the elution of peptido-LTs with excellent peak shape while the use of acetic acid usually results in poor peak symmetry (data not shown). Other investigators successfully used trifluoroacetic acid with similar results.[4]

DMSO added to solvent mixture A is used solely to counteract the effect of the increasing concentration of organic solvents on the UV photometer signal during the generation of the gradient (0–19 min of the elution program). DMSO has no significant effect on the separation of the various compounds and the amount added to solvent A can be modified as required to reduce the baseline drift.

[4] W. S. Powell, *Anal. Biochem.* **148,** 59 (1985).

Elution Program

The elution program (Table II) has been developed to take advantage of the ionic interaction occurring between the free silanols in non-end-capped C_{18} supports and the amino groups of the solutes.[5] The extreme sensitivity of capacity factors of peptido-LTs to changes in pH within the range of 3 and 5 in a variety of reversed-phase HPLC systems has been reported by several investigators and, in some cases, has been specifically utilized to increase the selectivity for the analysis of these compounds.[4,6,7] When the peptido-LTs are injected onto a non-end-capped C_{18} support at a pH value of ≈3, these compounds are strongly retained through ionic interactions with the support and the organic solvent content of the mobile phase can be increased to elute the solutes that interact with the support by reversed-phase mechanisms only (it is noteworthy that we have sometimes observed ionic interaction of the peptido-LTs on C_{18} support claimed to be end-capped by the manufacturers). The peptido-LTs can then be selectively eluted by either increasing the pH or ion strength of the mobile phase.[4,7] This is the strategy used in the development of the HPLC methodology reported herein.

The reversed-phase HPLC method as shown in Fig. 3 enables the separation of the main biologically active lipoxygenase products in less than 30 min. The major advantage of the procedure is to provide a group separation of the peptido-LTs; these compounds elute as very sharp peaks at much longer retention times than other lipoxygenase products retained by reversed-phase mechanisms only. Thus, the method offers excellent sensitivity and selectivity for the analysis of the peptido-LTs.

The difficulty encountered in the use of this procedure is that the capacity factor of LTC_4 progressively decreases with column use, up to a point where the LTC_4 peak will comigrate with or even elute before (as a broad peak) the sharp peak appearing at ≈22 min (see Fig. 3; this peak is caused by the change in pH of the mobile phase), making accurate and sensitive measurements of LTC_4 impossible. Although this problem will ultimately determine the column life for measurements of peptido-LTs, appropriate conditions for assay of LTC_4, D_4, and E_4 can be restored and the use of the column prolonged by replacing solvent C by a solvent mixture with a higher water content, such as 30/26/44, water/methanol/acetonitrile (v/v/v) and 0.06% H_3PO_4.

[5] R. J. Ruane and I. D. Wilson, *J. Pharm. Biomed. Anal.* **5,** 723 (1987).

[6] W. R. Mathews, J. Rokach, and R. C. Murphy, *Anal. Biochem.* **118,** 96 (1981).

[7] P. Borgeat, B. Fruteau de Laclos, H. Rabinovitch, S. Picard, P. Braquet, J. Hébert, and M. Laviolette, *J. Allergy Clin. Immunol.* **74,** 310 (1984).

TABLE II
HPLC PROGRAM

Time (min)	Solvent mixtures (% volume) A	B	C[a]	External events[b]
0	100	0	0	Pump 3 ON: flow 1.5 ml/min Pump 4 ON: flow 3.0 ml/min
3				Pumps 3 and 4 OFF Double 3-way valve in "elution" position
4	100	0	0	
4.1	85	15	0	
6	85	15	0	
6.5	55	45	0	
9	55	45	0	
11.5	30	70	0	
14	30	70	0	
14.3	0	100	0	
21	0	100	0	
21.1	0	0	100	
25				Double 3-way valve in "extraction" position Pump 4 ON: flow 3.0 ml/min
28	0	0	100	
28.1	100	0	0	
31				Pump 4 OFF Autosampler injection valve to load position and loading of a sample in autosampler loop
36.5				Autosampler injection valve to inject position and reactivation of HPLC controller program
38	Pumps 1 and 2: flow 0 ml/min			

[a] The left panel shows the composition (% volume of the solvent mixtures A, B, C, from pumps 1 and 2) of the mobile phase at the level of the dynamic mixer (DM, see Fig. 2) throughout the elution program. The combined flow of pumps 1 and 2 is maintained at 1.5 ml/min.

[b] The right panel shows the time table of other events occurring in the course of a complete HPLC run. The conditions of the apparatus between t = 31 and t = 36.5 min (final/initial conditions) are as follows: pumps 3 and 4 "OFF," double 3-way valve in "extraction" position, solvent mixture A is pumped through the analytical column (reequilibration) and the autosampler is in the process of loading a sample. When the desired volume of sample has been aspirated into the autosampler injection loop, the autosampler injection valve automatically switches to inject position which simultaneously activates a program on the HPLC controller (the chart mark signal of the autosampler is used for this purpose). These events occurring at t = 36.5 min also correspond to t = 0 since, at the time of injection, the autosampler is reset to t = 0 and reactivates the HPLC controller program. The autosampler and HPLC controller clocks

In its present form the elution program permits the separation and measurement of the 20-OH and 20-COOH metabolites of LTB_4 as the most polar compounds (retained through reversed-phase mechanism) (see Figs. 3 and 6). For analysis of samples containing more polar solutes, such as the recently described dinor derivative of 20-COOH-LTB_4,[8] minor modifications should be introduced to the elution program and to the composition of solvent mixture A (percentage water content should be increased) to enable adequate separation of these compounds.

Finally, we would like to emphasize that while the reversed-phase HPLC system reported here has been specifically developed for profiling of lipoxygenase products with group separation of LTC_4, D_4, E_4, the elution program can be shortened for analysis of samples containing no LTC_4, D_4, and E_4 by eliminating the use of the solvent mixture C (between t = 21.1 min and t = 28 min) which is used specifically for the elution of these compounds. Examples of analysis performed using a shortened elution program are shown below. It follows that for analysis of samples containing none of the peptido-LTs with free amino groups, the choice of the column (end-capped vs non-end-capped C_{18} support), the type of acid used to acidify or buffer the mobile phase, and the pH of the mobile phase are not critical since the capacity factors of hydroxy, hydroperoxy, car-

[8] T. W. Harper, M. J. Garrity, and R. C. Murphy, *J. Biol. Chem.* **261,** 5414 (1986).

are then almost synchronized (within 1 or 2 sec). The HPLC controller immediately turns pump 3 "ON" (at t = 0 min) by means of a programmed contact closure to move the sample on the EC; if necessary, the controller may be programmed to simultaneously turn pump 4 "ON" to dilute the sample (see under HPLC Procedures). When the sample transfer is completed at t = 3 min, pumps 3 and 4 are stopped, the position of the double 3-way value is switched from "extraction" to "elution" position (see Fig. 2), which connects the EC with the main pumps (1 and 2) and the analytical column, and the elution program described is executed. At t = 25 min when all solutes of interest have been eluted from EC, the double 3-way valve is switched back to "extraction" position and elution of solutes from the column continues until t = 28 min, by-passing GC and EC, while during that time the EC (and GC) is reequilibrated with water (pump 4). At t = 28.1 min, the column is reequilibrated with solvent mixture A until injection of the next sample (t = 36.5 min) or 38 min (see below). At t = 31 min, pump 4 is turned "OFF," and the autosampler, which was programmed for a run time of 31 min, switches to load position and starts loading to next sample (in the autosampler injection valve loop). With the instrument used, the loading of a 1.7-ml sample in the autosampler loop takes 5.5 min. Therefore, the HPLC controller program is interrupted and reactivated by the autosampler every 36.5 min until analysis of the last sample, in which case the HPLC controller program is not interrupted and stops pumps 1 and 2 flows at t = 38 min.

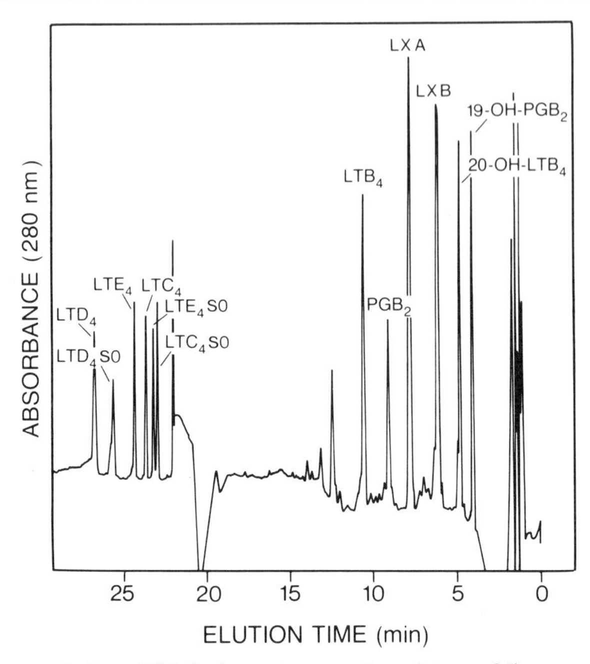

FIG. 3. Reversed-phase HPLC chromatogram of a mixture of lipoxygenase product standards. The amount injected of each compound varies from 10 to 20 ng. The two isomers of each of the peptido-LT sulfoxides (SO) are not separated in this HPLC system. The sharp peak appearing at ≈ 22 min elution time, 1 min before LTC_4SO is due to the change in the pH of the mobile phase. LXA, isomer (5*S*,6*S*,15*S*)-trihydroxy-(7*E*,9*E*,11*Z*,13*E*)-eicosatetraenoic acid; LXB, isomer (5*S*,14*R*,15*S*)-trihydroxy-(6*E*,8*E*,10*E*,12*E*)-eicosatetraenoic acid. The column used is the C_{18} Resolve (3.9 × 150 min, 5 μm particles), and the attenuation setting of the UV photometer is 0.01 od unit full scale. *N*-Acetyl derivatives of LTC_4, D_4, and E_4 (not shown on this figure) elute between 10 and 15 min, *N*-acetyl-LTC_4 eluting first, followed by *N*-acetyl-LTD_4, and *N*-acetyl-LTE_4. Reversed-phase HPLC chromatograms illustrating the separation of several other lipoxygenase products and obtained using a similar HPLC system are available from previous publications [P. Borgeat, B. Fruteau de Laclos, H. Rabinovitch, S. Picard, P. Braquet, J. Hébert and M. Laviolette, *J. Allergy Clin. Immunol.* **74,** 310 (1984); H. Salari, P. Borgeat, P. Braquet, and P. Sirois, *J. Allergy Clin. Immunol.* **77,** 720 (1986); M. E. Goldyne, G. F. Burrish, P. Poubelle and P. Borgeat, *J. Biol. Chem.* **259,** 8815 (1984)].

boxy, or oxo derivatives of unsaturated fatty acids as well as *N*-acetylated derivatives of peptido LTs are affected little by these parameters (data not shown).[3,4]

Assessments of Recoveries and Linearity of UV Signal

In this procedure the addition of the two internal standards PGB_2 and 19-OH-PGB_2 to the biological samples prior to any manipulation (denaturation, centrifugation, etc.) enables correction for loss of sample occurring prior to or during the injection by the autosampler, and allows for accurate comparisons of the amounts of various solutes in different samples. Furthermore, the use of the polar internal standard 19-OH-PGB_2, which elutes prior to the LTB_4 metabolite 20-COOH-LTB_4, allows the assessment of the efficiency of the on-line extraction.[1]

The determination of the recovery of different lipoxygenase products is, however, essential for absolute quantitation. Quantitation also relies on the linearity of the signal (peak area or peak height) over the mass range of lipoxygenase products measurable by UV photometry.

Recovery measurements are performed using three biologically active lipoxygenase products representative of compounds with low, intermediate, and high polarity, i.e., 5-HETE, LTB_4, and LTC_4, respectively. Experiments are performed as described under the section on Materials and Methods, using 25–40 pg of ^{3}H-labeled compounds. The percentage recovery (±S.E.M.) of 5-HETE, LTB_4, and LTC_4 are 86 ± 2%, 96 ± 1%, and 87 ± 4%, respectively. These results are identical whether the silica guard cartridge is present or not. When 50 ng of cold material are added as carrier to the labeled tracers, the percentage recoveries are unchanged, clearly indicating that the small loss of material could not be attributed to adsorption to surfaces in the system. In agreement with these findings, there is a linear relationship between the amounts of 20-OH-LTB_4, LTD_4, and LTE_4 injected and UV peak heights over the range of 0.85 to 42.5 ng injected (Fig. 4) (curves obtained for LTB_4 and LTC_4 are also linear and almost superimposable with those obtained for LTE_4

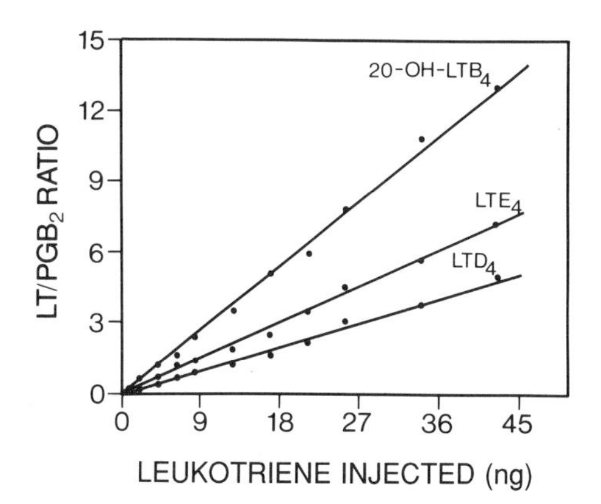

FIG. 4. Linearity of the UV signal in the analysis of LTs by RP-HPLC using the on-line extraction procedure. Analysis are performed on a C_{18} Resolve steel column (3.9 × 150 mm, 5 μm particles). Each injection contains 8.5 ng of PGB_2 and 19-OH-PGB_2 as internal standards and 0.85 to 42.5 ng of LTs. Each point is the mean of two determinations.

and LTD_4, respectively; data not shown); the correlation coefficient (*r*) is ≥ 0.993 for the five curves.

It is noteworthy that the slopes of curves as those shown in Fig. 4 will vary with the types of HPLC columns used, and will also vary somewhat for the same column used over prolonged periods of time. Consequently, a given set of curves can only be used as calibration curves for the assay of lipoxygenase products performed under identical HPLC conditions. However, given the linear relationship between the intensity of the UV signal and the mass of product injected, quantitation can be achieved from a single calibration point obtained by analysis of a mixture of known amounts (e.g., 10 ng each) of lipoxygenase products standards and internal standards. Thus, in our routine analysis of lipoxygenase products such a calibration mixture of standards is analyzed with every set of samples, and is used specifically for absolute quantitation of this set of samples.

Analysis of Biological Samples

Figures 5 to 8 illustrate some applications of the reversed-phase HPLC analysis of lipoxygenase products in biological media of various complexities. Figure 5A and B shows the profiles of lipoxygenase products in denatured suspensions of neutrophils. Such protein-free samples represent a simple type. Comparison of the two chromatograms clearly show that the zymosan-stimulated cells synthesized substantial amounts of LTB_4 (and its metabolites), as shown previously.[9] Identification of metabolites is based on comigration with authentic standards and specificity of absorption at 229 or 280 nm. This simple experiment illustrates the advantages of HPLC for studying lipoxygenase-mediated transformation of arachidonic acid. The profile obtained (Fig. 5B) reveals that LTB_4 was formed and metabolized by more than 90% into 20-OH- and 20-COOH-LTB_4 in a 5 : 1 ratio. Such detailed information could not be obtained using presently available immunoassay methods (unless the three compounds were isolated) and would have required considerable time and effort by MS techniques.

Figure 6A and B shows the reversed-phase HPLC analysis of guinea pig lung perfusates. Because of the large volume, the sample had to be concentrated by solid-phase extraction prior to analysis. Comparison of the two chromatograms shows that LTD_4 induced the release of (12*S*)-hydroxy-(5*Z*,8*E*,10*E*)-heptadecatrienoic acid (HHT) and 12-keto-(5*Z*,8*E*,10*E*)-heptadecatrienoic acid (12-keto-HT) by the perfused tissue, two cyclooxygenase products carrying UV chromophores (see Fig. 1) and

[9] P. E. Poubelle, S. Bourgoin, P. H. Naccache, and P. Borgeat, *Agents Actions* **27**, 388 (1989).

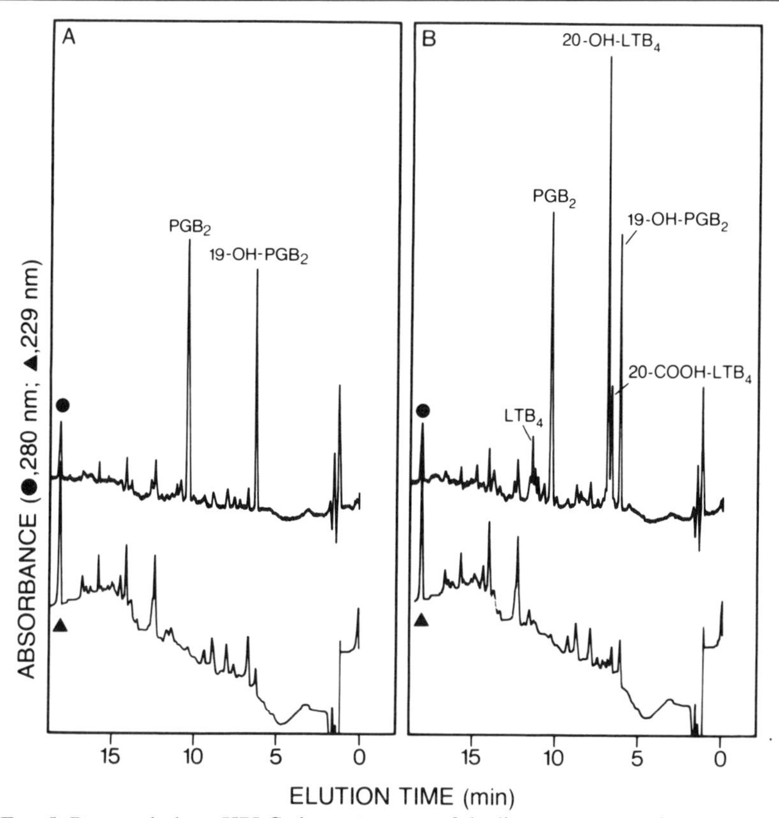

FIG. 5. Reversed-phase HPLC chromatograms of the lipoxygenase products generated by resting human neutrophils (A) or neutrophils incubated with zymosan (B). Neutrophils are prepared from human venous blood collected on sodium-EDTA and purified by treatment with dextran, centrifugation over Ficoll-Paque (Pharmacia, Sweden) cushions, and treatment with NH_4Cl as described previously [P. Borgeat and S. Picard, *Anal. Biochem.* **171,** 283 (1988)]. Aliquots of 1-ml of the final neutrophil suspension (10×10^6 cells/ml in Dulbecco's phosphate-buffered saline containing 0.5 m*M* $MgCl_2$ and 2 m*M* $CaCl_2$) are incubated 30 min at 37° in presence or absence of 3 mg/ml nonopsonized zymosan in polystyrene test tubes. At the end of the incubation period, the cell suspensions are denatured with 0.5 ml of a methanol/acetonitrile (1/1,v/v) mixture containing 12.5 ng each of PGB_2 and 19-OH-PGB_2 as internal standards. The denatured samples are centrifuged at 4° (600 *g*, 10 min) to remove the precipitated material, the supernatants are diluted to 2 ml with water, and 1.7 ml is injected. The total amounts of 20-OH- and 20-COOH-LTB_4 generated by the stimulated neutrophils are 12 and 3 ng, respectively. The column used is a C_{18} Resolve steel column (3.9×150 mm, 5 μm particles) and the attenuation settings of the UV photometers are of 0.01 and 0.02 od unit at 280 and 229 nm, respectively.

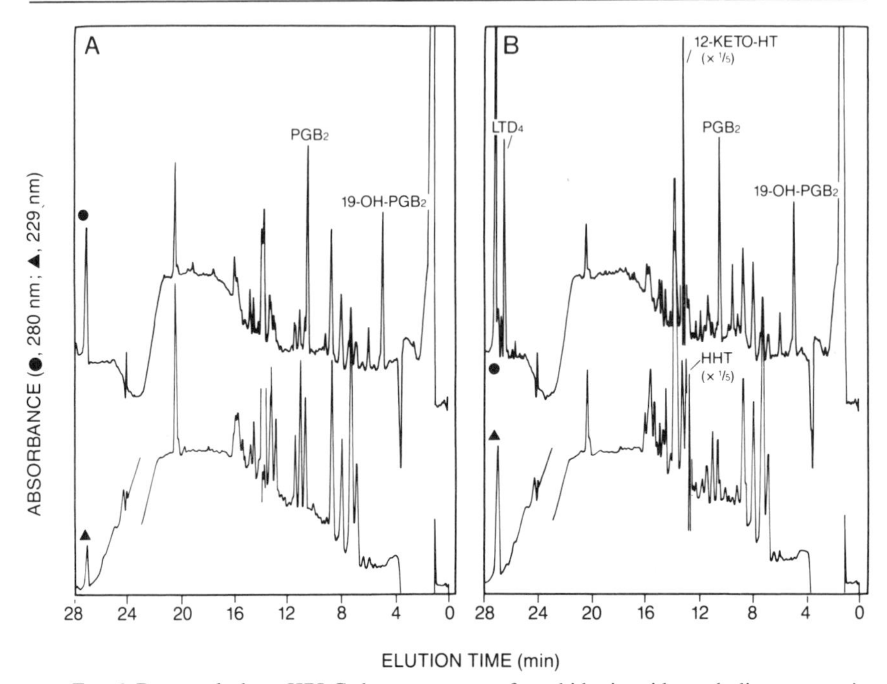

FIG. 6. Reversed-phase HPLC chromatograms of arachidonic acid metabolites present in a guinea pig lung perfusate before (A) and after (B) stimulation with LTD_4. A guinea pig lung is obtained and perfused with Hanks' buffered salt solution at 32 ml/min as described previously [H. Salari, P. Borgeat, P. Braquet, and P. Sirois, *J. Allergy Clin. Immunol.* **77,** 720 (1986)]. Following a 15-min washing period, a 15-ml aliquot of the perfusate is collected on ice and mixed with one volume of methanol containing 12.5 ng each of PGB_2 and 19-OH-PGB_2. The lung is then stimulated by a bolus injection of 50 ng of LTD_4 and 15 ml of the perfusate is collected and treated as described above. The samples are centrifuged to remove the precipitated material and extracted on Sep-Pak C_{18} cartridges (Waters, Millipore) as described previously [H. Salari, P. Borgeat, M. Fournier, J. Hébert, and G. Pelletier, *J. Exp. Med.* **162,** 1904 (1985)] to eliminate salts and reduce the volumes. The lipoxygenase and cyclooxygenase products are recovered from the Sep-Pak cartridges in 6 ml of water/methanol (10/90,v/v). The samples are evaporated to a volume of ≈ 2 ml under a nitrogen stream at 40° and 1.7 ml is injected. The column used is a C_{18} Resolve Radial Pak cartridge (5 × 100 mm, 10 μm particles); attenuation settings are as in Fig. 5 except for the peaks of HHT and 12-keto-HT which are shown at five times greater attenuation (0.1 and 0.05 od unit full scale, respectively). Amounts of the various metabolites (in B) are: HHT, ≈ 250 ng and 12-keto-HT, ≈ 125 ng. The analysis also indicates that 12 ng of LTD_4 out of the 50 ng injected are recovered in the perfusate.

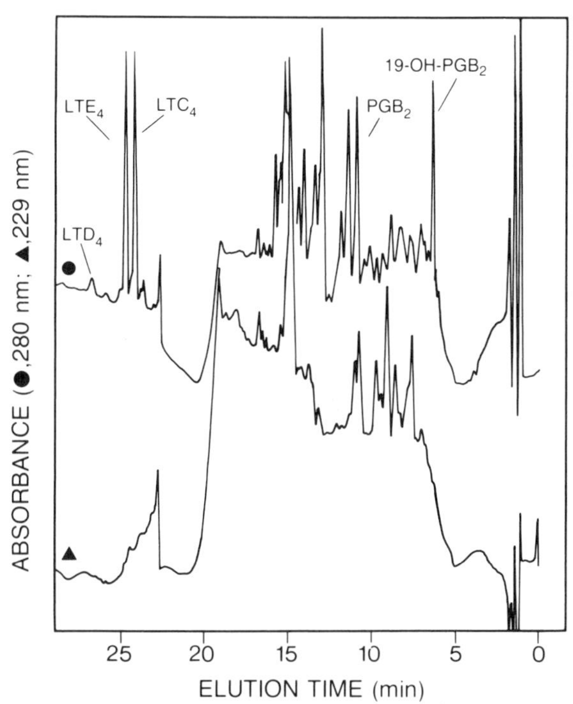

FIG. 7. Reversed-phase HPLC chromatogram of the lipoxygenase products present in an inflammatory mouse peritoneal exudate. A mouse is given an intraperitoneal injection of 1 mg of zymosan (in 0.5 ml of saline). The peritoneal exudate is collected 30 min later by washing the peritoneal cavity with 2 ml of saline as described previously [N. Doherty, P. E. Poubelle, P. Borgeat, T. H. Beaver, G. L. Westrich, and N. L. Schrader, *Prostaglandins,* **30,** 769 (1985)]. The exudate is denatured with 8 ml of ethanol and centrifuged. The clear supernatant is evaporated to dryness under reduced pressure at 40°. The residue is dissolved in 2 ml of methanol/acetonitrile/water (12.5/12.5/75,v/v/v) containing 25 ng each of PGB_2 and 19-OH-PGB_2 and 1.7 ml is injected. Total amounts of LTC_4, D_4, and E4 in the sample are 65,7, and 50 ng, respectively. LTB_4 is not detectable. The column used is as indicated in Fig. 6 legend; attenuation settings are 0.02 and 0.05 od unit at 280 and 229 nm, respectively.

reported previously as important arachidonic acid products in stimulated guinea pig lung.[10] HHT and 12-keto-HT were not detectable in the perfusate prior to stimulation with LTD_4 (Fig. 6A).

Figure 7 shows the profile of lipoxygenase products in a mouse peritoneal lavage fluid. Injection of zymosan into the peritoneal cavity induces an inflammatory reaction accompanied by plasma exudation. Because of the presence of plasma proteins in inflammatory exudates, such samples

[10] H. Salari, P. Borgeat, P. Braquet, and P. Sirois, *J. Allergy Clin. Immunol.* **77,** 720 (1986).

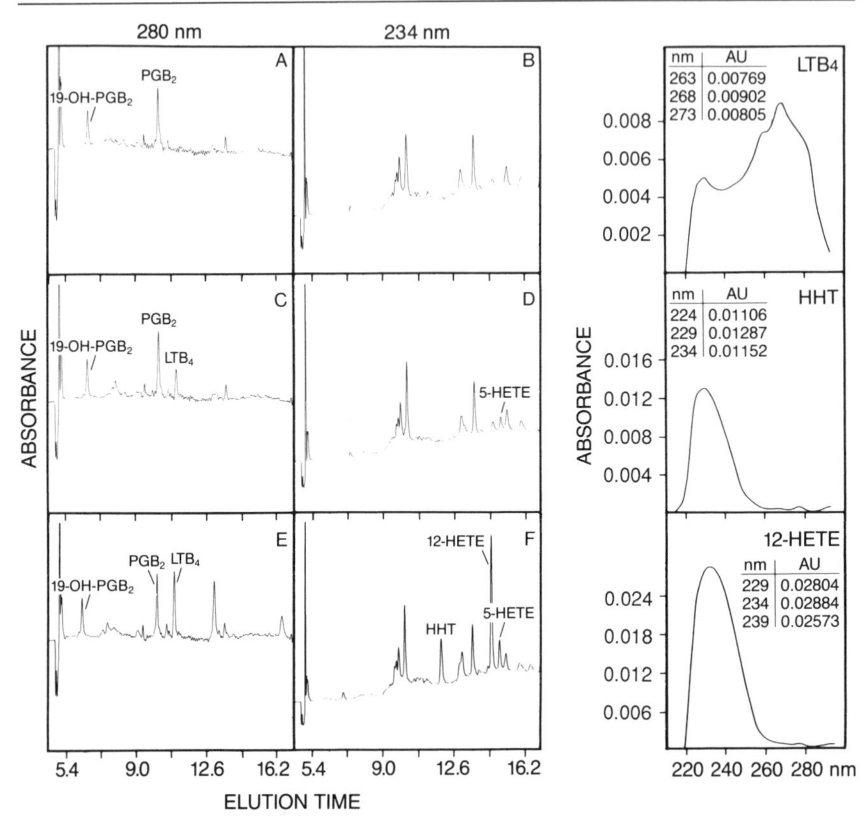

FIG. 8. Reversed-phase HPLC analysis (left panel) of lipoxygenase and cyclooxygenase products in the plasma of unstimulated (A,B), zymosan-stimulated (C,D), and ionophore A23187-stimulated human blood (E,F); right panel: UV spectra. Blood is obtained by venipuncture, collected on heparin, and aliquoted in polystyrene tubes (1 ml/tube). Blood aliquots are either stimulated with 2 mg/ml of nonopsonized zymosan for 30 min or with 50 μM ionophore for 10 min at 37°, or kept 30 min at 37° unstimulated (control). After the indicated incubation time, blood samples are centrifuged at 700 g, 15 min at 4°. Aliquots of 200-μl of plasma are collected and denatured with 800 μl of ethanol containing 25 ng/ml each of PGB_2 and 19-OH-PGB_2. The precipitated material is removed by centrifugation, the supernatants ($\approx$700 μl) are diluted to 1.2 ml with water, and 1-ml aliquots are injected. The column used is a C_{18} Nova-Pak cartridge (5 $\times$ 100 mm, 4 μm particles). Attenuation settings of the UV photometer (Polychrom model 9060 photodiode array detector, Varian, Sunnyvale, CA) are 0.025 and 0.05 od unit at 280 and 234 nm. Amounts of the various metabolites in chromatograms E and F are 25,150,32, and 50 ng for LTB_4, 12-HETE, 5-HETE, and HHT, respectively. Note that in these chromatograms the PGB_2 standard comigrated with a contaminant detectable at both 280 and 234 nm; therefore quantitation is based on the 19-OH-PGB_2 standard. The right panel shows the UV spectra of LTB_4, HHT, and 12-HETE recorded on-line from chromatograms E and F.

must be deproteinized by addition of ethanol or acetonitrile prior to HPLC analysis. The HPLC tracing clearly shows the presence of peptido-LTs in the inflammatory peritoneal fluid as reported before.[11] Because of the presence of many contaminants, it was not possible to confirm the presence of other metabolites.

The experiments shown in Figs. 6 and 7 constitute excellent examples of the usefulness of the group separation of the peptido-LTs on non-end-capped reversed-phase columns. Indeed, it was possible to quantify peptido-LTs in these samples containing a large number of unidentified solutes with UV absorption.

Figure 8 shows the analysis of arachidonic acid metabolites in plasma samples following stimulation of whole blood *ex vivo*. Metabolites are not detectable in plasma of unstimulated blood (A,B). Zymosan stimulation induces the formation of the 5-lipoxygenase products 5-HETE and LTB_4 (C,D). When blood is treated with the nonspecific activator of arachidonic acid metabolism, the ionophore A23187, both platelet and leukocyte metabolites could be identified in plasma; platelets release the 12-lipoxygenase product 12-HETE and the cyclooxygenase product HHT, whereas 5-HETE and LTB_4 are generated by phagocytes (E,F). Full UV spectra of three compounds recorded on-line using a photodiode array UV photometer support their identification as LTB_4, HHT, and 12-HETE (Fig. 8, right panel). The UV spectrum of 5-HETE (not shown) is similar to that of 12-HETE. These experiments clearly show that several arachidonic acid metabolites generated from endogenous substrate upon stimulation of blood *ex vivo* can be measured using reversed-phase HPLC in 100–300 μl plasma samples with minimal sample preparation procedures. LTB_4 synthesis in blood *ex vivo* was reported previously using immunoassay methods.[12,13]

Conclusions

Using several types of biological samples we have shown the usefulness of the methodology described herein for analysis of lipoxygenase products as well as some cyclooxygenase products. The major limitation of the method is the sensitivity, with a detection limit in the nano-

[11] N. Doherty, P. E. Poubelle, P. Borgeat, T. H. Beaver, G. L. Westrich, and N. L. Schrader, *Prostaglandins* **30,** 769 (1985).

[12] P. Gresele, J. Arnout, M. C. Coene, H. Deckmyn, and J. Vermylen, *Biochem. Biophys. Res. Commun.* **137,** 334 (1986).

[13] F. J. Sweeney, J. D. Eskra, and T. J. Carty, *Prostaglandins Leukotrienes Med.* **28,** 73 (1987).

gram range (using standard 4 to 5-mm diameter columns). Nevertheless, reversed-phase HPLC offers major advantages for lipoxygenase product analysis, since it provides a profile of compounds, as opposed to measurements of a single component, thereby supplying more information on the composition of the samples. In addition, sample preparation procedures are usually very simple, thus minimizing the possibility of sample loss and chemical alterations; this is a point of major concern considering that lipoxygenase products are polyunsaturated compounds susceptible to oxidation.

[13] Preparation of Antibodies Directed against Leukotrienes

By EDWARD C. HAYES

Introduction

The leukotrienes (LTC_4, LTD_4, LTE_4, and LTB_4) must be coupled to high-molecular weight carrier proteins such as keyhole limpet hemocyanin (KLH) or bovine serum albumin (BSA) to be immunogenic. Such coupling is possible because these leukotrienes contain a number of groups that can react directly with the carrier protein or with a conjugating reagent serving as a multicarbon spacer arm (see below and Table I). The use of such spacers will often lead to the generation of antibodies that are specific for the hapten rather than a determinant that is shared between the hapten and the carrier molecule. The site on the hapten that is used for coupling, the number of haptens per carrier (5–20), and the configuration of the hapten (whether it is the naturally occurring leukotriene or an analog) may have a profound effect on the antibody specificity. The influence of the carrier protein must also be considered. Although KLH is difficult to work with, it is one of the most immunogenic carriers and offers the additional advantage that antibodies directed against it do not cross-react with proteins found in most biological samples. A number of coupling procedures will be presented and the individual investigator must decide which are to be employed based on the titer and specificity of the antibody required as well as availability of chemical expertise.

TABLE I

PROPERTIES OF LEUKOTRIENE-SPECIFIC SERA

Immunogen	Titer[a]	$K_a (M^{-1})$	Sensitivity (pmol)[b]	Selectivity[c]
LTC$_4$–glut–BSA[d]	1 : 75	NR[e]	0.25	LTC$_4$: LTD$_4$, 100 : 16
LTC$_4$–(*N*-acetyl)-polyamino-BSA[f]	1 : 1000	NR	0.25	LTC$_4$: LTD$_4$: LTE$_4$, 100 : 0.08 : 0.07
LTC$_4$–maleimidohexanoic-S-KLH[g]	1 : 10,000	2×10^9	0.10	LTC$_4$: LTD$_4$: LTE$_4$, 100 : 43 : 8.3
LTC$_4$–(7-*cis*-hexahydro)maleimidohexanoic-S-KLH[h]	1 : 100	NR	0.15	LTC$_4$: LTD$_4$: LTE$_4$, 100 : 1.6 : 0.66
LTD$_4$–BSA[i]	1 : 40	2.8×10^9	0.5	LTD$_4$: LTC$_4$: LTE$_4$, 100 : 200 : 92
LTD$_4$–glut–thyroglobulin[j] + LTD$_4$–glut–KLH	1 : 5000	1.1×10^{10}	1×10^{-3}	LTD$_4$: LTC$_4$: LTE$_4$, 100 : 7 : 17
LTD$_4$–glut–KLH[j]	1 : 5000	1×10^9	9×10^{-3}	LTD$_4$: LTC$_4$: LTE$_4$, 100 : 8 : 48
LTB$_4$–BSA[k]	1 : 4000	NR	0.01	LTB$_4$: LTC$_4$: 5,12-di-HETE, 100 : 0.03 : 0.14
LTB$_4$–0 · CO–BSA[l] (12-oxy)	>1 : 25	3.2×10^9	0.3	LTB$_4$: 6-*trans*-LTB$_4$: 6-*trans*-12-*epi*-LTB$_4$: LTC$_4$, 100 : 30 : 3 : 5
LTB$_4$–hydrazidomaleimidohexanoic-S-KLH[m]	1 : 2000	5.8×10^8	0.1	LTB$_4$: 6-*trans*-LTB$_4$: 5,12-di-HETE : LTC$_4$, 100 : 6 : 7 : 0.02
LTB$_4$–BSA[n]	NR	4.9×10^{11}	7×10^{-3}	LTB$_4$: 6-*trans*-LTB$_4$: 6-*trans*-12-*epi*-LTB$_4$: 5,12-di-HETE : LTC$_4$, 100 : 60 : 7.5 : 3.4 : 0.02

[a] Approximate dilution employed in RIA.

[b] Lower limit of detection in competitive RIA.

[c] (Concentration of homologous cold ligand which displaces 50% of radiolabeled ligand in RIA)/(concentration of heterologous cold ligand which displaces 50% of radiolabeled ligand in RIA) × 100.

[d] U. Aehringhaus, R. H. Wolbling, W. Konig, C. Patrono, B. M. Peskar, and B. A. Peskar, *FEBS Lett.* **146,** 111 (1982).

[e] NR, Not reported.

[f] J. A. Lindgren, S. Hammerstrom, and E. J Goetzl, *FEBS Lett.* **152,** 83 (1983).

[g] R. N. Young, M. Kakushima, and J. Rokach, *Prostaglandins* **23,** 603 (1982).

[h] M. A. Wynalda, J. R. Brosher, M. K. Bach, D. R. Morton, and F. A. Fitzpatrick, *Anal. Chem.* **56,** 1862 (1984).

[i] L. Levine, R. A. Morgan, R. A. Lewis, K. F. Austen, D. A. Clark, A. Marfat, and E. J. Corey, *Proc. Natl. Acad. Sci. U.S.A.* **78,** 7692 (1981).

[j] B. C. Beaubien, J. R. Tippins, and H. R. Morris, *Biochem. Biophys. Res. Commun.* **125,** 97 (1984).

[k] J. A. Salmon, P. A. Simmons, and R. M. J. Palmer, *Prostaglandins* **24,** 225 (1982).

[l] R. A. Lewis, J. M. Mencia-Huerta, R. J. Soberman, D. Hoover, A. Marfat, E. J. Corey, and K. F. Austen, *Proc. Natl. Acad. Sci. U.S.A.* **79,** 7904 (1982).

[m] R. N. Young, R. Zamboni, and J. Rokach, *Prostaglandins* **26,** 605 (1983).

[n] F. Carey and R. A. Forder, *Prostaglandins Leukotrienes Med.* **22,** 57, (1986).

Preparation of LTC_4 Immunogens

Glutaraldehyde Coupling

Perhaps the most direct method of coupling LTC_4 to a carrier protein is via glutaraldehyde (glut).[1] LTC_4 (0.1 ml, 2 mg/ml) in water is added to 2.0 mg BSA in 0.1 ml phosphate buffer (0.2 *M*; pH 7.5) and then 0.1 ml of glutaraldehyde (0.21 *M*) is added dropwise with continuous stirring. The mixture is gently stirred at room temperature for 18 hr and protected from light during this incubation. Lysine (0.1 ml of 1.0 *M*, pH 7) is added to stop the reaction, the reaction volume brought up to 1 ml with phosphate buffer (0.1 *M*, pH 7.5), and the conjugate dialyzed extensively against phosphate-buffered saline (PBS).

Carbodiimide Coupling

Acetylated LTC_4 has been coupled to modified BSA (polyamino-BSA).[2] The LTC_4 is acetylated under argon by adding 10 μl of acetic anhydride to 100 μg LTC_4 dissolved in 0.3 ml pyridine. The sample is incubated at 22° for 15 min, evaporated to dryness with a stream of nitrogen, and then the product, acetylated LTC_4, is dissolved in 100 μl of methanol–water (1 : 1, v/v). The polyamino-BSA (PABSA) is prepared by dissolving 100 mg BSA (recrystallized), 30 mg triethylenetetraamine-HCl (recrystallized), and then 40 mg 1-ethyl-3-(3-dimethylamino-propyl)carbodiimide hydrochloride (EDC) in 10 ml water (pH 4.0 with HCl). The reaction is maintained at 22° for 16 hr, dialyzed against distilled water at 4°, and then lyophilized.

The acetylated LTC_4 (100 μg) is incubated (under argon) with 2.5 mg PABSA and 10 mg EDC in 0.3 ml distilled water (pH 4.0 with HCl) at 22° for 60 min and then at 4° for an additional 180 min. The immunogen is dialyzed against distilled water under nitrogen and used immediately to immunize rabbits. The conjugate is prepared fresh for each immunization.

Maleimidohexanoic Acid Chloride Coupling

Two procedures for the generation of an LTC_4 or LTC_4 analog immunogen have been published.[3–5] Either LTC_4 or the analog 7-*cis*-hexahydro-

[1] U. Aehringhaus, R. H. Wolbling, W. Konig, C. Patrono, B. M. Peskar, and B. A. Peskar, *FEBS Lett.* **146,** 111 (1982).

[2] J. A. Lindgren, S. Hammerstrom, and E. J. Goetzl, *FEBS Lett.* **152,** 83 (1983).

[3] R. N. Young, M. Kakushima, and J. Rokach, *Prostaglandins* **23,** 603 (1982).

[4] M. A. Wynalda, J. R. Brosher, M. K. Bach, D. R. Morton, and F. A. Fitzpatrick, *Anal. Chem.* **56,** 1862 (1984).

[5] E. C. Hayes, D. L. Lombardo, Y. Girard, A. L. Maycock, J. Rokach, A. S. Rosenthal, R. N. Young, R. W. Egan, and H. J. Zweerink, *J. Immunol.* **131,** 429 (1983).

LTC_4 is attached to thiolated KLH via a method that permits altering the hapten : carrier ratio.

The thiolated KLH is prepared by reaction with *S*-acetylmercaptosuccinic anhydride (SAMSA).[3] KLH (60 mg) is dissolved in 1.5 ml borate buffer (0.2 *M*, pH 8) and the insoluble material removed by centrifugation. The KLH concentration of the clarified solution is determined to be 24.6 mg by UV [E_{278} (1 mg/ml) = 1.36]. This solution is alternately purged with high purity nitrogen and subjugated to high vacuum several times to remove oxygen. Then 45 mg SAMSA is added in 5-mg aliquots over a 1-hr period during which time the pH is maintained at 8.0 with 1 *N* NaOH. The reaction is allowed to proceed for an additional hour and then *N*-ethylmaleimide (20 mg in 0.1 ml methanol) is added to block any free sulfhydryl groups. After an additional 90 min at room temperature, the thiolated KLH is clarified by centrifugation and separated from the low molecular weight material on a Sephadex G-50 column (100 ml bed volume) eluted with PBS-6.2 (0.1 *M* NaCl, 0.01 *M* sodium phosphate, pH 6.2). Approximately 17 mg protein is routinely recovered in the major peak eluting just after the void volume. The thiol analysis routinely yields 18 thiols per 100 kilodaltons (KDa) of protein.[6] LTC_4 or the LTC_4 analog (5 mg) is dissolved in anhydrous methanol (1 ml) containing triethylamine (80 μl) and maintained under nitrogen. Freshly prepared 6-*N*-maleimidohexanoic acid chloride[3] (25 μl of 10 mg in 100 μl anhydrous THF) is added and the reaction stirred under nitrogen at 22° for 10 min. Additional acid chloride (5 μl) is added and incubation continued for 10 min. The reaction mixture is concentrated to 0.2 ml with nitrogen diluted to 0.5 ml with sodium borate buffer (0.1 *M*, pH 7.2), and the remainder of the methanol removed under vacuum.

The derivatized LTC_4 is then conjugated with thiolated KLH as follows: The thiolated KLH (10 mg) in PBS-6.2 (4.5 ml) is deoxygenated (alternating vacuum and nitrogen), the pH brought to 11.5 with 1 *M* NaOH, and the reaction maintained at 23° under nitrogen for 1 hr. The pH is then adjusted to 7.2 (1 *N* HCl) and the derivatized LTC_4 (in borate) added. The mixture is incubated at 23° for 2 hr under nitrogen, NEM (1 mg in 10 μl methanol) added, and incubation continued for an additional hour. The conjugate is purified on a Sephadex G-50 column (1.5 × 75 cm) equilibrated with PBS-6.0 (0.1 *M* NaCl, 0.01 *M* phosphate, pH 6.0) and the pH of the eluted protein adjusted to 7.2 with 1 *M* NaOH. The conjugates prepared by this method contain between 7 and 10 LTC_4 residues per 100 KDa equivalents of KLH.

[6] G. L. Elleman, *Arch. Biochem. Biophys.* **82,** 70 (1959).

Preparation of LTD_4 Immunogens

Mixed Anhydride Coupling

LTD_4 has been coupled to BSA via the eicosanoid carboxyl.[7] The leukotriene moiety is a protected LTD_4, the dimethyl ester of *N*-trifluoroacetyl-LTD_4,[8,9] and its preparation has been described. Treatment of the derivitized LTD_4 with 2.5 molar equivalents of lithium hydroxide [in dimethoxyethane : water (4 : 1,v/v)] at 23° for 90 min removes the protecting methyl group from eicosanoid carboxyl moiety while leaving the glycine as the methyl ester. The product is purified by silica gel TLC and then incubated at −25° with 4 E of triethylamine and 2 E of isobutyl chloroformate in dry dimethoxyethane for 15 min. The product, a mixed anhydride of the glycine monomethyl ester of *N*-trifluoroacetyl-LTD_4 and the isobutyl ester of carbonic acid is held at −25° and BSA (40 mg/ml) in deionized water added to a 50 : 1 molar ratio (leukotriene : BSA). This solution is allowed to warm to 0° over 60 min and maintained at this temperature for an additional 90 min. Three volumes of methanol and 100 E of aqueous 0.15 *M* potassium carbonate (relative to leukotriene) are added and the reaction maintained at 23° for 60 min. This hydrolyzes the protective *N*-trifluoroacetyl and methyl ester functions and after neutralization with acetic acid the conjugate is purified on Sephadex G-25 equilibrated with 10% methanol in water. The molar ratio of LTD_4 : BSA obtained is 7.

Glutaraldehyde Coupling

LTD_4 has also been conjugated to thyroglobulin and to KLH via glutaraldehyde.[10] In one case, 300 μg thyroglobulin and 100 μg LTD_4 and, in the other case, 20 mg KLH and 0.75 mg LTD_4, are conjugated in the presence of 0.03% glutaraldehyde at 4° for 48 hr. In each reaction approximately 10^5 DPM of [^{3}H]LTD may be added to the reactions in order to estimate the efficiency of conjugation. The product is dialyzed extensively and then made up in physiological saline.

[7] L. Levine, R. A. Morgan, R. A. Lewis, K. F. Austen, D. A. Clark, A. Marfat, and E. J. Corey, *Proc. Natl. Acad. Sci. U.S.A.* **78,** 7692 (1981).

[8] E. J. Corey, D. A. Clark, G. Goto, A. Marfat, C. Mioskowski, B. Samuelsson, and S. Hammarstrom, *J. Am. Chem. Soc.* **102,** 1436 (1980).

[9] R. A. Lewis, J. M. Drazen, K. F. Austen, D. A. Clark, and E. J. Corey, *Biochem. Biophys. Res. Commun.* **96,** 271 (1980).

[10] B. C. Beaubien, J. R. Tippins, and H. R. Morris, *Biochem. Biophys. Res. Commun.* **125,** 97 (1984).

Preparation of LTB_4 Immunogens

Maleimidohexanoic Acid Chloride Coupling[11,12]

LTB_4 (8 mg of the free acid) is dissolved in 5 ml anhydrous diethyl ether and is then treated with 20 mg dicyclohexylcarbodiimide (DCC) at 0° for 24 hr under nitrogen. An additional 30 mg DCC is added and the mixture maintained at 0° for 48 hr by which time the LTB_4 is converted to the δ-lactone. The δ-lactone is dried under nitrogen, dissolved in 1 ml acetate : hexane (2 : 3), and purified by HPLC (Waters 10 μm, μ-Porasil) in ethyl acetate : hexane (1 : 2). The product (77% yield) eluted in 5.7 min at 4 ml/min. LTB_4 δ-lactone (4 mg) is then converted to the hydrazide by dissolving in tetrahydrofuran (THF, 1 ml) and hydrazine hydrate (0.5 ml) and mixing under nitrogen at 22° for 30 min. The hydrazide is extracted with diethyl ether (3 × 2 ml) and the combined ether extracts are dried over Na_2SO_4 and then evaporated to dryness under nitrogen. The LTB_4 hydrazide (2.5 mg) is then dissolved in dry methanol (1 ml) containing 20 μl triethylamine and mixed with 6-*N*-maleimidohexanoic acid chloride[3] (3.3 mg) in 0.1 ml anhydrous THF under nitrogen at 22°. This material is dried under nitrogen and then dissolved in deoxygenated methanol (1.2 ml). The product elutes from reversed-phase-HPLC [Waters, 10 μm μBondapak, C-18; eluted with methanol : water (75 : 25,v/v), retention time 4.5 min at 2 ml/min], however, this step is not necessary for purification but only for analysis.

The above product is coupled to thiolated KLH[3] basically as described above for the reaction of thiolated KLH with the 6-*N*-maleimidohexanoic acid derivative of LTC_4. In the case of LTB_4, however, the derivative of LTB_4 is added to the thiolated KLH in 1.2 ml methanol rather than borate and then incubated at 22° under nitrogen for 18 hr. *N*-Ethylmaleimide (5 mg in 0.1 ml methanol) is added and incubation continued for an additional hour. The methanol is evaporated with nitrogen and the conjugates purified on Sephadex G-50. Analysis of a typical conjugate yields 12 LTB_4 residues per 100 KDa equivalents of KLH.

Mixed Anhydride Coupling

LTB_4 has also been conjugated to the free lysine amino group of BSA through the 12-hydroxy function of the leukotriene.[13] The conjugate, as

[11] R. N. Young, R. Zamboni, and J. Rokach, *Prostaglandins* **26,** 605 (1983).

[12] J. Rokach, E. C. Hayes, Y. Girard, D. L. Lombardo, A. L. Maycock, A. S. Rosenthal, R. N. Young, R. Zamboni, and H. J. Zweerink, *Prostaglandins Leukotrienes Med.* **13,** 21 (1984).

[13] R. A. Lewis, J. M. Mencia-Huerta, R. J. Soberman, D. Hoover, A. Marfat, E. J. Corey, and K. F. Austen, *Proc. Natl. Acad. Sci. U.S.A.* **79,** 7904 (1982).

represented LTB_4-O-CO-NH(Lys)-BSA, contains a single carbon atom as a spacer. The starting material used in the generation of the immunogen is the 5-benzoate of LTB_4 methyl ester[14] (1.8 mg) which is mixed at 0° with *p*-nitrophenyl chloroformate (4 mg) in anhydrous pyridine (240 μl). The mixture is then brought to 23° for 10 min and then triethylamine (5 drops) is added to the reaction. This solution is applied to a silica gel column equilibrated with diethyl ether : hexane : triethylamine (33 : 66 : 1, v/v/v) at 0° and eluted with the same. This solvent is removed from fractions and yields of approximately 2.5 mg of the product, the 12-*p*-nitrophenoxycarbonyl derivative of LTB_4-5-benzoate methyl ester, should be obtained. This material should be stored at −20° under argon in dry benzene.

The LTB_4 derivative (2.5 mg) in benzene is evaporated onto the bottom of a 10-ml flask just before conjugation with BSA. Separately, BSA (33.3 mg) is dissolved in sodium borate buffer (170 μl, 0.2 *M*, pH 9.0) and then brought to 1 ml with dimethyl sulfoxide. An aliquot (450 μl) of this solution is added to the flask containing the LTB_4 derivative and this mixture maintained at 4° with stirring for 52 hr. The resulting yellow solution is mixed with 6 ml of methanol/0.15 *M* K_2CO_3 (1 : 3) at 4° and then stirred at 25° for 60 hr to remove the protective methyl ester and benzoate groups. The solution is neutralized (pH 7) with acetic acid (1 *M*), water (1 ml) added, and the LTB_4–BSA conjugate purified on Sephadex G-25 equilibrated with water : methanol (9 : 1,v/v). Appropriate fractions are concentrated under vacuum, combined, and dialyzed against distilled water (at 4°). The immunogen should have a LTB_4/BSA ratio of 10.9 (mol : mol).

A second method for preparing an LTB_4 immunogen coupled to BSA via a mixed anhydride has been employed.[15] A mixed anhydride of the carboxyl on LTB_4 and the isobutyl ester of carbonic acid is first prepared and this in turn, reacts, with free amino groups on BSA. LTB_4 (2.5 mg) is dissolved in 1,4-dioxane (0.25 ml) and then isobutyl chloroformate (2.5 μl) and triethylamine (4 μl) are added. BSA (7.5 mg) in 5% sodium bicarbonate (0.75 ml) is added and the reaction stirred at 4° for 1 hr. The conjugate is dialyzed extensively against sodium phosphate buffer (0.1 *M*, pH 7.4) and diluted to 0.5 mg protein/ml with PBS. The conjugate contains 9.5 mol LTB : 1 mol BSA by UV absorbance and 14.75 mol LTB_4 : 1 mol BSA as determined by adding [^{3}H]LTB_4 (100,000 cpm) to the starting LTB_4.

[14] E. J. Corey, A. Marfat, J. E. Monroe, K. S. Kim, P. B. Hopkins, and F. Brion, *Tetrahedron Lett.* **22,** 1070 (1981).

[15] J. A. Salmon, P. A. Simmons, and R. M. J. Palmer, *Prostaglandins* **24,** 225 (1982).

Carbodiimide Coupling

LTB_4 has been conjugated via carbodiimidazole (CDI).[16] LTB_4 (2.5 mg) is dissolved in dimethylformamide (DMF) (0.5 ml) and CDI (1.45 mg) added. The reaction is maintained at 22° for 1 hr and then cooled to 0°. BSA (7.15 mg in 0.715 ml deionized water) is added and the mixture incubated at 22° for 5 hr. The conjugate is dialyzed against DMF : water (1.3 : 2, v/v) at 4° for 16 hr, then extensively against water and lyophilized.

Immunization

The amount of leukotriene–protein conjugate employed to immunize rabbits has varied greatly (20 μg–2 mg per injection). In general, 100–500 μg of conjugate in approximately 1 ml of saline is emulsified with an equal volume of Freund's complete adjuvant (CFA). Rabbits are injected intradermally at multiple sites for primary immunization. The first boost, consisting of the same amount of conjugate in Freund's incomplete adjuvant (IFA) via the same route or intramuscular (im) is usually given 3 weeks later. Animals should be bled 7–10 days later and antibody levels determined. In those animals responding, antibodies of both high titer and high affinity are often obtained if these animals are rested for 3–6 months and then boosted again with the conjugate in IFA.

Radioimmunoassay

Several types of radioimmunoassays (RIAs) are suitable for use with leukotrienes and differ in the method of separating the free and immunoglobulin-bound ligand. These separation techniques include dextran-coated charcoal, second antibody precipitation, and protein A Sepharose immunoadsorption. In all of these methods the leukotriene-specific antiserum must be titrated to bind approximately 50% of the ^{3}H-labeled leukotriene. The labeled ligands employed in these assays are commercially available and have specific activities in the range of 20 to 100 Ci/mmol. The specific activity plays a major role in the sensitivity of the RIA. Approximately 5,000–10,000 cpm of ligand are incorporated in each tube of the RIA.

Dextran-coated charcoal (Dex-CC) may be utilized to bind free labeled ligand and therefore to separate it from antibody bound ligand. However, the Dex-CC should not remain in the reaction for any longer than is necessary to bind free ligand (1–5 min) as any ligand which disassociates from the antibody will adsorb to the Dex-CC.

[16] F. Carey and R. A. Forder, *Prostaglandins Leukotrienes Med.* **22,** 57, (1986).

To prepared Dex-CC suspend 30 g Norit A charcoal in 1 liter PET buffer (10 mM sodium phosphate, pH 7.4, 1 mM EDTA, and 0.25 mM thimerosal) containing 0.25% dextran T-70 (Pharmacia). Sediment the charcoal at 500 *g* and resuspend in 1 liter PET and repeat until the supernatant contains no particulates. Resuspend the final sediment in 200 ml PET. Dilute this stock 1 : 10 with PBS before use. The Dex-CC RIA is carried out as follows. The ^{3}H-labeled leukotriene is diluted in PBS to contain 10,000 cpm/25 μl. Twenty-five microliters of this is added to a 1.5-ml polypropylene microfuge tube containing 100 μl of PBS or 100 μl of unknown sample or 100 μl of the unlabeled competing ligand; 25 μl 20% horse serum in PBS or 25 μl 0.5% BSA in PBS; and 25 μl of immune serum (in PBS at the dilution which binds about 50% of the labeled ligand). The reaction is incubated at room temperature for 1 hr or overnight at 4° and 1 ml of the diluted Dex-CC added. After 2 min the charcoal is sedimented at 10,000 *g* for 1 min, an 0.8-ml aliquot of the supernatant removed, and the radioactivity determined.

Although the double-antibody RIA is carried out in a similar manner and gives very reproducible results, it is somewhat more time consuming. The [^{3}H]leukotriene in PBS (as above) is incubated for 1 hr with 25 μl of appropriately diluted antileukotriene serum; 100 μl of normal rabbit serum (diluted 1 : 25 in PBS); and 100 μl of PBS or 100 μl unknown sample or 100 μl of the unlabeled competing ligand. Goat anti-rabbit immunoglobulin at the equivalence point [800 μl of 1 : 8 dilution of Bio-Rad (No. 170-6010)] is added and the mixture maintained at 4° for 18 hr. The precipitates are centrifuged at 12,000 *g* for 3 min, supernatants removed, and the precipitates washed twice in PBS. The pellets are dissolved in 500 μl 0.1% SDS or 0.1 M NaOH and the radioactivity determined.

It is necessary to combine the RIAs with separation methods in order to evaluate the leukotrienes present in complex biological samples.[17,18] Interfering substances can be separated from the leukotrienes by organic extraction or Sep-Pak C_{18} (Waters, Milford, MA) adsorption. Samples containing a number of leukotrienes which cross-react with the antiserum must first be resolved on reversed-phase HPLC and the individual leukotrienes then determined by RIA.

[17] H. J. Zweerink, J. A. Limjuco and E. C. Hayes, *in* "Handbook of Experimental Pharmacology, Vol. 82" (C. Patrono and B. A. Peskar, eds.), p. 481. Springer-Verlag, Berlin and New York, 1987.

[18] S. P. Peters, E. S. Shulman, M. C. Liu, E. C. Hayes, and L. M. Lichtenstein, *J. Immunol. Methods* **64,** 335 (1983).

[14] Bioassay of paf-Acether by Rabbit Platelet Aggregation

By MARIE-JEANNE BOSSANT, EWA NINIO, DANIÈLE DELAUTIER, and JACQUES BENVENISTE

Three basic methods exist for accurate quantitation of chemical mediators: (1) measurement of their biological activity, (2) immunological identity, and (3) physicochemical properties, such as chromatographic mobility and mass spectrometric behavior. In the case of paf-acether (paf), our method of choice is a bioassay[1-9] since the immunoassays that are now being developed have not yet proved their usefulness in day-to-day use in the research laboratory. Physicochemical methods are hampered by the very low paf concentration typically encountered. The present biological method for quantitation of paf is based on the monitoring of platelet activation either assessing platelet aggregation in an aggregometer or monitoring the release of radiolabeled serotonin previously incorporated into platelets. Several criteria are used to distinguish paf from other platelet agonists: (1) conditions of platelet activation, (2) normal and reversed-phase chromatographic elution behavior identical to that of synthetic paf, (3) inactivation by phospholipases: A_2, C, and D, and resistance to lipase from *Rhizopus arrhizus,* and (4) when enough of the mediator can be recovered, gas chromatography and mass spectrometry characteristics. Other methods, less frequently used, have been described, e.g., the platelet-desensitizing activity of paf or, for some cellular species, the incorporation of [^{3}H]acetate into paf.

[1] J. Benveniste, P. M. Henson, and C. Cochrane, *J. Exp. Med.* **316,** 1356 (1972).
[2] J. Benveniste, *Nature (London)* **249,** 581 (1974).
[3] J. Benveniste, J. P. Le Couedic, J. Polonsky, and M. Tencé, *Nature (London)* **269,** 170 (1977).
[4] P. M. Henson, *J. Clin. Invest.* **60,** 481 (1977).
[5] J. P. Cazenave, J. Benveniste, and J. F. Mustard, *Lab. Invest.* **41,** 275 (1979).
[6] C. A. Demopoulos, R. N. Pinckard, and D. J. Hanahan, *J. Biol. Chem.* **254,** 9355 (1979).
[7] M. L. Blank, F. Snyder, L. W. Byers, B. Brooks, and E. E. Muirhead, *Biochem. Biophys. Res. Commun.* **90,** 1191 (1979).
[8] J. Benveniste, M. Tencé, P. Varenne, J. Bidault, C. Boullet, and J. Polonsky, *C.R. Acad. Sci. (Paris)* **289D,** 1037 (1979).
[9] J. Polonsky, M. Tencé, P. Varenne, B. C. Das, J. Lunel, and J. Benveniste, *Proc. Natl. Acad. Sci. U.S.A.* **77,** 7019 (1980).

Platelet Aggregation Method[1,5]

Preparation of Washed Rabbit Platelets

This method, a modification of that of Ardlie *et al.*,[10] is based on the differential centrifugation of whole blood since platelets have the lowest density and thus can be readily separated from leukocytes and erythrocytes.

Reagents

Ethylenediaminetetraacetate (EDTA), 0.2 *M*, pH 7.2

Tyrode's gelatin buffer devoid of Ca^{2+} (TG no Ca^{2+}-EGTA), pH 6.5, in m*M*: 2.6 KCl, 1.05 $MgCl_2$, 137 NaCl, 12.1 $NaHCO_3$, 5.6 glucose, 0.2 EGTA {[ethylenebis(oxyethylenenitrilo)]tetraacetic acid} and 0.25% gelatin (w/v)

Acetylsalicylic acid (Aspegic, Egic Lab, Amilly, France), 100 m*M*

Procedure. Adult New Zealand White rabbits are bled from the ear artery (40 ml) into a Falcon tube containing 1 ml of 0.2 *M* EDTA. Platelets are obtained using a modification of the method of Ardlie *et al.*[10] Blood is centrifuged at room temperature at 400 *g* for 20 min to allow separation into three layers: (1) the platelet-rich plasma, (2) the buffy coat, and (3) the erythrocyte and neutrophil layers. The platelet-rich plasma is recovered, treated with acetylsalicylic acid (0.1 m*M*, final concentration) for 30 min at room temperature,[5] and platelets are separated from plasma by centrifugation at 2000 *g* for 20 min. The platelet pellet is washed twice following resuspension in TG no Ca^{2+}-EGTA (pH 6.5) and centrifugation. Platelets are finally suspended in this same medium and kept at room temperature.

Aggregation of Platelets with paf

Platelet aggregation is monitored using an aggregometer, the principle of which is based on the measurement of light transmission through the platelet suspension (this physical phenomenon is termed turbidimetry). Aggregated platelets will allow proportionally higher light transmission than nonaggregated ones.

Reagents

Tyrode's gelatin buffer with calcium (TG), in m*M*: 2.6 KCl, 1 $MgCl_2$, 137 NaCl, 12.1 $NaHCO_3$, 5.6 glucose, 0.9 $CaCl_2$ and 4.2 HEPES (*N*-2-

[10] N. G. Ardlie, M. A. Packham, and J. F. Mustard, *Br. J. Haematol.* **19**, 7 (1970).

hydroxyethylpiperazine-*N'*-2-ethanesulfonic acid, Calbiochem, San Diego, CA), 0.25% gelatin (w/v), pH 7.4.

HEPES–BSA buffer, in m*M*: 4.2 HEPES, 137 NaCl, 2.6 KCl, 0.65 $CaCl_2$, 0.5 $MgCl_2$, and 0.25% fatty acid–free bovine serum albumin (BSA; Sigma Chemical Co., St. Louis, MO)

C16 : 0 PAF (1-hexadecyl-2-acetyl-*sn*-glycero-3-phosphocholine) or C18 : 0 PAF (1-octadecyl-2-acetyl-*sn*-glycero-3-phosphocholine) (Novabiochem, Clery en Vexin, France)

Creatine phosphate 133 m*M*/creatine phosphokinase 1333 U/ml (CP/CPK) (Sigma), stock solutions in 0.15 *M* NaCl are kept in aliquots (300 μl) at −20°

Procedure

PREPARATION OF STANDARD paf SOLUTIONS. Weigh 4.408 mg C18 : 0 paf or 4.192 mg C16 : 0 paf into a glass tube. Dissolve in 1 ml CH_2Cl_2 : CH_3OH (1 : 1), paf concentration: 8 m*M* (sample 1). Take 50 μl and evaporate under air stream. Add 2 ml CH_2Cl_2 : CH_3OH, 1 : 1 (v/v), paf concentration: 0.2 m*M* (sample 2). (It was empirically found that evaporation and resuspending paf in CH_2Cl_2 : CH_3OH, gives better yield than simply dissolving in the same solution.) Take 50 μl of solution 2 and evaporate. Add 2 ml CH_2Cl_2 : CH_3OH 1 : 1, paf concentration: 5 μ*M* (sample 3). Take 200 μl of solution 3 and evaporate. Add 1 ml NaCl 0.15 *M* with 0.25% BSA (w/v), paf concentration: 1 n*M* (sample 4). From sample 4, four other concentrations of standard paf are prepared by dilution with NaCl–BSA: 0.8, 0.6, 0.4, and 0.2 n*M*.

SAMPLE PREPARATION. Samples containing paf are prepared as follows. Cells with their medium (to assess the total formation of paf) or cells and medium separately (to assess cell-associated vs. released paf) are extracted with 80% ethanol (final concentration) and continuously mixed for 1 hr at 20°. The extracts are centrifuged, the pellets are discarded, and the supernatants brought to dryness under an air stream at 40°. Dry residues containing paf are then suspended in HEPES–BSA.

Tissues or *Escherichia coli*[11–14] are extracted according to a modified Bligh and Dyer technique.[15] Briefly, 1 volume CH_2Cl_2 and 2 volumes

[11] F. Snyder, *Med. Res. Rev.* **5,** 107 (1985).

[12] E. Ninio, *in* "New Horizons in Platelet Activating Factor Research" (L. M. Winslow and M. L. Lee, eds.), p. 27. Wiley, New York, 1987.

[13] J. Benveniste, *in* "Biological Membranes: Aberrations in Membrane Structure and Function" (M. L. Karnovsky, A. Leaf, and L. C. Bolis, eds.), p. 73. Liss, New York, 1988.

[14] Y. Denizot, E. Dassa, H.-Y. Kim, M. J. Bossant, N. Salem Jr., Y. Thomas, and J. Benveniste, *FEBS Lett.* **243,** 13 (1989).

[15] E. G. Bligh and W. J. Dyer, *Can. J. Biochem. Physiol.* **37,** 911 (1959).

CH_3OH are added to 1 volume aqueous cells solutions acidified at pH 3. This mixture is incubated for 24 hr at 4° with shaking. CH_2Cl_2 and H_2O 1 : 1 (v/v) are added to achieve phase separation and the lower phase is collected and evaporated to dryness under nitrogen.

Lyso-paf is quantified after chemical acetylation into paf.[16] Briefly, the dry residues of the ethanolic samples are treated overnight at room temperature with 100 μl acetic anhydride and 100 μl pyridine. After evaporation of the reagents, the dry residue is washed three times with 1 ml of CH_2Cl_2 and is then suspended in the HEPES–BSA described above; the paf is then assayed. The amount of lyso-paf is measured as the difference between the amounts of paf measured after and before acetylation of the samples. The yield of this method is 80%.[16]

The cell content of 1-*O*-alkyl-2-acyl-*sn*-glycero-3-phosphocholine is measured as paf after alkaline hydrolysis, followed by acetylation. The ethanolic samples are brought to dryness and are treated with 0.03 *N* NaOH in methanol for 2 hr at 22°. The pH is then adjusted to 7.4 with 1 *N* HCl and the mixture brought to dryness; the dry residue is then acetylated as described above, and then bioassayed for paf activity. The yield of alkaline hydrolysis is 99.1%.[16]

CALIBRATION OF PLATELET AGGREGATION. Platelets (2×10^9/ml) are stirred in 350 μl TG. To avoid the possibility that either arachidonic acid metabolites or ADP might be responsible for platelet aggregation, platelets are preincubated with acetylsalicylic acid during purification (see above) and the complex CP/CPK (300 μl of each, final concentration:1 m*M* and 10 U/ml, respectively) is added to 40 ml aggregation buffer.[5]

A calibration curve is obtained daily using 5 μl of the five dilutions of C16:0 or C18:0 paf. Linear regression is applied to the experimental data (Fig. 1) and the following equations are typically obtained: $y = 0.92\,x - 3.34$, $R = 0.98$ and $y = 0.43\,x - 2.41$, $R = 0.95$ for C16:0 and C18:0 paf, respectively. Sensitivity of platelets may vary from batch to batch and require a different range of paf dilutions to establish the calibration curve.

TESTING FOR PRESENCE OF PLATELET AGGREGATION INHIBITOR. Five microliters of the sample and 5 μl of paf standard solution (a concentration taken from the linear regression calibration curve that induces about 75–80% aggregation) are added to the platelet solution. If standard paf aggregation is not diminished, the sample can be quantified directly without purification. If it is decreased, purification (including thin-layer chromatography and/or liquid chromatography) is necessary.

[16] E. Jouvin-Marche, E. Ninio, G. Beaurain, M. Tencé, P. Niaudet, and J. Benveniste, *J. Immunol.* **133,** 892 (1984).

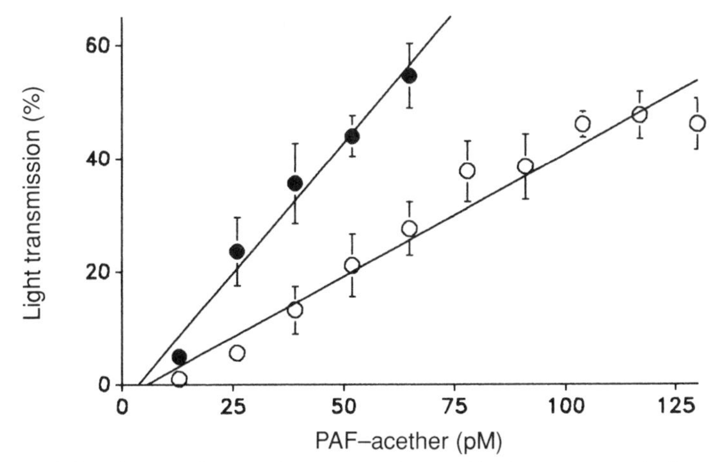

FIG. 1. Calibration plot for paf bioassay. Washed rabbit platelets were challenged with paf C16 : 0 (●) or paf C18 : 0 (○). Results in percentage of light transmission are means ± SEM of three to four determinations using independent platelet preparations.

CRITERIA FOR paf IDENTIFICATION.[17] To increase the specificity of this biological method, the aggregation is verified under different conditions: paf inactivation by phospholipases A_2, C, and D[3] or inhibition of its effects by paf antagonists such as BN 52021, kadsurenone, or CV 3988. By contrast, lipase A from *R. arrhizus* is devoid of any inactivation effect. It is also possible to analyze the molecular species of paf using reversed-phase liquid chromatography. Collected fractions corresponding to C16 : 0 or C18 : 0 or other paf analogs can be thus quantified using the bioassay.[18]

Conclusions

The platelet aggregation method is a very sensitive tool that can be used to detect as little as picogram quantities of paf. When using five different platelet preparations the variation coefficient was 23%. The bioassay specificity using crude biological extracts is obviously not very good but it reaches full specificity following prior purification of the samples and using stringent criteria for paf identification. The aggregation method has various advantages over the other method based on platelet activation: the

[17] J. Benveniste, J. Camussi, and J. Polonsky, *Monogr. Allergy* **12,** 138 (1977)

[18] L. Michel, Y. Denizot, Y. Thomas, F. Jean-Louis, C. Pitton, J. Benveniste, and L. Dubertret, *J. Immunol.* **141,** 948 (1988).

release of labeled serotonin. This latter technique uses expensive radioactive compounds which must be appropriately disposed of, necessitates a preincorporation step of the label, and, most importantly, does not yield immediate data. Off-range concentrations of paf cannot be adjusted in real time as is possible with aggregation techniques where results are immediately available, thus delaying the acquisition of valid results by several days. The aggregation method was and still is the basis for most of the studies conducted so far on this mediator, notably on the structure, the origin, the metabolism, and the physiological properties of paf. In addition, most of the paf antagonists have been developed using this test either as a screening procedure or for more complex pharmacological studies.

[15] Bioassay of Platelet-Activating Factor by Release of [^{3}H]Serotonin

By PETER M. HENSON

Introduction

Platelet-activating factor (PAF) is a highly potent cell communication mediator thought to be involved in a wide variety of pathophysiological processes. Three major approaches have been taken in assaying for the molecule: bioassays, including platelet aggregation, platelet secretion, and neutrophil activation; synthetic approaches, including incorporation of [^{3}H] acetate into the molecule; and physicochemical procedures, including mass spectrometry. Each has advantages and disadvantages. The bioassays are relatively simple, can be applied to a large number of samples, often require little workup, are highly sensitive, and by the use of appropriate standards, give reasonably quantitative results. They can be carried out with minimal equipment and are inexpensive. Their disadvantage, in general, is the potential lack of specificity (necessitating appropriate controls and/or purification), since many agents can activate platelets or neutrophils, the lack of quantitative precision that would be found in physical approaches, and the inability to obtain structural data from the assay.

Rabbit platelets are particularly sensitive to stimulation by PAF and are often used as indicator cells. Thus the assay is described for these cells, although it could be adapted to cells from humans. Platelet aggregation is more sensitive to low concentrations of the mediator than is secretion and

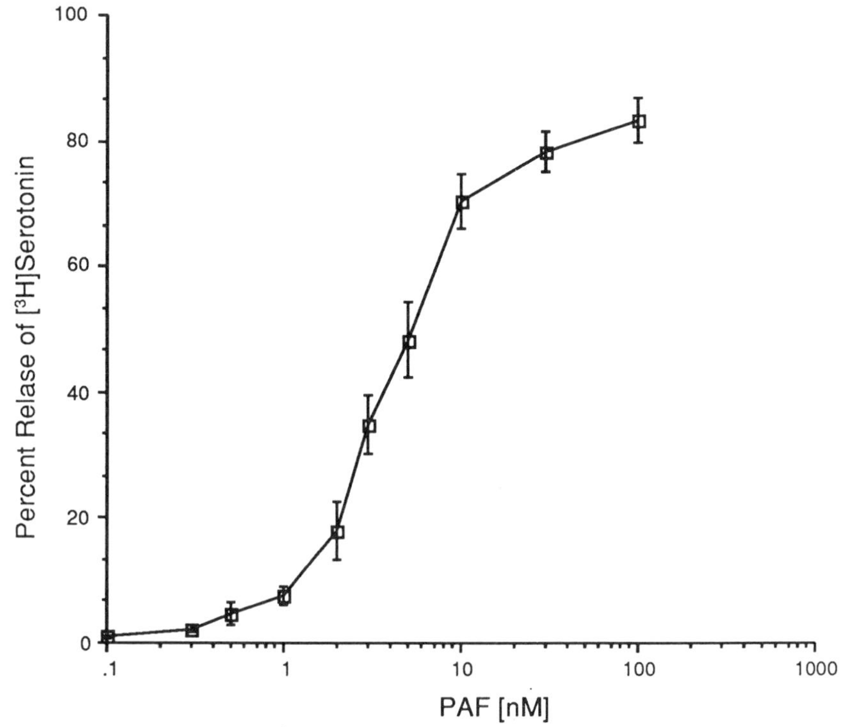

FIG. 1. Dose response of [^{3}H] serotonin release from platelets stimulated with increasing concentrations of hexadecyl-PAF. Mean of five standard curves ± SEM.

does not require separation of the platelets from the platelet-rich plasma, but does need specialized equipment (an aggregometer), can only be performed one or two assays at a time, and is less easily quantitated. Release of serotonin, on the other hand, is a simple assay, can be performed on multiple samples at once, and is linear over a wide range of PAF concentrations (see Fig. 1).

Reagents and Materials

Acid–citrate–dextrose (13.65 g citric acid, 25 g sodium citrate, 20 g dextrose in 1000 ml H_2O)[1]

Tyrode's buffer containing 0.25% w/v gelatin, but lacking calcium[2]

Tyrode's buffer containing gelatin with calcium

[1] R. H. Aster and J. H. Jandl, *J. Clin. Invest.* **43,** 843 (1964).

[2] N. G. Ardlie, M. A. Packham, and J. F. Mustard, *Br. J. Haematol.* **19,** 7 (1970).

[^{3}H]Serotonin binoxalate diluted to 1 μCi/μl (29.7 Ci/mmol, New England Nuclear, Boston, MA) PAF standards (hexadecyl-PAF)
EGTA, 0.1 *M* pH 7.2
9.25% Formaldehyde
10% Triton X-100
Scintiverse II Liquid scintillation cocktail

Equipment

Tabletop centrifuge
Liquid scintillation counter
37° water bath

Procedure (After Lynch et al.[3])

1. Collection and Labeling of Rabbit Platelets. Blood is collected free-flowing from the central ear artery of New Zealand White rabbits into a 1 : 6 volume (final) of acid–citrate–dextrose using a 19-gauge needle. It is centrifuged at room temperature at 300 *g* for 20 min to obtain platelet-rich plasma (PRP) which then is aspirated off using a transfer pipette. The PRP is incubated at 37° for 45 min with 1 μCi[^{3}H]serotonin binoxalate/ml. The platelets are pelleted by centrifugation at 1200 *g* for 20 min and washed.

2. Washing of Platelets.[2] Platelets are washed in Tyrode's gelatin buffer without calcium in the presence of 0.1 m*M* EGTA and pelleted at 1200 *g* for 20 min at room temperature. Platelets are washed once with Tyrode's gelatin buffer without calcium and no EGTA and again pelleted at 1200 *g* for 20 min. The platelet pellet is gently resuspended in Tyrode's gelatin buffer without calcium (3 ml for every 60 ml blood originally collected), and diluted to a concentration of 1.25×10^9 platelets/ml as determined from a hemocytometer or Coulter counter or a standard curve of absorbance of platelet suspensions in the spectrophotometer previously derived from known platelet concentrations.

3. [^{3}H]Serotonin Release Reaction. Polystyrene reaction tubes (75×12 mm Sarstedt tubes) containing replicate samples or PAF standards in a 0.45-ml volume of Tyrode's gelatin buffer with calcium, each receive a 0.05-ml aliquot (approximately 6.25×10^8) of the labeled platelets. After 90 sec of incubation at room temperature, the reaction is stopped by the addition of 0.02 ml of 9.25% formaldehyde. (Glutaraldehyde could also be used and acts more rapidly. Alternatively, if the tubes are cooled rapidly the cells can be centrifuged without the fixation step). The platelets are pelleted by centrifugation at 2500 *g* for 15 min at 4°.

[3] J. M. Lynch, G. Z. Lotner, S. J. Betz, and P. M. Henson, *J. Immunol.* **123,** 1219 (1979).

4. Quantitation. Aliquots of the supernatants (0.1 ml) are pipetted into 4-ml polypropylene scintillation vials. Three milliliters of Scintiverse II scintillation cocktail is added and the samples counted in a scintillation counter. The supernatant from the platelets in one set of reaction tubes containing only buffer are used to indicate the spontaneous release. Platelets lysed with 0.01 ml 10% Triton X-100 give the total [^{3}H]serotonin available. The release of serotonin is reported as a percentage of the total available. The amount of PAF present in the unknown samples is determined by comparing the percentage of serotonin released to that released by standard amounts of synthetic PAF. A standard curve of synthetic PAF is run with every assay. The results can then be expressed as nanogram equivalents of synthetic PAF.

Problems

This is a straightforward assay when the only platelet stimulant likely to be present in the sample is PAF. Thus it is particularly useful for examining PAF production from cells in serum-free culture. However, this is seldom the case and therefore appropriate controls and/or separatory procedures have to be employed. The assay is even less satisfactory when the PAF is present in complex biological fluids such as plasma or bronchoalveolar lavage fluid that may contain materials that themselves stimulate platelets, that interfere with the platelet stimulation, or that destroy the PAF. To confirm that the material being assayed is PAF it should be sensitive to destruction by phospholipase A_2. Much more effective, however, and essential for study of PAF in biological fluids or extracts of whole cells is some degree of purification before assay. Methods for isolating PAF are described by Hanahan in this volume [18]. Lipid extraction, TLC, and/or HPLC separations before assay ensure the accuracy of the results. In such cases, however, standards must be included for assessment of recovery through the separation steps. The degree of purification to employ depends on the nature of the material in which the PAF is being assayed. For extracts of cells, TLC is generally satisfactory. Since PAF is generally found in significant amounts within cells after stimulation (see Ref. 4), it is recommended that the cells be extracted and their PAF content also assayed. For blood or plasma, complete isolation through HPLC is recommended and even then the results may not match those obtained by mass spectrometric analysis (see [16] and [17] in this volume).

[4] D. L. Bratton and P. M. Henson, *in* "Platelet Activating Factor and Human Disease" (P. Barnes, C. Page, and P. Henson, eds.). Blackwell, Oxford, 1989.

Despite these caveats, the platelet serotonin assay for PAF is quick, simple, inexpensive, and has proved very useful in numerous studies of PAF production and its modulation.

Acknowledgments

This work was carried out in the F. L. Bryant Jr. Research Laboratory and was supported by National Institutes of Health Grant HL 34303.

[16] Quantitation of Platelet-Activating Factor by Gas Chromatography–Mass Spectrometry

By KEITH L. CLAY

Introduction

Quantitative analysis of platelet-activating factor (PAF, 1-*O*-alkyl-2-*O*-acetylglycero-3-phosphocholine) is complicated by the chemical nature of the molecule and by the necessity to measure it at very low concentrations (picomolar or less) in extremely complex biological mixtures. Measurement of PAF has frequently been accomplished by the use of its biological activities, through which this molecule was initially described.[1] Measurement of platelet aggregation[2] or of release of platelet serotonin content[3] has been used to estimate the amount of PAF present in biological mixtures. These procedures can give very valuable information and one should not underestimate the importance of biological assays in working with a substance that was initially described as a biological activity. Since the elucidation of the exact chemical nature of PAF, however, it has become possible to develop specific physical techniques for its measurement.[4]

Gas chromatography–mass spectrometry has been used as the primary method of measuring PAF. Procedures have been described for the measurement of this molecule in both the positive-ion electron-impact ionization mode[5] and in the negative-ion chemical ionization mode.[6] These procedures have the advantage over bioassay procedures of better

[1] J. Benveniste, P. M. Henson, and C. G. Cochrane, *J. Exp. Med.* **136,** 1356 (1972).
[2] G. Camussi, M. Aglietta, F. Malavasi, C. Tetta, W. Piacibello, F. Sanavio, and F. Bussolino, *J. Immunol.* **131,** 2397 (1983).
[3] J. M. Lynch, G. Z. Lotner, S. J. Betz, and P. M. Henson, *J. Immunol.* **123,** 1219 (1979).
[4] R. C. Murphy and K. L. Clay, *Am. Rev. Respir. Dis.* **136,** 207 (1987).
[5] K. Satouchi, M. Oda, K. Yasunaga, and L. Saito, *J. Biochem.* **94,** 2067 (1983).
[6] C. S. Ramesha, and W. C. Pickett, *Biomed. Environ. Mass Spectrom.* **13,** 107 (1986).

precision and accuracy and are less subject to false positive values due to cross-reactivity in the biological activities. GC–MS procedures, however, require much more sample preparation and are, therefore, much more labor intensive than the bioassays. A major advantage of GC–MS (or any mass spectrometry-based procedure) is that stable isotopically labeled variants of the analyte of interest can be added to the biological matrix at the earliest time of collection of the sample. These chemically identical, but physically distinguishable, molecules serve as almost ideal internal standards for the quantitation of the endogenous material. It should be noted that a major advantage of the use of stable isotopically labeled variants is that these molecules can serve as positive indicators of the adequacy of each individual analysis: if the internal standard is recovered and gives an adequate signal in the mass spectrometer, the analysis has been successful, even if none of the endogenous material has been detected. With other procedures, it is possible to obtain a false negative result due to inadequate recovery, but the use of the stable isotope-labeled internal standards gives a useful control on each individual sample and corrects for the variability of recoveries to be expected with biological matrixes of differing complexity.

Assay Procedure

Reagents

Acetyl-d_3 chloride, (Acetic anhydride)-d_6, Acetic acid-d_4, Hydrofluoric acid, 48%, Acetic acid (Aldrich, Milwaukee, WI)

Bis(trimethylsilyl)trifluoroacetamide (BSTFA), Silicic acid solid-phase extractor cartridges (Supelco, Bellefonte, PA)

Bis(*tert*-butyldimethylsilyl)trifluoroacetamide (MTBSTFA) (Pierce, Rockford, IL)

Silica gel G thin-layer chromatography plates (Analtech, Newark, DE)

Octadecylsilyl solid-phase extractor cartridges (Analytichem, Harbor City, CA)

L-Lysophosphatidylcholine-palmitoyl, stearoyl, and oleoyl (acyllyso-GPC). Phospholipase C(Type XIII, from *Bacillas cereus*). 1-*O*-Hexadecyl-2-*O*-acetylglycero-3-phosphocholine (hexadecyl-PAF) (Sigma, St. Louis, MO)

1-*O*-Octadecyl-2-*O*-acetylglycero-3-phosphocholine (Octadecyl-PAF); Hexadecyllysoglycero-3-phosphocholine and octadecyllysoglycero-3-phosphocholine (lyso-PAF) Bachem, Bubendorf, Switzerland)

Chloroform, dichloromethane, methanol, diethyl ether, hexane, perchloric acid, 70% (Fisher, Fair Lawn, NJ)

Internal Standards

Stable isotopically labeled variants of PAF and similar molecules are not commercially available, but preparation of excellent internal standards is quite simple and inexpensive. In essence, acetylation of the appropriate lysophospholipid with deuterated acetylating reagents will give the acetylated phosphatidylcholine with deuterium atoms in the acetyl moiety of the *sn*-2 position. The preparation of approximately 10 mg of internal standard by two different methods will be described.

Method 1

To 200 μl of perdeuteroacetic acid (acetic acid-d_4), are added 10 mg of lyso-PAF (or acyllyso-GPC) followed by 50 μl of acetyl-d_3 chloride. The solution is mixed by gentle agitation and left at room temperature for 30 min. A small amount of hydrogen chloride gas will be produced in the reaction, so it should be kept in a fume hood. At the end of 30 min, 2 ml of methanol are cautiously added, followed by 4 ml dichloromethane. Two phases are then produced by the addition of 2 ml water. The mixture should be gently shaken for several minutes, centrifuged to separate the phases, and the water (top) layer removed. The water layer should then be shaken with a fresh 2 ml portion of dichloromethane and the dichloromethane (bottom) layer combined with the initial extract. The dichloromethane phase will contain essentially all of the deuterated phosphatidylcholine and should be taken to dryness for further purification.

Method 2

To 200 μl of dichloromethane are added 10 mg of lyso-PAF (or acyllyso-GPC) and the solution sonicated to form a dispersion of the phospholipid. To this suspension is added 50 μl of (acetic anhydride)-d_6 and the suspension mixed. Perchloric acid (70%, 5 μl), is then added and the suspension mixed rapidly for 15 sec. Immediately on addition of the perchloric acid, the suspension should become clear as the reaction proceeds and the phospholipid goes into solution. After mixing for 15 sec, an additional 4 ml of dichloromethane is added, followed immediately by the addition of 2 ml of methanol and 2 ml water. The phases are mixed, separated, washed, and taken to dryness as described in method 1. It is critical to the success of this method that the material not be allowed to stand longer than a few seconds in an attempt to force completion. The reaction is essentially complete as soon as the reagents are mixed; any further delay in isolation of the products will result in oxidation and production of dark-colored products, and decreased yields.

Both of these procedures result in rapid and essentially complete reaction of the lysophospholipid with the acetylating reagent. For further purification, it is necessary only to separate the unreacted lysophospholipid from the acetylated product, although even that minor purification will not be necessary for some applications. Application of the dichloromethane layer as a streak on a preparative silica gel G TLC plate and development in chloroform/methanol/water (50 : 30 : 5, v/v/v) will result in clean separation of the acetylated from the nonacetylated lysophospholipid. Scraping and extraction of the appropriate area of the TLC plate will yield a solution suitable for use as internal standards for subsequent GC–MS analysis. It should be noted that the exact amount of trideuterated-PAF in this solution must be independently quantitated.

Principle of Method

PAF is too involatile to be directly analyzed by GC–MS. It must, therefore, be converted to a molecule that is amenable to gas phase analysis. The following steps are designed to: (1) isolate PAF from the biological mixture in a purity sufficient for subsequent GC–MS analysis; (2) hydrolyze the polar phosphocholine moiety to yield a diglyceride which can be derivatized to give a molecule with favorable properties for GC–MS; (3) prepare a derivative for positive-ion electron impact ionization–mass spectrometry for either estimation of the molecular species composition of the sample or for quantitative analysis of PAF and related molecules. The following listing presents two parallel schemes, the individual steps of which are interchangeable, depending on the required analysis.

1. Add deuterated internal standards to biological sample.
2. Lipid extraction (Bligh and Dyer solvent extraction or solid-phase ODS cartridge).
3. Isolation of PAF-enriched fraction (TLC, silica gel G or solid-phase silica cartridge).
4. Hydrolysis of phosphate ester (phospholipase C or hydrofluoric acid).
5. Isolation of diglyceride (TLC or solvent extraction).
6. Chemical derivatization (trimethylsilyl or *tert*-butyldimethylsilyl).
7. Gas chromatography–mass spectrometric analysis (selected-ion monitoring or full scan mass spectra).

Assay in Biological Fluids

The two parallel methods outlined above will be described, but it should be borne in mind that each of the steps of one method can be substituted for the analogous step in the other method. The choice of the optimum method will be dictated by the particular question and by the resources available in the laboratory.

The step common to both procedures, and the one that is critical for achievement of good accuracy and precision is addition of the deuterated internal standards. In general, it is desirable to add an amount of internal standard that is similar to the amount expected in the sample. Accuracy of addition of internal standards is essential for reliable quantitative results. Solutions of deuterated PAF and analogous compounds should be kept in ethanol solution, rather than chloroform or dichloromethane because of the difficulty of accurately pipetting the latter solvents. To ensure that the internal standard adequately corrects for losses or degradation which may occur in any isolation scheme, the internal standards should be added to the biological sample at the earliest possible time permitted by the experiment.

Method A

Lipids are extracted by the method of Bligh and Dyer.[7] One volume of the aqueous sample is mixed with equal volumes of methanol and chloroform. After thorough mixing, the solution is induced to separate into two phases by the addition of one volume each of water and chloroform. After mixing and low-speed centrifugation, the chloroform phase of this extraction is then taken to dryness and redissolved in a small amount of chloroform/methanol (1 : 1, v/v) for application to a silica gel G TLC plate. On a separate portionof the TLC plate, a standard solution of tritiated PAF should be applied to allow detection of the appropriate region of the plate by scanning with a radioactive TLC scanner (Berthold) after development. After drying, the plate is developed in a solvent system of chloroform/methanol/water (50 : 30 : 5, v/v/v). PAF will move to an R_f of approximately 0.3, but that value can change slightly with each analysis, and the exact position should be located by the use of radioactive standards. For extraction of this PAF fraction from the TLC plate, the silica gel scraped from the plate is shaken with 2 ml methanol/water (9 : 1, v/v). The silica gel is removed from the solution by filtration or centrifugation and the solution taken to dryness. When dry, the sample is subjected to phosphate ester hydrolysis with phospholipase C. Water (1 ml), diethyl ether (1 ml), and 5 units of phospholipase C are added to the sample and the suspension mixed by continuous rotation of the reaction tube for 2 hr. At the end of 2 hr, 1 ml hexane is added, the mixture shaken thoroughly, and then centrifuged to effect good phase separation. The upper organic layer is

[7] E. G. Bligh, and W. J. Dyer, *Can. J. Biochem, Physiol.* **37,** 911 (1959).

removed, taken to dryness with a gentle stream of nitrogen, and the diglyceride is purified by TLC on silica gel G with development in choloroform/methanol/acetic acid (98 : 2 : 1, v/v/v). Prior to analysis as the trimethylsilyl derivative, it is necessary to effect intramolecular rearrangement of the *sn*-2-acetyl group to the *sn*-3 position, since the naturaly occurring 1,2-isomer does not give useful spectra as the trimethylsilyl derivative. Isomerization is conveniently effected by simply allowing the diglyceride to remain on the TLC plate overnight, after which it is located by scanning with the radioactive TLC scanner, scraped, and extracted with diethyl ether containing 10% methanol. This ethereal solution is taken to dryness and the trimethylsilyl derivative prepared by heating for 15 min at 60° in a closed tube with 50 μl of BSTFA.

The trimethylsilyl derivative of the 1,3-isomer of the diglyceride derived from hexadecyl-PAF gives the mass spectrum illustrated in Fig. 1. For quantitative analysis, the ion at m/z 175 and its deuterated analog at m/z 178 are alternately measured by the GC–MS, and the ratio of the ion currents of these two ions used to measure the amount of PAF in the original sample. Figure 2 illustrates the chromatograms obtained when the analysis described was applied to measurement of PAF in the stimulated human neutrophil.

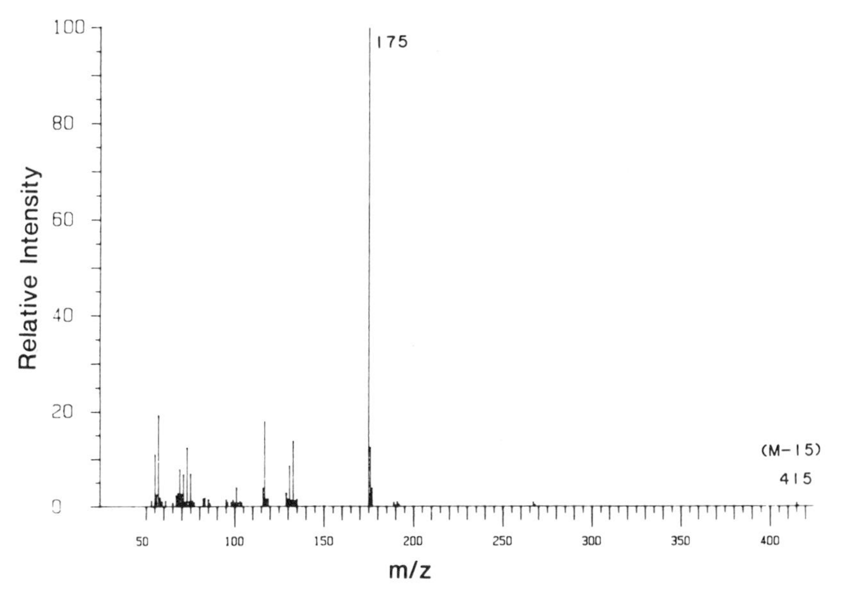

FIG. 1. Electron-impact ionization mass spectrum of the trimethylsilyl derivative of the 1,3-isomer of the hexadecylacetyl diglyceride derived from hexadecyl-PAF.

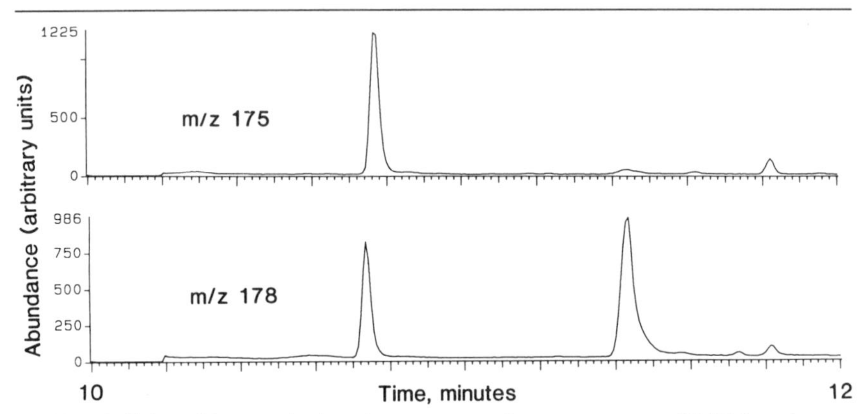

FIG. 2. Selected-ion monitoring chromatograms for measurement of PAF from human PMNs by GC–MS as the trimethylsilyl derivative. The top trace (*m/z* 175) is a measure of the amount of PAF in the sample; the bottom trace (*m/z* 178) is the ion current derived from the added deuterated internal standards, [$^2H_3$]hexadecyl-PAF and [$^2H_3$]octadecyl PAF.

Method B

In this procedure, the biological sample is mixed with 4 volumes of absolute ethanol to stop enzymatic activity and to extract PAF from the mixture. This ethanolic mixture is placed on ice for 1 hr to allow for good

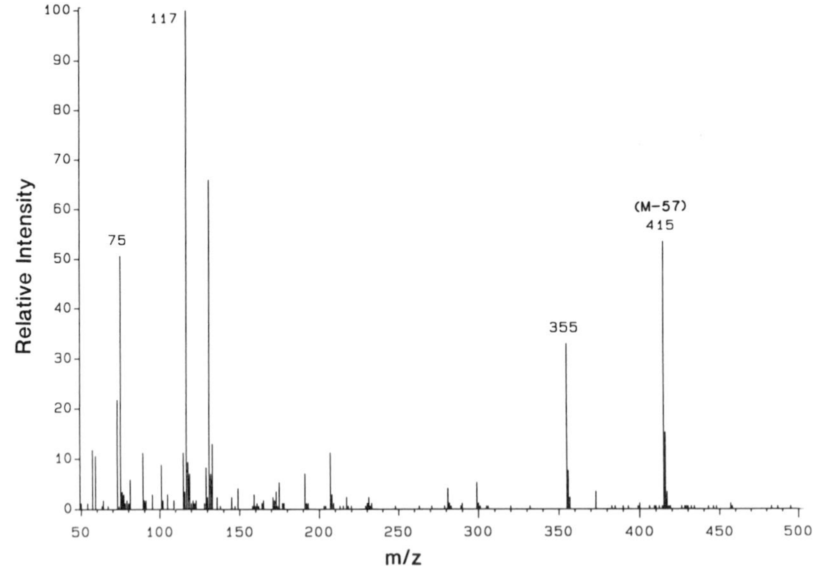

FIG. 3. Electron-impact ionization mass spectrum of *tert*-butyldimethylsilyl derivative of the diglyceride derived from hexadecyl-PAF.

protein precipitation and then centrifuged to separate the ethanolic solution from the precipitate. The ethanol is then diluted to approximately 50% by the addition of 3 volumes of water. This solution is then applied directly to a reversed-phase extractor column (100 mg silica-ODS, Bond-Elut, Analytichem) which has been activated with 5 ml ethanol and then washed with 5 ml water. The sample application can be conveniently performed either with the use of a vacuum manifold or with positive pressure from a syringe. After passing the 50% ethanolic solution through the ODS packing, the column is washed with 5 ml water followed by 5 ml 50% aqueous

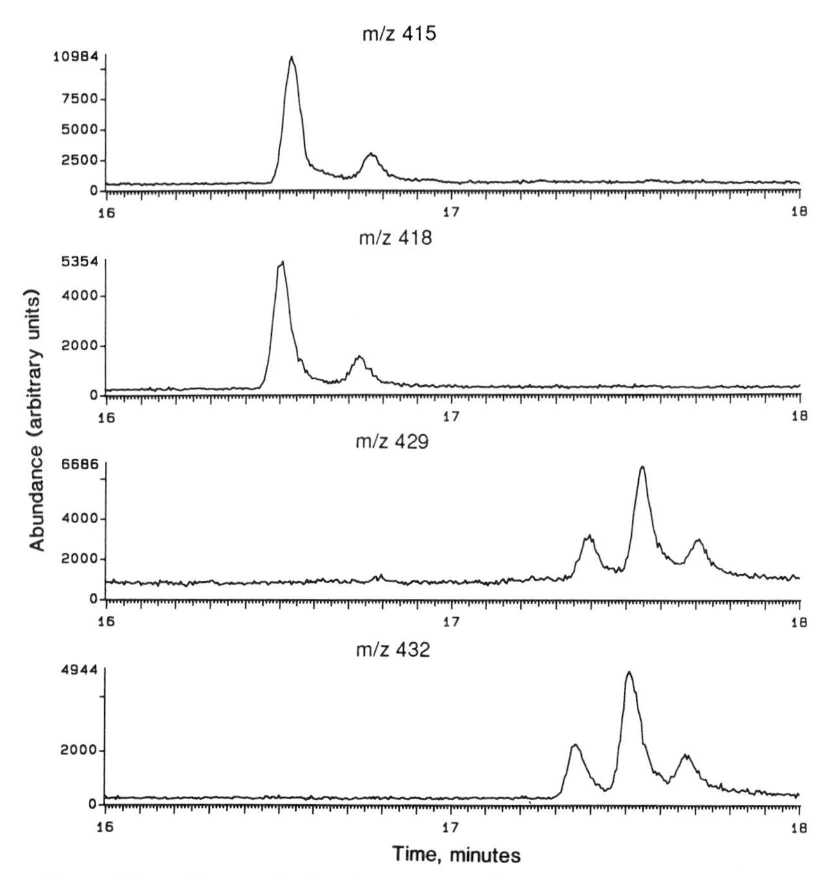

FIG. 4. Selected-ion monitoring chromatograms from an analysis of PAF derived from stimulated murine mast cells. The top two chromatograms (*m/z* 415 and *m/z* 418) are from hexadecyl-PAF and [$^2H_3$]hexadecyl-PAF. The bottom two traces are from the cellular hexadecanoylacetylphosphocholine (*m/z* 429) and its added internal standard [$^2H_3$]hexadecanoyl acetyl phosphocholine (*m/z* 432).

ethanol. A silicic acid solid-phase extractor cartridge (500 mg silica, Supelco) which has been washed with 5 ml ethanol is then connected directly to the end of the ODS cartridge with an adapter (Analytichem) and the PAF is eluted directly from the ODS cartridge onto the silica with 5 ml ethanol. The cartridges are uncoupled and the silica is washed with an additional 5 ml ethanol. The PAF fraction is then eluted into a polyethylene tube with 4 ml of methanol/water (3 : 1). This solution is then taken to dryness on a centrifugal vacuum dryer (Savant). The phosphate ester is cleaved with hydrofluoric acid (48%, 0.5 ml) which is added to the polyethylene tube containing the dried extract and left at room temperature for 4 hr. At the end of this time the diglycerides produced are extracted with 2 ml hexane. The hexane is washed once with water and then taken to dryness with a gentle stream of nitrogen. The dried extracts are then prepared for GC–MS analysis as the TBDMS derivative by reaction with 50 μl MTBSTFA for 15 min at 60°

The TBDMS derivative of PAF gives the mass spectrum illustrated in Fig. 3. This derivative gives an intense M−57 ion (*m/z* 415 for the hexadecylacetyl diglyceride from PAF) for each of the acetyl-containing diglycerides related to PAF. The combination of GC retention time and ion at M−57 gives good identification of the molecular species of compounds related to PAF. By alternately measuring the M−57 ion and its deuterated analog 3 mass units higher, accurate measurement of the amount of PAF can be made. Figure 4 illustrates measurement of hexadecyl-PAF (*m/z* = 415 and 418) and the palmitoylacetyl analog (*m/z* = 429 and 432) produced in response to calcium ionophore stimulus of murine mast cells. The two peaks for PAF and the three peaks observed for the palmitoyl analog are from all of the possible positional isomers of the acyl groups produced by intramolecular isomerization.

[17] Quantitative Analysis of Platelet-Activating Factor by Gas Chromatography–Negative-Ion Chemical Ionization Mass Spectrometry

By WALTER C. PICKETT and CHAKKODABYLU S. RAMESHA

Mass spectrometric analysis of the platelet-activating factor (PAF) has been severely limited due to the inherent involatility of this phospholipid. Although valuable structural information has been gathered with respect to the intact molecule utilizing fast atom bombardment mass

METHODS IN ENZYMOLOGY, VOL. 187

spectrometry[1-3] and to a lesser extent with thermo-spray[4] techniques, generally these methods have not been sensitive enough for the analysis of PAF derived from physiological fluids or tissue. Analysis of PAF as neutral derivatives has circumvented many of these problems by not only increasing volatility but also enhancing chromatographic characteristics. One of the most promising of these PAF derivatives is the pentafluorobenzoyl (PFB)-diglyceride. It is conveniently derived from PAF by the phospholipase C (PLC)-catalyzed hydrolysis to the diglyceride which is, in turn, esterified with pentafluarobenzoyl chloride. The PFB-diglyceride provides a highly sensitive and molecular species-selective method of PAF analysis.[5]

Reagents

1-*O*-Hexadecyl-2-acetyl-*sn*-glycero-3-phosphocholine (PAF)
1-*O*-Hexadecyl-2-lyso-*sn*-glycero-3-phosphocholine (lyso-PAF), primulin (Sigma, St. Louis, MO)
1-*O*-[^{3}H]Hexadecyl-2-acetyl-*sn*-glycero-3-phosphocholine (New England Nuclear, Boston, MA)
Perdeuteroacetic acid and trideuteroacetyl chloride (MSD Isotopes, St. Louis, MO)
Pentafluorobenzoyl chloride, PFB (Aldrich, Milwaukee, WI)
Phospholipse C, PLC (EC 3.1.4.3) from *Bacilus cereus* (CalBiochem, San Diego, CA)

Procedure

Synthesis of Deuterium-Labeled PAF.[5] One milligram of lyso-PAF is dissolved in 100 μl of perdeuteroacetic acid containing 25 mg of trideuteroacetyl chloride. After 1 hr at room temerature and under nitrogen the reaction is complete. [$^2H_3$]Acetyl-PAF is purified by thin-layer chromatography (see below) and stored in methanol. It is stable for over a year at −20°. As assessed by tracer lyso-PAF, recovery of lyso-PAF as [$^2H_3$]acetyl-PAF is greater than 98%.

Sample Preparation. The overall isolation and derivatization scheme is summarized in Fig. 1. Specifically, the lipids are extracted by a slight

[1] K. Clay, D. Stene, and R. C. Murphy, *Biomed. Mass. Spectrom.* **11,** 47 (1984).
[2] K. Clay, R. C. Murphy, J. Andres, J. Lynch, and P. Henson, *Biochem. Biophys. Res. Commun.* **121,** 815 (1984).
[3] S. T. Weintraub, J. Ludwig, G. Mott, L. McManus, C. Lear, and R. N. Pinckard, *Biochem. Biophys. Res. Commun.* **129,** *868 (1985).*
[4] H. Y. Kim and N. Salem, *Anal. Chem.* **59,** 722 (1987).
[5] C. S. Ramesha and W. C. Pickett, *Biomed. Mass. Spectrom.* **13,** 107 (1986).

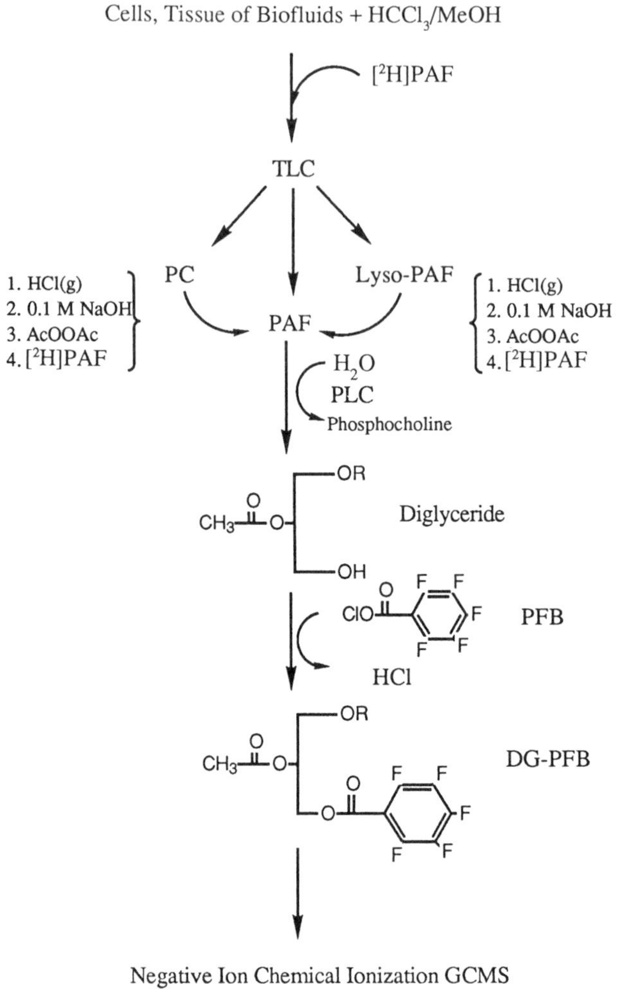

FIG. 1. Isolation and derivatization of PAF (lyso-PAF and PC) to the PFB-diglyceride.

modification of the Bligh–Dyer[6] procedure. To one volume of biological fluids such as urine, serum or cell suspensions, 3.5 volumes of methanol: chloroform 5 : 2 (v/v) are immediately added after collection or at specified times after stimulation. Tissue samples such as lung preparations are collected on dry ice, minced, and homogenized with a Polytron (Brinkman) in 5 : 2 (v/v) methanol : chloroform. The presence of solvent is needed to deactivate degradative enzymes particularly the acetylhydrolases. Acid inhibition of hydrolases is avoided especially in samples con-

[6] E. G. Bligh and W. J. Dyer, *Can. J. Biochem. Physiol.* **37,** 911 (1959).

taining erythrocytes due to poor PAF recoveries. Apparently this is due to the increase in the amount of pigments extracted at low pH.

At this point, 0.05–10 ng of [$^2H_3$]acetyl-PAF is added in 10 μl of MeOH as an internal standard. The samples are then stored or further processed by the addition of 1.5 and 1 volume additions of chloroform and water, respectively. From the resulting two phase system, the chloroform is removed and residual lipid remaining in the aqueous phase partitioned into an additional 2.5 ml of chloroform. PAF, PC, and lyso-PAF are isolated by TLC utilizing activated (100°, 30 min) silica G plates (500–1000 μm thick) developed with 65/25/0.5/2.5 (v/v/v/v) chloroform : methanol : acetic acid : water yielding R_s^f values of 0.35, 0.5 and 0.2, respectively. Plate thickness and bandwidth of silica receiving lipid extract are adjusted according to the amount of lipid applied in order to provide clear separation of PC, PAF, and lyso-PAF. Standards and sample PC (generally PAF and lyso-PAF) are identified under UV after lightly spraying with primulin (0.5 mg/ml in ethanol). Indentification of these lipids after exposure of the plate to I_2 vapors is strictly avoided due to the unexpectedly efficient depletion of unsaturated molecular species of PAF. The band corresponding to PAF and sometimes those also corresponding to standard PC and lyso-PAF are scraped, extracted[6] and stored (less than 2 days) in chloroform.

Enzymatic Hydrolysis to 1-O-Hexadecyl-2-acetylglycerol.[7] Purified PAF samples are placed in silanized 13 × 100 mm screw-cap vials with Teflon seals, the solvent removed under a stream of nitrogen, and suspended in 300 μl of 80 m*M* borate buffer (pH 8.0) and 1 ml of Et_2O. After the addition of 50 units of PLC, the samples are vigorously shaken 1 hr at 37° in tubes slanted 45° from vertical to increase the interfacial reaction area. The ether phase is retained and residual diglyceride is partitioned into two 1-ml portions of ether which are pooled with the original ether phase. Due to acetyl migration of 1,2-diglycerides to the thermodynamically favored 1,3-diglycerides, the samples are directly prepared for reaction with PFB without storage. Although borate stabilizes[8] the acetyl group during hydrolysis and purification prior to derivatization with PFB is not found necessary, it should be pointed out that silica catalyzes this acetyl migration and should be avoided. A chemical alternative (HF) to the PLC-catalyzed hydrolysis has been described.[9]

Synthesis of Pentafluorobenzoyl Derivatives. The ether extracts of 1-*O*-alkyl-2-acetylglycerol are immediately and rigorously dried (nitrogen stream with consecutive washes of ethanol and dry acetonitrile). To each tube is added 100 μl of PFB, which is then flushed with nitrogen, sealed,

[7] K. Satouchi and K. Saito, *Biomed. Mass Spectrom.* **6,** 396 (1979).

[8] Kito, H. Takamura, H. Narita, and R. Urade, *J. Biochem.* **98,** 327 (1985).

[9] P. Haroldsen, K. Clay, and R. C. Murphy, *J. Lipid Res.* **28,** 42 (1987).

and heated at 120° for 45° min. After removal of the excess reagent under a nitrogen stream, the derivative is dissolved in hexane and purified over a short silica-CC4 column (Pasteur pipette). The sample is quantitatively applied to the column with hexane, washed with an additional 4 ml, and eluted with 6 ml 25% diethyl ether in hexane. A 90% overall yield from aqueous sample to PFB derivative is obtained. For whole blood, plasma, or serum, the yield is reduced to 80%. The PFB derivative after evaporation is taken up and stored in tetradecane. Because of the high boiling point of tetradecane, reisolation of the PFB-diglyceride for purpoes of sample concentration requires additional chromatography.

Analysis of PAF Precursors and Metabolites. The above methodology is also applicable to the analysis of PC and lyso-PAF molecular species after conversion to PAF. Knowledge of these precursors is of great value in establishing substrate–product relationships necessary to determine specificity of those enzymes responsible for PAF homeostasis.[2] Specifically, PC or lyso-PC is isolated by TLC under the conditions described above, exposed to HCl gas for 5 min to remove plasmalogens, and then partitioned into chloroform.[6] After evaporation of the solvent, the extract (PC or lyso-PAF) is treated with mild base (0.1 *M* NaOH, 37° for 30 min) to remove *sn*-1 or *sn*-2 esters to yield 1-*O*-alkyl-2-lyso-*sn*-glycero-3-phosphocholine (lyso-PAF). After extraction, the lyso-PAF is converted to PAF by acetylation, which is complete overnight at 37° after dissolving the sample in 100 μl of acetic anhydride. Internal standard is included and the PC and lyso-PAF (now present as PAF) are derivatized to the PFB-DG exactly as described as above.

Mass Spectrometry. A Finnigan Model 4023-T modified with a PPINICI module interfaced with the INCOS data system is used. The mass spectrometry conditions are: Em 1.3 kV, emission current 0.44 mA, conversion dinode +3.1 kV, ionizer and analyzer temperatures 210° and 100°, respectively. The mass spectrometer is operated in the negative-ion chemical ionization mode with methane (0.3 torr) as the reagent gas. Spectra are obtained by scans of m/z from 100 to 650. In the selected-ion mode, the molecular anion of 1-*O*-hexadecyl-PAF, m/z 552 (or other molecular species) is quantitated by comparison with m/z 555, the deuterated internal standard with a 0.25–0.5 amu window.

Generation of a stable molecular anion with improved chromatographic characteristics is the major requirement for improving the quantitation of PAF. As shown in Fig. 2, the DG-PFB derivative is unusually stable, carrying greater than 92% of the ion current as the molecular anion. Apparently the DG-PFB molecular anion is more stable than analogous PFB adducts because characteristic anions such as $[M-PFB]^-$,pentafluorobenzoylate anion (m/z 211) and pentafluorotolu-

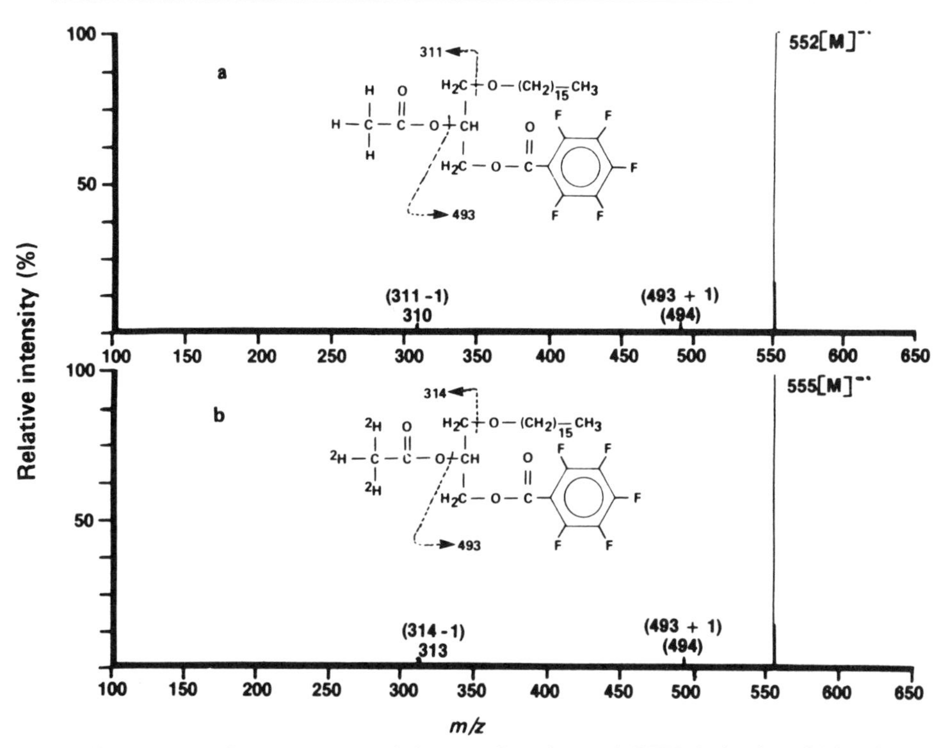

FIG. 2. The NCI mass spectra of the pentafluorobenzoyl (PFB) derivative of (a) 1-*O*-hexadecyl-2-acetylglycerol and (b) 1-*O*-hexadecyl-2-[$^2H_3$]acetylglycerol.

ene (*m*/*z* 181) are completely absent in the spectrum. Yet minor ions at *m*/*z* 311 and 493 provide useful structural information. Also indicated in Fig. 2 is the high degree of isotopic purity of the internal standard, [$^2H_3$]acetyl-PAF, necessary for maximal sensitivity.

Gas Chromatography. Despite the high molecular weight, the PFB derivatives are volatilized at temperatures compatible with moderately polar capillary columns. For this purpose, we employ DB5 fused silica columns (15 m, 0.25 mm ID; J & W Scientific, Rancho Cordova, CA) placed directly into the ion source. The column is held at an initial temperature of 220° for 1 min and programmed to 280° at 10°/min. The injection temperature is 290° and the GC–MS transfer lines are held between 285–300. Two to four microliters of sample are injected in the splitless mode with helium (9 psi) as the carrier gas.

Sensitivity. Under these conditions, a standard curve linear from greater than 10 pg to 100 fg of PAF is generated (Fig. 3). As mentioned

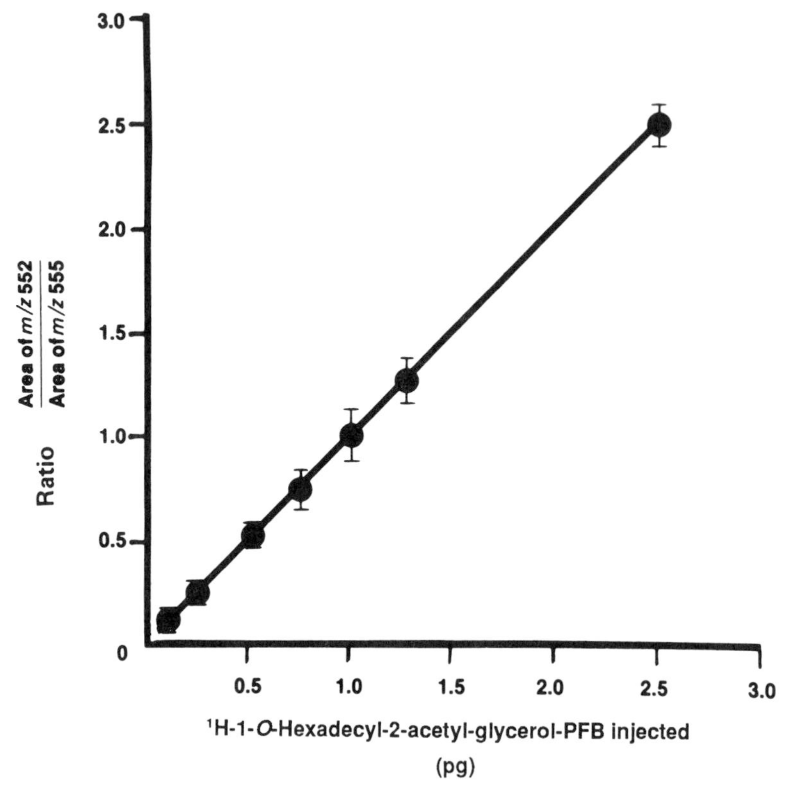

FIG. 3. Standard curve for 1-*O*-hexadecyl-2-acetylglycerol-PFB with 1 pg of 1-*O*-hexadecyl-2-[$^2H_3$]acetylglycerol as internal standard.

above, the increase in sensitivity is related to the stability of the molecular anion generated under these "soft" ionization conditions. Compared to other methods of PAF detection, this method is many orders of magnitudes more sensitive. For example, liquid chromatographic methods utilizing chromogenic[10] and fluorogenic[11] derivatives have detection limits of approximately 0.5 and 0.05 ng, respectively. Selective ion-monitoring assays of *tert*-butyldimethylsilyl derivatives of PAF have limits of approximately 0.25 ng[12] while serotonin release bioassays utilizing rabbit platelets have limits of 0.1 to 1 ng. The least sensitive but structuraly most informative is the FAB assay requiring greater that a nanogram of PAF.[1]

[10] M. L. Blank, M. Robinson, V. Fitzgerald, and F. Snyder, *J. Chromatog.* **298,** 473 (1984).
[11] C. S. Ramesha, W. C. Pickett, and D. Murthy, *J. Chromatog.* **491,** 37 (1989).
[12] K. Satouchi, M. Oda, K. Yasunaga, and K. Saito, *J. Biochem.* **94,** 2067 (1983).

Interfering Compounds. Acyl analogs of PAF (e.g., 1-*O*-hexadecyl-2-acetyl-*sn*-glycerophosphocholine) have been identified by Oda *et al.*[13] in cells stimulated with ionophore. Although biologically inactive, the "1-acyl-PAF" species can present an analysis problem. The *sn*-1 esters and those PAF molecular species with one additional carbon in the ether chain are isobaric. This is particularly problematic in the analysis of PAF derived from the guinea pig polymorphonuclear leukocytes (PMN) due to the high percentage of 1-*O*-heptadecyl molecular species. One must verify that the relatively large amount of the heptadecyl ether is not merely due to contamination with the hexadecanoyl ester. One of two approaches have been taken to solve this problem: (1) a mild hydrolysis followed by an acetylation step has been incorporated into the isolation scheme immediately following the initial TLC (Fig. 1); (2) mildly polar columns (DB5) have been utilized to resolve the alkyl and acyl species.[14] Although hydrolysis completely eliminates the acyl species, its also removes the internal standard and, therefore, requires a second incorporation of $[^2H_3]$acetyl-PAF partially through the isolation scheme.

Measurements of PAF Derived from Biological Sources. The high sensitivity of this assay has not only permitted the measurements of PAF from stimulated cells where PAF is abundant[14,15] but also from less abundant sources such as unstimulated cells[14] (Table I) or from tissue available only in small amounts such as psoriatic scales[16] (Table I).

As indicated in Table I, PAF can be measured in a variety of inflammatory cells and tissues ranging from 8 (guinea pig) to 32 (rat PMN) ng/10^7 cells. In all cases, the hexadecyl ether is the most abundant molecular species for PAF and alkyl-PC. The degree of heterogeneity increases in the basal versus the stimulated cells.

Molecular Species Analysis. Equaly important to the high sensitivity of this method is the ability to analyze molecular-species of PAF and lyso-PAF as well as the *sn*-1 chain length of alkyl-PC. Molecular species analysis permits identification of substrate–product relationships which reflect the specificity of key enzymes necessary to PAF homeostasis. As indicated in Fig. 4, analysis of selected anions corresponding to the various molecular species of PAF has revealed in the human PMN one major (1-*O*-hexadecyl, 96%) and a number of minor molecule species ranging in chain length from 14 to 19. The high degree of homogeneity is in agreement with Clay *et al.*[2] who concluded that highly specific biosynthetic enzymes

[13] M. Oda, K. Satouchi, K. Yasunaga, and K. Saito, *J. Immunol.* **134,** 1090 (1985).

[14] C. S. Ramesha and W. C. Pickett, *J. Biol. Chem.* **261,** 7592 (1986).

[15] C. S. Ramesha and W. C. Pickett. *J. Immunol.* **138,** 1558 (1987).

[16] C. S. Ramesha, N. Soter, and W. C. Pickett, *Agents Actions* **21,** 382 (1987).

TABLE I

AMOUNTS AND MAJOR MOLECULAR SPECIES OF PAF AND ALKYL-PC DERIVED FROM VARIOUS SOURCES

Source	Species	PAF			Alkyl-PC			Ref.
		ng/10^7 cells	Composition (%)	Major molecular species	μg/10^7 cells	Composition (%)	Major molecular species	
Psoriatic scales	Human	24[a]	51	C16				16
PMN	Rat							
	Stimulated[b]	32.5	96	C16				14,15
	EFAD[c] stimulated	5.3	79	C16				14,15
	Basal	0.28	76	C16	3.8	74	C16	14
	Mouse							
	Stimulated	10.4	80	C16				14
	Basal	0.64	75	C16	3.2	49	C16	14
	Guinea pig							
	Stimulated	8.4	35	C17				14
			35	C16				
	Basal	5.2	33	C16	5.1	74	C16	14
RBL[d]	Stimulated	18	80	C16				14

[a] pg/mg scales.
[b] With A23187.
[c] Essential fatty acid-deficient.
[d] RBL, Rat basophilic leukemia cell line.

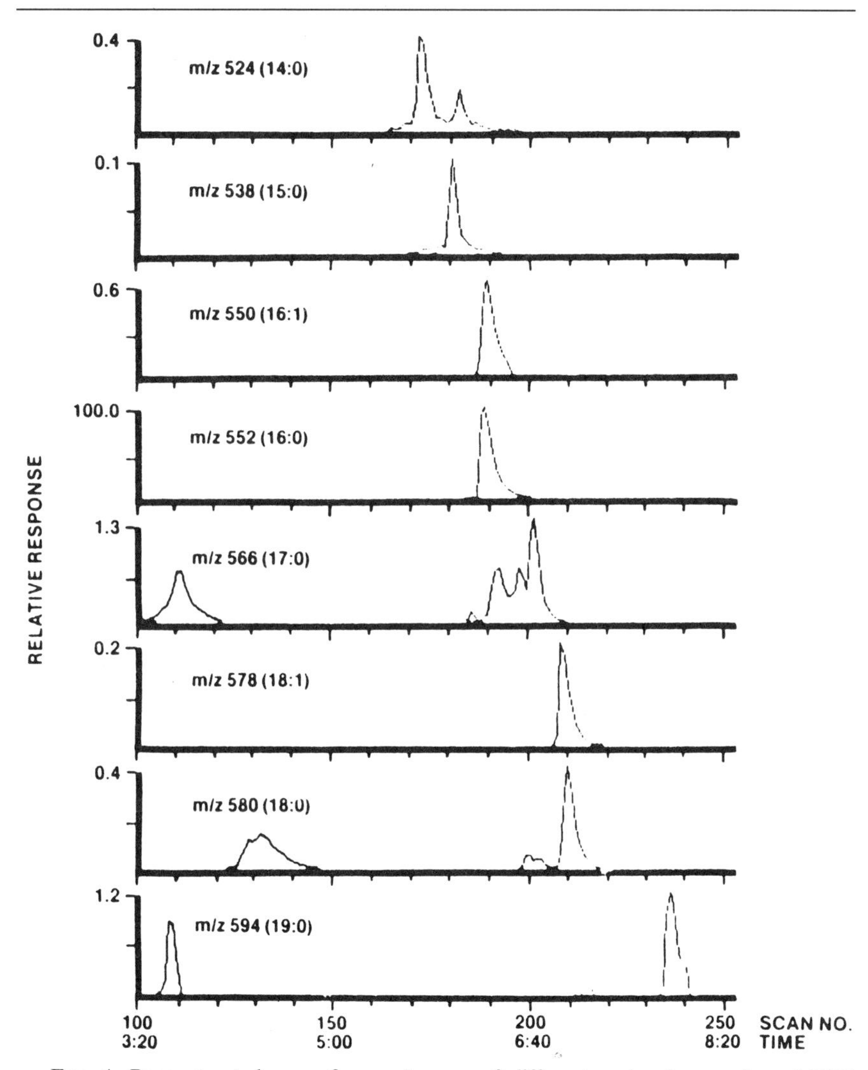

FIG. 4. Reconstructed mass fragmentogram of different molecular species of PAF synthesized by A23187-stimulated rat PMN. The PAF molecular species are analyzed as the PFB-diglyceride. The *m/z* represent the molecular anion of individual PAF species: the length and the number of double bonds of their 1-*O*-alkyl group are in parentheses.

must exist to yield a homogeneous product from a diverse substrate (alkyl-PC). We have since analyzed PAF molecular species in cells modified in essential fatty acids. Depleting cells of arachidonate inhibits PAF biosynthesis[15] (Table I). while supplementing with eicosapentaenoic acid has less of an effect.[17] These experiments lend additional support to the concept that PAF homeostasis is dependent on highly selective enzymes. Not only are few molecular species of PAF derived from many precursor molecules, but subtle changes in the fatty acid composition of key precursors affect substantial changes in PAF biosynthesis. The PFB-digylceride is a useful tool in the identification of these key intermediates and enzymes which, in turn, may provide targets for pharmacological intervention in the treatment of disease where PAF is an etioilogical component.

Conclusions

Analysis of PAF as the PFB-diglyceride is highly sensitive and molecular species selective method applicable to a wide variety of biological samples. The sensitivity is derived from the efficiency of ionization and the stabiity of the molecular anion. The method offers sensitivity and in some cases, selectivity over existing physicochemical methods of PAF analysis. Furthermore, this method is also more sensitive than bioassays for PAF, but more importantly, it obviates two important problems associated with PAF bioassays: (1) the differential response to various molecular species,[18] and (2) copurification of inhibitors with PAF[19] which may interfere with the bioassay.

[17] W. C. Pickett, D. Nytko, C. Dondero-Zahn, and C. Ramesha, *Prostaglandin Leukotriene Med.* **23,** 135 (1986).

[18] J. R. Surles, R. Wykle, J. O'Flaherty, W. Salzer, M. Thomas, F. Snyder, and C. Piantadosi, *J. Med Chem.* **28,** 73 (1985).

[19] M. Miwa, C. Hill, R. Kumat, J. Sugatani, M. Olson, and D. Hanahan, *J. Biol. Chem.* **262,** 527 (1984).

[18] Isolation of Platelet-Activating Factor and Purification by Thin-Layer Chromatography

By DONALD J. HANAHAN

Platelet-activating factor is a biologically active phosphoglyceride which is recognized as having an important role as an acute inflammatory substance and as a hypotensive agent in mammals. The most abundant and biologically potent naturally occurring form of this factor has been identified as 1-*O*-alkyl-2-acetyl-*sn*-glycero-3-phosphocholine (**1**) in which x can be 15 : 0, 17 : 0, and 18 : 1, with minor amounts of shorter chain length derivatives. 1-*O*-Acyl-2-acetyl-*sn*-glycero-3-phosphocholine (**2**) is much less abundant and has a significantly lesser biological activity. In structure

2 the acyl residue appears to be mainly the hexadecyl, octadecyl, and octadecenyl moieties. These two derivatives behave in a similar manner on thin-layer chromatography with indistinguishable R^f values. However, the low activity of the 1-*O*-acyl derivative toward rabbit platelets (approximately 300-fold less than the 1-*O*-alkyl form) and coupled with its very low concentrations, again compared to the alkyl derivative, makes it a minor consideration in any bioassay for platelet-activating factor.

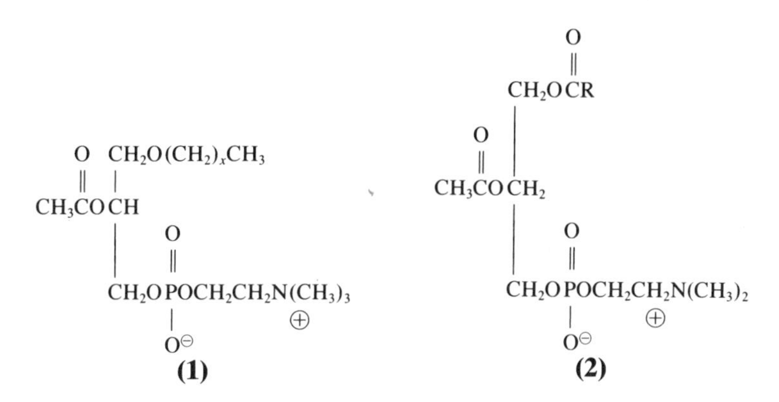

Extraction

As with most other phosphoglycerides, platelet-activating factor (PAF) does not form a covalent bond with cellular proteins. Although there is a significant binding of PAF with cellular proteins via the usual—hydrogen bonding,van der Waals forces—involving "bound" water, this phosphoglyceride is easily recovered from these complexes through the use of a combination of polar and nonpolar solvent systems. The most widely employed solvent system uses a mixture of methanol and chloroform. Although a number of different combinations have been proposed the most popular and effective one is basically that described by Bligh and Dyer.[1] Essentially using the (calculated) cell, tissue, or reaction mixture water as one component, sufficient methanol and chloroform are added to make a ratio of chloroform : methanol : water of 1 : 2 : 0.8. Subsequent recovery of the soluble component from such a treated reaction mixture will allow recovery of a homogeneous clear extract. Careful aspiration of this solution will allow for isolation of the lipids by addition of chloroform and water in volumes of 25% of the total extract and subjecting to a vortex mixer. Usually no more than two such mixings need to be done to achieve

[1] E. C. Bligh and W. J. Dyer, *Can. J. Biochem. Physiol.* **37,** 911 (1959).

an excellent separation of the PAF into the lower, chloroform-rich phase. If separation of phases does not occur within 2 to 3 min at room temperature, the sample should be centrifuged for 10 min at room temperature to achieve an excellent clean separation of phases. Finally, a word of caution regarding isolation of phospholipids and related compounds from cell or tissue preparations: there is always the potential for activation of lipolytic enzymes such as phospholipase A_2, phospholipase C, phospholipase D, and acetylhydrolase (an enzyme considered specific for PAF and related compounds) during solvent extraction. One precaution is to include a low pH buffer which would cause inactivation of acetylhydrolase.[2] Another approach is to conduct extractions at or below room temperature. Finally one can include high specific activity tritiated (usually in the alkyl residue or the acetate group) PAF as a monitor of the efficiency and effectiveness of the extraction procedure. The levels of radioactivity employed in the latter instance would not have any effect on any biological assay values.

Isolation of Platelet-Activating Factor

General Procedure

The total lipid extract,whether the chloroform-soluble material from a direct tissue or cell extract or the methanol eluate from a silicic acid column chromatographic separation, for example, are evaporated to dryness under nitrogen and dissolved in a small volume of chloroform/methanol (1 : 1, v/v). This sample can then be applied directly to silica gel G plates. When a total lipid sample is to be processed by thin-layer chromatography, in general, a sample corresponding to 200 to 300 μg total phosphorus is applied to a 500 μm plate. Overloading can cause serious problems since the usual repeated application of the sample to the origin area can create an apparent film of lipid that is not uniformly moved up the plate by the developing solvent.

Thin-Layer Chromatography

Silica gel G plates (500 μm, 10 $\times$ 20 cm) (Analtech, Newark, DE) are placed in tank containing a neutral solvent system of chloroform/methanol/water (65 : 35 : 6,v/v/v), the solvent allowed to migrate to the top, the plate removed, and air-dried. It is then activated by heating near 105° for 10 to 20 min. The plates are cooled to room temperature and then

[2] D. J. Hanahan and S. T. Weintraub, *in* "Methods of Biochemical Analysis" (D. Glick, ed.), Vol. 31, p. 195. Wiley (Interscience), New York, 1985.

divided into four lanes. A standard consisting of lysophosphatidylcholine (lysoPC), sphingomyelin, and phosphatidylcholine (PC) is applied to the first lane, an extracted sample (from above procedure) to the second, another extracted sample to the third and, finally, a standard again to the fourth lane. Plates are then developed in the same neutral solvent system. After development, only the standard lanes are sprayed with either 2-*p*-toluidinylnaphthylene 6-sulfonate (TNS) or phosphorus-detecting spray to visualize and mark standards. Silica gel in the sample lanes is scraped from an area immediately above lyso-PC to just below PC,using the standards as references. This PAF-containing fraction also includes any sphingomyelin present. A highly idealized separation of certain synthetic lipids is presented in Fig. 1. Experiments have shown that the amounts of sphingomyelin present do not affect PAF-induced aggregation of, or secretion from, washed rabbit platelets. The collected silica gel is mixed with chloroform/methanol/water (1 : 2 : 0.8,v/v/v) and allowed to stand for 30 min. Samples are then centrifuged at 800 *g* for 10 min at room temperature to pellet the silica gel and the supernatant (lipid extract) collected. Chloroform and water are then added so that proportions are chloroform/methanol/water (2 : 2 : 1.8) to effect phase separation. After vortex mixing, samples are centrifuged at 800 *g* for 10 min. The lower (chloroform-rich) phase is collected and evaporated under nitrogen, dissolved in chloroform/methanol (1 : 1), for subsequent chemical and biological testing.

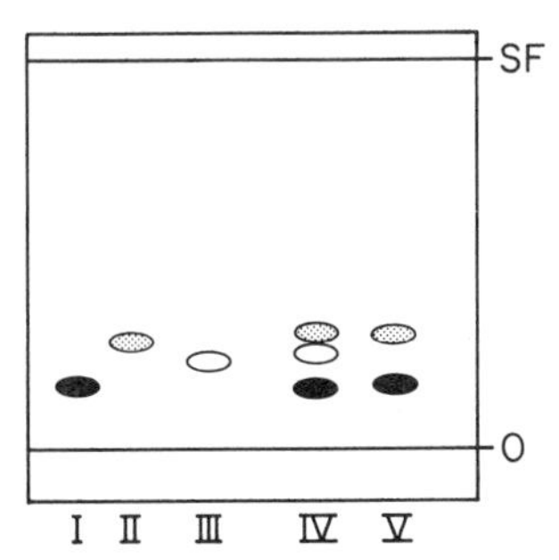

FIG. 1. Thin-layer chromatographic behavior of 1-*O*-hexadecyl-2-acetyl-*sn*-glycero-3-phosphocholine (a synthetic PAF). This chromatogram illustrates a typical separation of synthetic PAF from other phospholipids commonly associated with this type of activity in the usual isolation procedures (from tissues or cells). The plates were coated with silica gel G (250 μm) and the solvent system was chloroform–methanol–water (65 : 35 : 6,v/v/v). The numbered lanes contained the following compounds: (**I**) 1-*O*-hexadecyl-2-(lyso)-*sn*-glycero-3-phosphocholine; (**II**) sphingomyelin (from bovine erythrocytes); (**III**) 1-*O*-hexadecylacetyl-*sn*-glycero-3-phosphocholine; (**IV**) mixture of **I, II,** and **III;** (**V**) mixture of **I** and **II.** O, origin; SF, solvent front.

Comments on Procedure

The most reliable, reproducible, and effective support medium for separation of PAF is silica gel G. Obviously there are other types of support that could be employed, but this material has proved to be a consistent and reproducible media. It is, however, advisable to prerun commercially precoated plates in a solvent such as chloroform/methanol/water, (65 : 35 : 6,v/v/v), to move an organic material, commonly found in this type of plate, to the top of the plate. Such treatment allows a cleaner separation of applied lipids. One can expect to find PAF in the area above lysolecithin and running into the sphingomyelin region. It should be emphasized that the chromatogram in Fig. 1 represents a highly idealized separation. Normally, a lipid extract from a cellular or tissue preparation will show broader bands due to the mixture of fatty chain lengths present. This leads to overlapping adsorption isotherms and, hence, in the case of PAF, which is usually present in very low amounts,one can expect a trailing effect. As noted above, sphingomyelin causes no problem in the assay for biological activity of the isolated fraction. It is important to note, however, that the silica gel G removed from the thin-layer chromatograms should be placed in acid-washed tubes for extraction since past experience has shown loss of PAF if tubes are not pretreated with acid. Apparently there is sufficient alkali ions in the tubes to promote deacetylation of PAF.

It is not possible to use any spray reagent that would detect the levels of PAF usually found in cellular preparations. Hence, it is necessary to use reference compounds and to locate these with the TNS reagent noted above. This compound causes no difficulties in the biological assay for PAF since it is easily removed into the methanol water-rich phase in the isolation procedure noted above.

On occasion it is prudent to check on the degree of recovery of PAF from a thin-layer chromatogram. This can be easily achieved by inclusion of tritiated PAF (commercially available at 90 to 94 Ci/mmol) during lipid isolation from a tissue or cellular preparation, or addition to a sample just prior to application to a plate. Levels of radioactivity in the range of 1 to 2×10^4 or 10^5 cpm should be sufficient to establish the efficiency of the extraction technique. Recovery is in the range of 90 to 95%. Further, this amount of labeled PAF would not be detected in the biological assay.

Finally, it is important to determine whether all of the biological activity as isolated by the above technique is represented by 1-*O*-alkyl-2-acetyl-*sn*-glycero-3-phosphocholine. The most reasonable approach is to employ the well-established sensitivity of this lipid to base-catalyzed methanolysis during which acetate is cleanly removed leaving the biologically inactive lyso-PAF. Acetylation of this derivative should then give complete recovery of biological activity. If recoveries are lower than expected, it would be necessary to consider that the 1-*O*-acyl analog was also a component of the sample.

Summary

Isolation and purification of platelet-activating factor can be accomplished with reasonable ease and effectiveness given the precautions and suggestions outlined above. These experimental approaches represent a composite of those practiced in the author's laboratory combined with selected observations reported in the literature.[2–5]

[3] M. M. Billah and J. M. Johnston, *Biochem. Biophys. Res. Commun.* **113,** 51 (1983).
[4] J. Sugatani, K. Fujimura, M. Miwa, T. Mizuno, Y. Sameshina, and K. Saito, *FASEB J.* **3,** 65 (1989).
[5] J. H. Sisson, S. M. Prescott, T. M. McIntyre, and G. A. Zimmerman, *J. Immunol.* **138,** 3918 (1987).

[19] Separation and Characterization of Arachidonate-Containing Phosphoglycerides

By Floyd H. Chilton

Introduction

Over the years, research has shown that arachidonate and its oxygen-carrying metabolites are involved in several important cellular events. Arachidonic acid metabolism is initiated by the mobilization of esterified arachidonate from cellular phosphoglycerides by the activation of intracellular phospholipases (for a review, see 1). There are a diverse group of phosphoglyceride molecular species that contain arachidonate in mammalian cells (Fig. 1). In general, arachidonate is located at the *sn*-2 position of these arachidonate-containing phosphoglyceride and is usually associated with ethanolamine-, choline-, and inositol-linked phosphoglycerides. Phosphoglyceride composition studies of several cell types indicate that a great deal of variation exists with respect to the quantity of arachidonate at the *sn*-2 position associated with different radyl(acyl, alkyl, alk-1-enyl) substituents at the *sn*-1 position. For example, in the inflammatory cells,

[1] R. F. Irvine, *Biochem. J.* **204,** 3 (1982).

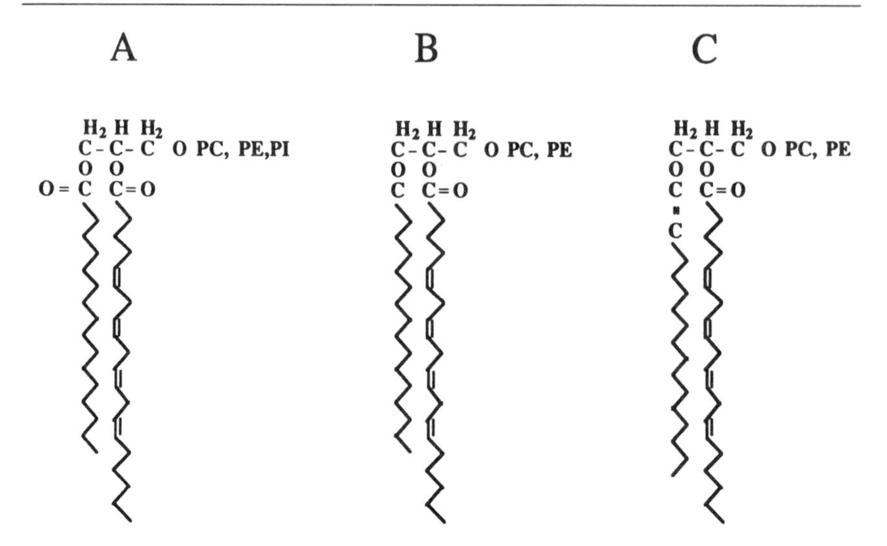

FIG. 1. Major arachidonate-containing phosphoglyceride subclasses. (A) 1-acyl-2-arachidonoyl-GPC, -GPE, -GPI. (B) 1-alkyl-2-arachidonoyl-GPC, -GPE. (C) 1-alk-1-enyl-2-arachidonoyl-GPC, -GPE.

such as the neutrophil and macrophage, the bulk of arachidonate within choline phosphoglycerides is associated with 1-alkyl-linked species.[2–4] In contrast, 1-acyl-2-arachidonoylphosphocholine contains most of the arachidonate found in choline-containing phosphoglycerides of cells such as the platelet and endothelial cell.[5,6] In all of these cells, the majority of arachidonate found in ethanolamine-linked phospholipids is associated with 1-alk-1-enyl molecular species.

A number of recent studies suggest that the *sn*-1 position linkage of archidonate-containing phosphoglycerides is an important component that determines the biochemical pathways available to these molecules. For example, labeling studies in a number of cells indicate that exogenously added arachidonic acid is taken up by 1-acyl-, 1-alkyl-, and 1-alk-1-enyl-linked phosphoglycerides at nonuniform rates and by distinct mecha-

[2] H. W. Mueller, J. T. O'Flaherty, D. G. Green, M. P. Samuel, and R. L. Wykle, *J. Lipid Res.* **25,** 284 (1984).

[3] F. H. Chilton and T. R. Connel, *J. Biol. Chem.* **263,** 5260 (1988).

[4] T. Sugiura, M. Nakajima, N. Sekiguichi, Y. Nakagawa, and K. Waku, *Lipids* **18,** 125 (1983).

[5] H. W. Mueller, A. D. Purdon, J. B. Smith, and R. L. Wykle, *Lipids* **18,** 814 (1983).

[6] M. L. Blank, A. A. Spector, T. L. Kaduce, and F. Snyder, *Biochim. Biophys. Acta* **877,** 211 (1986).

nisms.[7,8] In the case of choline-linked phosphoglycerides, arachidonic acid appears to be initially incorporated into 1-acyl-linked molecular species by a CoA-dependent acyltransferase followed by its transfer to 1-alkyl-linked molecular species utilizing a CoA-independent transacylase.[9–12] The fact that there are a wide variety of arachidonate-containing phosphoglyceride molecular species in some inflammatory cells has led to questions as to which classes, subclasses, or molecular species are important sources of arachidonic acid utilized for eicosanoid biosynthesis. The relative degradation of different arachidonate-containing phosphoglyceride molecular species has recently been examined in several inflammatory cells. Studies in the neutrophil and macrophage indicate that all major phosphoglyceride subclasses (1-acyl, 1-alkyl, and 1-alk-1-enyl) provide arachidonate during cell activation.[13–16] However, on a mole basis, 1-ether-linked phosphoglycerides clearly provide the bulk of arachidonate released from phosphoglycerides in the human neutrophil.[3] By contrast, arachidonate is released primarily from 1-acyl-linked phosphoglycerides in the platelet.[17] All of these studies suggest that the 1-radyl linkage of arachidonate-containing phosphoglycerides may play an important role in determining the metabolic fate of these molecules.

It is clear from the aforementioned studies that the ability to separate phosphoglyceride classes, subclasses, and molecular species is crucial to studies that examine the incorporation or release of arachidonic acid into or from cellular phosphoglycerides. Over the years, several TLC and HPLC systems have been reported to have achieved resonable resolution of intact phosphoglycerides. Other systems have also been developed to separate phosphoglyceride subclasses or molecular species as diglyceride derivatives. This chapter focuses on high-performance liquid chromatography (HPLC) and thin-layer chromatography (TLC) methods used in

[7] F. H. Chilton and R. C. Murphy, *J. Biol. Chem.* **261,** 7771 (1986).
[8] T. Sugiura, O. Katayama, J. Fukui, Y. Nakagawa, and K. Waku, *FEBS Lett.* **165,** 273 (1984).
[9] M. Robinson, M. L. Blank, and F. Snyder, *J. Biol. Chem.* **260,** 7789 (1985).
[10] R. M. Kramer, G. M. Patton, C. R. Pritzker, and D. Deykin, *J. Biol. Chem.* **259,** 13316 (1984).
[11] T. Sugiura and K. Waku, *Biochem. Biophys. Res. Commun.* **127,** 384 (1985).
[12] F. H. Chilton, J. S. Hadley, and R. C. Murphy, *Biochim. Biophys. Acta* **917,** 48 (1987).
[13] F. H. Chilton, J. M. Ellis, S. C. Olson, and R. L. Wykle, *J. Biol. Chem.* **259,** 12014 (1984).
[14] C. L. Swendsen, J. M. Ellis, F. H. Chilton, J. T. O'Flaherty, and R. L. Wykle, *Biochem. Biophys. Res. Commun.* **113,** 72 (1983).
[15] D. H. Albert and F. Snyder, *Biochim. Biophys. Acta* **796,** 92 (1984).
[16] F. H. Chilton, *Biochem. J.* **258,** 327 (1989).
[17] A. D. Purdon and J. B. Smith, *J. Biol. Chem.* **260,** 12700 (1985).

our laboratory to achieve separation of arachidonate-containing phosphoglyceride classes, subclasses, and molecular species.

Reagents

[5,6,8,9,11,12,14,15-^{3}H]Arachidonic acid (83 Ci/mmol) was purchased from DuPont-New England Nuclear (Boston, MA). All solvents were HPLC grade from Fisher (Springfield, NJ). Lipid standards used in TLC were purchased from Avanti Polar Lipids (Birmingham, AL). Essentially fatty acid-free human serum albumin (HSA) was purchased from Sigma (St. Louis, MO). Hanks' balanced salt solution (HBSS) was purchased from Gibco (Grand Island, NY). *N*-(*tert*-butyldimethylsilyl)-*N*-methyltrifluoroacetamide for preparing the *tert*-butyldimethylsilyl derivative was purchased from Aldrich (Milwaukee, WI). Octadeuteroarachidonic acid was a generous gift from Dr. Howard Sprecher (Ohio State University). This compound is commercially available from Biomol Research Laboratories, Inc. (Plymouth Meeting, PA).

Methods

Preparation of Cells. Human neutrophils are prepared from venous blood obtained from healthy human donors immediately before the neutrophil isolation as previously described.[18] Murine mast cells (PT-18) were obtained from Dr. D. H. Pluznik (National Institutes of Health, Bethesda, MD). This cell line was established from antigenically stimulated C3H.SW mouse spleen cells grown in RPMI-1640 medium supplemented with supernatant from concanavalin A-stimulated mouse spleen cells. The PT-18 subline was characterized as a mast cell according to morphological and biochemical criteria.[19] The cells were grown at 37° in a 5% CO_2 atmosphere in RPMI containing 1 m*M* L-glutamine, 25 m*M* HEPES, and 50 μM 2-mercaptoethanol supplemented with 10% fetal calf serum (FCS), 0.1% minimum essential medium containing amino acids, 0.1% minimum essential medium containing nonessential amino acids, 0.5% mouse interleukin-3, and 1% penicillin–streptomycin.

Incorporation of [^{3}H]Arachidonic Acid into Cellular Glycerolipids. In labeling studies, which examine the incorporation of exogenous arachidonic acid into phosphoglyceride classes, subclasses, and molecular species, cells are generally pulse labeled with [^{3}H]arachidonic acid ($<0.1\ \mu M$) complexed with HSA (5.0 mg/ml) for a short period of time (5–15 min) at 37°. The cells are then removed from the supernatant fluid by centri-

[18] L. R. DeChatelet and P. S. Shirley, *Infect. Immun.* **35,** 206 (1982).

[19] D. H. Pluznik, N. S. Tare, M. M. Zatz, and A. L. Goldstein, *Exp. Hematol.* **10,** 211 (1982).

fugation (225 *g*, 8 min, 4°), washed (twice) with buffer containing HSA, and then suspended in buffer. Esterified [^{3}H]arachidonate is then allowed to equilibrate among all pools by allowing the reaction mixture to incubate at 37° for various periods of time. In experiments where the phosoph-glyceride distribution of endogenous arachidonate is to be determined or experiments where the changes in the content of endogenous arachidonate are quantitated during cell activation, cell glycerolipids are labeled with [^{3}H]arachidonic acid for 30 min in order to obtain retention times for individual glycerolipid classes as well as to determine recoveries of arachidonate-containing glycerolipids through the extraction and chromatography procedures. In all experiments, lipids are extracted from the cell pellets by the method of Bligh and Dyer.[20]

HPLC Separation of Glycerolipid Classes. Glycerolipid classes are separated by normal-phase HPLC using a modified method of Patton *et al.*[21] In this system, cellular glycerolipids are placed on an Ultrasphere-Si column (4.6 mm × 250 mm; Rainin Instrument Co., Woburn, MA) and eluted with 2-propanol/25 m*M* phosphate buffer (pH 7.0)/hexane/ethanol/acetic acid (245.0 : 15.0 : 183.5 : 50.0 : 0.3, v/v/v/v/v) at 1.0 ml/min for 5 min. After 5 min, the solvent is changed to 2-propanol/25 m*M* phosphate buffer (pH 7.0)/hexane/ethanol/acetic acid (245.0 : 25.0 : 183.5 : 40.0 : 0.3, v/v/v/v/v) over a 10-min period. The retention time of the glycerolipid classes in this system is very sensitive to the composition of phosphate buffer in the running solvent. Our method differs from the original method[8] in the fact that we utilized a gradient with phosphate buffer from 3.0 to 5.0%. This is found to be necessary to resolve early eluting glycerolipid classes such as the neutral lipids, phosphatidylethanolamine (PE), and phosphatidic acid and important for maintaining the resolution for normal-phase columns which had been used for sometime.

The eluting solvent in this system is prepared in the following manner: First, 2-propanol is added to 25 m*M* potassium phosphate (pH 7.0) followed by the addition of hexane and ethanol. The solvent is then filtered through an FH Millipore filter (Millipore Corp., Bedford, MA). Finally, glacial acetic acid is then added to the filtered solvent.

In general, separated glycerolipid classes are collected in 1-min fractions. A portion of the fraction is then counted for radioactivity using liquid scintillation counting. One of the major advantages that we have found with this HPLC technique is the high recoveries (>85%) of glycerolipids from the column. This is crucial when trying to determine small changes in mass or radioactivity. Once the fractions have been collected, those that

[20] E. G. Bligh and W. J. Dyer, *Can. J. Biochem. Physiol.* **37,** 911 (1959).

[21] G. M. Patton, J. M. Fasulo, and S. J. Robins, *J. Lipid Res.* **23,** 190 (1982).

contain individual glycerolipid classes are combined and a portion of the isolated phosphoglyceride is prepared for GC–MS analysis. The mole quantity of unlabeled arachidonic acid is determined by hydrolyzing the glycerolipids with 2 *N* KOH in methanol/water (75 : 25, v/v), for 30 min at 60° to liberate free arachidonic acid. Octadeuteroarachidonic acid (250 ng) as an internal standard is added to the reaction mixture. After 30 min, additional water is added and the pH adjusted to 3.0 with 6 *N* HCl. The free arachidonic acid is extracted with hexane. The solvents are then removed and arachidonic acid is converted to the *tert*-butyldimethylsilyl ester by the addition of *N-tert*-butyldimethylsilyl-*N*-methyltrifluoroacetamide/acetonitrile (1 : 1, v/v) at 60° for 30 min.[23] The solvents are removed from the sample with a stream of nitrogen and the sample suspended in hexane. The *tert*-butyldimethylsilylarachidonate is analyzed by GC–MS using selected-ion monitoring techniques to monitor the M−57 ion at *m/z* 361 and the *tert*-butyldimethylsilyloctadeuteroarachidonate M−57 ion at *m/z* 369. GC–MS is carried out on a HP selective-mass detection system (HP 5790). The gas chromatography is performed using a cross-linked methylsilicone column threaded directly into the mass spectrometric ion source. The initial column temperature is 80° and programmed to 220° at 30°/min. At that point, the temperature is increased 10°/min to 260°. The injector temperature is 250° and the gas chromatographic mass spectrometric interface line is at 280°. Helium is used as a carrier gas and 1–2 μl of the samples are injected in the splitless mode.

Illustrated in Fig. 2 is a radioactive and mass profile of arachidonate in the various glycerolipid classes of the murine mast cell maintained in culture with [^{3}H]arachidonic acid for 24 hr. These data point out that by 24 hr, the quantity of labeled arachidonate within each glycerolipid class is similar to that of the endogenous mass of arachidonate found within that class. This suggests that the radioactivity within the cells has reached equilibrium among phosphoglyceride classes by 24 hr of culture.

TLC Separation of Choline- and Ethanolamine-Linked Subclasses. Quite often in analyzing arachidonate-containing phosphoglycerides, it is desirable to separate them into 1-acyl-, 1-alkyl-, and 1-alk-1-enyl-linked subclasses. This is less complex than analyzing the numerous phosphoglyceride molecular species which make up each phosphoglyceride subclass. In our laboratory, purified choline- and ethanolamine-linked phosphoglycerides (from normal-phase HPLC, see above) are further separated into 1-acyl, 1-alkyl, and 1-alk-1-enyl subclasses and analyzed for their content of labeled and unlabeled arachidonate. In this procedure, phosphoglycerides are dissolved in 2.0 ml of ethyl ether. One milliliter of 0.1 *M* Tris buffer (pH 7.4) containing 20 units of phospholipase C (*Bacillus cereus*, Sigma Chemical Company) is added to the ethyl ether. The reac-

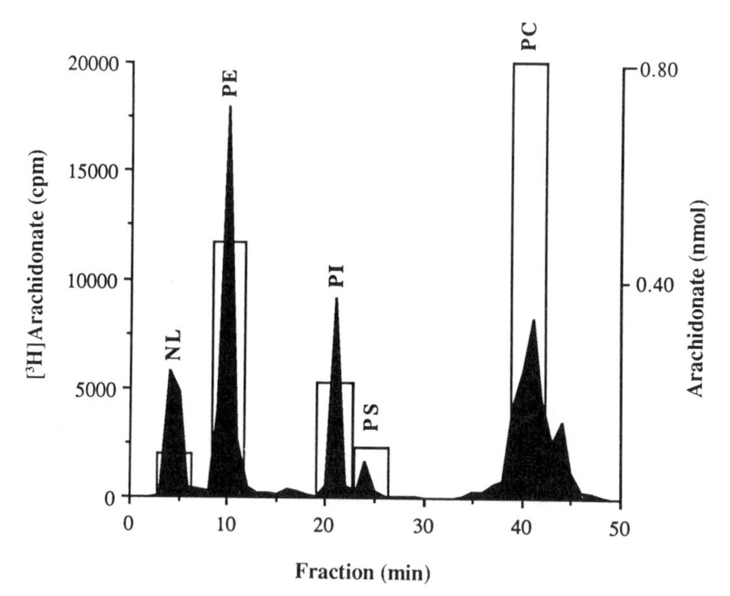

FIG. 2. Radioactive and mass profile of arachidonate in glycerolipid classes. Mast cells are maintained in culture for 24 hr with [^{3}H]arachidonic acid and analyzed as described. The label and mass of the arachidonate are represented by filled and open peaks, respectively. In the case of the mass, fractions 4–7, 8–13, 19–22, 23–26, and 34–48 are combined and analyzed for arachidonate in NL, PE, PI, PS, and PC, respectively.

tion mixture is then shaken at 25° for 2 or 7 hr for choline- and ethanolamine-linked phosphoglycerides, respectively. The diradylglycerols are then extracted and converted into 1,2-diradyl-3-acetylglycerols by incubating them with acetic anhydride and pyridine (4 : 1, v/v) for 12 hr at 37°. The 1,2-diradyl-3-acetylglycerols are then separated into 1-acyl, 1-alkyl, and 1-alk-1-enyl subclasses by TLC on layers of silica gel G (Analtech, Newark, DE) developed in benzene/hexane/ethyl ether (50 : 45 :4 v/v). The distribution of labeled arachidonic acid in these subclasses is then monitored utilizing a TLC radioactivity scanner (Bioscan Inc., Washington, D.C.). Each of the separated sublcasses are then extracted from the silica gel with ethyl ether/methanol (9 : 1, v/v). The amount of label in each of the glycerolipids is determined by liquid scintillation counting. The mole quantity of unlabeled arachidonic acid is determined by alkaline hydrolysis of the arachidonate from the 1,2-diradyl-3-acetylglycerols followed by GC–MS as described above.

Figure 3 shows the distribution of labeled arachidonate in choline-linked subclasses of the human neutrophil pulse-labeled for 15 min. Good resolution of all three major subclasses (1-acyl, 1-alkyl, and 1-alk-1-enyl)

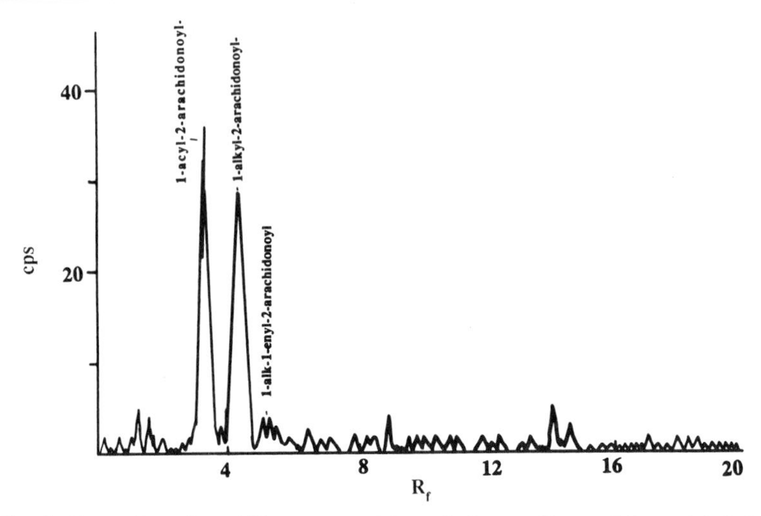

FIG. 3. Separation of arachidonate-containing subclasses. Neutrophils are labeled with [^{3}H]arachidonic acid for 15 min and choline-containing phosphoglycerides isolated. These are then converted to diglyceride acetates and separated as described.

of choline-linked phosphoglycerides can be obtained using the aforementioned procedures. In this particular labeling experiment, labeled arachidonate is distributed primarily in 1-acyl-2-arachidonoyl-GPC while the bulk of the mass of arachidonate is found in 1-alkyl-2-arachidonoyl-GPC (data not shown) indicating that the label in the subclasses is not at equilibrium with the mass of arachidonate.

HPLC Separation of Arachidonate-Containing Phosphoglyceride Molecular Species. Several HPLC systems have been developed to separate phosphoglyceride molecular species. The analysis of phosphoglyceride molecular species has been performed on intact phosphoglycerides or the derivatives of the diradylglycerides obtained from the hydrolysis of intact phosphoglycerides. In general, those HPLC systems that separate diradylglyceride derivatives of phosphoglyceride molecular species accomplish better resolution of the individual molecular species than those separating intact phosphoglycerides. In fact, Blank and coworkers[22] and Haroldsen and Murphy[23] have described reversed-phase HPLC systems that resolve derivative diglycerides of all major phosphoglyceride molecular species either as benzoates or dinitrobenzoates, respectively. In these systems, phosphoglyceride classes are isolated

[22] M. L. Blank, M. Robinson, V. Fitzgerald, and F. Snyder, *J. Chromatog.* **298,** 473 (1984).
[23] P. E. Haroldsen and R. C. Murphy, *Biomed. Environ. Mass Spectrom.* **14,** 573 (1987).

and the polar head group removed by phospholipase C. Once the diradylglycerides are formed, their polarity is further reduced and their ultraviolet absorption increased by the addition of benzoate or dinitrobenzoate derivatives. Finally, these diglyceride derivatives must be further separated into 1-acyl-, 1-alkyl-, and 1-alk-1-enyl-linked subclasses by normal-phase HPLC or TLC before they are partitioned into molecular species by reversed-phase HPLC.

In some instances, the resolution of all the major molecular species is not required or the information lost from hydrolysis of the intact phosphoglycerides is undersirable. Using a modified reversed-phase system described by Patton *et al.*,[21] we have been able to resolve a number of the major arachidonate-containing phosphoglycerides. This HPLC system has a major advantage over systems for diradylglyceride derivatives in that it requires much less sample preparation prior to HPLC. On the other hand, a disadvantage of this system is that it does not allow for baseline resolution of all major phosphoglyceride molecular species. Prior to reversed-phase HPLC, glycerolipid classes are separated by normal-phase HPLC as described above. The various molecular species of choline-, ethanolamine-, and inositol-linked phosphoglycerides are then separated by reversed-phase HPLC using an Ultrasphere ODS column (4.6 × 250 mm; Rainin, Woburn, MA) eluted with methanol/water/acetonitrile (905 : 70 : 25, v/v/v) and 20 m*M* choline chloride at 2ml/min at 40°. The running solvent is prepared by mixing choline chloride dissolved in water (filtered with a 0.45 μm filter, Millipore) with acetonitrile and methanol.

Using this system, we have examined the initial incorporation and subsequent remodeling of the major arachidonate-containing phosphoglycerides in the human neutrophil. In the experiment illustrated in Fig. 4, neutrophils are pulse-labeled for 5 min, washed, and allowed to incubate for an additional 120 min. During this 120-min incubation, [^{3}H]arachidonic acid is distributed into many major and minor molecular species of choline-, ethanolamine-, and inositol-linked phosphoglycerides. Nine, six, and two major arachidonate-containing peaks in phosphatidylcholine (PC), phosphatidylethanolamine (PE), and phosphatidylinositol (PI), respectively, are identified. Each of the major label-containing peaks have been analyzed by fast atom bombardment–mass spectrometry (FAB–MS) and shown to have strong positive ions for the indicated arachidonate-containing molecular species. As shown in Fig. 4, arachidonic acid is incorporated into predominantly 1-acyl- and 1-alkyl-linked PC molecular species within 2 hr. In contrast to PC, 1-alk-1-enyl-linked molecular species contain a major portion of labeled arachidonate found within PE at 2 hr. Only 1-acyl-linked species of PI were labeled during the 2-hr incubation. FAB^{M}S indicated no 1-alkyl- or 1-alk-1-enyl-linked PI species with

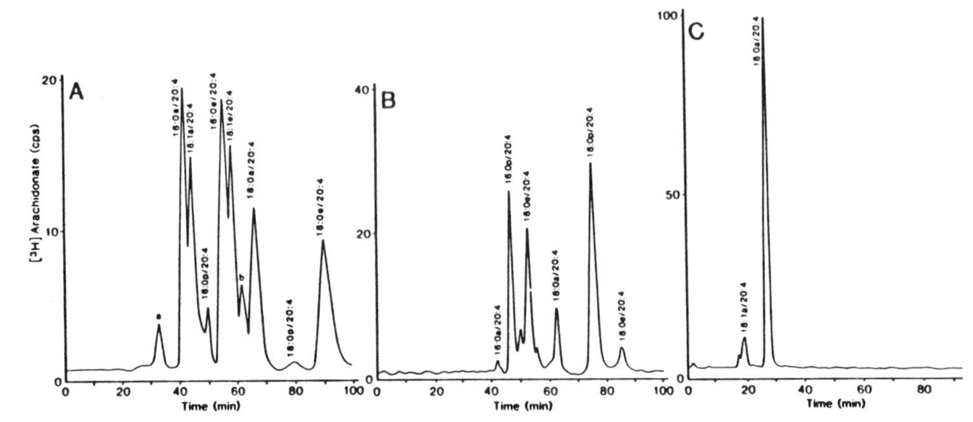

FIG. 4. Separation of arachidonate-containing phosphoglyceride molecular species. (A) PC; (B) PE; (C) PI. Neutrophils are labeled for 2 hr as described. Choline-, ethanolamine-, and inositol-linked phosphoglycerides are then isolated and separated into molecular species by reversed-phase HPLC. The identification of molecular species is established by FAB–MS and abbreviated by *sn*-1-radyl carbon chain: number of double bonds at *sn*-1 position/arachidonate at the *sn*-2 position. The radyl group at the *sn*-1 is either acyl- (a), alkyl- (e), or alk-1-enyl- (p) linked.

arachidonate at the *sn*-2 position exist in the neutrophil. In addition to arachidonate-containing phosphoglycerides, there are also a number of other ions present in the FAB–MS mass spectra of these peaks that correspond to other phosphoglyceride molecular species. Therefore, as previously mentioned, a disadvantage of this system as compared to diglyceride derivative systems is that it does not resolve arachidonate-containing molecular species from all other phosphoglyceride molecular species. Nevertheless, FAB–MS analysis did indicate that this system separates all the major arachidonate-containing molecular species from each other using intact phosphoglycerides. The complete FAB–MS analysis of each of the arachidonate-containing peaks in the neutrophil has recently been characterized.[24]

Utilizing this HPLC system, it is also possible to isolate specific peaks and measure the mass or radioactivity of arachidonate in a given phosphoglyceride molecular species as described above. We have now employed this approach to measure the [^{3}H]arachidonate/arachidonate ratio of the major arachidonate-containing phosphoglyceride molecular species after labeling neutrophils or mast cells for various times with exogenous arachidonic acid. These studies have emphasized the difficulty of obtaining

[24] F. H. Chilton and R. C. Murphy, *Prostaglandins Leukotrienes Med.* **23,** 141 (1986).

equilibrium labeling of all major arachidonate-containing phosphoglyceride molecular species within mammalian cells. Furthermore, they also pointed out how major discrepancies can arise when comparing mass and label to determine sources of arachidonate used for eicosanoid biosynthesis.

Acknowledgment

Work discussed in this chapter was supported by National Institutes of Health Grants AI24985 and AI26771.

[20] High-Performance Liquid Chromatography Separation and Determination of Lipoxins

By CHARLES N. SERHAN

The lipoxins are a series of biologically active, acyclic eicosanoids which contain a conjugated tetraene structure as a characteristic feature.[1-3] The two main compounds are positional isomers: one is designated lipoxin A_4 (LXA_4) and the other lipoxin B_4 (LXB_4) (Fig. 1). Multiple pathways exist for lipoxin biosynthesis that are substrate-, cell type-, and species-specific.[4-9] One route, which involves the transformation of 15-HETE and formation of a 5(6)-epoxytetraene by human leukocytes, is outlined in Fig. 1. Other routes can involve the transcellular metabolism of

[1] C. N. Serhan, M. Hamberg, and B. Samuelsson, *Proc. Natl. Acad. Sci. U.S.A.* **81,** 5335 (1984).

[2] B. Samuelsson, S.-E. Dahlén, J. Å. Lindgren, C. A. Rouzer, and C. N. Serhan, *Science* **237,** 1171 (1987).

[3] C. N. Serhan, K. C. Nicolaou, S. E. Webber, C. A. Veale, S.-E. Dahlén, T. J. Puustinen, and B. Samuelsson, *J. Biol. Chem.* **261,** 16340 (1986).

[4] H. Kühn, R. Wiesner, L. Alder, B. J. Fitzsimmons, J. Rokach, and A. R. Brash, *Eur. J. Biochem.* **169,** 593 (1987).

[5] N. Ueda, S. Yamamoto, B. J. Fitzsimmons, and J. Rokach, *Biochem. Biophys. Res. Commun.* **144,** 996 (1987).

[6] P. Walstra, J. Verhagen, M. A. Vermeer, J. P. M. Klerks, G. A. Veldink, and J. F. G. Vliegenthart, *FEBS Lett.* **228,** 167 (1988).

[7] C.N. Serhan, U. Hirsch, J. Palmblad, and B. Samuelsson, *FEBS Lett.* **217,** 242 (1987).

[8] B. K. Lam, C. N. Serhan, B. Samuelsson, and P. Y.-K. Wong, *Biochem. Biophys. Res. Commun.* **144,** 123 (1987).

[9] C. Edenius, J. Haeggström, and J. Å. Lindgren, *Biochem. Biophys. Res. Commun.* **157,** 801 (1988).

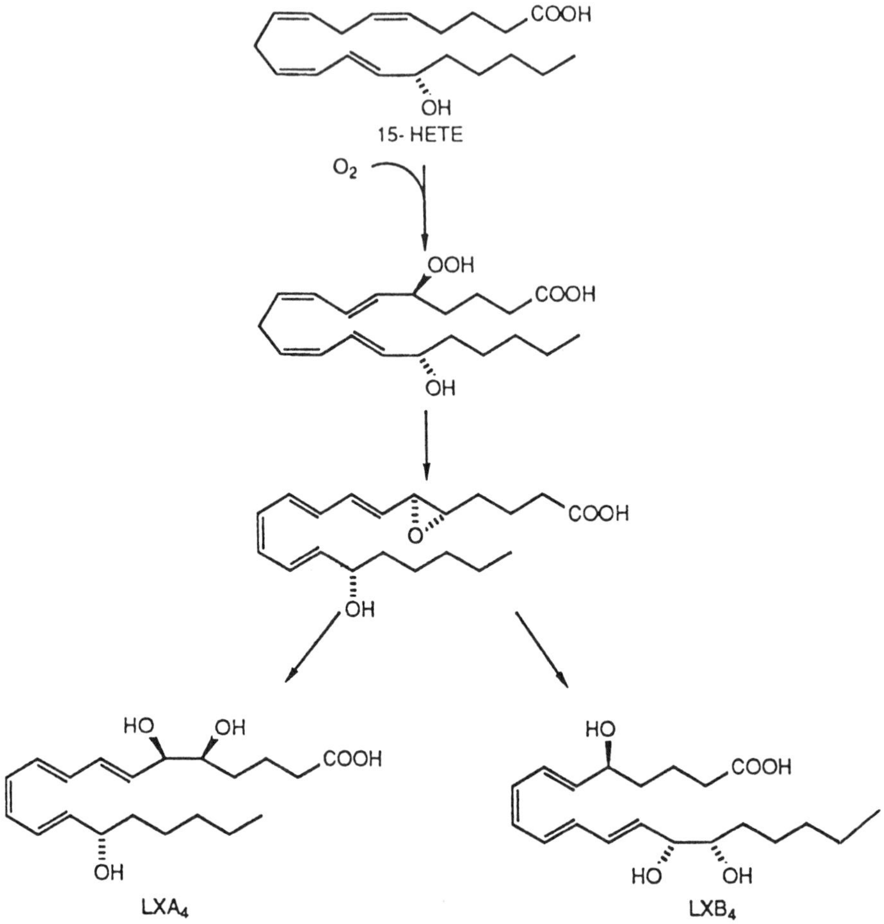

FIG. 1. One biosynthetic pathway for lipoxin formation (transformation of 15-HETE by activated human leukocytes). 15-HETE, (15*S*)-hydroxy-5,8,11-*cis*-13-*trans*-eicosatetraenoic acid; LXA_4, (5*S*,6*R*,15*S*)-trihydroxy-7,9,13-*trans*-11-*cis*-eicosatetraenoic acid; LXB_4, (5*S*,14*R*,15*S*)-trihydroxy-6,10,12-*trans*-8-*cis*-eicosatetraenoic acid.

various substrates including leukotriene A_4, arachidonic acid, 5,15-DHETE, and 5-HETE.[4–9] LXA_4 and LXB_4 each display biological activities that can be distinguished from those of other eiscosanoids.[10] Several isomers of LXA_4 and LXB_4 have been identified, including a novel 7-*cis*-

[10] S.-E. Dahlén, L. Franzén, J. Raud, C. N. Serhan, P. Westlund, E. Wikström, T. Björck, H. Matsuda, S. E. Webber, C. A. Veale, T. Puustinen, J. Haeggström, K. C. Nicolaou, and B. Samuelsson, *Adv. Exp. Med. Biol.* **229,** 107 (1988).

11-*trans*-LXA_4.[3,11,12] Since LXA_4, LXB_4, and their isomers differ in both biological actions and potencies, their identification within biologically derived materials is of interest. This chapter describes the isolation, reversed-phase high-performance liquid chromatography (RP-HPLC) separation, and determination of lipoxins of the four series.

Procedure

Materials

HPLC-grade solvents were from American Scientific Products, Burdick and Jackson (Muskegon, MI). Methyl formate was from Sigma Chemical Company (St. Louis, MO). Sep-Pak C_{18} cartridges were from Waters Associates (Milford, MA). Synthetic LXA_4, LXB_4, LTB_4, and other eicosanoids used as reference materials were from Biomol Research Laboratories (Philadephia, PA). Synthetic 7-*cis*-11-*trans*-LXA_4 as well as the all-*trans* isomers of LXA_4 and LXB_4 were prepared[3,12] and were provided by K. C. Nicolaou, Department of Chemistry, University of Pennsylvania (Philadelphia, PA).

Extraction of Lipoxins

The lipoxins can be extracted along with the prostaglandins (PG) and leukotrienes (LT) from *in vitro* incubations of various cell types suspended in phosphate-buffered saline (PBS) by utilizing a combination of techniques.[3,13] (Trivial names used in this article are consistent with the recent nomenclature proposal.[14]) To terminate the incubations, ethanol (2 volumes to that of the incubation volume) and PGB_2 (as internal standard, or radiolabeled eicosanoids can be used) are added to the cell suspensions. The mixture is allowed to stand at 4° for at least 30 min. Following centrifugation (1200 *g*, 15 min), the supernatants are removed and saved. The pellets are then suspended in methanol (2 volumes). This step is repeated twice and the resulting ethanol- and methanol-containing fractions are pooled and dried by rotoevaporation under reduced pressure in a round-bottom flask.

Materials coating the round-bottom flask are next suspended in methanol : water (1 : 45, v/v) by vortexing (~1–2 min) and then transferred into a

[11] J. Adams, B. J. Fitzsimmons, Y. Girard, Y. Leblanc, J. F. Evans, and J. Rokach, *J. Am. Chem. Soc.* **107,** 464 (1985).

[12] K. C. Nicolaou, B. E. Marron, C. A. Veale, S. E. Webber, S.-E. Dahlén, B. Samuelsson, and C. N. Serhan, *Biochim. Biophys. Acta* **1003,** 44 (1989).

[13] W. S. Powell, *J. Biol. Chem.* **259,** 3082 (1984).

[14] C. N. Serhan, P. Y.-K. Wong, and B. Samuelsson, *Prostaglandins* **34,** 201 (1987).

glass syringe. The samples are rapidly acidified with HCl to pH 3.5 and loaded onto cartridges containing ODS silica (C_{18} Sep-Pak). Since both LXA_4 and LXB_4 can undergo isomerization to their corresponding all-*trans* isomers,[3,12] the acidification, cartridge loading, and washing with 10 ml water to obtain pH 6–7 are performed rapidly (i.e., within 60 sec of the addition of acid). To ensure that appropriate pH values are achieved, eluates from each sample are checked with colorpHast (EM Science, Cherry Hill, NJ). Next, the cartridges are eluted with hexane (10 ml) followed by methyl formate (10 ml).[13] The lipoxins, dihydroxyeicosatetraenoic acids, and tri-HETEs (i.e., LTB_4, (5*S*,12*S*)-DHETE as well as their ω products) are eluted with methyl formate. Materials eluting in this fraction can either be concentrated with a stream of argon suspended in mobile phase and injected directly on reversed-phase HPLC (*vide infra*) or they can be treated with diazomethane before reversed-phase HPLC. Chromatography of the methyl esters will enable separation of the all-*trans*-LXB_4 isomers [8-*trans*-LXB_4 and (14*S*)-8-*trans*-LXB_4].[3]

As in the case of 5-HETE, the lipoxins can easily form 1,5-γ-lactones, which drastically changes their chromatographic behavior. Therefore, the UV absorbance of the extracted materials should be examined prior to HPLC in order to provide quantitative data on recovery of the compounds following analysis.

High-Performance Liquid Chromatography

The lipoxins display strong absorbance in UV because of their conjugated tetraene structure.[1] The presence of the tetraene chromophore renders these compounds well suited for reversed-phase HPLC separation coupled with photodiode array rapid spectral detection (see later Fig. 3). An isocratic system which enables the separation of LXA_4, LXB_4, and 7-*cis*-11-*trans*-LXA_4 from their all-*trans* isomers employs an Altex Ultrasphere-ODS (4.6 mm × 25 cm) column eluted with methanol : water : acetic acid (70 : 30 : 0.01) as mobile phase (flow rate 0.7 ml/min). A representative chromatogram of the lipoxins obtained from human neutrophils is shown in Fig. 2.

If the resolution of diene-, triene-, and tetraene-containing eicosanoids is needed within individual samples, the extracted materials can be injected, for example, into a gradient reversed-phase HPLC system equipped with a photodiode array detector. The column, an Altex Ultrasphere-ODS (4.6 mm × 25 cm), is eluted with a gradient solvent controller (LKB, Bromma, Sweden) using methanol : water : acetic acid (65 : 35 : 0.01, v/v/v) as phase one (injection t_0 = 20 min) and a linear gradient with methanol : acetic acid (99.99 : 0.01, v/v) as phase two (30–50

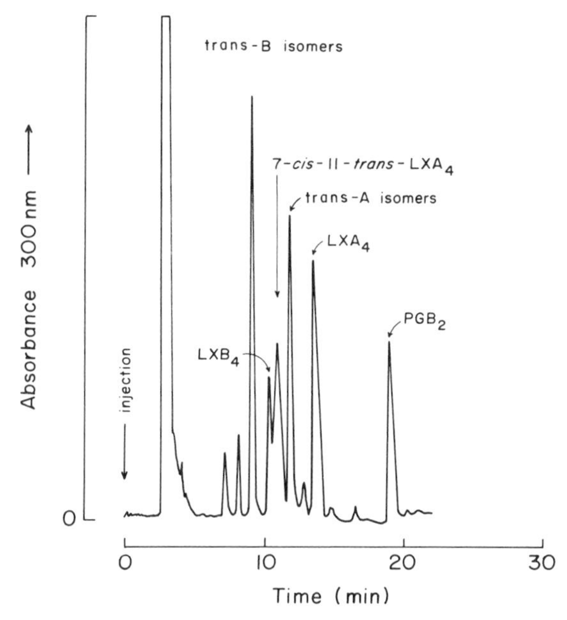

FIG. 2. Reversed-phase HPLC chromatogram of lipoxins obtained from neutrophils. Neutrophils (30 = 10^6 cells; 1 ml total incubation volume) were exposed to 15-HETE (30 μM) and ionophore A23187 (2.5 μM; 20 min, 37°). The incubation was stopped by addition of alcohol and the products were extracted and chromatographed as described in the text.

min). The flow rate is 1.0 ml/min with an initial pressure of 110 bar. Representative chromatograms are shown in Fig. 3A and a topogram of the same spectral data is given in Fig. 3B.

The HPLC system consisted of an LKB dual-pump gradient HPLC equipped with a solvent controller and photodiode array rapid spectral detector linked to an AT&T PC 6300. Post-HPLC run analyses were performed utilizing either a 2140-202 Wavescan program or Wavescan EG 2146-002 program (Bromma, Sweden) and a Nelson Analytical 3000 series chromatography data system (Paramus, NJ). The lipoxins can be quantitated by computer-aided manipulation from their UV spectra recorded on-line following HPLC. Standard curves are generated for compounds from each series utilizing known quantities of synthetic standards. This system permitted detection of ~0.1 ng of lipoxin/10^7 human neutrophils following computer-aided analysis at a single wavelength (300 nm).

The two trans isomers of LXB_4, 8-*trans*-LXB_4 and (14*S*)-*trans*-LXB_4, coelute on HPLC as free acids, while they can be separated as methyl esters on reversed-phase HPLC. The 7-*cis*-11-*trans*-LXA_4 can be separated from the all-*trans*-LXA_4 isomers as its free acid. However, these

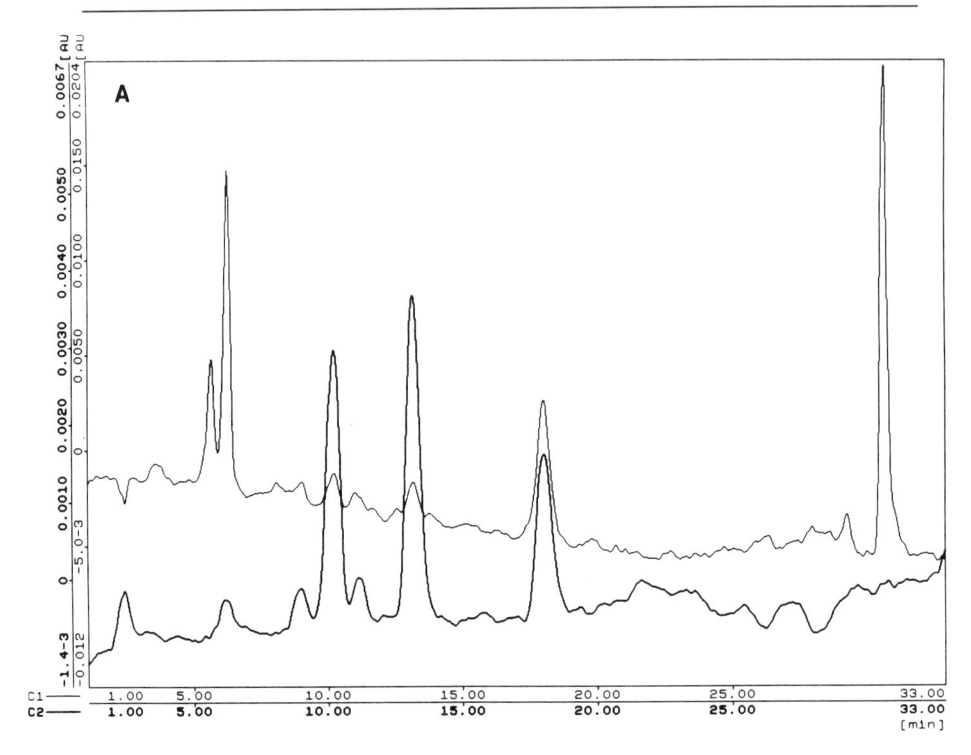

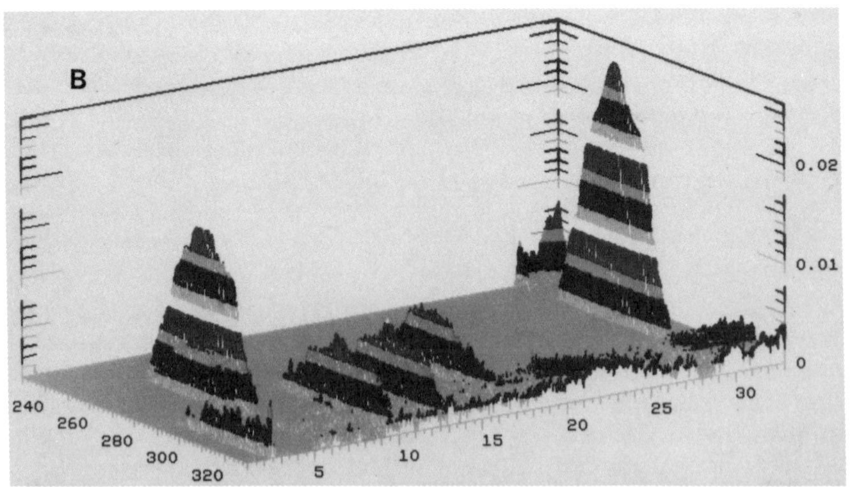

FIG. 3. (A) Reversed-phase HPLC chromatograms. A representative profile of authentic standards following a single injection of a mixture of leukotrienes and lipoxins into a gradient system (see text for description). The profiles obtained at 270 nm (upper chromatogram [C1]) and 300 nm (lower chromatogram [C2]) were recalled from stored UV spectral data moni-

three compounds [7-*cis*-11-*trans*-LXA_4, (6*S*)-11-*trans*-LXA_4, and 11-*trans*-LXA_4] coelute when chromatographed as the methyl esters (see Table I). Although some of these products may coelute in different HPLC systems, and they share similar prominent ions in their mass spectra, determination of the *C* value (equivalent chain length) of their methyl ester trimethylsilyl derivatives has proved to be a useful criterion for identifying the individual isomers.[3,12] In general, the lipoxins that contain all-trans geometry of their conjugated tetraene display higher *C* values than either LXA_4, LXB_4, or other cis-containing compounds such as 7-*cis*-11-*trans*-LXA_4 (Table I). Determination of the *C* value is particularly useful, since LXA_4 and its isomers give very similar mass spectra, as does LXB_4 and its isomers.[1,3,12] This situation is unlike that observed with the mass spectra of LTB_4 and its isomers [i.e., (5*S*, 12*S*)-DHETE, 6-*trans*-, and 12-epi-6-*trans*-LTB_4]. The *C* values for the lipoxins were determined by the method of Bergstrom *et al.*[15] as originally described for prostaglandins and related factors. Briefly, mixtures of methyl esters of fatty acids (20–26 carbons) are injected in the GC–MS and their retention times plotted on a logarithmic scale as a function of the number of carbon atoms present in each fatty acid on a normal scale. This procedure is useful because it provides a calibration plot for various GC columns and temperature programs or offers an index for eicosanoids and related products.

Concluding Remarks

As is the case when identifying other eicosanoids, lipoxins should be identified from biological sources by fulfilling several criteria that may include: (1) coelution of the product with authentic materials in more than one HPLC system (retention time); (2) coelution of the derivatized material with that of authentic standard (determination of *C* value); (3) UV spectral analyses of the isolated material; (4) the presence of diagnostic ions in the mass spectra of several derivatives (i.e., *n*-butyl boronate-

[15] S. Bergström, R. Ryhage, B. Samuelsson, and J. Sjövall, *J. Biol. Chem.* **238,** 3555 (1963).

tored by scanning between 220–340 nm ± 0.9 nm; integration time was 0.5 sec. Post-HPLC analyses were performed with the Wavescan EG 2140-002 program and a Nelson Analytical 3000 series chromatography data system. The elution times were: 5.60 min (20-COOH-LTB_4), 6.11 min (20-OH-LTB_4), 9.00 min (all-*trans*-LXB_4), 10.20 min (LXB_4), 11.22 min (all-*trans*-LXA_4), 13.27 min (LXA_4), 17.87 min (PGB_2), and 30.65 min (LTB_4). (B) Topogram of HPLC spectral data. The stored spectral data presented in (A) is plotted here from the Wavescan EG 2140-002 program. Wavelength range 230–330 nm; absorbance range 0.00–0.03 *A*, and time 1.0–33.00 min. It can be seen that the lipoxin region is clearly separated from LTB_4 and its ω oxidation products.

TABLE I
PROPERTIES OF LIPOXIN A_4, B_4, AND THEIR ISOMERS

	HPLC Retention time[a] (min : sec)			GC–MS[c]	
Parent compound	Free acids	Methyl esters	UV γ_{max} MeOH[b]	*C* value[d]	Prominent ions[e]
LXA_4	12:53	11:59	287, 300, 315	24.1	171, 173, 203 (100%), 582M
LXB_4	10:13	9:86	288, 300, 315	24.0	173 (100%), 203, 582M
7-*cis*-11-*trans*-LXA_4	10:35	11:06	288, 302, 316	25.3	171, 173, 203 (100%), 582M
11-*trans*-LXA_4	11:45	11:06	287, 301, 316	27.7	173, 203 (100%), 582M
(6*S*)-11-*trans*-LXA_4	11:45	11:06	287, 301, 316	27.5	173, 203 (100%), 582M
8-*trans*-LXB_4	9:29	9:14	288, 301, 316	28.3	173 (100%), 203, 582M
(14*S*)-8-*trans*-LXB_4	9:29	8:46	288, 301, 316	28.0	173 (100%), 203, 582M

[a] The retention times were determined for each compound under identical conditions. The column was eluted with methanol : water : acetic acid (65 : 35 : 0.01, v/v/v) as mobile phase with a flow rate of 1.0 ml/min for chromatography of the compounds as free acids. The methyl esters were chromatographed without acid in the mobile phase using methanol : water (65 : 35).

[b] Spectra were recorded in methanol.

[c] GC–MS was performed with a Hewlett-Packard 5988A MS equipped with an HP 59970A work station, software, and 5890 GC. A fused silica capillary SE-30 (Supelco, Inc., Bellefonte, PA) column 2-4004, 30 m, 0.25 mm id, 0.25 μM *df* was employed with a temperature program. The splitless on time was 0.90; initial temperature was 150° (1 min), followed by 230° (4 min), 240° (8 min), and 245° (12.0 min) with a 12.0 min solvent delay time. The retention times of standard fatty acid methyl esters were 10.69 ± 0.30 (C_{20}), 15.35 ± 0.10 (C_{22}), 22.47 ± 0.19 (C_{24}), and 34.19 ± 0.30 (C_{26}) [$n = 6$, min : sec, $\overline{X} \pm SE$].

[d] *C* values (equivalent chain length) were determined using retention times obtained with 20–26 carbon methyl esters of fatty acid standards as described by Bergström *et al.*[15]

[e] Fragmentation patterns were obtained with the methyl esters of the trimethylsilyl derivatives of each parent compound.

trimethylsilyl derivatives); and (5) bioassay of the isolated products and comparison with authentic materials.

Acknowledgments

The author thanks Ms. Kelly Sheppard for technical assistance and Mary Halm Small for skillful preparation of the manuscript. This work was supported in part by NIH grants #AI26714 and GM38765. C.N.S. is a recipient of the J.V. Satterfield Arthritis Investigator Award from the National Arthritis Foundation and is a Pew Scholar (1988).

[21] Quantitation of Epoxy- and Dihydroxyeicosatrienoic Acids by Stable Isotope-Dilution Mass Spectrometry

By JOHN TURK, W. THOMAS STUMP, WENDY CONRAD-KESSEL, ROBERT R. SEABOLD, and BRYAN A. WOLF

Arachidonic acid may be converted to four regionally isomeric epoxyeicosatrienoic acids (EET)[1] by the action of a microsomal cytochrome-P-450 monooxygenase in the presence of NADPH.[2] These compounds are hydrolyzed to the corresponding vicinal diols (dihydroxyeicosatrienoic acids or DHET)[3] by a cytosolic epoxide hydrolase.[4] These metabolites are produced from exogenous arachidonate by intact hepatocytes[5] and may be endogenous components of hepatocyte membranes.[6] Exogenous EET have been reported to influence the function of endocrine,[7,8] vascular,[9] and renal[10] tissues, suggesting a possible mediator role for these compounds.

[1] E. Oliw, F. Guengerich, and J. Oates, *J. Biol. Chem.* **257,** 3771 (1982).
[2] M. Laniado-Schwartzman, K. L. Davis, J. C. McGill, R. D. Levere, and N. G. Abraham, *J. Biol. Chem.* **263,** 2536 (1988).
[3] E. Oliw, J. Lawson, A. Brash, and J. Oates, *J. Biol. Chem.* **256,** 9929 (1981).
[4] N. Chacos, J. Capdevila, J. Falck, S. Manna, C. Martin-Wixtrom, S. Gill, B. Hammock, and R. Estabrook, *Arch. Biochem. Biophys.* **223,** 639 (1983).
[5] E. Oliw and P. Moldeus, *Biochim. Biophys. Acta* **721,** 135 (1982).
[6] J. Capdevila, B. Pramanik, J. Napol, S. Manna, and J. Falck, *Arch. Biochem. Biophys.* **231,** 511 (1984).
[7] G. Snyder, J. Capdevila, N. Chacos, S. Manna, and J. Falck, *Proc. Natl. Acad. Sci. U.S.A.* **80,** 3504 (1983).
[8] J. Capdivila, N. Chacos, J. R. Falck, S. Manna, A. Negro-Vilar, and S. Ojeda, *Endocrinology* **113,** 421 (1983).
[9] K. G. Proctor, J. R. Falck, and J. Capdevila, *Circ. Res.* **60,** 50 (1987).
[10] L. Lapuerta, N. Chacos, J. R. Falck, H. Jacobson, and J. H. Capdevila, *Am. J. Med. Sci.* **295,** 275 (1988).

Useful in the evaluation of such possible mediator functions is the ability to quantitate the endogenous EET and DHET content of cells and changes in their levels induced by physiological signals. Described here is a method for measurement of picomolar amounts of EET and DHET compounds by gas chromatography–mass spectrometry. The method involves preparation of $^2H/^3H$-labeled standards of the four regionally isomeric EET compounds and the four corresponding DHET from $^2H/^3H$-labeled arachidonate. These labeled compounds are then used as internal standards in an assay which involves their addition in organic solvent to terminate cell incubations, extraction, high-performance liquid chromatographic separation of the regionally isomeric standards (located by virtue of their ^3H content), conversion to the pentafluorobenzyl ester (PFBE) derivatives (EET) or to the trimethylsilyl (TMS)-PFBE derivatives (DHET), and gas chromatographic–mass spectrometric analysis in the negative-ion (methane) chemical ionization mode[11] with selected-ion monitoring. The endogenous EET or DHET compounds are then quantitated relative to the ^2H-labeled standards.

Procedure

Preparation of Standards

[$^2H_8/^3H_8$]Arachidonic Acid and Unlabeled Epoxyeicosatrienoic Acids. [$^3H_8$]Arachidonate (200 Ci/mmol) is obtained from NEN Research products (Boston, MA). [$^2H_8$]Arachidonate was prepared from 5,8,11, 14-eicosatetrynoic acid (obtained from Dr. James Hamilton, Hoffman LaRoche, Nutley, NJ) with $^2H_2$ gas (MG Scientific Gases, Chicago, IL) using nickel acetate as a catalyst.[12] It may now also be obtained from commercial sources (Cayman Chemical, Ann Arbor, MI or BIOMOL Research Laboratories, Philadelphia, PA); unlabeled EET compounds may also be obtained from these suppliers.

5,6-[$^2H_8/^3H_8$]Epoxyeicosatrienoic Acid and 5,6-Dihydroxy[$^2H_8/^3H_8$]eicosatrienoic Acid. The [$^2H_8/^3H_8$]-5,6-EET is prepared from [$^2H_8/^3H_8$]arachidonate via 6-iodo-5-hydroxy[$^2H_8/^3H_3$]eicosatetraenoic acid, δ-lactone[5,13] as follows: Dissolve $KHCO_3$ (60 mg) and KI (160 mg) in water (4 ml). Add I_2 crystals (460 mg) and tetrahydrofuran (6 ml). (The tetrahydrofuran in this and all subsequent steps must be freshly distilled from

[11] I. A. Blair, S. E. Barrow, K. A. Waddell, P. J. Lewis, and C. T. Dollery, this series, Vol. 86, p. 467.

[12] D. F. Taber, M. A. Phillips, and W. C. Hubbard, *Prostaglandins* **22,** 349 (1981).

[13] E. J. Corey, H. Niwa, and J. R. Falck, *J. Am. Chem. Soc.* **101,** 1586 (1979).

$LiAlH_4$.) The resultant iodine reagent solution will be dark brown and opaque. Chill this solution on ice. Concentrate [$^2H_8$/$^3H_8$]arachidonate (5 mg) to a pure oil under nitrogen in a separate vessel (1 ml Reacti-vial, Pierce, Rockford, IL). Add 0.14 ml of the iodine reagent solution to the [$^2H_8$/$^3H_8$]arachidonate and vortex. Maintain on ice for 30 min. At the end of this period, add a saturated solution of Na_2SO_3 in water to the reaction vessel dropwise with vortexing until the solution becomes colorless. Extract twice with hexane (0.5 ml). Wash the hexane extracts with water (0.2 ml) and then dry over Na_2SO_4. The solution will contain virtually pure 6-iodo-5-hydroxy [$^2H_8$/$^3H_8$]eicosa-8,11,14-trienoic acid, δ-lactone. This compound is stable for at least 6 months at −70° in hexane and exhibits an R_f of 0.52 on Analtech (Newark, DE) silica gel GF TLC plates (2.5 × 10 cm, 250 μm) in solvent system A [ethyl acetate/hexane/acetic acid (25/75/0.1,v/v/v)]. To convert the 6-iodo-5-hydroxy[$^2H_8$/$^3H_8$]eicosatrienoic acid, δ-lactone to the [$^2H_8$/$^3H_8$]-5,6-EET, concentrate the hexane solution to dryness under nitrogen and add 0.2 ml of a solution of tetrahydrofuran (3 ml) and 0.2 *N* LiOH (2 ml). Vortex and maintain on ice for 30 min. Acidify carefully to pH 4.0 with 0.1 *N* HCl. Do NOT overacidify. Extract twice with 0.5 ml of hexane. Wash the extract with 0.2 ml water and dry over Na_2SO_4. This crude preparation of [$^2H_8$/$^3H_8$]-5,6-EET must be purified by normal-phase HPLC on column I (10 μm, 3.9 mm × 30 cm, waters, μPorasil, Milford, MA) in solvent system B [hexane/2-propanol/acetic acid (100/0.3/0.1, v/v/v)] Retention volume is about 54 ml under these conditions, and the compound may be located by liquid scintillation counting of aliquots (10 μl) of the eluant (4-ml fractions). The material is then concentrated to dryness and stored at −70° in hexane with 5% pyridine. Decomposition occurs at the rate of about 1% a day. Since the iodolactone intermediate is substantially more stable, it is desirable to store that compound and to generate the [$^2H_8$/$^3H_8$]-5,6-EET just before use.

The [$^2H_8$/$^3H_8$]-5,6-EET is converted to the corresponding vicinal diol ([$^2H_8$/$^3H_8$])-5,6-DHET) in a 1 ml Reacti-vial. Add 0.4 ml CH_2Cl_2 and vortex. Add 0.1 ml 0.5 *M* HCl and vortex. Allow to stand 12 hr at room temperature. Aspirate the aqueous (upper) layer, and wash the organic (lower) layer three times with 0.4 ml water. The product is the δ-lactone of [$^2H_8$/$^3H_8$]-5,6-DHET. This material is converted to the free acid by alkaline hydrolysis. Concentrate the CH_2Cl_2 to dryness under nitrogen. Add 0.05 ml of a solution of triethylamine/pyridine/water (1/10/10, v/v/v), vortex, and allow to stand 30 min at room temperature. Reconstitute in 0.05 ml of solvent C [hexane/2-propanol/acetic acid (100/1.5/0.2, v/v/v)] and purify by normal-phase HPLC on column I. The retention volume of the [$^2H_8$/$^3H_8$]-5,6-DHET is 104 ml (flow rate 4 ml/min).

$^2H_8/^3H_8$-labeled Standards of Other, Regionally Isomeric Epoxyeicosatrienoic Acids (14,15-EET, 11,12-EET, and 8,9-EET) and Dihydroxyeicosatrienoic Acids (14,15-DHET, 11,12-DHET, and 8,9-DHET). These epoxides are prepared from [$^2H_8/^3H_8$]arachidonate with *m*-chloroperoxybenzoic acid.[3,5] Add 1 mg of [$^2H_8/^3H_8$]arachidonate (50 mg/ml in ethanol) to a 1 ml Reacti-vial and concentrate to a pure oil under nitrogen. Add 0.2 ml of a solution (2.86 mg/ml) of *m*-chloroperoxybenzoic acid in CH_2Cl_2 and vortex. Allow to stand at room temperature for 12 hr. Concentrate to dryness under nitrogen. Reconstitute in 0.1 ml of methanol, and achieve partial separation of the regionally isomeric epoxyeicosatrienoic acids by reversed-phase HPLC[3,5] on column II (Waters, μBondapak C_{18}, 10 μm, 3.9 mm $\times$ 30 cm) in solvent D [methanol/water/acetic acid (73/27/0.2, v/v/v)]. The retention volumes of the regional isomers in this system are approximately 72 ml (14,15-EET), 88 ml (11,12-EET and 8,9-EET), and 96 ml (5,6-EET), but the isomers are incompletely separated and must be further purified by normal-phase HPLC[14] on column I in solvent B. The retention volumes of the regional isomers in this system are approximately 23 ml (14,15-EET), 25 ml (11,12-EET), 34 ml (8,9-EET), and 54 ml (5,6-EET). Yields of the 5,6-EET are low with this procedure, and the 5,6-EET should be prepared instead via the iodolactone as described above.

The [$^2H_8/^3H_8$]EET regional isomers are converted to the corresponding vicinal diols (14,15-DHET, 11,12-DHET, and 8,9-DHET) by a procedure involving acid hydrolysis of the methyl esters.[15] Add 20 μg of the [$^2H_8/^3H_8$]EET isomer to a 1 ml Reacti-vial as a CH_2Cl_2 solution and concentrate to dryness under nitrogen. Add 0.02 ml methanol and vortex. Add 0.2 ml of an ethereal solution of diazomethane, cap, and vortex. This converts the EET to its methyl ester derivative (EET-ME). (The methyl esters are substantially more stable than the free acids under conditions of acid hydrolysis.) Concentrate to dryness under nitrogen. Add 0.2 ml of a solution of perchloric acid in tetrahydrofuran. (To prepare this solution, use tetrahydrofuran freshly distilled from $LiAlH_4$. Dilute 0.1 ml of 70% perchloric acid with 0.6 ml water. Add 1.9 ml of tetrahydrofuran.) Vortex and allow to stand at 4° for 12 hr. This converts the EET-ME to the DHET-ME. Add 0.1 ml water and vortex. Add 0.5 ml CH_2Cl_2 and vortex. Use a tabletop centrifuge to clarify the CH_2Cl_2 (lower) layer. Aspirate the aqueous (upper) layer, and wash the CH_2Cl_2 phase twice with water.

[14] N. Chacos, J. R. Falck, C. Wixtrom, and J. Capdevila, *Biochem. Biophys. Res. Commun.* **104,** 916 (1982).

[15] S. Manna, J. R. Falck, N. Chacos, and J. Capdevila, *Tetrahedron Lett.* **24,** 33 (1983).

Concentrate to dryness under nitrogen. Add 0.5 ml of dimethoxyethane and vortex. Add 0.1 ml of 3 *N* LiOH, vortex, and add a magnetic stir bar. Flush with nitrogen and cap tightly. Stir magnetically at 60° for 90 min. This converts the DHET-ME to the DHET-free acid. Cool to room temperature and transfer the contents of the Reacti-vial to a 10-ml beaker. Wash vial three times with 1 ml water and add the washes to the beaker. Adjust to pH 4 with 0.5 *N* HCl. Transfer to a 7-ml screw-cap vial, and extract with 3 ml CH_2Cl_2. Aspirate the aqueous (upper) phase, and wash the CH_2Cl_2 (lower) phase twice with 1 ml water. Concentrate to dryness under nitrogen, reconstitute in 0.1 ml solvent C and purify by normal-phase HPLC on column I. The retention volumes for the [$^2H_8$/$^3H_8$]DHET regional isomers are approximately 28 ml (14,15-DHET), 34 ml (11,12-DHET), 64 ml (8,9-DHET), and 104 ml (5,6-DHET).

Derivatization and Analysis by Gas Chromatography–Negative-Ion Chemical Ionization–Mass Spectrometry

Both EET and DHET compounds are analyzed as the pentafluorobenzyl ester (PFBE) derivatives which is formed as follows: The analyte is concentrated to dryness under nitrogen in a 1-ml Reacti-vial, to which is added 0.02 ml of solution E [*N,N*-dimethylacetamide/tetramethylammonium hydroxide/methanol (8/5/15,v/v/v)] and 0.2 ml of solution F [pentafluorobenzyl bromide/dimethylacetamide (1/3, v/v)]. The vial is vortexed and allowed to stand at room temperature for 30 min. The contents are then concentrated to dryness under nitrogen and reconstituted in 0.1 ml water. Extraction is performed with 0.5 ml CH_2Cl_2. The aqueous (upper) layer is aspirated and discarded. The CH_2Cl_2 (lower) layer is washed a second time with 0.1 ml water and concentrated to dryness under nitrogen. The EET-PFBE are reconstituted in heptane (0.02 ml) and analyzed by GC–NCI–MS. The DHET-PFBE are first converted to the bistrimethylsilyl ether (TMS) derivatives by treatment with pyridine (0.01 ml) and *N,O*-bis(trimethylsilyl)trifluoroacetamide (0.01 ml) for 30 min at room temperature, followed by concentration to dryness under nitrogen and reconstitution in heptane (0.02 ml). The resultant DHET-PFBE, TMS derivative is then analyzed by GC–NCI–MS.

Gas chromatography was performed on a Hewlett Packard 5840A gas chromatograph interfaced with a Hewlett Packard 5985B mass spectrometer. The capillary column employed was a Hewlett Packard Ultra-performance (cross-linked methylsilicone, id 0.31 mm, film thickness 0.17 μm, 25 m length) and was operated with a Grob-type injector in the splitless mode. The carrier gas was helium (inlet pressure 15 lb/in^2). Injector and interface temperatures were 250° and 260°, respectively. GC oven temperature was programmed to remain at 85° for 0.5 min after injection and then to rise at a rate of 30° min to a final temperature of 225° (for EET-PFBE) or to 240° for (DHET-PFBE, TMS). The mass spectrometer was operated in the negative-ion chemical ionization (NCI) mode with methane as reagent gas at a source pressure of 1 torr. Absolute retention times varied with the age and condition of the GC column. The retention times listed below are representative values and illustrate the relative retention times of regional isomers. Under the conditions described above, the following retention times of you [$^2H_8$]EET-PFBE regional isomers were observed: 13.1 min (5.6-EET), 13.0 min (8,9-EET), 12.9 min (11,12-EET), and 12.9 min (14,15-EET). The corresponding unlabeled EET-PFBE compounds eluted 0.07 min after the $^2H_8$-labeled compounds. The [$^2H_8$]DHET-PFBE, TMS isomers under the conditions described above exhibited the following retention times: 12.4 min (14,15-DHET), 12.1 min (11,12-DHET), 12.1 min (8,9-DHET), and 12.2 min (5,6-DHET). The EET regional isomers are insufficiently well resolved from each other to permit their distinction by GC retention time alone. This was also true, in general, for the DHET compounds, although 11,12-DHET and 14,15-DHET were resolved to baseline.

The NCI mass spectra of both the EET-PFBE and of the DHET-PFBE, TMS derivatives consisted essentially of a single ion generated by the loss of the pentafluorobenzyl moiety (M−181). For the [$^2H_8$]EET-PFBE compounds this ion is m/z 327. The analogous ion in the spectrum of the unlabeled EET-PFBE is m/z 319. For the [$^2H_8$]DHET-PFBE, TMS compounds the [M−181] ion is m/z 489 and for the unlabeled compounds is m/z 481. Quantitation of endogenous EET compounds therefore involved selected-ion monitoring of the ion pair m/z 319 and m/z 327. Quantitation of the DHET compounds involved selected-ion monitoring of the ion pair m/z 481 and m/z 489. NCI mass spectrometric monitoring confers great sensitivity for quantitative analyses of both classes of compounds but does not permit distinction among the regional isomers, which must be separated from each other by liquid chromatography before GC–NCI–MS analysis. Although electron-impact (EI) mass spectrometry is capable, in principle, of achieving structural distinctions between the regional isomers, the quantitative sensitivity with NCI–MS is more than two orders of magnitude greater than that with EI–MS. With standard materials the sensitivity of the GC–MCI–MS method extends from below 40 pg and remains linear over several orders of magnitude. This is illustrated in Fig. 1, which is a representative standard curve for quantitation of 5,6-DHET by the procedures described above.

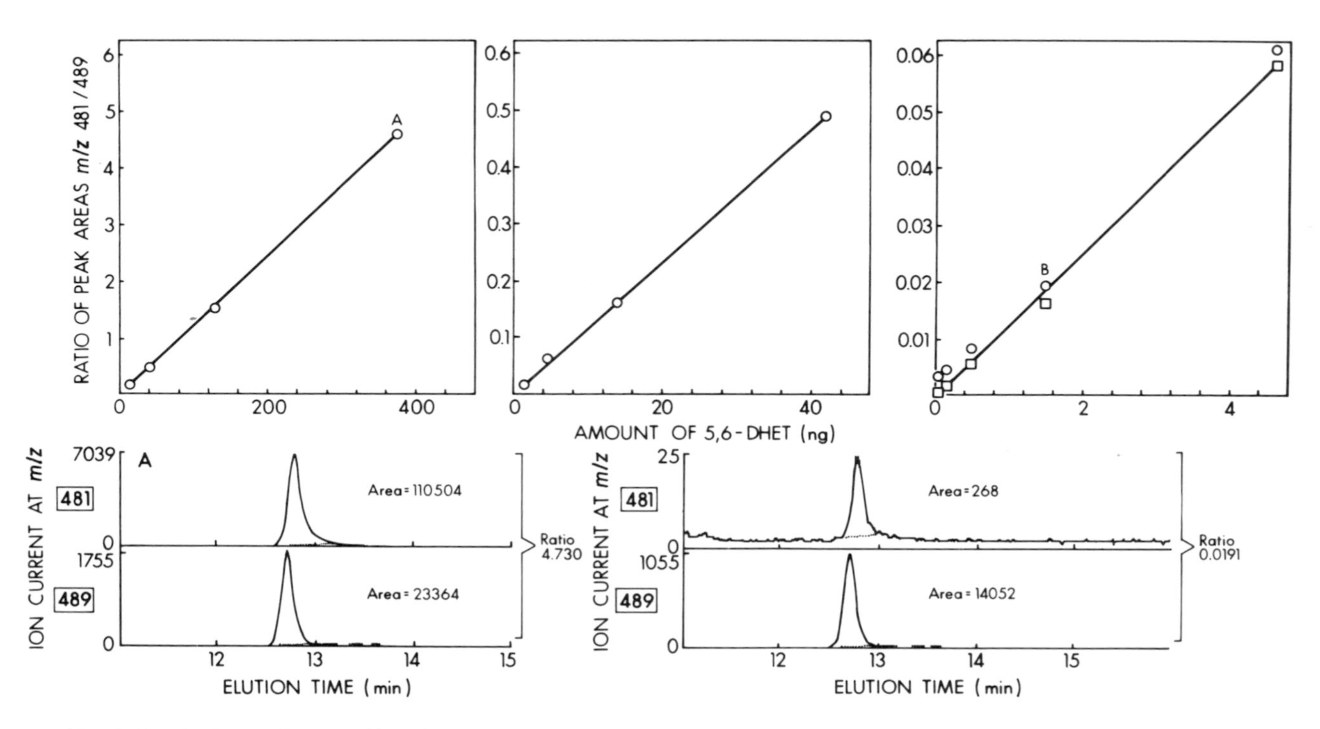

FIG. 1. Standard curve for quantities of 5,6-dihydroxyeicosatrienoic acid by stable-isotope dilution gas chromatography–negative-ion chemical ionization mass spectrometry.

Measurement of Production of Epoxyeicosatrienoic Acids and Dihydroxyeicosatrienoic Acids by Rat Liver Microsomes with Stable Isotope-Dilution Gas Chromatography–Negative-Ion Chemical Ionization Mass Spectrometry

Rat liver microsomes were prepared by placing fresh liver in ice-cold 0.15 *M* saline, mincing, and determining the approximate volume of the minced tissue. Four volumes of buffer (0.15 *M* NaCl, 0.05 *M* Tris-HCl, 0.001 *M* EDTA, pH 7.4) were added, and the tissue homogenized by four strokes of the Teflon pestle (800 rpm) of a Potter–Elvehjem tissue grinder. The homogenate was centrifuged (15 min, 17,000 *g*, 4°), and the pellet discarded. The supernatant was then centrifuged (80 min, 4°, 100,000 *g*). The resulting pellet was washed once with buffer and resuspended in a volume of buffer equal to 20% of the original volume of minced tissue. Protein content was determined by the method of Lowry, *et al.*[16]

Microsomes (final protein concentration 0.8 mg/ml) were incubated with NADPH (1 m*M*, added as a 0.94 mg/ml solution in water and arachidonate [30 μM, added as a 7.5 $\mu g/\mu l$ solution in dimethyl sulfoxide/0.1 *M* Na_2CO_3 (1/1, v/v)] for 15 min at 37° with stirring. In some cases arachidonate, NADPH, or the microsomes themselves were omitted, or an inhibitor was added. Inhibitors included BW755C [500 μM, added as a 5 $\mu g/\mu l$ solution in dimethyl sulfoxide/0.15 *M* NaCl (11/178, v/v)], nordihydroguaiaretic acid (NDGA) [50 μM, added as a 1.75 $\mu g/\mu l$ solution in dimethyl sulfoxide/0.15 *M* NaCl (143/428, v/v)], metyrapone [50 μM, added as a 2.3 $\mu g/\mu l$ solution in dimethyl sulfoxide/Krebs–Ringer bicarbonate (ref. 17) (43/387, v/v)], or eicosa-5,8,11,14-tetrynoic acid (ETYA) (20 μM, added as a 0.6 $\mu g/\mu l$ solution in 0.1 M Na_2CO_3).

Incubations were terminated by addition of 1 volume of methanol containing 75 ng of $^2H_8/^3H_8$-labeled standards of each EET or DHET compound to be measured. Tubes were centrifuged (tabletop centrifuge, 2000 rpm, 5 min at room temperature) to remove particulate matter. The supernatant was acidified to pH 4.0 with 0.5 *N* HCl and extracted with CH_2Cl_2 (0.5 ml twice). The extract was washed once with water, concentrated to dryness under nitrogen, and reconstituted in HPLC solvent. EET compounds were then analyzed on column I, solvent B, and DHET compounds were analyzed on column I, solvent C, as described above. The isolated EET and DHET isomers were then collected, concentrated, derivatized and quantitated by GC–NCI–MS as described above. [Ade-

[16] O. H. Lowry, N. J. Rosebrough, A. L. Farr, and R. J. Randal, *J. Biol. Chem.* **193,** 265 (1951).

[17] J. Turk, B. A. Wolf, P. G. Comens, J. Colca, B. Jakschik, and M. L. McDaniel, *Biochim. Biophys. Acta* **835,** 1 (1985).

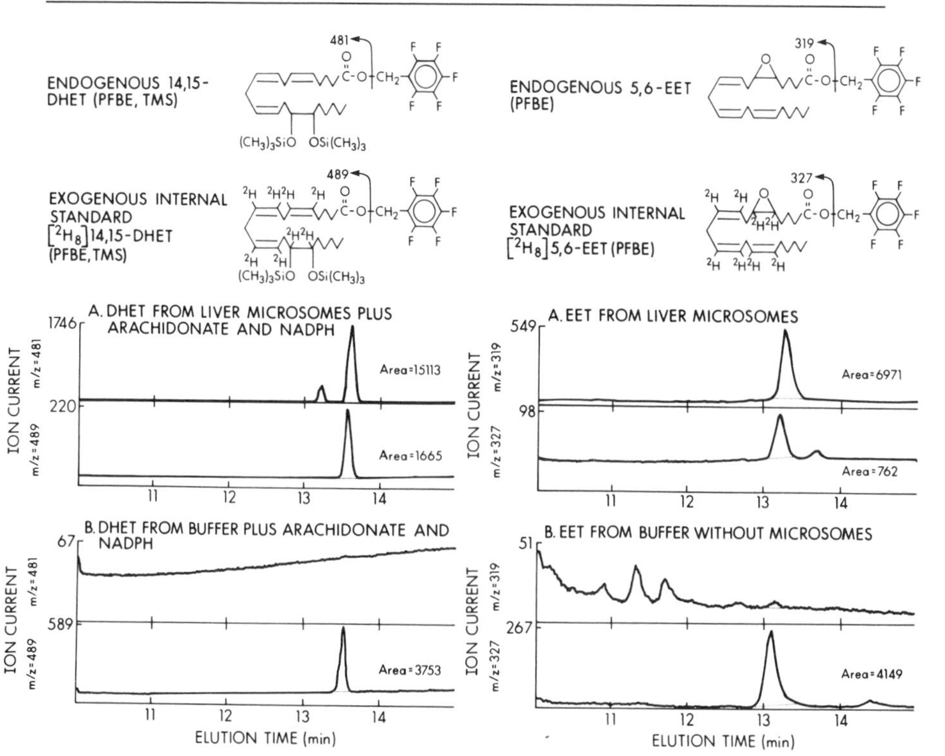

FIG. 2. Demonstration of NADPH-dependent oxygenation of arachidonic acid by stable-isotope dilution gas chromatography–negative-ion chemical ionization mass spectrometry in liver microsomes.

quate recovery of the 5,6-DHET compound required the application of a delactonization procedure before HPLC and before derivatization. This procedure involved treating the sample with triethylamine/pyridine/water (1/10/10, v/v/v) for 30 min at room temperature.] Figure 2 illustrates quantitation of a DHET (left-hand side) compound and an EET (right-hand side) compound from liver microsomes using this method.

Table I shows that DHET production is dependent on the presence of microsomes and NADPH in this system and is amplified by arachidonate. Table II indicates that in the presence of NADPH and arachidonate, liver microsomes produce measurable amounts of all four DHET and all four EET regional isomers. The disproportionately low production of 5,6-DHET relative to 5,6-EET may reflect the resistance of the latter compound to epoxide hydrolase, as reported by Chacos *et al.*[4] Table II indicates that BW755C and metyrapone, but not ETYA or NDGA, inhibit

TABLE I
SYNTHESIS OF CYTOCHROME-*P*-450 MONOOXYGENSE METABOLITES OF ARACHIDONIC ACID BY LIVER MICROSOMES IN THE PRESENCE OF NADPH[a]

Entry	Liver microsomes	Arachidonic acid	NADPH	Amount of 14,15-DHET (μg)	Percentage of control
1	+	+	+	6.30	100
2	+	+	−	0.06	<1
3	+	−	−	0.04	<1
4	+	−	+	2.27	36
5	−	+	+	<0.01	0
6	−	+	−	<0.01	0
7	−	−	+	<0.01	0
8	−	−	−	<0.01	0

[a] Microsomes are prepared from rat liver and incubated in the presence or absence of NADPH (1 m*M*) and exogenous arachidonic acid (100 μM) for 15 min at 37°. Microsomal protein concentration is 0.8 mg/ml. Incubations are terminated by addition of 1 volume of methanol containing 75 ng of $^2H/^3H$-labeled DHET or EET compounds as internal standards. Products are extracted, analyzed by normal-phase- HPLC (column II, solvent C), derivatized, and analyzed by GC–NCI–MS

NADPH-dependent oxygenation of arachidonate to DHET compounds at the tested concentrations in liver microsomes.

Preparation of $^{18}O_2$-Labeled Epoxyeicosatrienoic and Dihydroxyeicosatrienoic Acids

An alternative approach to preparation of heavy isotope-labeled analogs of epoxyeicosatrienoic acids is to perform $^{18}O_2$ exchange of the commercially available, unlabeled epoxyeicosatrienoic acids (Biomol Laboratories, Philadelphia, PA; or Cayman Chemical, Ann Arbor, MI) with butyrylcholinesterase and $H_2{}^{18}O$ using procedures described by Wescott *et al.*[18] The $^{18}O_2$-labeled compounds could then, in principle, be employed in stable isotope dilution GC–NCI–MS assays similar to those described above. This approach has been examined with each of the four regionally isomeric epoxyeicosatrienoic acids. The 5,6-EET was found to be too unstable in aqueous solution (half-life about 1 hr) to permit completion of the 30-hr exchange reaction. The procedure was applied to each of the other three regionally isomeric EET compounds and involved addition of

[18] J. Y. Westcott, K. L. Clay and R. C. Murphy, *Biomed. Mass Spectrom.* **12,** 714–718 (1985).

TABLE II
A. REGIONAL ISOMERIC COMPOSITION OF CYTOCHROME-*P*-450 MONOOXYGENASE METABOLITES OF ARACHIDONIC ACID FROM RAT LIVER MICROSOMES AND B. INFLUENCE OF PHARMOCOLOGICAL INHIBITORS ON THEIR PRODUCTION

periment	Compound	Amount (μg)	B. Inhibitor	Amount of 14,15-DHET (μg)	Percentage of control amount of DHET	Significance of difference from control
1 and 2	14,15-DHET	5.37 ± 3.09	None	2.38 ± 0.83	100 ± 0	n.a.[a]
	11,12-DHET	13.06 ± 0.70	(Control)			
	8,9-DHET	5.93 ± 3.95	BW755C	0.44 ± 0.25	19 ± 6	$0.01 < p < 0.02$
	5,6-DHET	0.6 ± 0.02	(500 μM)			
			ETYA	1.73 ± 0.43	84 ± 16	$0.50 < p < 1.00$
3	14,15-EET	3.17	(20 μM)			
	11,12-EET	0.77	NDGA	2.84 ± 1.32	114 ± 22	$0.05 < p < 1.00$
	8,9-EET	0.58	(50 μM)			
	5,6-EET	1.83	Metyrapone	1.45 ± 0.60	56 ± 13	$0.02 < p < 0.05$
			(50 μM)			

[a] n.a., Not applicable.

0.2 ml of $H_2{}^{18}O$ (MSD Isotopes, St. Louis, MO) and 120 units of Sigma Type XI butyrylcholinesterase to 1 μg of the EET in a 1 ml Reacti-vial and incubation for 30 hr at 37°. This is followed by acidification to pH 3.5 with 1 *N* HCl in $H_2{}^{18}O$ and extraction with 0.5 ml CH_2Cl_2. The resultant $[^{18}O_2]$EET compounds exhibited the following isotopic disturbtion upon derivatization and GC–NCI–MS analysis: 14,15-EET (4.3% $^{16}O/^{16}O$, 31.8% $^{16}O^{18}O$, 63.9% $^{18}O/^{18}O$); 11,12-EET (19.3% $^{16}O/^{16}O$, 45.7% $^{16}O/^{18}O$, 35% $^{18}O/^{18}O$); 8,9-EET (21.3% $^{16}O/^{16}O$, 47.9% $^{16}O_2/^{18}O$, 30.8% $^{18}O/^{18}O$). Similar exchange reactions were performed with 14,15-DHET (0.51% $^{16}O/^{16}O$, 4.5% $^{16}O/^{18}O$, 95% $^{18}O/^{18}O$) and with 5,6-DHET (27.9% $^{16}O/^{16}O$, 34.2% $^{16}O/^{18}O$, 37.9% $^{18}O/^{18}O$). A greater degree of isotopic purity could likely be achieved by a second exchange. Unlike $^2H_8$-labeled compounds, which lose ^2H under conditions of catalytic hydrogenation, the heavy isotope label in $[^{18}O_2]$carboxylate-labeled compounds is stable under such conditions. To perform catalytic hydrogenation, the EET compounds or their methyl esters (5 μg) are dissolved in 0.15 ml of ethanol with 100 μg of platinum oxide in a 1-ml Reacti-vial. Hydrogen is bubbled through the solution for 1 to 5 min. Water (0.35 ml) is then added, and the solution acidified to pH 3.5 with 1 *N* HCl (for free acids) or extracted directly (for EET methyl esters) with 0.5 ml ethyl acetate. Products are analyzed by GC–MS. Hydrogenation of the methyl esters of each of the four regionally isomeric EET compounds is complete at 5 min and results in an increase in the C value from 20.9 to 21.4 as reported.[1] The free acids of all of the EET compounds are unstable under conditions of catalytic hydrogenation and decomposed completely by reaction times of 1 min.

Discussion

The reported insulin and glucagon secretagogue properties of exogenous epoxyeicosatrienoic acids has been taken to imply that these compounds may participate in secretion of hormones from pancreatic islets in response to physiological stimuli.[19] The stable isotope dilution GC–NCI–MS method described above has, therefore, been used to quantitate production of epoxyeicosatrienoic acids and of dihydroxyeicosatrienoic acids from endogenous arachidonate by intact islets and from exogenous arachidonate by islet microsomes.[17] Islet microsomes were found to produce less than 3% of the amounts of these compounds produced by liver microsomes per unit protein mass. Production of these compounds by intact islets was less than 3% of the amount of 12-hydroxy-5,8,10,14-eicosatetraenoic acid (12-HETE) produced in the same incubations.[17] Metyrapone at a concentration (50 μM) which clearly suppressed production of epoxyeicosatrienoic acids by liver microsomes did not influence glucose-induced insulin secretion. NDGA (50 μM) and ETYA (20 μM) at concentrations which suppressed both insulin secretion and islet 12-HETE biosynthesis did not influence production of epoxyeicosatrienoic acids in liver microsomes. These observations indicate that epoxyeicosatrienoic acids are unlikely to participate in glucose-induced insulin secretion and illustrate the importance of measuring production of these compounds from endogenous substrate in intact cells responding to physiological stimuli.

Acknowledgment

Supported by Public Health Service Grants R01-DK-34388, K04-DK-01553, RR-00954, and S10-RR-04693

[19] J. Falck, S. Manna, J. Moltz, N. Chacos, and J. Capdevila, *Biochem. Biophys. Res. Commun.* **114,** 743 (1983).

[22] High-Performance Liquid Chromatography for Chiral Analysis of Eicosanoids

By ALAN R. BRASH and DAN J. HAWKINS

The advantages of chiral-phase columns for analysis of enantiomeric composition are so compelling that they are considered here first and foremost. Although an alternative approach may give better resolution, and this can be advantageous for preparative purposes, for analytical work, the chiral phase methods are clearly the method of choice. The only serious disadvantage is the current high cost of the columns. Other methods are a useful back-up if the chiral columns fail to resolve the enantiomers. Most of the examples given here are of separation of hydroxyeicosatetraenoic acids (HETEs), but the same approaches are straightforward to adapt for other eicosanoids. The methods sections are prefaced with some brief comments on the two systems for assignment of chirality.

Nomenclature: *R*/*S* and *D*/*L* Systems

The procedure for assigning *R* or *S* (Cahn–Ingold–Prelog system) is straightforward and applicable to any molecule and is described in any college chemistry textbook. The designation of *R* or *S* is not directly related to the orientation of the chiral substituents in space. Rather, assignment is based on the "priority order" of substituents surrounding the chiral center. This can lead to a change in the *R* or *S* designation, for example, on addition or removal of a double bond (as in arachidonate/eicosapentaenoate analogs, or before and after hydrogenation of an HETE). Specific examples were given in a review of lipoxygenase stereochemistry.[1]

The Fischer D/L designation *is* directly related to the physical orientation of the chiral center. Although it is not always straightforward to make the D or L assignment in a complex molecule, the procedure is simple to apply to a linear molecule such as arachidonic acid (Fig. 1). The D/L designations have a real physical meaning—"L" is on the left, and does not change on hydrogenation of surrounding double bonds or on addition of new functional groups; this is a very helpful characteristic of the Fischer

[1] H. Kühn, T. Schewe, and S. M. Rapoport, *Adv. Enzymol.* **58,** 273 (1986).

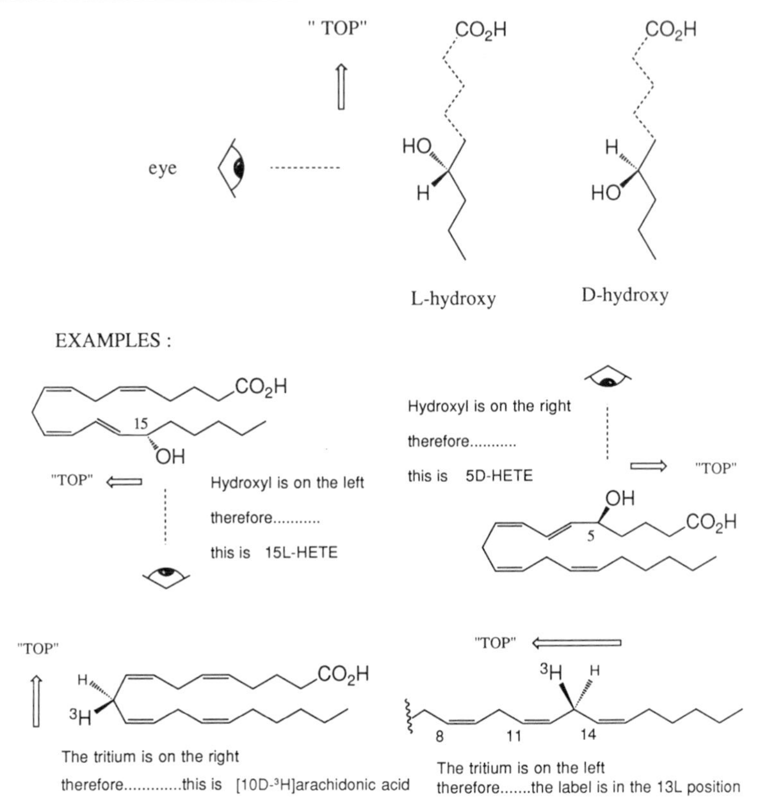

FIG. 1. Assignment of the D/L configuration of hydroxy fatty acids. To assign the hydroxyl configuration, imagine the fatty acid held by the CO_2H group and allowed to dangle like a string. When looked at as shown, the hydroxyl is either on the left (L) or on the right (D).

nomenclature. Sometimes the D/L and the *R*/*S* assignments are both given: e.g., $5D_s$-HETE, $8L_s15L_s$-DiHPETE, [$13L_s$-^{3}H]arachidonic acid.

Resolution on Chiral-Phase HPLC Columns

Aside from the obvious ease of use, the important advantages of chiral-phase columns are the sensitivity (analyses can be carried out on small amounts of material, e.g., 0.1 μg of HETE), and the accuracy (no derivatization is required, and, therefore, interfering peaks from reaction by-products are absent; also inaccuracy resulting from chirally impure resolving agents is eliminated). Figure 2 shows the chiral phases of the two types of columns discussed below.

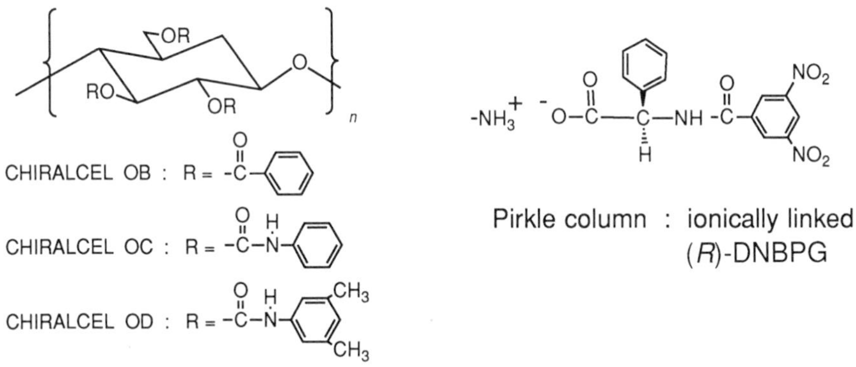

CHIRALCEL OB : R =

CHIRALCEL OC : R =

CHIRALCEL OD : R =

Pirkle column : ionically linked (*R*)-DNBPG

FIG. 2. Chiral stationary phases of Chiralcel and (*R*)-DNBPG columns.

Chiralcel Columns

Chiralcel columns give the best resolution of racemic HETEs that has been reported to date. The complete resolution of 12(*RS*)-HETE methyl ester on a Chiralcel OC column was demonstrated by Kitamura *et al.*,[2] and the method has also been used by Takahashi *et al.*[3] Using a solvent system of hexane/2-propanol (100 : 1, v/v) and a flow rate of 1 ml/min, the enantiomers eluted in the order 12*R* (retention volume ~21 ml) followed by 12*S* (retention volume ~ 23.5 ml).[2] Takahashi *et al.* used the same column with the solvent modified to 2% 2-propanol; the retention times are halved with no sacrifice in resolution.

On investigating these columns, we found that the Chiralcel OB column gives excellent resolution of the methyl esters of 15(*RS*)-HETE and 8(*RS*)-HETE (Fig. 3). Note that the two enantiomers elute from the Chiralcel column with different peak shapes; however, the two peaks are identical in area.

Use of the Chiralcel column method has revealed the very high stereofidelity of certain lipoxygenase reactions. We found an aliquot of coral 8-lipoxygenase-derived 8(*R*)-HETE to be 99.4% 8*R* enantiomer. A sample of 15(*S*)-HETE prepared using the soybean lipoxygenase was comprised of 99.7% 15*S* enantiomer (only 0.3% 15*R*). In the past, when we analyzed the enantiomeric composition of soybean lipoxygenase products using resolving agents, we typically obtained ratios of 97 : 3 (*S* : *R*). In

[2] S. Kitamura, T. Shimizu, I. Miki, T. Izumi, T. Kasama, A. Sato, H. Sano, and Y. Seyama, *Eur. J Biochem.* **176,** 725 (1988).

[3] Y. Takahashi, N. Ueda, and S. Yamamoto, *Arch. Biochem. Biophys.* **266,** 613 (1988).

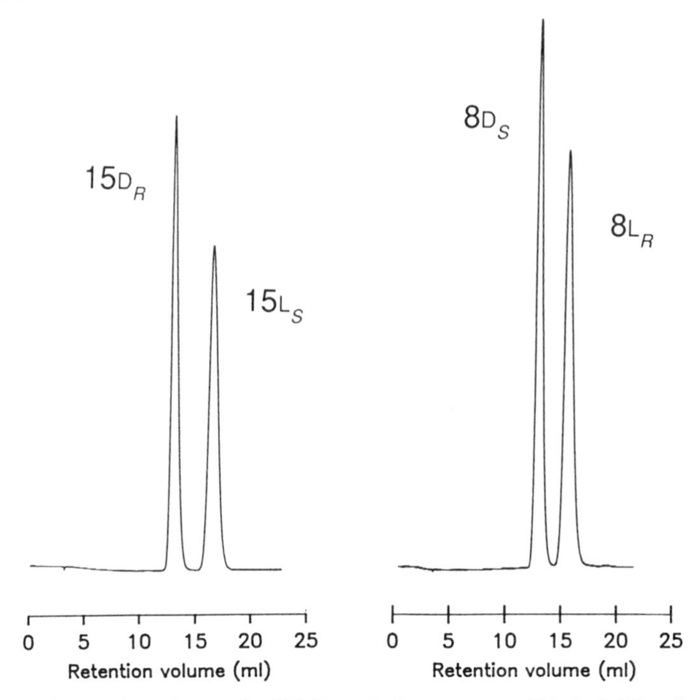

FIG. 3. Separation of racemic HETE methyl esters on a Chiralcel OB column. The chromatograms are of racemic 15-HETE methyl ester (left) and racemic 8-HETE methyl ester (right). Column: Chiralcel OB (250 × 4.6 mm) (J. T. Baker, Inc., Phillipsburg, NJ). Solvent: hexane/2-propanol 100 : 2 (v/v). Flow rate 0.5 ml/min. UV detection at 235 nm.

retrospect, it would appear that use of the resolving agents may introduce inaccuracies which can be attributed to chirally impure resolving agent, racemization during derivatization, or the presence of interfering peaks from the reagents.

All regioisomers of racemic epoxyeicosatrienoates (EETs) can be completely resolved on Chiralcel OB or OD columns.[4] For example, the *cis*-epoxide 14,15-EET methyl ester is resolved on a Chiralcel OB column using a solvent of 0.014% 2-propanol in hexane, and a flow rate of 1.1 ml/min; the 14(*R*),15(*S*)-epoxide elutes at 46.5 min and its enantiomer at 55 min.[4]

Chiralcel columns also can be used with reversed-phase solvents, and some racemates are resolved only in the reversed-phase mode. For exam-

[4] T. D. Hammonds, I. A. Blair, J. R. Falck, and J. H. Capdevila, *Anal. Biochem.* **182,** 300 (1989).

ple, optimal separation of the enantiomers of 5,6-EET methyl ester is obtained on a Chiralcel OB column using a solvent of 30% water/ethanol.[4] The ethanol/water mixtures which are recommended for reversed-phase separations on Chiralcel columns are particularly viscous, and there is the danger of overpressurizing the column bed. This can result in a physical collapse of the structure of the column, thus generating extremely high pressures and rendering the column useless. Use of low flow rates ($\leq$0.5 ml/min) is advisable in the reversed-phase mode.

As a service to potential customers, the manufacturers of Chiralcel columns (Diacel Chemical Industries, Tokyo, Japan) will investigate the separation of a racemic sample on the complete range of their chiral-phase columns.

Pirkle Columns

Silica derivatized with the substituent (*R*)-(–)-*N*-3,5-dinitrobenzoylphenylglycine (DNBPG) is one of the commonly used chiral phases developed by W. H. Pirkle. Partial resolution of racemic HETE methyl esters is achieved on these columns.[5] This degree of separation has proved adequate for establishing whether a product is either a racemic mixture or is predominantly of the *R* or *S* configuration.[6–8] The 9(*RS*)- and 13(*RS*)-hydroxylinoleates can be completely resolved on the DNBPG column.[5]

It is worth mentioning that there is considerable variation in the resolving power of individual HPLC columns prepared with the DNBPG phase. Unfortunately, there appears to be an element of luck involved in obtaining a "good" column. The DNBPG phase may be covalently or ionically linked to the silica. In our experience, either of these types experiences a noticeable loss of resolution after extensive use; chromatographic efficiency is maintained but the separation factors (relative elution volumes) diminish. Theoretically, the ionically linked phase can be regenerated, and this is available as a service from some manufacturers.

The resolution of HETE enantiomers on the DNBPG column is improved when the alcohol moiety of the HETE methyl ester is converted to a benzoyl or naphthoyl ester (Fig. 4).[9] The new group is not in itself chiral, and the improved separation relies on an enhanced interaction of the

[5] H.Kühn, R. Wiesner, V. Z Lankin, A. Nekrasov, L. Alder, and T. Schewe, *Anal. Biochem.* **160,** 24 (1987).

[6] D. J. Hawkins and A. R. Brash, *J. Biol. Chem.* **262,** 7629 (1987).

[7] J.-P. Falqueyret, Y. Leblanc, J. Rokach, and D. Riendeau, *Biochem. Biophys. Res. Commun.* **156,** 1083 (1988).

[8] E. H. Oliw and H. Sprecher, *Biochim. Biophys. Acta* **1002,** 283 (1989).

[9] D. J. Hawkins, H. Kühn, E. H. Petty, and A. R. Brash, *Anal. Biochem.* **173,** 456 (1988).

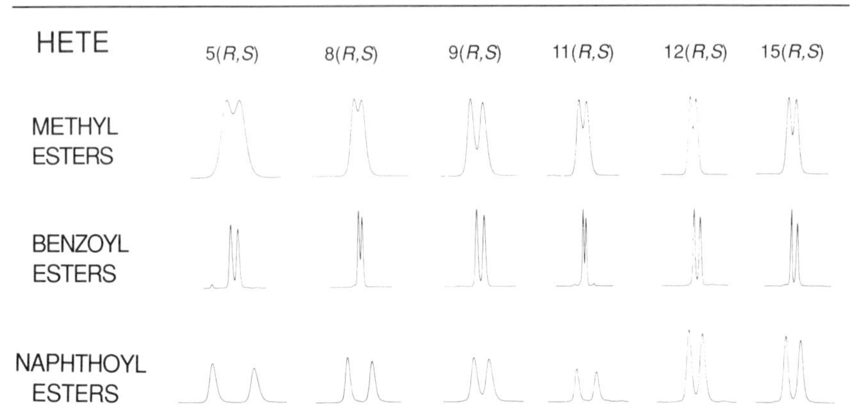

FIG. 4. Resolution of HETE methyl esters and aromatic derivatives on the DNBPG column. Partial UV (235 nm) chromatograms of enantiomers of HETE methyl esters and their aromatic ester derivatives are separated on the DNBPG chiral phase HPLC column [Regis Chemical Co., Morton Grove, IL; (*R*)-(−)-*N*-3,5-dinitrobenzoyl-α-phenylglycine (DNBPG), ionically linked 5 μm, modified Spherisorb; 25 cm × 4.6 mm]. All products are eluted at 1.2 ml/min; racemic HETE methyl esters are eluted with hexane:2-propanol 100 : 0.5 (v/v) and aromatic esters with hexane:2-propanol 100 : 0.25 (v/v). Retention times are given in the original publication.[9] (Figure reproduced from Ref. 9 by permission.)

aromatic ester with the DNBPG phase. To form these derivatives, the HETE methyl ester is dissolved in 30 μl of anhydrous pyridine and reacted with 3 μl of either benzoyl chloride or 1-naphthoyl chloride (Aldrich, Milwaukee, WI) for 15 min at room temperature. The reaction mixture is then evaporated under nitrogen and the residue extracted with hexane. Hexane-soluble products are subsequently purified by reversed-phase HPLC and then chromatographed on the DNBPG chiral column.

This derivatization protocol is quite time-consuming. Furthermore, in our experience, the naphthoyl chloride is apt to degrade on storage. Although the separations are much better than for underivatized HETEs on the DNBPG columns, the results offer little improvement over the direct analyses on Chiralcel columns (Fig. 5). Overall, the advent of chiral phases with the resolving power of the Chiralcel series has now superceded the earlier advantages of the derivatization methods. The derivatization does increase resolution, and this may offer some advantage for preparative separations.

Conversion to Diastereomer

Reaction with Resolving Agents

Another approach to the measurement of enantiomeric composition involves reaction of the racemate with a chiral derivatizing reagent (resolv-

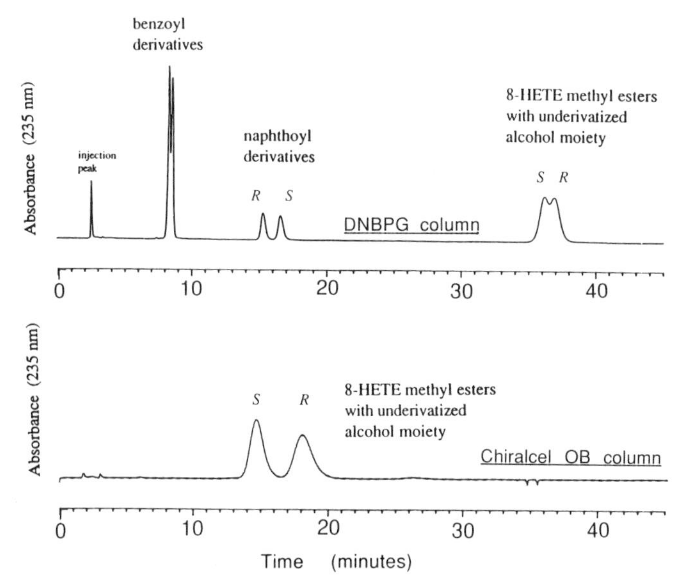

FIG. 5. Comparison between DNBPG and Chiralcel OB columns. UV HPLC chromatogram for 8(*R*,*S*)-HETE methyl esters and aromatic ester derivatives. Products are eluted from the DNBPG column at 1.2 ml/min with hexane : 2-propanol, 100 : 0.5 (v/v). 8-HETE methyl ester enantiomers are eluted from the Chiralcel OB column at 2 ml/min with hexane : 2-propanol, 100 : 0.75 (v/v). (Top portion of figure adapted from Ref. 9 by permission.)

ing agent), thus giving a mixture of diastereomers, which, in principle, can be resolved by conventional chromatography. As with the chiral column methods, it is difficult to predict if resolution can be accomplished. It is quite common to find that the diastereomers do not resolve.

Resolving agents which have been used for the separation of racemic eicosanoids by HPLC include the isocyanate of dehydroabietylamine,[10–12] menthyl chloroformate,[13–15] and α-methoxy-α-trifluoromethylphenacetyl chloride.[16]

The derivatives of dehydroabietylamine usually give the best resolution. But this derivative suffers from the disadvantages that the reagent is

[10] J. R. Falck, S. Manna, H. R. Jacobson, R. W. Estabrook, N. Chacos, and J. Capdevila, *J. Am. Chem. Soc.* **106,** 3334 (1984).

[11] A. R. Brash, C. D. Ingram, and T. M. Harris, *Biochemistry* **26,** 5465 (1987).

[12] P. M. Woollard, *Biochem. Biophys. Res. Commun.* **136,** 169 (1986).

[13] A. R. Brash, A. T. Porter, and R. L. Maas, *J. Biol. Chem.* **260,** 4210 (1985).

[14] A. R. Brash, S. W. Baertschi, C. D. Ingram, and T. M. Harris, *J. Biol. Chem.* **262,** 15829 (1987).

[15] S. W. Baertschi, C. D. Ingram, T. M. Harris, and A. R. Brash, *Biochemistry* **27,** 18 (1988).

[16] J. C. Andre and M. O. Funk, *Anal. Biochem.* **158,** 316 (1986).

not commercially available (preparation is described in Ref. 17), and the derivatization reaction is extremely sluggish, typically taking several days for reaction to reach completion.

Formation of the menthoxycarbonyl derivative is a simple procedure which utilizes commercially available reagents. Typically, 10–100 μg of the methyl ester of the HETE is taken to dryness, and then mixed with 100 μl dry toluene, 10 μl dry pyridine, and 5 μl menthylchloroformate (Aldrich). Reaction of unhindered hydroxyl groups will be complete in 30 min at room temperature. The derivatization of vicinal diols requires more forcing conditions: 100° for 1 to 2 hr under nitrogen or argon. Subsequently, the solvents are evaporated to dryness and the menthoxycarbonyl derivatives are extracted from the residue with hexane. The original enantiomers (now diastereomers) may be resolved by SP-HPLC. A solvent of hexane/2-propanol (100 : 0.2, v/v) is appropriate for the derivative of 5(*RS*)-HETE, with the solvent changed to 0.1% 2-propanol for 8(*RS*)-HETE or 11(*RS*)-HETE. The menthoxycarbonyl derivatives of 9-HETE, 12-HETE, and 15-HETE cannot be resolved using this system. The menthoxycarbonyl derivative of 15-HETE methyl ester is resolved on an Ag^+-loaded cation-exchange column.[18]

We found that the menthoxycarbonyl derivatives of vicinal diols (benzyl esters) tailed very badly on SP-HPLC; resolution of the original enantiomers was accomplished by reversed-phase HPLC using 100% acetonitrile as solvent. Interestingly, CH_3CN was superior to methanol as the reversed-phase solvent: retention volumes of the menthoxycarbonyl derivatives were greater using CH_3CN and impurities from the derivatization reaction were not retained. The result was good resolution and "clean" chromatograms (UV detection at 210 nm).[14,15]

A potential disadvantage of the derivatization with a chiral reagent is the possibility of unequal rates of reaction with the two enantiomers of the analyte. It is imperative to demonstrate that a racemic standard is resolved into two peaks of equal area.

Reaction with Lipoxygenase

In some cases this may be a useful alternative method for the introduction of a new chiral center. Thus, the analysis of chirality of an HETE can be accomplished by enzymatic conversion to a DiH(P)ETE. A special advantage of this approach is that the diastereomers are usually well

[17] E. J. Corey and S. Hashimoto, *Tetrahedron Lett.* **22,** 299 (1981).

[18] M. Claeys, M.-C. Coene, A. G. Herman, G. H. Jouvenaz, and D. H. Nugteren, *Biochim. Biophys. Acta* **713,** 160 (1982).

resolved by conventional HPLC. This allows a precise determination of the percentage abundance of the minor enantiomer.

Yoshimoto *et al.*[19] used the method to assign the chirality of the 12-lipoxygenase of porcine leukocytes. Reaction of authentic 5(*S*)-HETE with the leukocyte 12-lipoxygenase gave a single product which after reduction of the 12-hydroperoxide was shown to be indistinguishable from 5(*S*), 12(*S*)-DiHETE. This effectively demonstrated that the leukocyte enzyme is a 12*S*-lipoxygenase.

Similarly, reaction of 8-HETE with the soybean lipoxygenase was used to assign the chirality of the 8-lipoxygenases of starfish oocytes[20] and of the coral *Plexaura homomalla*.[14] Enzymatic 15*S*-oxygenation of racemic 8-HETE gives [after reduction of the 15(*S*)-hydroperoxide] equal amounts of 8*S*,15*S*-DiHETE and 8*R*,15*S*-DiHETE. These diastereomers are easily resolved on reversed-phase or SP-HPLC.[14,20]

Acknowledgments

This work was supported by NIH Grants DK-35275, HD-05797, and GM-15431.

[19] T. Yoshimoto, Y. Miyamoto, K. Ochi, and S. Yamamoto, *Biochim. Biophys. Acta* **713,** 638 (1982).

[20] L. Meijer, A. R. Brash, R. W. Bryant, K. Ng, J. Maclouf, and H. Sprecher, *J. Biol. Chem.* **261,** 17040 (1986).

[23] Extraction of Phospholipids and Analysis of Phospholipid Molecular Species

By GEORGE M. PATTON and SANDER J. ROBINS

Introduction

The phospholipids (PLs) are a family of molecules which, with the exception of sphingomyelin (SM), are characterized by the presence of a phosphate group esterified to the *sn*-3 position of glycerol and 1 (lyso-PLs) to 4 (cardiolipin) acyl residues attached through an ester bond at the *sn*-2 position of glycerol and either an ester (acyl), ether (alkyl), or vinyl ether (alkenyl) bond at the *sn*-1 position of glycerol. Sphingomyelin, a derivative of serine, contains one acyl residue attached by an amide bond to the amino group at the 2-position and a phosphorylcholine at the 1-position of a long-chain base. Except for phosphatidic acid (PA), the PLs also contain a "base" esterified to the phosphate. It is these "head groups" (phosphate + base) that define the various PL classes, and it is differences in the properties of the head groups that permit the chromatographic resolution of the PLs into separate classes. Under certain circumstances, the purified PL classes can be further separated into subclasses based upon the type of chemical bond between the acyl group and the *sn*-1 position of glycerol. The purified PL classes or subclasses can then be further separated into molecular species based on the acyl composition of the molecules. For any particular PL class or subclass it is possible to resolve as many as 60 molecular species. Typically, however, 90% or more of the total mass of a PL is confined to only a few (8–12) major molecular species. Arachidonic acid [20 : 4 (*n*–6)],[1] which is the focus of this volume, is found predominantly in four major molecular species, namely 18 : 0–20 : 4, 16 : 0–20 : 4, 18 : 1(*n*–9)–20 : 4, and 18 : 1(*n*–7)–20 : 4.

The separation of a particular PL class into its various molecular species can be readily accomplished by reversed-phase chromatography of either the intact PL or after the PL has been converted to an appropriate derivative. There are advantages and disadvantages to both procedures. The separation of intact molecular species is quick, simple, and is essential if radioactivity in either the phosphate or the base is to be determined. However, relatively large amounts of material are required, the resolution is not particularly good, quantitation is difficult and, at present, there is no satisfactory procedure for separating the intact PL classes into diacyl, alkylacyl, and alkenylacyl subclasses. The molecular species separation of derivatized PLs is a more complex procedure, but it affords much better resolution, greater sensitivity, and ease of quantitation. Moreover, derivatization is essential if the diacyl, alkylacyl, and alkenylacyl subclasses are to be separated.

A major concern when fractionating lipids which contain 20 : 4 or other polyunsaturated fatty acids is oxidation of the double bonds. Antioxidants like 2,6-di-*tert*-butyl *p*-cresol (BHT) or tocopherol are useful in preventing oxidation, but they are not necessarily sufficient protection since they are frequently separated from the lipids during the fractionation. Moreover, in large quantities the antioxidants and their oxidation products can interfere with the fractionation procedures, especially when the fractionation is

[1] Fatty acids are designated by the notation XX : Y(*n*–*z*). XX is the carbon number of the acyl chain; Y is the number of double bonds; and (*n*–*z*) indicates the location of the double bond, i.e., *n* is the carbon number of the fatty acid and *z* is the location of the first double bond counting from the methyl end of the fatty acid.

being monitored by absorbance in the 195–214 nm range. Whether or not antioxidants are used, a few general precautions are necessary. Always keep the lipids sealed under an inert atmosphere (nitrogen or argon gas) in screw-capped culture tubes with Teflon-lined caps. Ground glass-stoppered tubes are not air tight. Never store lipids dry, and whenever possible, store lipid samples dissolved in chloroform. The oxidation of fatty acids proceeds by a free radical mechanism which allows for the possibility of a polymerization reaction especially with dry or highly concentrated polyunsaturated lipids. Chloroform affords some protection against polymerization.

Extraction of Tissue Lipids

The first step in the fractionation of PLs is to extract the lipids from the tissue. This is accomplished by extracting the tissue with a combination of organic solvents in which (1) the lipids are chemically stable, (2) any enzymes that might metabolize the lipids are denatured and (3) where miscibility with water is sufficient so that a single phase is formed (in order to ensure an efficient extraction). Many methods have been described for the extraction of lipids from tissues (see Ref. 2 for a review of these methods), but only a few of these procedures are in common use. Two of these, Folch *et al.*[3] and Bligh and Dyer,[4] use chloroform/methanol, while one, which has been described by Radin,[5] uses hexane/2-propanol. The choice of methods depends on the source of the lipids. The Folch procedure is the preferred method for the extraction of lipids from tissue, but when extracting larger volumes, the method of Bligh and Dyer is frequently more convenient because it results in a smaller final volume of extract. The hexane/2-propanol system has the advantage of being less toxic, and it can be adapted for the extraction of both tissues and of larger aqueous samples; but to date the procedure has not received wide acceptance. An evaluation of its performance with any specific tissue, therefore, is not yet possible. The method is thoroughly discussed in an earlier volume of this series.[5] None of the above procedures quantitatively extract the polyphosphoinositides and specialized procedures have been developed for this purpose.[6]

[2] N. S. Radin, *in* "Lipids and Related Compounds" (A. A. Boulton, G. B. Baker, and L. A. Horrocks, eds.), p. 61. P. I. Humanan Press, Clifton, NJ 1988.

[3] J. Folch, M. Lees, and G. H. Sloane Stanley, *J. Biol. Chem.* **226,** 497 (1957).

[4] E. G Bligh and W. J. Dyer, *Can. J. Biochem. Physiol.* **37,** 911 (1959).

[5] N. S. Radin, this series, Vol. 72, p. 5.

[6] J. Schacht, this series Vol. 72, p. 626.

Folch Extraction

One gram of tissue is homogenized in a blade- or sonic probe-type homogenizer with 19 ml of chloroform/methanol (2 : 1). (A mortar and pestle or glass/Teflon homogenizer is usually not suitable for homogenization in chloroform/methanol.) Alternatively, the tissue can be homogenized in water or buffer after which 1 ml of the homogenate is extracted with 19 ml of chloroform/methanol (2 : 1). The sample is sealed under nitrogen and allowed to sit for several hours at room temperature to permit complete extraction of the lipids. The extract is then partitioned by the addition of 0.2 volume (4 ml) of saline (0.9% NaCl). The tubes are vigorously shaken and then centrifuged to separate the phases. The upper phase (methanol/water) is removed and discarded. The lower chloroform phase, which contains the PL, is removed with a Pasteur pipette being careful to leave behind the protein that remains at the interface. Although it is not generally necessary (unless there is concern about contamination with radioactive precursors), the lower phase, either before or after transfer, can be washed with 0.4 volume (8 ml) of theroretical upper phase [chloroform/methanol/saline (3 : 48 : 47)] or methanol/saline (1 : 1). The lower phase is then dried under nitrogen and stored in chloroform.

The extracts can be filtered (preferably through a sintered glass filter) to remove particulate material before partitioning. The nonlipid residue and the filter are then washed several times with small amounts of chloroform/methanol (2 : 1). The filtrate and washes are combined before partitioning with 0.2 volume of saline. A final ratio of 1 g tissue to 19 ml chloroform/methanol (2 : 1) must be maintained.

Bligh and Dyer Extraction

For each 2 ml of water, 7.5 ml of chloroform/methanol (1 : 2) is added. After vigorous mixing, an additional 2.5 ml of chloroform is added and, after further mixing, another 2.5 ml of water is added. The extract is again vigorously mixed and filtered if necessary. The extract (or filtrate) is centrifuged to separate the phases and the upper phase discarded. The lower phase is transferred to a clean tube and dried under nitrogen. If the extract is filtered, the nonlipid residue and funnel are washed with chloroform/methanol (1 : 1). The water content of the final filtrate should be adjusted to maintain the ratio of chloroform/methanol/water (5 : 5 : 4.5).

Separation of PL Classes

Neutral lipids (NL) and all the major PL classes, i.e., phosphatidylethanolamine (PE), phosphatidylinositol (PI), phosphatidylserine (PS), cardiolipin (CL), phosphatidylcholine (PC), sphingomyelin (SM), and lyso-PC can be separated by normal-phase (silica) chromatography on a 4.0 mm × 25 cm, 5 μm, LiChrospher Si 100 column (E. Merck, Darmstadt, FRG) with a mobile phase consisting of hexane/2-propanol/ethanol/25 m*M* potassium phosphate (pH 7.0)/acetic acid (485 : 376 : 100 : 56 : 0.275). This mobile phase is saturated with potassium phosphate, therefore it is necessary to filter the mobile phase through a 0.45 μm or finer filter to remove precipitated potassium phosphate.

For most tissues the separation of the major PL classes is straightforward because of the limited number of lipids present and the favorable distribution of the lipid mass among the PL classes. For some tissues, of which the brain is an example, the separation is more difficult because of an unfavorable distribution of the lipid classes (very large amounts of neutral glycosphingolipids and PS) and an unusually wide variety of molecular species within a given PL class which results in broad peaks. However, as can be seen in Fig. 1, this system completely resolves the major PL classes of both brain and liver, but only about one-half as much brain extract as compared to liver extract could be applied to the column and still achieve complete separation of the lipid classes. Several minor, but potentially interesting, PLs, phosphatidylglycerol, lyso-PE, and *N*-monomethyl-PE, all elute together between PE and PI. The retention time of PA is highly variable depending on the particular column and the precise amount of acetic acid and water in the mobile phase. Under the conditions described in Fig. 1, PA elutes just after PS. *N,N*-Dimethyl PE (retention time = 0.4 × PC) elutes after PA.

The order in which the PL classes elute depends primarily on the polarity of the head groups. Therefore, small changes in the composition of the mobile phase, with the exception of the acetic acid, have little effect on the elution sequence. The retention time (or elution volume) depends primarily on the amount of water in the mobile phase. As the amount of water in the mobile phase increases, the retention time of all the PL classes decreases, but the relative position of the PLs, with the exception of PA, is not affected. However, the lipid classes do not resolve as well. Generally, the retention time of PA does not decrease as much as that of the other PL classes as the concentration of water in the mobile phase increases, especially if the water is added after filtering the mobile phase.

A small change in the concentration of acetic acid in the mobile phase, on the other hand, has a profound effect on the retention time of CL. An increase in the amount of acetic acid results in an increase in the retention time of CL, but does not materially affect the retention time of the other major PL classes. Since there are differences in the selectivity of silica columns, even from the same manufacturer, it is generally necessary to

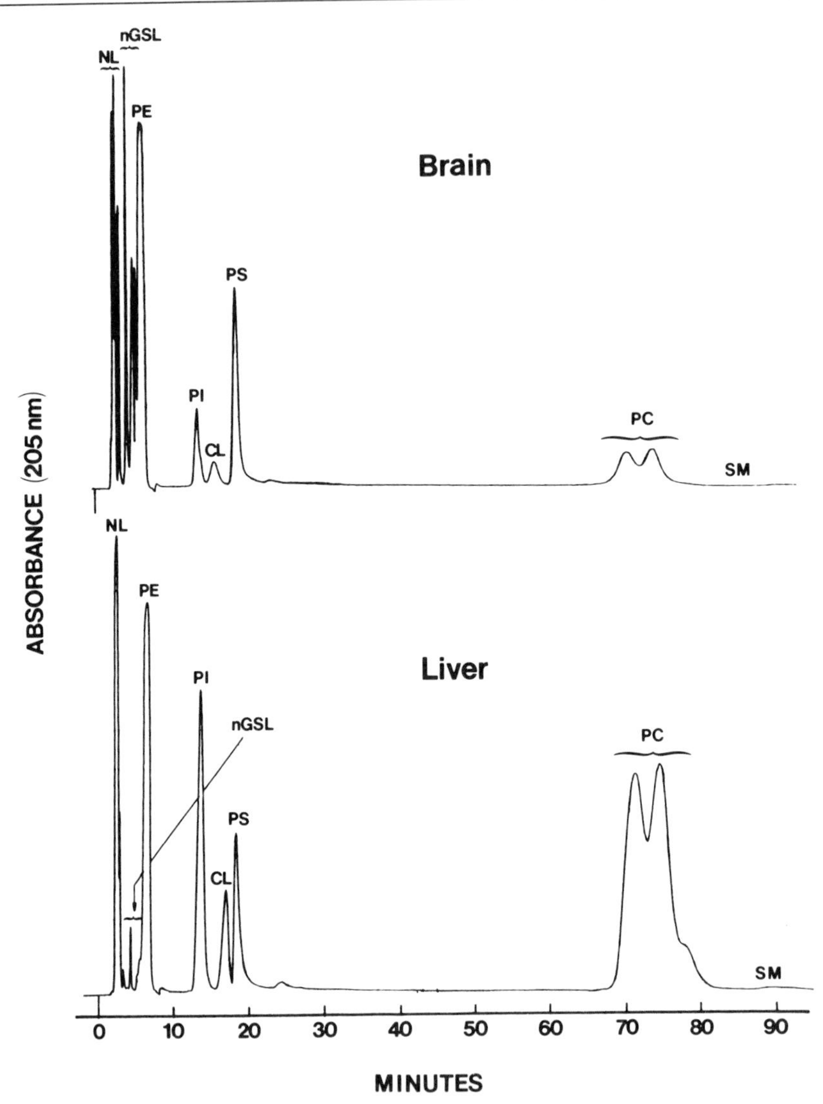

FIG. 1. Separation of PL classes by normal-phase HPLC. One milligram of rat brain PLs or 2 mg of rat liver PLs are dissolved in 100 μl of the mobile phase and injected onto a 4.0 mm $\times$ 25 cm LiChrospher Si 100 (5 μm) column. The column is eluted with hexane/2-propanol/ethanol/25 m*M* potassium phosphate (pH 7.0)/acetic acid (485 : 367 : 100 : 56 : 0.275) at a flow rate of 1.0 ml/min. Detection is by absorbance at 205 nm with an attenuation of 2.56 absorbance units full scale (AUFS). NL, neutral lipids; nGSL, neutral glycosphingolipids; PE, phosphatidylethanolamine; PI, phosphatidylinositol; CL, cardiolipin; PS, phosphatidylserine; PC, phosphatidylcholine; SM, sphingomyelin. Lyso-PC (not shown) has a retention time 2.6 times that of PC.

make small adjustments in the acetic acid concentration with each column in order to maximmize the resolution of PI, CL, and PS. The effect of acetic acid concentration on the retention time of CL can also be used to obtain pure NL, PE, PC, and either PI or PS in a short time by increasing the amount of water in the mobile phase by 10–20 ml/liter and then adjusting the acetic acid concentration such that CL coelutes with either PI or PS. The fraction containing CL plus either PI or PS can then be resolved by rechromatography in a mobile phase with a different acetic acid concentration. A more thorough discussion of factors affecting the elution of the PL classes can be found in Ref. 7.

Separation of Intact Molecular Species

The purified PL classes can be separated into molecular species on a C18 reversed-phase column with a mobile phase composed of 0.35% choline chloride in methanol/water/acetonitrile (90 : 7.5 : 2.5). Although the purified PL fractions from the silia column can be applied to the reversed-phase column directly, it is preferable to Folch partition the purified PLs to remove the potassium phosphate, especially if the samples are to be stored before further fractionation.

The separation of a PL class into molecular species results primarily from the differential solubility of the acyl chains of the PL in the stationary phase, i.e., the hydrocarbon chain of the octadecyl silane. For molecules with no head group or only a slightly polar head group, the chromatography is efficient, and reasonably narrow symmetrical peaks are obtained. As the polarity of the head group increases (PE → PC), the efficiency of the chromatography decreases, i.e., the peaks become wider and less symmetrical (tailing). The decreased efficiency is caused by interactions between the polar groups of molecules immobilized in the stationary phase and those dissolved in the mobile phase as well as between the polar groups of the PL and any exposed silanols (silica) on the reversed-phase column. The addition of a salt, choline chloride, to the mobile phase helps to ameliorate this problem, but it is only partially effective. Moreover, for those PLs with a free carboxyl group (PS) or one or more free phosphate groups (PA and the polyphosphoinositides), the interaction of the polar head groups is so intense that tailing obliterates the molecular species separation. This excessive tailing can be corrected by acidifying the mobile phase with 1 ml of phosphoric acid per liter of mobile phase. However, this highly acidic mobile phase can result in the hydrolysis of one or both of

[7] G. M. Patton and S. J. Robins, *in* "Chromatography of Lipids in Biomedical Research and Clinical Diagnosis" (A. Kuksis, ed.), p. 311. Elsevier, Amsterdam, 1987.

the acyl chains. Therefore, if the intact molecular species are to be recovered for further analysis, it is necessary to immediately neutralize the phosphoric acid with sodium bicarbonate before storing or attempting to concentrate the samples.

Because the major factor governing reversed-phase chromatography of PL is the solubility of the acyl chains in the stationary phase, the retention time of any particular molecular species of either PE, PA, PS, or PC is approximately the same. Because PI and the polyphosphoinositides have a large number of polar sites on the head group, which makes them more soluble in the mobile phase, they have a retention time about one-half that of the other PLs. CL, with four acyl chains, is so soluble in the stationary phase that it effectively does not elute with methanol. Therefore, a less polar mobile phase is required.[8] Rat brian PE, purified as described above, will be used to illustrate a number of the HPLC procedures involved in separating arachidonic acid containing molecular species.

The separation of intact brain PE into individual molecular species is shown in Fig. 2A. The location of the major arachidonyl-containing molecular species as well as some of the other major molecular species are indicated by number in the figures and identified in Table I. Brain PE, like PE from most other tissues, is predominantly a mixture of the diacyl- and alkenylacyl-PE with only traces of alkylacyl-PE. As can be seen (Table I), the diacyl-PE elute before the corresponding alkenylacyl-PE and the alkenylacyl-PE elute before the corresponding alkylacyl molecular species. The major arachidonyl molecular species of brain PE are the diacyl and alkenylacyl 18 : 0–20 : 4 with lesser but still considerable amounts of both diacyl and alkenylacyl 16 : 0–20 : 4 and 18 : 1–20 : 4. PE is unique among the PL classes in that it contains large amounts of molecular species with two unsaturated acyl groups. Because one of the acyl groups is usually 18 : 1, the 18 : 1–22 : 6, 18 : 1–20 : 4, and 18 : 1–18 : 1 molecular species are relatively major components. Furthermore, in brain as well as many other tissues, there is little, if any, diacyl or alkenylacy-PE 16 : 0–18 : 2 which would coelute with the 18 : 1–20 : 4 molecular species of the same subclass. Therefore, the 18 : 1–20 : 4 peaks are reasonably pure. In contrast, both the 16 : 0–20 : 4 and 18 : 0–20 : 4 molecular species are heavily contaminated with 18 : 1–22 : 6 and 18 : 1–18 : 1, respectively. Brian PE is also unusually rich in 22 : 4(*n*–6) and, consequently, 18 : 0–20 : 4 is also heavily contaminated with 16 : 0–22 : 4, which is the major 22 : 4-containing molecular species. In most other PL classes, the opposite would be true, i.e., the 16 : 0–20 : 4 and 18 : 0–20 : 4 peaks would be relatively pure. The 16 : 0–20 : 4 peak would be contaminated with only trace

[8] J. I. Teng and L. L. Smith, *J. Chromatogr.* **339**, 35 (1985).

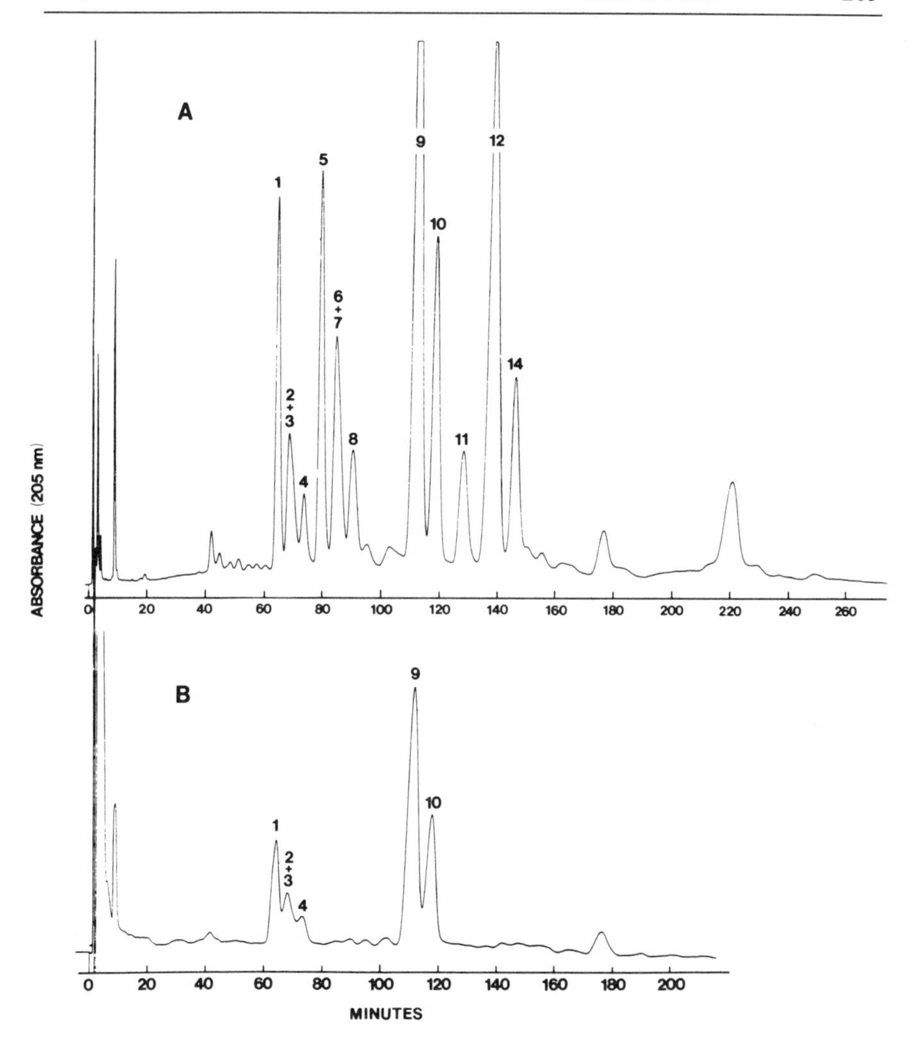

FIG. 2. Molecular species separation of intact brain PE by reversed-phase chromatography. (A) Total PE (50 μg); (B) diacyl-PE (20 μg), dissolved in 100 μl of methanol is injected onto a 2 mim × 25 cm Ultrasphere ODS (5 μm) column and eluted with 0.35% choline chloride in methanol/water/acetonitrile (90 : 7.5 : 2.5) at a flow rate of 0.3 ml/min. Detection is by absorbance at 205 nm with an attenuation of 0.16 AUFS. The peaks are identified by number in Table I.

amounts of 18 : 1–22 : 6, 16 : 0–16 : 1, and 18 : 1–16 : 1, and the 18 : 0–20 : 4 peak would contain only trace amounts of 18 : 1–18 : 1 and 16 : 0–22 : 4, but the 18 : 1–20 : 4 peak would be heavily contaminated with 16 : 0–18 : 2. In addition to the major diacyl and alkenylacyl arachidonyl species (16 : 0–

TABLE I
MAJOR MOLECULAR SPECIES SEPARATED FROM INTACT BRAIN PHOSPHATIDYLETHANOLAMINE

Peak no.	Major components[a]	Notation		
1	Diacyl	16:0–22:6		
2	Diacyl	18:1–22:6		
3	Diacyl	16:0–20:4		
4	Diacyl	18:1–20:4		
5	Alkenylacyl	16:0–22:6		
6	Alkenylacyl	18:1–22:6		
7	Alkenylacyl	16:0–20:4		
8	Alkenylacyl	18:1–20:4		
9	Diacyl	18:0–22:6	16:0–18:1	
10	Diacyl	18:0–20:4	16:0–22:4	
11	Unidentified			
12	Alkenylacyl	18:0–22:6		
13	Alkenylacyl	16:0–18:1		
14	Alkenylacyl	18:0–20:4	16:0–22:4	18:1–18:1
15	Unidentified			
16	Diacyl	18:0–18:1		
17	Alkylacyl	18:0–20:4		

[a] The precise distribution of the fatty acid between the *sn*-1 and *sn*-2 positions of glycerol are not known, but since 18:0, 16:0, and 18:1 are the major fatty acid at the *sn*-1 position, the molecular species will be referred to as though the least unsaturated fatty acid is in the *sn*-1 position.

20:4, 18:1–20:4, and 18:0–20:4), there are also a number of minor alkenylacyl arachidonyl molecular species (18:2–20:4, 14:0–20:4, 15:0–20:4, 16:1–20:4) which elute in the vincinity of the diacyl 16:0–20:4 and 18:1–20:4 species. Likewise, to the extent that such molecular species exist, the minor alkylacyl-PE will coelute with alkenylacyl 16:0–20:4 and 18:1–20:4. In addition, the 16:0–20:4 and 18:1–20:4 alkylacyl species will elute with the diacyl 18:0–20:4. The problem of coelution of the subclasses is most acute with PE, because in most tissues it is particularly rich in both diacyl and alkenylacyl molecular species. However, depending on the source of the lipids, significant amounts of alkylacyl and alkenylacyl lipids can also occur in the other PL classes, particularly PC. Thus, with intact PLs it is difficult to obtain truly pure arachidonyl molecular species, both because of contamination with other minor arachidonyl species from a different subclass as well as other non-arachidonyl species from the same or different subclass.

It is possible to reposition the molecular species to some extent by increasing the concentration of acetonitrile in the mobile phase. Acetonitrile forms a complex with the double bonds of the fatty acids, making the fatty acid (and the molecular species) more soluble in the mobile phase, thereby decreasing the retention time of a molecular species in proportion to the number of double bonds in its acyl groups. In general, however, increasing the proportion of acetonitrile in the mobile phase is not a satisfactory solution to the problem of contamination of the intact arachidonyl molecular species. Although the 18 : 1–22 : 6 will move away from the 16 : 0–20 : 4 and the 16 : 0–20 : 4 will move away from the other main contaminants (16 : 0–16 : 1 and 16 : 1–18 : 1), the 18 : 1–20 : 4 will move into the 16 : 0–20 : 4. Moreover, while the 18 : 0–20 : 4 will move away from its main contaminant, 18 : 1–18 : 1, it moves into the 16 : 0–18 : 1 and if the amount of acetonitrile is increased further it will eventually move into the 16 : 0–16 : 0. Furthermore, 16 : 0–22 : 4, the other major contaminant of 18 : 0–20 : 4, will continue to elute with the 18 : 0–20 : 4. In addition, the alkenylacyl 16 : 0–22 : 6 will move into the diacyl 16 : 0–20 : 4, and the alkenylacyl 18 : 0–22 : 6 will move into the diacyl 18 : 0–20 : 4. Moreover, as the proportion of acetonitrile in the mobile phase increases, tailing becomes an increasing problem.

While it is not possible to separate the intact PL classes into subclasses, it is possible to selectively degrade the alkenylacyl subclass thereby yielding relatively pure diacyl-PL. (The alkylacyl subclass is usually, but not always, only a minor component of a PL class). Figure 2B shows the separation of brain diacyl- and alkylacyl-PE molecular species after the alkenylacyl-PE was decomposed by exposure to HCl gas. Exposure of dried alkenylacyl lipids to HCl vapors for 10 min results in cleavage of the vinyl ether bond with little or no degradation of the diacyl or alkylacyl lipid. The breakdown products of the alkenylacyl-PE elute near the solvent front. With the alkenylacyl-PE eliminated, the minor diacyl molecular species which coelute with the alkenylacyl 16 : 0–20 : 4 and 18 : 1–20 : 4 can be seen. Among these are the diacyl 16 : 0–20 : 3 and 18 : 1–20 : 3 as well as the 16 : 0–22 : 5(*n*–6) and 18 : 1–22 : 5(*n*–6) molecular species which can arise by elongation and desaturation of arachidonic acid. Likewise, the alkenylacyl 18 : 0–20 : 4 is contaminated with diacyl 18 : 0–20 : 3(*n*–6) and 18 : 0–22 : 5(*n*–6).

Separation of Derivatized Molecular Species

Because the resolution of intact molecular species is not always adequate, especially with lipids that contain a mixture of subclasses, it is frequently advantegous to convert the PL into a nonpolar derivative and then separate the lipid class into its subclasses. The first step in the derivatization procedure is the conversion of the PL into diacylglycerides

(DG) by hydrolysis with phospholipase C. The DG are then converted into a suitable nonpolar derivative. This procedure has the added advantage of improving the resolution of the molecular species and, if the DG are derivatized with a UV-absorbing chromophore, of permitting the quantitation of each subclass and/or molecular species by integration of the area under the peaks as they elute from an appropriate column.

Many methods have been described for the hydrolysis of lipids with phospholipase C, but not all of the methods are suitable for this application since some procedures yield a mixture of 1,2- and 1,3-DG. The hydrolysis of the common glycero-PLs and lyso-PLs (with the exception of the inositol-containing lipids) can be accomplished by enzymatic hydrolysis with phospholipase C as described below. The inositol-containing lipids are converted to DG by hydrolysis with PI-specific phospholipase C. Sphingomyelin can be converted to ceramide by hydrolysis with sphingomyelinase[9] and the resulting ceramides can then be derivatized by the same procedures as the DG.

Phospholipase C Hydrolysis of PL

The lipids to be hydrolyzed (PE, PA, PS, PC, CL,[10] or their corresponding lyso-PL) are dried under nitrogen in a 15 ml screw-capped culture tube. Two milliliters of diethyl ether and 6 units of *Bacillus cereus* phospholipase C (Boehinger Mannheim, type I) dissolved in 0.5 ml of 50 m*M* potassium phosphate buffer (pH 7.0) are added to the tube. The tube is sealed under nitrogen and vigorously shaken for 1 min on a vortex mixer. (If large amounts of PA are being hydrolyzed it may be necessary to shake for 2 or 3 min.) The reaction mixture is then extracted three times with 4 ml of hexane. The hexane phases, which contain the DG, are pooled and dried under nitrogen at room temperature. Heating the hexane extract while drying can promote the rearrangement of the DG.

PI-Specific Phospholipase C Hydrolysis of PI

PI is dried under nitrogen in a 15-ml screw-capped culture tube. Two milliliters of diethyl ether/benzene (1 : 1), and 0.25 units of PI-specific phospholipase C (Boehringer Mannheim) dissolved in 0.5 ml of 25 m*M* potassium phosphate buffer (pH 7.0) are added to the tube. The tube is sealed under nitrogen and vigorously shaken for 2 min and then extracted three times with 4 ml of hexane.

[9] H. Ikezawa, M. Mori, T. Ohyabu, and R. Taguchi, *Biochim. Biophys. Acta* **528,** 247 (1978).

[10] The phospholipase C hydrolysis of CL results in the formation of two molecules of DG for each molecule of CL. Thus, the actual molecular species of CL cannot be determined by this procedure.

Purification of DG

If the relative amount of the derivatized subclasses is to be quantitated by UV absorbance, then it is necessary to purify the DG before derivatization in order to remove impurities that interfere with the quantitation of the alkylacyl subclass. Otherwise, the DG can be derivatized directly. The DG can be purifed by chromatography on a LiChrospher Si 100 column with hexane/tetrahydrofuran/acetic acid (500 : 50 : 0.1) as the mobile phase (Fig. 3). The DG elute as two clusters of peaks. The first cluster, which is the 1,2-DG, contains both the alkylacyl- and alkenylacyl-DG. The second cluster contains the diacyl-DG which, under the conditions of the phospholipase C hydrolysis, have undergone an intramolecular rearrangement and are recovered as 1,3-DG. If only one or the other of the DG is required or if there is no need to quantitate the subclasses, then the 1,2- and 1,3-DG can be collected separately. If, however, it is necessary to quantitate the subclasses, then it is best to collect all the DG together.

Benzoylation of DG

The DG can be converted to any one of a number of derivatives depending on the requirements of a particular experiment. However, since the alkenylacyl-DG decomposes in acidic organic solvents, it is necessary to avoid any derivatization procedure which is acidic or which generates unneutralized acid. Moreover, if it is necessary to separate a lipid class into subclasses, then it is also necessary to derivatize the DG with a nonpolar chromophore. The separation of the subclasses is based on differences in the polarity of the bonds at the *sn*-1 position of glycerol and is accomplished by silica gel chromatography with a nonpolar mobile phase. With a polar chromophore, the polarity of the mobile phase would have to be increased in order to elute the lipids. With a polar mobile phase, the small differences in the polarity of the bonds at the *sn*-1 position of glycerol may not be sufficient to permit separation of the subclasses. Nakagawa and Horrocks,[11] who initially described the separation of the subclasses, used DG acetates, but the benzoyl derivative works equally as well.

The DG can be converted to the benzoyl ester as described by Ullman and McCluer[12] for the perbenzoylation of neutral glycosphingolipids.[13] The DG is dried under nitrogen in a small (8 ml) screw-capped culture tube and 0.5 ml of 10% benzoyl chloride in dry pyridine is added. The tube is

[11] Y. Nakagawa, and L. A. Horrocks, *J. Lipid Res.* **24,** 1268 (1983).

[12] M. D. Ullman, and R. H. McCluer, *J. Lipid Res.* **48,** 371 (1977).

[13] When using this procedure to derivatize the ceramide resulting from the hydrolysis of sphingomyelin with sphingomyelinase, both the free hydroxyl group and the amide nitrogen of the ceramide are benzoylated.

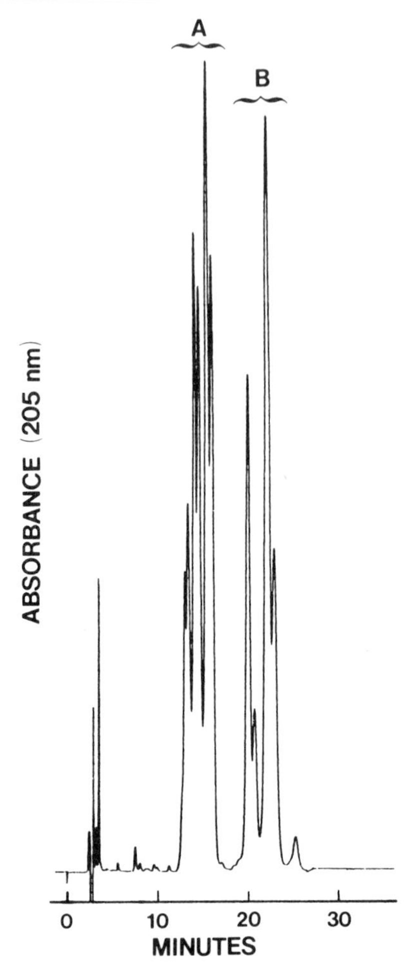

FIG. 3. Purification of DG by normal-phase chromatography. DG, obtained by phospholipase C hydrolysis of rat brain PE, are dissolved in 100 μl of hexane and injected onto a 4.0 mm $\times$ 25 cm LiChrospher Si 100 (5 μm) column and eluted with hexane/tetrahydrofuran/acetic acid (500 : 50 : 0.1) at a flow rate of 1.0 ml/min. Detection is by absorbance at 205 nm with an attenuation of 1.28 AUFS. (A) 1,2-DG; (B) 1,3-DG.

sealed under nitrogen, mixed thoroughly, and incubated at 37° for 16 hr. The excess pyridine and benzoyl chloride are then evaporated under nitrogen and 3 ml of hexane are added to the residue followed by 2 ml of methanol/water (80 : 20) saturated with sodium carbonate. After thorough mixing, the tube is centrifuged to separate the phases and the lower phase (methanol/water) is removed and discarded. The hexane phase is washed

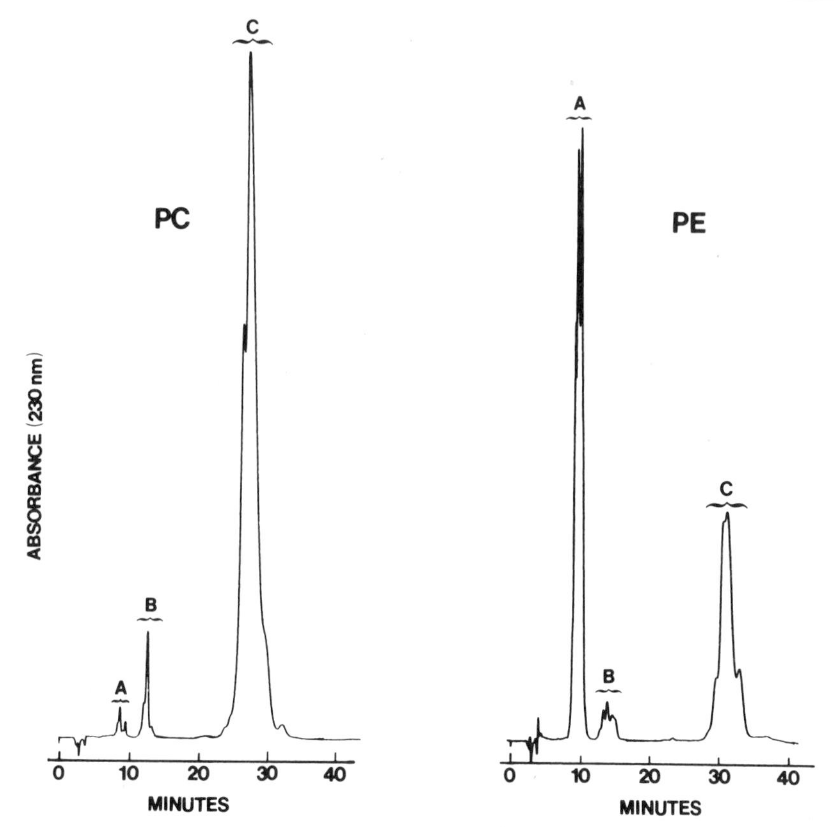

FIG. 4. Separation of benzoyl-DG into subclasses by normal-phase chromatography. Benzoyl-DG derived from rat brain PC and PE is dissolved in 100 μl of hexane, and injected onto a 4.0 mm × 25 cm LiChrospher Si 100 (5 μm) column and eluted with cyclohexane/hexane/methyl *tert*-butylether/acetic acid (375 : 125 : 20 : 0.1) at a flow rate of 1.0 ml/min. Detection is by absorbance at 230 nm with an attenuation of 1.28 AUFS. (A) Alkenylacylbenzoyl-DG; (B) alkylacylbenzoyl-DG; and (C) diacylbenzoyl-DG.

three more times with 2 ml of the sodium carbonate saturated methanol/water and then twice with 2 ml of methanol/water (80 : 20). The hexane layer which contains the benzoyl-DG is removed to another tube and dried under nitrogen.

Separation of Benzoyl-DG into Subclasses

The separation of the benzoylated PLs into the diacyl, alkenylacyl, and alkylacyl subclasses is performed essentially as described by Nakagawa and Horrocks.[11] Figure 4 shows the separation of the benzoylated subclasses of brain PC and PE by chromatography on a 4.0 mm × 25 cm, 5 μm, LiChrospher Si 100 column eluted with cyclohexane/hexane/methyl *tert*-butylether/acetic acid (375 : 125 : 20 : 0.1). On a column packed with 5 or 10 μm spherical particles, each subclass resolves into a cluster of peaks. On a column packed with either 10 or 30 μm irregular particles, each subclass elutes as a single peak. If the relative amount of each subclass is to be determined by integration, then the amount of material applied to the column should be reduced such that the absorbance for the largest peak is no more than 0.2 *A* in order to assure linearity of the detector response.

Molecular Species Separation of Benzoyl-DG

Finally, the purified benzoylated subclasses can be separated into molecular species by reversed-phase chromatography with methanol/water/acetonitrile (946.5 : 38.5 : 15) as the mobile phase. The separation of total brain PE (A) and the three individual subclasses (B, diacyl; C, alkenylacyl; D, alkylacyl) is shown in Fig. 5. When examining the molecular species pattern of the subclasses, it is clear that all three subclasses contain essentially the same major molecular species, although the relative amounts of the molecular species differ among the subclasses (see ref. 11 for a detailed analysis of the brain PE molecular species). The factors which affect the separation of the derivatized DGs are essentially the same as those which affect the separation of the intact molecular species, i.e., the composition of the acyl chains. However, since there are no charged groups, there is no need for choline chloride in the mobile phase and, since the benzoyl-DG are much less polar than the intact PLs, the amount of water in the mobile phase is reduced. The resolution of the benzoylated DG is significantly better than the resolution obtained with intact PE (Fig. 2). In particular, the 16 : 0–20 : 4 is clearly resolved from 18 : 1–22 : 6. Likewise, the 18 : 1–20 : 4 resolves from the 16 : 0–18 : 2 (the minor peak eluting immediately after 18 : 1–20 : 4). However, derivatization still does not permit the separation of 16 : 0–22 : 4 from 18 : 0–20 : 4 and, with this mobile phase, 18 : 0–20 : 4 is still contaminated with 18 : 1–18 : 1. Even though the resolution of the arachidonyl molecular species appears reasonably good, except for the contamination of the 18 : 0–20 : 4 peak, there are, in fact, a large number of extremely minor components contaminating the peaks. The extent of this contamination can be estimated by examining Fig. 6, which shows a portion of a high-resolution molecular species separation of total benzoylated PE, diacyl-PE and alkenylacyl-PE from rat brain. At this level of resolution, the 18 : 1–20 : 4 peak (peaks 4 and 8) partially resolves into its two major components, the 18 : 1(*n*–7)–20 : 4 (first peak) and the 18 : 1(*n*–9)–20 : 4. Moreover, 18 : 1–18 : 1 (front peak)

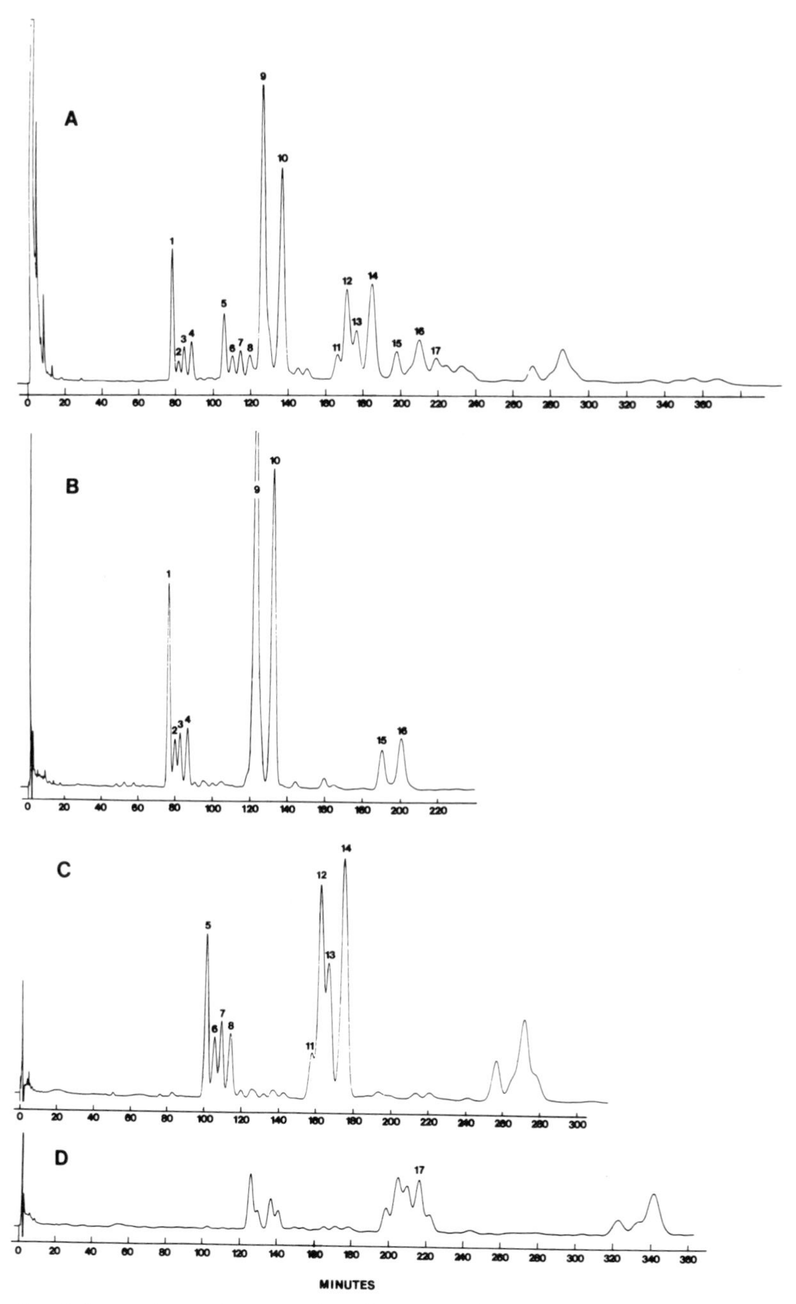

FIG. 5. Molecular species separation of benzoylated brain PE by reversed-phase chromatography. Total (A), diacyl-(B), alkenylacyl-(C), and alkylacyl-(D) benzoyl-DG are dissolved in methanol and 100 μl is injected onto a 2.0 mm × 25 cm Ultrasphere ODS (5 μm) column and eluted with methanol/water/acetonitrile (946.5 : 38.5 : 15.0) at a flow rate of 0.3 ml/min. Detection is by absorbance at 230 nm with an attenuation of 0.08 AUFS. The numbered peaks are identified in Table I.

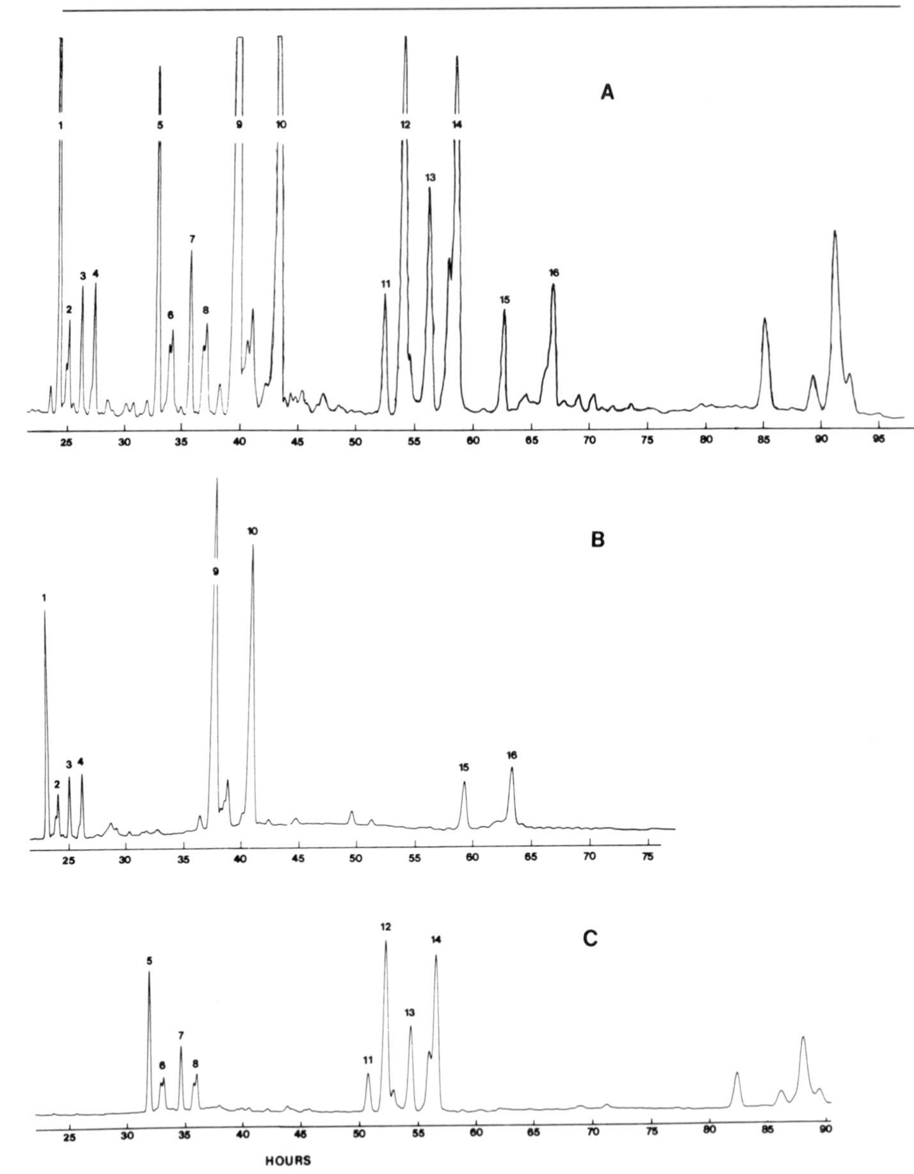

FIG. 6. High-resolution molecular species separation of benzoylated brain PE by reversed-phase chromatography. Total (A), diacyl- (B), and alkenylacyl- (C) benzoyl-DG are dissolved in methanol and 100 μl is injected onto six, 2 mm × 25 cm, Ultrasphere ODS (5 μm) columns connected in series and eluted with methanol/water/acetonitrile (946.5 : 38.5 : 15.0) at a flow rate of 0.1 ml/min. Detection is by absorbance at 230 nm with an attenuation of 0.16 AUFS. The numbered peaks are identified in Table I.

partially resolves from 18 : 0–20 : 4. However, 16 : 0–22 : 4 still cannot be separated from the 18 : 0–20 : 4. There are also a large number of minor unidentified molecular species that elute almost continuously throughout the chromatogram. Based on the fatty acid analysis of tissue lipids, the presence of these minor moleculr species would be expected, since in addition to the relatively small number of major fatty acids [16 : 0, 18 : 0, 18 : 1(*n*–9), 18 : 2(*n*–6), 20 : 3(*n*–6), 20 : 4(*n*–6), 20 : 5(*n*–3), 22 : 4(*n*–6), 22 : 5(*n*–6), 22 : 6(*n*–3)], there is also a very large number of minor fatty acids (at least 50) including odd-carbon number fatty acids, branched-chain fatty acids, and numerous positional isomers of the unsaturated fatty acids with both cis and trans double bonds.[14] While these minor molecular species are of little concern when determining the amount of material present in a major archidonyl peak, they are of potential concern when collecting peaks for determining radioactivity, since many of these peaks will contain arachidonic acid or metabolites of arachidonic acid, i.e., 22 : 4(*n*–6) and 22 : 5(*n*–6).

Comments about Methods

1. Virtually all modern HPLC equipment when functioning properly is more than adequate for all of the chromatography described here. While much of the modern HPLC equipment is gradient capable, all the chromatography systems described here are isocratic, but they are readily adaptable to gradient chromatography with two restrictions. (a) The PL class separation is not readily adaptable to a gradient system if the resolution of PI, PS, and CL is required, because of the inordinate amount of time required to completely equilibrate the column after a change in the composition of the mobile phase. Changes in the composition of the mobile phase alter the partitioning of the acetic acid between the column and the mobile phase, and PI, PS, and CL cannot be separated reliably unless the column is completely equilibrated with acetic acid. (b) Gradient elution is also not recommended if the amount of material is to be quantitated by integration, because gradient elution continually alters the size of the peaks for a given amount of material. Although it is possible to compensate for this effect, in practice greater precision is obtained with isocratic elution.

2. Every silica column is unique and the differences between columns from different manufacturers can be quite significant in the separation of the PL classes. Although some silica columns other than LiChrospher Si 100 columns will undoubtedly resolve the PL classes, others will not. Thus, if purchasing a new column, a LiChrospher Si 100 (or some other

[14] G. M. Patton, S. Cann, H. Brunengraber, and J. M. Lowenstein, this series, Vol. 72, p. 8.

column that is known to work) is strongly recommended for that application. For the other normal-phase separations, virtually any good quality silica column is adequate. Likewise, although there are slight differences in selectivity among reversed-phase columns, virtually any reversed-phase column will resolve the molecular species, especially the benzoyl-DG. The major characteristic to consider in choosing a reversed-phase column is the efficiency of the column, i.e., the number of theoretical plates/column (plates/meter × length). Thus a good 25 cm, 5 μm particle size column will have 15,000 plates (i.e., 60,000 plates/m × 0.25 m) which is about the same number of plates as a good 15 cm, 3 μm particle column (100,000 plates/m × 0.15 m). For the separation of intact molecular species (but not the derivatized molecular species), the carbon load (i.e., percentage carbon) of the reversed-phase column is also a consideration. A higher carbon load generally implies fewer free silanols. Free silanol groups are a major cause of tailing in that application. The silica columns can all be stored virtually indefinitely in any of the mobile phases with which the column is being used. Likewise, with the exception of the highly acidic (phosphoric acid) mobile phase used for the molecular species separation of the PS, PA, and the polyphosphoinositides, the reversed-phase columns can be stored in the mobile phase. For the chromatography systems described here, a guard column is not necessary and is not recommended, but a screen-type filter is highly recommended.

3. When injecting a sample onto an HPLC column, it is best to dissolve the sample in the mobile phase or a solvent very similar to the mobile phase. This practice avoids two problems. (a) If the sample is only sparingly soluble in the mobile phase (as is the case with benzoylated DG) and it is injected in a solvent in which it is very soluble, some of the sample will come out of solution when the injection solvent is diluted by the mobile phase. This results in a pure component eluting as two peaks or a peak with a front shoulder. (b) With whole lipid extracts (either Folch or Bligh and Dyer, but not with the Radin procedure), there is nonlipid material (mostly proteins) which is soluble in chloroform but which is not soluble in the mobile phase. If this material is injected onto the HPLC column in chloroform, the nonlipid material precipitates on the column. This causes a rapid increase in the back-pressure of the column and, as the protein accumulates on the column, alters the chromatographic characteristics of the column.

4. Quantitation of the PL classes is generally accomplished by phosphorus analysis of the total lipid extract after partitioning, and of the purified PL classes. If the PL classes are to be quantitated by phosphorus analysis it is necessary to Folch partition the purified lipid classes and wash the Folch lower phase three times to remove all the phosphate in the

mobile phase. Likewise, if intact molecular species from the reversed-phase column are to be quantitated by phosphorus analysis, it is necessary to Folch partition the fractions to reduce the amount of choline chloride because large amounts of this salt interfere with the phosphorus determination. If, on the other hand, the PL classes are going to be derivatized, then quantitation can be accomplished by adding an internal standard to the initial lipid extract before filtering or partitioning. This is particularly helpful with small amounts of material since the phosphorus determination requires a relatively large amount of material.

Since suitable internal standards are not available for all the PL classes, an indirect method of quantitation is required. As an example, an internal standard (di–14 : 0–PC) is added to the initial lipid extract, and after the PL classes are separated a second internal standard (di–16 : 1–PC) is added to each PL class, including the PC. (Since the purified PL classes will be hydrolyzed, the second internal standard need not be of the same class as the PL being quantitated.) When the PC molecular species are analyzed, the ratio of the first internal standard (di–14 : 0) to total C indicates the amount of PC originally present in the lipid extract. When the molecular species of the other PL classes (and PC) are analyzed, the second internal standard (di–16 : 1) permits quantitation of the individual classes relative to the amount of PC. If the relative amount of the subclasses is to be quantitated by normal-phase chromatography and integration, then, after the molecular species separation, the area of the diacyl class must be corrected for the amount of internal standard added. If an internal standard is added to PI, then both phospholipase C- and PI-specific phospholipase C must be included in the PI hydrolysis reaction.

5. Extreme care must be exercised when using [5,6,8,9,11,12,14,15-^{3}H] 20 : 4 in studies of the metabolism of arachidonyl molecular species. Those molecular species containing the heavily tritiated 20 : 4 elute significantly before the corresponding unlabeled molecular species. Thus, depending on the efficiency of the chromatography system, the ^{3}H-labeled 20 : 4 molecular species elute as a separate peak (benzoylated DG) or at the leading edge of the unlabeled peak (intact molecular species). [^{14}C]arachidonic acid does not exhibit this behavior.

Acknowledgments

This work was supported by the General Medical Research Service of the Veterans Administration and National Institute of Health Grant AM28640.

[24] Macrophage Phospholipase A_2 Specific for *sn*-2-Arachidonic Acid

By CHRISTINA C. LESLIE

The availability of free arachidonic acid is thought to be the rate-limiting step in the production of the eicosanoids. Because most arachidonic acid is esterified at the *sn*-2 position of phospholipid, phospholipase A_2 is thought to play a central role in providing arachidonic acid for subsequent metabolism. In addition, phospholipase A_2-mediated hydrolysis of arachidonic acid from the ether-linked membrane phospholipid, 1-*O*-alkyl-2-arachidonoylglycerophosphocholine (−GPC), is the first step in the production of the potent phospholipid mediator platelet-activating factor (PAF) in inflammatory cells. The lyso-PAF formed is then acetylated to PAF. Compared to diacyl-GPC, the 1-*O*-alkyl-linked species is enriched in arachidonic acid in neutrophils and macrophages. Consequently, the eicosanoids and PAF can be coordinately produced from a common phospholipid precursor through the initial action of a phospholipase A_2. Recently, a few studies have described phospholipase A_2 enzymes in macrophages,[1,2] neutrophils,[3] and platelets,[4] that exhibit relative specificity for *sn*-2-arachidonic acid. The assay of this arachidonoyl-specific phospholipase A_2 from the macrophage cell line, RAW 264.7 and the preparation of the 1-*O*-hexadecyl-2-[^{3}H]arachidonoyl-GPC substrate will be described.

Assay Method: Hydrolysis of [^{3}H]Arachidonic Acid from 1-*O*-Hexadecyl-2-[^{3}H]Arachidonoyl-GPC

Assay Mixture

Final concentrations in the assay are Tris-HCl buffer, 50 m*M*, pH 8.0; 1-*O*-Hexadecyl-2-[^{3}H]arachidonoyl-GPC, 30 μM (100,000 dpm); $CaCl_2$, 1 n*M*–10 m*M*; Enzyme: 50 μg of the 100,000 *g* crude soluble fraction or less depending on the stage of purification. Final assay volume is 100 μl when crude enzyme is used and 50 μl for more purified preparations.

[1] C. C. Leslie, D. R. Voelker, J. Y. Channon, M. W. Wall, and P. T. Zelarney, *Biochim. Biophys. Acta* **963,** 476 (1988).
[2] I. Flesch, B. Schmidt, and E. Ferber, *Z. Naturforsch.* **40c,** 356 (1985).
[3] F. Alonso, P. M. Henson, and C. C. Leslie, *Biochim. Biophys. Acta* **878,** 273 (1986).
[4] D. K. Kim, I. Kudo, and K. Inoue, *J. Biochem.* **104,** 492 (1988).

Procedure

The radiolabeled 1-*O*-hexadecyl-2-[^{3}H]arachidonoyl-GPC (prepared as described below) is adjusted to 100,000 dpm/3 nmol with unlabeled 1-*O*-hexadecyl-2-arachidonoyl-GPC (Biomol Research Laboratories, Philadelphia, PA) and enough for several assays is evaporated under nitrogen to remove the solvent (chloroform/methanol, 90/10, v/v). Tris-HCl (50 m*M*, pH 8.0) is added to make a concentration of lipid that is three to five times the final concentration in the assay and the lipids are solubilized by sonication for 4 min at 4° using a microprobe (Braun Instruments, Burlingame, CA). Aliquots of the sonicated lipids are counted to determine the percentage liposome formation (usually 90–95%). The radiolabeled substrate and calcium solutions are aliquoted into assay tubes (disposable, 13 × 100 mm screw-cap tubes with Teflon caps), the tubes are purged with nitrogen, capped, and then placed in a 37° water bath. The reaction is started by the addition of enzyme and incubated for 1 min with shaking. The reaction is stopped with 2.5 ml of Dole[5] reagent (2-propanol/heptane/0.5 *M* H_2SO_4, 20/5/1, v/v/v) which is followed by the addition of 1.5 ml heptane and 1 ml water. Unlabeled arachidonic acid (20 μg) is also added as cold carrier. After vortexing the extraction mixture, the hydrolyzed [^{3}H]arachidonic acid is separated from 1-*O*-hexadecyl-2-[^{3}H]arachidonoyl-GPC by silicic acid column chromatography. Column reservoirs (1 ml) with frits (Analytichem International, Harbor City, CA) containing approximately 300 mg silicic acid (100–200 mesh Unisil, Clarkson, Williamsport, PA) are used in a Visiprep solid-phase extraction vacuum manifold (Supelco, Bellefonte, PA). The upper phase of the Dole extraction is passed through the column and then the column washed with 2 ml of chloroform (diethyl ether can also be used). The eluent is collected in scintillation vials, dried, and then counted by liquid scintillation spectrometry.

Preparation of 1-O-Hexadecyl-2-[^{3}H]arachidonoyl-GPC

Procedure. To form 1-*O*-hexadecyl-2-[^{3}H]arachidonoyl-GPC, 1-*O*-hexadecyllyso-GPC (Bachem Feinchemikalien, Bubendorf, Switzerland) is incubated with [5,6,8,9,11,12,14,15-^{3}H]arachidonic acid (95 Ci/mmol, E.I. duPont, Boston, MA) in the presence of rat liver microsomes. Unlabeled arachidonic acid (25 mg) and 50 μCi of [^{3}H]arachidonic acid are added together into a 13 × 100 disposable glass screw-cap tube and the solvents evaporated under nitrogen. NaOH (100 μl of a 0.1 *N* solution) is then added and vortexed to make the sodium salt. 1-*O*-Hexadecyllyso-

[5] V. P. Dole and H. Meinertz, *J. Biol. Chem.* **235,** 2595 (1960).

GPC is evaporated under nitrogen, resuspended in 0.1 *M* Tris-HCl, pH 7.4, and then 150 μg added to the sodium salt of [^{3}H]arachidonic acid. The following components are then added (final concentrations in a 1.0 ml assay volume): 5 m*M* ATP, 2 m*M* $MgCl_2$, 0.15 m*M* coenzyme A, 0.1 *M* Tris-HCl, pH 7.4, and 500 μg microsomal protein. The reaction mixture is purged with nitrogen, vortexed, and then incubated for 6 to 8 hr at 28°. The reaction mixture is then extracted according to Bligh and Dyer.[6] The 1-*O*-hexadecyl-2-[^{3}H]arachidonoyl-GPC is isolated by reversed-phase HPLC using an Ultrasphere ODS Altex column (5 μm, 4.6 mm id × 25 cm) (Rainin Instrument Co., Emeryville, CA) maintained at 30° as described by Patton *et al.*[7] A mobile phase of methanol/acetonitrile/water (90 : 5 : 5,v/v/v) containing 20 m*M* choline chloride is used at a flow rate of 2.2 ml/min. All solvents are HPLC grade and filtered and deaerated before use. Absorbance is monitored at 206 nm. The column is washed thoroughly overnight with methanol (500 ml) followed by mobile phase (250 ml) before injecting the sample. The sample is dried down and resuspended in mobile phase (500 μl) and injected. The 1-*O*-hexadecyl-2-[^{3}H]arachidonoyl-GPC fraction is identified and quantitated by comparing the retention time and peak height to known amounts of standard 1-*O*-hexadecyl-2-arachidonoyl-GPC, which elutes at approximately 29–31 min (Fig. 1). Routinely, 90–95% of the [^{3}H]arachidonic acid is converted to 1-*O*-hexadecyl-2-[^{3}H]arachidonoyl-GPC resulting in a specific activity of approximately 0.3 Ci/mmol. With this high degree of conversion the 1-*O*-hexadecyl-2-[^{3}H]arachidonoyl-GPC is the major lipid peak observed. The other lipid peaks are derived from the liver microsomes. The fractions containing the greatest amount of 1-*O*-hexadecyl-2-[^{3}H]arachidonoyl-GPC are extracted and stored at −70° under nitrogen.

Preparation of Rat Liver Microsomes

A fresh rat liver is placed in a beaker with approximately 30 ml of cold STE buffer (0.25 *M* sucrose, 10 m*M* Tris-HCl, pH 7.4, 1.0 m*M* EDTA), minced with scissors, and then homogenized in a Polytron for 15 sec (twice). The homogenate is centrifuged at 1500 *g* for 10 min at 4°. The supernatant is filtered through gauze and the filtrate centrifuged at 16,000 *g* for 15 min at 4° to remove mitochondria. The supernatant is centrifuged at 100,000 *g* for 45 min at 4° to obtain a microsomal pellet. The microsomes are washed twice in STE buffer by centrifugation at 100,000 *g* for 45 min at 4°. After the last wash the microsomes are resuspended in STE buffer and frozen in small aliquots at −20°. Aliquots are thawed only once and then

[6] E. G. Bligh and W. J. Dyer, *Can. J. Biochem. Physiol.* **39,** 911 (1959).

[7] G. M. Patton, J. M. Fasulo, and S. J. Robins, *J. Lipid Res.* **23,** 190 (1982).

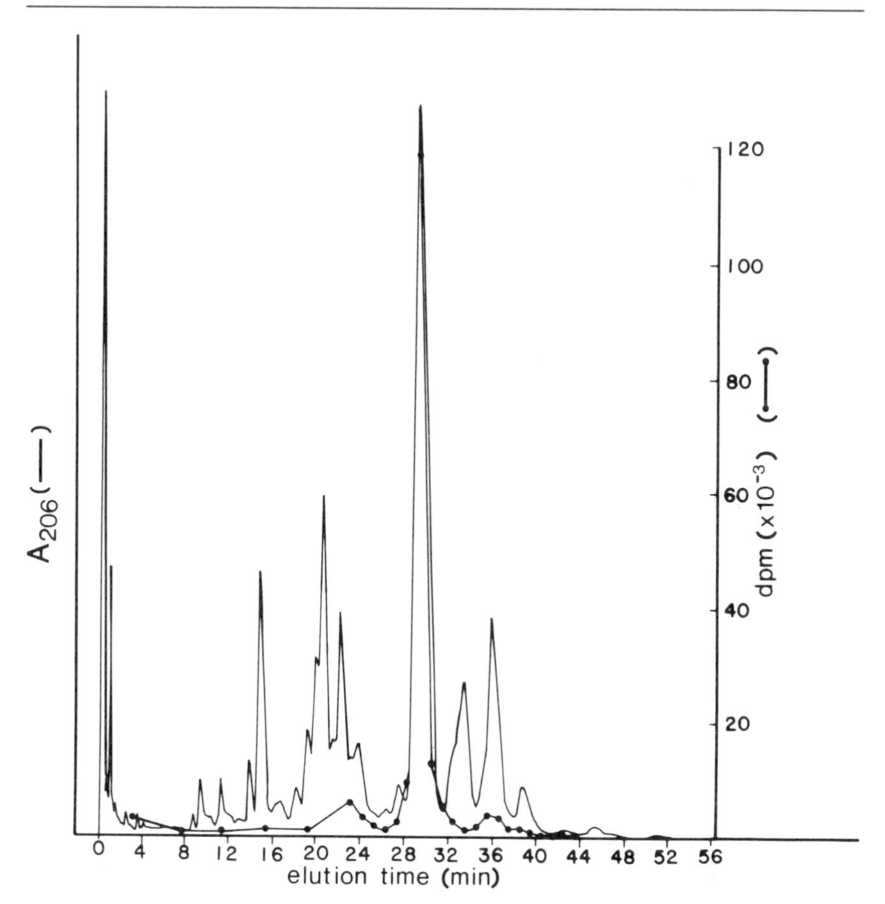

FIG. 1. Isolation of 1-*O*-hexadecyl-2-[^{3}H]arachidonoyl-GPC by reversed-phase HPLC. Fractions are collected every 4 min after injection up to 24 min and then every minute thereafter. Aliquots from each fraction are counted by liquid scintillation spectrometry.

discarded. Microsomes that have been frozen for up to 6 months have been used without appreciable loss of activity.

Preparation of Calcium Solutions

Since the intracellular arachidonoyl-hydrolyzing phospholipase A_2 enzyme requires calcium, it is of interest to monitor the activity of the enzyme at intracellular levels of free calcium (0.1–1.0 μM). Calcium solutions at this level are prepared using the fluorescent calcium indicator indo-1 (Molecular Probes, Inc., Eugene, OR) as described by Grynkiewicz *et al.*[8] The dye, indo-1, when bound to Ca^{2+} fluoresces at an excitation wavelength of 355 nm and emits at 400 nm. When indo-1 is free of Ca^{2+} (in the presence of excess EGTA), it fluoresces at 355 nm but now emits at a peak of 480 nm. To determine the amount of Ca^{2+} bound to the dye, which would represent the available free Ca^{2+} in the solution, the ratio of the fluorescence value at each emission wavelength ($R = F^{400}/F^{480}$) is used in Eq. (1):

$$[Ca^{2+}](nM) = K_d \frac{S_{f2}}{S_{b2}} \frac{R - R_{min}}{R_{max} - R} \tag{1}$$

where K_d equals 250 nM, which is the dissociation constant of indo-1 for Ca^{2+}; S_{f2}, fluorescence value at 480 nm for the free dye at zero Ca^{2+} (measured in excess EGTA); S_{b2}, fluorescence value at 480 nm for the Ca^{2+}-bound dye (measured in 1 mM Ca^{2+}); R_{min}, ratio of the fluorescence values F_{400}/F_{480} measured at zero Ca^{2+}; and R_{max}, ratio of the fluorescence values F_{400}/F_{480} measured at saturating Ca^{2+}. A Hitachi F-4010 fluorescence spectrophotometer is used for all measurements.

The exact ratios of the components used in the phospholipase A_2, assay are used in the cuvette to measure fluorescence of the calcium solutions. For example, for a 50-μl phospholipase A_2 assay the following components are used: 20 μl of enzyme, 25 μl substrate, and 5 μl of calcium solution. Consequently, in a 3.0-ml cuvette, 60 times each of the solutions is used.

Enzyme Solution (1.2 ml). Whether the crude cytosolic fraction or the enzyme at various stages of purification is used, it is dialyzed against two 2-liter volumes of 50 mM Tris-HCl, pH 8.0, containing 10% glycerol. Since large amounts of the partially purified enzyme are not available, the final dialyzate buffer is used in place of the actual enzyme since it should be the same ionic composition as the dialyzed enzyme. When the actual enzyme solution has been tested with a known calcium solution, no significant effects on the calcium concentration have been observed.

Substrate (1.4 ml). The actual liposomes are not used to prepare the calcium solution because of the large amount that would be required. Rather, the buffer (Tris-HCl, pH 8.0) in which the dried lipids are sonicated is used. When the actual liposomes have been tested with a known calcium solution, no significant effects on the calcium concentration have been observed.

Indo-1 (90 μl). A 60 μM stock solution is made in 50 mM Tris-HCl buffer, pH 8.0, aliquoted into 1.0-ml volumes and stored in the dark at $-20°$.

[8] G. Grynkiewicz, M. Poenie, and R. Y. Tsien, *J. Biol. Chem.* **260,** 3440 (1985).

Calcium Solutions (300 μl). For preparing calcium solutions that will result in concentrations of 0.1 to 1.0 μ*M* calcium in the phospholipase assay, 50 m*M* Tris-HCl, pH 8.0, containing 100 μ*M* EGTA is used. After first establishing R_{max} and R_{min} with this Tris/EGTA buffer (see below), increments of a stock $CaCl_2$ solution are then added to the Tris/EGTA buffer, starting with 10 m*M* $CaCl_2$ to 1 m*M* $CaCl_2$ (for fine tuning). After each increment of calcium is added to the Tris/EGTA buffer and mixed, 300 μl are removed and added to the other components of the cuvette. It is necessary to thoroughly mix the calcium solution with the other solutions in the cuvette since they contain glycerol. The calcium solution is mixed in with a disposable Berel plastic transfer pipette and the solution is also stirred with a magnetic stir bar during fluorescence readings. It is also important to thoroughly rinse the cuvette and magnetic stirrer with distilled water between readings which is done in a cuvette washer and only handling the stir bar with plastic tongs.

To establish R_{min}, all of the above components are added to the cuvette and a fluorescence scan performed. To ensure that this mixture represents zero calcium, additional EGTA can be added and another scan performed; however, in our experience this should not be necessary. Calcium (30 μl of a 100 m*M* $CaCl_2$ stock) is then added directly to the cuvette, mixed thoroughly, and another scan performed to established R_{max}.

Preparation of Macrophage Phospholipase A_2

The RAW 264.7 macrophage cell line (ATCC, Rockville, MD) is used as a source for the arachidonoyl-hydrolyzing phospholipase A_2. This macrophage cell line, originally established from a murine tumor induced by Abelson's leukemia virus, secretes lysozyme, is phagocytic, and has receptors for complement and immunoglobulin.[9] In addition, the RAW 264.7 cells release arachidonic acid in response to zymosan and endotoxin.[1] Consequently, they are a useful macrophage model for studying an arachidonoyl-hydrolyzing phospholipase A_2 and they can be grown in large quantities in suspension, which is necessary for enzyme purification. The cells are cultured at 37° in a humidified incubator containing air/CO_2 (19:1) in a 1:1 mixture of Dulbecco's modified Eagle's medium (high glucose) and RPMI 1640 (both from Hazelton Biologics, Inc., Lenexa, KS). The medium is supplemented with 2 m*M* glutamine, 100 units/ml penicillin, 100 μg/ml streptomycin, and 10% iron-supplemented bovine calf serum (HyClone Laboratories Inc., Logan, UT). Since the RAW 264.7 cells are growth arrested by low levels of endotoxin (0.5 ng/ml), all cell

[9] W. C. Raschke, S. Baird, P. Ralph, and I. Nakoinz, *Cell* **15,** 261 (1978).

culture reagents contain less than 0.5 ng/ml endotoxin as determined by a *Limulus* amoebocyte lysate assay kit (Associates of Cape Cod, Woods Hole, MA). The RAW 264.7 cells are routinely screened for mycoplasma contamination using a Gen-Probe mycoplasma T. C. II rapid detection system (Gen-Probe, San Diego, CA). The RAW 264.7 cells can be grown either adherent or in suspension, when large quantities of cells are needed for purification of the phospholipase A_2. Approximately 12 liters of suspension cells are harvested weekly, however, the suspension cultures are always started from adherent cells every week rather than continuously passaging suspension cells. The RAW 264.7 cells are grown to confluence in 18, 175-cm^2 tissue culture flasks. This takes 1 week by starting with one 175-cm^2 flask and subculturing three times during the week. For starting suspension cultures, the cells are scraped with a rubber policeman and cells from two 175-cm^2 flasks are seeded into each of eight, 1-liter microcarrier spinner flasks (Bellco Glass Inc., Vineland, NJ) and brought to approximately 600 ml with complete medium. Fresh medium is added two more times during the week to a final volume of 1.5 liters. The cells from eight Spinner flasks are harvested by centrifugation at 4° in 1-liter bottles using a Damon/IEC rotor (#981) operated at 280 g for 10 min. The cell pellets are combined and washed twice with homogenization buffer (0.34 *M* sucrose, 10 m*M* HEPES, 1 m*M* EDTA, 1 m*M* EGTA, 1 m*M* phenylmethylsulfonyl chloride, 1 μg/ml leupeptin, pH 7.4) by centrifugation at 4° in 50-ml conical tubes at 460 *g* for 10 min. A 20–25 ml cell pellet is obtained from 12 liters of cells in suspension. The cells are resuspended in a volume of homogenization buffer that is 40% of the cell pellet volume, and glycerol is then added to a final concentration of 30% (w/v). The cells can be used fresh (in which case it is not necessary to add the glycerol) or frozen at −70° in 10-ml aliquots for up to 1 month. Generally, 2–3 weeks worth of cells are combined to start a purification.

To prepare cell-free homogenates from a large quantity of cells (to start a purification), the cells are homogenized by nitrogen cavitation after equilibration at 600 psi for 30 min at 4° using a 920-ml cell disruption bomb (Parr Instrument Co., Moline, IL). For smaller amounts of fresh cells (to study properties of the crude enzyme), sonication can be used with comparable recovery of enzyme activity. For sonication, the cells are suspended in homogenization buffer at 100–200 $\times$ 10^6 cells/ml and sonicated on ice using a microprobe (B. Braun Instruments, Burlingame, CA) for 30 to 60 sec, depending on the volume of the cell suspension. Cell disruption is monitored by light microscopy. The homogenized cells are centrifuged at 100,000 *g* for 1 hr to separate the soluble and particulate fractions. The phospholipase A_2 activity is found almost exclusively in the high-speed supernatant when the cells are homogenized in the presence of chelators,

TABLE I
PARTIAL PURIFICATION OF PHOSPHOLIPASE A_2 FROM RAW 264.7 MACROPHAGES

Step	Protein (mg)	Total[a] activity (milliunits)	Specific activity (milliunits/mg)	Yield (%)	Purification (fold)
Crude enzyme 100,000 *g* supernatant	1070	890	0.8		
$(NH_4)_2SO_4$	400	580	1.5	65	1.9
Sepharose 6B	38	278	7.3	31	9.0
Mono Q	2	96	58.0	11	73.0

[a] One unit (U) of enzyme activity is the amount of enzyme that hydrolyzes 1 μmol of substrate per minute at 37°.

routinely resulting in a specific activity of 1 to 2 nmol/min/mg protein. If chelators are not present, there is a dramatic decrease in the recovery of phospholipase A_2 activity in the soluble fraction. We have shown that this phospholipase A_2 exhibits calcium-dependent association with membrane.[10]

Partial Purification of Macrophage Phospholipase A_2

A procedure for extensive purification of the RAW 264.7 phospholipase A_2 has been published in detail elsewhere.[1] This highly purifed form of the enzyme is relatively unstable and limited quantities are available. Consequently, for routine studies on the biochemical properties of the enzyme a partially purified preparation is used. A typical purification is summarized in Table I. For purification, cells are always homogenized by nitrogen cavitation in homogenization buffer containing chelators, protease inhibitors, and glycerol as described above. The 100,000 *g* supernatant, which contains the phospholipase A_2, is adjusted to 60% $(NH_4)_2SO_4$ with a saturated solution of $(NH_4)_2SO_4$ and stirred for 1 hr on ice. The precipitated solution is centrifuged at 100,000 *g* for 45 min and the precipitate solubilized in 10 m*M* Tris-HCl buffer, pH 8.0, containing 10% glycerol (Buffer A). The solubilized pellet is centrifuged at 12,000 *g* for 10 min to remove the small amount of insoluble material. The amount of protein precipitated by $(NH_4)_2SO_4$ is variable and ranges from 40 to 60% of the protein in the 100,000 *g* supernatant. The final volume of the ammonium sulfate-precipitated material is adjusted to 2–5% of the subsequent

[10] J. Y. Channon and C. C. Leslie, *J. Biol. Chem.* **265,** in press (1990).

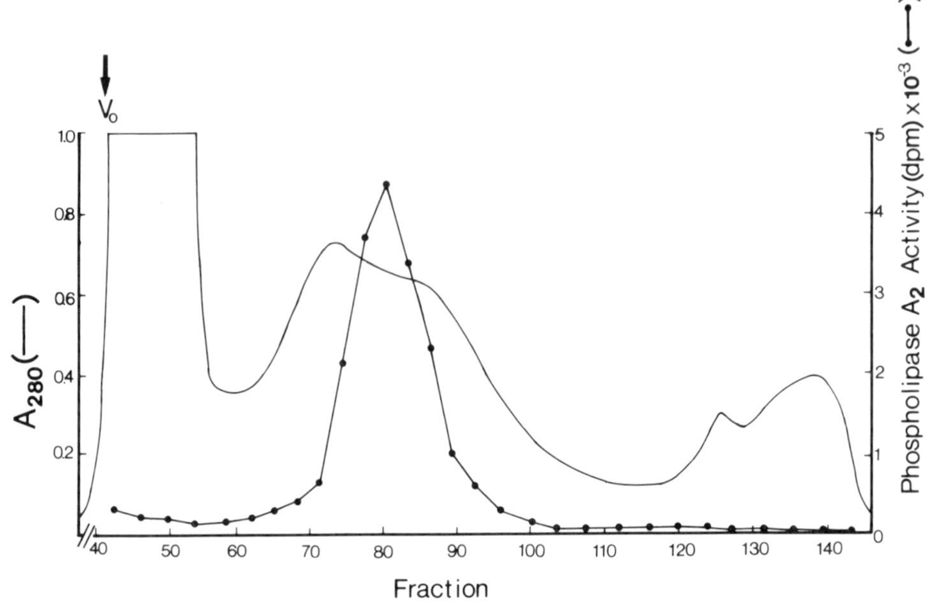

FIG. 2. Sepharose 6B chromatography of the solubilized ammonium sulfate-precipitated phospholipase A_2.

Sepharose 6B column (5 × 80 cm), which is equilibrated with Buffer A (Fig. 2). Fractions (10.0-ml) are eluted at a rate of 120 ml/hr with Buffer A and 30-μl aliquots are assayed for phospholipase A_2 activity. The phospholipase elutes as a single peak of activity at a molecular weight between 150,000–200,000. Active fractions are pooled and adsorbed onto a Mono Q column (HR 10/10) previously equilibrated with Buffer A containing 0.2 *M* NaCl (Fig. 3). Fractions are eluted in a linear salt gradient (160 ml) from 0.2 to 0.375 *M* NaCl in Buffer A at 2 ml/hr. The phospholipase-containing fractions eluting at 0.27–0.29 *M* NaCl are pooled and dialyzed against two 2-liter volumes of Buffer A and then frozen in aliquots at −20°. The frozen enzyme is used for up to 4 weeks with little loss in activity. The presence of glycerol is essential for maintaining enzyme activity.

Properties of Macrophage Phospholipase A_2

The properties of the macrophage phospholipase A_2 are very similar in the crude cytosolic fraction (100,000 *g* supernatant) and in more purified preparations.[1] The enzyme requires calcium for activity, is optimally active at pH 9.0, and rapidly exhibits nonlinear kinetics at incubation times longer than 1 min. The macrophage phospholipase A_2 exhibits apparent

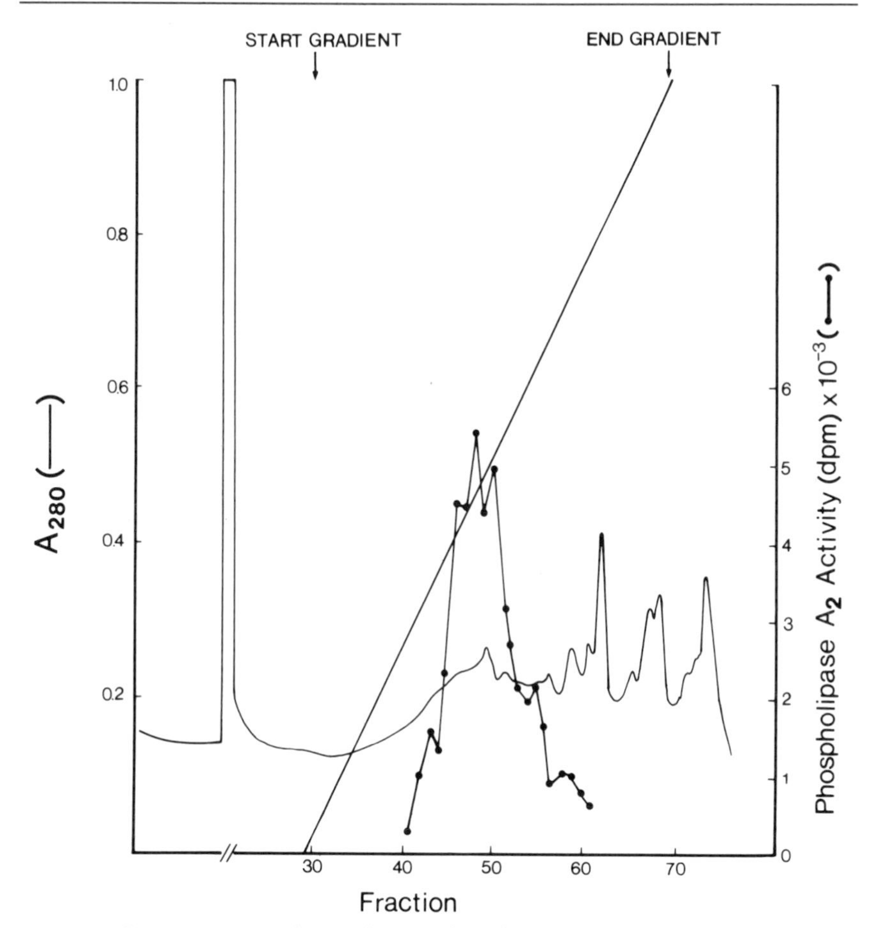

FIG. 3. Chromatography of the active peak from Sepharose 6B over a Mono Q column.

specificity for phospholipid substrates containing *sn*-2-arachidonic acid. Phospholipids containing oleic or linoleic acids at the *sn*-2 position are very poorly hydrolyzed. The enzyme does not exhibit base-group preference, since arachidonic acid is hydrolyzed similarly from choline- or ethanolamine-containing phospholipid substrates. The enzyme also hydrolyzes arachidonic acid equally well from 1-acyl- or 1-alkyl-linked, choline-containing phospholipid substrates.

Acknowledgments

This work was supported by NIH grant HL34303.

[25] Measurement of Phosphoinositide-Specific Phospholipase C Activity

By JOHN E. BLEASDALE, JAMES C. MCGUIRE, and GREGORY A. BALA

Phosphoinositide-specific phospholipases C (EC 3.1.4.10 and EC 3.1.4.11) comprise a class of phosphodiesterases that catalyze the hydrolysis of either phosphatidylinositol or one of its derivatives to produce diacylglycerol and inositol phosphates. These phospholipases C have multiple involvement in the production of eicosanoids. They are involved not only in the mobilization (from phosphatidylinositol) of arachidonic acid used to fuel eicosanoid synthesis, but also in Ca^{2+}-dependent regulation of arachidonic acid mobilization that involves phospholipase C-dependent production of the intracellular second messenger inositol 1,4,5-trisphosphate (IP_3). Since assays for phosphoinositide-specific phospholipase C were last discussed in this series,[1] there have been several important developments that are relevant to consideration of appropriate assay methods. These include (*i*) confirmation of dissimilar isozymic forms of phospholipases C,[2] (*ii*) recognition of G-protein involvement in the regulation of activity of some phospholipases C,[3] (*iii*) improved methods for the analysis of phospholipase C-catalyzed reaction products, especially inositol phosphates,[4] (*iv*) the use of Ca^{2+} buffers to define calcium requirements for phospholipase C activity,[5] and (*v*) the commercial availability of a variety of radiolabeled substrates and products of phospholipase C. One consequence of the diversity of the phospholipases C is that no single assay is appropriate for all. The methods described here are restricted to, (*i*) a two-step procedure for purification to near homogeneity of the M_r 86,000 isoform of phospholipase C from human amnion, (*ii*) a general radiochemical procedure for measurement of phospholipase C-catalyzed hydrolysis of phosphoinositides added in suspension, and (*iii*) a method for the measurement of IP_3 generated from endogenous substrates by membranes isolated from human polymorphonuclear neutrophils.

[1] S. E. Rittenhouse, this series, Vol. 86, p. 3.
[2] S. G. Rhee, P.-G. Suh, S.-H. Ryu, and S. Y. Lee, *Science* **244,** 546 (1989).
[3] S. Cockcroft, *Trends Biochem. Sci.* **12,** 75 (1987).
[4] R. F. Irvine, E. E. Anggard, A. J. Letcher, and C. P. Downes, *Biochem. J.* **229,** 505 (1985).
[5] D. Renard, J. Paggioli, B. Berthon, and M. Claret, *Biochem. J.* **243,** 391 (1987).

Isolation of Major Phospholipase C from Human Amnion

Human placentas are obtained on ice within 2 hr of delivery at term of gestation (phospholipase C activity changes with gestational age but not mode of delivery).[6] Amnion membrane is peeled away from the underlying chorion and decidua, washed four times in saline solution (0.9% w/v), and then spread (mucus side up) over a glass plate on ice. Mucus is removed by gentle scraping with a glass microscope slide and the tissue is washed twice more with saline solution. Each cleaned amnion (approximately 12 g) is minced finely with scissors and then homogenized (Polytron homogenizer, 3 × 20 sec at 90% max, 4°) in 30 ml of saline solution (0.9% w/v) that contains phenylmethylsulfonyl fluoride (1.5 m*M*), sodium tetrathionate (5m*M*), and EGTA (6.5 m*M*). The homogenate is centrifuged (3000 g_{av} for 20 min) and the resulting supernatant fluid is recentrifuged (105,000 g_{av} for 60 min). The resulting supernatant fluid (approximately 1 mg protein/ml) is dialyzed overnight at 4° against Tris-succinate (50 m*M*, pH 8.0) that contains dithiothreitol (1 m*M*) using dialysis tubing with a nominal molecular weight cut-off of M_r 50,000 (Spectrum Medical Industries, Los Angeles, CA). Dialyzed supernatant fluid is subjected to anion-exchange high-performance liquid chromatography on an LKB Ultrochrom GTi chromatograph using an LKB DEAE 5PW (8 × 75 mm) column eluted with a gradient of Tris-succinate (50 m*M*), pH 8.0–4.0, with the gradient profile depicted in Fig. 1, at 1 ml/min. A pH gradient, rather than a salt gradient, is necessary because activity of this phospholipase C is irreversibly lost at salt concentrations greater than 0.2 *M*. Column effluent is collected in 1-ml fractions into tubes that contain 0.25 ml of glycerol and phospholipase C activity in 0.05-ml samples of the enzyme/glycerol mixture is measured as the rate of hydrolysis of phosphatidyl[^{3}H]inositol using the procedure described below. Fractions containing phospholipase C activity (Fig. 1) are pooled and exchanged into potassium phosphate buffer (10 m*M*, pH 7.0) by use of centrifugal ultrafiltration (Centriprep 30 concentrator tubes, Amicon Inc., Danvers, MA). This material is subjected to high-performance hydroxyapatite chromatography using a Bio-Rad HPHT Bio-Gel 7.8 × 100 mm column eluted at 1 ml/min with a gradient of potassium phosphate (10 m*M*, pH 7.0) to potassium phosphate (500 m*M*, pH 7.0) having the profile depicted in Fig. 1. Column effluent is collected in 1-ml fractions into glycerol (0.25 ml) and phospholipase C activity is assayed in the presence of bovine serum albumin (3 mg/ml).

In each chromatographic procedure essentially all recovered phospholipase C activity is in a single peak. A final specific activity of approximately 100 μmol/min/mg protein is obtained. This represents an approximately 1000-fold purification with an estimated recovery of 35%. (The

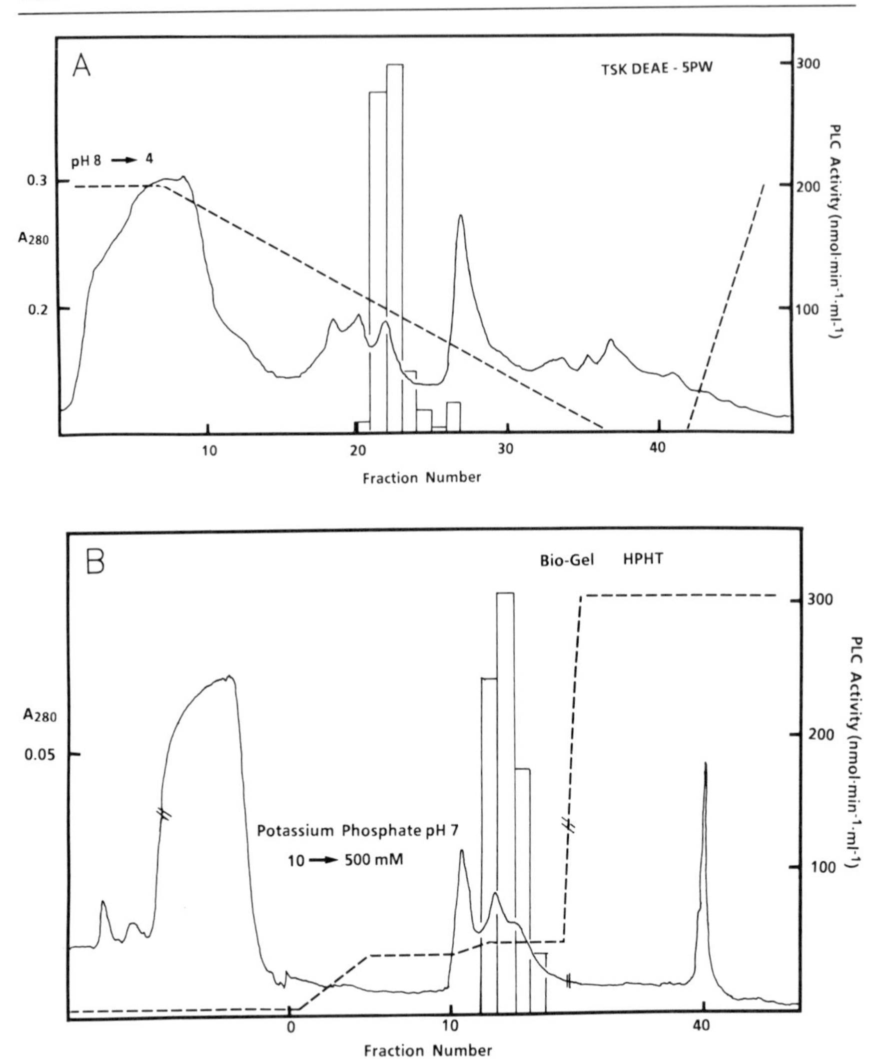

FIG. 1. Isolation of phospholipase C from human amnion. Typical elution profiles from (A) anion-exchange chromatography (TSK DEAE-5PW) and (B) hydroxyapatite chromatography (Bio-Rad HPHT BioGel). Absorbance measured at 280 nm is depicted as a solid line, phospholipase C activity as the open bars, and the gradient profile as the broken line. Fractions without bars were essentially devoid of phospholipase C activity. (C) 7.5% SDS–PAGE of amnion phospholipase C at three stages of purification: (1) 105,000 *g* supernatant fraction, (2) fraction with greatest specific activity of phospholipase C after anion-exchange chromatography, and (3) fraction with greatest specific activity of phospholipase C after hydroxyapatite chromatography (stained with Coomassie blue).

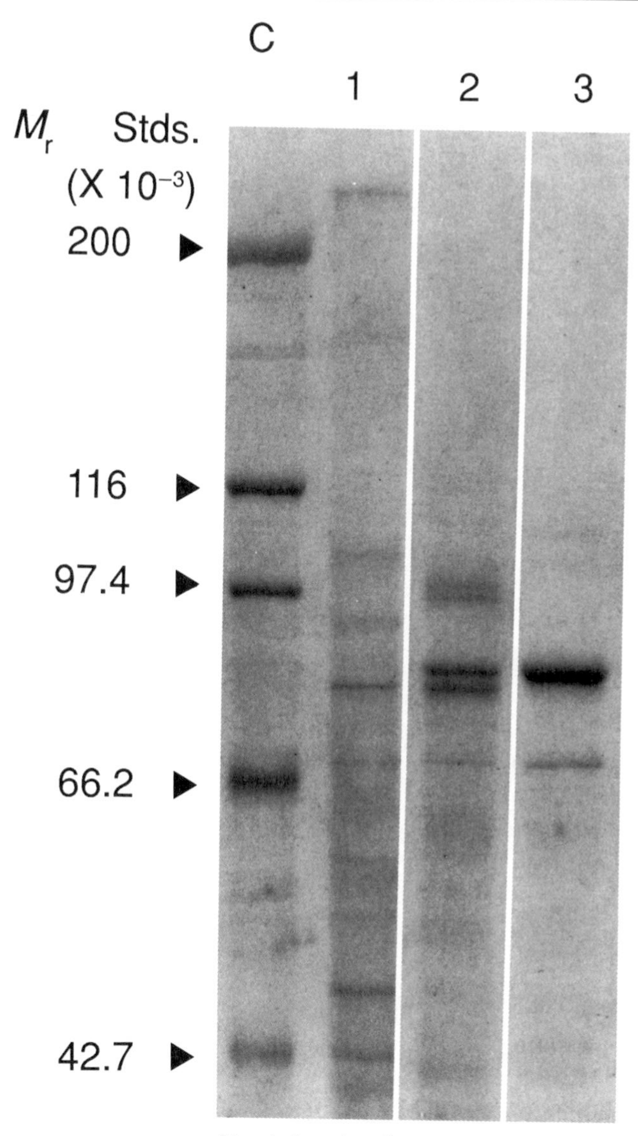

FIG. 1. (*continued*)

presence of phosphate in enzyme fractions after hydroxyapatite chromatography necessitates the addition of extra calcium to assays and complicates exact estimates of recovery.) SDS-PAGE of amnion phospholipase C purified by these procedures reveals a predominant band at M_r 86,000 with only small amounts of other proteins (Fig. 1).

Measurement of Phospholipase C-Catalyzed Hydrolysis of Phosphoinositides in Suspension

Reagents

Phosphatidylinositol from liver (Serdary Research Laboratories, London, Ontario, Canada or Avanti Polar Lipids, Birmingham, AL)
Phosphatidyl[2-^{3}H]inositol (New England Nuclear, Boston, MA or Amersham Corporation, Arlington Heights, IL)
HEPES–NaOH buffer (50 m*M*, pH 7.0)
$CaCl_2$[40 m*M* in HEPES–NaOH buffer (50 m*M*, pH 7.0)]
Chloroform (Analytical Reagent Grade)
Chloroform/methanol (1 : 2, v/v)
HCl (1 *M*)

The method described uses phosphatidyl[2-^{3}H]inositol as substrate but with minor modification (below) can be employed successfully to measure the rate of hydrolysis of phosphatidyl[2-^{3}H]inositol 4,5-bisphosphate (from Sigma Chemical Company, St. Louis, MO and New England Nuclear). Methods for the purification of phosphatidylinositol from tissues and for the enzymatic preparation of phosphatidyl[2-^{3}H]inositol have been described in detail[1,6,7] and are not further discussed because these materials are now available commercially. Radiolabeled substrate is prepared at concentrations 2.5-times that present in the assay mixture as follows. Phosphatidylinositol and phosphatidyl[2-^{3}H]inositol (in chloroform) are mixed to achieve a specific radioactivity of approximately 50 mCi/mol lipid P and transferred to a glass test tube. Chloroform is removed by evaporation under nitrogen, HEPES-NaOH buffer (50 m*M*, pH 7.0) is added (1 ml/5 μmol lipid P), and the substrate is dispersed by use of ultrasound (Bransonic 220 bath-type sonicator, 125 W, Bransonic Company, Shelton, CT) for 5 min. Standard assay mixtures consist of HEPES-NaOH buffer (50 m*M*, pH 7.0), $CaCl_2$ (4 m*M*), phosphatidyl[2-^{3}H]inositol (2 m*M*), and enzyme [20–100 μg protein when using crude tissue extracts or 1 μg of purified phospholipase C stabilized with bovine serum albumin (3 mg/ml)] in a total volume of 0.25 ml. Reactions are initiated with the addition of either enzyme or substrate and conducted at 37° for 15 min in a shaking water bath. Reactions are terminated with 1 ml of chloroform/methanol (1 : 2, v/v), followed by chloroform (0.3 ml), and 0.3 ml of HCl (1 *M*). Samples are mixed vigorously and then centrifuged (500 g_{av} for 5 min at room temperature) to yield two phases. The water-soluble prod-

[6] G. C. Di Renzo, J. M. Johnston, T. Okazaki, J. R. Okita, P. C. MacDonald, and J. E. Bleasdale, *J. Clin. Invest.* **67,** 847 (1981).

[7] V. G. Mahadevappa and B. J. Holub, *in* "Chromatography of Lipids in Biomedical Research and Clinical Diagnosis" (A. Kuksis, ed.), p. 225. Elsevier, Amsterdam, 1987.

ucts (a mixture of [^{3}H]inositol 1-phosphate and [^{3}H]inositol 1,2-cyclic phosphate) are contained in the upper phase. A sample (0.75 ml) of the upper phase (total volume of 1.2 ml) is transferred to a scintillation vial and mixed with scintillation fluid (ACS, Amersham Corporation, Arlington Heights, IL) for the measurement of ^{3}H by liquid scintillation spectrometry (with correction for quenching). Blank values are obtained from samples in which enzyme is added after quenching with chloroform/methanol.

Comments

If phospholipases C exhibit preference for molecular species of phosphatidylinositol with particular fatty acyl substituents, appropriate *in vitro* assay conditions for the demonstration of such specificity have not been described. In general, the fatty acid composition of the phosphatidylinositol does not greatly influence phospholipase C activity *in vitro*. For instance, the fatty acid composition of the diacylglycerol generated from pig liver phosphatidylinositol (mixed molecular species) by amnion phospholipase C is almost identical to that of the substrate.[6] The lack of demonstrable preference of phospholipase C for phosphatidylinositol molecules that contain arachidonic acid does not argue against a function of this enzyme in arachidonic acid mobilization because in most mammalian tissues more than half of the phosphatidylinositol molecules contain arachidonic acid.[8]

Dithiothreitol or mercaptoethanol when added to the assay mixture does not affect amnion phospholipase C activity but may be required for *in vitro* activity of other phospholipases C.[9] Phenylmethylsulfonyl fluoride, added to amnion homogenates to inhibit proteolysis during purification, does not inhibit phospholipase C activity, however, other phospholipases C may be at least partially inhibited by phenylmethylsulfonyl fluoride.[10] The large apparent calcium requirement for phospholipase C activity in this *in vitro* assay likely reflects binding of calcium to assay components other than phospholipase C because the calcium requirement decreases greatly as more purified enzyme is used and increases as the substrate concentration is increased.[6,11] When Ca^{2+} is buffered at 1 μM by use of a Ca/EGTA buffer,[12,13] the rate of hydrolysis of phosphatidylinositol

[8] D. A. White, *in* "Form and Function of Phospholipids" (G. B. Ansell, J. N. Hawthorne, and R. M. C. Dawson, eds.), p. 441. Elsevier, Amsterdam, 1973.

[9] R. S. Atherton and J. N. Hawthorne, *Eur. J. Biochem.* **4,** 68 (1968).

[10] R. Walenga, J. Y. Vanderhoek, and M. B. Feinstein, *J. Biol. Chem.* **255,** 6024 (1980).

[11] N. Sagawa, J. E. Bleasdale, and G. C. Di Renzo, *Biochim. Biophys. Acta* **752,** 153 (1983).

[12] J. Raaflaub, *in* "Methods of Biochemical Analysis" (P. Glick, ed.), Vol. 3, p. 301. Wiley (Interscience), New York, 1986.

[13] T. Bartfai, *in* "Advances in Cyclic Nucleotide Research" (G. Brooker, P. Greengard, and G. A. Robison, eds.), Vol. 10, p. 219. Raven Press, New York, 1979.

(0.2 mM) by amnion phospholipase C is only 20% of maximum. The rate of hydrolysis of phosphatidylinositol 4,5-bisphosphate (0.2 mM) is maximal at Ca^{2+} (1 μM) and is inhibited 80% at Ca^{2+} (4 mM).

Assays similar to that described here have been reported with the major differences being in choice of assay buffer, inclusion of detergent, and the procedure for extraction of products. If Ca/EGTA buffers are to be used, care should be taken to avoid a buffer system that binds Ca^{2+}. Sodium deoxycholate is usually the detergent employed in phospholipase C assays and, although the activities of some phospholipases C are greatly affected by deoxycholate while others are only slightly affected, in most cases maximal phospholipase C activity is observed at 1 mg of deoxycholate per milliliter and pronounced inhibition is observed at higher concentrations. The most commonly employed extraction procedures employed in phospholipase C assays are the chloroform/methanol/HCl system described here, chloroform/methanol/HCl/EGTA,[14] chloroform/methanol/KCl,[6] and chloroform/butanol/HCl.[15] Since one of the products of phospholipase C-catalyzed hydrolysis of phosphoinositides is an acid-labile cyclic-1,2-phosphate ester, acid extraction procedures should be avoided if initial products are to be identified.[4] Inclusion of EGTA in the extraction system has no effect on the partitioning of either substrate or products. When KCl is substituted for HCl, blank values for assays of phosphatidyl[2-^{3}H]inositol hydrolysis are approximately doubled and counting efficiency is reduced. In contrast, however, the high blank values observed in assays of phosphatidyl[2-^{3}H]inositol 4,5-bisphophate hydrolysis when chloroform/methanol/HCl is used for extraction are greatly reduced when KCl is substituted for HCl. Chloroform/butanol/HCl extraction yields blank values for phosphatidylinositol hydrolysis that are greater than those with chloroform/methanol/HCl and is inferior to chloroform/methanol/KCl in assays of phosphatidylinositol 4,5-bisphosphate hydrolysis. Butanol extraction procedures, however, reduce the recovery into the upper phase of lysophosphatidylinositol[15] that may be formed by phospholipase A_2-dependent hydrolysis of phosphatidylinositol when crude sources of enzyme are used.

Measurement of Phospholipase C-Catalyzed Production of Inositol Trisphosphate by Membranes from Polymorphonuclear Neutrophils

Phospholipase C-catalyzed formation of IP_3 by isolated membranes from either endogenous or exogenously added phosphatidylinositol 4,5-

[14] S. L. Hofmann and P. W. Majerus, *J. Biol. Chem.* **257,** 6461 (1982).
[15] S. Rittenhouse-Simmons, *J. Clin. Invest.* **63,** 580 (1979).

bisphosphate has been measured but the data is influenced by the mode of substrate presentation.[16] Hydrolysis of endogenous substrate is preferable but complicated by artifacts that may accompany the necessary radiolabeling of the endogenous phosphatidylinositol 4,5-bisphosphate either before or after the membranes are isolated.[17] The method described here employs modifications of established procedures[18–21] and allows the measurement of IP_3 production by freshly isolated neutrophil membranes without the requirement of radiolabeling endogenous phosphatidylinositol 4,5-bisphosphate. The method depends upon a competitive binding assay for IP_3 utilizing a binding protein (microsomes) either prepared from bovine adrenal cortex[21] or obtained commercially (Amersham Corporation).

Reagents

Dextran solution [6%, w/v in saline (0.9% w/v)]
Hanks' balanced salt solution (without calcium and magnesium)
Lymphocyte Separation Medium (Litton Bionetics Inc., Kensington, MD)
Incubation buffer[20] [HEPES-NaOH buffer (10 mM, pH 7.0) that contains KCl (115 mM), KH_2PO_4 (5 mM), EGTA (2 mM), and $MgSO_4$ (0.9 mM)]
Saline solutions (0.9% and 3.6%, w/v)
Sucrose solution (41% w/v in assay buffer)
Lysis buffer [incubation buffer that contains soybean trypsin inhibitor (50 μg/ml), leupeptin (30 μM), benzamidine (0.5 mM), phenylmethylsulfonyl fluoride (0.2 mM), and ATP (1 mM)] (all components from Sigma)
$CaCl_2$ (21.2 mM and 150 mM in assay buffer)
Guanosine 5′-*O*-(3-thiotriphosphate) lithium salt (from Sigma or Boehringer Mannheim Biochemicals, Indianapolis, IN)
Trypan blue (0.2% w/v in 0.9% saline)
Assay buffer [Tris-HCl (0.1 M, pH 9.0), EDTA (4 mM), and bovine serum albumin (4 mg/ml)]

One-half unit of human blood withdrawn into acid-citrate-dextrose[22] is centrifuged (300 g_{av} for 20 min at room temperature) in 50-ml plastic tubes (Falcon). The resulting platelet-rich plasma (upper layer) is removed and

[16] H. Sommermeyer, B. Behl, E. Oberdisse, and K. Resch, *J. Biol. Chem.* **264,** 906 (1989).
[17] C. D. Smith and R. Snyderman, this series, Vol. 141, p. 261.
[18] A. Bøyum, *Scand. J. Clin. Lab. Invest.* **21** (*Suppl. 97*), 77 (1968).
[19] T. Maeda, K. Balakrishnan and S. Q. Mehdi, *Biochim. Biophys. Acta* **731,** 115 (1983).
[20] J. L. Boyer, C. P. Downes, and T. K. Harden, *J. Biol. Chem.* **264,** 884 (1989).
[21] S. Palmer, K. T. Hughes, D. Y. Lee, and M. J. O. Wakelam, *Cell. Signal.* **1,** 147 (1989).
[22] R. H. Aster and J. H. Jandl, *J. Clin. Invest.* **43,** 843 (1964).

retained. The lower layer of cells (5 volumes) is mixed gently with 1 volume of dextran solution (6%, w/v in 0.9% saline), diluted to 50 ml with saline solution (0.9%, w/v), and transferred to 50-ml polypropylene-graduated cylinders for 1 hr at room temperature. During this time, the platelet-rich plasma is centrifuged (2500 g_{av} for 15 min) and the resulting supernatant fluid (platelet-poor plasma) is retained. After 1 hr, the top layer of fluid in each graduated cylinder (approximately 25 ml) is removed (taking care not to disturb erythrocytes below) and centrifuged (275 g_{av} for 15 min). Each cell pellet is suspended in 8 ml of platelet-poor plasma/0.9% saline (1 : 3, by volume) and layered over 3 ml of Lymphocyte Separation Medium in 15-ml polypropylene centrifuge tubes and centrifuged (750 g_{av} for 25 min at room temperature). The supernatant fluid is discarded and the cell pellet suspended in 4 ml of saline solution (0.9%, w/v). The cell suspension is mixed rapidly with 20 ml of distilled deionized water (to lyse erythrocytes) and, after 20 sec, 8 ml of saline solution (3.6%, w/v) is added rapidly with mixing. The cell suspension is centrifuged (275 g_{av} for 5 min). The cell lysis step is repeated if the cell pellet is visibly contaminated with erythrocytes, otherwise, the cell pellet is resuspended in 25 ml of Hanks' balanced salt solution without calcium or magnesium. A small portion is diluted (1 : 20) with saline solution (0.9% w/v), mixed with an equal volume of Trypan blue (0.2% w/v), and cell number and viability determined using a hemocytometer (approximate yield is 0.5–1.0 $\times$ 10^9 cells/half unit of blood).

The cell suspension is centrifuged (275 g for 10 min) and the cell pellet (approximately 8 $\times$ 10^8 cells) is suspended in 30 ml of lysis buffer, placed in a 100-ml polypropylene beaker, and subjected to nitrogen cavitation (500 psi for 30 min while stirring on ice). The cell lysate is homogenized (Polytron homogenizer, 2 $\times$ 10 sec at 75% max power at 4°) and then layered over sucrose solution (41%, w/v) (2 $\times$ 10 ml) in two Ultra-Clear centrifuge tubes (Beckman Instruments Inc.) that were soaked overnight in EGTA (1 m*M*) and then rinsed extensively with deionized water.[19] The tubes are centrifuged (95,000 g_{av} for 60 min) in a swinging-bucket rotor and the layer of material migrating to the interface above the sucrose solution collected, diluted with 10 vol of ice-cold incubation buffer, and centrifuged (95,000 g_{av} for 20 min). The membrane pellet is washed once by resuspension in 30 ml of incubation buffer followed by centrifugation (95,000 g_{av} for 20 min). The washed membranes are resuspended in 5 ml of incubation buffer and protein content is measured (BCA protein reagent, Pierce Chemical Company, Rockford, IL, using bovine serum albumin for calibration; approximate yield is 1 mg protein/8 $\times$ 10^8 cells). Membranes are kept on ice until used (within 2 hr of preparation). Standard reaction mixtures contain membranes (20 μg of protein), guanosine 5′-(3-

thiotriphosphate) (GTPγS) (10 μM), and $CaCl_2$ (added to achieve a particular buffered Ca^{2+} concentration[12,13]) in a total volume of 0.25 ml of incubation buffer in plastic centrifuge tubes (15-ml Falcon). Incubations are initiated with the addition of membranes, conducted at 37° for 15–600 sec at 37° in a shaking water bath and terminated with the addition of 80 μl of ice-cold trichloroacetic acid (20%, w/v) followed by 10 μl of bovine serum albumin (100 mg/ml). After 10 min on ice, samples are centrifuged (1200 *g* for 10 min) and a portion of each supernatant fluid (0.25 ml) extracted twice with 10 volumes of water-saturated diethyl ether. Ether extracts are discarded and traces of ether are removed from the aqueous phase, first, by aspiration after freezing the aqueous phase at −70° and, second, by centrifugation of the thawed aqueous phase for 2 min *in vacuo* (Speed Vac Concentrator, Savant Instruments Inc., Farmingdale, NY). A portion of the aqueous extract (0.1 ml) is mixed with 10 μl of $NaHCO_3$ (0.1 *M*) in a 1.5-ml Eppendorf plastic centrifuge tube. All subsequent operations are at 4°. Known amounts of inositol 1,4,5-trisphosphate [Sigma or Amersham Corporation (0.2 to 50 pmol in 0.1 ml water)] are added to a separate series of Eppendorf tubes to construct a calibration curve. Assay buffer (0.1 ml)is added to each sample followed by 0.1 ml of carrier-free aqueous [^{3}H]inositol 1,4,5-trisphosphate (1-5 Ci/mmol, Amersham Corporation, approximately 7 nCi per sample). Binding protein (0.1 ml) (from Amersham Corporation or prepared as described by Palmer *et al.*[21]) sufficient to bind approximately 45% of the [^{3}H]inositol 1,4,5-trisphosphate is added and samples incubated at 4° for 15 min before centrifugation (1500 g_{av} for 7 min). Supernatant fluids are aspirated and discarded and the tip of each tube (containing the binding protein) cut off and collected in 0.75 ml of water in a scintillation vial. The binding protein is released into the water by mixing vigorously *before* the addition of 10 ml of scintillation fluid (ACS, Amersham Corporation). Bound [^{3}H]inositol 1,4,5-trisphosphate is measured by liquid scintillation spectrometry and the amount of IP_3 in samples computed using a calibration curve of $\%B/B_O$ versus amount of IP_3 (0.2 to 50 pmol) {$\% B/B_o = 100 \times$ [Standard or unknown disintegrations per minute (dpm) minus nonspecific binding dpm]/[zero standard dpm minus nonspecific binding dpm]}.

Comments

The assay is valid for the measurement of IP_3 production in the range of 0.5 to 15 pmol (Fig. 2). Recovery of IP_3 during the extraction procedure (assessed as recovery of added carrier-free [^{3}H]inositol 1,4,5-trisphosphate to samples containing 1–10 pmol IP_3) is greater than 90%. The binding protein is relatively specific for inositol 1,4,5-trisphosphate

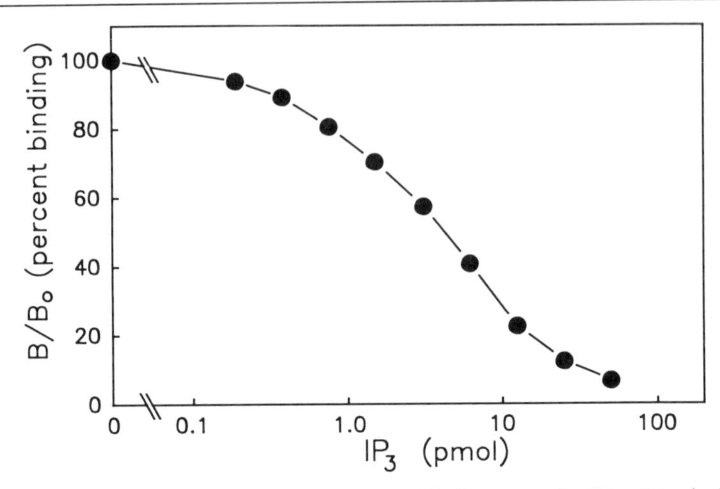

FIG. 2. Calibration curve for the competitive binding assay for IP_3. A typical curve describing the displacement of $[^3H]IP_3$ by known amounts of IP_3 is illustrated. A superimposable calibration curve is obtained when IP_3 standards are diluted not with water but with a neutralized extract of neutrophil membranes (as described in the text) provided that the pH is carefully controlled.

with only inositol 1,3,4,5-tetrakisphosphate exhibiting significant cross-reactivity (approximately 7%). When IP_3 (15 pmol) is added to neutrophil membranes incubated at 37° in the presence of Ca^{2+} (1 μM), the recovery of IP_3 declines over time (80% recovered after 5 min). Loss of added IP_3 is accelerated in the presence of Ca^{2+} (1 mM). Neutrophil membranes incubated in the absence of Ca^{2+} (plus EGTA) exhibit no production of IP_3. Addition of Ca^{2+} (1 μM) elicits a small production of IP_3 that is stimulated greatly by the coaddition of GTPγS (10 μM) (approximately 60 pmol/min/mg protein) and is further enhanced when the chemotactic peptide formylmethionylleucylphenylalanine (200 nM) is added. The rate of GTPγS-elicited IP_3 production subsides rapidly even though the addition of Ca^{2+} (1 mM) still elicits a large production of IP_3 [>10 times the rate of IP_3 elicited by GTPγS and Ca^{2+} (1 μM)]. Addition of ATP (1 mM) to membranes incubated with GTPSγS (10 μM) and Ca^{2+} (1 μM) stimulates IP_3 production. This effect of ATP is consistent with the suggestion of replenishment by endogenous kinases of the pool of phosphatidylinositol 4,5-bisphosphate available to phospholipase C.[17] Alternatively, it may reflect ATP-induced production of IP_3 via the activation of a P_2-purinergic receptor.[23] The mechanism by which receptor activation

[23] D. S. Cowen, H. M. Lazarus, S. B. Shurin, S. E. Stoll, and G. R. Dubyak, *J. Clin. Invest.* **83,** 1651 (1989).

leads to stimulation of phospholipase C activity in this and similar systems[20] remains to be defined. This assay may be useful for investigating receptor-coupled activation of phospholipase C since perturbation of the membrane is minimized.

Acknowledgment

We are grateful to D. K. Piper for preparing this manuscript.

[26] Solubilization of Arachidonate-CoA Ligase from Cell Membranes, Chromatographic Separation from Nonspecific Long-Chain Fatty Acid CoA Ligase, and Isolation of Mutant Cell Line Defective in Arachidonate-CoA Ligase

By MICHAEL LAPOSATA

An acyl-CoA synthetase which utilizes all long-chain fatty acids was first described by Kornberg and Pricer in 1953.[1] Several purifications of this nonspecific long-chain acyl-CoA synthetase enzyme (long-chain-fatty-acid-CoA ligase, EC 6.2.1.3) from a variety of sources including rat liver microsomes,[2] rat liver mitochondria,[3] and *Candida lipolytica*[4] have been reported. An acyl-CoA synthetase specific for arachidonate and other eicosanoid precursor fatty acids was more recently described in 1982.[5] Two important pieces of evidence suggested the existence of an arachidonate-specific acyl-CoA synthetase which was different from the previously characterized acyl-CoA synthetase showing broad specificity for long-chain fatty acids. First, when crude homogenates of platelets were heated to 45° for 30 min, arachidonoyl-CoA-forming activity was lost four times faster than nonspecific long-chain acyl-CoA synthetase activity.[5] Second, a mutant cell line defective in arachidonate incorporation into phospholipids was isolated and found to lack arachidonoyl, but not nonspecific acyl-CoA synthetase activity.[6]

[1] A. Kornberg and W. E. Pricer, *J. Biol. Chem.* **204,** 329 (1953).
[2] T. Tanaka, K. Hosaka, M. Hoshimaru, and S. Numa, *Eur. J. Biochem.* **98,** 165 (1979).
[3] D. P. Philipp and P. Parsons, *J. Biol. Chem.* **254,** 10776 (1979).
[4] K. Hosaka, M. Mishina, T. Tanaka, and S. Numa, *Eur. J. Biochem.* **93,** 197 (1979).
[5] D. B. Wilson, S. M. Prescott, and P. W. Majerus, *J. Biol. Chem.* **257,** 3510 (1982).
[6] E. J. Neufeld, T. E. Bross, and P. W. Majerus, *J. Biol. Chem.* **259,** 1986 (1984).

However, the conclusive proof for the existence of an arachidonate-specific acyl-CoA synthetase was provided by chromatographic separation of this enzyme activity from the nonspecific long-chain acyl-CoA synthetase.[7] Arachidonoyl-CoA synthetase (arachidonate-CoA ligase, EC 6.2.1.15) can be solubilized from cell membranes using 1% Nonidet P-40 (NP-40) and 10 m*M* EDTA, and it can be separated from the nonspecific enzyme using hydroxylapatite chromatography. The fatty acid substrate specificity of arachidonoyl-CoA synthetase includes all fatty acids that can subsequently be converted by cyclooxygenase or lipoxygenase to eicosanoids.[8] Deficiency of arachidonoyl-CoA synthetase is associated with decreased uptake of arachidonate into cells, enhanced turnover of arachidonoylphosphatidylcholine, and decreased agonist-induced arachidonate release.[9]

Enzyme and Protein Assays

Nonspecific long-chain acyl-CoA synthetase and arachidonoyl-CoA synthetase activities are assayed as described in complete detail earlier in this series.[10] For the enzyme assays, unlabeled oleic acid and arachidonic acid can be obtained from Nu-Chek Prep., Inc. (Elysian, MN). [1-^{14}C]Oleic acid (55 μCi/μmol) and [1-^{14}C]arachidonic acid (55 μCi/umol) are available from Du-Pont-New England Nuclear (Boston, MA). Protein determinations on samples containing NP-40 are precipitated on ice with acetone, centrifuged, and the precipitate resuspended in 0.1 ml of 0.2 *N* NaOH and boiled for 10 min. The Bio-Rad protein assay can be used to quantitate protein in these treated samples.[7] Protein standards in 1% NP-40 must be processed identically for use in the Bio-Rad assay.

Solubilization of Arachidonoyl-CoA Synthetase

Arachidonoyl-CoA synthetase has been solubilized and separated from nonspecific long-chain acyl-CoA synthetase using as starting material calf brain, cultured mouse fibrosarcoma cells, and human platelets. Arachidonoyl-CoA synthetase is solubilized from a particulate fraction of calf brain using the nonionic detergent NP-40 as described below and outlined in Table I. The enzyme is recovered in a high-speed supernatant fraction

[7] M. Laposata, E. L. Reich, and P. W. Majerus, *J. Biol. Chem.* **260,** 11016 (1985).

[8] E. J. Neufeld, H. Sprecher, R. W. Evans, and P. W. Majerus, *J. Lipid Res.* **25,** 288 (1984).

[9] P. W. Majerus, E. J. Neufeld, and M. Laposata, *in* "Inositol and Phosphoinositides: Metabolism and Regulation" (J. E. Bleasdale, J. Eichberg, and G. Hauser, eds.), p. 443. Humana Press, Clifton, NJ, 1985.

[10] M. Laposata and P. W. Majerus, this series, Vol. 141, p. 350.

TABLE 1
SOLUBILIZATION OF ARACHIDONOYL-CoA SYNTHETASE FROM CALF BRAIN

Step 1.	Frozen calf brain shredded and homogenized in a blender
Step 2.	Homogenate of calf brain centrifuged 1000 *g*, 5 min, 4° (pellet of this spin discarded)
Step 3.	Supernatant centrifuged at 30,000 *g*, 60 min, 4° (supernatant of this spin discarded)
Step 4.	Resuspended pellet incubated with 1% NP-40 and 10 m*M* EDTA, 2 hr, 4°, to solublize enzymes from cell membranes
Step 5.	Detergent-extracted membranes centrifuged at 30,000 *g*, 60 min, 4°. The supernatant of this spin contains arachidonoyl-CoA synthetase activity solubilized from cell membranes. The results of four separate preparations showed only a very slight purification of enzyme activity from Steps 1 to 5, with a 1.10 ± 0.06-fold (mean ± S.E.M) increase in specific activity and a yield of 31.75 ± 2.69% (mean ± S.E.M.)

after NP-40 extraction with a 25–40% yield. A variety of other detergents, including 3-[(3-cholamidopropyl)dimethylammonio]-1-propane sulfonate (CHAPS) and various extraction conditions do not improve this result.

Fresh calf brain obtained from a local slaughterhouse is cut into sections and suspended in a buffer containing 0.32 *M* sucrose, 10 m*M* HEPES, pH 7.5, 10 m*M* 2-mercaptoethanol, 20 μg/ml phenylmethylsulfonyl fluoride, 1 m*M* benzamidine, and 0.01% soybean trypsin inhibitor. The brain sections are stored frozen at −70°. Within 3 months of collection, frozen brains are shredded in a Hobart commercial vegetable shredder at 4° and the homogenate diluted with the above buffer to 5 mg of protein per milliliter. The preparation is further homogenized at 4° in a Waring blender with three 30-sec bursts and then centrifuged at 1000 *g* for 5 min at 4°. The supernatant is collected, centrifuged at 30,000 *g* for 60 min at 4°, and the membrane pellet resuspended at 10 mg protein per milliliter in 20 m*M* phosphate buffer, pH 7.4, containing 10 m*M* 2-mercaptoethanol. The mixture is brought to 1% NP-40 and 10 m*M* EDTA, stirred 2 hr at 4°, and centrifuged for 1 hr at 30,000 *g*. The supernatant contains arachidonoyl-CoA synthetase and nonspecific long-chain acyl-CoA synthetase among numerous solubilized enzymes.

Platelets are obtained from normal human volunteers and isolated as described earlier in this series.[11] The platelets are suspended at 10^9/ml in buffer containing 26.2 m*M* sodium phosphate and 6.8 m*M* potassium phosphate, pH 6.5, 118 m*M* NaCl, 5.6 m*M* glucose, and 10 m*M* 2-mercaptoethanol. After sonication on ice with three 30-sec bursts at 100 W with a Biosonik IV sonicator (Bronwill, Rochester, NY), membranes are

[11] N. L. Baenziger and P. W. Majerus, this series, Vol. 31, p. 149.

collected by centrifugation at 50,000 *g* for 1 hr at 4°. The platelet particulate fraction from 35 × 10^9 platelets is resuspended by sonication in 6 ml of 20 m*M* phosphate buffer, pH 7.4, containing 1% NP-40, 10 m*M* EDTA, and 10 m*M* 2-mercaptoethanol. The mixture is stirred at 4° for 1 hr and then centrifuged at 50,000 *g* for 45 min. The supernatant contains the solubilized enzymes. Platelets contain 5–10 times as much arachidonoyl-CoA synthetase as nonspecific acyl-CoA synthetase activity, the highest ratio reported among a variety of cells and tissues tested.[7]

Confluent monolayers of cells in tissue culture are harvested by trypsinization and centrifugation. After neutralizing the trypsin, the cells are suspended in Dulbecco's phosphate-buffered saline containing 10 m*M* 2-mercaptoethanol. The cells are then disrupted by sonication on ice with two 30-sec bursts of 100 W and the membranes sedimented by centrifugation at 50,000 *g* for 1 hr at 4°. The membrane pellet is resuspended by sonication in 20 m*M* phosphate buffer, pH 7.4, 10 m*M* 2-mercaptoethanol, and 1% NP-40. The mixture is stirred for 1 hr at 4° and then centrifuged at 50,000 *g* for 1 hr. The supernatant contains the solubilized enzymes.

Separation of Acyl-CoA Synthetases by Hydroxylapatite Chromatography

Arachidonoyl-CoA synthetase can be separated from the nonspecific acyl-CoA synthetase by chromatography on hydroxylapatite at 4°. While there is no purification of arachidonoyl-CoA synthetase activity achieved by this step, this enzyme is separated from the nonspecific acyl-CoA synthetase activity.

The supernatant solution, containing the solubilized enzyme from 2 to 3 calf brains, is added to a 1.2 × 12 cm column of hydroxylapatite equilibrated with 20 m*M* potassium phosphate, pH 7.4, 1% NP-40, and 10 m*M* 2-mercaptoethanol. Elution is carried out with 135 ml of 80 m*M* potassium phosphate followed by a 270 ml linear gradient from 80 to 300 m*M* phosphate at a flow rate of 30 ml/hr. All buffers must include 1% NP-40 and 10 m*M* 2-mercaptoethanol. Fractions contain 6 ml. Only arachidonoyl-CoA synthetase activity is eluted by the phosphate gradient. However, a significant amount of arachidonoyl-CoA-forming activity elutes with 80 m*M* potassium phosphate in the same fractions as the nonspecific enzyme. We have determined that most of this activity represents nonspecific acyl-CoA synthetase using arachidonate as substrate and not from contamination of these fractions by the arachidonate-specific enzyme.

The supernatant, which contains the solubilized enzyme from 20–40 × 10^9 platelets, is added to a 1.2 × 3.8 cm column of hydroxylapatite, equilibrated as described for calf brain. Elution is carried out with 45 ml of

80 mM potassium phosphate followed by 90 ml of a linear gradient from 80 to 300 mM phosphate at a flow rate of 30 ml/hr. Fractions contain 2 ml.

The supernatant of the detergent-treated membrane preparation containing the solubilized enzyme from 8 to 12 150-mm diameter petri dishes of cultured cells is added to an 0.7×6.5 cm column of hydroxylapatite, which is equilibrated as described for calf brain. Elution is carried out with 20 ml of 80 mM potassium phosphate, pH 7.4, 1% NP-40, and 10 mM 2-mercaptoethanol, followed by a linear gradient of 40 ml from 80 to 300 mM phosphate at a flow rate of 10 ml/hr. Fractions contain 1 ml.

Production of Mutant Cell Line Defective in Arachidonoyl-CoA Synthetase

The rationale for isolation of a mutant in arachidonoyl-CoA synthetase is based on the assumption that this enzyme activity is essential for arachidonate uptake into cells, and that selected clones of mutant cells defective in arachidonate uptake will be defective in arachidonoyl-CoA synthetase activity. In the radiation suicide technique described below,[12] actively metabolizing cells which can incorporate highly radioactive [^{3}H]arachidonate ultimately die from prolonged radiation exposure during the 2–4 months they are stored frozen in liquid nitrogen. Conversely, uptake defective mutants survive because they cannot accumulate an intracellular pool of highly radioactive [^{3}H]arachidonate. Mouse fibrosarcoma $HSDM_1C_1$ cells are grown to approximately one-half confluence in 75-cm^2 flasks. Ethylmethane sulfonate (Sigma, St. Louis, MO), 0.025% (v/v), is added as a mutagen for 18 hr. The mutagen is removed, fresh medium added, and the cells are allowed to grow until nearly confluent. [5,6,8,9,11,12,14,15-^{3}H]Arachidonate (75 mM, 110 Ci/mmol) in 10 ml delipidated medium (Ham's F-10 containing 4.2% delipidated horse serum) is added to the cells for 2 hr. The labeling medium is aspirated and the cells washed twice with PBS, trypsinized, centrifuged, and resuspended in two parts cryoprotective medium (M.A. Bioproducts, Walkersville, MD) containing 20% horse serum and one part delipidated medium. After 1–4 months, the cells are thawed and returned to control medium (Ham's F-10 containing 12.5% horse serum and 2.5% fetal calf serum). Over the first 2 weeks in culture, most of the suicide-labeled cells die, leaving 20–100 colonies of survivors per 10^6 cells plated. The method of Raetz *et al.*[13] is used to identify clones of cells defective in arachidonate uptake which

[12] J. Pouyssegur, A. Franchi, J. C. Salomon, and P. Silvestre, *Proc. Natl. Acad. Sci. U.S.A.* **77,** 2698 (1980).

[13] C. R. H. Raetz, M. M. Wermuth, T. M. McIntyre, J. D. Esko, and D. C. Wing, *Proc. Natl. Acad. Sci. U.S.A.* **79,** 3223 (1982).

survive [^{3}H]arachidonate suicide selection. The petri dishes of cells with survivors develop geographically isolated clones on the surface of the dish. These cultures are allowed to grow and are fed 8 and 15 days after plating, without disturbing an overlying polyester filter placed on top of the cells to permit replica plating. The replica filters are carefully removed 2–3 weeks after plating the cells and the master plates returned to the incubator after being refed with 10 ml of conditioned medium [a 1 : 1 mixture of control medium and medium which had been exposed to a monolayer of $HSDM_1C_1$ cells for 24 hr and then filtered to remove viable cells, supplemented with 20 μg/ml bovine insulin (Sigma)]. Each filter is washed twice with PBS to remove loose cells and then incubated for 2 hr in control medium containing 1 μCi/ml [^{3}H]arachidonate (~10 n*M*). Each filter is then washed three times with PBS to remove excess label, and phospholipids and protein from the cells are precipitated onto the filter by incubating the filter in 10% trichloroacetic acid for 1 hr at room temperature. Finally, each filter is stained with Coomassie blue and then destained to detect cell colonies as described by Esko and Raetz.[14] The filter is dried, sprayed with ENHANCE fluor (DuPont-New England Nuclear), and used to expose autoradiographs on preflashed X-Omat AR5 film (Kodak, Rochester, NY) for 2–4 days at −70°. Autoradiographs are compared to their corresponding filters stained with Coomassie blue. Colonies (Coomassie blue spots) without corresponding autoradiograph spots are tentatively identified as arachidonate uptake mutants to be tested for deficiency of arachidonoyl-CoA synthetase activity. These colonies are collected from the master plate with trypsin using an 0.8-cm cloning cylinder (Bellco, Vineland, NJ). Each tentatively identified uptake defective colony is then grown up in larger cell numbers and directly assayed for arachidonate uptake and arachidonoyl-CoA synthetase enzyme activity. Approximately 1% of the survivors of suicide selection are defective in both uptake and enzyme activity.

Acknowledgments

This research was supported by Grant DK37454 from the National Institutes of Health. I would like to acknowledge the significant contribution of Dr. Elizabeth Reich Barry in the completion of these studies.

[14] J. D. Esko and C. R. H. Raetz, *Proc. Natl. Acad. Sci. U.S.A.* **77,** 5192 (1980).

Section II

Biosynthesis, Enzymology, and Chemical Synthesis

A. Prostaglandins
Articles 27 and 28

B. Leukotrienes
Articles 29 through 38

C. Platelet-Activating Factor
Article 39

D. Other Oxidative Products
Articles 40 through 43

[27] Preparation of Prostaglandin H_2: Extended Purification/Analysis Scheme

By DUANE VENTON, GUY LE BRETON, and ELIZABETH HALL

Prostaglandin H_2 (PGH_2) is routinely prepared from ram seminal vesicles and isolated by gravity flow silicic acid chromatography.[1,2] In our hands these procedures produce PGH_2 which is contaminated with small amounts of HHT, HETEs, $PGF_{2\alpha}$, PGE_2, PGD_2, and other unidentified nonarachidonic acid substances, and which has a purity that rarely exceeds 80%. The bonded-phase rechromatography methods described herein provide homogeneous PGH_2 as assayed by high-performance liquid chromatography (HPLC) and ammonia, direct chemical ionization–mass spectrometry (DCI-MS).[3]

Procedure

Chromatography

A fundamental limit to any technique used in the purification of PGH_2 resides in the reactivity of this molecule with protic solvents and silica-based adsorbents (presumably via the acidic Si–OH bond). A study evaluating various chromatography systems and solvents has shown that cyano-bonded, stationary-phase columns eluted with a hexane–2-propanol gradient is an effective means for further purification of PGH_2. 2-Propanol, used as the polar organic modifier in the mobile phase, appears to cause less PGH_2 degradation than other alcohols that were tested.[3]

Gradient HPLC can be performed using a Waters (Milford, MA) system equipped with a Rheodyne Model 7125 injector (100 μl loop), and two Model 510 pumps controlled by a Waters Model 680 automated gradient controller. The elution profiles can be followed using a Waters Model 441 UV spectrophotometer, equipped with a zinc lamp for 214 nm detection.

The column for the semipreparative PGH_2 purification is 36 $\times$ $\frac{1}{8}$ inches, and is packed dry with light vibration using a cyano-bonded stationary phase (Spherisorb, 8 μm, Alltech Associates, Chicago, IL). The gradient

[1] G. Graff, this series, Vol. 86, p. 376.

[2] K. Green, M. Hamberg, B. Samuelsson, M. Smigel, and J. C. Frolich, *Adv. Prostaglandin Thromboxane Res.* **5,** 39 (1978).

[3] I. M. Zulak, M. L. Puttemans, A. B. Schilling, E. R. Hall, and D. L. Venton, *Anal. Biochem.* **154,** 152 (1986).

used for the semipreparative work is hexane/2-propanol (98/2, v/v) for the first 25 min followed by a linear gradient (1% 2-propanol/min) to a final concentration of 20% 2-propanol at a flow rate of 1.5 ml/min. The flow rates, mobile phases, and gradients for the analytical cyano-bonded column (IBM Instruments, 250 × 4.5 mm, particle size 5 μm) are the same except that the gradient is started after 12, rather than 25, min. All chromatography solvents may be purchased from Alltech Associates, are HPLC grade, and have an extra low cut-off in the UV.

Mass Spectrometry

We have found ammonia DCI-MS to be a particularly useful method for directly analyzing prostaglandin preparations, including those containing the labile PGH_2 molecule.[4,5] The technique has fairly good sensitivity (100 ng) and produces little fragmentation. The most intense ammonia DCI-MS ion adduct for the prostaglandin related compounds is $[M + NH_4]^+$, i.e., M + 18. When concentrations are high the $[M + NH_4–H_2O]^+$ and $[M + N_2H_7]^+$ ions may also be seen, corresponding to molecular weights of M and M + 35. Since prostaglandins E_2 and D_2 are simply rearrangement products of PGH_2, they have the same molecular weight as PGH_2 and cannot be differentiated from PGH_2 in the ammonia DCI mass spectra. However, when deuterated ammonia (ND_3) is used as the reagent gas, prostaglandins differing in the number of hydroxyl groups may be differentiated. The base peak ND_3 adduct ions are PGH_2, 376, and $PGE_2(D_2)$, 377. The DCI technique volatilizes all organic material in the sample, and is, therefore, also useful in assessing non-prostaglandin contaminants such as lipids, proteins, and other potential residues frequently found in biological isolates.

Mass spectrometry analyses can be obtained using a Finnigan MAT 4510 mass spectrometer with INCOS Data System. Chemical ionization (CI) spectra may be obtained with a Finnigan direct chemical ionization (DCI) probe, using ammonia (Linde UHP, Limox Specialty Gases, Hillside, IL) at a source pressure of 0.4 torr and a source temperature of 60 to 80°. The DCI controller start and stop settings are 0 and 1350 mA, and the ramp rate is 50 mA/sec. The PGH_2 sample is dissolved in 1–3 μl of benzene, applied directly onto the probe tip (rhenium filament), and the solvent allowed to evaporate. The spectrum presented in Fig. 3 (see later) is from scans (0.95/sec) corresponding to the first, and major, burst of ions recorded in the reconstructed ion current (i.e., 200–300 mA), which are summed and displayed as the spectra for the sample.

[4] S. R. Cepa, E. R. Hall, and D. L. Venton, *Prostaglandins* **27,** 645 (1984).

[5] A. B. Schilling, I. M. Zulak, M. L. Puttemans, E. R. Hall, and D. L. Venton, *Biomed. Environ. Mass Spectrom.* **13,** 545 (1986).

Prostaglandin H_2: Biosynthesis, Purification, and Analyses

Unless stated otherwise, all biochemical reagents are from Sigma Chemical Co. (St. Louis, MO). The following vesicular microsome preparation, PGH_2 biosynthesis, and initial silicic acid purification step are taken with little modification from the original work of Graff[1] and Green *et al.*[2] Throughout the chromatography processes, care is taken to keep preparations as cold as possible. Solvents are removed from all chromatographic fractions with a rotoevaporator attached to a vacuum pump carefully trapped with two efficient vapor traps at dry ice–acetone temperature. Bumping is avoided by beginning the process with precooled samples. The heat of vaporization maintains the samples at a very low temperature as the solvents are removed and appears to reduce reactions leading to the degradation of PGH_2.

Preparation of Ram Seminal Vesicle Microsomes

Ovine seminal vesicles, obtained from freshly slaughtered mature noncastrated rams, are trimmed of excess fat and fibrous tissue, and immediately frozen at −80° for storage. The tissue (120 g) is thawed as needed in sucrose (250 ml, 0.2 *M*) EDTA buffer (0.001 *M*, pH 7.4, 4°), minced with scissors, and homogenized with a Polytron homogenizier for 1 min at 4°. The homogenate is centrifuged at 10,000 *g* for 20 min and the supernatant decanted through cheesecloth at 4°. The filtrate is then centrifuged at 100,000 *g* for 60 min at 4°. The resulting pellet is resuspended in potassium phosphate buffer (120 ml, 50 m*M*, pH 7.4) containing bovine serum albumin (3 mg/ml, fatty acid-free) and again centrifuged at 100,000 *g* for 60 min at 4°. The pellet is resuspended and centrifuged three times in the same buffer without albumin and the final pellet suspended in a minimal volume of buffer at 4° and homogenized in a Potter–Elvehjem hand homogenizer. This suspension is then dispersed in phosphate buffer (final volume, 20 ml), frozen, and stored at −80° until use. The fatty acid free-albumin in the first wash helps to remove endogenous arachidonic acid and is used in place of the organic solvent extraction of the microsomes used by Graff and others. The three subsequent washes are without albumin so that when the arachidonic acid is added it is available for the reaction and not bound to the albumin.

PGH_2 Biosynthesis

Frozen vesicle microsomes (approximately 20 g) are thawed and added to Tris buffer (80 ml, 0.1 *M*, pH 8.0) to which has been added hemoglobin (7.5 mg), isoproterenol (25 mg), and tryptophan (100 mg). Lands and coworkers have shown that full advantage of the peroxidase activity requires a phenolic peroxidase cosubstrate (isoproterenol) and a prosthetic heme in the incubation medium.[6] The tryptophan is first dissolved in a small amount of buffer with the use of a few drops of NaOH (2 *M*), then added to the bulk solution, and the pH checked and adjusted as needed with HCl (1 *M*). Oxygen is bubbled through this solution for 5 min with shaking, followed by the addition of *p*-hydroxymercuribenzoate (100 mg) and an additional 5-min period of oxygenation. The reaction is initiated by addition of 5,8,11,14-eicosatetraenoic (arachidonic) acid (10 mg, Nu-Chek Prep) and may contain [1-^{14}C]arachidonic acid (0.032 μCi/μmol, Amersham, Arlington Heights, IL) in ethanol (0.5 ml) as a tracer for arachidonic acid metabolites. The reaction is terminated after 75 sec with a NaCl (2 *M*) and citric acid (2 *M*) solution (10 ml) and immediately poured into a 1-liter separatory funnel containing ethyl ether (100 ml) at $-80°$ (dry ice–acetone bath). The funnel is inverted once and the ether decanted into a beaker setting in a dry ice–acetone bath that contains magnesium sulfate (10 g). Another aliquot of ether (100 ml) is added to the funnel followed by agitation and removal of the ether by decanting into the beaker. The combined ether fractions are vacuum-filtered through a scintered glass funnel into a 500-ml round bottom flask. The ether is removed on a rotoevaporator and the residue dissolved in hexane/petroleum ether/ethyl ether (40/40/20, v/v/v, 2 ml) and stored at $-80°$, or used directly for the silicic acid chromatography.

PGH_2 *Purification by Silicic Acid Chromatography*

Silicic acid (100–200 mesh, Unicil, Clarkson Chemicals, Williamsport, PA) is dried for 3 hr at 110° before use. The silicic acid (9.6 g) is added to Celite 545 (1.5 g, Fisher, Itasca, IL) and hexane/petroleum ether/ethyl ether (40/40/20, v/v/v, 10 ml). The slurry is poured into a 300 × 20 mm jacketed column and a mixture of dry ice and acetone added to the jacket. Celite is added to improve the column flow characteristics, which, in turn, allows the application of very light nitrogen pressure to increase the flow rate of the mobile phase. The reaction products from the PGH_2 biosynthesis are applied to the column and three fractions are collected: fraction I is eluted with hexane/petroleum ether/ethyl ether (40/40/20, v/v/v, 100 ml), fraction II is eluted with hexane/ethyl ether (60/40, v/v, 100 ml), and fraction III, containing the majority of PGH_2, is eluted with hexane/ethyl ether (40/60, v/v,100 ml). Solvents are removed, without heating, from fraction III on a rotoevaporator under pump vacuum, and the residue dissolved in dry benzene and stored at $-80°$ until use. The amount of PGH_2 produced is calculated based on the specific activity of the starting arachidonic acid. A typical yield of PGH_2 from this procedure is 2.0 mg (18%),

[6] M. E. Hemler, C. G. Crawford, and W. E. M. Lands, *Biochemistry* **17,** 1772 (1978).

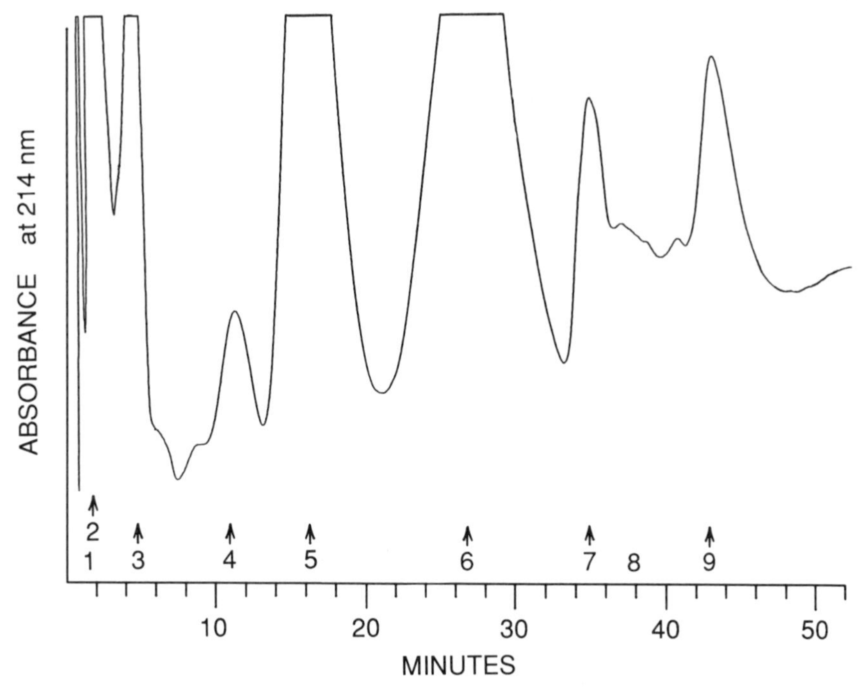

FIG. 1. HPLC chromatography of PGH_2 fraction III from silicic acid column on a semipreparative cyano-bonded column. Flow, 1.5 ml/min; UV detection, 214 nm; solvent, hexane/2-propanol 98/2 (v/v) for 25 min, followed by a linear gradient (1% 2-propanol/min) to a final concentration of 20% 2-propanol. Peaks: Nos. 1–3 non-arachidonic acid-related contamination; No. 4, HETEs; No. 5, HHT; No. 6, PGH_2; No. 7, PGD_2; No. 8, PGE_2; No. 9, $PGF_{2\alpha}$. See text for further discussion. Reproduced from Zulak *et al.*[3]

whose purity, as judged by chromatographic and mass spectral analyses, rarely exceeds 80%.

Bonded-Phase HPLC Purification of PGH_2

Aliquots (ca. 200 μg) of the PGH_2 preparation, isolated by silicic acid chromatography, are taken to dryness with a stream of nitrogen and redissolved in 2-propanol/hexane (<100 μl, 50/50, v/v) for injection on the self-packed preparative cyano-bonded column. Figure 1 shows a typical chromatogram with the absorption scale enlarged to emphasize the impurities. Total arachidonic acid metabolite recovery is typically 75–80% (i.e., percentage applied radioactivity). The largest peak (No. 6, 28 min) is PGH_2, which is collected and the solvents removed with a rotoevaporator attached to a pump. The residue, 50–120 μg PGH_2 of >97% purity, is dissolved in dry benzene and stored at $-80°$.

Other arachidonic acid metabolites (as indicated by radiochemical detection) may be seen as peaks Nos. 4, 5, 7, 8, and 9. These have been assigned as HETEs, HHT, PGD_2, PGE_2, and $PGF_{2\alpha}$, respectively.[3] In addition, non-arachidonic acid contaminants may be seen as peaks in the first 7 min of elution (peaks Nos. 1, 2, and 3) and as nonresolved adsorption between peak No. 7 and peak No. 9. Apparently because of non-prostaglandin impurities remaining in the silicic acid chromatographed samples, only about five injections are made before the column must be repacked or reversed and washed with hexane–2-propanol (50/50, v/v) overnight. It should be noted that different workers have found varying retention times with amount of column use and packing techniques. Consequently, care should be exercised in identifying the PGH_2-containing fractions from the semipreparative cyano-bonded column. This may be conveniently done by the usual bioassays or by the use of ammonia DCI–MS described herein.

Ammonia DCI mass spectra of the PGH_2 preparation, whose chromatogram is shown in Fig. 1, shows ions at 352, 370, and 387 for PGH_2 and its isomers PGE_2 and PGD_2. Additional peaks at 280, 298, and 315 are identified as HHT and its ammonia ion adducts, at 338 as HETEs, and at 354 as $PGF_{2\alpha}$. Using deuterated ammonia as the reagent gas, the ion currents corresponding to PGH_2 and the E_2 and D_2 series prostaglandins are separated and their intensity is found to compare well with the ratio of these metabolites as determined by chromatography (data not shown).

Benzene is removed at pump pressure from an aliquot of the PGH_2 (ca. 50 μg) as isolated from the semipreparative cyano-bonded column, the residue redissolved in hexane–2-propanol (50/50, v/v) and rechromatographed on the analytical cyano-bonded column. Figure 2 shows a typical chromatogram indicating that significant purification of the PGH_2 has been obtained. However, there still remain small ($<3\%$), but measurable, amounts of HHT (peak No. 2 at 7 min) and $PGD(E_2)$ (peak No. 3 at 18–19 min) as assigned by DCI–MS. Similar mass spectral analysis of peaks near the solvent front (largely injected 2-propanol) did not indicate the presence of any arachidonic acid. The increased background, beginning at about 16 min, is due to the 2-propanol gradient. The peak containing PGH_2 (No. 1, Fig. 2) is collected, the solvents removed under vacuum, and the PGH_2 redissolved in dry benzene and stored at $-80°$ for later use.

Figure 3 shows the ammonia DCI mass spectrum of PGH_2 as isolated from the analytical cyano-bonded column. Control spectra indicate that the peak at 111 amu is from the MS source and/or solvent used, and appears only when sample concentrations are low (ca. 20 ng). The peaks at 371, 372, 352, and 387 appear to arise only from PGH_2, with the major adduct ion at 370 showing a linear response between 40 and 1000 ng

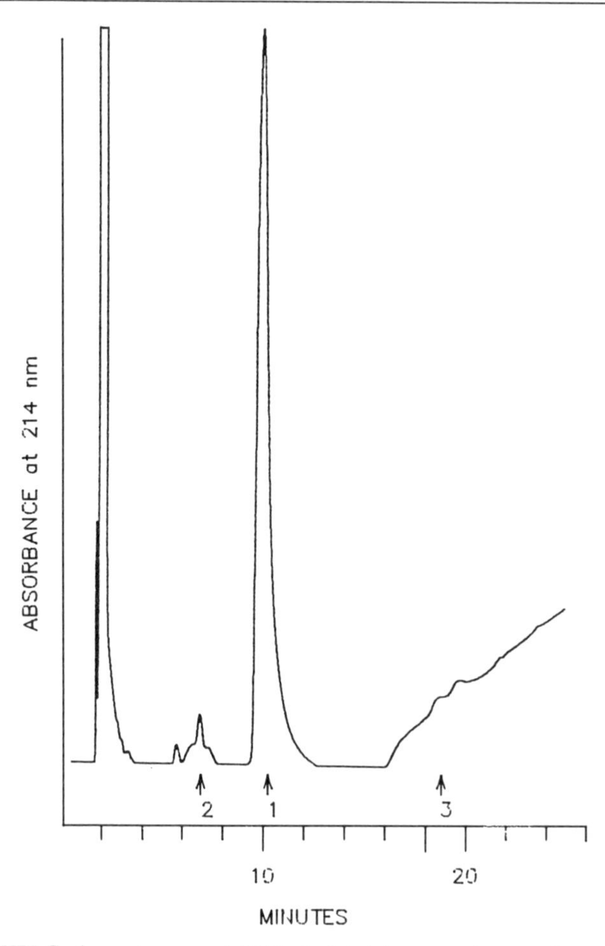

FIG. 2. HPLC chromatogram of PGH_2 isolated from semipreparative cyano-bonded column (see Fig. 1), reinjected on analytical cyano-bonded column. Chromatography conditions the same as given in Fig. 1 except that the gradient is started after 12, rather than 25, min. Peak No. 1, PGH_2; No. 2, HHT; No. 3, $PGD_2(E)_2$. See text for further discussion. Reproduced from Zulak *et al.*[3]

(r = 0.994, n = 6) with the lower limit of detection being about 3 ng. We interpret the peak at 298 amu as HHT arising from the decomposition of PGH_2 on the probe tip and/or from a gas phase reaction in the source. Thus, rechromatography of the sample remaining after MS on the analytical cyano-bonded column did not indicate any detectable $PGE_2(D_2)$ or HHT. In addition, ND_3 DCI–MS, which differentiates between PGH_2 and $PGE_2(D_2)$, supports this finding. Taken together, these studies suggest

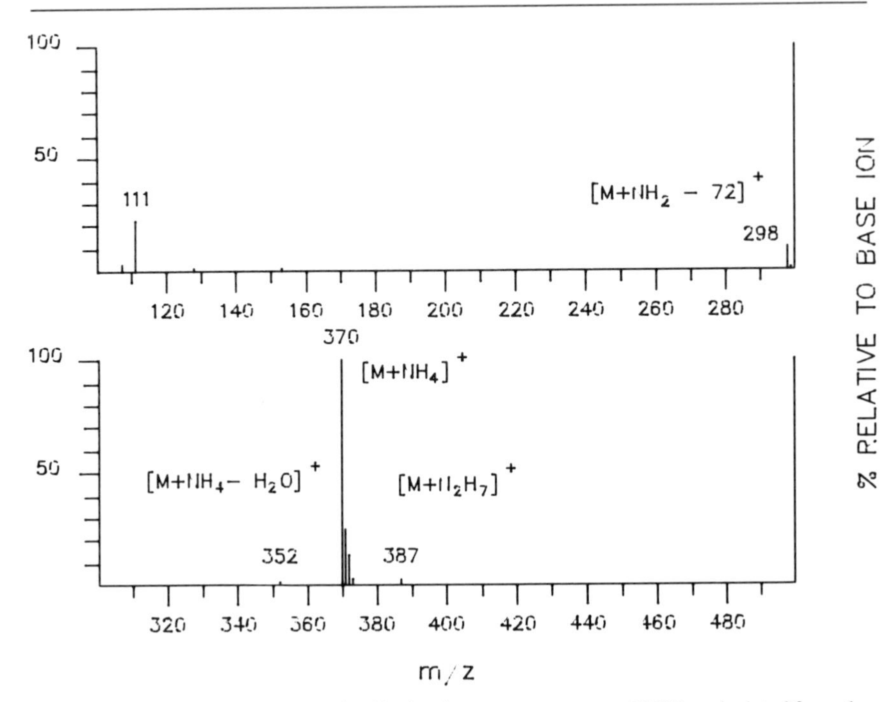

FIG. 3. Ammonia direct-chemical ionization mass spectrum of PGH_2 as isolated from the analytical cyano-bonded column (see Fig. 2, peak No. 1). See text for further details. Reproduced from Schilling *et al.*[5]

that the hexane–2-propanol, polar-bonded-phase chromatography system is nondestructive to the PGH_2 molecule. The PGH_2 prepared by these techniques appears to be homogenous and remains free of detectable decomposition products for 2 week (500 ng/μl dry benzene, a dry glass vial, −80°). After several weeks, some decomposition (<10%) may be detected by the techniques outlined herein. In our hands, silanized glass vials did not retard this degradation.

PGH_2 of approximately 97% purity as isolated by the semipreparative cyano-bonded column is likely to be sufficient for most biological tasks. When PGH_2 of greater purity is required, rechromatography of this preparation using an analytical cyano-bonded column can be used to produce essentially homogeneous PGH_2.

[28] Purification and Properties of Pregnancy-Inducible Rabbit Lung Cytochrome P-450 Prostaglandin ω-Hydroxylase

By A. SCOTT MUERHOFF, DAVID E. WILLIAMS, and BETTIE SUE SILER MASTERS

Pulmonary microsomes from pregnant rabbits or female rabbits pretreated with progesterone or human chorionic gonadotropin catalyze the ω-hydroxylation of prostaglandins and related eicosanoids, an activity which is very low or absent in nonpregnant or control rabbits.[1,2] Williams *et al.*[3] described the isolation of a unique cytochrome P-450 (P-$450_{PG\omega}$) from lung microsomes of pregnant rabbits (day 25–28 of gestation) which catalyzes the regiospecific hydroxylation of prostaglandins of the E, F, and A series at the ω- or 20-position. This P-450 does not catalyze the ω- or (ω-1)-hydroxylation of lauric acid or the N-demethylation of benzphetamine—activities which are associated with distinct rabbit lung cytochromes P-450.[3,4] A similar, if not identical, cytochrome P-450 [P-$450_{p\text{-}2}$ or P-450IVA4 (Ref. 5)] has been purified from pulmonary microsomes of female rabbits pretreated with progesterone and has been shown to catalyze the ω-hydroxylation of medium- and long-chain fatty acids and prostaglandins.[6] The procedures described herein for the purification of the pregnancy-inducible rabbit lung cytochrome P-450 prostaglandin ω-hydroxylase follow the methods of Williams *et al.*[3] with some modifications.

Purification and Properties of P-$450_{PG\omega}$

Principle. The purification of the cytochrome P-$450_{PG\omega}$ from rabbit pulmonary microsomes is based on a procedure originally described by

[1] W. S. Powell, and S. Solomon, *J. Biol. Chem.* **253,** 4609 (1978).

[2] W. S. Powell, *J. Biol. Chem.* **253,** 6711 (1978).

[3] D. E. Williams, S. E. Hale, R. T. Okita, and B. S. S. Masters, *J. Biol. Chem.* **260,** 14600 (1984).

[4] C. J. Serabjit-Singh, C. R. Wolf, I. G. C. Robertson, and R. M. Philpot, *J. Biol. Chem.* **254,** 9901 (1979).

[5] D. W. Nebert, D. R. Nelson, M. Adesnik, M. J. Coon, R. W. Estabrook, F. J. Gonzalez, F. P. Guengerich, I. C. Gunsalus, E. F. Johnson, B. Kemper, W. Levin, I. R. Phillips, R. Sato, and M. R. Waterman, *DNA* **8,** 1 (1989).

[6] S. Yamamoto, E. Kusunose, K. Ogita, M. Kaku, K. Ichihara, and M. Kusunose, *J. Biochem.* (*Tokyo*) **96,** 593 (1984).

Guengerich,[7] as modified by Williams *et al.*[3] The procedure involves the initial isolation of total rabbit lung microsomal cytochrome *P*-450 by aminooctyl-Sepharose chromatography followed by separation of the rabbit lung cytochromes *P*-450 by chromatography on DEAE-Sepharose. Removal of detergent from the enzymes is then performed by chromatography on hydroxylapatite or by hydrophobic absorption chromatography. The procedure described by Guengerich[7] has been used for the copurification of NADPH–cytochrome-*P*-450 reductase and cytochrome b_5. However, we have preferred the methods of Yasukochi and Masters[8] and Strittmatter *et al.*[9] for the purification of NADPH–cytochrome-*P*-450 reductase and cytochrome b_5, respectively.

Procedures

Materials. Whole lungs from pregnant rabbits at 25–28 days of gestation are obtained from Pel Freez Biologicals (Rogers, AR). The lungs are perfused with saline and then frozen in liquid nitrogen prior to shipment on dry ice. The lungs are stored at −80° until used for preparation of microsomes. Sepharose 4B and DEAE-Sepharose CL-6B are from Pharmacia-LKB Biotechnology, Inc. (Piscataway, NJ). Hydroxylapatite is prepared by the method of Tiselius *et al.*[10] Aminooctyl-Sepharose 4B is prepared in the laboratory by the method of Imai and Sato.[11] Cholic acid (Aldrich Chemicals, Milwaukee, WI) is recrystallized from ethanol prior to use.

Preparation and Detergent Solubilization of Microsomes

Microsomes are prepared from 36 pairs of lungs from 25 to 28-day pregnant rabbits as described by Williams *et al.*[3] The protein concentration of the microsomal suspension is determined by the method of Lowry *et al.*[12] and the total *P*-450 content is determined by the method of Remmer *et al.*[13] using 91 mM^{-1} cm^{-1} as the extinction coefficient for the 450–490 nm absorbance difference.[14] Microsomes are diluted to 2 mg/ml in cold 100 m*M* potassium phosphate buffer, pH 7.25, containing 20% glycerol, (v/v) 1 m*M* EDTA, 1 m*M* dithiothreitol (DTT), and 0.1 m*M* phenylmethylsulfonyl fluoride (PMSF). Recrystallized cholate [sodium salt, 20% stock

[7] F. P. Guengerich, *Mol. Pharmacol.* **13,** 911 (1977).
[8] Y. Yasukochi and B. S. S. Masters, *J. Biol. Chem.* **251,** 5337 (1976).
[9] P. Strittmatter, P. Fleming, M. Connors, and D. Concoran, this series, Vol. 52, p. 97.
[10] A. Tiselius, S. Hjerten, and O. Levin, *Arch. Biochem. Biophys.* **65,** 132 (1956).
[11] Y. Imai and R. Sato, *J. Biochem.* **75,** 689 (1974).
[12] O. H. Lowry, N. J. Rosebrough, A. L. Farr, and R. J. Randall, *J. Biol. Chem.* **193,** 265 (1951).
[13] H. Remmer, H. Greim, J. B. Schenkman, and R. W. Estabrook, this series, Vol. 10, p. 703.
[14] T. Omura and R. Sato, *J. Biol. Chem.* **239,** 2379 (1964).

solution in distilled water (w/v), pH 7.0] is then added dropwise, over a period of 15–20 min while stirring at 4°, to a final concentration of 0.6%. The solution is stirred vigorously, but without foaming, for an additional 30 min and the unsolubilized material is removed by centrifugation at 106,000 *g* for 90 min at 4°. Following centrifugation, the tubes are carefully removed from the rotor so as not to disturb the lipid layer which often forms on top of the supernatant. This layer is removed by aspiration prior to recovery of the supernatant.

Aminooctyl-Sepharose 4B Chromatography

The solubilized microsomal protein solution is applied to an aminooctyl-Sepharose (AOS) column (2.5 × 25 cm) equilibrated previously in 100 m*M* potassium phosphate buffer, pH 7.25, containing 0.7% cholate (w/v), 20% glycerol (v/v), 1 m*M* EDTA, 1 m*M* DTT, and 0.1 m*M* PMSF at a flow rate of 30–40 ml/hr. During the loading procedure, the cytochrome *P*-450 binds to the top of the column as a reddish-brown band which gradually lengthens during the course of the sample application but does not move down the column. Cytochrome b_5 elutes from the column in the void volume. The AOS column is then washed with 4–5 column volumes of 100 m*M* potassium phosphate buffer, pH 7.25, containing 0.46% cholate (w/v), 20% glycerol (v/v), 1 m*M* EDTA, 1 m*M* DTT, and 0.1 m*M* PMSF at 30 to 40 ml/hr. The cytochrome *P*-450 bound to the column becomes a deeper red color during the wash step but remains bound to the top of the column.

The bound cytochrome *P*-450 is eluted using 100 m*M* potassium phosphate buffer, pH 7.25, containing 0.37% cholate (w/v), 0.08% Lubrol PX (w/v), 20% glycerol (v/v), 1 m*M* EDTA, 1 m*M* DTT, and 0.1 m*M* PMSF. Fractions collected during elution are monitored at 280 and 416 nm as a measure of protein and heme content, respectively. Under these conditions, a single hemoprotein peak is eluted from the column. The peak fractions are pooled and the *P*-450 concentration determined from the CO-difference spectrum.[13] Typical yields of *P*-450 from the AOS column are 45–55% of the total rabbit lung microsomal cytochrome *P*-450 (see Table I).

The *P*-fraction obtained from the AOS column is concentrated approximately 4-fold (but not more than 5-fold) using an Amicon Ultrafiltration Device and a PM-30 ultrafiltration membrane (Amicon Corp., Danvers, MA). The concentrated sample is dialyzed for 2 hr at 4° against 1 liter of distilled water containing 20% glycerol (v/v), 1 m*M* EDTA, 1 m*M* DTT, and 0.1 m*M* PMSF and then overnight against 10 m*M* potassium phosphate buffer, pH 7.7, containing 20% glycerol, 0.1 m*M* EDTA, 0.1 m*M* DTT, 0.1 m*M* PMSF, 0.1% cholate (w/v), and 0.2% Lubrol PX (w/v) (buffer A) in preparation for ion-exchange chromatography.

TABLE I
PURIFICATION OF THREE FORMS OF CYTOCHROME *P*-450 FROM RABBIT LUNG MICROSOMES

Purification step	Protein (mg/ml)	Volume (ml)	Total protein (mg)	*P*-450 nmol/ml)	Total *P*-450 (nmol)	Specific content (nmol/mg)	Recovery (% microsomal *P*-45[illegible]
Microsomes	12.5	100	1250	5.78	578	0.46	100
AO-Sepharose	3.0	50	150	6.04	302	2.01	52.2
DEAE-Sepharose 4B							
Peak I (form 2)	0.76	95	72.2	1.72	164	2.26	28.4
Peak II (PGω)	0.58	47	27.3	1.33	62.5	2.29	10.8
Peak III (form 5)	0.54	16	8.6	0.54	8.6	1.00	1.5
Totals		158	108		235		40.7
Hydroxylapatite							
Form 2	8.30	2.0	16.6	66.0	132	8.0	22.8
PGω	0.82	1.5	1.23	4.6	6.9	5.6	1.2
Form 5	0.46	1.0	0.46	4.5[b]	4.5	n.d.[c]	n.d.
Extracti-Gel D							
PGω	0.97	8.0	7.8	3.0	24.0[d]	3.1	4.1
Total PGω recovery			9.0		30.9		5.3

[a] Prepared from 36 pairs of lungs from 25- to 28-day pregnant rabbits.
[b] Based on 416 nm absorbance, CO-difference spectrum was not obtainable with this particular enzym[e] preparation.
[c] n.d., Not determined.
[d] The *P*-450 $_{PG\omega}$ was first applied to the hydroxylapatite column; that which did not bind was applied to t[he] Extracti-Gel D column.

The aminooctyl-Sepharose is regenerated according to the method of Guengerich[7] and can be reused up to three times without significant loss of binding capacity or resolution; with further use, however, losses in recovery become noticeable. The gel can be stored for several months in distilled water containing 0.02% sodium azide; however, extensive washing with 3–4 liters of distilled water is recommended prior to the next use.

Ion-Exchange Chromatography on DEAE-Sepharose

The separation of rabbit lung *P*-450 isoforms is performed on a DEAE-Sepharose CL-6B column (1.5 × 24 cm), equilibrated in buffer A. The dialyzed sample is applied to the ion-exchange column at a flow rate of 15 to 20 ml/hr while collecting 3 to 4-ml fractions. The column is washed with buffer A at which time the first hemoprotein band separates from the bound material. As this band reaches the bottom of the column, the two remaining *P*-450s are eluted with a linear gradient of 0 to 100 m*M* NaCl (in buffer A, 500 ml total volume). The 280 and 416 nm absorbances of the

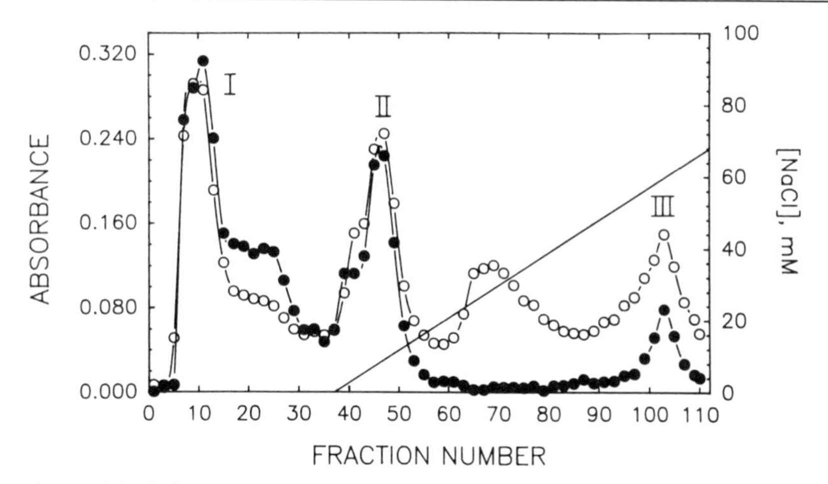

FIG. 1. DEAE-Sepharose elution profile of cytochromes P-450 from lung microsomes of pregnant rabbits. The fractions containing cytochrome P-450 obtained from the AOS column are pooled, concentrated, dialyzed, and then loaded onto a DEAE-Sepharose column and the P-450s eluted as described in the text. The absorbances at 416 nm (filled circles) and 280 nm (open circles) of the fractions are shown. The NaCl gradient is indicated by the solid line.

fractions are shown in Fig. 1. Peak I corresponds to cytochrome P-450 form 2 [P450IIB4 (Ref. 5)] which catalyzes the hydroxylation of various drugs and xenobiotics; peak II corresponds to P-$450_{PG\omega}$, which catalyzes the ω-hydroxylation of eicosanoids; and peak III corresponds to P-450 form 5 [P450IVB1 (Ref. 5)], which catalyzes the ω- and (ω-1)-hydroxylation of lauric acid[3] as well as the oxidation of 2-aminofluorene[15] and 4-ipomeanol.[16] Prior to pooling fractions from each of the peaks, and in order to ensure homogeneity of the pooled samples, aliquots of every other fraction are analyzed for purity by SDS–PAGE[17] on 10% polyacrylamide gels (data not shown). The gels revealed that the trailing shoulder on peak I contains small amounts of P-$450_{PG\omega}$ and the leading shoulder on peak II contains small amounts of P-450 form 2; therefore, only those peak fractions which appeared to be homogeneous by SDS–PAGE were pooled. The recovery of all forms of rabbit lung cytochrome P-450 from the ion-exchange column is typically 30–40% relative to the P-450 present in the microsomal fraction (see Table I).

[15] I. G. C. Robertson, R. M. Philpot, E. Zeiger, and C. R. Wolf, *Mol. Pharmacol.* **20,** 662 (1981).

[16] C. R. Wolf, C. N. Statham, M. G. McMenamin, J. R. Bend, M. R. Boyd, and R. M. Philpot, *Mol. Pharmacol.* **22,** 738 (1982).

[17] U. K. Laemmli, *Nature (London)* **227,** 680 (1970).

TABLE II
PGE_1 ω-HYDROXYLASE ACTIVITIES ASSOCIATED WITH SEVERAL CYTOCHROME $P\text{-}450_{PG\omega}$ PREPARATIONS[a]

$P\text{-}450_{PG\omega}$ preparation	PGE_1 ω-hydroxylation (nmol/min/nmol P-450)
1	2.32[b]
2	2.74[c]
3	1.94
4	1.94
5	2.11[d]
5	1.25[e]

[a] Reconstitutions were performed as described in the Methods except where noted. Values represent the average of duplicate determinations.

[b] The ratio of ionic to nonionic detergent used during ion-exchange chromatography was 2 : 1.

[c] The ratio of ionic to nonionic detergent used during ion-exchange chromatography in preparations 2–5 was 1 : 2. Hydroxylapatite was used for detergent removal in preparations 2–4.

[d] Detergent removal was by hydrophobic absorption chromatography.

[e] Didecanoylphosphatidylcholine (DDPC) was substituted for DLPC in the reconstitution system.

The separation of these three forms of rabbit lung cytochromes P-450 by ion-exchange chromatography has been achieved using a ratio of ionic (cholate) to nonionic (Lubrol PX) detergent of 2 : 1, in contrast to the 1 : 2 ratio described herein. With a ratio of 2 : 1, slightly lower yields of $P\text{-}450_{PG\omega}$ are obtained; however, the yield of P-450 form 5 is increased by as much as 50%. The cytochrome $P\text{-}450_{PG\omega}$ isolated under either of these conditions does not display any apparent difference in reaction rate with PGE_1 (see Table II).

Detergent Removal

Principle. The presence of detergents in the purified P-450 sample can have detrimental effects on reconstitution reactions by potentially disrupting the interaction between cytochrome P-450, NADPH–cytochrome-P-450 reductase, and/or cytochrome b_5, as well as affecting the solubility and/or accessibility of the substrate. Therefore, detergent is removed from the DEAE-isolated proteins by one of two methods: hydroxylapatite chromatography or hydrophobic absorption chromatography.

Hydroxylapatite Chromatography. The pooled fractions obtained from the DEAE column from peaks I (P-450 form 2) and II (P-$450_{PG\omega}$) are adjusted to pH 7.25 and applied to 1×5 cm and 1×3 cm hydroxylapatite columns, respectively, which have been equilibrated with 10 mM potassium phosphate buffer, pH 7.25, containing 20% glycerol, 0.1 mM EDTA, 0.1 mM DTT, and 0.1 mM PMSF (buffer B). The pooled fractions from peak III (P-450 form 5) are diluted 3-fold with buffer B, adjusted to pH 7.25, and applied to a 1×2 cm hydroxylapatite column. The hydroxylapatite columns are washed with 250 ml of buffer B to remove detergent from the hydroxylapatite-bound enzymes. The enzymes are then eluted with 0.30 M potassium phosphate buffer, pH 7.25, containing 20% glycerol, 0.1 mM EDTA, 0.1 mM DTT, and 0.1 mM PMSF. The eluted P-450s are dialyzed overnight at 4° against 50 mM potassium phosphate buffer, pH 7.4, containing 20% glycerol, 0.1 mM EDTA, 0.1 mM DTT, and 0.1 mM PMSF. CO-difference spectra of the detergent-free cytochromes P-450 are measured and used to calculate the enzyme concentration.[13] The three enzymes exhibit unique absorption maxima in their CO-difference spectra: P-450 form 2, 451 nm; P-$450_{PG\omega}$, 450 nm; and P-450 form 5, 449 nm. The protein concentration of the purified P-450s is determined by the Bio-Rad Protein Assay (Bio-Rad, Richmond, CA).

Hydrophobic Absorption Chromatography. The pooled fractions from peak II (P-$450_{PG\omega}$) obtained from the DEAE column are applied to an Extracti-Gel D detergent removal column (Pierce Chemical, Co., Rockford, IL) which had been equilibrated in buffer B. This gel has a high binding capacity for detergents such as cholate and CHAPS (50 mg/ml gel) and Lubrol PX (106 mg/ml gel). The enzyme sample is collected in a single fraction as it passes through the column and is then concentrated using an Amicon Ultrafiltration Device and a PM-30 membrane, if necessary. A CO-difference spectrum of the detergent-free P-450 is obtained and used to calculate the enzyme concentration.

There are advantages and disadvantages to the two detergent removal techniques with respect to the particular form of rabbit lung P-450. P-450, form 2, binds very efficiently to hydroxylapatite under low ionic strength conditions and is readily eluted by the 0.30 M potassium phosphate buffer; thus, the recovery is very high (see Table I). In addition, the purity of P-450, form 2, is improved following the procedure, as evidenced by the 3-fold increase in specific content (see Table I). P-$450_{PG\omega}$ does not bind well to hydroxylapatite (various ionic strength and pH conditions have been examined—all resulting in inefficient binding), thus, the recovery of P-450$_{PG\omega}$ by this method is low, although that which is obtained exhibits a 2.5-fold increase in specific content over the DEAE-isolated enzyme. Hydrophobic absorption chromatography on Extracti-Gel D of P-450$_{PG\omega}$ is much faster than hydroxylapatite chromatography and produces better recovery of enzyme although the increase in specific content is minimal. Therefore, this method does not offer a means for further purification and is used for detergent removal only. Rabbit lung cytochrome P-450 form 5 also binds poorly to hydroxylapatite (see Table I) and, therefore, detergent removal via Extracti-Gel D (or a similar procedure) is the method of choice for this form. It is important to note that P-450$_{PG\omega}$ subjected to either of these detergent removal procedures displays no significant differences in the rates of PGE_1 ω-hydroxylation (see Table II). The enzyme is stored for up to 7 months at $-80°$ without loss of activity.

Analysis of the purified, detergent-free rabbit lung cytochromes P-450 by discontinuous SDS–PAGE[17] followed by silver staining of the gel[18] demonstrates their differences in relative molecular weight as well as their high degree of purity (see Fig. 2). The gel indicates that the enzymes isolated by this procedure are greater than 95% pure. The apparent purity demonstrated by SDS–PAGE suggests that the relatively low specific content of the P-450s may be due to the loss of heme during the purification procedure.

Measurement of Prostaglandin ω-Hydroxylase Activity

Principle. The conversion of radiolabeled prostaglandins and other eicosanoids to their ω-hydroxy derivatives by the purified rabbit lung cytochrome P-450$_{PG\omega}$ is determined by integration of peak areas as detected by an on-line radioactivity flow detector following separation of radiolabeled substrate and product by reversed-phase high-performance liquid chromatography (RP-HPLC).[19,20]

Procedures

Reagents

[5,6(n)-^{3}H]PGE_1 (Amersham, Arlington Heights, IL) and PGE_1 (unlabeled; commercially available from Cayman Chemicals, Ann

[18] W. Wray, T. Boulikas, V. P. Wray, and R. Hancock, *Anal. Biochem.* **118,** 197 (1981).

[19] R. T. Okita, R. J. Soberman, J. M. Bergholte, B. S. S. Masters, R. Hayes, and R. C. Murphy, *Mol. Pharmacol.* **32,** 706 (1987).

[20] M. T. Leithauser, D. L. Roerig, S. M. Winquist, A. Gee, R. T. Okita, and B. S. S. Masters, *Prostaglandins* **36,** 819 (1988).

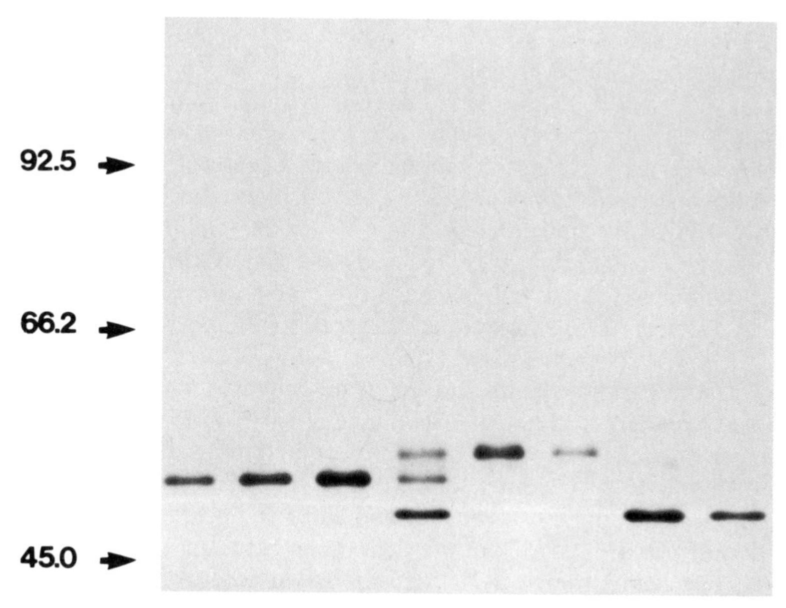

FIG. 2. SDS–PAGE of purified rabbit lung cytochromes P-450. Samples are diluted with sample buffer (1:2) containing 50 mM Tris, pH 6.8, 10% glycerol (v/v), 2 mM EDTA, 1% 2-mercaptoethanol (v/v), and 0.05% bromphenol blue (w/v) and heated at 100° for 2 min. The samples are then electrophoresed through a gel containing a 3% polyacrylamide stacking gel and a 10% separating gel according to the method of Laemmli.[17] Lanes 1–3 contain 0.25, 0.50, and 0.75 μg of P-450$_{PG\omega}$, respectively; lane 4, a mixture of 0.25 μg each of the three rabbit lung P-450s; lanes 5 and 6, 0.50 and 0.25 μg P-450 form 5, respectively; lanes 7 and 8, 0.50 and 0.25 μg P-450 form 2, respectively. The molecular weights ($\times 10^{-3}$) of the standards are shown at the left.

Arbor, MI) are mixed together to prepare a final stock solution of 10 mM [^{3}H]PGE$_1$ (2-4 μCi/μmol) in 100% methanol

Other radiolabeled substrates are from Amersham; unlabeled eicosanoids are available from Cayman Chemicals.

100 mM potassium phosphate buffer, pH 7.4, containing 5 mM $MgCl_2$, and 1 mM EDTA

Purified rabbit lung cytochrome P-450$_{PG\omega}$

Detergent-solubilized NADPH–cytochrome-P-450 reductase is purified by the method of Yasukochi and Masters[8] and cytochrome b_5 is purified by the method of Strittmatter *et al.*[9]

Dilauroylphosphatidylcholine (DLPC), 1 mg/ml, in buffer
Isocitrate, 0.20 *M*, in buffer
Isocitrate dehydrogenase, Type IV from bovine heart
(Sigma Chemical Co., St. Louis, MO)
NADPH, 50 m*M*, in buffer
Ethyl acetate (HPLC grade)

Assay. [^{3}H]PGE_1 is added to a glass reaction vial (typically 16 × 125 mm round-bottom, screw-cap tubes are used) to a final concentration of 50 μM and the methanol evaporated under a gentle stream of nitrogen. The following components are then added in the order indicated: 50 μg DLPC (sonicated immediately prior to use), equimolar NADPH–cytochrome-*P*-450 reductase, cytochrome b_5, and cytochrome $P\text{-}450_{PG\omega}$. These components are combined such that a single drop is formed on the bottom of the tube. The mixture is incubated for 10 min at room temperature. An NADPH-regenerating system consisting of 20 m*M* isocitrate (final concentration) and 0.026 units of isocitrate dehydrogenase is then added followed by buffer to obtain a final volume of 1.0 ml. The reaction mixture is preincubated for 2 min at 37° and the reaction initiated by the addition of NADPH to a final concentration of 1 m*M*. The reconstituted system is incubated at 37° for 20 to 40 min and the reaction terminated by the addition of 1 *N* HCl such that the final pH is between 3 and 4 (typically 90 μl). The reaction mixture is then extracted twice with 3 ml of ethyl acetate. The organic layers are combined in a silanized, conical glass tube and the ethyl acetate is evaporated under nitogen. The sample can be stored in the capped tube at −20° for several weeks prior to analysis for the presence of hydroxylated products by reversed-phase HPLC. Reactions involving polyunsaturated fatty acids such as arachidonic acid should be analyzed within 1 to 2 days.

Microsomal prostaglandin ω-hydroxylase activity is determined essentially as described above except that the microsomes replace the NADPH-cytochrome *P*-450 reductase, pure *P*-450, and cytochrome b_5 in the reaction mixture. Reactions containing microsomal concentrations over 0.50 mg/ml are extracted with ethyl acetate as described above and then back-extracted with 3 ml of distilled water in order to remove excess protein. The aqueous layer is then discarded. Typical specific activities obtained with pulmonary microsomes prepared from 25 to 28-day pregnant rabbits are 2.5–3.0 nmol of ω-hydroxy-PGE_1 formed per minute per milligram microsomal protein.

Reversed-Phase HPLC. The dried sample is resuspended in 0.20 ml of the HPLC solvent (39.8% CH_3CN containing 0.2% benzene/59.8% water with 0.2% acetic acid) by vortexing and 0.06 ml of the sample is injected onto the reversed-phase column (Bio-Rad Bio-Sil ODS-5S (C_{18}), 5μm particle size, 150 × 4 mm). Elution is performed isocratically using the HPLC solvent described above at a flow rate of 0.7 ml/min using a Varian Model 5000 Liquid Chromatograph with simultaneous monitoring of radioactivity by a Radiomatic Radioactive Flow Detector (Radiomatic Instruments and Chemical Co., Inc., Tampa, FL). Under these conditions, unmetabolized PGE_1 elutes with a retention time of 7.0 min and ω-hydroxy-PGE_1 elutes with a retention time of 3.1 min (see Fig. 3A). The integrated areas of the peaks, as determined by the microprocessor of the radioactive flow detector, are used to determine the percentage conversion of substrate to product, which, when multiplied by the number of nanomoles of substrate used in the reaction, divided by the total reaction time, will give the rate of the reaction in nanomoles per minute.

The above *in vitro* reconstitution procedure is used to examine the ability of purified P-$450_{PG\omega}$ to ω-hydroxylate products of the cyclooxygenase or lipoxygenase pathways as well as fatty acids. However, the reversed-phase HPLC solvent systems and elution procedures utilized to analyze the reaction products vary depending upon the substrate used.

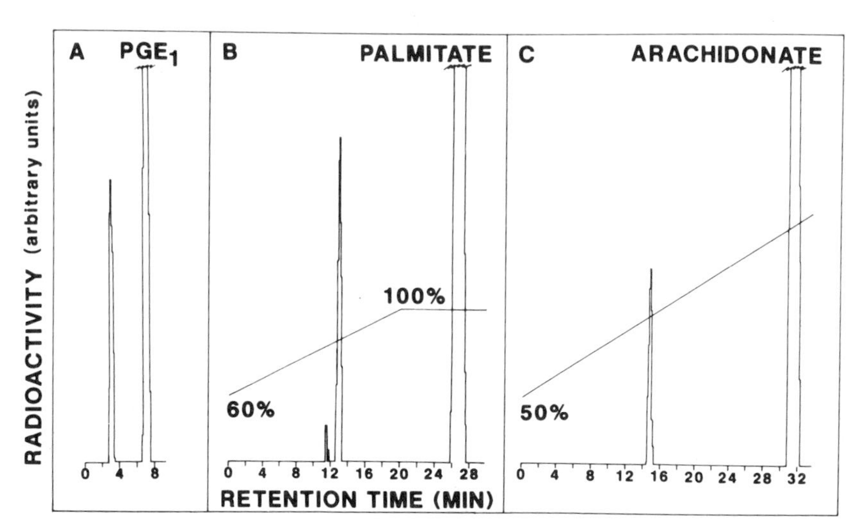

FIG. 3. RP-HPLC elution profiles of oxidation products formed upon incubation of PGE_1, palmitate, or arachidonate with cytochrome P-$450_{PG\omega}$, in a reconstituted system. Cytochrome P-$450_{PG\omega}$ (25 pmol) is reconstituted *in vitro* with equimolar NADPH–cytochrome-P-450 reductase and cytochrome b_5, 50 μg of DLPC, an NADPH-regenerating system, and either [^{3}H]PGE_1 (25 μM final concentration), [^{14}C]palmitate (50 μM final concentration), or [^{14}C] arachidonate (50 μM final concentration) with specific activities of 1–2 μCi/μmol. The reactions are incubated at 37° for 30 min, terminated by addition of 1 N HCl (final pH 3.5), and then extracted with ethyl acetate and analyzed by reversed-phase HPLC as described in the Methods. Retention times of the metabolites are given in the text.

The analysis of the palmitate reaction products involves resuspension of the sample in 60% CH_3CN/40% water and elution of the substrate and products from the reversed-phase column with a linear gradient from 60 to 100% acetonitrile (2%/min).[21] Under these conditions, the retention times of palmitate and the product of the reaction, which is presumed to be ω-hydroxypalmitate, are 26.6 and 13.0 min, respectively (see Fig. 3B). Analysis of arachidonic acid reaction products[22] involves resuspension of the sample in 49.8% CH_3CN containing 0.20% benzene/49.8% water with 0.20% acetic acid and elution with a linear gradient from 50 to 100% CH_3CN over 40 min (1.25% min). Unmetabolized arachidonic acid elutes with a retention time of 31.5 min and the product of the reaction, which is presumed to be ω-hydroxyarachidonate, elutes at 15.0 min (see Fig. 3C). Analysis of lauric acid reaction products is by the method of Okita and Chance.[23]

ω-Hydroxylase Activities of P-450$_{PG\omega}$

The rates of ω-hydroxylation of products of the cyclooxygenase and lipoxygenase pathways and various fatty acids by the purified, detergent-free rabbit lung P-450$_{PG\omega}$ are shown in Table III. The enzyme exhibits highest turnover rates with PGE_1 as the substrate in the reconstitution system described above and has a K_m for this prostaglandin of 3 μM.[3] Lung microsomes isolated from pregnant rabbits have been shown to catalyze the ω-hydroxylation of 15-hydroxyeicosatetraenoic acid (15-HETE) but not of 5- or 12-HETE or leukotriene B_4.[19] The reconstituted P-450$_{PG\omega}$, purified from lung microsomes of pregnant rabbits, catalyzes the ω-hydroxylation of 15-HETE at rates comparable to those observed with prostaglandins. The turnover rates with palmitate or myristate are significantly lower than that observed with PGE_1. This is in contrast to the findings of Yamamoto *et al.*[6] who demonstrated that P-450$_{p\text{-}2}$, purified from lung microsomes of progesterone-treated female rabbits, catalyzes the ω-hydroxylation of palmitate at rates 3–5 times greater than that observed with prostaglandins. Although the reasons for these differences in reaction rates are not clear, possible explanations include differences in assay procedure or purification methods which could fractionate distinct but closely related P-450s. P-450$_{PG\omega}$ also oxidizes arachidonic acid to a more polar product which elutes as a single peak from the reversed-phase

[21] R. T. Okita, unpublished observations

[22] J. Capdevila, Y. R. Kim, C. Martin-Wixtrom, J. R. Falck, S. Manna. and R. W. Estabrook, *Arch. Biochem. Biophys.* **243,** 8 (1985).

[23] R. T. Okita, and C. Chance, *Biochem. Biophys. Res. Commun.* **121,** 304 (1984).

TABLE III
ω-HYDROXYLASE ACTIVITY OF RECONSTITUTED RABBIT LUNG CYTOCHROME $P\text{-}450_{PG\omega}$ TOWARD VARIOUS SUBSTRATES[a]

Substrate	ω-Hydroxylase activity (nmol/min/nmol/P-450)
PGE_1	2.88
PGA_1	2.15
PGD_2	1.08
15-HETE	1.68
Arachidonate	1.05[b]
Stearate	nd[c]
Palmitate	0.41
Myristate	0.27
Laurate	nd

[a] Reactions were performed as described in the Methods. The substrates were present in the reaction mixture at 40–50 μM except for PGE_1 and 15-HETE which were present at 25 and 16 μM, respectively. Metabolites of stearate and myristate were detected as described in the Methods for palmitate. Metabolites of 15-HETE were detected by reversed-phase-HPLC as previously described.[19] Values represent the averages of duplicate determinations.

[b] This value represents the rate of conversion of arachidonic acid to the metabolite(s) present in the single peak isolated by reversed-phase-HPLC (see Fig. 3c).

[c] nd, Not detected.

column (see Fig. 3c). The retention times of this metabolite, as determined by reversed-phase and normal phase HPLC by Capdevila, are identical to that of authentic 20-hydroxyarachidonate identified by mass spectrometry according to the procedures of Capdevila *et al.*[22]

N-Terminal Sequence Analysis of P-450 Form 5 and $P\text{-}450_{PG\omega}$

Principle. N-terminal sequence analysis of the purified rabbit lung P-450s provides useful information for determination of the similarity between various P-450 forms and is also a means for the determination of the purity of the protein preparation.

TABLE IV
N-TERMINAL SEQUENCES OF SEVERAL CYTOCHROMES *P*-450

	N-terminal sequence																
P-450	1	2	3	4	5	6	7	8	9	10	11	12	13	14	15	16	Ref.
Form 2	N-Met	Glu	Phe	Ser	Leu	Leu	Leu	Leu	Leu	Ala	Phe	Leu	Ala	Gly	Leu	Leu	24
Form 5	N-Met	Leu	Gly	Phe	Leu	Ser	Arg	Leu	Gly	Leu	Trp	Ala	Ser	Gly	Leu	Ile	24
Form 5[a]	N-Met	Leu	Gly	Phe	Leu	Ser	-X-	-X-	-X-	-X-	Trp	Ala	Ser	Gly	Leu	Ile	this chapter
PGω	N-Ala	Leu	Ser	Pro	Thr	Arg	Leu	Pro	Gly	Ser	Leu	Ser	Gly	Leu	Leu	Gln	this chapter
p-2	N-Ala	Leu	Ser	Pro	Thr	Arg	Leu	Pro	Gly	Ser	Leu	Ser	Gly	Leu	Leu	Gln	25
LAω[b]	Ala	Leu	Ser	Ser	Thr	Arg	Phe	Thr	Gly	Ser	Ile	Ser	Gly	Phe	Leu	Gln	26

[a] Residues 7–10 were not identified.
[b] The first four amino acids were not included in order to facilitate alignment with the rabbit lung cytochrome $P\text{-}450_{PG\omega}$, namely, N-Met-Ser-Val-Ser- (26).

Methods and Results

The P-450s are prepared for sequencing by dialysis against 4 liters of distilled water containing 0.01% acetic acid in a multiple-well minidialysis apparatus. Dialysis of the sample is necessary in order to remove glycerol, PMSF, and DTT which may interfere with sequencing reactions. The microsequence analysis was conducted using an Applied Biosystems Model 447A Pulsed Liquid Phase Peptide Sequencer at the Protein and Nucleic Acid Shared Research Facility at the Medical College of Wisconsin, Milwaukee, WI. The results of triplicate analyses of P-$450_{PG\omega}$ were identical and the data obtained did not demonstrate the simultaneous sequencing of multiple proteins, indicating the homogeneity of the enzyme preparation (data not shown). The N-terminal sequences obtained for P-450 form 5 and P-$450_{PG\omega}$ are compared (Table IV) with those of rabbit lung cytochrome P-450 form 2 and form 5 reported by Parandoosh *et al.*,[24] rabbit lung P-$450_{p\text{-}2}$ reported by Matsubara *et al.*,[25] and the rat liver clofibrate-inducible lauric acid ω-hydroxylase (P-$450_{LA\omega}$).[25] The amino terminal sequence of rabbit lung P-450 form 5 obtained by Parandoosh *et al.*[24] is identical to those residues identified for cytochrome P-450 form 5 isolated from lung microsomes of pregnant rabbits. Cytochrome P-$450_{PG\omega}$ shares complete sequence homology through the first 16 amino acids with cytochrome P-$450_{p\text{-}2}$, isolated from either progesterone-treated or pregnant animals, the N-terminal amino acid sequences of which were determined by Matsubara *et al.*[25] This similarity in N-terminal sequence is suggestive, but not definitive proof, of the identity of the progesterone-inducible and the pregnancy-inducible enzymes. It is important to note in this regard that a lauric acid-hydroxylating cytochrome P-450 [P-450IVA5 (Ref. 5)] has been expressed in monkey kidney cells (COS-1 cells) which shares the identical N-terminal sequence with P-$450_{PG\omega}$ and cytochrome P-$450_{p\text{-}2}$ of Matsubara *et al.*,[25] but exhibits no prostaglandin ω-hydroxylase activity.[27] Therefore, caution must be taken when in interpreting these N-terminal sequences.

Acknowledgments

This work was supported by NIH GM-31296 to BSSM. We gratefully acknowledge Dr. Richard T. Okita for his valuable advice and many helpful suggestions throughout the course of this work.

[24] Z. Parandoosh, V. S. Fujita, M. J. Coon, and R. M. Philpot, *Drug. Met. Dispos.* **15,** 59 (1987).

[25] S. Matsubara, S. Yamamoto, K. Sogawa, N. Yokotani, Y. Fujii-Kuriyama, M. Haniu, J. E. Shively, O. Gotoh, E. Kusunose, and M. Kusunose, *J. Biol. Chem.* **262,** 13366 (1987).

[26] J. P. Hardwick, B. J. Song, E. Huberman, and F. J. Gonzalez, *J. Biol. Chem.* **262,** 801 (1987).

[27] E. F. Johnson and B. S. S. Masters, unpublished observations.

[29] Purification of Arachidonate 5-Lipoxygenase from Potato Tubers

By PALLU REDDANNA, J. WHELAN, K. R. MADDIPATI, and C. CHANNA REDDY

Lipoxygenases (EC 1.13.11.12) are a group of closely related enzymes that appear to be widely distributed in plants and mammalian tissues.[1] They catalyze the dioxygenation of polyunsaturated fatty acids (PUFA) containing all *cis*-methylene-interrupted double bonds. The nomenclature for lipoxygenases is based on the site of insertion of molecular oxygen onto the fatty acid molecule, in this case, arachidonic acid. Accordingly, they have been classified as 5-, 8-, 9-, 11-, 12-, and 15-lipoxygenases and are found in plants and animals.[2,3] One of the most physiologically important lipoxygenases and the subject of greater interest in recent years is 5-lipoxygenase, which catalyzes the first committed step in leukotriene (LT) biosynthesis.[4-6] Leukotrienes are critical mediators of inflammation.[5] 5-Lipoxygenase has been purified from a number of mammalian cells, including human[7] and porcine[8] leukocytes, and from plants.[9-11] The purified enzyme from mammalian sources requires Ca^{2+} and ATP for its maximal activity.[7] Recently, some information on the primary structure of the mammalian 5-lipoxygenase has been derived from cDNA clones of mRNAs isolated from various sources.[12-14] Nevertheless, information on

[1] H. Kühn, T. Schewe, and S. M. Rapoport, *Adv. Enzymol.* **58,** 274 (1986).

[2] T. Galliard and H. W.-S. Chan, "The Biochemistry of Plants" (P. K. Stumpf, ed.), p. 131. Academic Press, New York, 1980.

[3] T. Schewe, S. M. Rapoport, and H. Kühn, *Adv. Enzymol.* **58,** 191 (1986).

[4] P. Borgeat, M. Hamberg, and B. Samuelsson, *J. Biol. Chem.* **251,** 7816 (1976); correction, *J. Biol. Chem.* **252,** 8772 (1977).

[5] B. Samuelsson, *Science* **220,** 568 (1983).

[6] B. Samuelsson, S.-E. Dahlen, J. A. Lindgren, C. A. Rouzer, and C. N. Serhan, *Science* **237,** 1171 (1987).

[7] C. A. Rouzer, T. Matsumoto, and B. Samuelsson, *Proc. Natl. Acad. Sci. U.S.A.* **83,** 857 (1986).

[8] N. Ueda, S. Kaneko, T. Yoshimoto, and S. Yamamoto, *J. Biol. Chem.* **261,** 7982 (1986).

[9] T. Shimizu, O. Radmark, and B. Samuelsson, *Proc. Natl. Acad. Sci. U.S.A.* **81,** 689 (1984).

[10] P. Reddanna, K. R. Maddipati, and C. Channa Reddy, *FEBS Lett.* **193,** 39 (1985).

[11] P. Reddanna, J. Whelan, P. S. Reddy, and C. Channa Reddy, *Biochem. Biophys. Res. Commun.* **157,** 1348 (1988).

[12] T. Matsumoto, C.-D. Funk, O. Radmark, J.-O. Hoog, H. Jornvall, and B. Samuelsson, *Proc. Natl. Acad. Sci. U.S.A.* **85,** 26 (1988); and correction, *Proc. Natl. Acad. Sci. U.S.A.* **85,** 3406 (1988).

the mechanistic details of this enzyme-catalyzed reaction is scant due to limitations in obtaining sufficient quantity of the purified enzyme in stable form from mammalian sources. Therefore, potato tubers are frequently employed as the source of large amounts of 5-lipoxygenase. This chapter briefly describes the assay, the purification, and the properties of archidonate 5-lipoxygenase (EC 1.13.11.34) from potato tubers.

Enzyme Assay

Principle. The mechanism of dioxygenation of PUFA catalyzed by lipoxygenases involves three main steps: (1) abstraction of a hydrogen atom from the double allylic methylene carbon atom (the first and rate-limiting step), (2) conjugation of double bonds followed by rearrangement of the radical electron, and (3) addition of molecular oxygen (this occurs at a diffusion-controlled rate under ambient O_2 concentrations). Based on this mechanism, it is possible to determine the lipoxygenase activity by measuring: (1) absorbance of conjugated diene at 235 nm, (2) O_2 uptake by means of a Clark oxygen electrode, and (3) product formation by means of HPLC or TLC. Although each one of these methods can be used for monitoring enzyme activity during the course of purification of 5-lipoxygenase from potato tubers, the polarographic method is found to be the most convenient and reproducible.

Reagents

Incubation buffer: 0.15 *M* potassium phosphate buffer, pH 6.3, kept at 25°

Arachidonic acid stock solution (40 m*M*): 12.2 mg arachidonic acid per milliliter absolute ethanol stored under argon at −20°; a portion of the stock solution is allowed to warm to room temperature prior to assay.

Incubation Procedure. The lipoxygenase activity is measured at 30° by determining the O_2 consumption with a Yellow Springs Instrument (Yellow Springs, OH) Model 53 oxygen monitor equipped with a thermostat incubation cell (20 × 68 mm) and a YSI Clark oxygen probe covered with a standard Teflon membrane (YSI 5539). A suitable strip chart recorder or other data acquisition device is used to record $[O_2]$ (as percentage of O_2 in air-saturated buffer vs. time). The assay mixture consists of 2.0 ml potassium phosphate buffer, 10 μl of arachidonic acid stock solution, enzyme

[13] R. A. F. Dixon, R. E. Jones, R. E. Diehl, C. D. Bennett, S. Kargman, and C. A. Rouzer, *Proc. Natl. Acad. Sci. U.S.A.* **85,** 416 (1988).

[14] J. M. Balcarek, T. W. Theisen, M. N. Cook, A. Varrichio, S.-M. Hwang, M. W. Strohsacker, and S. T. Crooke, *J. Biol. Chem.* **263,** 13937 (1988).

solution (10–100 μg of protein) and water in a total volume of 3.0 ml. Buffer, enzyme solution, and water are added into the cell which is closed with the O_2 electrode after stirring to ensure air saturation and temperature equilibration. The contents are stirred continuously and the $[O_2]$ is recorded to establish a straight baseline prior to starting the reaction by injecting arachidonic acid stock solution through the capillary bore of the probe. The change in $[O_2]$ is recorded for several minutes in order to determine the lipoxygenase activity as the rate of O_2 consumption during the linear portion of the reaction. Assuming air-saturated buffer at 30°, and 1 atm pressure, full-scale (100%) deflection of the recorder corresponds to a dissolved $[O_2]$ of 0.23 μmol/ml. The lipoxygenase activity is calculated using the following equation

$$\frac{V_T \times [O_2]_s \times \Delta\%[O_2]}{Enz_{vol} \times 100} = \frac{\mu\text{mol } O_2 \text{ incorporated/min}}{\text{ml of enzyme}}$$

where V_T is the total volume of reaction mixture (ml); $[O_2]_s$, concentration of O_2 in air-saturated buffer at 30° (μmol/ml); $\Delta\%[O_2]$, maximum percentage change per minute in $[O_2]$ taken during the linear portion of the reaction ($\%[O_2]$/min); and Enz_{vol}, volume of enzyme added (ml).

Definition of Enzyme Unit

One unit of lipoxygenase activity is equal to one micromole of O_2 consumed per minute. The specific activity is reported in units per milligram.

Purification Procedures

Fresh red potato tubers are the preferred source of 5-lipoxygenase. The white potato varieties we have tested, as well as the red potatoes from long-term commercial storage, usually have much lower levels of 5-lipoxygenase activity. The enzyme activity is also significantly lower in potatoes stored at −20° for more than 1 week than in fresh potatoes. We have observed that potato tubers contain significant amounts of α-linolenic acid which serves as a substrate for 5-lipoxygenase in cell-free extracts. The enzyme is inactivated during substrate conversion by a self-catalyzed mechanism. Therefore, inclusion of EDTA, sodium metabisulfite and ascorbic acid in the homogenizing buffer is necessary to protect the enzyme from the self-catalyzed inactivation.

Preparation of Crude Extract

All purifications are performed at 0–4°. One kilogram of red potato tubers is minced and homogenized in a Waring blender for 30 sec with 2 volumes of 100 m*M* potassium phosphate buffer, pH 6.3, containing 2 m*M* sodium metabisulfite, 2 m*M* ascorbic acid, and 1 m*M* EDTA. The homogenate is filtered through two layers of cheesecloth and the filtrate is centrifuged at 10,000 *g* for 20 min. The pellet is discarded and the supernatant is subjected to $(NH_4)_2SO_4$ fractionation.

$(NH_4)_2SO_4$ Fractionation

Solid $(NH_4)_2SO_4$ is added under continuous stirring to the enzyme solution obtained from the previous step to give a final concentration of 15% saturation. After all of the $(NH_4)_2sO_4$ is dissolved, the pH of the solution adjusted to 6.3 with 1 *M* NH_4OH, and the contents stirred for 1 hr. The precipitated protein is removed by centrifugation at 15,000 *g* for 15 min and discarded. The resulting supernatant is brought to 45% saturation by further addition of solid $(NH_4)SO_4$, stirred for 1 hr, and then centrifuged at 15,000 *g* for 15 min. The pellet is resuspended in 40 m*M* potassium phosphate buffer, pH 6.3 (hereafter referred to as buffer A), to give a final volume of about 250 ml. This suspension is dialyzed for 24 hr against 4 liters of buffer A with two changes of buffer.

DEAE-Cellulose (Whatman DE-52) Column Chromatography

The dialyzed enzyme solution is centrifuged at 20,000 *g* for 30 min and the supernatant applied to a DE-52 column (2.5 × 40 cm) previously degassed and equilibrated with buffer A. Using the same buffer as eluant, and a flow rate of 1.5 ml/min, the column is washed until the absorbance at 280 nm has dropped to below 0.2. No significant lipoxygenase activity is detected in the flow-through fractions. The eluting buffer is then changed to a linear KCl gradient prepared using 1 liter of buffer A and 1 liter of buffer A containing 0.25 *M* KCl, and the effluent collected in 10-ml fractions. The lipoxygenase activity is eluted at approximately 0.13 *M* KCl. For further purification of 5-lipoxygenase, the fractions containing lipoxygenase activity are pooled, concentrated by ultrafiltration to a final volume of ~10 ml, and dialyzed against 100 volumes of buffer A overnight. The dialyzed enzyme solution is centrifuged and the supernatant passed through a 0.2 μm Millipore filter prior to HPLC injection.

Hydrophobic Interaction HPLC

The lipoxygenase preparation (~10 ml) from the previous step is further purified by HPLC using a Supelco (Bellefonte, PA) LC-HINT hydrophobic interaction column (4 × 100 mm). The HPLC column is equilibrated with buffer A containing 2*M* $(NH_4)_2SO_4$ and the enzyme solution (1 ml) is injected into the column. The protein is eluted with a linear

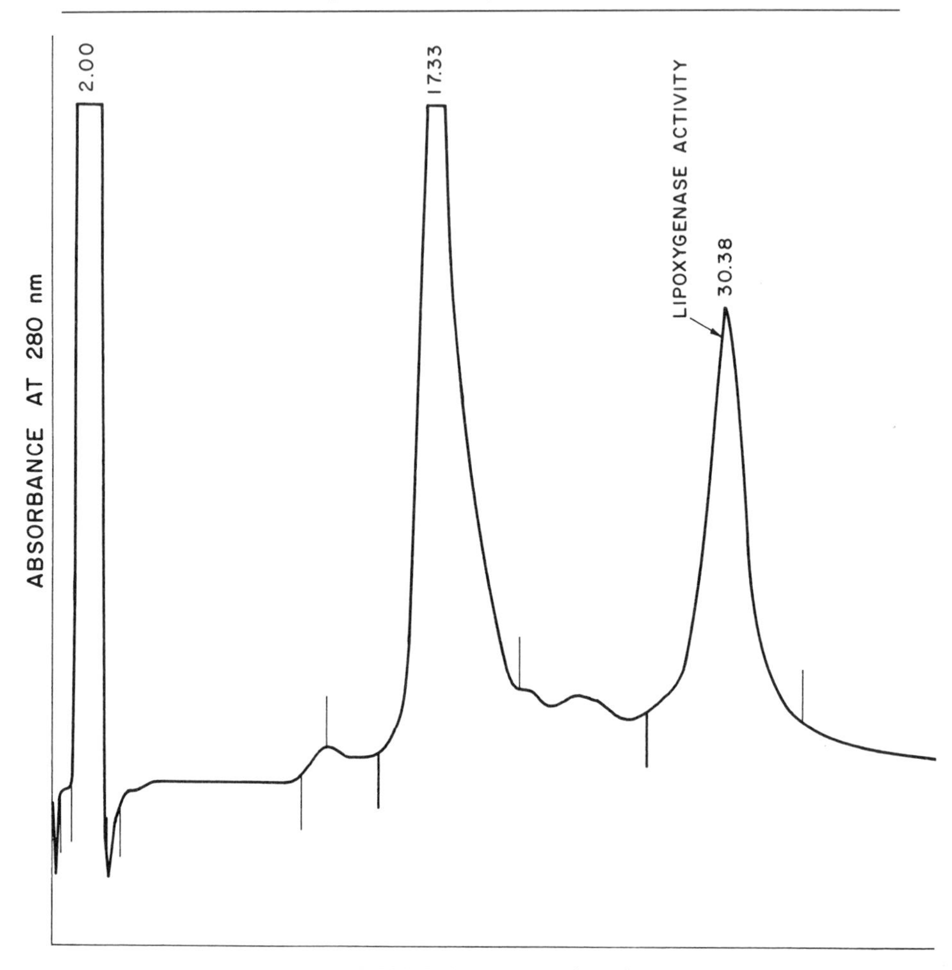

FIG. 1. HPLC purification on a Supelco LC-HINT column of a 5-lipoxygenase preparation obtained from chromatography on a DEAE-cellulose column. The HPLC effluent is monitored continuously for protein at 280 nm.

$(NH_4)_4SO_4$ gradient in buffer A at a flow rate of 1 ml/min. The gradient is programmed in such a way that the concentration of $(NH_4)_2SO_4$ is linearly reduced from 2.0 to 0.0 *M* in 1 hr. As illustrated in Fig. 1, all the lipoxygenase activity is eluted as a single peak at approximately 30 min. The remaining enzyme solution from the previous step is repeatedly injected into the HPLC column and the peaks containing the lipoxygenase activity are pooled. At this stage, SDS-gel electrophoresis indicates that the enzyme preparation is about 90–95% pure. It is further purified by anion-exchange chromatography on HPLC.

Anion-Exchange HPLC

The lipoxygenase recovered from hydrophobic interaction HPLC is concentrated and dialyzed against 50 m*M* sodium acetate buffer, pH 5.6. The dialyzed enzyme is applied to a Waters DEAE-5PW HPLC column (7.5 mm × 7.5 cm) (Milford, MA) previously equilibrated with 50 m*M* sodium acetate buffer, pH 5.6. The absorbed enzyme is eluted with a linear pH gradient (5.6 → 3.0) formed from 50 m*M* sodium acetate and 50 m*M* acetic acid at a flow rate of 1 ml/min for 30 min. As illustrated in Fig. 2, two peaks of lipoxygenase activity are well separated from impurities. The peaks at 22.30 and 25.70 min with lipoxygenase activity are designated as L_1 and L_2, respectively. The L_1 peak (Fig. 2) has a specific activity of ~14 μmol/min/mg and accounts for 80% of the total lipoxygenase activity obtained from this purification procedure whereas the L_2 peak has a specific activity of 6 μmol/min/mg and accounts for 20% of the total lipoxygenase activity. As indicated in the next section, L_2 appears to be a degradation product of the L_1 form.

Properties of the Enzyme

A summary of a typical 5-lipoxygenase purification from red potato tubers is given in Table I. The procedure described above has proved to be convenient for obtaining electrophoretically pure lipoxygenase preparations for kinetic and mechanistic studies. Interestingly, the profile of arachidonic acid oxidation products, especially the hydroperoxyeicosatetraenoic acids (HPETEs), from the purified enzyme is identical to

TABLE I
PURIFICATION OF ARACHIDONATE 5-LIPOXYGENASE FROM POTATO TUBERS

Purification step	Total protein (mg)	Total activity (units)	Specific activity (units/mg)	Yield (%)
Crude extract	27,669	2,967	0.1	100
15–45% $(NH_4)_2SO_4$ fraction	4,360	2,552	0.6	86
DEAE-cellulose chromatography	1,238	2,144	1.7	72
Hydrophobic interaction HPLC	248	1,721	6.5	58
Anion-exchange HPLC				
L_1 fraction	52	710	13.7	24
L_2 fraction	30	169	5.7	6

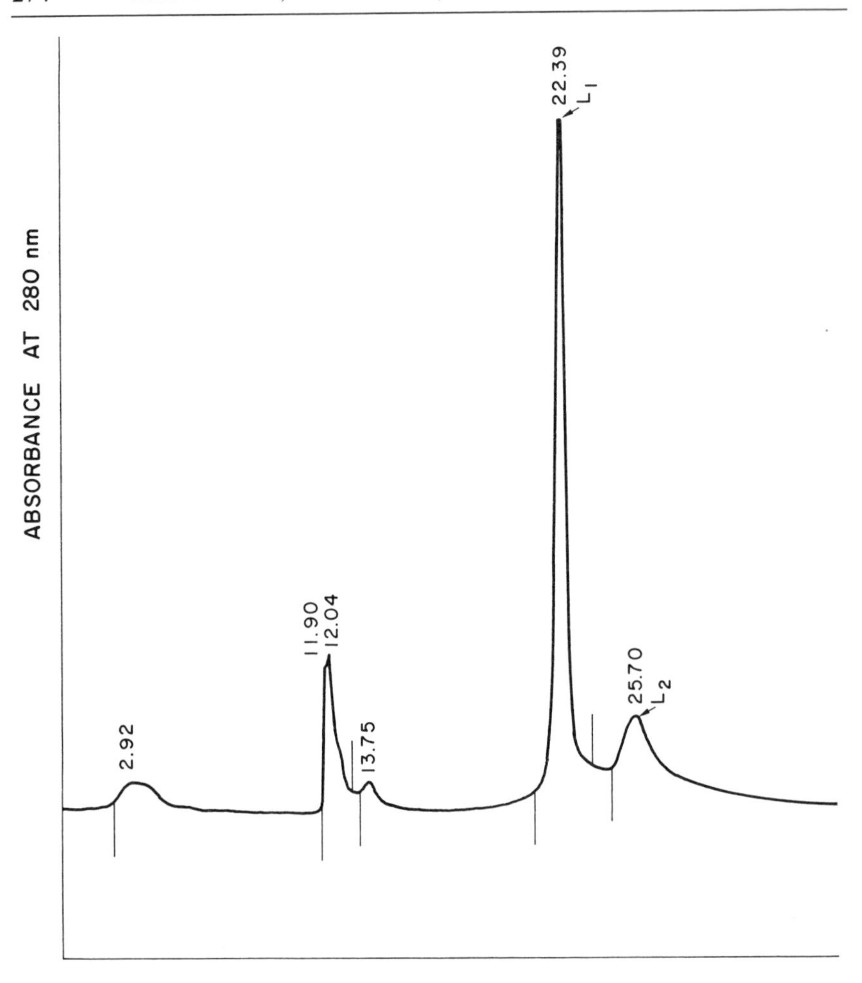

FIG. 2. Anion-exchange chromatography on a Waters DEAE-5PW HPLC column of the 5-lipoxygenase peak from the LC-HINT column. The HPLC effluent is monitored at 280 nm.

that from the partially purified enzyme preparation obtained after DEAE-cellulose column chromatograph. Therefore, one can conveniently use the latter enzyme source for the preparation of HPETEs in multimilligram quantities. Indeed, the 15–45% $(NH_4)_2SO_4$ fraction can also be used for the preparation of HPETEs, especially the 5-HPETE. However, the residual oxidation products from endogenous α-linolenic acid still present in this enzyme preparation complicate purification of some HPETEs. The enzyme preparation is free of endogenous substrate-derived products after the DEAE-cellulose chromatographic step.

Purity and Physical Properties

The freshly prepared L_1 form of 5-lipoxygenase (from the anion-exchange HPLC step) elutes as a single symmetrical peak when analyzed by rechromatography on an anion-exchange HPLC column, and the specific activity is the same in all the protein fractions. Only one protein band is seen on SDS-gel electrophoresis, even when the gels are overloaded. Similar results are observed with the L_2 form of 5-lipoxygenase obtained after anion-exchange HPLC. The molecular masses for L_1 and L_2, determined by SDS-gel electrophoresis, are estimated to be 85,000 and 35,000 Da, respectively. Gel filtration of these two enzyme forms along with marker proteins on a Sephadex G-150 column yields the same molecular mass estimates. The isoelectric points for L_1 and L_2, determined by isoelectric focusing, are 4.5 and 4.25, respectively. Interestingly, rabbit antiserum fo L_1 fails to cross-react with L_2 but does cross-react with a component, possibly the 5-lipoxygenase, of PMNL cells in Western blot analysis.

Storage and Stability

When the enzyme is stored as a suspension in 40 mM potassium phosphate buffer, pH 6.3, containing 2 M $(NH_4)_2SO_4$, the L_1 form is stable for months at $-20°$. However, at 4°, the L_1 form is gradually transformed into the L_2 form with a significant reduction in specific activity from 14 to 6 μmol/min/mg protein. The L_2 form eventually loses all of its activity over a few days at 4° and more rapidly in the absence of $(NH_4)_2SO_4$. It is suggested that L_1 form should be employed for all the kinetic and mechanistic studies, since our data indicate that L_2 is derived from L_1.

Catalytic Properties

The L_1 form of potato 5-lipoxygenase has an apparent K_m of 33 μM and an apparent V_{max} of 14 μmol/min/mg protein with arachidonic acid as substrate. This enzyme catalyzes the dioxygenation of arachidonic acid at all six possible positions (Fig. 3). When arachidonic acid is incubated with the purified L_1 form, 5-HPETE is the most abundant primary oxygenation product, representing 60% of the total HPETEs formed, followed by 8-, 9-, 11-, 12-, and 15-HPETEs in decreasing order of abundance. An analysis of the products of a control incubation mixture employing a heat-denatured L_1 preparation and archidonic acid reveals no HPETE formation, indicat-

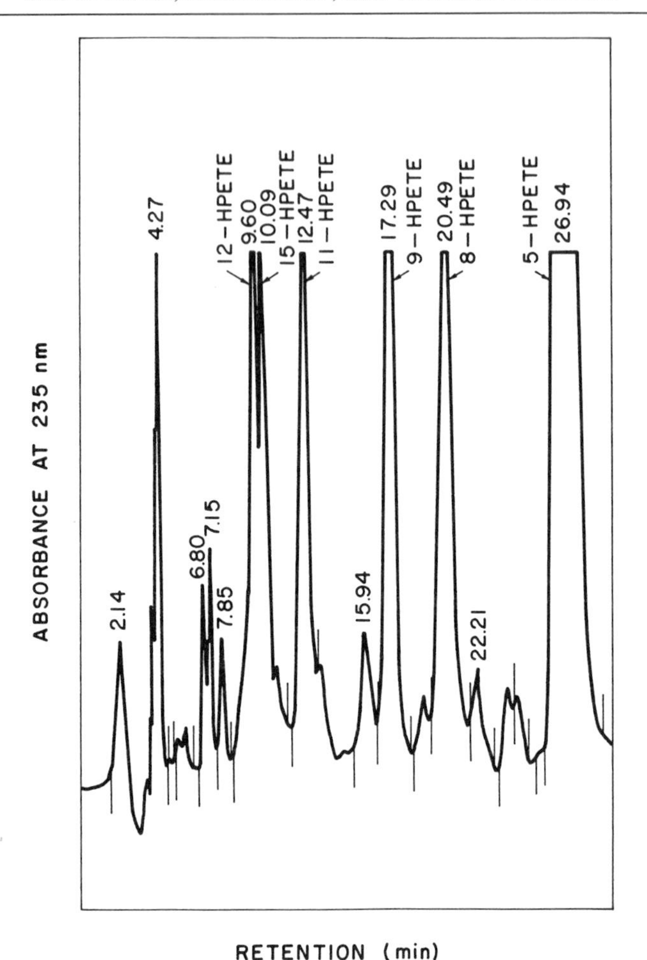

FIG. 3. SP-HPLC analysis of monohydroperoxy products of arachidonic acid generated by the purified L_1 form of potato arachidonate 5-lipoxygenase. The chromatogram is developed isocratically with hexane/2-propanol/acetic acid (993/6/1, v/v/v) at a flow rate of 2 ml/min and the column effluent is monitored at 235 nm.

ing that the HPETEs have been generated enzymatically. Stereochemical analysis by HPLC on a chiral column indicates that the 5-, 8-, and 11-HPEPEs have the *S* configuration, whereas the 9-, 12-, and 15-HPETEs are mixtures of both *S* and *R* (the proportions of *S* and *R* isomers in the latter are not equal). In addition to six possible HPETEs generated from arachidonic acid, small amounts of corresponding hydroxy fatty acids are recovered from the potato enzyme-catalyzed reaction, which suggests that the purified enzyme exhibits peroxidase activity. This is further confirmed by incubating 5-HPETE with the purified enzyme under anerobic conditions where a significant amount of 5-HETE formation is observed in the reaction mixture. The pH of the reaction has a profound influence on the product profile. At pH values below 6.5, the ratio between HPETEs and DiH(P)ETEs (judged by ratio of absorption at 235 and 268 nm) is ~2.5, whereas at pH 7.3 the DiH(P)ETE formation is significantly reduced. At pH 9.0, the lipoxygenase activity is only about 1% of the activity at pH 6.3, but the formation of 8-HPETE exceeds that of 5-HPETE, while the other HPETEs drop to undetectable levels.

In addition to the six HPETEs identified during L_1-catalyzed oxidation of arachidonic acid, all possible diH(P)ETEs are detected, with each exhibiting a distinct absorption spectrum. The diH(P)ETEs include those that are directly generated by dual lipoxygenase activity and those that result from the nonenzymatic hydrolysis of LTA_4s. Furthermore, when 5(OOH), 15-(OH)-, or 5(OH),15(OOH)-arachidonic acid are incubated with the L_1 form, all possible isomers of lipoxin A and lipoxin B are isolated from the reaction. We have demonstrated further that the potato lipoxygenase can act at all possible positions of eicosapentaenoic acid and docosahexaenoic acid to give rise to their respective hydroperoxy compounds.[15,16] Given its ease of preparation and the wide range of compounds it can produce, potato arachidonate 5-lipoxygenase is useful in the preparation of many experimentally interesting metabolites of arachidonic acid and other PUFAs.

[15] J. Whelan, P. Reddanna, G. Prasad, M. K. Rao, and C. C. Reddy, *Ann. N.Y. Acad. Sci.* **524,** 391 (1988).

[16] J. Whelan, P. Reddanna, G. Prasad, and C. C. Reddy, *in* "Proceedings of the Short Course on Polyunsaturated Fatty Acids and Eicosanoids" (W. E. M. Lands, ed.), p. 468. AOCS Press, Champaign, IL, 1987.

[30] Leukotriene Metabolism by Isolated Rat Hepatocytes

By Michael A. Shirley and Danny O. Stene

Leukotrienes are potent biologically active molecules, yet the roles played by these substances as mediators of inflammation remain unclear. One approach to the study of eicosanoid involvement in physiological and pathological processes has been to assess the production of eicosanoids in association with particular normal or disease states. A crucial part of such an approach for leukotrienes must focus on their metabolic fate *in vivo*. Metabolites which may accumulate in urine or bile might serve as indicators of leukotriene synthesis at remote sites such as the lung or skin. Recently, as a model of *in vivo* metabolism, isolated rat hepatocytes have been shown to metabolize leukotrienes to polar metabolites.[1] This chapter will report on the methods used for the production of leukotriene metabolites by isolated rat hepatocytes.

Isolation and Purification of Hepatocytes

Animals

Male Sprague-Dawley rats fed *ad libitum* on lab chow and tap water, and ranging from 200 to 650 g, are used. Neither the size of the animal, nor the time of day at which cells are isolated, appears to affect the results of leukotriene metabolism studies. However, smaller animals tend to give better cell yields and viabilities than larger animals.

Buffers

The buffers used are a perfusion buffer (buffer A) containing 116 m*M* NaCl, 6 m*M* KCl, 0.74 m*M* KH_2PO_4, 0.6 m*M* $MgSO_4$, 12 m*M* $NaHCO_3$, 15 μM glucose, pH 7.1–7.4, equilibrated with 95% O_2, 5% CO_2 (O_2/CO_2); a wash buffer (buffer B) containing 126 m*M* NaCl, 5.2 m*M* KCl, 3 m*M* Na_2PO_4, 0.9 m*M* $MgSO_4$, 0.12 m*M* $CaCl_2$, 10 μM glucose, pH 7.1–7.4, equilibrated with O_2/CO_2; and an incubation buffer (buffer C) prepared by adding Tris [2-amino-2-(hydroxymethyl)-1,3-propanediol] and $CaCl_2$ to buffer B, to the final concentrations of 10 m*M* Tris, and 1 m*M* $CaCl_2$, pH 7.4.[2]

Reagents for Metabolite Purification

Leukotriene E_4 [LTE_4, (5*S*)-hydroxy-(6*R*)-*S*-cysteinyl-(7*E*,9*E*,11*Z*,14*Z*)-eicosatetraenoic acid] was received as a gift from Dr. J. Rokach (Merck-Frosst, Canada Inc.), or is purchased from BioMol (Philadelphia, PA). Leukotriene B_4 [LTB_4, (5*S*,12*R*)-dihydroxy-(6*Z*,8*E*,10*E*,14*Z*)-eicosatetraenoic acid] is purchased from Cayman Chemical Co. (Ann Arbor, MI). [^3H]LTE_4 [(5*S*)-hydroxy-(6*R*)-*S*-cysteinyl-(7*E*,9*E*,11*Z*,14*Z*)-[14,15-$^3H_2$]eicosatetraenoic acid, 39 Ci/mmol] and [^3H]LTB_4 [(5*S*,12*R*)-dihydroxy-(6*Z*,8*E*,10*E*,14*Z*)-[5,6,8,9,11,12,14,15-$^3H_8$]eicosatetraenoic acid,

[1] D. O. Stene and R. C. Murphy, *J. Biol. Chem.* **263,** 2773 (1988).

[2] M. J. Garrity, E. P. Brass, and E. P. Robertson, *Biochim. Biophys. Acta* **796,** 136 (1984).

200 Ci/mmol] are purchased from New England Nuclear (Boston, MA). Supplies for chromatography include octadecylsilyl (ODS) flash chromatography cartridges (Sep-Pak C_{18}) from Waters Associates (Milford, MA), reversed-phase high-performance liquid chromatography (HPLC) cartridge columns from Jones Chromatography (Columbus, OH), and HPLC solvents from Fisher (Fairlawn, NJ).

Surgery and Cell Isolation

The surgical procedure described by Wagle and Ingebretsen, as revised by Brass, is used.[3,4] The perfusion apparatus used in these studies is improvised from standard lab glassware. It allows the perfusion buffer to be warmed and gassed, and maintained at a perfusion pressure of approximately 17 cm water. Unlike the apparatus described by Wagle and Ingebretsen, the buffer is gassed directly, and the apparatus is not enclosed.

Following intraperitoneal injection of sodium pentobarbital (60 mg/kg), the anesthetized rat is placed in a dissecting tray and the abdomen opened by a midline scissor incision from just above the bladder to the xiphoid process. Anterolateral incisions are made on each side from the level of the bladder to the level of the kidneys and the resulting tissue flaps pulled aside to expose the peritoneal cavity. After exposing the posterior peritoneum by pulling the intestines to the right of the animal, ligatures are placed around the portal vein and inferior vena cava as described,[3] and the portal vein is cannulated with an 18-gauge 5.1 cm Quik-Cath (Travenol) attached to the outlet of the perfusion apparatus with intravenous tubing (Travenol, Deerfield, IL). After securing the portal vein cannula, the liver is perfused with buffer A, which is maintained at 37° and equilibrated with O_2/CO_2 by the perfusion system. The perfusion system outlet is 40 cm above the dissecting tray. As the liver perfusion is started, the inferior vena cava is cut well below the kidney to provide the necessary fluid exit.

The ligamentum teres hepatis and diaphragm are then cut, and the anterior thoracic wall removed by making parallel cuts of the rib cage along the midclavicular line. A loose ligature is placed around the inferior vena cava just superior to the diaphragm, as is another just inferior to the right atrium. A small incision is made in the vessel between the two ligatures and an 18-gauge catheter (with the stylet removed and cut to about 1 cm) is inserted and secured. The ligature around the inferior vena cava in the abdomen is then tied.

The liver is dissected from the animal by first freeing the lower margin of the abdominal vena cava just below the ligature. The perfusion line is

[3] S. R. Wagle and W. R. Ingebretsen, Jr., this series, Vol. 35, p. 579.

[4] E. Brass, personal communication, 1986.

then moved to the thoracic catheter, and the portal vein and liver are dissected free from the surrounding mesentery. The perfusion line is then returned to the portal vein, and the liver dissected free from the posterior abdominal wall. The tubing connecting the liver to the perfusion apparatus is removed and the liver is attached to the outlet of the perfusion apparatus by way of the portal vein catheter.

The perfusion buffer is adjusted to 1 m*M* $CaCl_2$ by addition of 2 ml 100 m*M* $CaCl_2$, and 3 units/ml collagenase (Type II, Cooper Biomedical, Freehold, NJ) is added. The liver is perfused with this collagenase buffer for 10 min. via the portal vein, 5 min via the inferior vena cava, and then 5 min with collagenase-free/Ca^{2+}-free buffer via the portal vein. The liver is then disconnected from the perfusion system and placed into a 250-ml silanized glass beaker containing about 30 ml of buffer A warmed to 37°. [All glassware used during the purification of hepatocytes is silanized by rinsing with neat dimethyldichlorosilane (Pierce, Rockford, IL) followed by rinsing with dry methanol.] Hepatocytes are teased from the liver into the buffer using forceps. The resulting cell suspension is filtered through nylon mesh (70 mesh, Nytex nylon) into a 200-ml Erlenmeyer flask and placed into a water bath for 10 min at 37° with gentle shaking and continuous O_2/CO_2 ventilation. Cells are poured from the flask through the nylon mesh into 50-ml plastic centrifuge tubes chilled on ice, the tubes are filled with buffer B (4°), and then are centrifuged at 65 *g* for 3 min. After removing the supernatant, the pellet is very gently resuspended in chilled buffer B followed by recentrifugation. This is repeated, and the washed hepatocytes are gently resuspended in chilled buffer C. The cells are counted, and the viability evaluated by the trypan blue exclusion test.[5] Cells used for metabolism studies are always >85% viable.

Generation and Purification of Leukotriene Metabolites

Incubation Conditions

Incubations with LTE_4 are conducted with 3×10^6 cells/ml in a gently shaking water bath at 37° with O_2/CO_2 ventilation. For studies of substrate concentration effects, [^{3}H]LTE_4, at concentrations of 0.05, 0.5, 10, 25, and 50 μM, is incubated with 2 or 5 ml of cells in a 25-ml plastic tissue culture flask for 30 min. For studies of the time course of LTE_4 metabolism, 15 ml of cells in a 50-ml flask are incubated with 5 μM [^{3}H]LTE_4 for up to 60 min. Incubations with LTB_4 (3 μM) are conducted with 5×10^6 cells/ml for 45 min as above.

[5] P. O. Seglen, *Methods Cell Biol.* **13,** 29 (1976).

Metabolite Purification

Hepatocyte incubations are stopped by the addition of 4 volumes chilled ethanol. After storage overnight at −70°, precipitated proteins are pelleted by centrifugation at room temperature. For the purification of LTE_4 metabolites, the ethanolic supernatant is diluted by addition of 10 volumes of reversed-phase HPLC solvent A (see below), and partially purified by solid-phase extraction on an ODS Sep-Pak. The Sep-Pak cartridges are washed with solvent A and eluted with methanol. The methanol eluent is evaporated under vacuum and redissolved in 80% solvent A : 20% methanol for analysis by reversed-phase HPLC. The radioactivity content of the sample before and after evaporation is determined to quantitate the production of volatile counts. The recovery of tritium-labeled metabolites from the Sep-Paks averages 85% throughout these experiments.

The ethanolic supernatant containing LTB_4 metabolites is taken to dryness under vacuum and the residue dissolved in water (acidified to pH 3.5 with acetic acid). This sample is taken for direct analysis by reversed-phase HPLC.

LTE_4 metabolites are subjected to reversed-phase HPLC using a linear gradient from 80% solvent A to 100% solvent B. Solvent A is water with 16.6 m*M* ammonium acetate, pH 4.5, and 132 μM tetrasodium ethylenediaminetetraacetic acid (EDTA); solvent B is 90% methanol, 10% solvent A, 0.1% acetic acid, and 132 μM EDTA. For LTB_4 metabolites, the same reversed-phase HPLC system is used except that ammonium formate is the buffer, the gradient is started at 90% solvent A, and EDTA is not included. The reversed-phase HPLC effluent is monitored by a photodiode array detector (HP-1040A, Hewlett-Packard, Palo Alto, CA), and/or a radioactivity detector (FLO-ONE Beta, model IC radioactive flow detector, Radiomatic Instruments, Tampa, FL). The UV wavelengths monitored are at 270 nm (for LTB_4 metabolites), 280 nm, or a broad-band UV signal which sums the absorbance from 260 to 340 nm (for LTE_4 metabolites). One assumption made for these experiments is that, during purification and reversed-phase HPLC of leukotriene metabolites, there are no differences in the recoveries of the various metabolites. Total recovery of leukotriene metabolites from reversed-phase HPLC ranges from 50 to 80%. The recoveries of LTE_4 metabolites are best when the columns are washed prior to use with water/methanol (90 : 10, v/v) and 132 m*M* EDTA as described.[6]

[6] F. S. Anderson, J. Y. Westcott, J. A. Zirrolli, and R. C. Murphy, *Anal. Chem.* **55,** 1837 (1983).

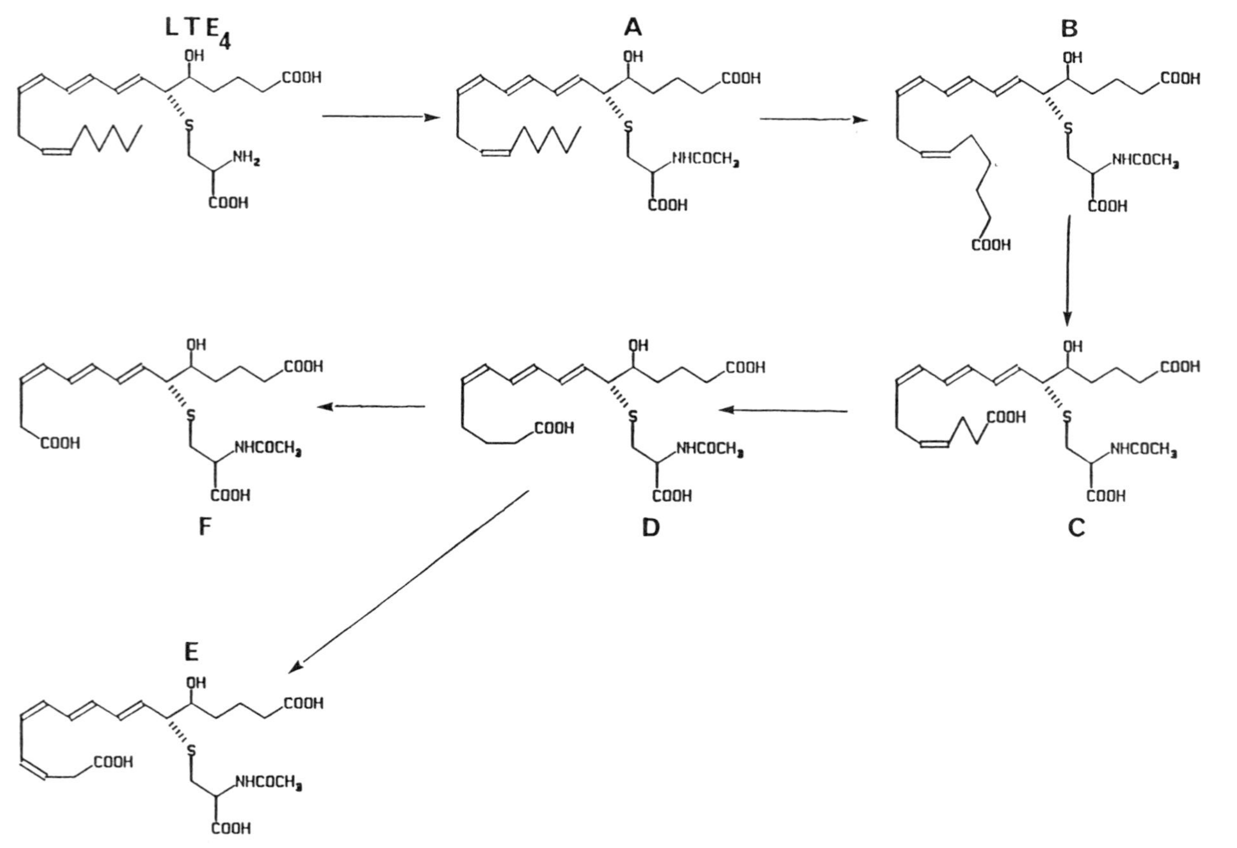

FIG. 1. LTE_4 metabolites from isolated rat hepatocytes. (A) *N*-Acetyl-LTE_4; (B) 20-carboxy-*N*-acetyl-LTE_4; (C) 18-carboxy-19,20-dinor-*N*-acetyl-LTE_4; (D) 16-carboxy-17,18,19,20-tetranor-14,15-dihydro-*N*-acetyl-LTE_4; (E) 16-carboxy-17,18,19,20-tetranor-Δ^{13}-*N*-acetyl-LTE_4; and (F) 14-carboxy-15,16,17,18,19,20-hexanor-*N*-LTE_4.

LTE_4 Metabolism

LTE_4 has been shown to be metabolized by isolated rat hepatocytes to six polar metabolites (Fig. 1).[1] The initial *N*-acetylation of LTE_4 is followed by ω-oxidation and β-oxidation from the ω terminus. All of these metabolites retain the typical leukotriene UV chromophore with absorption maxima at 280 nm, except for Metabolite E which is a conjugated tetraene with a UV absorption maximum at 309 nm.

The metabolism of LTE_4 is dependent on the concentration of LTE_4 exposed to the isolated hepatocytes (Figs. 2 and 3). When 50 μM LTE_4 is used, the only metabolite observed is *N*-acetyl-LTE_4. However, as the concentration of LTE_4 decreases, the conversion of LTE_4 to ω/β-oxidized metabolites increases. When 25 μM LTE_4 is used, ω/β-oxidized metabolites appear, but *N*-acetyl-LTE_4 is still the dominant metabolite. At low substrate concentrations (less than 5 μM), only ω/β-oxidized metabolites are present after 30-min incubation. Loss of the tritium label is up to 45% of the total indicating metabolism of LTE_4 beyond the 16-carboxydihydro

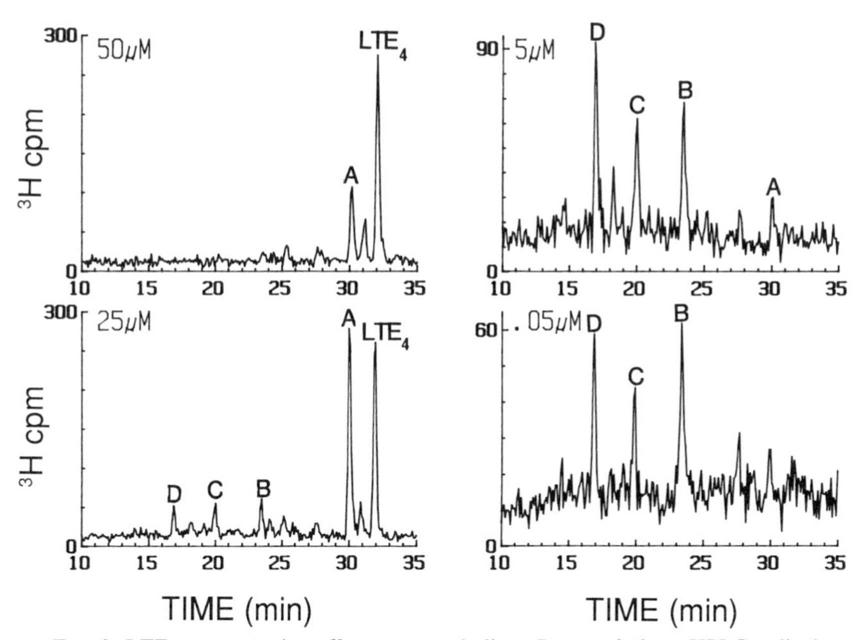

FIG. 2. LTE_4 concentration effects on metabolism. Reversed-phase HPLC radiochromatograms of [^{3}H]LTE4 metabolites from isolated rat hepatocytes (3×10^6 cells/ml, 30 min) at the LTE_4 concentrations indicated. The identities of the metabolites are determined by comparison with biologically derived and synthetic standards (see Fig. 1). Reversed-phase HPLC conditions are as noted in the text.

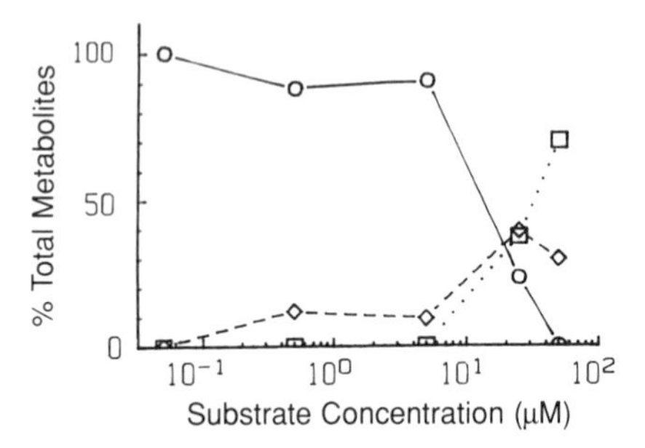

FIG. 3. LTE_4 concentration effects on metabolism. The effect of substrate concentration on the extent of LTE_4 metabolism by isolated rat hepatocytes is plotted. Cells are incubated with from 0.05 to 50 μM LTE_4 for 30 min as described in the text. Data points represent, for the ω/β-oxidized metabolites (○), the sum of the reversed-phase HPLC UV peak heights of the metabolites divided by the sum of the peak heights for all metabolites and LTE_4. For LTE_4 (□) and *N*-acetyl-LTE_4 (◇), the data points represent the reversed-phase HPLC peak heights divided by the sum of the peak heights for all metabolites and LTE_4.

metabolite. When the relative amount of ω/β-oxidized metabolites is plotted versus the LTE_4 concentration, it is apparent that at substrate concentrations about 10 μM, the rate of LTE_4 metabolism by isolated rat hepatocytes is nonlinear.

When LTE_4 is incubated with hepatocytes at 5 μM, the extent of LTE_4 metabolism increases as a function of time. In order to determine the extent to which metabolism continues past the formation of Metabolite D, the broad-band monitoring capability of the photodiode array detector is used to generate a signal derived from monitoring 260 to 340 nm. Although the extinction coefficient of the conjugated tetraene chromophore of Metabolite E is not known, it is estimated to be approximately 80,000 at 309 nm based values reported in the literature.[7] The extinction coefficient for LTE_4 has been determined to be 40,000 at 280 nm.[8] The relative changes in the broad band signals are assumed to be linear for all metabolites, and the peak heights for Metabolite E are divided by 2 as a correction for the higher extinction coefficient of this metabolite. Using this approach, the relative amounts of Metabolites C, D, and E change as a function of time (Fig. 4; LTE_4, and Metabolites A and B are not shown). Clearly, Metabolite E is a major LTE_4 metabolite at 60 min. It is also apparent that the relative amount of volatile tritium increases over time, and this increase parallels the increases in Metabolites D and E.

[7] G. A. J. Pitt and R. A. Morton, *Prog. Chem. Fats Lipids* **4,** 227 (1957).

[8] R. A. Lewis, J. M. Drazen, K. F. Austen, D. A. Clark, and E. J. Corey, *Biochem. Biophys. Res. Commun.* **96,** 271 (1980).

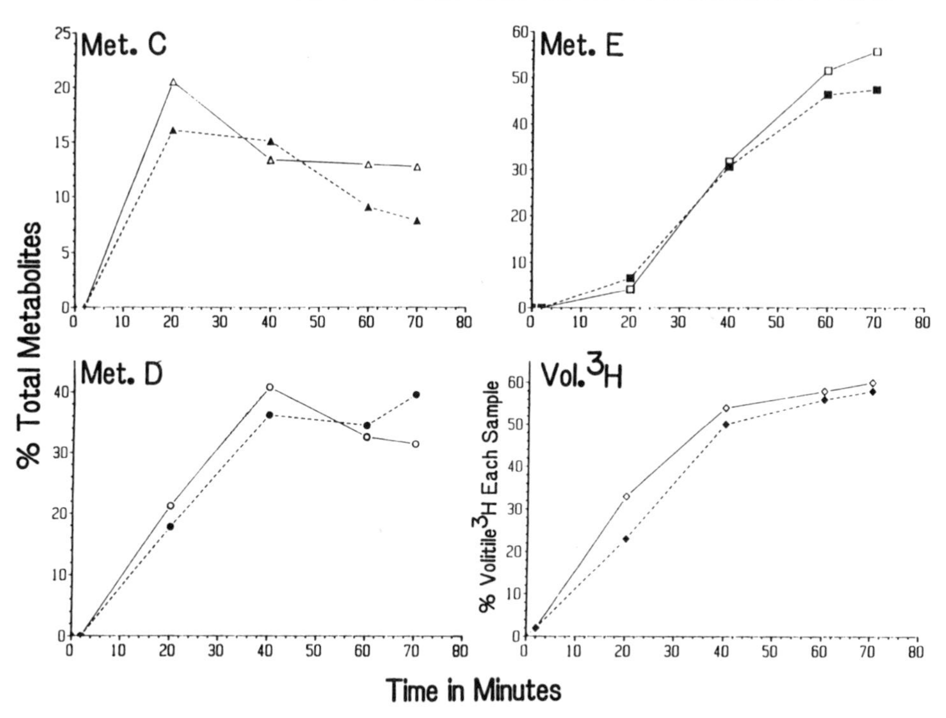

FIG. 4. Metabolism of LTE_4 as a function of time. The amount of Metabolite E increases relative to other metabolites as a function of time during metabolism of LTE_4 by isolated rat hepatocytes (5 μM LTE_4, 3×10^6 cells/ml). A broad-band UV signal from 260 to 340 nm is used to measure metabolites by reversed-phase HPLC. The relative amount of volatile tritium increases in parallel with the increases in Metabolites D and E. The data are from two experiments, the open points from one experiment, and the closed points from another (see the text for an explanation of how the data is calculated).

LTB_4 Metabolism

Rat hepatocyte metabolism of LTB_4 results in three major metabolites as assessed by reversed-phase HPLC (Fig. 5). As with LTE_4 metabolism, LTB_4 is also ω- and β-oxidized from the ω terminus. The formation of 20-COOH-LTB_4 and 18-COOH-LTB_4 has been reported previously.[9] Preliminary mass spectral data indicate that 16-COOH-14,15-dihydrotetranor-LTB_4 is also formed. Volatile products account for approximately 30% of the starting radiolabel. One or more highly polar metabolites, representing 11% of the nonvolatile radiolabel, elute with the

[9] T. W. Harper, M. J. Garrity, and R. C. Murphy, *J. Biol. Chem.* **261,** 5414 (1986).

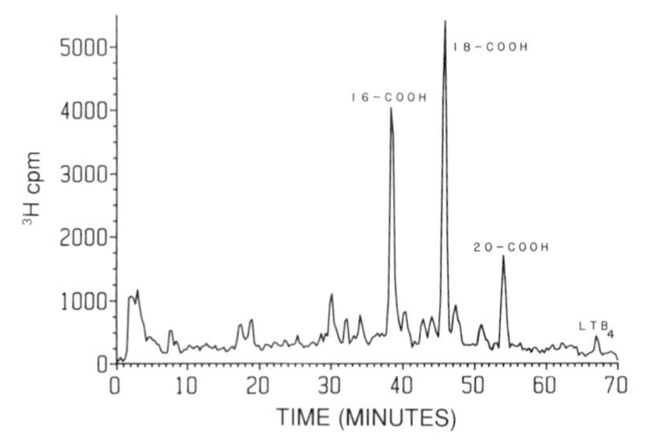

FIG. 5. LTB_4 metabolism by isolated rat hepatocytes. The reversed-phase HPLC radiochromatogram of [^{3}H]LTB_4 metabolites formed by hepatocytes (5 ml cells, 5×10^6 cells/ml) incubated with LTB_4 (3 μM) for 45 min. The identity of 20-COOH-LTB_4 is determined by coelution with a synthetic standard. 18-COOH-LTB_4 and 16-COOH-LTB_4 are identified by electron impact mass spectrometry. Reversed-phase HPLC conditions are as outlined in the text.

solvent front of the gradient reversed-phase HPLC system. Further studies on the metabolism of LTB_4 by isolated rat hepatocytes are in progress.

The relevance of the isolated rat hepatocyte model of LTE_4 metabolism to the fate of leukotrienes *in vivo* has been supported by the identification of 20-COOH-LTE_4 (Metabolite B) and 16-COOH-dihydro-LTE_4 (Metabolite D) in rats injected with LTE_4.[10] This hepatocyte system serves as a biosynthetic source of these metabolites.

[10] P. Perrin, J. Zirrolli, D. O. Stene, J. P. Lellouche, J. P. Beaucourt, and R. C. Murphy, *Prostaglandins* **37,** 53 (1989).

[31] Purification and Characterization of Human Lung Leukotriene A_4 Hydrolase

By NOBUYA OHISHI, TAKASHI IZUMI, YOUSUKE SEYAMA, and TAKAO SHIMIZU

Leukotriene (LT) A_4 hydrolase (EC 3.3.2.6) catalyzes enzymatic hydration of LTA_4 [5(*S*)-*trans*-5,6-oxido-7,9-*trans*-11,14-*cis*-eicosatetraenoic acid] to LTB_4 [5(*S*), 12 (*R*)-dihydroxy-6,14-*cis*-8,10-*trans*-eicosatetraenoic acid], a potent chemotactic substance. This reaction on the epoxide differs from that of epoxide hydrolases (EC 3.3.2.3). The latter results in the synthesis of vicinal alcohols (glycol), while the former synthesizes a compound possessing two hydroxy groups at C-5 and C-12 with a conjugated triene in between. LTA_4 hydrolase activity was previously reported mainly in blood cells, especially polymorphonuclear leukocytes and erythrocytes, and the enzymes from these sources were purified.[1-3] The enzyme activity is, however, ubiquitously distributed in the cytosolic fraction of various organs of the guinea pig, and higher activities are observed in small intestine, lung, and aorta.[4] To compare the lung enzyme with those found in blood cells, and for further characterization, the enzyme was purified from the human lung.[5] Recently, this enzyme has also been purified from the guinea pig liver[6] and lung.[7]

Preparation and Purification of LTA_4 Methyl Ester

LTA_4 methyl esters are synthesized according to Ohkawa and Terao[8] and Gleason *et al.*[9] These are obtained as mixtures of several geometric isomers, and can be separated by straight-phase HPLC. The HPLC conditions are as follows: column—ChemcoPak Nucleosil 50-5, 1×30 cm (Chemco, Osaka); solvent—*n*-hexane/ethyl acetate/triethylamine (100/1.2/0.4, v/v/v); flow rate—3 ml/min; column temperature—4°; UV monitor—280 nm. The column temperature is critical for good resolution and recovery. Under these conditions, four major peaks representing various isomers of LTA_4 methyl ester appear at retention times of between 40 and 60 min, the second-appearing peak being LTA_4 methyl ester (Fig. 1). Gas chromatography–mass spectrometry and proton magnetic resonance studies are necessary for structural identification.

[1] O. Rådmark, T. Shimizu, H. Jörnvall, and B. Samuelsson, *J. Biol. Chem.* **259,** 12339 (1984).

[2] J. F. Evans, P. Dupuis, and A. W. Ford-Hutchinson, *Biochim. Biophys. Acta* **840,** 43 (1985).

[3] J. McGee and F. Fitzpatrick, *J. Biol. Chem.* **260,** 12832 (1985).

[4] T. Izumi, T. Shimizu, Y. Seyama, N. Ohishi, and F. Takaku, *Biochem. Biophys. Res. Commun.* **135,** 139 (1986).

[5] N. Ohishi, T. Izumi, M. Minami, S. Kitamura, Y. Seyama, S. Ohkawa, S. Terao, H. Yotsumoto, F. Takaku, and T. Shimizu, *J. Biol. Chem.* **262,** 10200 (1987).

[6] J. Haeggström, T. Bergman, H. Jörnvall, and O. Rådmark, *Eur. J. Biochem.* **174,** 717 (1988).

[7] H. Bito, N. Ohishi, I. Miki, M. Minami, T. Tanabe, T. Shimizu, and Y. Seyama, *J. Biochem.* **105,** 261 (1989).

[8] S. Ohkawa and S. Terao, *J. Takeda Res. Lab.* **42,** 13 (1983).

[9] J. G. Gleason, D. B. Bryan, and C. M. Kinzig, *Tetrahedron Lett.* **21,** 1129 (1980).

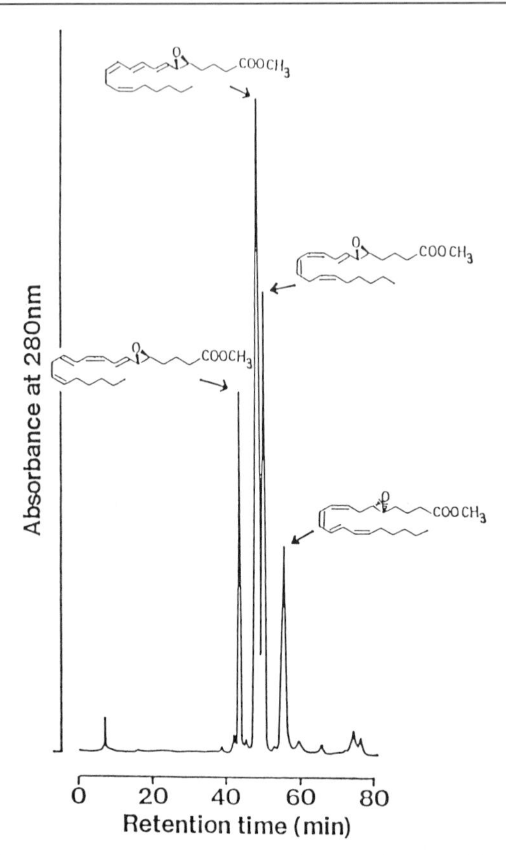

FIG. 1. Straight-phase HPLC separation of LTA_4 methyl ester and its geometric isomers. Four major peaks elute at retention times around 40–60 min and the structure of each peak is illustrated.

Saponification of LTA_4 Methyl Ester

LTA_4 is unstable in aqueous buffer solutions at acidic and neutral pH, and is stabilized in alkaline organic solutions at a lower temperature. LTA_4 is synthesized as the methyl ester form and is stored in alkaline organic solution such as benzene/triethylamine (100/2, v/v) in a nitrogen atmosphere at $-70°$. When in use, LTA_4 methyl ester is saponified in tetrahydrofuran (400 μl) plus 0.1 *M* lithium hydroxide in water (100 μl) under conditions of vigorous mixing overnight at room temperature. The aliquot is then dried under a stream of nitrogen (complete dryness should be avoided), and dissolved with an appropriate amount of ethanol. The concentration of LTA_4 is determined from UV absorption, using a molar extinction coefficient of 40,000 $M^{-1}cm^{-1}$ at 280 nm.

Assay of LTA_4 Hydrolase

The standard incubation mixture (50 μl in 1.5 ml sampling tube) contains the enzyme in 0.1 *M* Tris-HCl buffer (pH 7.8). After preincubation for several minutes at 37°, 1 μg of LTA_4 in 1 μl of ethanol is added. After 1 min of incubation, 117 μl of stopping solution (0.1% acetic acid in methanol containing 0.3 nmol of prostaglandin B_2 as an internal standard) is added. After the mixture is kept at −20° for at least 20 min followed by centrifugation at 10^4 *g* for 10 min, a 50-μl aliquot of the supernatant is directly injected onto HPLC. The conditions are as follows: column—TSK-ODS $80T_M$, 0.46 × 15 cm (Tosoh, Tokyo); solvent—methanol/water/acetic acid (70/30/0.05, v/v/v); flow rate—1 ml/min; column temperature—35°; UV monitor—270 nm. PGB_2 and LTB_4 elute at about 7 and 11 min, respectively, and the amount of LTB_4 formed is calculated from the peak area ratio LTB_4/PGB_2.

Purification Procedure

All procedures other than column chromatography are done at 4°. Column chromatography is performed using a fast protein liquid chromatography (FPLC) system (Pharmacia, Uppsala), at room temperature. All centrifugations are carried out at 10,000 *g* at 4° for 20 min, unless otherwise indicated.

Step 1. Human lung (about a 50 g autopsy specimen with no inflammatory and other disease processes) is minced with scissors in 3 volumes of ice-cold 20 m*M* potassium phosphate buffer containing 5 m*M* EDTA and 0.15 *M* NaCl (pH 7.4). After homogenization using a Polytron-type homogenizer (output dial 3, 1 min, 3 times), the sample is centrifuged.

Step 2. The 10,000 *g* supernatant is subjected to ammonium sulfate fractionation. The precipitates between 40 and 70% saturation are collected, dissolved in 50 ml of 20 m*M* Tris-HCl buffer (pH 8), and then dialyzed against two changes of 40 volumes of the same buffer.

Step 3. After centrifugation and filtration through 0.45 μm filters, the dialyzed sample (one-third or one-fourth of total volume per one cycle of chromatography) is applied to a Mono Q HR 10/10 column (1 × 10 cm, Pharmacia) preequilibrated with 20 m*M* Tris-HCl buffer (pH 8) at a flow rate of 2 ml/min. The column is washed with the same buffer (about 20 ml), and the adsorbed proteins are eluted with a 75 ml linear gradient between 0–0.15 *M* KCl, in the same buffer. The enzyme activity elutes at a KCl concentration of around 0.1 *M*.

Step 4. Active fractions from the previous step are collected (10–15 ml), and the same volume of 2 *M* ammonium sulfate solution (adjusted to pH 7.2 with 2 *M* Tris) is added. The ammonium sulfate solution must be added dropwise very slowly while stirring, otherwise insoluble materials may appear. The sample (one-half of total sample per one cycle of chromatography) is then applied to a phenyl-Superose HR 5/5 column (0.5 × 5 cm, Pharmacia) preequilibrated with 0.1 *M* Tris-HCl buffer containing 1 *M* ammonium sulfate (pH 7.2). After washing with the same buffer at a flow rate of 0.5 ml/min, the column is eluted with a 20 ml linear gradient between the starting buffer and the terminating buffer (0.1 *M* Tris-HCl buffer, pH 7.6). The enzyme activity elutes at the ammonium sulfate concentration of around 0.35 *M*.

Step 5. Active fractions from the previous step are collected (about 6 ml) and concentrated to 1 ml with a Centricon-10 (Amicon, Danvers, MA). The sample is applied to a Superose 12 HR 16/50 column (1.6 × 50 cm, Pharmacia) and eluted with 20 m*M* Tris-HCl buffer containing 0.1 *M* KCl (pH 8.0) at a flow rate of 1 ml/min. The enzyme activity is eluted at a retention volume of around 55–60 ml.

Step 6. The final step of the purification is hydroxyapatite column chromatography. In some cases, depending on the starting materials, this step can be omitted. After dialysis against 100 volumes of 5 m*M* potassium phosphate buffer (pH 6.3), active fractions from the gel filtration chromatography are applied to a hydroxyapatite column (KB column, 0.78 × 13.5 cm, KOKEN, Tokyo), preequilibrated with the same buffer, and eluted with a 50-ml linear gradient of 5 to 200 m*M* potassium phosphate buffer (pH 6.3). The enzyme activity elutes at around 60 m*M* potassium phosphate concentration.

A representative result of the purification is summarized in Table I. The purified enzyme has a specific activity ranging from 0.2 to 0.4 μmol LTB_4/mg·min under standard assay conditions. (For the comparison with V_{max}, refer to "kinetic properties.")

Although we initially attempted to purify the enzyme from the human lung using the method described for human leukocytes,[1] a major impurity with a molecular weight of about 80,000 could not be removed. Because this impurity is successfully removed by phenyl-Superose column chromatography, this is a key step in the purification of the enzyme from the human lung. The yield in this step is, however, low (about 30–40%).

Physical Properties

Molecular weights of the purified LTA_4 hydrolase from the human lung are determined as 68,000–71,000 on SDS–PAGE, values in good accord

TABLE I
PURIFICATION OF LTA_4 HYDROLASE FROM HUMAN LUNG

Step	Total protein (mg)	Total activity (nmol/min)	Specific activity (nmol/mg · min)	Yield (%)
1. 10,000 *g* Supernatant	1,300	440	0.34	100
2. 40–70% $(NH_4)_2SO_4$	470	310	0.66	70
3. Mono Q	20	250	13	57
4. Phenyl-Superose	1.2	—[a]	—[a]	—[a]
5. Superose 12	0.28	59	210	13
6. Hydroxyapatite	0.17	38	220	8.6

[a] In this step, ammonium sulfate inhibits the enzyme activity; therefore, precise determination of the activity is not feasible.

with those noted for the enzyme from other sources.[1,2,6,7] except for that from human erythrocytes (54,000 ± 1000 on SDS–PAGE).[3] The p*I* value is determined to be 5.1–5.3 on a Mono P chromatofocusing column. The N-terminal amino acid sequence of the human lung enzyme is Pro-Glu-Ile-Val-Asp-Thr-Xaa-Ser-Leu-Ala-Ser-Pro-Ala-Ser-Val,[10] which is identical with findings in human leukocytes.[1] The N-terminal sequence for the guinea pig liver[6] and lung[7] enzyme is identical to that of the human enzyme, except for the conservative substitution of two amino acid residues, Ile-3 and Ser-14 (human) to Val and Thr (guinea pig), respectively.

This enzyme is stable at 4° for at least 1 month, but is sensitive to freezing and thawing, one cycle of which decreases activity by about 50%.

Kinetic Properties

The pH optima of the enzyme reaction ranges between 7.8 and 9.0. Neither divalent cations (Ca^{2+}, Mg^{2+}, Mn^{2+}) nor divalent cation chelators (EDTA, EGTA) affect the enzyme activity. The enzyme activity is reduced by 20% in the presence of 1 *M* KCl and by 50% in the presence of 0.2 *M* ammonium sulfate. 2-Mercaptoethanol (5 m*M*) or dithiothreitol (0.5 m*M*) does not affect the enzyme activity.[11] On the other hand, SH-blocking reagents such as *N*-ethylmaleimide, *p*-chloromercuribenzoic acid, and $HgCl_2$ inhibit the enzyme activity (IC_{50} 3 m*M*, 0.7 μ*M*, 0.5 μ*M*,

[10] Xaa at N-7 amino acid is determined as Cys by the molecular cloning of LTA_4 hydrolase cDNA (see footnotes 16 and 17).

[11] SH-reducing reagents affect the elution profile of the enzyme on Mono Q and Mono P column chromatography, when included in the purification procedure of the enzyme from guinea pig lung.[7]

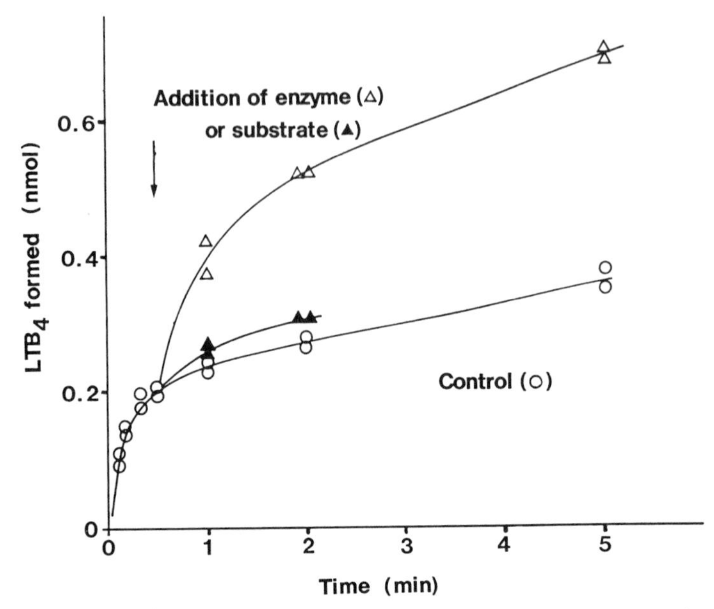

FIG. 2. Time course of the enzymatic hydration of LTA_4 by the human lung LTA_4 hydrolase in the presence of albumin. The purified enzyme (0.44 μg) is incubated in 50 μl reaction mixture containing LTA_4 (80 μM), BSA (2 mg/ml), and 0.1 M Tris-HCl buffer (pH 7.8) at 37° for 5 sec to 5 min (○). After 30-sec incubation (indicated by vertical arrow), 0.44 μg of enzyme (△) or 1.1 μg of LTA_4 (▲) is added and the reaction halted at the time indicated.

respectively, when 0.44 μg of enzyme is preincubated with these reagents for 5 min at 37°).

In determining kinetic parameters, two points should be taken into consideration. First, LTA_4 is unstable in aqueous buffer solutions at neutral pH (half-life less than 10 sec at 25° at pH 7.8) and nonenzymatically hydrolyzed to 6-*trans*-LTB_4 (major products) and 5,6-dihydroxy acids (minor products). In fact, under the standard assay condition, the reaction ceases within 20 sec. Second, the enzyme reaction follows a suicide-type one, namely, the enzyme is inactivated by the substrate, LTA_4. Figure 2 shows a typical time course of the reaction in the presence of bovine serum albumin which stabilizes LTA_4.[12] Again, the rate of reaction declines after 30 sec, and is fully recovered when the enzyme, but not the substrate, is added. Since LTB_4 (8 μM) or 6-*trans*-LTB_4 (38 μM) does not inhibit the enzyme reaction, the substrate, LTA_4, may inactivate the enzyme. For the above two reasons, the initial velocity must be determined from reactions

[12] F. A. Fitzpatrick, D. R. Morton, and M. A. Wynalda, *J. Biol. Chem.* **257,** 4680 (1982).

TABLE II
SUBSTRATE AND INACTIVATOR SPECIFICITIES OF LTA_4 HYDROLASE
Required Structures:

Substrate: R, O, COOH

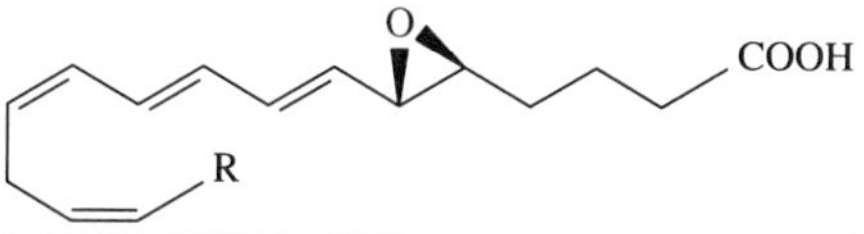

Inactivator: R′, O, R″

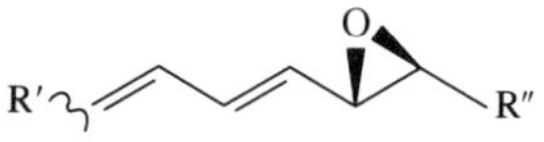

Compound	Structure	Substrate	Inactivator
	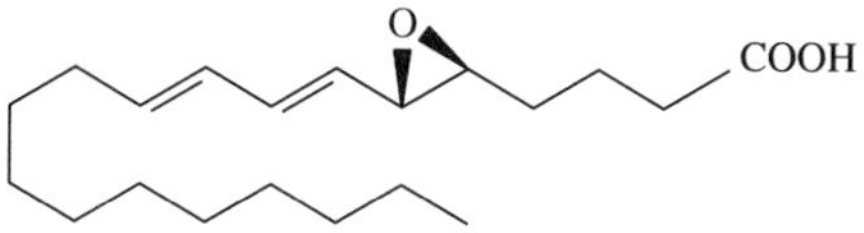 O, COOH	−	+
LTA_3	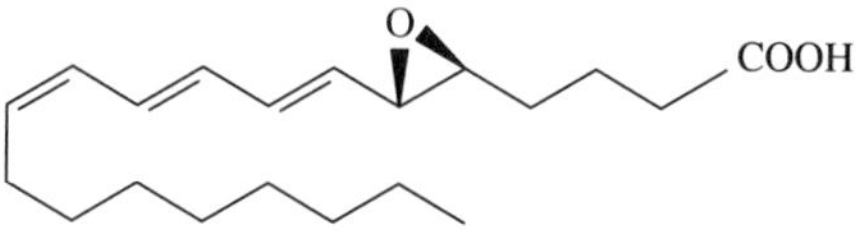 O, COOH	−	+
LTA_4	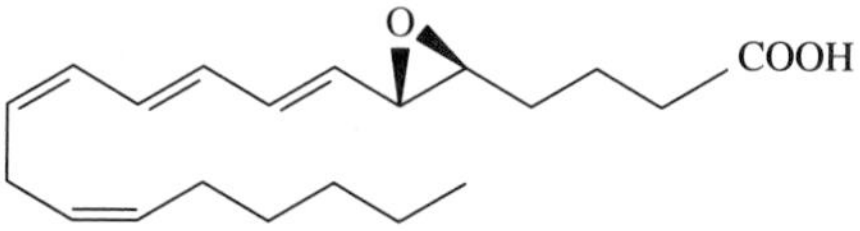 O, COOH	+	+
LTA_5	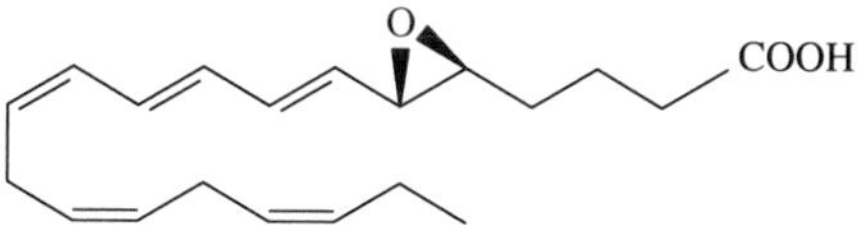 O, COOH	+	+
LTA_4-Me	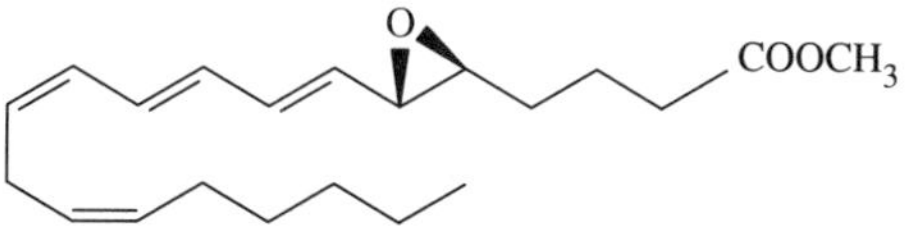 O, $COOCH_3$	−	+
	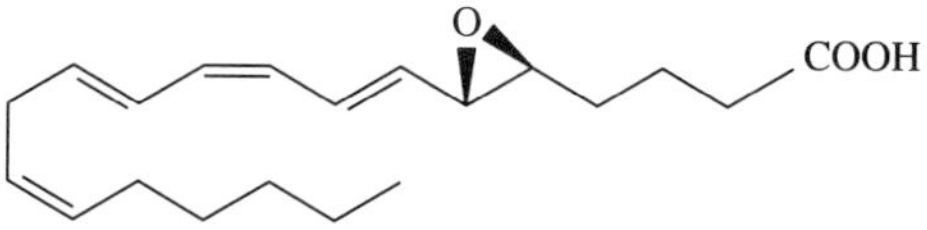 O, COOH	−	+
	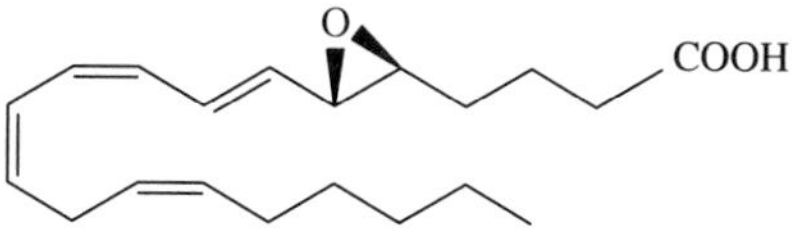 O, COOH	−	+

(*continued*)

TABLE II (*continued*)

Compound	Structure	Substrate	Inactivator
	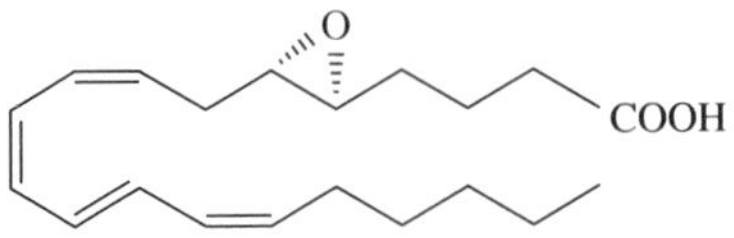 COOH	–	–
14,15-LTA_4	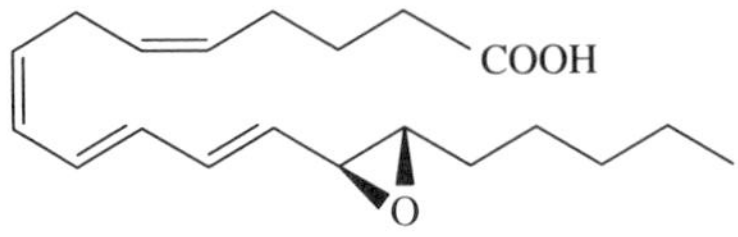 COOH, O	–	+
11,12-EET	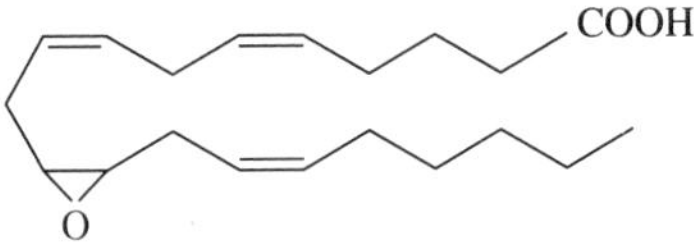 COOH, O		–
14,15-EET	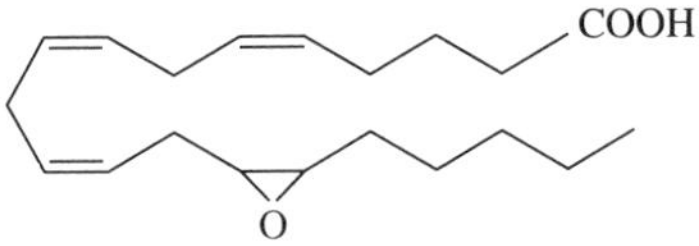 COOH, O		–
Styrene oxide	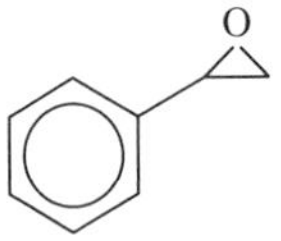 O	–	–

with as short an incubation time as practically possible. Thus, V_{max} and K_m values obtained from 10-sec incubation are 2–3 μmol LTB_4/mg·min and 12–13 μM, respectively. This V_{max} value is about 10-fold greater than that obtained with a 1-min incubation of the standard assay condition (0.2–0.4 μmol/mg·min).

Substrate and Inactivator Specificities

LTA_4 methyl ester, geometric isomers of LTA_4 (7,11-*trans*-9-*cis* form and 7-*trans*-9,11-*cis* form), and 14,15-LTA_4 do not serve as a substrate but inactivate the lung enzyme, the concentration of 50% inactivation with 0.8 μg of enzyme being less than 10 μM. On the other hand, styrene oxide and 5(*S*)-*trans*-5,6-oxido-8,10,14-*cis*-12-*trans*-eicosatetraenoic acid (an isomer of LTA_4 without allylic epoxide, see Fig. 1) neither serve as a substrate nor do they inactivate the lung enzyme. Neutrophil LTA_4 hydrolase also converts LTA_5 to LTB_5, albeit less efficiently than LTA_4 to LTB_4.[13] 5(*S*)-*trans*-5,6-oxido-7,9-*trans*-Eicosadienoic acid and LTA_3 fail to serve as a substrate, but they inactivate the neutrophil enzyme.[14,15] Finally, 11, 12-oxide-5,8,14-*cis*-eicosatrienoic acid (EET) and 14,15-EET do not inactivate the erythrocyte enzyme.[3] All these results are summarized in Table II. Taken together, LTA_4 hydrolase shows a strict substrate specificity, only for LTA_4 and for LTA_5, and may be susceptible to inactivation by 1,2-oxide-3,5-hexadiene structure which includes highly reactive allylic epoxide. The covalent binding of LTA_4 to the enzyme has been reported,[15] but the relationship between this binding and suicide-type inactivation is not well understood.

Human LTA_4 hydrolase cDNA has been cloned and the complete primary structure elucidated.[16,17] This clone is expressed in *Escherichia coli* as a fusion protein, with full enzyme activity.[18]

Acknowledgment

This work is supported by grant-in-aid from the Ministry of Education, Science and Culture of Japan.

[13] D. J. Nathaniel, J. F. Evans, Y. Leblanc, C. Léveillé, B. J. Fitzsimmons, and A. W. Ford-Hutchinson, *Biochem. Biophys. Res. Commun.* **131,** 827 (1985).

[14] J. F. Evans, D. J. Nathaniel, S. Charleson, C. Léveillé, R. Zamboni, Y. Leblanc, R. Frenette, B. J. Fitzsimmons, S. Leger, P. Hamel, and A. W. Ford-Hutchinson, *Prostaglandins Leukotrienes Med.* **23,** 167 (1986).

[15] J. F. Evans, D. J. Nathaniel, R. J. Zamboni, and A. W. Ford-Hutchinson, *J. Biol. Chem.* **260,** 10966 (1985).

[16] M. Minami, S. Ohno, H. Kawasaki, O. Rådmark, B. Samuelsson, H. Jörnvall, T. Shimizu, Y. Seyama, and K. Suzuki, *J. Biol. Chem.* **262,** 13873 (1987).

[17] C. D. Funk, O. Rådmark, J. Y. Fu, T. Matsumoto, H. Jörnvall, T. Shimizu, and B. Samuelsson, *Proc. Natl. Acad. Sci. U.S.A.* **84,** 6677 (1987).

[18] M. Minami, Y. Minami, Y. Emori, H. Kawasaki, S. Ohno, K. Suzuki, N. Ohishi, T. Shimizu, and Y. Seyama, *FEBS Lett.* **229,** 279 (1988).

[32] Potato Arachidonate 5-Lipoxygenase: Purification, Characterization, and Preparation of 5(*S*)-Hydroperoxyeicosatetraenoic Acid

By TAKAO SHIMIZU, ZEN-ICHIRO HONDA, ICHIRO MIKI, YOUSUKE SEYAMA, TAKASHI IZUMI, OLOF RÅDMARK, and BENGT SAMUELSSON

Introduction

Synthesis of biologically active leukotrienes (LTs) is initiated by 5-lipoxygenation of arachidonic acid to yield 5(*S*)-hydroperoxy-6-*trans*-8, 11, 14-*cis*-eicosatetraenoic acid (5-HPETE), which is further transformed by dehydration to LTA_4, a key intermediate in the formation of LTs.[1,2] The reaction mechanism and the enzyme responsible for the conversion of 5-HPETE to LTA_4 (originally termed as dehydratase) remained to be elucidated. Using a unique lipoxygenase from potato tubers, Shimizu *et al.*[3] in 1984 demonstrated that a single enzyme (arachidonate 5-lipoxygenase) catalyzes both 5-lipoxygenation of arachidonic acid, and the subsequent conversion of 5-HPETE to LTA_4. Later, four groups independently purified Ca^{2+}-requiring 5-lipoxygenase from several mammalian tissues (human leukocytes,[4] porcine leukocytes,[5] murine mast cells,[6] and rat basophilic leukemia cells[7]). They demonstrated that these enzymes had dual enzyme activity, as observed in case of potato lipoxygenase. In addition to facilitating a better understanding of the mechanism of multiple lipoxygenase reactions, the enzyme is useful for preparation of arachidonate metabolites such as 5-HPETE and 5(*S*),12(*S*)-dihydroxy-6,10-*trans*-8,14-*cis*-eicosatetraenoic acid [5(*S*),12(*S*)-diHETE] as well as for the initial-phase screening of 5-lipoxygenase inhibitors.

We describe herein the purification and characterization of potato lipoxygenase, and detailed procedures for the syntheses of 5-HPETE and 5(*S*),12(*S*)-diHETE.

[1] B. Samuelsson, *Science* **220,** 568 (1983).
[2] T. Shimizu, *Int. J. Biochem.* **20,** 611 (1988).
[3] T. Shimizu, O. Rådmark, and B. Samuelsson, *Proc. Natl. Acad. Sci. U.S.A.* **81,** 689 (1984).
[4] C. A. Rouzer and B. Samuelsson, *Proc. Natl. Acad. Sci. U.S.A.* **82,** 6040 (1985).
[5] N. Ueda, S. Kaneko, T. Yoshimoto, and S. Yamamoto, *J. Biol. Chem.* **261,** 7982 (1986).
[6] T. Shimizu, T. Izumi, Y. Seyama, K. Tadokoro, O. Rådmark, and B. Samuelsson, *Proc. Natl. Acad. Sci. U.S.A.* **83,** 4175 (1986).
[7] G. K. Hogaboom, M. Cook, J. F. Newton, A. Varrichio, R. G. L. Shorr, H. M. Sarau, and S. T. Crook, *Mol. Pharmacol.* **30,** 510 (1986).

Potato Selection

Enzyme activity varies significantly from potato to potato. Some lots of potatoes have no detectable enzyme activity, presumably due to storage conditions (duration and temperature, radiation, etc.) rather than to species differences. Before starting purification of the enzyme, or preparation of 5-HPETE, first the enzyme activity should be checked using the ammonium sulfate fraction (see below). The use of fresh potatoes obtained directly from farms provides the most successful results.

Assay of Potato Lipoxygenase

Like other lipoxygenases, the potato lipoxygenase activity can be determined by four different assay methods: (1) spectrophotometry, (2) oxygen monitoring, (3) thin-layer chromatography (TLC), and (4) high-performance liquid chromatography (HPLC). Table I summarizes the characteristics of each method. For every assay, one unit of the enzyme activity is defined as the amount that synthesizes 1 μmol of 9-hydroperoxy-10-*trans*-12-*cis*-octadecadienoic acid from linoleic acid for 1 min at 24°. The specific activity is expressed as units per milligram of protein. The protein concentration is measured by the method of Lowry *et al.*[8] with bovine serum albumin (Sigma, St. Louis, MO) as a standard.

Spectrophotometric Assay[3]

The standard assay mixture contains 0.1 *M* potassium phosphate buffer, pH 6.3, and enzyme in a total volume of 1 ml. The quartz cuvettes should be maintained at 24°. The reaction is initiated by the addition of 100 μM linoleic acid (Sigma) (10 μl of 10 m*M* stock solution in ethanol), and the increase in the absorbance at 234 nm (typical for the conjugated diene structure) is continuously monitored. To the reference cuvette is added 10 μl of ethanol before starting the reaction. Usually, the time-course curve exhibits a sigmoidal shape (lag time before the maximal velocity). The sharpest slope in the middle of the reaction is taken as the initial velocity using 28,000 M^{-1} cm^{-1} as a molar extinction coefficient.

Oxygen Monitor Assay

The composition of the assay mixture is the same as for the spectrophotometric assay. The reaction volume depends on the oxygen monitor instrument. After calibrating zero and 100% (equivalent to 220 μM of O_2

[8] O. H. Lowry, N. J. Rosebrough, A. L. Farr, and R. J. Randall, *J. Biol. Chem.* **193,** 265 (1951).

TABLE I
COMPARISON OF LIPOXYGENASE ASSAY METHODS

Parameter	Method			
	Spectrophotometry	Oxygen monitoring	TLC	HPLC
Used for crude enzyme	No	Yes	Yes	Yes
Interference				
With turbid sample	Yes	No	No	No
By UV substances	Yes	No	No	Yes; sometimes
Measures initial velocity	Yes	Yes	No	No
Sensitivity	High	Low	Very high	High
Specificity	Low	Low	High	High
Radioisotope	No use	No use	Necessary	Not necessary
Speed	Fast Real time	Fast Real time	30 min for TLC; overnight for autoradiography	30 min for each sample[a]

[a] Use of an automatic injector saves time.

concentration) with sodium dithionite, and air, respectively, the substrate [linoleic acid or arachidonic acid (Nu-Chek Prep, Elysian, MN)] is added to initiate the reaction, and the initial velocity of the oxygen consumption determined. The major advantage of this method is that it is suited to turbid solutions, such as crude extracts. This method is also useful for screening of various inhibitors possessing UV absorbance; otherwise, these compounds interfere with the spectrophotometric or, sometimes, HPLC assay.

Thin-Layer Chromatographic Assay

1-^{14}C-Labeled fatty acid (10 μM, 100,000 dpm) is incubated with enzyme in 100 mM potassium phosphate buffer, pH 6.3, in a total volume of 100 μl. The incubation is carried out for 5 min at 24° and terminated by the addition of 0.3 ml of an ice-cold mixture of diethyl ether/methanol/0.2 M citric acid (30 : 4 : 1, v/v/v). The organic layers (100 μl) are spotted on a silica gel glass plate (10 cm × 10 cm, HPTLC, Merck) at 4°. The solvent system and R_f values of fatty acid metabolites are described elsewhere in this volume.[9]

HPLC Analysis

A mixture consisting of 100 mM potassium phosphate buffer, pH 6.3, and enzyme in a total volume of 0.05 ml is preincubated for 5 min at 24° using an Eppendorf tube (1.5 ml). The reaction is initiated by the addition of 100 μM of linoleic acid (5 nmol dissolved in 2 μl of ethanol, 2.5 mM stock solution). After 5 min, the reaction is terminated by the addition of 150 μl of acidic methanol (0.1% acetic acid in methanol). The mixture is kept at −20° for at least 30 min, followed by centrifugation at 10,000 g for 10 min at 4°. The supernatant (100 μl) is directly applied to an ODS column (TSK gel ODS 80, 0.46 × 15 cm, Tosoh, Tokyo), which is eluted with a solvent system of methanol/water/acetic acid (75 : 25 : 0.01, v/v/v) at a flow rate of 1 ml/min. The column temperature is kept constant at 35° in a column oven. Absorbance at 234 nm is monitored. The retention time of both 9-hydroperoxyoctadecadienoic and 9-hydroxyoctadecadienoic acid is 14 min, whereas that of the 13-hydro(pero)xy derivatives is 13 min. When arachidonic acid is used as a substrate, the products and their retention times are as follows: 15-hydro(pero)xyeicosatetraenoic acid [15-H(P)ETE, minor product], 14 min; 11-H(P)ETE (minor product), 15 min; and 5-H(P)ETE (major product), 19 min. The recovery of these monohydroperoxy acids exceeds 85%.

[9] N. Ueda and S. Yamamoto, this volume [38].

Purification of Lipoxygenase from Potato Tubers

The potato lipoxygenase is purified to homogeneity, with several modifications of the methods originally described by Sekiya *et al.*[10] All procedures are carried out at 0°–4°, and centrifugation at 10,000 *g* for 15 min, unless otherwise indicated.

Step 1. Ammonium Sulfate Fractionation. After peeling, the potatoes (500 g) are cut into small slices with a knife, and homogenized in a Waring blender with 0.6 volume of freshly prepared 0.1 *M* acetate buffer (pH 4.5) containing 2 m*M* ascorbic acid and 2 m*M* $Na_2S_2O_5$. The homogenate is filtered through four layers of gauze and centrifuged. The supernatant is subjected to ammonium sulfate fractionation. The precipitate formed between 25 and 45% saturation is dissolved in 40 ml of 50 m*M* potassium phosphate buffer (pH 6.8) (Buffer A). The sample is dialyzed overnight against two changes of 1 liter of Buffer A.

Step 2. Hydroxyapatite Chromatography. After centrifugation, the sample is applied to a hydroxyapatite column (Hypatite C, Clarkson, Williamsport, NY, 3 × 10 cm), preequilibrated with Buffer A. The enzyme passes through the column without adsorption.

Step 3. DEAE-Sephadex Column Chromatography. The eluate is applied to a DEAE-Sephadex A-50 column (4 × 20 cm, Pharmacia), which has been equilibrated with Buffer A. After washing the column with about 1 liter of Buffer A, a linear gradient elution is performed 0 and 0.4 *M* NaCl in Buffer A (400 ml each). The enzyme elutes at around a 0.1 *M* concentration of NaCl.

Step 4. Gel Filtration on Sephacryl S-300. The active fraction is concentrated to about 8 ml with the aid of Diaflo membrane YM30 (Amicon, Danvers, MA) and applied to a Sephacryl S-300 column (3 × 80 cm, Pharmacia) which is equilibrated with 20 m*M* Tris-HCl buffer/0.1 *M* NaCl (pH 7.2, Buffer B). Elution is carried out with Buffer B at a flow rate of 0.6 ml/min. The enzyme appears at around 300 ml of the retention volume.

Step 5. Mono Q Column Chromatography. The active fraction from the previous step is directly applied to a Mono Q column (1 × 10 cm, Pharmacia), previously equilibrated with Buffer B. After washing the column with the same buffer, a linear gradient elution is carried out with 50 ml each of Buffer B and 20 m*M* Tris-HCl buffer/1 *M* NaCl (pH 7.2) at a flow rate of 2 ml/min.

Step 6. Mono P Chromatofocusing Column. The active fractions are combined and dialyzed against two changes of 100 volumes of 25 m*M*

[10] J. Sekiya, H. Aoshima, T. Kajiwara, T. Togo, and A. Hatanaka, *Agric. Biol. Chem.* **41,** 827 (1977).

piperazine-HCl buffer (pH 5.3), each for 1 hr, and applied to a Mono P column (0.5 × 20 cm, Pharmacia) that is equilibrated with the same piperazine buffer. The enzyme is eluted with 30 ml of 10% Polybuffer 74–HCl buffer (pH 4). The enzyme elutes at pH about 4.5, and is defined as the purified enzyme. Column procedures with Mono Q and Mono P are carried out with an FPLC system (Pharmacia) at room temperature. The pH of the purified enzyme should be adjusted immediately to 6.3 by the addition of 1 *M* K_2HPO_4. The specific activity of the purified enzyme is 43 units/mg of protein at 24°, when 100 *μM* linoleic acid is used as a substrate. Yields are 10–15%. The enzyme is kept frozen at −70° without any detectable loss of activity for at least 1 month. Recently, an improved procedure of the purification was reported by Mulliez *et al.*[11] They added low concentrations of detergents and glycerol, and obtained an enzyme with a higher specific activity (140–160 units/mg). DEAE-Sephadex and Sephacryl S-300 can be replaced by equivalent columns such as Q-Sepharose HR (Pharmacia) and Superose 12 Prep grade (Pharmacia), respectively.

General Properties of the Enzyme

The potato lipoxygenase is a monomeric protein with a molecular weight of 95,000. It is within a reasonable variation of those reported by other workers (100,000,[10] and 92,000[11]). The p*I* of the enzyme is estimated to be 4.5 by Mono P chromatofocusing, which is significantly lower than the value obtained by others (p*I* 4.9), determined by analytical isoelectric focusing.[11] The one mole of nonheme iron is described as a prosthetic group by other investigators.[10,11]

The enzyme shows a broad substrate specificity toward various fatty acids. Linoleic acid is the best substrate, followed by arachidonic acid, and bishomo-γ-linolenic acid. When arachidonic acid is used as a substrate, in addition to the major product (5-HPETE), 11-HPETE and 15-HPETE are also produced. In addition, 5(*S*),12(*S*)-dihydroperoxy-6,8,10,14-eicosatetraenoic acid, 6-*trans*-LTB_4, and 12-*epi*-6-*trans*-LTB_4 are produced. (The ratio of these products are 0.8 : 1 : 1, depending on the reaction conditions.) The latter two diHETE are nonenzymatic hydrolysis products of LTA_4.[12] $^{18}O_2$ experiments as well as the trapping of the intermediate by methanol, suggested that the purified enzyme synthesizes LTA_4.[3] Thus, the purified potato lipoxygenase catalyzes not only the 5-lipoxygenation of arachidonic acid, but further conversion of 5-HPETE to LTA_4 and

[11] E. Mulliez, J.-P. Leblanc, J.-J. Girerd, M. Rigaud, and J.-C. Chottard, *Biochim. Biophys. Acta* **916,** 13 (1987).

[12] P. Borgeat and B. Samuelsson, *Proc. Natl. Acad. Sci. U.S.A.* **76,** 3213 (1979).

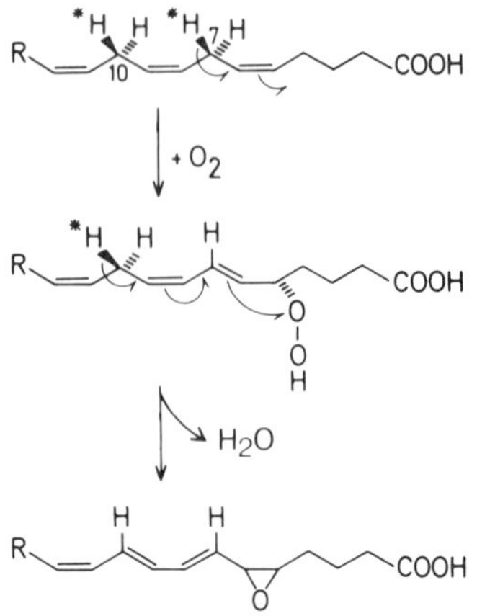

FIG. 1. Proposed reaction mechanism for the formation of LTA_4 from arachidonic acid by successive eliminations of D-hydrogens at C-7 and C-10 (see Ref. 3).

5(*S*),12(*S*)-diHPETE. The incubation of the purified enzyme with bishomo-γ-linolenic acid results in the syntheses of 8-hydroperoxy acid (major product),[13] with a small amount of 11- and 12-hydroperoxy acids. K_m values for linoleic acid, arachidonic acid, and bishomo-γ-linolenic acid are 200, 30, and 50 μM, respectively. The *S* configuration of 5-HPETE was originally reported by Corey *et al.*[14] The chirality at C-8 of 8-hydroxy-9,11,14-eicosatrienoic acid was determined by derivatization with menthyl chloroformate, followed by oxidative ozonolysis. For details, refer to Shimizu *et al.*[3] and Hamberg.[15] The proposed reaction mechanism of LTA_4 formation is illustrated in Fig. 1 (see Ref. 3). The abstraction of D-hydrogen at C-10 was proved later using a purified 5-lipoxygenase from murine mast cells.[6] The mechanism of the formation of 5(*S*),12(*S*)-diHPETE remains unclear.

Preparation of 5-Hydroperoxyeicosatetraenoic Acid

Small-Scale Preparation (Pilot Experiment)

The reaction mixture contains 0.1 *M* potassium phosphate buffer (pH 6.3), 0.03% Tween 20, 300 μM arachidonic acid, and varying amounts of

[13] T. Shimizu, O. Rådmark, and B. Samuelsson, *Adv. Prostaglandin Thromboxane Leukotriene Res.* **15,** 177 (1985).

[14] E. J. Corey, J. O. Albright, A. E. Barton, and S. Hashimoto, *J. Am. Chem. Soc.* **102,** 1436 (1980).

[15] M. Hamberg, *Anal. Biochem.* **43,** 515 (1971).

potato lipoxygenase in a total volume of 0.1 ml. The reaction is started by the addition of arachidonic acid. The incubation is carried out at 24° for varying times (usually 10 min), and is terminated by the addition of 0.2 ml of a mixture (acetonitrile/methanol/acetic acid, 350 : 150 : 1, v/v/v) containing 2 nmol of 13-hydroxy-9,11-octadecadienoic acid [internal standard, prepared by the incubation of linoleic acid with soybean lipoxidase (Sigma), or purchased from Oxford Biomedical Research Inc.]. The mixture is kept at −20° for 30 min, followed by centrifugation at 10,000 *g* for 5 min at 4°. Aliquots of the supernatants are applied to HPLC (TSK ODS 80, 0.46 × 15 cm, Tosoh, Tokyo). HPLC conditions are as follows: solvent system—acetonitrile/methanol/water/acetic acid (350 : 150 : 250 : 1, v/v/v/v); column temperature—35°; flow rate—1 ml/min; retention times—13-hydroxy-9,11-octadecadienoic acid, 12 min, 5-HETE, 16 min, and 5-HPETE, 17 min. Optimal conditions (especially, the amount of enzyme and the time of incubation) required for preparing 5-HPETE should be determined before a large-scale incubation. The longer incubation time allows the synthesis of more 5,12-diHETE. The excess amount of the enzyme is often inhibitory for producing both 5-HPETE and 5(*S*),12(*S*)-diHETE.

Large-Scale Preparation

Reagents and Materials

Arachidonic acid (Nu-Chek, Elysian, MN)
Tween 20 (Sigma)
Partially purified potato lipoxygenase (dialyzed sample after ammonium sulfate fractionation)
Distilled diethyl ether (600 ml)
Distilled water (800 ml)
Methanol (200 ml)
Iatrobeads silica gel 6RS-8060 (2 g, Iatron, Tokyo, kept in 95% ethanol[16])
Anhydrous Na_2SO_4 (200 g)
6 *N* HCl
Hexane/diethyl ether (9 : 1, v/v, 150 ml)
Hexane/diethyl ether (6 : 4, v/v, 100 ml)

Equipment and Instruments

Kjeldahl flask, short neck (1 liter, 300 ml, 50 ml)

[16] S. Kitamura, T. Shimizu, I. Miki, T. Izumi, T. Kasama, A. Sato, H. Sano, and Y. Seyama, *Eur. J. Biochem.* **176,** 725 (1988).

O_2 gas
Rotary evaporator
Separating funnel (2 liter)
Glass column (5 × 20 cm) for anhydrous Na_2SO_4

Procedure

STEP 1. INCUBATION. About 10 mg of arachidonic acid (ethanolic solution) are added to [100-*X*] ml of 50 m*M* potassium phosphate buffer (pH 6.3) containing 0.03% Tween 20 (*X* is the volume of enzyme solution, see below) in a 300-ml flask. The mixture is placed for 1 min in a sonication bath. An appropriate amount (*X* ml) of the enzyme (as determined by the small-scale incubation) is added to initiate the reaction. Incubations are carried out for 10 min (or the time determined as best from the small-scale incubation) at 24° under oxygen spraying, and the reaction terminated by adding 2 volumes of ice-cold methanol.

STEP 2. PARTITION AND COLUMN CHROMATOGRAPHY. The mixture is kept at −20° for 30 min, followed by centrifugation. The supernatant is acidified to pH 3 with 6 *N* HCl. The extraction is based on the method described by Borgeat *et al.*[17,18] The supernatant (about 300 ml) is poured into a 2-liter separating funnel, and 600 ml (6 volumes of the reaction volume) of diethyl ether and 400 ml (4 volumes of the reaction volume) of distilled water are added. After vigorous mixing, the aqueous layer is discarded. Two hundred milliliters of distilled water are then added, mixed well, and the water discarded. Final washing is done using 50 ml (half-volume of the reaction volume) of distilled water. The diethyl ether layer is passed through a column of anhydrous Na_2SO_4 (5 × 10 cm) to reduce the water content. The organic layer is dried with a rotary evaporator (the bath temperature should be kept below 30°). Usually, about 700 ml of organic solution is dried in a 1-liter flask to about 20 to 30 ml, which is transferred to a small-size (50-ml) flask. After evaporation, the residue is dissolved in about 5 ml of hexane/diethyl ether (9 : 1, v/v).

STEP 3. SILICA GEL COLUMN CHROMATOGRAPHY. The sample is applied to a silica gel column (Iatrobeads, 2 g), which has been maintained with 95% ethanol,[16] washed with 60 ml of ethanol (99.5%), and then equilibrated with 60 ml of hexane/diethyl ether (9 : 1). After application of the sample, the column is washed with 60 ml of hexane/diethyl ether (9 : 1), eluting unreacted arachidonic acid. 5-HPETE elutes with 60 ml of hexane/diethyl ether (6 : 4). 5-HETE and 5-keto compound appear in the hexane/diethyl

[17] P. Borgeat, M. Hamberg, and B. Samuelsson, *J. Biol. Chem.* **251,** 7816 (1976).
[18] P. Borgeat and B. Samuelsson, *Proc. Natl. Acad. Sci. U.S.A.* **76,** 2148 (1979).

ether (4 : 6) fraction. More polar compounds such as 6-*trans*-LTB_4 or 5(*S*),12(*S*)-diHPETE elute with 60 ml of ethyl acetate (see below).

STEP 4. PURIFICATION OF 5-HPETE BY HPLC. Two methods can be used to purify 5-HPETE by HPLC. (1) Straight-phase HPLC. The hexane/diethyl ether (6 : 4) fraction is evaporated to dryness with a rotary evaporator, and applied to a straight-phase HPLC (Nucleosil 50-5, Macherey-Nagel, Duren, 1 × 30 cm, 4 ml/min) with a solvent system of hexane/2-propanol/acetic acid (97 : 3 : 0.01, v/v). 5-HPETE appears at a retention time of about 13 min, before the peak of 5-HETE. The solvent is dried under a nitrogen stream, and the dried residue reconstituted in ethanol (99.5%). 5-HPETE can be kept at −20° for 1 month. (2) Reversed-phase HPLC. The column (TSK ODS 120 A, 0.78 × 30 cm, Tosoh, Tokyo) is eluted with the same solvent as for the small-scale preparation at a flow rate of 3 ml/min. The 5-HPETE fraction is combined, and extracted twice with about 4 volumes of diethyl ether. Before evaporation, the organic phase is washed with dilute ammonia water for neutralization.

Usually, the yield of 5-HPETE from arachidonic acid is about 8 to 20%. The chirality of 5-HPETE can be determined using a chiral-phase HPLC[16] (Chiralcel OC, 0.46 × 25 cm, Daisel, Tokyo) after it is reduced to 5-HETE.

Preparation of 5(*S*),12(*S*)-diHETE

Optimal conditions should be determined in small-scale experiments. The primary product [5(*S*),12(*S*)-diHPETE] is assayed by reversed-phase HPLC using acetonitrile/methanol/water/acetic acid (300 : 100 : 300 : 6, v/v/v/v) containing 0.05% EDTA (disodium salt)[19] as a solvent. HPLC conditions are as follows: column—TSK ODS 80 (0.46 × 15 cm, Tosoh, Tokyo); temperature—35°; flow rate—1 ml/min. The absorbance at 270 nm is monitored. The product appears at a retenion time of 12 min. Under the same conditions, 6-*trans*-LTB_4 and its 12-epimer appear at retention times 8.5 and 9 min, respectively. After incubation under oxygen spraying, 2 volumes of ice-cold methanol are added to halt the reaction. Proteins are precipitated at −20° for 30 min, and removed by centrifugation. Solid $NaBH_4$ is added to the supernatant, until the pH of the solution exceeds 10. It is kept in an ice bath for 30 min, and the pH brought to 3 by the addition of 6 *N* HCl. Extraction with diethyl ether, evaporation, and silicic acid column chromatography are performed, as described above. The ethyl acetate fraction of the Iatrobeads column is concentrated with a rotary evaporator, and purified by reversed-phase HPLC (TSK

[19] T. Izumi, T. Shimizu, Y. Seyama, N. Ohishi, and F. Takaku, *Biochem. Biophys. Res. Commun.* **135,** 139 (1986).

ODS 120 A, 0.78 × 30 cm) in a solvent system of acetonitrile/methanol/water/acetic acid (300 : 100 : 300 : 6, v/v/v/v) containing 0.05% EDTA (disodium salt) at a flow rate of 4 ml/min. 5(*S*),12(*S*)-diHETE appears with a retention time of about 15 min, shortly after LTB_4 standard.

5(*S*),12(*S*)-diHETE is extremely unstable, especially on exposure to light and oxygen. It is recommended, therefore, that this compound be stored at −80° after saturating the solution with argon gas.

Acknowledgment

This work was supported by a grant-in-aid from the Ministry of Education, Science and Culture of Japan, the Swedish Medical Research Council, and the Japan–Sweden Foundation.

[33] Leukotriene C_4 Synthase: Characterization in Mouse Mastocytoma Cells

By MATS SÖDERSTRÖM, BENGT MANNERVIK, and SVEN HAMMARSTRÖM

Leukotriene C_4 (LTC_4) is formed by conjugation of leukotriene A_4 (LTA_4) with glutathione (GSH). In biological systems, the reaction is catalyzed by a membrane-bound enzyme, leukotriene C_4 synthase (EC 2.5.1.37). In addition, cytosolic glutathione transferases, in particular, members of the class Mu, have been shown to catalyze formation of LTC_4.[1] The most efficient isoenzymes are transferase 6-6 isolated from rat brain,[2] transferase 4-4 from rat liver,[3] and transferase μ from human liver.[4]

Nomenclature

The process catalyzed is chemically an LTA_4 : glutathione epoxide transferase reaction, formally similar to conjugation reactions involving

[1] B. Mannervik, P. Ålin, C. Guthenberg, H. Jensson, M. K. Tahir, M. Warholm, and H. Jörnvall, *Proc Natl. Acad. Sci. U.S.A.* **82,** 7202 (1985).

[2] S. Tsuchida, T. Izumi, T. Shimizu, T. Ishikawa, I. Hatayama, K. Satoh, and K. Sato, *Eur. J. Biochem.* **170,** 159 (1987).

[3] B. Mannervik, H. Jensson, P. Ålin, L. Örning, and S. Hammarström, *FEBS Lett.* **175,** 289 (1984).

[4] M. Söderström, B. Mannervik, L. Örning, and S. Hammarström, *Biochem. Biophys. Res. Commun.* **128,** 265 (1985).

glutathione transferases (EC 2.5.1.18) (see Mannervik and Danielson[5] for a recent review). Members of the latter family of enzymes also catalyze the formation of LTC_4 (and its C-1 methyl ester), but do not appear to be involved in the main pathway of leukotriene biosynthesis. The name leukotriene C_4 synthase, used for the enzyme described in this chapter, has been adopted to distinguish the enzyme from the above glutathione transferases, which display broad substrate specificity. Further, the form "synthase" rather than "synthetase" should be used, since the latter designation is synonymous to "ligase" and refers to synthetic reactions coupled with the hydrolysis of a phosphate ester bond.[6]

Reports from three groups of investigators have shown that LTC_4 formation in rat basophilic leukemia cells is catalyzed by a membrane-bound enzyme.[7-9] Leukotriene C_4 synthase activity has been described and an enzyme partially purified from the microsomal fraction of guinea pig lung.[10,11] The formation of LTC_4 is especially high in mouse mastocytoma cells, the source from which LTC_4 was first isolated.[12] The partial purification of leukotriene C_4 synthase from this source is described here.

Assay Method

Principle. Enzyme activity is determined by measuring LTC_4 formation by means of reversed-phase HPLC on a C_{18} column.

Reagents

Sodium phosphate buffer, 0.1 *M*, pH 7.0, containing 1.0 m*M* EDTA

Glutathione (GSH), 130 m*M* in deionized water

LTA_4 lithium salt, 1.0 m*M*, containing an addition of labeled material equivalent to 2.5 μM [14,15-$^3H_2$]LTA_4 (40 Ci/mmol, in 99.5% ethanol).

[5] B. Mannervik and U. H. Danielson, *CRC Crit. Rev. Biochem.* **23,** 283 (1988).

[6] "Enzyme Nomenclature," Recommendations (1984) of the Nomenclature Committee of the International Union of Biochemistry, Academic Press, San Diego, 1984.

[7] M. K. Bach, J. R. Brashler, and D. R. Morton, Jr., *Arch. Biochem. Biophys.* **230,** 455 (1984).

[8] B. A. Jakschik, T. Harper, and R. C. Murphy, *J. Biol. Chem.* **257,** 5346 (1982).

[9] T. Yoshimoto, R. J. Soberman, R. A. Lewis, and K. F. Austen, *Proc. Natl. Acad. Sci. U.S.A.* **82,** 8399 (1985).

[10] T. Izumi, Z. Honda, N. Ohishi, S. Tsuchida, K. Sato, S. Kitamura, T. Shimizu, and Y. Seyama, *Biochim. Biophys. Acta* **959,** 305 (1988).

[11] T. Yoshimoto, R. J. Soberman, B. Spur, and K. F. Austen, *J. Clin. Invest.* **81,** 866 (1988).

[12] R. C. Murphy, S. Hammarström, and B. Samuelsson, *Proc. Natl. Acad. Sci. U.S.A.* **76,** 4275 (1979).

Hydrolysis of LTA_4 Methyl Ester

Stock solutions of LTA_4 methyl ester in hexane containing 5% triethylamine (1.2 mM) should be kept at $-80°$. Before the start of an assay, unlabeled LTA_4 methyl ester mixed with $[14,15\text{-}^3H_2]LTA_4$ methyl ester is brought to near dryness under a stream of argon and redissolved in 20–50 μl of ethanol to give a concentration of 2 mM. The solution is kept at $-20°$ for a few minutes before the addition of 1 volume (20–50 μl) of 1.0 M LiOH. The hydrolysis is allowed to proceed at room temperature for 30 min and the mixture then kept at 0° until used (within 1 hr).

Procedure. Buffer (50–95 μl) and enzyme sample (5–50 μl) are added to a test tube. The solution is preincubated for 1 min at 30° before the addition of 4 μl of GSH to give a final concentration of 5 mM. After 1 min the reaction is started by the addition of 3.8 μl of LTA_4 lithium salt. The reaction is terminated by the addition of 100 μl of methanol after a specified time (0.5–5 min). The incubation solution is kept at $-20°$ for about 1 hr in order to allow protein to precipitate and is then centrifuged at 3000 g for 10 min. An aliquot (150-μl) of the resulting supernatant fraction is injected onto a reversed-phase HPLC Nucleosil C_{18} column. The mobile phase is prepared by adding 0.07% acetic acid and 0.03% phosphoric acid to a mixture of methanol/water, 7 : 3. The pH of the mobile phase is adjusted to 5.6 with NH_4OH. The chromatography is performed isocratically at a flow rate of 1 ml/min. Eluted compounds are detected at 280 nm utilizing a UV absorbance detector. The retention time for LTC_4 is determined by means of a synthetic standard. The amount of radioactivity corresponding to LTC_4 is measured using a liquid scintillation counter or by using a radioactivity detector equipped with a flow cell. The product formed is identified by its characteristic UV spectrum[12] with a maximum at 280 and shoulders at 270 and 290 nm, and by the retention time from HPLC determined with the standard sample of LTC_4. Figure 1 shows a representative chromatogram of an assay mixture.

In cases where the chromatograms are not well resolved in the region containing LTC_4, an extraction procedure may be introduced for enrichment of the reaction product before chromatography. Poor resolution may be caused by the spontaneous hydrolysis of the substrate LTA_4, which gives rise to products that interfere with the determination of LTC_4. This is particularly likely to arise in kinetic experiments in which initial rates are desired. When the extraction procedure is included, the incubation is terminated by the addition of 200 μl of ethyl acetate, preacidified to pH 3 by equilibrating with 1 M HCl. The organic phase, which contains LTC_4, is removed and LTC_4 is reextracted with 100 μl of 0.1 M $NaPO_4$ buffer, pH 7.0, and the final aqueous phase is injected onto the HPLC column as described above.

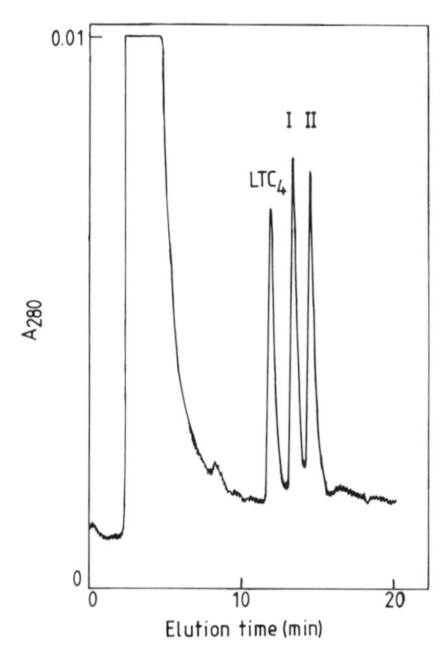

FIG. 1. Chromatogram of a reaction mixture from an incubation of leukotriene C_4 synthase with LTA_4 and GSH, analyzed by reversed-phase HPLC on a Nucleosil C_{18} column. Peaks labeled I and II are nonenzymatically formed hydrolysis products of LTA_4.

A radiometric assay based on the use of the methyl ester of LTA_4 has also been used for leukotriene C_4 synthase.[3] This method, however, is less specific for leukotriene C_4 synthase, since glutathione transferases, which generally have higher activity with LTA_4 methyl ester than with LTA_4, may interfere when present.

Definition of Units and Specific Activity. A unit of enzyme activity is defined as the amount of enzyme that catalyzes the formation of 1 μmol of LTC_4 per minute at 30° using 5 mM concentration of GSH and 35 μM concentration of LTA_4 lithium salt. Specific activity is defined as units per milligram of protein. Protein concentration is determined according to the method of Peterson[13] with bovine serum albumin as a standard.

Tumor Cell Propagation

Mouse mastocytoma cells (CXBGABMCT-1) are propagated intramuscularly in the hindlegs of male or female CB_6Fl (C57 bl × BALB/c)

[13] G. L. Peterson, *Anal. Biochem.* **83,** 346 (1977).

mice, aged 6 weeks to 6 months. When the tumors reach a size of 0.5 to 1.5 cm^3 (17–20 days after injection), the animal is killed by excessive ether anesthesia. The tumor from each leg is excised and minced in 5 ml of phosphate-buffered saline (150 m*M* NaCl/37 m*M* KH_2PO_4/11 m*M* Na_2HPO_4, pH 7.2) by using the blunt end of a Pasteur pipette. Large tissue fragments are allowed to sediment. The cell suspension is then injected (0.1 ml per leg) into a new group of mice.

Preparation of Leukotriene C_4 Synthase from Mouse Mastocytoma Cells

Step 1. Preparation of Microsomal Fraction. Tumors from 20 animals are excised and homogenized in 0.25 *M* ice-cold sucrose, using four up-and-down strokes at 400 rpm in a Potter-Elvehjem homogenizer. The homogenate is adjusted to 20% (w/v) and centrifuged at 10,000 *g* for 15 min. The supernatant fraction is centrifuged at 105,000 *g* for 60 min. The resulting microsomal pellet is suspended in 0.1 *M* $NaPO_4$ buffer, pH 8.0, and recentrifuged at 105,000 *g* for 60 min; all centrifugations made at 5°.

Step 2. Solubilization of Leukotriene C_4 Synthase. The microsome fraction is suspended in 0.1 *M* $NaPO_4$ buffer, pH 8.0. Glutathione, dissolved in water, is added to a final concentration of 1 m*M*. The suspension is stirred slowly at 5° using a magnetic stirrer and a 10% (w/v) solution of CHAPS (3-[(3-cholamidopropyl)-dimethylammonio]-1-propanesulfonate) is added dropwise within a period of 10 min to a final concentration of 1%. The suspension is stirred for an additional 20 min at 5° and is then centrifuged for 60 min at 105,000 *g* in order to remove undissolved material.

Step 3. Chromatography on Sephadex G-25. The solubilized material from step 2 is passed through a Sephadex G-25 column (4 × 15 cm), equilibrated with 10 m*M* $NaPO_4$ buffer, pH 7.0, 0.1% CHAPS, and 1 m*M* GSH (buffer A).

Step 4. Chromatography on Hydroxyapatite. The pooled material from step 3 is applied to a hydroxyapatite column (4 × 10 cm) equilibrated with buffer A. The column is washed with 200 ml of buffer A and 200 ml of 70 m*M* $NaPO_4$ buffer, pH 7.0, 0.1% CHAPS, and 1 m*M* GSH (buffer B). The leukotriene C_4 synthase activity is eluted with 0.3 *M* $NaPO_4$ buffer, pH 7.0, 0.1% CHAPS, and 1 m*M* GSH (buffer C).

Step 5. Chromatography on Sephadex G-25. The pooled material from step 4 is desalted on a Sephadex G-25 column (4 × 25 cm), equilibrated with 0.1 *M* $NaPO_4$ buffer, pH 7.0, 0.1% CHAPS, and 1 m*M* GSH. The eluted enzyme is concentrated by ultrafiltration on an Amicon (Danvers, MA) concentration cell, using a PM 10 filter.

The resulting partially purified leukotriene C_4 synthase is enriched 4-fold in comparison with the microsomal fraction. The specific activity is in the range of 30 milliunits per milligram protein (Table I).

TABLE I
PARTIAL PURIFICATION OF MURINE LEUKOTRIENE C_4 SYNTHASE

Step	Volume (ml)	Total protein (mg)	Total activity (units)	Specific activity (mU/mg)
1. Microsomes	16.3	220	1.75	7.95
2. Solubilized microsomes	15.5	67	1.70	25.4
3. Sephadex G-25	38	68.4	1.82	26.6
4. Hydroxyapatite	50	52.5	1.63	31.0
5. Sephadex G-25	70	52.5	1.60	30.5

Properties of Leukotriene C_4 Synthase

Molecular and Kinetic Properties

Leukotriene C_4 synthase from mouse mastocytoma cells has an isoelectric point at pH 6. The enzyme shows activity with LTA_4 and LTA_4 methyl ester. In distinction from the less substrate-specific glutathione transferases, leukotriene C_4 synthase displays no activity with 1-chloro-2,4-dinitrobenzene, 1,2-dichloro-4-nitrobenzene, ethacrynic acid, 4-hydroxynonenal, or bromosulfophthalein as electrophilic substrates.[10,14,15]

Sensitivity toward Inhibitors

Leukotriene C_4 synthase is inhibited by Rose Bengal with an IC_{50} value of 50 μM in the standard assay system for LTA_4 methyl ester.[14] Indomethacin and triphenyltin chloride are poor inhibitors of leukotriene C_4 synthase (IC_{50} = 1000 μM and >100 μM, respectively).[14]

Distinction from Cytosolic and Microsomal Glutathione Transferases

Microsomal glutathione transferase, purified from mouse liver, has no detectable activity with LTA_4 or LTA_4 methyl ester and is consequently distinct from leukotriene C_4 synthase.[14] The substrate specificity profiles and the sensitivities to various inhibitors distinguish the cytosolic transferases from leukotriene C_4 synthase, even though these enzymes display some activity with LTA_4 and (in most cases even higher) with LTA_4 methyl ester. Antibodies raised against the microsomal and the cytosolic

[14] M. Söderström, S. Hammarström, and B. Mannervik, *Biochem. J.* **250,** 713 (1988).

[15] M. Söderström, S. Hammarström, and B. Mannervik, unpublished result (1988).

glutathione transferases neither inhibit nor precipitate leukotriene C_4 synthase from mouse mastocytoma cells.

Acknowledgments

This work was supported by grants from the National Institutes of Health, Bethesda, MD (RO1 HL 33258-01-03), from the Swedish Medical Research Council (03X-5914), and from the Swedish Natural Science Research Council. M.S. was a recipient of a fellowship from the Swedish Medical Research Council.

[34] Leukocyte Arachidonate 5-Lipoxygenase: Isolation and Characterization

By CAROL A. ROUZER and BENGT SAMUELSSON

Assay Method

Principles. There are a number of issues that complicate the assay of leukocyte arachidonate 5-lipoxygenase (EC 1.13.11.34). First, the enzyme catalyzes two reactions, including the oxidation of arachidonic acid to 5-hydroperoxy-6,8,11,14-eicosatetraenoic acid (5-HPETE) followed by the dehydration of 5-HPETE to leukotriene (LT) A_4 (5,6-oxido-7,9,11,14-eicosatetraenoic acid).[1–3] In crude cell-free systems, the efficiency of the second reaction is poor such that a mixture of metabolites is obtained from the interaction of 5-lipoxygenase with arachidonic acid. These may include 5-HPETE, its reduction product, 5-hydroxyeicosatetraenoic acid (5-HETE), the nonenzymatic hydrolysis products of LTA_4 (isomers of 5,12-dihydroxyeicosatetraenoic acid), and enzymatically derived LT, as well. Second, the kinetics of the 5-lipoxygenase reaction are complex, including a lag phase and the first-order irreversible inactivation of the enzyme during the course of the reaction.[4,5] Third, many sources of leukocyte 5-lipoxygenase may also contain arachidonate 12- and 15-lipoxygenases which cause problems in specificity for the assay of crude samples as well

[1] C. A. Rouzer, T. Matsumoto, and B. Samuelsson, *Proc. Natl. Acad. Sci. U.S.A.* **83,** 857 (1986).
[2] N. Ueda, S. Kaneko, T. Yoshimoto, and S. Yamamoto, *J. Biol. Chem.* **261,** 7982 (1986).
[3] G. K. Hogaboom, M. Cook, J. F. Newton, A. Varrichio, R. G. L. Shorr, H. M. Sarau, and S. T. Crooke, *Mol. Pharmacol.* **30,** 510 (1986).
[4] D. Aharony and R. L. Stein, *J. Biol. Chem.* **261,** 11512 (1986).
[5] C. A. Rouzer, N. A. Thornberry, and H. G. Bull, *Ann. N.Y. Acad. Sci.* **524,** 1 (1988).

as competition for substrate.[6] Finally, maximal activity of purified leukocyte 5-lipoxygenase requires Ca^{2+}, ATP, and phosphatidylcholine micelles or a cell membrane preparation. In the case of the human leukocyte enzyme, other, as yet, undefined cellular components have also been shown to stimulate enzyme activity.[2-7]

This enzyme, therefore, does not readily lend itself to routine enzymological techniques. However, investigators have developed assay methods that have provided adequate information for basic purification and characterization studies. In most cases, samples containing enzyme are allowed to react with excess arachidonic acid until the reaction ceases due to enzyme inactivation, and the products are measured by high-performance liquid chromatography (HPLC) or thin-layer chromatography (TLC). This assay provides the specificity required to distinguish contaminating lipoxygenases, and although it is not a direct measure of initial velocity, there is usually a good linear correlation between enzyme amount and product formation over a reasonable range of enzyme concentrations. Since, in most cell-free preparations the ratio of 5-HPETE plus 5-HETE to LT is 8–10 : 1, and varies little, many investigators measure only 5-HPETE plus 5-HETE formation.[1-3] For kinetics studies, a spectrophotometric assay has been devised, and the use of hydroperoxide stimulators and special substrates has been proposed to simplify the reaction.[4,5,8] This chapter describes only an HPLC assay which will suffice for most practical applications.

HPLC Assay. Incubation mixtures contain 0.1 *M* Tris-HCl, 2 m*M* ATP, 2.0 m*M* EDTA, 3 m*M* $CaCl_2$, 1.8 m*M* dithiothreitol (DTT) plus desired stimulatory factors (see below) and the sample to be assayed in a total volume of 1 ml at pH 7.5. Note that 2 m*M* EDTA is present in most buffers used for the isolation of 5-lipoxygenase. For this reason, we include 2 m*M* EDTA in all assay reagents in order to maintain a constant ratio of Ca^{2+} : EDTA in the final reaction mixtures. Since 5-lipoxygenase is unstable in the presence of Ca^{2+}, the EDTA also helps to preserve enzyme activity. The $CaCl_2$ is not added to the reaction mixture until immediately before the incubations are to begin. The samples are warmed at 37° for 5 min, and then the reaction is initiated by the rapid addition of 5 μl of a solution of 20 m*M* arachidonic acid plus 0.4 μM 15-hydroperoxy-11,13-eicosadienoic acid in ethanol. After 10 min of incubation, the samples are placed on ice and 1 ml of 1 μM 13-hydroxyoctadecadienoic acid (13-HOD) is added to serve as internal standard. The samples (100 μl) may then be

[6] C. A. Rouzer and B. Samuelsson, *Proc. Natl. Acad. Sci. U.S.A.* **82,** 6040 (1985).

[7] C. A. Rouzer, T. Shimizu, and B. Samuelsson, *Proc. Natl. Acad. Sci. U.S.A.* **82,** 7502 (1985).

[8] J. F. Navé, B. Dulery, C. Gaget, and J. B. Ducep, *Prostaglandins* **36,** 385 (1988).

injected directly onto the HPLC after removal of precipitated protein by centrifugation (10,000 *g*, 10 min, 4°). Alternatively, to improve sensitivity, the samples may be acidified with 10 μl of formic acid, and extracted twice with 1 ml of chloroform. The extracts are evaporated to dryness under a stream of nitrogen and the residues are then taken up in 100 μl of methanol for injection (25 μl) onto the HPLC. The 5-lipoxygenase products are separated on a column (3.9 mm × 15 cm) of Nova-Pak C_{18} (Waters, Milford, MA) eluted isocratically at 1 ml/min with methanol/water/trifluoroacetic acid/triethylamine (75/25/0.2/0.1, v/v/v/v).[9] In this system, 5-HPETE and 5-HETE coelute at 14 min and are quantitated by the ratio of the area of the corresponding peak to that of the 13-HOD peak (retention time, 9 min). One unit of 5-lipoxygenase activity is defined here as the amount of enzyme that produces 1 nmol of product under these reaction conditions.[5,6,10,11]

Purification of 5-Lipoxygenase

The purification procedure described here was developed for the 5-lipoxygenase from human peripheral blood leukocytes.[6] A very similar scheme has been reported for the isolation of 5-lipoxygenase from rat basophilic leukemia cells.[3,12] Alternatively, the 5-lipoxygenase of porcine leukocytes has been purified by immunoaffinity chromatography using monoclonal antibodies.[2]

Step 1. Isolation of Leukocytes. Human leukocyte concentrates (buffy coat) are obtained from local blood collection centers. For one enzyme preparation, 80–100 units (3–4 liters total volume) of buffy coat are utilized. The leukocyte concentrates are combined with an equal volume of a cold (4°) solution of 2% (w/v) dextran T-500 (Pharmacia) plus 0.9% (w/v) NaCl in water. The resulting cell suspension is allowed to stand at room temperature in 2-liter plastic graduated cylinders to allow sedimentation of erythrocytes. When the sedimentation interface has reached one-half the total volume of the suspension, the supernatant containing leukocytes and plasma is removed and the cells are recovered by centrifugation at 500 *g* for 12 min at 4°. The cell pellets are resuspended in 400 ml of 20 m*M* Tris-HCl, pH 7.4, containing 6.75 g/liter NH_4Cl. After incubation for 10 min at 37°, during which time lysis of remaining erythrocytes occurs, the leukocytes are again recovered by centrifugation and the lysis step is repeated. The cells are then washed once in 200 ml of Dulbecco's

[9] J. D. Eskra, M. J. Pereira, and M. J. Ernest, *Anal. Biochem.* **154,** 332 (1986).

[10] C. A. Rouzer and B. Samuelsson, *FEBS Lett.* **204,** 293 (1987).

[11] C. A. Rouzer and S. Kargman, *J. Biol. Chem.* **263,** 10980 (1988).

[12] A. M. Goetze, L. Fayer, J. Bouska, D. Bornemeier, and G. W. Carter, *Prostaglandins* **29,** 689 (1985).

phosphate-buffered saline (PBS) and resuspended at a concentration of 200×10^6/ml in KPB-1 (50 m*M* potassium phosphate, pH 7.1, containing 0.1 *M* NaCl, 2 m*M* EDTA, 1 m*M* DTT, and the protease inhibitors 0.5 m*M* phenylmethylsulfonyl fluoride, and 60 μg/ml soybean trypsin inhibitor).

The leukocytes are homogenized by sonication using a Cole Parmer (Chicago, IL) 4710 Series ultrasonic homogenizer with an output control setting of 3 and a percentage duty cycle setting of 70. Six pulsatile bursts of 20 sec each separated by 1 min are delivered to the samples (100–200 ml) maintained at 4°. The sonicate is then subjected to centrifugation at 10,000 *g* for 15 min, and the resulting supernatant used for subsequent purification steps.

Step 2. Ammonium Sulfate Fractionation. To the 10,000 *g* supernatant is added 0.176 g/ml of solid ammonium sulfate with gentle stirring at 4°. The pH is maintained at 7.2–7.6 through the dropwise addition of 10 *N* NaOH. After 1 hr, the precipitated protein is removed by centrifugation (10,000 *g*, 15 min) and discarded. To the supernatant is added 0.198 g/ml of ammonium sulfate as described above. In this case, however, the precipitated protein (30–60% ppt) is redissolved in 30 ml of KPB-2 (20 m*M* potassium phosphate, pH 7.1, plus 2 m*M* EDTA and 1 m*M* DTT). This precipitate may be used immediately in step 3, or it may be dialyzed against KPB-3 (KPB-2 plus 0.1 *M* NaCl and 20% glycerol) and stored for several months at −70° for use in later purifications.

Step 3. Gel Filtration Chromatography on AcA 44. All chromatography steps are carried out in a cold cabinet maintained at 4°. Freshly prepared or frozen and thawed 30–60% precipitate (30–50 ml) is subjected to centrifugation at 10,000 *g* for 10 min and then loaded on a column (5.0 × 91 cm) of AcA 44 (LKB) packed in Tris-1 (50 m*M* Tris HCl, pH 7.9, containing 2 m*M* EDTA and 1 m*M* DTT). The column is eluted upward at a rate of 2.45 ml/min with Tris-1, and 11-ml fractions are collected. The elution volume of 5-lipoxygenase is 840–1070 ml, corresponding to an apparent molecular weight of 76,000–38,000. This column may be run overnight and the fractions allowed to stand until morning for completion of the procedure. From steps 4 through 6, however, the enzyme becomes increasingly unstable and best results are obtained if these procedures are completed in succession without delay. It is extremely important that all buffers in subsequent steps are thoroughly deoxygenated. A continuous helium sparge is recommended.

Step 4. HPLC on Mono Q HR 10/10. The eluate (230 ml) from the AcA 44 column is applied directly to a Mono Q HR 10/10 column equilibrated in Tris-2 (Tris-1 plus 20% glycerol). The column is washed at a flow rate of 2 ml/min with Tris-2 and then eluted with a gradient Tris-2 to Tris-2 plus 0.3 *M* KCl totaling 240 ml. Fractions of 4-ml are collected. 5-Lipoxygenase elutes at fractions 28–32. Note that the direct application of the AcA 44 eluate to the Mono Q column without concentration or dialysis is a modification over the previously reported method that results in a 2- to 5-fold improvement in recovery.[6] Under these conditions the enzyme elutes from the Mono Q column as a single peak of activity as opposed to the bifid peak observed with the older method.

Step 5. HPLC on Hydroxyapatite (BioGel HPHT). The eluate (20 ml) from step 4 is concentrated to 7.5 ml by ultrafiltration in a 50-mm cell using a YM10 membrane (Amicon, Danvers, MA). The sample is then subjected to chromatography according to manufacturer's directions on 3 PD-10 columns (Pharmacia, Uppsala, Sweden) equilibrated and eluted with KPB-4 [10 m*M* potassium phosphate, pH 7.3, 0.3 m*M* $CaCl_2$, 1 m*M* DTT, 20 μM ferrous ammonium sulfate (FAS), 20% glycerol]. The resulting sample (10.5 ml) is then applied to a BioGel HPHT column (Bio-Rad, Richmond, CA). Note that, due to technical difficulties with the guard column supplied by Bio-Rad for the HPHT column, we have replaced it with a Mono S HR 5/5 column (Pharmacia). Using Mono S as a guard column provides adequate protection for the HPHT column and does not significantly alter the chromatographic behavior of the sample.

The HPHT column is washed at 0.3 ml/min with KPB-4 and then developed with a gradient of KPB-4 to 10% KPB-5 (500 m*M* potassium phosphate, 6 μM $CaCl_2$, 1 m*M* DTT, 20 μM FAS, 20% glycerol) totaling 10 ml, and from 10 to 50% KPB-5 totaling 50 ml. Fractions of 0.9-ml are collected. 5-Lipoxygenase elutes at fractions 26–31.

After this step, the enzyme is quite unstable and can no longer be recovered in good yield from concentration or desalting procedures. Care should be taken to avoid vigorous mixing and prolonged exposure to oxygen.

Step 6. HPLC on Mono Q HR 5/5. The eluate (5.4 ml) from the HPHT column is immediately and gently mixed with 4.6 ml of TEA-1 (25 m*M* triethanolamine acetate, pH 7.8, containing 2 m*M* EDTA, 1 m*M* DTT, 20 μM FAS, and 25% glycerol). This sample is then applied to a Mono Q HR 5/5 column (Pharmacia) that has been equilibrated with TEA-2 (TEA-1 at pH 7.25). The column is washed with TEA-2 and then eluted with a gradient (40 ml) of TEA-2 to TEA-2 plus 0.4 *M* sodium acetate. Fractions of 1 ml are collected. 5-Lipoxygenase elutes in fractions 18 through 22. The purified enzyme can be stored for several months at $-70°$ if the final elution buffer is made 50% in glycerol.

Preparation of 5-Lipoxygenase Stimulatory Factors from Human Leukocytes

Unlike the 5-lipoxygenase of rat and pig, purification of the human leukocyte 5-lipoxygenase has revealed a number of cellular fractions that stimulate 5-lipoxygenase product formation in the routine HPLC assay.[6] The preparation of these fractions follows below.

100,000 g Pellet. Human leukocyte 10,000 *g* supernatants prepared as described above are subjected to centrifugation at 100,000 *g* for 60 min at 4°. The supernatants are discarded and the pellets gently resuspended in KPB-2 using a glass homogenizer with a tightly fitting Teflon pestle. The resulting suspension (5 mg of protein/ml) is added (50 to 200 μl/ml) to 1-ml reaction mixtures in the HPLC assay. It can be stored for up to several months at −70°. NOTE: This factor may be replaced in the assay by phosphatidylcholine micelles.[13]

60–90% Precipitate. To the final supernatant from the ammonium sulfate fractionation of 5-lipoxygenase (see step 2) is added 0.227 g/ml $(NH_4)_2SO_4$ on ice with gentle stirring. After 1 hr, the precipitated protein is removed by centrifugation at 10,000 *g* for 15 min, and dissolved in 30 ml of KPB-2. Following dialysis against KPB-3 as described in step 2, this material (12 mg of protein/ml) can be stored at −70°. In the HPLC assay, it is stimulatory at concentrations of 20 to 100 μl/ml.

MQ-PTF. In previous reports, a 5-lipoxygenase stimulatory activity was found in the nonadherent protein (pass-through fraction) from the Mono Q HR 10/10 column (step 4).[6] Details for the preparation of this factor (designated MQ-PTF) were described at that time. In the purification procedure outlined here, modifications of the Mono Q HR 10/10 step have been made to improve 5-lipoxygenase recovery. Under these conditions, the presence of a stimulatory activity in the MQ-PTF has been inconsistent; however, adequate activity of purified enzyme can be obtained in the assay in the absence of this material.

Properties of 5-Lipoxygenase

The procedure described here yields a protein that is 85 to 95% pure. As shown in Table I, the yield of enzyme is 3–6% (300–600 mg) with an approximately 400-fold increase in specific activity if the assay is performed in the presence of the leukocyte stimulatory factors (100,000 *g* pellet and 60–90% precipitate), as well as Ca^{2+} and ATP. Activity of the purified protein decreases rapidly at 4° under room atmosphere in the presence of the final chromatography buffer.

Sequence data obtained from the purified enzyme have provided the necessary information to allow the molecular cloning of a cDNA for human leukocyte 5-lipoxygenase. The cDNA encodes for a protein of

[13] T. Puustinen, M. M. Scheffer, and B. Samuelsson, *Biochim. Biophys. Acta* **960,** 261 (1988).

TABLE I
PURIFICATION OF HUMAN LEUKOCYTE 5-LIPOXYGENASE[a]

Step	Total protein (mg)	Total activity (units)	Specific activity (units/mg)	Purification (-fold)	Yield (%)
10,000 *g* Supernatant	3,400	31,500	9.25		
30–60% Precipitate	997	17,800	17.8	1.92	56.5
AcA 44	316	18,300	57.9	6.26	58.1
Mono Q HR 10/10	16.6	5,770	348	37.6	18.3
HPHT	1.52	3,040	2,000	216	9.6
Mono Q HR 5/5	0.447	1,714	3,830	414	5.4

[a] Assays were performed in the presence of the leukocyte 60–90% precipitate and 100,000 *g* pellet stimulatory factors.

78,000 Da, composed of 674 amino acids.[13,14] The deduced amino acid sequence is consistent with a cytosolic protein possessing some hydrophobic regions. There are two consensus sequences for Ca^{2+} binding sites, either of which may be involved in the Ca^{2+} stimulation of 5-lipoxygenase activity. There is also a region of the enzyme that bears a homology to the interface binding domains of lipoprotein lipase and hepatic lipase. This portion may be involved in the interaction of 5-lipoxygenase with its substrate, or with stimulatory membranes and phospholipids. Finally, there are significant regions of homology between the human leukocyte 5-lipoxygenase, the rat leukocyte 5-lipoxygenase, the human reticulocyte 15-lipoxygenase, and the lipoxygenase from soybeans.[14–17]

Due to limited availability and poor stability, very few detailed kinetic studies have been performed with purified leukocyte 5-lipoxygenase. Investigations performed with partially purified samples suggest a deviation from Michaelis–Menten behavior, with substrate inhibition observed.[4,8] As has been shown for the lipoxygenase from soybeans, leukocyte 5-lipoxygenase displays a kinetic lag phase that is eliminated by fatty acyl hydroperoxides, and exacerbated by reducing agents such as sulfhydryl compounds.[4,5,10] This suggests that, like the soybean enzyme, leukocyte

[14] R. A. F. Dixon, R. E. Jones, R. E. Diehl, C. D. Bennett, S. Kargman, and C. A. Rouzer, *Proc. Natl. Acad. Sci. U.S.A.* **85,** 416 (1985).

[15] T. Matsumoto, C. D. Funk, O. Rådmark, J. O. Hoog, H. Jörnvall, and B. Samuelsson, *Proc. Natl. Acad. Sci. U.S.A.* **85,** 26 (1988); *ibid.*, p. 3406 (1988).

[16] J. M. Balcarek, T. Theisen, M. Cook, A. Varrichio, A. Hwang, S. M. Stroschaker, and S. T. Crooke, *J. Biol. Chem.* **263,** 13937 (1988).

[17] E. Sigal, C. S. Craik, E. Highland, D. Grunberger, L. L. Costello, R. A. F. Dixon, and J. A. Nadel, *Biochem. Biophys. Res. Commun.* **157,** 457 (1988).

5-lipoxygenase may undergo a product-mediated oxidative activation step; however, the chemical mechanism of such a step remains to be elucidated. As previously mentioned, 5-lipoxygenase undergoes first-order irreversible inactivation during the course of its reaction.[4,5] Interestingly, a similar enzyme inactivation occurs upon the exposure of the enzyme to fatty acid hydroperoxide in the absence of substrate turnover. It is possible, therefore, that the product-activated form of the enzyme is unstable.

Studies have been performed on the effects of various stimulatory factors on human 5-lipoxygenase kinetics.[5] These have shown that ATP, the 100,000 *g* pellet, and the 60–90% precipitate all stimulate the initial velocity of the reaction, and that the increased total product synthesis induced by none of these factors is due to a decreased rate of enzyme inactivation. Interestingly, the 60–90% precipitate actually increases the enzyme inactivation rate as well as the initial velocity. The effects of ATP are not influenced by temperature, but those of both the 100,000 *g* pellet and the 60–90% precipitate are augmented with increasing temperature. Much remains to be determined concerning the regulation and activation of 5-lipoxygenase in the intact leukocyte. However, recent evidence indicating that a Ca^{2+}-dependent association of the enzyme with membrane occurs in ionophore-challenged leukocytes suggests possible roles for both Ca^{2+} and membrane in the regulation of 5-lipoxygenase activity.[10,18]

Acknowledgments

The authors acknowledge the contributions of many friends and colleagues at the Karolinska Institute and the Merck Frosst Centre for Therapeutic Research. We also gratefully acknowledge the support of the Swedish Medical Research Council (Grant # 03X-217), and we thank Ms. Barbara Pearce for excellent secretarial services.

[18] C. A. Rouzer and B. Samuelsson, *Proc. Natl. Acad. Sci. U.S.A.* **84,** 7933 (1987).

[35] Cytochrome P-450_{LTB} and Inactivation of Leukotriene B_4

By ROY J. SOBERMAN

Introduction

5(S),12(R)-Dihydroxy-6,14-*cis*-8,10-*trans*-eicosatetraenoic acid [leukotriene B_4 (LTB_4)] is generally considered the most potent chemotactic compound yet described for human polymorphonuclear leukocytes (PMN). The biological functions of chemotaxis and receptor binding of LTB_4 are inactivated by progressive oxidation of C : 20.[1-4] The ω-oxidation of this pathway is initiated by the enzymatic action of P-450_{LTB}, a cytochrome P-450 mixed-function oxidase located, apparently exclusively, in human PMN.[5-10] Cytochrome P-450_{LTB} is distinct from the ω-hydroxylases which ω-oxidize prostaglandins and lauric acid by both its substrate specificity[9] and lack of immunological cross-reactivity with polyclonal antibody to both these enzymes.[9] Cytochrome P-450_{LTB} shows a highly restricted substrate specificity, requiring the "correct" chirality of both the 5- and 12-hydroxyl group configuration and the cis and trans configuration of the double bonds and hydroxyl groups for maximal substrate activity. Cytochrome P-450_{LTB} catalyzes the following reactions:

$$LTB_4 + O_2 + NADPH + H^+ \rightarrow 20\text{-OH-}LTB_4 + NADP^+ + H_2O \quad (1)$$

$$20\text{-OH-}LTB_4 - (R\text{-}\overset{\displaystyle H}{\underset{\displaystyle H}{\overset{|}{\underset{|}{C}}}}\text{-OH}) + NADPH + O_2 \rightarrow 20\text{-CHO (20-aldehyde)} - LTB_4 \; (R\text{-}\overset{\displaystyle O}{\overset{\|}{C}}\text{-H}) + NADP^+ + H_2O \quad (2)$$

The overall reaction catalyzed by this enzyme is

$$LTB_4 + 2NADPH + 2O_2 \rightarrow 20\text{-CHO-}LTB_4 + 2NAD^+ + 2H_2O$$

The second reaction, the apparent direct enzymatic oxidation of 20-OH-LTB_4 by a cytochrome P-450, actually results from the enzymatic addition of a second hydroxyl group followed by the spontaneous dehydration of a dihydroxylated intermediate to an aldehyde. 20-CHO-LTB_4 can also be hydroxylated to 20-COOH-LTB_4 by cytochrome P-450_{LTB}, but this reaction is relatively slow.

[1] G. Hansson, J. Å. Lindgren, S.-E. Dahlén, P. Hedqvist, and B. Samuelsson, *FEBS Lett.* **130,** 107 (1981).

[2] W. Jubiz, O. Rådmark, C. Malmsten, G. Hansson, J. Å. Lindgren, J. Pålmblad, A.-M. Udén, and B. Samuelsson, *J. Biol. Chem.* **257,** 6106 (1982).

[3] S. Shak and I. M. Goldstein, *J. Biol. Chem.* **259,** 10181 (1984).

[4] W. S. Powell, *J. Biol. Chem.* **259,** 3082 (1984).

[5] H. Sumimoto, K. Takeshige, H. Sakai, and S. Minakami, *Biochem. Biophys. Res. Commun.* **125,** 615 (1984).

[6] S. Shak and I. M. Goldstein, *Biochem. Biophys. Res. Commun.* **123,** 475 (1984).

[7] S. Shak and I. M. Goldstein, *J. Clin. Invest.* **76,** 1218 (1985).

[8] R. J. Soberman, T. W. Harper, R. C. Murphy, and K. F. Austen, *Proc. Natl. Acad. Sci. U.S.A.* **82,** 2292 (1985).

[9] R. J. Soberman, R. T. Okita, B. Fitzsimmons, J. Rokach, B. Spur, and K. F. Austen, *J. Biol. Chem.* **262,** 12421 (1987).

[10] R. J. Soberman, J. P. Sutyak, R. T. Okita, D. F. Wendelborn, L. J. Roberts II, and K. F. Austen, *J. Biol. Chem.* **263,** 7996 (1988).

Principle. The conversion of LTB_4 to 20-OH-LTB_4 and the conversion of 20-OH-LTB_4 to 20-CHO-LTB_4 and 20-COOH-LTB_4 is measured by reversed-phase high-performance liquid chromatography (RP-HPLC) at 280 nm and quantitated by integrated absorbance.

Procedures

Reagents. LTB_4 and 20-OH LTB_4 can be purchased from either Cayman Chemical Company Corporation (Ann Arbor, MI) or Biomol Corporation (Philadelphia, PA). NADP and NADPH are available from Sigma (St. Louis, MO). Methanol HPLC (grade) is obtained from Burdick and Jackson. Acetic acid (reagent grade) and ammonium acetate (HPLC grade) are purchased from Fischer Chemical Corporation. PMN and PMN microsomes are prepared exactly as described by Soberman *et al.*[9,10] and in this series.[11]

Stock Solutions

Reaction buffer, 50 mM Tris-HCl, pH 8.0
NADPH, 1.0 mM in Buffer A
Substrate, 150 μM LTB_4 or 20-OH-LTB_4 in ethanol
PMN microsomes, generally 0.5–1.0 mg/ml

Preparation of 20-CHO LTB_4

Approximately 500 units of equine alcohol dehydrogenase are diluted to 200 ml with 50 mM Tris-HCl buffer (pH 7.0) and dialyzed against 6 liters of this buffer for 48 h to remove the ethanol present in the commercial enzyme preparation. 20-OH-LTB_4 (100 μg) in methanol is then evaporated to dryness under nitrogen in a 250-ml beaker and then resuspended in 80 ml of 0.2 M glycine buffer (pH 10.0; the optimal buffer for alcohol dehydrogenase). NAD^+ is then added to a concentration of 500 μM, and 54 units of the dialyzed alcohol dehydrogenase added to initiate the reaction. The reaction is allowed to proceed for 90 min at room temperature and then terminated by the addition of 100 ml of stop solution. Twenty 10-ml aliquots are removed, diluted to a volume of 50 ml with cold (4°) distilled water (pH 5.6), and separately concentrated by application onto a single reversed-phase Sep-Pak column (Waters Associates, Milford, MA) previously prepared by washing with 20 ml of methanol, followed by 20 ml of distilled water. The Sep-Pak column is then washed with 20 ml of water, adjusted to pH 5.6 with 0.6 N NaOH, and substrate and products eluted with 2 ml of methanol. The methanol eluate fraction is evaporated to a

[11] R. J. Soberman and R. T. Okita, this series, Vol. 163, p. 349.

volume of 0.75 ml under nitrogen, mixed with an equal volume of 50 m*M* ammonium acetate (pH 5.6), and injected onto a 5-μm reversed-phase HPLC column (0.46 × 25 cm) equilibrated in a solvent of methanol/50 m*M* ammonium acetate (pH 5.6) (1 : 1, v/v). No guard column is attached. 20-OH-LTB_4 (retention time = 19.1 ± 1.6 min; mean ± SE, n = 5), 20-CHO-LTB_4 (retention time = 17.6 ± 1.3 min; mean ± SE, n = 5), and 20-COOH-LTB_4 (retention time = 9.3 ± 0.5 min; mean ± SE, n = 5 (Ref. 10) are eluted in this solvent at a flow rate of 1 ml/min. The 20-CHO-LTB_4 peak is collected by hand. Reinjection of a sample of this fraction in the same solvent system demonstrates a single peak. The yield of pure 20-CHO-LTB_4 is 22.9 ± 8.3% (mean ± SD, n = 5).[10]

Assay. PMN microsomes (generally 30–75 μl comprising 12.5–75 μg) are incubated in either a plastic Eppendorf or a borosilicate centrifuge tube, with 25 pmol NADPH and 750 pmol (5 μl of substrate stock) substrate reaction buffer in a final volume of 500 μl. The reaction mixture is then incubated at 37° for 20 min and the reaction terminated by the addition of 500 μl of stop solution. The reaction is then terminated by centrifugation at maximal speed for 5 min in an Eppendorf centrifuge at maximal speed. The supernatant is removed for direct analysis by reversed-phase HPLC.

Reversed-Phase HPLC Analysis. Reversed-phase HPLC analysis is performed on a 0.46 × 25-cm, 5 μm analytical column coupled to a 10 μm reversed-phase guard column. The initial solvent is 53% methanol, 47% 50 m*M* ammonium acetate, pH 5.6 (v/v). The column is equilibrated at a flow rate of 1.0 ml/ml in this buffer. 20-COOH-LTB_4 elutes first at 8.0 min, followed by 20-CHO-LTB_4 at 16.5 min and 20-OH-LTB_4 at 18 min (Fig. 1). After 20 min, the solvent composition is altered to 100% methanol. LTB_4 is then eluted at 25 min in a sharp peak. On-line monitoring at 270 nm is performed. The relative areas of the peaks are used to calculate the percentage conversion of substrate to product. When multiplied by the amount of substrate added (750 pmol) the amount of product formed is calculated.

Comments

This system allows the measurement of LTB_4 and the products of all three ω-oxidations catalyzed by cytochrome P-450_{LTB}. The K_m for the conversion of LTB_4 to 20-OH-LTB_4 and 20-OH-LTB_4 to 20-CHO-LTB_4 have all been reported to be in the range of 0.2–2.0 μm,[7–10] indicating that these reactions are equally efficient. The V_{max} for both these reactions are reported to be ≃400 pmol/min/mg. When 20-CHO is used as a substrate

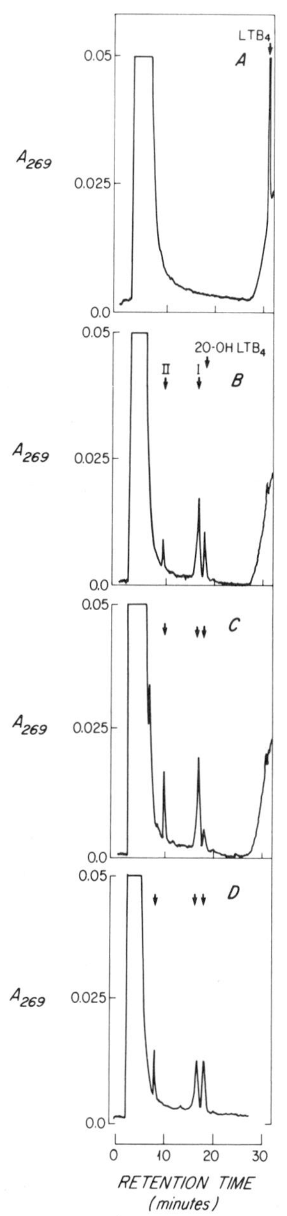

FIG. 1. Metabolism of LTB_4 and 20-OH-LTB_4 by PMN microsomes. LTB_4 (0.6 μM) is incubated alone (A) or with 67.2 (B) or 134.4 (C) μg of PMN microsomes together with 100 μM NADPH for 40 min at 37° in 50 mM Tris-HCl buffer (pH 8.0), and the reaction terminated and analyzed by reversed-phase HPLC in solvent A. 20-OH-LTB_4 (1.0 μM) is incubated with 180.0 μg from a second preparation of microsomes (D) for 40 min at 37° in the same buffer, and the reaction terminated and analyzed as described above. Arrows indicate the retention times of chemically synthesized standards and compounds I and II. (Reprinted by permission of the *Journal of Biological Chemistry*.)

the K_m is lower (0.08 μM) but the V_{max} is markedly less (5 pmol $\times$ min^{-1} $\times$ mg^{-1}).

Acknowledgment

This work was supported by Grants AI-22563, AR-38633 from the National Institutes of Health and in part by a grant-in-aid from the American Heart Association, Massachusetts Affiliate, Inc.

[36] Cytosolic Liver Enzymes Catalyzing Hydrolysis of Leukotriene A_4 to Leukotriene B_4 and 5,6-Dihydroxyeicosatetraenoic Acid

By JESPER Z. HAEGGSTRÖM

The unstable allylic epoxide leukotriene (LT) A_4 [5(*S*)-*trans*-5,6-oxido-7,9-*trans*-11,14-*cis*-eicosatetraenoic acid] is a key intermediate in the biosynthesis of the biologically active leukotrienes.

Enzymatic hydrolysis of LTA_4 may be catalyzed by two different enzymes with several functional and structural differences.[1] Thus, LTA_4 hydrolase (EC 3.3.2.6) converts the epoxide into LTB_4, whereas cytosolic epoxide hydrolase (EC 3.3.2.3) generates 5(*S*),6(*R*)-dihydroxy-7,9-*trans*-11,14-*cis*-eicosate-traenoic acid (5,6-DiHETE).[2] Epoxide hydrolases, microsomal or cytosolic, are believed to be involved in detoxification of various harmful xenobiotic epoxides.

Although this chapter is primarily concerned with LTA_4 hydrolase and cytosolic epoxide hydrolase from guinea pig and mouse liver, respectively, the methods have a broader application since extended studies with synthetic LTA_4 and xenobiotic epoxides have shown that both enzymatic activities have a widespread occurrence in mammalian tissues.[3–5]

[1] J. Haeggström, J. Meijer, and O. Rådmark, *J. Biol. Chem.* **261,** 6332 (1986).

[2] J. Haeggström, A. Wetterholm, M. Hamberg, J. Meijer, R. Zipkin, and O. Rådmark, *Biochim. Biophys. Acta* **958,** 469 (1988).

[3] T. Izumi, T. Shimizu, Y. Seyama, N. Ohishi, and F. Takaku, *Biochem. Biophys. Res. Commun.* **135,** 139 (1986).

[4] J. F. Medina, J. Haeggström, M. Kumlin, and O. Rådmark, *Biochim. Biophys. Acta* **961,** 203 (1988).

[5] S. S. Gill and B. D. Hammock, *Biochem. Pharmacol.* **29,** 389 (1980).

Preparation of Liver Homogenates and Subcellular Fractions

The liver of an anesthetized and heparinized guinea pig or mouse is perfused with cold 0.9% NaCl. After cautious removal of the gall bladder, the liver is homogenized in three parts (v/w) 50 m*M* potassium phosphate buffer, pH 7.4, utilizing a Potter–Elvehjem homogenizer. If the liver will be used for purification of LTA_4 hydrolase, the homogenization should be carried out in 10 m*M* Tris/HCl, pH 8, 5 m*M* phenylmethylsulfonyl fluoride, and 2 m*M* EDTA. A simple subcellular fractionation is performed by sequential centrifugation at 20,000 *g* for 30 min followed by 105,000 *g* for 60 min at 4°. The resulting pellets should be washed with buffer and recentrifuged 2–3 times prior to incubations.

General Procedures for Incubations and Extractions

Typically, 500-μl aliquots of homogenates or subcellular fractions are incubated at 37° for 10 min with LTA_4 lithium salt (10–20 μM) added as an ethanol solution. To retain an unimpaired activity of cytosolic epoxide hydrolase it is essential not to incubate with LTA_4 dissolved in tetrahydrofuran, a solvent frequently used in saponification of LTA_4 methyl ester. The formation of 5,6-DHETE from LTA_4 is inhibited at low concentrations (1–2%) of tetrahydrofuran.[6] The reactions are terminated by the addition of 3 to 5 volumes of methanol, containing a defined amount of internal standard, e.g., prostaglandin (PG) B_1. Precipitated proteins are removed by centrifugation or filtration. After acidification of the samples to an apparent pH $\approx$ 3 with 0.1 *M* HCl, products may be conveniently extracted either with diethyl ether or on Sep-Pak C_{18} cartridges (Waters Associates, Milford, MA) eluted with methyl formate after previous washings.[7] In our hands, the former method gives more reproducible recoveries of LTA_4 transformation products. The extract is evaporated under a stream of nitrogen and the residue reconstituted in an appropriate mobile phase for further analysis by HPLC.

Characterization of Enzymatic Hydrolysis Products

The basis for identification of LTB_4 and 5,6-DHETE is reversed-phase HPLC. Free acids are separated by a Nucleosil C_{18} column (Maeherey-Nagel, Düren, FRG) (250 × 4.5 mm) eluted with a mixture of methanol/water/acetic acid (70 : 30 : 0.01, v/v/v) at 1 ml/min. The ultraviolet absorbance is monitored at 270 nm. A typical chromatogram is depicted in

[6] J. Meijer and J. W. DePierre, *Eur. J. Biochem.* **150,** 7 (1985).

[7] W. S. Powell, *Prostaglandins* **20,** 947 (1980).

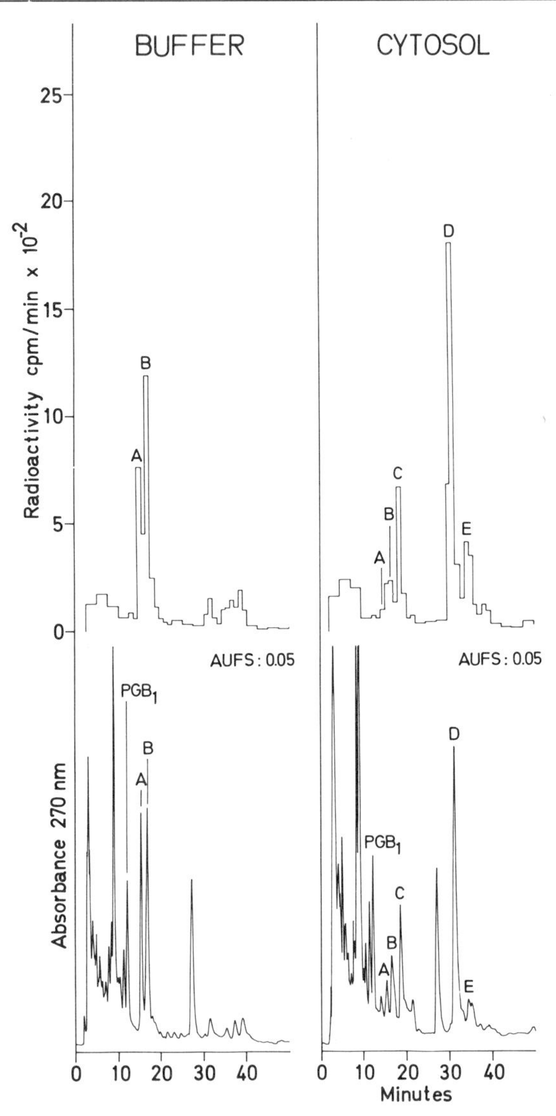

FIG. 1. Reversed-phase HPLC analysis of products formed in an incubation of mouse liver cytosol (10 mg of protein/ml) with 20 μM [^{3}H]LTA$_4$ (37°, 10 min) as compared to a control incubation with 50 mM potassium phosphate buffer, pH 7.4. To each incubate (500 μl), 420 ng of PGB$_1$ was added as internal standard. The column (Nucleosil C$_{18}$, 250 × 4.5 mm) was eluted with methanol/water/acetic acid (70 : 30 : 0.01, v/v/v) at 1 ml/min. Upper panels show the distribution of tritium in each collected fraction as determined by liquid scintillation counting. Lower panels show the continuous recordings of ultraviolet absorption at 270 nm. Peak A: Δ^6-*trans*-LTB$_4$; peak B: 12-epi-Δ^6-*trans*-LTB$_4$; peak C: LTB$_4$; peak D: 5(S),6(R)-dihydroxy-7,9-*trans*-11,14-*cis*-eicosatetraenoic acid; peak E: Unidentified isomers of 5,6-DHETE.

Fig. 1. In this chromatographic system, LTB_4 and 5,6-DHETE elute approximately in 19 and 32 min, respectively.

Both compounds may be further purified and examined by straight-phase HPLC, UV spectroscopy, and gas chromatography coupled to mass spectrometry (GC–MS). The resolution of various isomers of 5,6-DHETE (methyl esters) is poor in straight-phase HPLC, which has limited our use of this technique. In ultraviolet spectroscopy, with methanol as the solvent, 5,6-DHETE appears as a conjugated triene with maximal absorbance at 272 nm and with shoulders at 263 and 284 nm. For 5,6-DHETE, we use an extinction coefficient of $4.0 \times 10^4 M^{-1} \times cm^{-1}$ (at 272 nm) due to the structural similarity with LTA_4.

For GC–MS, the free acids of LTB_4 and 5,6-DHETE are converted to methyl ester trimethylsilyl ethers and analyzed on a capillary column (SE-30, fused silica, 25 m × 0.25 mm) coupled directly to a mass spectrometer (VG 7070E). The gas chromatograph is operated isothermally (240°–250°) with helium as the carrier gas. The analysis of 5,6-DHETE by GC–MS consistently results in two peaks with *C* values of 23.8 and 24.7 and with similar mass spectra (for details see Ref. 1). The latter peak most probably reflects an 11-cis to 11-trans isomerization of the parent compound during the analytical procedure.[4]

Subcellular Distribution of Epoxide Hydrolase and LTA_4 Hydrolase in Liver Tissue

The distribution of enzymatic activity in subcellular fractions of mouse liver homogenates differ between cytosolic epoxide hydrolase and LTA_4 hydrolase. Typically, formation of LTB_4 is almost exclusively detected in the high-speed supernatant whereas formation of 5,6-DHETE is seen in both the 105,000 *g* supernatant (cytosol) and the 20,000 *g* pellet. The activity in the pellet fraction most probably reflects the presence of mitochondrial/peroxisomal epoxide hydrolase, an enzymatic activity similar to cytosolic epoxide hydrolase with regard to molecular weight, substrate specificity, and antigenic properties.[8]

Substrate Specificity

Cytosolic (and microsomal) epoxide hydrolase has a broad substrate specificity and has frequently been characterized with stable aromatic and aliphatic xenobiotic epoxides. Besides LTA_4, a number of other epoxides derived from arachidonic acid, e.g., 15(*S*)-*trans*-14,15-oxido-5,8-*cis*-10,12-*trans*-eicosatetraenoic acid (14,15-LTA_4), may serve as substrates

[8] J. Meijer, G. Lundqvist, and J. W. DePierre, *Eur. J. Biochem.* **167,** 269 (1987).

for cytosolic epoxide hydrolase.[9] The microsomal enzyme (from rat liver) has no detectable catalytic activity toward LTA_4.[1]

In contrast, the substrate specificity of LTA_4 hydrolase seems to be very narrow.[10] Two isomers of LTA_4, namely LTA_3 and LTA_5, may be enzymatically hydrolyzed by this enzyme.[11,12] Thus, LTA_5 is transformed into LTB_5, although with lower efficiency as compared to hydrolysis of LTA_4 into LTB_4, whereas LTA_3 is a very poor substrate.

Purification of Cytosolic Epoxide Hydrolase from Mouse Liver Cytosol

Cytosolic epoxide hydrolase can be purified to apparent homogeneity from mouse liver by a procedure that involves column chromatography on DEAE-cellulose, phenyl-Sepharose, and hydroxyapatite, utilizing xenobiotic substrate in the assay of enzyme activity.[13]

Purification of LTA_4 Hydrolase from Guinea Pig Liver Cytosol

Step 1. Precipitate nucleic acids by adding streptomycin sulfate in water (10% w/v) to cytosol with continuous stirring on ice. After 30 min, remove the precipitate by centrifugation (10,000 *g*, 15 min, 4°). Proceed with ammonium sulfate precipitation (on ice) and collect the 40–80% saturated fraction by centrifugation. Dissolve the pellet in 50 m*M* Tris/HCl, pH 8.0, to give a protein concentration of 25 mg/ml.

Step 2. The 40–80% ammonium sulfate fraction is further purified by molecular exclusion chromatography on a column packed with AcA 44 (LKB-produkter, Bromma, Sweden) or equivalent, equilibrated with 50 m*M* Tris/HCl, pH 8.0. For good resolution, the sample volume should be adjusted to less than 2% of the bed volume.

Step 3. Pool the active fractions from step 2 and treat the sample with 2 m*M* dithiothreitol for 30 min. Load this sample onto DEAE-cellulose (Whatman DE-52) preequilibrated with 10 m*M* Tris/HCl, pH 8.0. When nonadsorbing proteins are eluted, apply a linear gradient of KCl (50–250 m*M*) that increases with approx. 0.25 m*M*/ml. Active fractions will appear in the range of 90 to 130 m*M* KCl. In this chromatographic step, most of the cytosolic epoxide hydrolase activity will be separated from the LTA_4 hy-

[9] A. Wetterholm, J. Haeggström, M. Hamberg, J. Meijer, and O. Rådmark, *Eur. J. Biochem.* **173,** 531 (1988).

[10] N. Ohishi, T. Izumi, M. Minami, S. Kitamura, Y. Seyama, S. Ohkawa, S. Terao, H. Yotsumoto, F. Takaku, and T. Shimizu, *J. Biol. Chem.* **262,** 10200 (1987).

[11] J. F. Evans, D. J. Nathaniel, R. J. Zamboni, and A. W. Ford-Hutchinson, *J. Biol. Chem.* **260,** 10966 (1985).

[12] D. J. Nathaniel, J. F. Evans, Y. Leblanc, C. Léveillé, B. J. Fitzsimmons, and A. W. Ford-Hutchinson, *Biochem. Biophys. Res. Commun.* **131,** 827 (1985).

[13] J. Meijer and J. W. DePierre, *Eur. J. Biochem.* **148,** 421 (1985).

drolase activity. The former enzyme elutes in the flow-through peak and early fractions of the gradient.

Step 4. Active fractions from step 3 are pooled and concentrated by ultrafiltration (Amicon PM10, Danvers, MA) in the presence of 2 m*M* dithiothreitol. The reducing agent should be added at this stage in order to prevent the appearance of multiple peaks of activity in the following chromatography. Anion-exchange separation with fast protein liquid chromatography (FPLC) is then performed on a column, Mono Q HR 5/5 (Pharmacia Fine Chemicals, Uppsala, Sweden), attached to the FPLC system and equilibrated at 10°–12° with 10 m*M* Tris/HCl, pH 8.0, supplemented with 0.1 m*M* dithiothreitol. The sample buffer is exchanged to the column equilibration buffer by molecular exclusion chromatography (PD-10, Pharmacia Fine Chemicals). Inject an aliquot of the enzyme pool, await the peak of nonadsorbing proteins, and then apply a gradient of KCl (0–200 m*M*) in a total volume of 50 ml. Enzyme activity is recovered in fractions between 40–85 m*M* KCl.

Step 5. Adsorption chromatography on hydroxyapatite is performed on a column (BioGel HPHT, Bio-Rad, Richmond, CA) attached to the FPLC system and equilibrated with 10 m*M* potassium phosphate buffer, pH 7.0, supplemented with 0.3 m*M* $CaCl_2$ and 0.1 m*M* dithiothreitol. Concentration and desalting of collected fractions from step 4 are conveniently carried out by repeated dilution and ultrafiltration of the sample with the column equilibration buffer. The sample is applied to the column at a flow rate of 0.25 ml/min and after elution of nonadsorbing proteins, the flow is increased to 0.5 ml/min and a linear gradient of phosphate in two steps (10–160 m*M* in 50 ml followed by 160–310 m*M* in 5 ml) is developed by mixing with 500 m*M* potassium phosphate buffer, pH 7.0, supplemented with 6 μM $CaCl_2$ and 0.1 m*M* dithiothreitol. Enzymatic activity is collected between 60–85 m*M* potassium phosphate.

Step 6. The final purification is achieved by chromatofocusing on a Mono P column (HR 5/20, Pharmacia Fine Chemicals) equilibrated with 25 m*M* triethanolamine, the pH adjusted to 8.3 with iminodiacetic acid, and supplemented with 0.1 m*M* dithiothreitol. The sample buffer from step 5 is exchanged to the alkaline starting buffer by repeated concentration and dilution as described above. To achieve maximal resolution in the chromatography, the column should be washed with 1–2 ml of 2 *M* sodium iminodiacetate and reequilibrated prior to each sample injection. Proteins are eluted with a pH gradient (8–5) by changing the buffer to a mixture of Polybuffer 74 and 96 (70 : 30 by volume, Pharmacia), the pH adjusted to 5.0 with iminodiacetic acid, and supplemented with 0.1 m*M* dithiothreitol, at a flow rate of 0.5 ml/min. Immediately after each chromatographic run, the pH should be measured in collected fractions and restored to an alkaline pH with 1 *M* Tris/HCl, pH 8.0, to prevent inactivation of the enzyme. In

TABLE I
PURIFICATION OF LTA_4 HYDROLASE FROM GUINEA PIG LIVER

Fraction	Volume (ml)	Total protein (mg)	Total[a] activity (U)	Specific activity (U/mg)	Yield (%)	Purification (-fold)
Cytosol	102	1835	0.73	0.0004	—	—
Precipitations	40	975	1.42	0.0015	100	1
AcA 44[b]	265	500	1.45	0.0029	102	2
DEAE-cellulose	80	62	0.86	0.014	61	9
Mono Q[c]	21	18	0.80	0.044	56	29
Hydroxyapatite	13.5	2.6	0.44	0.17	31	114
Mono P[d]	1.1	0.15	0.27	1.8	19	1200

[a] Aliquots of the enzyme from each step of purification were incubated with LTA_4 (100 μM) at 37° for 1 min. Analysis and quantitation of LTB_4 were performed with reversed-phase HPLC as described in the text. Enzymatic activity was expressed as micromoles of LTB_4 formed per minute. In crude cytosol, lower amounts of activity were detected as compared to following steps most probably due to the presence of competing enzyme activities. Therefore, the activity in the ammonium sulfate precipitate was set to 100%.

[b] Molecular exclusion chromatography.

[c] FPLC, anion-exchange separation.

[d] FPLC, chromatofocusing.

this system, guinea pig liver LTA_4 hydrolase is consistently recovered at a pH equal to p*I* of 6.2.

By this procedure the protein is purified more than 1000-fold to near homogeneity with a yield of about 20% (Table I). The total activity found in the 40–80% ammonium sulfate precipitate is usually higher than in crude cytosol which contains other enzymes, e.g., cytosolic epoxide hydrolase and glutathione transferases, that could compete with LTA_4 hydrolase for the substrate. In general, results of enzyme activity determinations vary considerably depending on the incubation conditions, especially temperature and substrate solvent. Thus, the activity seems to be somewhat reduced when LTA_4 is added in tetrahydrofuran and incubations on ice may give more product than incubations at 37°, probably due to increased substrate stability. It should be noted that the purified enzyme is inactivated upon freezing.

Enzyme Kinetics

Performing enzyme kinetic experiments with LTA_4 confronts the investigator with the problem of substrate instability. At 25° in a buffer of pH 7.4 without any organic solvent, the half-life of LTA_4 is <10 sec, but the stability increases at higher pH and lower temperature. In addition, LTA_4

hydrolase is inactivated to various degrees when exposed to its substrate.[10,11,14,15] To minimize the influence of these factors on apparent kinetic constants for LTA_4 hydrolase, we developed a method with short-time (only 5 sec) incubations at three low temperatures that allow the construction of an Arrhenius plot.[15] Kinetic experiments may also be performed at only one higher temperature but would, in our opinion, give less reliable results.

Below we outline two experimental protocols that we have used in determinations of apparent values of K_m and V_{max} for mouse liver cytosolic epoxide hydrolase and guinea pig liver LTA_4 hydrolase, respectively. The protocols are not interchangeable since cytosolic epoxide hydrolase is efficiently inhibited at low temperatures[6] and seems not susceptible to inactivation by LTA_4.[2]

Determination of Apparent K_m and V_{max} for Cytosolic Epoxide Hydrolase

Dilute purified mouse liver cytosolic epoxide hydrolase in argon-saturated 50 m*M* Tris/HCl, pH 8.0, to a concentration of about 40 μg/ml. Incubate aliquots (100 μl) of the enzyme at 37° with LTA_4 at concentrations ranging from 5 to 100 μ*M*. The substrate should be added as an ethanol solution and the final concentration of solvent should be equal in all samples and not exceed 2%. Stop the reactions after 10 sec by the addition of 5 volumes of methanol. Add a defined amount of internal standard, e.g., PGB_1, and extract the samples as outlined above. Quantitation of 5,6-DHETE is carried out by reversed-phase HPLC analysis of the sample combined with standard injections of known amounts of authentic compound.[2] At each concentration of substrate, a control incubation with buffer alone has to be included. Over the range of concentrations used (5–100 μ*M*) the two nonenzymatic diastereoisomers of 5,6-DHETE[16] appear in a constant ratio. Therefore, the nonenzymatic formation of 5,6-DHETE can be estimated from the peak height of the other diastereoisomer. This background is subtracted from the total formation of 5,6-DHETE to give a more accurate determination of the enzymatic product. Losses during the analytical procedure are calculated and corrected for by means of the internal standard. The effect of substrate concentration on initial reaction velocity may be plotted in a linear fashion according to Eadie[17] and Hofstee[18] where the slope of the line and its

[14] J. McGee and F. Fitzpatrick, *J. Biol. Chem.* **260,** 12832 (1985).

[15] J. Haeggström, T. Bergman, H. Jörnvall, and O. Rådmark, *Eur. J. Biochem.* **174,** 717 (1988).

[16] P. Borgeat and B. Samuelsson, *Proc. Natl. Acad. Sci. U.S.A.* **76,** 3213 (1979).

[17] G. S. Eadie, *J. Biol. Chem.* **146,** 85 (1942).

[18] B. H. J. Hofstee, *Nature (London)* **184,** 1296 (1959).

intercept with the ordinate gives the apparent values of K_m and V_{max}, respectively. In our hands, this method yielded values of 5 μM (K_m) and 550 nmol $\times$ mg^{-1} $\times$ min^{-1} (V_{max}) for mouse liver cytosolic epoxide hydrolase.[2]

Determination of Apparent K_m and V_{max} for LTA_4 Hydrolase

Dilute purified guinea pig liver LTA_4 hydrolase to a concentration of about 3 μg/ml in 50 mM Tris/HCl, pH 8.0, containing 30% glycerol to prevent freezing at low temperatures. Incubate aliquots (100 μl) of the enzyme with LTA_4 at concentrations between 5 and 100 μM at three (or more) different temperatures, $-10°$, 0°, and $+10°$. The substrate is added in an equal volume of ethanol (final concentration $<2\%$). Stop the reactions after only 5 sec by the addition of 5 volumes of methanol followed by the internal standard (PGB_1) and extract the samples as outlined above. Quantitation of LTB_4 is based on the peak height ratio between LTB_4 and PGB_1 in reversed-phase HPLC.[19] The effect of substrate concentration on the initial reaction velocity is plotted in a linear fashion according to Eadie[17] and Hofstee[18] (see previous section) for the three temperatures, respectively. The temperature dependence of the maximal initial reaction velocity can then be tested according to Arrhenius where log V_{max} is plotted against $1/T$.[20] An example is shown in Fig. 2. If the determinations have been carefully performed, this plot will show a high degree of linearity. By extrapolation to $T = 310$ K (37°) the apparent value of V_{max} at this temperature can be calculated. We have deduced an apparent value of V_{max} at 37° of 68 μmol $\times$ mg^{-1} $\times$ min^{-1} for guinea pig liver LTA_4 hydrolase with this method and the turnover number (k_{cat}) was calculated to 80 sec^{-1}.[15] Likewise, the temperature dependence of the Michaelis constant may be tested by plotting pK_m against $1/T$ which, by extrapolation, will give the apparent value of K_m at 37° (27 μM).[15,20] From values of k_{cat} and K_m the apparent second-order rate constant (k_{cat}/K_m) for guinea pig liver LTA_4 hydrolase was calculated to 3×10^6 $M^{-1} \times$ sec^{-1}, indicating a high catalytic efficiency.[15]

Besides the reduction of error due to substrate instability and inactivation of the enzyme, this protocol also has the advantage of permitting an estimation of the activation energy (E_A) required for the enzymatic catalysis (~43 kJ/mol for guinea pig liver LTA_4 hydrolase). This value is calculated from the slope of the line in the Arrhenius plot (log V_{max} vs. $1/T$) and

[19] O. Rådmark, T. Shimizu, H. Jörnvall, and B. Samuelsson, *J. Biol. Chem.* **259**, 12339 (1984).

[20] M. Dixon and E. C. Webb, *in* "Enzymes," 3rd Ed., pp. 164–182. Longman Group Limited, London, 1979.

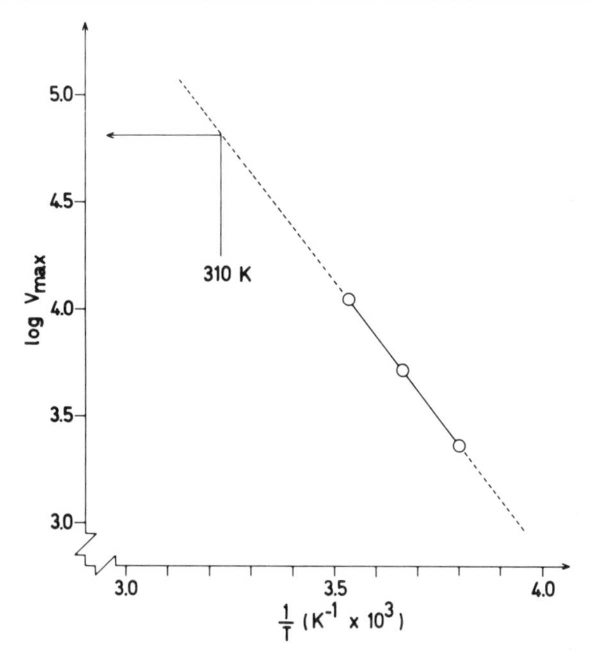

FIG. 2. Arrhenius plot (log V_{max} vs. $1/T$) of kinetic data obtained at three different temperatures (+10°, 0°, and −10°). The experiment was performed with freshly prepared guinea pig liver LTA_4 hydrolase. Aliquots of purified enzyme (0.33 μg dissolved in 100 μl 50 mM Tris/HCl, pH 8.0, with 30% glycerol) were incubated with LTA_4 (5–90 μM) for 5 sec. Reactions were stopped with 5 volumes of methanol. Internal standard PGB_1 (150 ng/sample) was added prior to acidic ether extraction and analysis by reversed-phase HPLC. Values of maximal initial reaction velocities were obtained from Eadie–Hofstee plots (V vs. V/[S]).

may also be obtained with impure preparations of enzyme.[20] However, it should be noted that the slope of the Arrhenius plot may be altered at some point between +10° and +37°, reflecting a sudden shift to a different activation energy at a certain critical temperature.[20]

Other investigators have usually employed alternative methods in determinations of kinetic constants for LTA_4 hydrolase,[19,21] i.e., short-time (10 sec) incubations at 37° (similar to the ones described above for cytosolic epoxide hydrolase) or incubations in buffer containing albumin, a protein known to stabilize LTA_4.[22] Albumin most likely competes with LTA_4 hydrolase for the substrate and would thus influence the kinetic data obtained with the latter method.[15]

[21] J. F. Evans, P. Dupuis, and A. W. Ford-Hutchinson, *Biochim. Biophys. Acta* **840,** 43 (1985).

[22] F. A. Fitzpatrick, D. R. Morton, and M. A. Wynalda, *J. Biol. Chem.* **257,** 4680 (1982).

TABLE II
COMPARISON OF GUINEA PIG LIVER LTA_4 HYDROLASE WITH MOUSE LIVER CYTOSOLIC EPOXIDE HYDROLASE

Parameter	LTA_4 hydrolase[a]	Cytosolic epoxide hydrolase[b]
Quaternary structure	Monomer	Dimer
M_r	67,000–71,000	120,000
Stokes radius (nm)	3.0	4.2
pH optimum	~8	~7
p*I*	6.2	5.5
Inactivation by LTA_4	+	–
K_m (LTA_4)	27	5
(μm)	(7)[c]	
V_{max} (LTA_4)	68	0.5
(μmol/mg/min)	(10)[c]	
Reaction with		
anti-LTA_4	+	–
hydrolase AB		
Reaction with		
anticytosolic epoxide	–	+
hydrolase AB		

[a] Haeggström *et al.*,[15] and J.Y. Fu, J. Haeggström, P. Collins, J. Meijer, and O. Rådmark, *Biochim. Biophys. Acta* **1006,** 121 (1989).

[b] Meijer and DePierre,[6,13] Haeggström *et al.*,[2] and J. Y. Fu, J. Haeggström, P. Collins, J. Meijer, and O. Rådmark, *Biochim. Biophys. Acta* **1006,** 121 (1989).

[c] Values within parentheses were obtained with a method similar to that used for cytosolic epoxide hydrolase, i.e., short-time (10 sec) incubations at 37°.

For a comparison, some enzyme characteristics for mouse liver cytosolic epoxide hydrolase and guinea pig liver LTA_4 hydrolase are listed in Table II.

Acknowledgments

This work was supported by grants from the Swedish Medical Research Council (Nos. 03X-217 and 03X-07467) and O. E. and Edla Johanssons Foundation.

[37] Purification and Properties of Leukotriene C_4 Synthase from Guinea Pig Lung Microsomes

By ROY J. SOBERMAN

Leukotriene C_4 synthase is the enzyme that catalyzes the conjugation of leukotriene (LT) A_4 with reduced glutathione (GSH). This enzyme is a novel member of the family of glutathione *S*-transferases and is distinct from other cytosolic glutathione *S*-transferases, and from the microsomal glutathione *S*-transferase that utilizes 1-chloro-2,4-dinitrobenze as a substrate. It has been partially purified and characterized from rat basophilic leukemia (RBL-1) cell microsomes[1,2] and from guinea pig lung microsome.[3,4]

Principle. The conjugation of LTA_4 and GSH to form LTC_4 is monitored by on-line monitoring at 280 nm and quantitated by integrated absorbance.

Reagents

LTA_4-methyl ester (me) can be purchased from either Biomol (Philadelphia, PA) or Cayman Chemical Corp. (Ann Arbor, MI). GSH, prostaglandin (PG) B_2, Triton X-102, and glycerol are obtained from Sigma (St. Louis, MO).

Reaction Buffer: 50 m*M* HEPES, pH 7.6

Substrate stock solutions: (a) 100 m*M* GSH; (b) 200 μM LTA_4-me or LTA_4 (prepared by hydrolysis of LTA_4-me as previously described[2])

Stop solution: Methanol : acetonitrile : acetic acid 50 : 50 : 1 (v/v/v) containing 100 μM PGB_2/0.5 ml

Reversed-phase HPLC Buffer: (A) methanol : acetonitrile : water : acetic acid, pH 5.6 (13 : 5 : 8.1 : 0.9, v/v/v/v) (for LTA_4-me); or (B) methanol : acetonitrile : water : acetic acid, pH 5.6 (5 : 5 : 8.1 : 0.9, v/v/v/v) (for LTA_4)

Reversed-phase HPLC column (C_{18}): 5 μM, 4.1 × 250 mm with a 10-μm guard column

[1] M. K. Bach, J. R. Brashler, and D. R. Morton, Jr., *Arch. Biochem. Biophys.* **230,** 455 (1984).

[2] T. Yoshimoto, R. J. Soberman, R. A. Lewis, and K. F. Austen, *Proc. Natl. Acad. Sci. U.S.A.* **82,** 8399 (1986).

[3] T. Yoshimoto, R. J. Soberman, B. Spur, and K. F. Austen, *J. Clin. Invest.* **81,** 866 (1988).

[4] T. Izumi, Z. Honda, N. Ohishi, S. Kitamura, S. Tsuchida, K. Sato, T. Shimuzu, and Y. Seyama, *Biochim. Biophys. Acta* **959,** 305 (1988).

Assay Method

Enzyme, 10 nmol LTA_4 or LTA_4-methyl ester are incubated for 10 min in reaction buffer with 5 μmol GSH and enzyme in a 0.5 ml volume at 22°. Stop solution, 0.5 ml, is added to terminate the reaction and the mixture centrifuged in an Eppendorf microfuge tube. For LTA_4-me, the supernatant is analyzed in buffer A at a flow rate of 1 ml/min with on-line monitoring at 280 nm. The retention times of PGB_2 and LTC_4-me are 8.2 min and 12.2 min, respectively.[3] The total amount of LTC_4-me formed is quantified by comparing the ratio of LTC_4-me to PGB_2 with a standard curve. When LTC_4 analysis is performed, the ratio of the integrated area of LTC_4 (9.1 min) to PGB_2 (16.4 min) is calculated.

Purification

Preparation of Guinea Pig Lung Microsomes. Lungs are collected in ice-cold 20 m*M* Tris-HCl buffer, pH 8.0, containing 1 m*M* ethylenediaminetetraacetic acid (EDTA) and 5 m*M* 2-mercaptoethanol. They are then cut into small pieces with scissors. The fragments are rinsed once in the same buffer and then resuspended in 5-volume wet weight of buffer and disrupted using a Tekmar Tissuemizer model SDT-1810. The homogenate is centrifuged at 1000 *g* for 10 min to remove nondisrupted tissue and the supernatant filtered through cheesecloth. The filtrate is then centrifuged at 10,000 *g* for 10 min and the resulting supernatant removed and centrifuged at 105,000 *g* for 60 min. All centrifugations are carried out at 4° and homogenization is performed on ice. The supernatant (cytosol) is removed, the microsomes resuspended and washed once in the same buffer, centrifuged at 105,000 *g* for 1 hr, and resuspended in an equal volume of the same buffer.

Glycerol is added to microsomes to achieve a concentration of 20% (w/v). To solubilize the enzyme, Triton X-102 and deoxycholate are added dropwise to concentrations of 0.4% each and then stirred for 1 hr at 4°. The mixture is then centrifuged at 105,000 *g* for 1 hr. Approximately 50% of the activity is recovered in the supernatant.

Approximately 35 ml of solubilized enzyme, is applied to a Sepharose CL-4B column (4.4 × 50 cm) equilibrated with 20 m*M* Tris-HCl buffer at pH 8.0, 1 m*M* EDTA, 5 m*M* 2-mercaptoethanol, and 10% glycerol containing 0.1% Triton X-102 and 0.1% deoxycholic acid. This mixture of detergent is required to prevent the formation of large aggregates. The column is run at a flow rate of ≃45 ml/hr. Microsomal LTC_4 synthase is eluted at an apparent size of ≃55,000 distinguishing it from the microsomal glutathione *S*-transferase which eluted at a molecular weight of ≃34,000.

The peak fractions containing LTC$_4$ synthase activity are pooled and applied to a DEAE-Sephacel column (2.5 × 4 cm) equilibrated with the same buffer. The enzymatic activity is recovered in the flow-through. The pH of the column should be kept below 8.0, as significant amounts of activity will adhere at pH values greater than 8.2. LTC$_4$ synthase appears in the effluent. An approximately 30-fold purification is achieved in this step.

The DEAE-Sephacel flow-through is then applied to an agarose-butylamine column (2.5 × 10 cm) previously equilibrated with the same buffer. The column is washed with two column volumes of buffer and then eluted with the same buffer; 6-ml fractions are collected.

The pooled enzyme preparation is injected onto a DEAE-3SW HPLC column (2.15 × 15 cm) equilibrated with 50 m*M* potassium phosphate buffer at pH 8.0 containing 1 m*M* EDTA, 5 m*M* 2-mercaptoethanol, 10% glycerol, and 0.1% Triton X-102. The column is washed at a flow rate of 3 ml/min and then eluted in a linear gradient (120 ml) of 0 to 0.2 *M* NaCl and 0 to 0.1% deoxycholic acid in the same buffer. LTC$_4$ synthase elutes at a salt concentration of ≃0.16 *M* and a deoxycholate concentration of 0.08%. Peak active fractions are pooled, frozen at −70°, and used as the source of purified LTC$_4$ synthase. We have previously achieved an overall purification of ≃91-fold with a 3% yield.[3]

Properties

The enzyme has somewhat lower K_m for LTA$_4$ (3 μM) than for LTA$_4$-me (15 μM), although the V_{max} for LTA$_4$ (108 μmol/3 min/mg) is significantly lower than for LTA$_4$-me (420 μmol/min/mg) allowing the latter to be used as a substrate for detection during purification. Both substrates complete with each other, indicating that they are substrates for the same enzyme. LTC$_4$ is an excellent inhibitor of LTC$_4$-me synthase activity. With LTA$_4$-me as a substrate, the IC$_{50}$ for LTC$_{50}$ is 2.1 μM. The actual subunit size and composition awaits complete purification.

Acknowledgment

This work was supported by grants AI22563, AR38638 and by a grant in aid from the American Heart Association with funds contributed in part by the Massachusetts Chapter.

[38] Immunoaffinity Purification of Arachidonate 5-Lipoxygenase from Porcine Leukocytes

By NATSUO UEDA and SHOZO YAMAMOTO

Arachidonate 5-lipoxygenase was highly purified from the cytosol fraction of porcine leukocytes by immunoaffinity chromatography using a monoclonal anti-5-lipoxygenase antibody as a ligand.[1] The purified enzyme oxygenated arachidonic acid at the C-5 position to produce 5-hydroperoxy-6,8,11,14-eicosatetraenoic acid (5-HPETE). The same enzyme preparation also converted the 5-HPETE to leukotriene (LT)A_4. Enzymological evidence confirmed that a single enzyme protein catalyzed both the 5-oxygenase and the LTA synthase activities.[1]

Assay of Radiolabeled Arachidonate Transformation

Reagents

[1-^{14}C]Arachidonic acid mixed with nonradioactive arachidonic acid (50,000 cpm/5 nmol/5 μl in ethanol), 25 μM
Tris-HCl, 50 mM, pH 7.4, or potassium phosphate buffer, 50 mM, pH 7.4
$CaCl_2$, 2 mM
ATP, 2 mM
Enzyme solution, less than 75 μl
All in total 200 μl.

Procedure. The reaction is started by the addition of [1-^{14}C]arachidonic acid, continued with shaking for 3 min at 24°, and terminated by the addition of 0.3 ml of a mixture of ethyl ether/methanol/1 M citric acid (30 : 4 : 1, v/v/v). The content of the tube is mixed by a vortex mixer for 15 sec, and then centrifuged at 1500 g for 5 min at 4°. The following step is performed in a cold room at 4°. The ethereal extract (the upper layer) is spotted on a precoated silica gel glass plate (Merck 60 F254, 20 cm × 20 cm, 0.25 mm thickness) in a 1.5-cm width at the origin 2 cm above the bottom. The solvent is completely evaporated with cold air. For the routine enzyme assay, the plate is developed for 50 min in a solvent mixture of ethyl ether/petroleum ether/acetic acid (85 : 15 : 0.1, v/v/v). The glass tank containing the solvent and the plate is placed in a deep-freeze at −10° to

[1] N. Ueda, S. Kaneko, T. Yoshimoto, and S. Yamamoto, *J. Biol. Chem.* **261,** 7982 (1986).

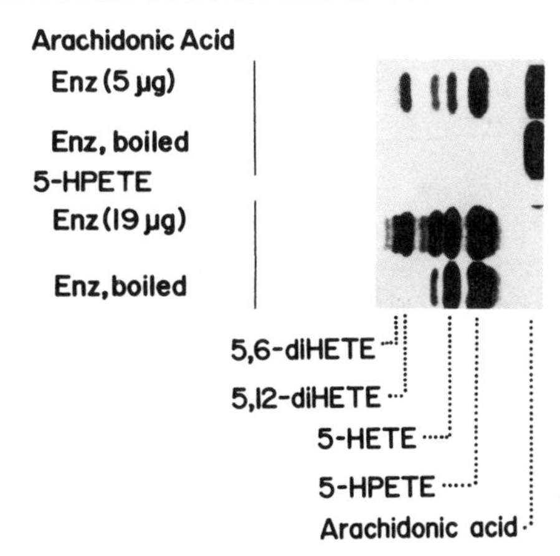

FIG. 1. Thin-layer chromatography of the reaction products from arachidonic acid and 5-HPETE. The purified 5-lipoxygenase is allowed to react with [1-^{14}C]arachidonic acid or [1-^{14}C]5-HPETE. Thin-layer chromatography is carried out using a solvent mixture of the organic phase of isooctane/ethyl acetate/water/acetic acid (50 : 110 : 100 : 20, v/v/v/v).

prevent the hydroperoxy product from nonenzymatic degradation at room temperature. Distribution of radioactivity on the plate is detected by a radiochromatoscanner or by autoradiography.[2] For a better separation of polar products, the organic phase of a mixture of isooctane/ethyl acetate/water/acetic acid (50 : 110 : 100 : 20, v/v/v/v) is used as a solvent system, and the plate is developed at 20° for 70 min. A typical chromatogram using the latter solvent is presented in Fig. 1. Compounds containing conjugated triene (6-*trans*-LTB_4) or diene (5-HPETE) can be visualized by illumination of the silica gel plate containing fluorescent indicator using a UV lamp. LTA synthase assay is performed under the same condition except that [1-^{14}C]5-HPETE (50,000 cpm/0.5 nmol/5 μl in ethanol) is used as a substrate. [1-^{14}C]5-HPETE is prepared with the aid of the purified 5-lipoxygenase[1] or potato lipoxygenase.[3]

Spectrophotometric Assay

Reagents

Arachidonic acid (12.5 nmol/5 μl in ethanol), 25 μM
Tris-HCl, 50 mM, pH 7.4
$CaCl_2$, 0.4 mM

[2] S. Yamamoto, this series, Vol. 86, p. 55.
[3] T. Shimizu, Z. Honda, I. Miki, Y. Seyama, T. Izumi, O. Rådmark, and B. Samuelsson, this volume [32].

ATP, 0.2 m*M*
Phosphatidylcholine (12 μg/2 μl in ethanol), 24 μg/ml
Enzyme solution, less than 100 μl
All in total 500 μl.

Procedure. The reaction is carried out in a 1.4-ml cuvette (10-mm light path), and initiated by the addition of the enzyme to the assay mixture.[4] Increase in the absorption at 235 nm derived from a conjugated diene is continuously monitored at 22° using a spectrophotometer. The concentration of 5-HPETE is determined on the basis of a molecular coefficient, $\varepsilon_{235} = 23{,}000/M/\text{cm}$.[5]

Enzyme Purification

Preparation of Cytosol Fraction of Porcine Leukocytes

Ten liters of fresh blood are collected from several pigs at a local slaughterhouse, and mixed immediately with 2 liters of a mixture of 0.14 *M* citric acid, 0.20 *M* trisodium citrate, and 0.22 *M* glucose.[6] After the addition of 2 liters of 6% dextran in saline, the mixture is kept at room temperature for 1 to 2 hr to sediment erythrocytes. The upper layer is centrifuged at 350 *g* for 15 min. The resulting precipitate from a 450 ml-portion of the upper layer is suspended in 20 ml of phosphate-buffered saline at pH 7.4, and subjected to hypotonic treatment with 200 ml of distilled water for 30 sec to remove contaminating erythrocytes. After the addition of 200 ml of 1.8% NaCl, the cell suspension is centrifuged at 200 *g* for 10 min, and the sedimented leukocytes washed three times with phosphate-buffered saline (pH 7.4), each followed by centrifugation at 200 *g* for 10 min at room temperature. About 60 to 65 g wet weight of leukocytes are obtained from 10 liters of whole blood.

The cells are suspended in 7 times volume of 20 m*M* Tris-HCl buffer (pH 7.4) containing 0.5 m*M* EDTA and 0.5 m*M* dithiothreitol, and subjected to sonic disruption at 20 kHz for 15 sec. The sonicate is centrifuged at 10,000 *g* for 10 min at 4°, and the supernatant is further centrifuged at 105,000 *g* for 60 min at 4°. The high-speed supernatant (cytosol fraction, 350 ml, 8 mg protein/ml) can be stored at −70° for several months without a significant loss of the 5-lipoxygenase activity. Since the cytosol fraction contains a very high activity of 12-lipoxygenase, this fraction metab-

[4] D. Riendeau, J.-P. Falgueyret, D. J. Nathaniel, J. Rokach, N. Ueda, and S. Yamamoto, *Biochem. Pharmacol.* **38,** 2313 (1989).

[5] M. J. Gibian and P. Vandenberg, *Anal. Biochem.* **163,** 343 (1987).

[6] T. Yoshimoto, Y. Miyamoto, K. Ochi, and S. Yamamoto, *Biochim. Biophys. Acta* **713,** 638 (1982).

olizes arachidonic acid predominantly to 12-hydroperoxy-5,8,10,14-eicosatetraenoic acid (12-HPETE), part of which is further degraded.[6] Compounds derived from the 5-lipoxygenase reaction are not readily detected.

Preparation of Monoclonal Anti-5-Lipoxygenase Antibodies

The cytosol fraction of porcine leukocytes is subjected to ammonium sulfate fractionation at 30–50% saturation and DEAE-cellulose chromatography.[7] The partially purified enzyme (300 μg protein) is emulsified with an equal volume of Freund's complete adjuvant, and the emulsion injected into the peritoneal cavity of C3H/He mice. After 2 weeks, the booster is performed with the same amount of the enzyme emulsified with incomplete adjuvant. After 2 more weeks, only the enzyme (300 μg protein) is injected. Three days after the last injection, the spleen is removed from each mouse under sterile conditions. The spleen cells are fused with myeloma cells (SP2/0-Ag14) in the presence of 50% polyethylene glycol 1000.[8] Hybridoma cells are selected with "HAT medium,"[8] and the cells producing anti-5-lipoxygenase antibody cloned in soft agar.[9] The anti-5-lipoxygenase antibody is titrated as described previously.[7] The cloned cells (about 10^7 cells) are grown in the peritoneal cavity of a C3H/He mouse. After about 2 weeks, ascites fluid is collected, and the antibody purified to an IgG fraction by ammonium sulfate fractionation at 50% saturation and by protein A–Sepharose chromatography. About 30 mg of IgG are obtained from one mouse.

Conjugation of Antibody to Agarose Gel

One bottle (25 ml) of Affi-Gel 10 (active ester agarose; Bio-Rad, Richmond, CA) is washed with 2-propanol and then with distilled water. A species of monoclonal anti-5-lipoxygenase antibody termed as *5Lox-6* is dialyzed against 0.1 *M* HEPES buffer at pH 7.5. The antibody (18 mg IgG/13 ml) is incubated with 25 ml of Affi-Gel 10 at 4° for 4 hr with occasional stirring. More than 98% of the added antibody is conjugated to the gel. After removing HEPES buffer, the remaining active ester of the gel is blocked by incubation with 0.1 *M* ethanolamine-HCl, pH 8.0, at 20° for 1 hr. The antibody-gel conjugate is then washed with a large amount of 0.1 *M* HEPES buffer at pH 7.5 and stored at 4°.

[7] S. Kaneko, N. Ueda, T. Tonai, T. Maruyama, T. Yoshimoto, and S. Yamamoto, *J. Biol. Chem.* **262,** 6741 (1987).

[8] J. W. Goding, *J. Immunol. Methods* **39,** 285 (1980).

[9] R. H. Kennett, *in* "Monoclonal Antibodies" (R. H. Kennett, T. J. McKearn, and K. B. Bechtol, eds.), p. 372. Plenum, New York, 1980.

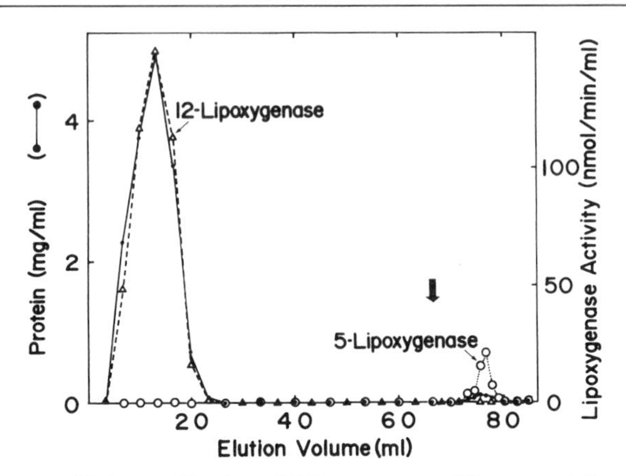

FIG. 2. Immunoaffinity purification of 5-lipoxygenase. The cytosol fraction (10 ml) of porcine leukocytes is applied to the standard immunoaffinity column. 5-Lipoxygenase is eluted as indicated by an arrow.

Immunoaffinity Chromatography

Immunoaffinity chromatography is performed on the immobilized antibody at 28° (shown in Fig. 2).[1] A column (10 × 100 mm) is packed with 5.5 ml of the immobilized antibody and preequilibrated with 20 m*M* Tris-HCl buffer at pH 7.4 containing 0.5 m*M* EDTA and 0.5 m*M* dithiothreitol. The cytosol fraction of porcine leukocytes (20 ml) is applied to the column at a flow rate of 40 ml/hr. After the column is washed with 45 ml of the same buffer at a flow rate of 75 ml/hr, 5-lipoxygenase is eluted with 50 m*M* sodium carbonate buffer at pH 10.0 containing 0.2% deoxycholic acid, 1 m*M* EDTA, and 0.5 m*M* dithiothreitol at a flow rate of 40 ml/hr. One-milliliter fractions are collected and immediately adjusted to pH 8 by the addition of 0.2 ml of 0.5 *M* Tris-HCl buffer at pH 7.4. Active fractions (about 0.3 mg protein) are combined and stored at −70°. The immobilized antibody can be used repeatedly (more than 30 times) although the purity of the eluted enzyme gradually decreases.

Properties of Purified Enzyme

The immunoaffinity-purified 5-lipoxygenase shows a major band corresponding to a molecular weight of about 72,000 when analyzed by 7.5% polyacrylamide gel electrophoresis in the presence of 0.1% sodium dodecyl sulfate.[1] The specific activity of the purified enzyme is about 1 μmol/min/mg protein at 24°. The enzyme activity is recovered from the cytosol fraction in about 50% yield. Most of the several minor bands of contaminating proteins are removed by rechromatography using the same procedure although the specific enzyme activity is not improved due to instability of the enzyme. Half-life of the activity of the purified enzyme is about 24 hr when kept at 2°. However, an appreciable loss of the activity is not observed for several weeks when stored at −70°.

The reaction time course shows a typical self-catalyzed inactivation.[1,4] The enzyme reaction at 24° almost ceases within 3 min. Optimal pH is 7.5–8.0. The purified enzyme is activated by both ATP and calcium ion at millimolar concentration.[1] Lipid peroxides abolish the initial lag phase which is observed when the enzyme is incubated with arachidonic acid. Activation by phosphatidylcholine is also found.[4]

When the purified enzyme is incubated with [1-^{14}C]arachidonic acid and the radioactive products analyzed by thin-layer chromatography, a major product is 5-HPETE (Fig. 1). The minor more polar products are 5-hydroxy-6,8,11,14-eicosatetraenoic acid (5-HETE) and hydrolytic products of LTA_4 (6-*trans*-LTB_4 and its 12-epimer and 5,6-dihydroxy-7,9,11,14-eicosatetraenoic acid). Exogenously added 5-HPETE is also converted to LTA_4 by the same enzyme preparation, indicating that the purified enzyme has both LTA synthase and 5-oxygenase activities. The LTA_4 synthesis is also activated by ATP and calcium ion, and blocked by 5-lipoxygenase-specific inhibitors.[1] Heat stability and the reaction time course of the LTA synthase activity are similar to those of the 5-oxygenase activity. These results show that a single enzyme protein has both 5-oxygenase and LTA synthase activities.

The purified 5-lipoxygenase is active with various HPETEs and HETEs. The enzyme catalyzes 5-oxygenation of 15-HPETE,[1] 12-HPETE,[1] and 15-HETE,[10] and 6*R*-oxygenation of 5-HPETE and 5-HETE.[11] The 5-lipoxygenase is also involved in the lipoxin synthesis from 5,15-diHPETE via 5,6-epoxide.[10]

[10] N. Ueda, S. Yamamoto, B. Fitzsimmons, and J. Rokach, *Biochem. Biophys. Res. Commun.* **144,** 997 (1987).

[11] N. Ueda and S. Yamamoto, *J. Biol. Chem.* **263,** 1937 (1988).

[39] Platelet-Activating Factor Acetylhydrolase from Human Plasma

By DIANA M. STAFFORINI, THOMAS M. MCINTYRE, and STEPHEN M. PRESCOTT

Platelet-activating factor (1-*O*-alkyl-2-acetyl-*sn*-glycero-3-phosphocholine, PAF) is a phospholipid that has been shown to possess potent biological activity. PAF induces hypotension, leukopenia, and thrombocytopenia,[1] and increases vascular permeability.[2] PAF also activates platelets, neutrophils, and macrophages. A variety of cells, e.g., neutrophils, macrophages, and endothelial cells, synthesize PAF on appropriate stimulation.[3] Monocytes and macrophages secrete most of their PAF, and this phospholipid has been identified in human blood, saliva, urine, and amniotic fluid.[4-6] Thus, the rate of its removal could be an important means of regulating its bioactivity.

The degradation of PAF occurs by removal of the acetoyl group esterified at the *sn*-2 position of the glycerol backbone. The products lack the bioactive properties of PAF. Farr *et al.*[7] were the first to demonstrate the occurrence, in mammalian plasma, of an enzyme that catalyzed the reaction shown in Eq. (1).

$$\text{PAF} \xrightarrow{\text{PAF acetylhydrolase}} \text{lyso-PAF} + \text{Acetate} \quad (1)$$

[1] M. Halonen, J. D. Palmer, I. C. Lohman, L. M. McManus, and R. N. Pinckard, *Annu. Rev. Respir. Dis.* **122,** 915 (1980).

[2] J. Bjork and G. Smedegard, *J. Allergy Clin. Immunol.* **71,** 145 (1983).

[3] S. M. Prescott, G. A. Zimmerman, and T. M. McIntyre, *Proc. Natl. Acad. Sci. U.S.A.* **81,** 3534 (1984).

[4] K. E. Grandel, R. S. Farr, A. A. Wanderer, T. C. Eisenstadt, and S. I. Wasserman, *N. Engl. J. Med.* **313,** 405 (1985).

[5] C. P. Cox, M. L. Wardlow, R. Jorgensen, and R. S. Farr, *J. Immunol.* **127,** 46 (1981).

[6] M. M. Billah and J. M. Johnston, *Biochem. Biophys. Res. Commun.* **113,** 51 (1983).

[7] R. S. Farr, C. P. Cox, M. L. Wardlow, and R. Jorgensen, *Clin. Immunol. Immunopathol.* **15,** 318 (1980).

The PAF acetylhydrolase activity in human plasma is selective for PAF since a short acyl residue at the *sn*-2 position is required for hydrolysis to occur; long-chain phosphatidylcholines are not hydrolyzed by this enzyme. The cellular sources of the human plasma PAF acetylhydrolase are not known, but likely candidates include liver cells as well as macrophages. A PAF acetylhydrolase activity is present in the cytosolic fraction of a variety of mammalian tissues.[8] This activity seems to belong to a family of intracellular PAF acetylhydrolases that differ from the plasma enzyme by a variety of criteria, although they all require a short-chain residue at the *sn*-2 position and are calcium independent.

Assay for PAF Acetylhydrolase

Principle. The most convenient assay for the determination of PAF acetylhydrolase activity utilizes 2-[*acetyl*-^{3}H]PAF as substrate; the [^{3}H]acetate generated by hydrolysis is quickly and efficiently separated from labeled substrate since the product is soluble in water, while the substrate is a lipid. We separate them by reversed-phase column chromatography on disposable octadecylsilica gel cartridges; the substrate binds to the column, while the radioactive acetate passes through. Thus, the amount of radioactivity released represents the amount of enzymatic activity in a given fraction, and can be conveniently quantified by liquid scintillation spectrometry.[9]

Reagents

[*acetyl*-^{3}H]PAF: 4 ml of a 0.1 m*M* solution are prepared by first mixing 400 nmol of PAF (supplied in chloroform, Avanti Polar Lipids, Birmingham, AL), with 4.5 μCi of hexadecyl-2-acetyl-*sn*-glyceryl-3-phosphorylcholine, 1-*O*-[*acetyl*-^{3}H(N)] (supplied in ethanol, New England Nuclear, Boston, MA). After evaporation of the solvents (by a stream of nitrogen), 4 ml of HEPES buffer are added and the solution then sonicated for 5 min at 4° and 100 watts, using a 4-mm needle probe in a Braun Sonicator (model 1510). This solution should be stored frozen to avoid nonenzymatic hydrolysis; it can be reused for at least 1 week. The sonication step is repeated each time the substrate is thawed. Duplicate aliquots should be counted to determine the specific radioactivity of the substrate prepared each time; our working solutions are 100 μ*M* with 10,000 cpm/nmol.

HEPES buffer: 0.1 *M*; adjust the pH to 7.2 with 1 *M* potassium hydroxide. Prepare 100 ml.

[8] M. L. Blank, T-c. Lee, V. Fitzgerald, and F. Snyder, *J. Biol. Chem.* **256,** 175 (1981).

[9] D. M. Stafforini, S. M. Prescott, and T. M. McIntyre, *J. Biol. Chem.* **262,** 4223 (1987).

Acetic acid: prepare 100 ml of a 10 *M* solution. Slowly add 57.1 ml of glacial acetic acid to 42.9 ml of distilled water.

Sodium acetate: 0.1 *M*; prepare 500 ml.

Octadecylsilica gel cartridges; purchased from Baker Chemical Co. (Phillipsburg, NJ). Each assay requires an individual column which can be reused several times if they are properly regenerated. Wash each column with 3 ml of chloroform : methanol (1 : 2), followed by 3 ml of 95% ethanol, and, finally, 3 ml of water.

Standard Procedure. Aliquots of 10 μl of the samples to be assayed (diluted in HEPES buffer, if necessary) are mixed with 40 μl of 0.1 m*M* [*acetyl*-^{3}H]PAF in polypropylene tubes and then incubated for 30 min at 37°. Glass tubes should be avoided to minimize substrate binding to the glass surface. After incubation, 50 μl of acetic acid are added followed by 1.5 ml of sodium acetate solution. Each reaction mixture is then passed through a C^{18} gel cartridge and the filtrates collected in 15-ml scintillation vials. Each assay tube is then washed with an additional 1.5 ml of sodium acetate solution and the wash also passed through the cartridge and combined with the original effluent. Ten milliliters of Opti-Fluor (Packard Instruments, Downers Grove, IL) are added to the vials and the amount of radioactivity determined in a liquid scintillation counter.

When there are many samples, it is convenient to use a multiplace vacuum manifold to allow several samples to be processed simultaneously. If only a few samples need to be assayed, one can obtain satisfactory results by manually pushing the product of the reaction through a syringe attached to a C_{18} cartridge.

Enzyme Activity. The amount of enzymatic activity present is expressed in micromoles per milliliter per hour, after correction for quenching, incubation time, dilution factors, and the amount of enzyme present in the assay.

Comments

We have also used another C_{18} silica gel cartridge, the Waters Sep-Pak (Milford, MA), but have found the Baker product to be more suitable for this assay as the Waters Sep-Pak gave more variable results. However, both types of cartridges can be reused at least ten times without loss of binding capacity.

We have used this assay for the determination of hydrolase activity in serum and plasma samples, and found that both sources contain the same amount of activity. The collection of plasma in EDTA, citrate, or heparin resulted in equal activities.

The amount of protein permissible in the solution to be applied to the cartridges should be determined, since large amounts of protein will prevent binding of PAF to the resin. This results in what appears to be high activity since the product becomes contaminated with substrate. This should be examined by carrying out mock assays terminated at zero time. The Baker cartridge has a higher capacity than the Waters cartridge, since only 3.4% of the labeled substrate leached from the column in the presence of 10 μl of plasma, in contrast to the Waters Sep-Pak, which allowed 10.1% of the label to leach. These considerations are particularly important when crude, protein-rich samples such as plasma, lipoproteins, and tissue extracts are assayed for PAF acetylhydrolase activity.

Following any change in procedure, one should verify that the apparent product is not contaminated with substrate. The effluent should be extracted into $CHCl_3$ and examined by TLC[10] or HPLC.[11]

Other Assays for PAF Acetylhydrolase

Other assays have been described for the determination of PAF acetylhydrolase activity. Farr *et al.*[7] used an assay that consisted of incubation of PAF with serum, a phase-partitioning step, and, finally, measurement of platelet aggregation. This assay is cumbersome, and lacks sensitivity since it detects the loss of substrate rather than the appearance of product. In addition, the PAF concentrations where platelet aggregation is linear are in the nanomolar range, while the PAF acetylhydrolase requires micromolar concentrations of substrate to achieve maximal initial rates. To ensure a linear response, one is forced to dilute the samples, and, consequently, increase the source of error.

Blank *et al.*[8] employed 2-[*acetyl*-^{3}H]PAF as a substrate and measured the release of radiolabeled acetate. The separation of released product from the lipid substrate was by phase–phase partitioning and measurement of radioactivity in the aqueous phase. This assay is slower than the one described above and care must be taken to remove all $CHCl_3$ to avoid quenching of the liquid scintillation counting.

Determination of Half-Life of PAF

It is occasionally useful to determine the ability of a given biological sample to hydrolyze subsaturating concentrations of PAF, as discussed below.

Reagents

[*alkyl*-^{3}H]PAF: prepare 100 μl of a PAF solution with a final concentration of 10^{-8} to 10^{-6} M by mixing unlabeled PAF (Avanti Polar

[10] H. W. Mueller, J. T. O'Flaherty, and R. L. Wykle, *J. Biol. Chem.* **258,** 6213 (1983).

[11] A. R. Brash, C. D. Ingram, and T. M. Harris, *Biochemistry* **26,** 546 (1987).

Lipids) with hexadecyl-2-acetyl-*sn*-glyceryl-3-phosphorylcholine, 1-*O*-[*hexadecyl*-1′, 2′-^{3}H(N)] (New England Nuclear). Dry the solvents under a stream of nitrogen. Then add 100 μl of 50 m*M* Tris-HCl buffer (pH 7.5) and briefly sonicate this solution at 4°.

Procedure. Add 50 μl of the [*alkyl*-^{3}H]PAF solution to 250 μl of the biological sample to be tested. Remove 50 μl-aliquots after 0, 2, 4, 6, 9, and 15 min at room temperature and place each one in a microcentrifuge tube containing 62.5 μl of chloroform and 125 μl of methanol to stop the reaction. This results in the formation of a Bligh–Dyer monophase.[12] Add approximately 40 nmol of PAF and lyso-PAF as carriers and then break the phases by adding 62.5 μl of water and 62.5 μl of chloroform. Remove the lower phase after a brief centrifugation step and dry it under nitrogen. Then, resuspend the extracted phospholipids in 50 μl of chloroform : methanol (9 : 1). Spot the samples on thin-layer chromatography plates that had been previously activated by heating at 100° for 1 hr. Develop the plates using chloroform : methanol : acetic acid : water (50 : 25 : 8 : 4).[10] Scrape the spots corresponding to PAF, lyso-PAF, and the rest of the lane and determine the amount of radioactivity present in the scrapings after adding 5 ml of Opti-Fluor (Packard Instruments).

Expression of Results. The amount of PAF hydrolyzed is determined by calculating the percentage of counts present in the PAF spot for all the time points tested. The percentage of PAF remaining is then plotted on a log scale versus time (linear scale). In human plasma, this usually gives pseudo-first-order kinetics and the half-life can then be calculated from this line.

Properties of PAF Acetylhydrolase

Effect of Metals and Chelators

The plasma acetylhydrolase does not require the presence of divalent cations for maximal activity. The addition of EDTA to whole blood as an anticoagulant does not alter enzymatic activity, and the activity of the nearly homogeneous enzyme is not altered by the addition of calcium.[9]

Association with Lipoproteins

The plasma PAF acetylhydrolase activity is found exclusively in the lipoprotein fraction. About two-thirds of the activity is associated with low-density lipoproteins (LDL) and one-third with high-density lipo-

[12] E. G. Bligh and W. F. Dyer, *Can. J. Biochem. Physiol.* **37,** 911 (1959).

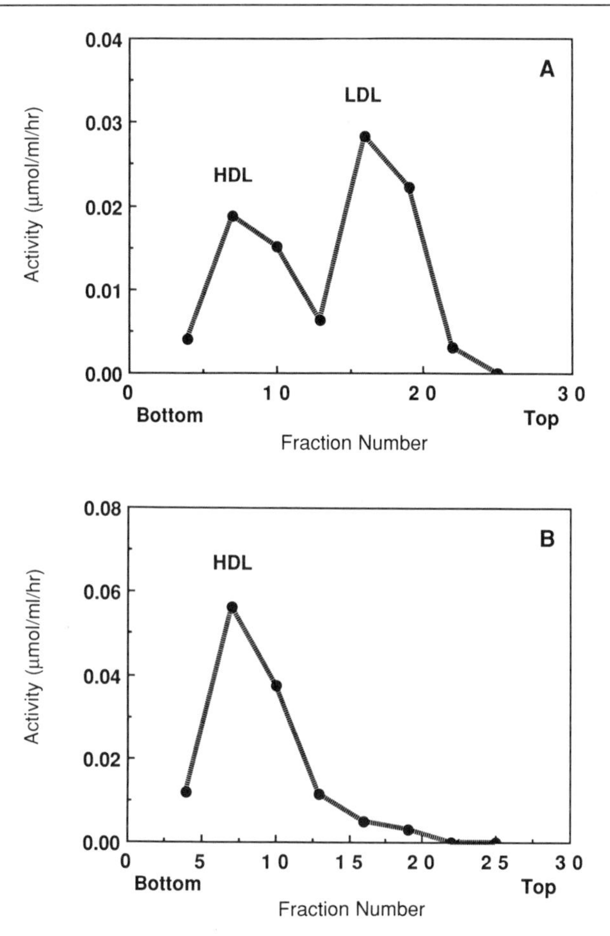

FIG. 1. Distribution of plasma PAF acetylhydrolase activity at two pH values. Samples of fresh human plasma (0.6 ml) are dialyzed overnight against 25 m*M* potassium phosphate buffer at (A) pH 7.4 or (B) pH 6.0. Each sample is then subjected to ultracentrifugation in a KBr density gradient. From Stafforini *et al.*[14]

proteins (HDL). This was first shown by subjecting human plasma to ultracentrifugation in a discontinuous NaCl/KBr density gradient. A typical profile of enzymatic activity is shown in Fig. 1A. The localization of PAF acetylhydrolase to lipoprotein particles may be a mechanism to concentrate PAF in the same particle as the enzyme; in addition, the lipoprotein particle provides an appropriate microenvironment, a lipid/solvent interface, for the efficient hydrolysis of PAF. These issues will be further discussed in the following section.

The distribution of PAF acetylhydrolase in HDL and LDL is puzzling since these two lipoprotein particles do not have a common metabolic pathway. Several lines of evidence, including immunological cross-reactivity, demonstrate that the activities associated with each particle are due to the same enzyme.[13] In addition, the enzymatic activity can be specifically transferred between lipoproteins in a pH-dependent manner. For example, after overnight dialysis at pH 6, the LDL-associated activity transfers to HDL (Fig. 1B). Conversely, at high pH the HDL-associated activity migrates to LDL.[14] The basis for this is not known, but the phenomenon has proved to be useful in a variety of studies.[13,15]

Surface Dilution Kinetics

The ability of an interface to stimulate the hydrolysis of PAF by the acetylhydrolase was first demonstrated in an experiment in which the hydrolysis of PAF by detergent-free purified acetylhydrolase was enhanced by inactive LDL, at substrate concentrations below the critical micellar concentration of PAF.[9] PAF associates with LDL while the purified enzyme does not. This strongly implies that the enhancement of hydrolysis is due to an effect of LDL on the presentation of the substrate, rather than to an altered environment of the enzyme.[9]

To further test whether PAF hydrolysis occurs more effectively at a lipid interface, we examined the reaction in a mixed micellar system in which the concentration of PAF or a nonionic detergent (Triton X-100) were systematically varied either alone or in tandem. The mixed system can distinguish reactions that occur at the micelle interface from those that occur in the bulk solvent. When the detergent concentration was increased while maintaining the amount of PAF per unit volume constant, the relative abundance of PAF at the micelle interface was decreased. In turn, a diminished rate of hydrolysis was observed. This occurs because the reaction is affected by the abundance of substrate at the micellar surface and not by its total concentration in the assay.

To rule out a direct effect of Triton X-100 on the acetylhydrolase, we performed experiments in which both PAF and Triton were increased proportionately, i.e., the relative abundance of PAF at the micelle surface was kept constant. This resulted in no inhibition by the detergent. Dennis and co-workers have observed that the cobra venom (*Naja naja naja*)

[13] D. M. Stafforini, M. E. Carter, G. A. Zimmerman, T. M. McIntyre, and S. M. Prescott, *Proc. Natl. Acad. Sci. U.S.A.* **86,** 2393 (1989).

[14] D. M. Stafforini, T. M. McIntyre, M. E. Carter, and S. M. Prescott, *J. Biol. Chem.* **262,** 4215 (1987).

[15] K. E. Stremler, D. M. Stafforini, S. M. Prescott, G. A. Zimmerman, and T. M. McIntyre, *J. Biol. Chem.* **264,** 5331 (1989).

TABLE I
EFFECT OF LIPOPROTEIN ENVIRONMENT ON THE CATALYTIC EFFICIENCY OF PAF ACETYLHYDROLASE[a]

Acetylhydrolase tested	Half-life (min)
Plasma	7.8
HDL	>15
LDL	5.1
R-HDL[b]	>15
R-LDL[c]	6.5

[a] Reproduced with permission from Stafforini *et al.*[13]

[b] R-HDL: An HDL particle that has been treated with diisopropylfluorophosphate to inactivate its PAF-acetylhydrolase activity, and that was then repleted with PAF acetylhydrolase from LDL.

[c] R-LDL: An LDL particle that has been treated with diisopropylfluorophosphate to inactivate its PAF-acetylhydrolase activity, and that was then repleted with PAF acetylhydrolase from HDL.

phospholipase A_2 exhibits a similar behavior which they termed "surface dilution kinetics."[16]

Effect of Lipoproteins on Catalytic Behavior of PAF Acetylhydrolase

When we examined the PAF acetylhydrolase at substrate concentrations well below saturation, we found that there are marked differences in the efficiency of hydrolysis catalyzed by the enzyme in the two different lipoprotein particles. In this type of experiment we measured the half-life of PAF at 10^{-9}–10^{-8} *M*, i.e., conditions that are likely to represent the situation *in vivo*, where only low concentrations of PAF have been detected. As shown in Table I, when the enzyme is in an LDL particle, it is much more efficient in degrading these subsaturating concentrations of PAF than when the same activity is localized in an HDL particle.[13]

Thus, the environment in which the PAF acetylhydrolase functions strongly influences its catalytic activity. This kind of experiment is useful

[16] E. A. Dennis, *in* "The Enzymes" (P. D. Boyer, ed.), 3rd Ed., Vol. 16, p. 307. Academic Press, New York, 1983.

when trying to assess the ability of a given system to hydrolyze low concentrations of substrate. It is of importance to notice, however, that this type of measurement cannot replace the true enzyme assay, where enzyme is rate-limiting and substrate is present in great excess.

Purification of PAF Acetylhydrolase

Isolation of LDL[17]

Reagents

Normal human plasma: 1 liter, anticoagulated with EDTA or sodium citrate
Sodium phosphotungstate: prepare 100 ml of a 4% solution. Adjust the pH to 7 with NaOH
$MgCl_2$: prepare 100 ml of a 2 *M* solution
Sodium citrate: prepare 1 liter of a 0.2% unbuffered solution
Buffer A: 10 m*M* potassium phosphate buffer (pH 6.8) containing 0.1% Tween 20.

Procedure. To 1 liter of plasma, add 100 ml of the sodium phosphotungstate solution and 25 ml of the $MgCl_2$ solution. Turbidity should develop almost instantaneously. Centrifuge the resulting solution at 6000 *g* for 10 min; resuspend the precipitate in 1 liter of sodium citrate solution with the aid of a rubber policeman. When the precipitate, which contains LDL and VLDL, is completely redissolved, add 10 g of NaCl and then 30 ml of the $MgCl_2$ solution. Repeat the centrifugation step and discard the supernatant. Wash the precipitate by redissolving it in sodium citrate, add NaCl and $MgCl_2$ as above, and wash the precipitate a second time with sodium citrate as above. The final LDL–VLDL precipitate should be resuspended in 1 liter of buffer A and stirred for 2 hr at 4° to allow solubilization to occur. From this point on, the enzyme preparation should be kept in a detergent-containing solution (ideally, in buffer with 0.1% Tween 20). Failure to do so can result in nonspecific binding to surfaces and loss in activity.

Chromatographic Steps

Reagents

DEAE-Sepharose CL-6B: 400 ml
Sephacryl S-300: 450 ml
Hydroxyapatite: 10 ml
Sigmacell (Sigma, St. Louis, MO)

[17] M. Burstein and H. R. Scholnick, *Adv. Lipid Res.* **11,** 67 (1973).

Procedure. To the LDL-VLDL precipitate in 1 liter of buffer A, add 300 ml of DEAE-Sepharose CL-6B equilibrated in buffer A. This mixture should be stirred for several hours or until less than 5% of the total PAF acetylhydrolase activity is found in an aliquot of the supernatant. Pack the slurry in a column of wide diameter (we use a 5 × 60 cm glass column), allow most of the buffer A to run through, and wash the column with an additional 500–600 ml of buffer A. Then, add four 125-ml washes of KCl (0.3 *M*) in buffer A. The second and third washes contain most of the PAF acetylhydrolase activity. Dilute the active washes 5-fold with buffer A and place the preparation on a 2.5 × 22 cm column packed with DEAE-Sepharose CL-6B in buffer A. The PAF acetylhydrolase activity can be recovered by applying a KCl gradient (0–0.3 *M*, total volume = 500 ml). It is occasionally necessary to "chase" the column with an additional 100–200 ml of 0.3 *M* KCl in buffer A to maximize recovery.

The pooled fractions are concentrated to 25–30 ml in an Amicon (Danvers, MA) ultrafiltration cell, using a PM10 filter. This preparation should be subjected to a gel permeation step using Sephacryl S-300 (we use a 2.5 × 90 cm column equilibrated in buffer A). The active fractions should then be concentrated by ultrafiltration, and placed on an 80% hydroxylapatite: 20% Sigmacell column (0.9 × 14 cm) in buffer A; the cellulose increases flow rate and prevents clogging. Elution is achieved using a 100-ml linear gradient of potassium phosphate buffer, pH 6.8 (0.1–0.3 *M*), containing 0.1% Tween 20. The active fractions should be concentrated to approximately 15 ml by ultrafiltration, as above.

Nondenaturing Polyacrylamide Gel Electrophoresis. Dialyze the hydroxylapatite effluent (15 ml) for 16–18 hr against two changes of 1 liter of buffer A, and add sucrose to a final concentration of 10%. Place this solution on a 3-mm slab gel of 7% polyacrylamide; use riboflavin as the catalyst. Electrophorese for about 8 hr at a constant current of 30 mA. All solutions should contain 0.1% Tween 20, since the purified enzyme tends to stick to surfaces in the absence of detergents. After electrophoresis, slice the gel into 0.5-cm horizontal strips, force each slice through a syringe, and combine the smashed slices with 5 ml of buffer A. Allow this solution to incubate overnight at 4°. It may be necessary to divide the preparation and run two 3-mm gels to improve resolution. Enzymatic activity is usually present in slices corresponding to an R_f of 0.22. A summary of the entire purification procedure is shown in Table II. The SDS–PAGE pattern of the purified preparation is shown in Fig. 2.

Comments

If desired, the enzyme can be further purified by chromatography on heparin-agarose columns or HPLC on Mono Q and/or C_4 columns. In

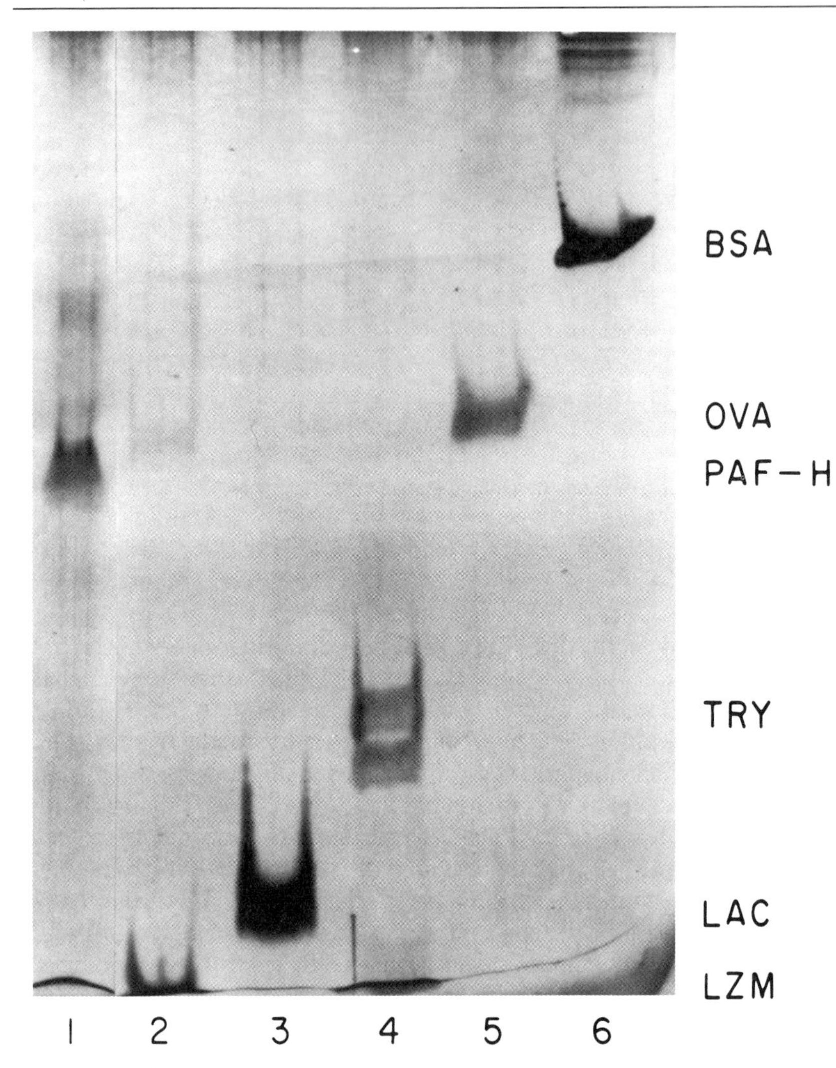

FIG. 2. SDS–PAGE of purified PAF acetylhydrolase. A sample of enzyme that had been purified 25,000-fold is subjected to SDS–PAGE and stained with silver (lane 1). Molecular weight standards: lysozyme (LZM, lane 2, M_r 14,300), β-lactoglobulin (LAC, lane 3, M_r 18,400), trypsinogen (TRY, lane 4, M_r 24,000), ovalbumin (OVA, lane 5, M_r 45,000), bovine serum albumin (BSA, lane 6, M_r 66,000). PAF-H, PAF acetylhydrolase. From Stafforini *et al.*[9]

TABLE II
PURIFICATION OF PAF ACETYLHYDROLASE FROM HUMAN PLASMA[a]

Fraction	Units (μmol/hr)	Protein (mg)	Specific activity (units/mg)	Purification (-fold)	Recovered (%)
Plasma	1,642	72,462	0.023		100
LDL/VLDL	1,426	1,340	1.064	46	87
Batch DEAE	1,282	715	1.793	78	78
Column DEAE	672	26	25.58	1,124	41
Sephacryl S-300	399	4.1	97.32	4,231	24
Hydroxylapatite	166	1.0	166.0	7,217	10
Preparative PAGE	159	0.28	567.8	24,689	10

[a] Reproduced with permission from Stafforini *et al.*[9]

some cases the enzyme may be loaded in the absence of detergents and then eluted using Tween 20 or CHAPS (Bio-Rad, Richmond, CA). We have often found that the enzyme sticks irreversibly to some hydrophobic gels.

Substrate Specificity

The molecular features of PAF necessary for its recognition as a substrate of PAF acetylhydrolase are summarized in Tables III and IV. The presence of an alkyl group is not necessary for hydrolysis, since the 1-acyl analog of PAF is an equally good substrate (not shown). In addition, the number of methyl groups in the phosphocholine moiety, does not seem to play a critical role in substrate recognition (Table III).

TABLE III
SUBSTRATE SPECIFICITY OF PAF ACETYLHYDROLASE[a]: EFFECT OF ETHANOLAMINE DERIVATIVES ON ACTIVITY

Substrate	Concentration (μM)	[^{3}H] Acetate (%)	Inhibition type
1-*O*-Alkyl-2-acetyl-*sn*-glycero-3-phosphodimethylethanolamine	10	71	Competitive
	200	9	($K_i = 12\ \mu M$)
1-*O*-Alkyl-2-acetyl-*sn*-glycero-3-phosphomonomethylethanolamine	10	74	Not determined
	200	11	
1-*O*-Alkyl-2-acetyl-*sn*-glycero-3-phosphoethanolamine	10	77	Competitive
	200	19	($K_i = 22\ \mu M$)

[a] Reproduced with permission from Stafforini *et al.*[9]

TABLE IV
SUBSTRATE SPECIFICITY OF PAF ACETYLHYDROLASE[a]: EFFECT OF *sn*-2-ACYL CHAIN LENGTH ON RATE OF HYDROLYSIS OF 1-ALKYL-2-ACYLGLYCEROPHOSPHOCHOLINE SUBSTRATES[b]

Substrate	Concentration (n*M*)	Velocity (nmol/hr/mg)
1-*O*-[alkyl-^{3}H]-2-Acetyl-GPC	0.60	24.0
1-*O*-[alkyl-^{3}H]-2-Propionyl-GPC	0.66	1.4
1-*O*-[alkyl-^{3}H]-2-Butyroyl-GPC	0.53	1.3
1-*O*-[alkyl-^{3}H]-2-Hexanoyl-GPC	2.95	0.7

[a] Reproduced with permission from Stafforini *et al.*[9]
[b] GPC, Glycerophosphocholine.

In contrast to the lack of specificity of the enzyme for the type of linkage or substitutions at the *sn*-1 and *sn*-3 positions, the enzyme recognizes only phospholipids with short acyl groups at the *sn*-2 position. This preference for short acyl residues at the *sn*-2 position is most apparent at subsaturating concentrations of substrate. For example, propionyl, butyroyl, and hexanoyl analogs were hydrolyzed at much slower rates (about 1/20 of the hydrolytic rate with PAF; Table IV).

We have also established that the purified plasma hydrolase is capable of utilizing oxidized phospholipids with short acyl chains at the *sn*-2 position of the glycerol backbone.[15] We synthesized 1-palmitoyl-2-oxovaleroylglycerophosphocholine (5-oxovaleroyl-PC) and determined its kinetic properties using both plasma and the purified enzyme (Table V). The enzyme has a high affinity for this oxidized phospholipid, as evidenced by the similar K_m values for PAF and 5-oxovaleroyl-PC. The rate of hydrolysis of the 5-oxovaleroyl-GPC is somewhat slower than that for PAF, under

TABLE V
HYDROLYSIS OF OXIDIZED PHOSPHOLIPIDS BY PAF ACETYLHYDROLASE[a]

Substrate	Concentration (*M*)	Half-life (min)	K_m (μM)	V_{max} (μmol/mg/hr)
PAF	3.3×10^{-6}	10.2	N/A[b]	N/A
5-Oxovaleroyl-GPC	3.3×10^{-6}	21.8	N/A	N/A
PAF	$0\text{-}4 \times 10^{-5}$	N/A	12.5	167
5-Oxovaleroyl-GPC	$0\text{-}4 \times 10^{-5}$	N/A	11.3	100

[a] Adapted from Stremler *et al.*[15]
[b] N/A, Not applicable.

conditions where substrate was near saturation and in half-life experiments carried out at subsaturating levels of substrate (Table II). This is presumably due to the longer oxoacyl group esterifying the *sn*-2 position, in agreement with our findings with the valeroyl analog of PAF. Additional experiments, including antibody cross-reactivity, have allowed us to establish that in plasma, PAF and oxidized phospholipids are hydrolyzed by the same phospholipase, i.e., the PAF acetylhydrolase.

[40] Synthesis of Epoxyeicosatrienoic Acids and Heteroatom Analogs

By J. R. FALCK, PENDRI YADAGIRI, and JORGE CAPDEVILA

Various studies have established an alternative route for the enzymatic generation of biologically active, oxygenated eicosanoids.[1] This pathway is known as the "epoxygenase" branch of the arachidonate cascade and is mediated by cytochrome *P*-450. Several epoxygenase metabolites have been identified as endogenous constituents of mammalian systems,[2–6] but their physiological significance is presently unclear. A comprehensive review of the current status of the epoxygenase pathway with a critical discussion of its implications has appeared.[7]

The most characteristic and extensively studied epoxygenase metabolites are the four regioisomeric *cis*-epoxyeicosatrienoic acids (EETs) (Fig. 1). They are relatively stable to hydrolysis at physiological pH [e.g., leukotriene A_4 (LTA_4) and hepoxilin A_3].[8] The exception is 5,6-EET

[1] J. Capdevila, G. Snyder, and J. R. Falck, *in* "Microsomes and Drug Oxidations" (A. R. Boobis, J. Caldwell, F. DeMatteis, and C. R. Elcombie, eds.), p. 84. Taylor and Francis, Ltd., London and New York, 1985.

[2] J. Capdevila, B. Pramanik, J. L. Napoli, S. Manna, and J. R. Falck, *Arch. Biochem. Biophys.* **231,** 511 (1984).

[3] R. Toto, A. Siddhanta, S. Manna, B. Pramanik, J. R. Falck, and J. Capdevila, *Biochim. Biophys. Acta* **919,** 132 (1987).

[4] J. R. Falck, V. J. Schueler, H. R. Jacobson, A. K. Siddhanta, B. Pramanik, and J. Capdevila, *J. Lipid Res.* **28,** 840 (1987).

[5] R. C. Murphy, J. R. Falck, S. Lumin, P. Yadagiri, J. A. Zirrolli, M. Balazy, J. L. Masferrer, N. G. Abraham, and M. L. Schwartzman, *J. Biol. Chem.* **263,** 17197 (1988).

[6] P. M. Woollard, *Biochem. Biophys. Res. Commun.* **136,** 169 (1986).

[7] F. A. Fitzpatrick and R. C. Murphy, *Pharmacol. Rev.* **40,** 229 (1989).

[8] N. Chacos, J. Capdevila, J. R. Falck, S. Manna, C. Martin-Wixtrom, S. S. Gill, B. D. Hammock, and R. W. Estabrook, *Arch. Biochem. Biophys.* **223,** 639 (1983).

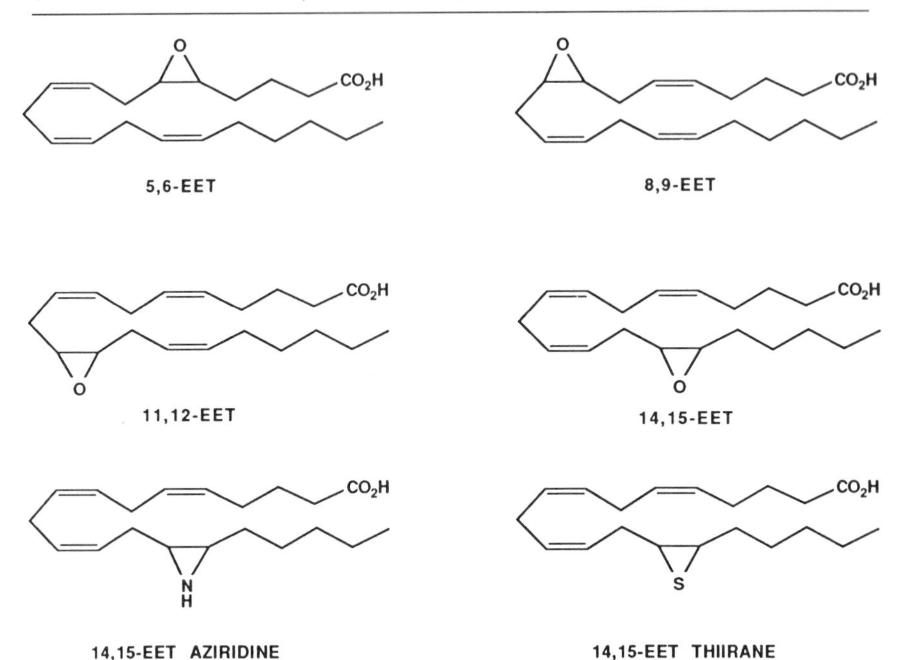

FIG. 1. Structure of epoxyeicosatrienoic acids and heteroatom analogs.

which, in its free-acid form, readily decomposes to 5,6-dihydroxyeicosatrienoic acid and/or the corresponding δ-lactone.[9] The absolute configuration of the EETs has been determined using material obtained from *in vitro* incubation of arachidonic acid with the major phenobarbital-inducible form of rat liver microsomal cytochrome *P*-450.[10] Recent results suggest, however, that the enantiomeric composition is highly dependent on the identity of the cytochrome *P*-450 isozyme.[11] This may be significant since in some instances biological activity is sensitive to EET stereochemistry.[12]

Chiral syntheses of all the EET enantiomers have been achieved (Table I). In general, these syntheses are impractical for laboratories not equipped for multistep organic synthesis. Most investigators have instead

[9] D. Schlondorff, E. Petty, J. A. Oates, M. Jacoby, and S. D. Levine, *Am. J. Physiol.* **253,** F464 (1987).

[10] J. R. Falck, S. Manna, H. R. Jacobson, R. W. Estabrook, N. Chacos, and J. Capdevila, *J. Am. Chem. Soc.* **106,** 3334 (1984).

[11] J. Capdevila, unpublished results, 1989.

[12] F. Fitzpatrick, M. Ennis, M. Baze, M. Wynalda, J. McGee, and W. Liggett, *J. Biol. Chem.* **261,** 15334 (1986).

TABLE I
ASYMMETRIC SYNTHESES OF EPOXYEICOSATRIENOIC ACIDS AND ANALOGS

EET isomer	Overall yield (%)	Chirality source	Source[a]
5(*S*),6(*R*)	5	2-Deoxy-D-glucose	1
5(*R*),6(*S*)	7	2-Deoxy-D-glucose	1
5(*S*),6(*R*)-20-OH	11	Dimethyl L-malate	2
8(*S*),9(*R*)	15–20	Dimethyl L-malate	3
8(*R*),9(*S*)	15–20	Dimethyl D-malate	3
11(*S*),12(*R*)	15–20	Dimethyl D-malate	3
11(*S*),12(*R*)	1–3	Methyl 15(*S*)-HETE	4
11(*R*),12(*S*)	15–20	Dimethyl L-malate	3
14(*S*),15(*R*)	2–6	Methyl 15(*S*)-HETE	4
14(*S*),15(*R*)	15	Sharpless epoxidation	5
14(*R*),15(*S*)	3	Methyl 15(*S*)-HETE	4
14(*R*),15(*S*)	13	2-Deoxy-D-glucose	1
14(*R*),15(*S*)	15	Sharpless epoxidation	5
trans-14(*S*),15(*S*)	—[b]	Sharpless epoxidation	5
trans-14(*R*),15(*R*)	—[b]	Sharpless epoxidation	5
trans-14(*R*),15(*R*)	33	Methyl 15(*S*)-HETE	4
14(*R*),15(*S*)-20-OH	11	Dimethyl L-malate	2

[a] Key to references: (1) C. A. Moustakis, J. Viala, J. Capdevila, and J. R. Falck, *J. Am. Chem. Soc.* **107,** 5283 (1985); (2) S. Lumin, P. Yadagiri, J. R. Falck, J. Capdevila, P. Mosset, and R. Gree, *J. Chem. Soc. Chem. Commun.* p. 389 (1987); (3) P. Mosset, P. Yadagiri, S. Lumin, J. Capdevila, and J. R. Falck, *Tetrahedron Lett.* **27,** 6035 (1986); (4) J. R. Falck, S. Manna, and J. Capdevila, *Tetrahedron Lett.* **25,** 2443 (1984); (5) M. D. Ennis and M. E. Baze, *Tetrahedron Lett.* **27,** 6031 (1986).

[b] Incomplete data.

relied on racemic EETs prepared by Corey's site-specific oxidation methodology[13,14] or, more conveniently, by nonselective peracid epoxidation of arachidonic acid.[15] This chapter describes procedures based on the latter reaction that have proved reliable in the authors' laboratories for the production and purification of nanomole–millimole amounts of EETs and their methyl esters. In light of the increasing interest in EET heteroatom analogs, protocols[16] for converting EETs to *cis*-thiiranes and *cis*-aziridines (Fig. 1) are also included.

[13] E. J. Corey, H. Niwa, and J. R. Falck, *J. Am. Chem. Soc.* **101,** 1586 (1979).

[14] E. J. Corey, A. Marfat, J. R. Falck, and J. O. Albright, *J. Am. Chem. Soc.* **102,** 1433 (1980).

[15] S.-K. Chung and A. I. Scott, *Tetrahedron Lett.* p. 3023 (1974).

[16] J. R. Falck, S. Manna, J. Viala, A. K. Siddhanta, C. A. Moustakis, and J. Capdevila, *Tetrahedron Lett.* **26,** 2287 (1985).

Experimental Section

General Procedures. Dichloromethane is distilled from calcium hydride. Tetrahydrofuran and ether are distilled from sodium benzophenone ketyl. All other solvents and solutions are peroxide free and sparged with argon to remove dissolved oxygen immediately prior to use. All reactions and distillations are conducted under an inert atmosphere of nitrogen or argon. Methyl arachidonate (>99%) and arachidonic acid (>99%) are obtained from Nu-Chek-Prep (Elysian, MN). Radiolabeled arachidonate (^{14}C or ^{3}H) is purchased from DuPont-New England Nuclear (Boston, MA) or Amersham (Arlington Heights, IL) and mixed with unlabeled arachidonate to give the desired specific activity prior to epoxidation. 3-Chloroperoxybenzoic acid (80–85%, technical grade) and all other chemicals which are reagent grade or better are obtained from Aldrich (Milwaukee, WI). Woelm neutral alumina W-200 (activity grade I) is obtained from ICN Pharmaceuticals (Cleveland, OH). Eicosanoids are stored at −78° under argon in glass containers with fluorocarbon-faced caps. Extracts are dried over anhydrous sodium sulfate. Solvent removal refers to evaporation under reduced pressure on a rotary evaporator.

Thin-layer chromatography (TLC) is performed on silica gel 60 F-254 plates (20 × 20 cm, 0.25 mm thickness) from E. Merck (Darmstadt, FRG) in glass tanks preequilibrated with solvent. Products are visualized by transillumination of the plate with a bright, white light, by exposure to iodine vapor, or by spraying with 5% ethanolic phosphomolybdic acid and heating briefly at 170°. Silica gel is removed from the TLC plate with a razor blade and extracted with methanol/dichloromethane (CH_2Cl_2) (1 : 10). Radioactivity of HPLC eluants is measured on-line with a Ramona-LS (Raytest Gmbh).

Epoxidation of Arachidonic Acid

3-Chloroperoxybenzoic acid (19 mg, 0.11 mmol) is added portionwise to a stirring, 0° solution of arachidonic acid (31 mg, 0.10 mmol) in CH_2Cl_2 (8 ml). After 12 hr at 0°, methyl sulfide (Me_2S) (50 μl) is added and the mixture maintained at ambient temperature for 20 min to quench any remaining peracid. The reaction is diluted with CH_2Cl_2 (15 ml), transferred to a separatory funnel, washed with water (3 × 15 ml), then with saturated NaCl solution (15 ml), and dried. Solvent removal gives a viscous oil which is purified immediately.

Normal-phase HPLC purification (Fig. 2) yields unreacted arachidonic acid (8 mg), 14,15-EET (7 mg), 11,12-EET (5 mg), 8,9-EET (3 mg), and 5,6-EET (1.5 mg). Comparable results are obtained using TLC and methanol/CH_2Cl_2 (5 : 95, 5 elutions) as eluant: arachidonic acid ($R_f \approx$

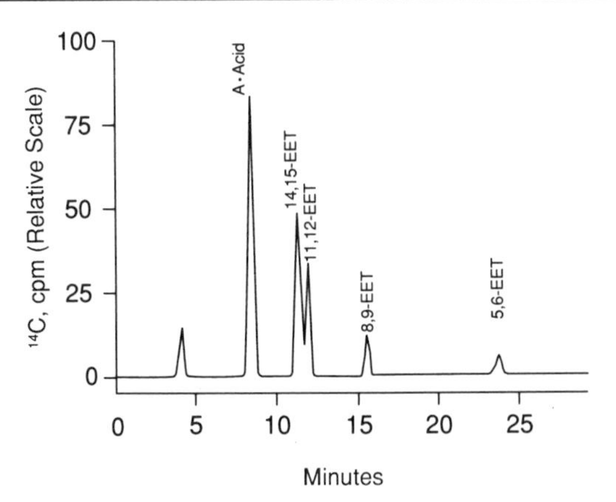

FIG. 2. Normal-phase HPLC analysis of the reaction products from epoxidation of arachidonic acid. The crude product is resolved on a Beckman (Fullerton, CA) Ultrasphere-Si (5 μm, 25 $\times$ 0.46 cm) using 0.4% 2-propanol/0.1% acetic acid/99.5% hexane at a flow rate of 2 ml/min: arachidonic acid ($R_t \approx$ 8.14 min), 14,15-EET ($R_t \approx$ 11.13 min), 11,12-EET ($R_t \approx$ 11.47 min), 8,9-EET ($R_t \approx$ 15.40), and 5,6-EET ($R_t \approx$ 23.44 min). Radioactivity is monitored on-line with a Ramona-LS.

0.76), 14,15-EET ($R_f \approx$ 0.70), 11,12-EET ($R_f \approx$ 0.68), 8,9-EET ($R_f \approx$ 0.60), 5,6-EET ($R_f \approx$ 0.54), and by-products ($R_f \approx$ 0.47–0.40).

Epoxidation of Methyl Arachidonate

3-Chloroperoxybenzoic acid (25 mg, 0.14 mmol) is added portionwise to a stirring, 0° mixture of methyl arachidonate (43 mg, 0.135 mmol) and anhydrous Na_2CO_3 (17 mg, 0.16 mmol) in CH_2Cl_2 (4 ml). After stirring for 15 hr at 0°, Me_2S (50 μl) is added and the mixture maintained at ambient temperature for 20 min to quench any remaining peracid. The reaction is diluted with CH_2Cl_2 (15 ml), transferred to a separatory funnel, washed with water (3 $\times$ 10 ml), then with saturated NaCl solution (10 ml), and dried. Solvent removal gives a viscous oil (43 mg).

Reversed-phase HPLC purification (Fig. 3) affords unreacted methyl arachidonate (12 mg), a coeluting mixture of three methyl EETs (i.e., methyl 5,6-, 8,9-, and 11,12-EET), methyl 14,15-EET (6 mg), and polar by-products. Resolution of the mixed methyl EET fraction by normal-phase HPLC (Fig. 4) furnishes methyl 5,6-EET (3 mg), methyl 8,9-EET (5 mg), and methyl 11,12-EET (5 mg). TLC purification of the crude reaction product using ethyl ether (Et_2O)/CH_2Cl_2/hexane (1 : 1 : 8, 5 elutions) gives incomplete resolution of the regioisomeric methyl EETs: methyl arachidonate ($R_f \approx$ 0.88), methyl 14,15-, and 11,12-EET ($R_f \approx$

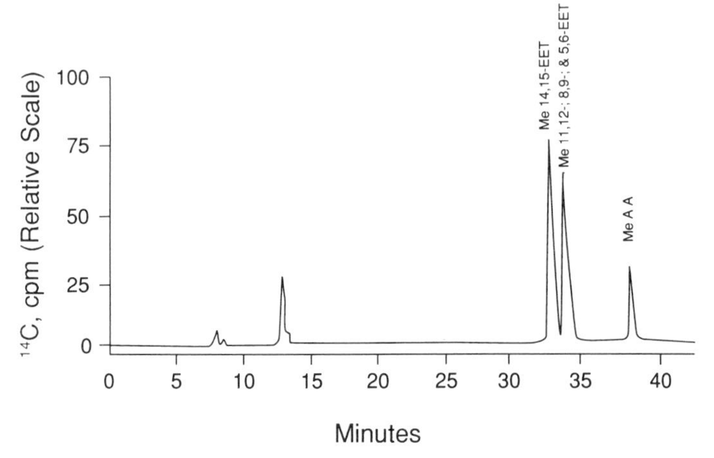

FIG. 3. Reversed-phase HPLC analysis of the reaction products from epoxidation of methyl arachidonate. The crude product is resolved partially on a Rainin Microsorb C_{18} (5 μm, 25 × 0.46 cm) using a linear gradient of 50% solvent A/50% solvent B to 100% solvent B over 40 min at a flow rate of 1 ml/min: methyl arachidonate ($R_t \approx$ 39.2 min), methyl 5,6-, 8,9-, and 11,12-EET coeluted ($R_t \approx$ 34.1 min), methyl 14,15-EET ($R_t \approx$ 32.7 min). Solvent A is 0.1% acetic acid/99.9% H_2O; solvent B is 0.1% acetic acid/99.9% CH_3CN. Radioactivity is monitored on-line with a Ramona-LS.

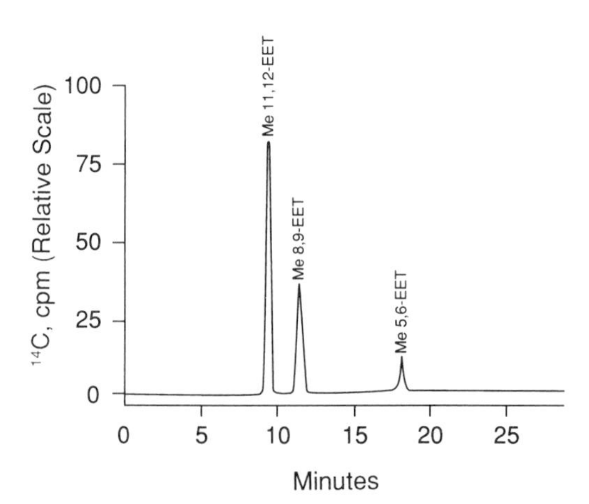

FIG. 4. Normal-phase HPLC analysis of methyl 5,6-, 8,9- and 11,12-EET. The mixed methyl EET fraction from Fig. 3 is resolved on a Waters (Milford, MA) μPorasil (5 μm, 25 × 0.46 cm) using a linear gradient from 0.15% 2-propanol/99.85% hexane to 0.4% 2-propanol/ 99.6% hexane over 30 min at a flow rate of 2 ml/min: methyl 5,6-EET ($R_t \approx$ 18.0 min), methyl 8,9-EET ($R_t \approx$ 11.5 min), and methyl 11,12-EET ($R_t \approx$ 9.6 min). Radioactivity is monitored on-line with a Ramona-LS.

0.55), methyl 8,9-EET ($R_f \approx 0.51$), methyl 5,6-EET ($R_f \approx 0.42$), and polar by-products ($R_f \approx 0.16$).

Saponification of Methyl 8,9-, 11,12-, or 14,15-EET: General Procedure

To a stirring, 0° solution of methyl EET (16.7 mg, 0.05 mmol) in MeOH (4.5 ml) is slowly added 0.33 *N* aqueous NaOH (1.5 ml). The homogeneous reaction is stirred at ambient temperature and monitored by TLC analysis until complete (≈12 hr). After cooling to 0°, the reaction is diluted with water (5 ml) and carefully acidified with 0.25 *M* aqueous oxalic acid (1.0 ml) to pH 4–4.5. The solution is transferred to a separatory funnel and extracted with Et_2O (3 × 25 ml). The combined ethereal extracts are washed with water until the aqueous layer is neutral, then washed with saturated NaCl solution (30 ml), and dried. Solvent removal affords EET-free acid in essentially quantitative yield.

The products can be purified by HPLC as described in Fig. 2. TLC analysis using methanol/CH_2Cl_2 (5 : 95) as eluant shows: 8,9-EET ($R_f \approx$ 0.17), 11,12-EET ($R_f \approx 0.18$), and 14,15-EET ($R_f \approx 0.18$).

Saponification of Methyl 5,6-EET

To a stirring, 0° solution of methyl 5,6-EET (16.7 mg, 0.05 mmol) in tetrahydrofuran (4.5 ml) is slowly added 0.41 *N* aqueous NaOH (1.5 ml). The homogeneous mixture is stirred at ambient temperature until the hydrolysis is judged complete (≈12 hr) by TLC analysis (ethyl acetate/hexane, 1 : 2, ester $R_f \approx 0.49$). The reaction is diluted with ice chips (15 g) and transferred to a separatory funnel containing ice-cold water (30 ml), Et_2O (30 ml), 0.25 *M* aqueous oxalic acid (1.1 ml), and glacial acetic acid (100 μl). After vigorous mixing, the upper organic layer is separated. The aqueous layer is extracted again with fresh Et_2O (2 × 20 ml) and the combined organic extracts are washed with water, until the washings are neutral, then with saturated NaCl solution (25 ml), and dried. Solvent removal affords 5,6-EET-free acid accompanied by a variable amount of the corresponding 5,6-diol. The products are somewhat labile and should be used without delay following purification by HPLC (Fig. 2) or by TLC using methanol/CH_2Cl_2 (1 : 10) as eluant: 5,6-EET ($R_f \approx 0.38$), 5,6-diol ($R_f \approx 0.17$).

Thiirane (Episulfide) Heteroatom Analog

Preparation of the *cis*-14,15-thiirane (episulfide) heteroatom analog[16] from methyl 14,15-EET is representative.

A mixture of methyl 14,15-EET (25 mg, 0.075 mmol) and anhydrous KSCN (145 mg, 1.5 mmol) in dry methanol (10 ml) is heated to reflux for

72 hr. After cooling to 0°, the reaction is diluted with water (30 ml), carefully acidified to pH 4.5 with dilute hydrochloric acid, and extracted in a separatory funnel with Et_2O/hexanes (1 : 1, 3 × 20 ml). The combined organic extracts are washed with water (3 × 20 ml), then with saturated NaCl solution (30 ml), and dried. Solvent removal gives an oily residue which is purified by TLC eluted with Et_2O/hexanes (1 : 2) to furnish methyl 14,15-thiirane (20.5 mg, $R_f \approx 0.55$) and 14,15-thiirane-free acid (2 mg, $R_f \approx 0.24$). The methyl ester is hydrolyzed as described for methyl 14,15-EET.

Aziridine Heteroatom Analog

Preparation of the *cis*-14,15-aziridine heteroatom analog[16] from methyl 14,15-EET is representative.

A solution of sodium azide (45 mg, 0.69 mmol) in water (0.6 ml) is added with stirring to benzene (1 ml) followed by concentrated H_2SO_4 (17 μl). *Caution: hydrazoic acid is toxic and should be handled in a hood while wearing gloves and eye protection.* After 10 min, the upper organic layer is removed by decantation and dried. According to the method of Posner and Rogers,[17] the resultant benzene solution of hydrazoic acid is added to a well-stirred suspension of Woelm neutral alumina W-200 (600 mg) in dry Et_2O (3 ml). A solution of methyl 14,15-EET (25 mg, 0.075 mmol) in dry Et_2O (1 ml) is added after 5 min and the stirring continued until the reaction is judged complete (≈10 min) by TLC analysis. The reaction is diluted with methanol (20 ml), stirred vigorously for 15 min, filtered through a Celite (diatomaceous earth) pad, and the filter cake washed thoroughly with methanol. Solvent removal provides the regioisomeric methyl 14,15-azidohydrins (23 mg) as an oil which are used directly in the next step. TLC analysis utilizing Et_2O/hexanes (1 : 2) as eluent gives the following: methyl 14,15-EET ($R_f \approx 0.40$), methyl 14,15-azidohydrins ($R_f \approx 0.26$ and 0.28).

After azeotropic drying with benzene, the above crude product is dissolved in dry Et_2O (4 ml). To this is added triphenylphosphine (Ph_3P) (27 mg, 0.10 mmol) and the mixture is stirred at ambient temperature for 18 hr, then at reflux for 12 hr. Solvent removal and rapid chromatography of the residue on a column of silica gel (2 g) using CH_3CN as eluant gives methyl 14,15-aziridine (17 mg). TLC analysis using CH_3CN : $R_f \approx 0.14$. The ester is hydrolyzed as described for methyl 14,15-EET.

Acknowledgments

Work in the authors' laboratories was supported by grants from the Robert A. Welch Foundation, Juvenile Diabetes Foundation, Kroc Foundation, NATO, and the USPHS NIH.

[17] G. H. Posner and D. Z. Rogers, *J. Am. Chem. Soc.* **99,** 8208 (1977).

[41] Isolation of Rabbit Renomedullary Cells and Arachidonate Metabolism

By MAIREAD A. CARROLL, ELIZABETH D. DRUGGE, CATHERINE E. DUNN, and JOHN C. MCGIFF

Novel arachidonic acid (AA) metabolites arising from the so-called third pathway are generated by cells of the medullary thick ascending limb of Henle's loop (mTALH). The cells of the mTALH contain high concentrations of Na^+,K^+-ATPase,[1] which plays a key role in the active translocation of sodium and potassium across the basal membrane and is modulated in this activity by these AA metabolites. Thus, the regulation of extracellular fluid volume depends in large part on mechanisms operating within the renal medulla, particularly the mTALH which transports NaCl in excess of water and also establishes the medullary solute gradient.

We have shown that mTALH cells are invested primarily with the cytochrome *P*-450 monooxygenase pathway[2,3] of AA metabolism; their capacity to generate prostaglandins is limited, if not negligible. This pathway requires the presence of NADPH and molecular oxygen for the oxidative metabolism of AA to 5,6-, 8,9-, 11,12-, and 14,15-epoxyeicosatrienoic acids (EETs) and their corresponding dihydroxyeicosatrienoic acids (DHTs), as well as to several hydroxyeicosatetraenoic acids (HETEs).[4,5]

The cytochrome *P*-450-dependent AA metabolites, unlike the metabolites of cyclooxygenase or lipoxygenase, do not demonstrate characteristic UV absorbance. Further, radioimmunoassays have not yet been developed. When using thin-layer chromatography (TLC) or high-performance liquid chromatography (HPLC), these metabolites are detected by the conversion of radiolabeled AA. Evidence for the conversion of AA through the cytochrome *P*-450 pathway can be attained by: (1) the absolute dependency of this pathway, in broken cell preparations, on the presence of NADPH; (2) inhibition of product formation by SKF 525A and other

[1] P. L. Jorgensen, *Physiol. Rev.* **60,** 864 (1980).

[2] N. R. Ferreri, M. Schwartzman, N. G. Abraham, P. N. Chander, and J. C. McGiff, *J. Pharmacol. Exp. Ther.* **231,** 441 (1984).

[3] M. Schwartzman, N. R. Ferreri, M. A. Carroll, E. Songu-Mize, and J. C. McGiff, *Nature (London)* **314,** 620 (1985).

[4] J. Capdevila, N. Chacos, J. Werringloer, R. A. Prough, and R. W. Estabrook, *Proc. Natl. Acad. Sci. U.S.A.* **78,** 5362 (1981).

[5] E. H. Oliw, F. P. Guengerich, and J. A. Oates, *J. Biol. Chem.* **257,** 3771 (1982).

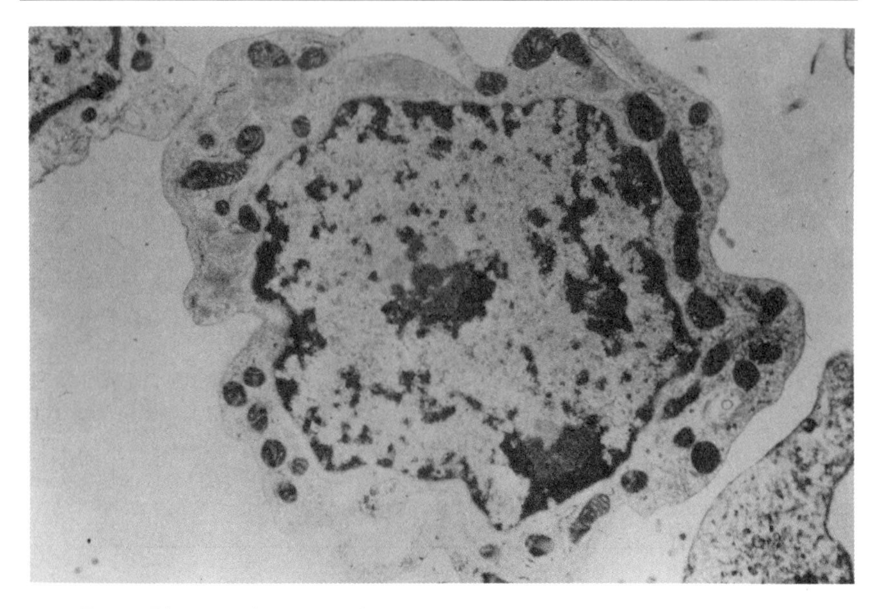

FIG. 1. Electron micrograph of an isolated TALH cell. (Magnification: ×8960.)

monooxygenase inhibitors[6]; and (3) the failure of indomethacin[6] to affect product formation. In our hands, nordihydroguariaretic acid, a lipoxygenase inhibitor, also reduces cytochrome *P*-450 activity. Further evidence of cytochrome *P*-450-dependent metabolism of AA, although behond the scope of this chapter, can be obtained by the manipulation of cytochrome *P*-450 content with inducers or depleters that produce corresponding changes in conversion of AA metabolites arising from this pathway.[7]

In addition to high Na^+,K^+-ATPase activity, mTALH cells obtained from rabbits have distinct morphological and histochemical characteristics.[2] These epithelial cells are morphologically recognized as being 8–10 μm in diameter, with large nuclei, numerous mitochondria, and few apical microvilli (Fig. 1). Histochemically mTALH cells, unlike proximal tubular cells, have low alkaline phosphatase activity and express Tamm–Horsfall glycoprotein, a specific marker for TALH and distal tubular cells.

There are various methods for isolating mTALH cells. Primary cultures have been generated from microdissected explants of rabbit medul-

[6] J. Capdevila, L. Gill, M. Orellana, L. J. Marnett, J. I. Mason, P. Yadagiri, and J. R. Falck, *Arch. Biochem. Biophys.* **261,** 259 (1988).

[7] M. L. Schwartzman, N. G. Abraham, M. A. Carroll, R. D. Levere, and J. C. McGiff, *Biochem. J.* **238,** 283 (1986).

lary thick ascending limbs.[8] Cultures have also been established from mTALH cells isolated by Ficoll density-gradient centrifugation.[9] More recently, cortical and medullary TALH cells have been isolated with an immunodissection technique, utilizing the specific cell-surface Tamm–Horsfall protein.[10] We have found an efficient method of isolating mTALH cells by centrifugal elutriation, a procedure that separates cells according to their sedimentation rate. This method provides a more highly purified suspension (85–95%) of mTALH cells, with sufficient yield (20–30 × 10^6) required for the biochemical studies, than can be achieved through microdissection techniques. Further, it requires less time (3–4 hr) for cell isolation, essential for preservation of cytochrome *P*-450 activity, than density gradient centrifugation.

By using centrifugal elutriation to separate and obtain mTALH cells, we have been able to study AA metabolism by these cells[2,3,11] in relative homogeneity, and also to establish the conditions necessary to grow them and preserve cytochrome *P*-450 activity in culture.[12]

Procedure for mTALH Cell Isolation

Freshly Isolated mTALH Cells from Adult Rabbits

New Zealand White male rabbits (3.0–3.2 kg) are anesthetized with ketamine-HCl (50 mg/kg) and xylazine (13 mg/kg). After laparotomy, the renal arteries are cannulated and the kidneys flushed with 0.9% saline (150 ml per kidney). The kidneys are excised and sectioned through the longitudinal axis, and the inner stripe of the outer medulla (red medulla) is carefully separated from the cortex and inner medulla. The inner stripe is cut into pieces and trypsinized for 20 min at room temperature in trypsin–EDTA (0.05% trypsin/0.53 m*M* EDTA; Gibco, Grand Island, NY) plus 0.1% bovine serum albumin. The tissue is mechanically disrupted using two microscope slides in RPMI 1640 culture media containing excess bovine serum albumin (1.0%) to stop trypsinization. The resultant cell suspension is filtered through a 30-μm nylon mesh to remove intact tubules, and then washed (700 *g;* 5 min) in Dulbecco's phosphate-buffered

[8] M. B. Burg, N. Green, S. Schraby, R. Steele, and J. Handler, *Am. J. Physiol.* **242,** C229 (1982).

[9] D. M. Scott, K. Zierold, E. Linne-Saffran, and R. Kinne, *Pfluegers Arch.* **402,** R7 (1984).

[10] M. G. Allen, A. Nakao, W. K. Sonnenburg, M. Burnatowska-Hledin, W. S. Spielman, and W. L. Smith, *Am. J. Physiol.* **255,** F704 (1988).

[11] M. A. Carroll, M. L. Schwartzman, M. Baba, M. J. S. Miller, and J. C. McGiff, *Am. J. Physiol.* **255,** F151 (1988).

[12] E. D. Drugge, M. A. Carroll, and J. C. McGiff, *Am. J. Physiol.* **256,** C1070 (1990).

saline (PBS) (Gibco, Grand Island, NY), without calcium (to reduce cell clumping), containing 0.2% bovine serum albumin. The final cell pellet, an outer medullary cell population, is resuspended in 10 ml of PBS plus 0.2% bovine serum albumin and separated into several fractions by centrifugal elutriation. The Beckman (San Remos, CA) JE-6B elutriator rotor and Beckman J2-21 centrifuge are used to separate different cell types of the inner stripe. Cells derived mainly from the mTALH are eluted at a flow rate of 19 ml/min at 2000 rpm and 20° after cells with slower sedimentation rates have been eluted. The recovered mTALH cells are washed (700 *g* for 10 min) with PBS and calcium (1 m*M*) to remove excess bovine serum albumin. The cells are counted using light microscopy (magnification: ×100) and the mTALH cell population is divided into 3×10^6 aliquots in preparation for AA metabolism studies. Approximately $20–30 \times 10^6$ mTALH cells are isolated, from each pair of rabbit kidneys.

Preparation of Cultured mTALH Cells from Neonatal Rabbits

Medullary TALH cells are isolated, as described above, under sterile conditions from neonatal New Zealand White male rabbits (0.5–0.7 kg). After isolation, mTALH cells are resuspended and cultured in a medium which was found to support attachment and growth of these cells as well as to allow the expression of cytochrome *P*-450 activity. The media consists of Dulbecco's modified essential medium: Ham's F12 medium (1/1), 5% fetal bovine serum, epidermal growth factor (20 ng/ml), streptomycin/penicillin (100 U/ml), Fungizone (1 μg/ml), and hemin (1×10^{-6} *M*). Plating density is 1×10^6 cells/ml. Two milliliters of media are spread over the bottom of a 60-mm tissue culture dish containing a Millipore filter. Media is removed after 5 days and replaced with 4 ml of fresh media. This step is repeated every 3 days thereafter. After 3 weeks, the polygonal cells form confluent monolayers and exhibit contact inhibition. These cultures are 90–95% pure mTALH cells as shown by Tamm–Horsfall immunofluorescence.

Tamm–Horsfall Protein Identification

Two million freshly isolated mTALH cells are washed with PBS plus calcium. The 3-week-old mTALH cell cultures are grown on coverslips (2 × 2 cm) and washed three times with PBS + calcium, then covered with methanol (20°) for 30 min. Methanol is removed and cells are washed with PBS. All washes are repeated three times for 3 minutes each. Using an indirect immunofluorescent method, 200 μl of goat antiserum to human Tamm–Horsfall protein [anti-Tamm–Horsfall protein (Orgamon Teknika

Corp., West Chester, PA) 1/5 dilution of reconstituted stock] is incubated with the cells for 30 min at 37°. Controls are incubated with 200 μl of rabbit anti-goat IgG (1 mg/ml). The cells are washed and incubated with 200 μl of fluorescein-conjugated rabbit antiserum to goat IgG (1/16 dilution of reconstituted stock) for 30 min at 37°. This second fluorescent antibody is used to tag the specific Tamm–Horsfall antibody. Those cells with an abundance of fluorescence (relative to controls) assessed by fluorescent microscopy are judged to be mTALH cells.

Arachidonic Acid Metabolism in Freshly Isolated and Cultured mTALH Cells

Freshly isolated mTALH cells (3×10^6/ml) are incubated immediately with AA in PBS with calcium. Three-week-old cultures are washed twice with RPMI 1640 media and scraped from culture dishes. Cells are centrifuged for 5 min (700 *g*), counted, and resuspended in PBS with calcium at a density of 3×10^6 cells/ml.

Arachidonic Acid Metabolism in Whole Cells

Either freshly isolated or cultured mTALH cells (3×10^6 cells/tube) are incubated in a shaking water bath at 37° for 30 min with 0.4 μCi ^{14}C-labeled AA (7 μM; specific activity 56 mCi/mmol) using uncapped glass test tubes. In experiments using inhibitors, agents are preincubated with the cells 10 min before the addition of ^{14}C-labeled AA. SKF-525A (100 μM) is dissolved in water and diluted with PBS. Indomethacin (10 μM) is dissolved in 4.2% $NaHCO_3$ and diluted with PBS. The reaction is terminated by acidification to pH 3.5–4.0, and acidic lipids are extracted at 90% efficiency by 2 volumes of ethyl acetate. The extracts are dried under nitrogen and kept at −70°. Product formation is assessed by TLC on silica gel G plates using a solvent system of ethyl acetate : isooctane : acetic acid : water (110/50/20/100, v/v/v/v). The approximate R_f value for AA is 0.79. Radioactive zones are visualized by autoradiography, cut, and evaluated in a liquid scintillation counter. The results are expressed as micrograms of AA converted/milligram protein/30 min. mTALH cells convert 0.1–0.3 μg AA/mg protein to cytochrome *P*-450-dependent metabolites.

Arachidonic Acid Metabolism in Broken Cell Preparations

Cells are resuspended in 200 μl of distilled water and disrupted by freezing and thawing for three cycles. The broken cells are then divided into samples (0.8–1.0 mg protein/sample) and resuspended in 1 ml PBS and calcium, pH 7.4. Samples are incubated with the NADPH-generating system (glucose 6-phosphate, 0.1 mM, NADP$^+$ 0.4 μM, glucose-6-phosphate dehydrogenase, 1 unit) and ^{14}C-labeled AA (0.4 μCi/sample, 7 μM, specific activity 56 mCi/mmol) for 30 min in a shaking water bath at 37°. Inhibitors are added 10 min prior to the addition of ^{14}C-labeled AA and the NADPH-generating system. The reaction termination and radioactive product extraction are performed as previously stated.

Separation of Arachidonic Acid Metabolites by Reversed-Phase HPLC

Following extraction, samples are resuspended in 300 μl methanol and ^{14}C-labeled AA metabolites separated by reversed-phase HPLC on a μBondapak C_{18} column (Whatman, Clifton, NJ), using a linear solvent gradient ranging from water: acetonitrile: acetic acid (1/1/0.01, v/v/v) to acetonitrile: acetic acid (1/0.01, v/v); the rate of change is 1.25%/min at a flow rate of 1 ml/min. Radioactive products are detected using an on-line radioactive flow detector (Radioflowmatic, Tampa, FL). The retention times of peaks I and II are 18.2 (range 18–19) and 15.7 (range 15–16) min, respectively (Fig. 2). Separation by HPLC reveals that the arachidonate metabolites of peaks I and II have retention times different from those of

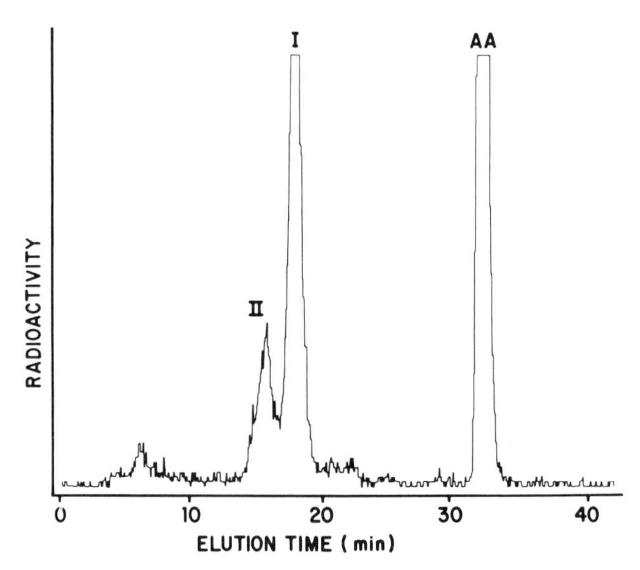

FIG. 2. Reversed-phase HPLC separation of arachidonic acid (AA) metabolites formed by mTALH cells. mTALH cells (3 × 10^6) are incubated with 0.4 μCi ^{14}C-labeled AA in PBS for 30 min at 37°. The supernate is extracted and radioactive metabolites separated using a μBondapak C_{18} column. Peaks I and II have retention times of 18.2 and 15.7 min, respectively.

arachidonate products of lipoxygenases, the mono- and diHETEs. Further, the lack of UV absorbance at 234 nm and 276 nm indicates the absence of conjugated diene or triene structures, respectively, and is further evidence that the oxygenated metabolites of AA are formed by a cytochrome *P*-450-dependent system.

Stimulated Release of mTALH Cell Arachidonic Acid Metabolites

Segmentation of nephron responsiveness to hormones may be expressed through adenylate cyclase linked to specific receptors, presumably via regulatory binding proteins.[13] Arginine vasopressin (AVP)-stimulated adenylate cyclase activity is restricted to the collecting ducts and mTALH. Calcitonin (SCT) also stimulates adenylate cyclase in the mTALH, whereas adrenocorticotropic hormone, parathyroid hormone, and isoprenaline do not. Hormonal stimulation of AA metabolism in mTALH cells is studied by prelabeling with ^{14}C-labeled AA (0.4 μCi per 3×10^6 cells) for 90 min at 37°. Within 90 min, 80–90% of ^{14}C-labeled AA is incorporated into phospholipids, chiefly phosphatidylcholine and phosphatidylethanolamine, and triglycerides. The cell suspension is centrifuged and the cells are then resuspended in fresh PBS and calcium and incubated for an additional 10 min with the following peptide hormones: AVP (4×10^{-10}–4×10^{-7} *M*), and SCT (2×10^{-10}–2×10^{-7} *M*). The reaction is stopped by acidification, cells are removed by centrifugation, and the media extracted with ethyl acetate. AA metabolites are separated by TLC as described above; radioactive zones corresponding to peaks I and II are cut out and counted using a liquid scintillation counter. The percentage increase in the formation of peaks I and II compared with the unstimulated control value is plotted. The control value of 0.060 ± 0.007 μg/mg protein/10 min is considerably lower than the values obtained when the 90–min period of prelabeling is omitted, as it represents release of bound radiolabeled AA which was esterified to lipids during prelabeling.

AVP and SCT, when added to mTALH cells in concentrations ranging from 10^{-10}–10^{-7} *M*, stimulate release of the oxygenated arachidonate metabolites of peaks I and II by as much as 2-fold, depending on the dose of the peptides. In contrast, concentrations of bradykinin and angiotensin II that are three orders of magnitude greater (10^{-6} *M*) have much lesser effects on the release of oxygenated metabolites of AA than either AVP or SCT. The phosphodiesterase inhibitor, 1-isobutyl-3-methylxanthine (10 μ*M*), increases formation of peaks I and II almost 2-fold, as does dibutyryl-cAMP (1 m*M*). Parthyroid hormone and isoprenaline, which do

[13] F. Morel, *Am. J. Physiol.* **240,** F159 (1984).

not affect adenylate cyclase in the mTALH segment, have no effect on AA metabolite formation by mTALH cells at concentrations of 10^{-10} to 10^{-7} *M*.

Concluding Remarks

The centrifugal elutriation method to isolate mTALH cells described here is applicable to the separation of many cell types, especially cells of epithelial origin and blood cells. This technique is reproducible in terms of the purity of the mTALH cell fraction. We have, however, noticed considerable variation in the quantity of mTALH cells isolated, and this is generally associated with a reduction in AA metabolism. This variation is of a seasonal nature, more apparent during the summer months, and may be related to dehydration and/or endemic diseases of the rabbits.

Although we are able to retain and express cytochrome *P*-450 activity in cultured mTALH cells, different NADPH-dependent metabolites are formed compared to peaks I and II produced by freshly isolated cells. The cell culture conditions may have induced a significant change in the expression of the multiple cytochrome *P*-450 isozymes.

Acknowledgments

We wish to thank Jennifer Jones for typing this manuscript and Sallie McGiff for editorial assistance. This work was supported by NHLBI Grants 5 RO1 HL25394 and 5 PO1 HL34300; American Heart Association Grant-in-Aid, 86 11 22 and National Research Service Award, 5 F32 DK 07788.

[42] Ocular Cytochrome *P*-450 Metabolism of Arachidonate: Synthesis and Bioassay

By MICHAL LANIADO SCHWARTZMAN and NADER G. ABRAHAM

The microsomal cytochrome *P*-450 monooxygenases represent the third metabolic pathway of arachidonic acid in animal tissues. The enzyme system is composed of three components: cytochrome *P*-450 hemoprotein, NADPH–cytochrome-*P*-450 (*c*) reductase as the flavoprotein, and phospholipids. The content and activities of the cytochrome *P*-450 enzymes can be regulated by their substrates as inducers as well as by hormones and heme availability. In the presence of NADPH and molecular oxygen, microsomal cytochrome *P*-450 converts arachidonate to several oxygenated metabolites: (1) epoxyeicosatrienoic acids (EETs), which can undergo hydrolysis by epoxide hydrolase to form the corresponding diol derivatives (DHTs); (2) monohydroxyeicosatetraenoic acids (HETEs), and (3) ω- and ω-1-hydroxylated compounds.[1] The formation of these metabolites is inhibited by carbon monoxide, SKF-525A, and antibodies to NADPH–cytochrome-*P*-450 (*c*) reductase and is reduced after induction of heme oxygenase with heavy metals.[2,3] Heme oxygenase is the key enzyme in heme degradation, thus regulating heme availability to several hemoproteins including cytochrome *P*-450. A decrease in microsomal heme levels brought about by heme degradation is a signal for proteolytic degradation of several cytochrome *P*-450 isozymes as well as for modulation of cytochrome *P*-450 gene expression.[4]

Although several ocular tissues such as the retinal pigment epithelium and the iris ciliary body contain relatively high concentrations of the cytochrome *P*-450 hemoprotein and NADPH–cytochrome-*P*-450 (*c*) reductase, the corneal epithelium has the highest specific activity of the cytochrome *P*-450 arachidonate-metabolizing enzymes.[5,6] Microsomal cytochrome *P*-450 of human, bovine, and rabbit corneal epithelium convert arachidonate to four major metabolites separated by HPLC and initially designated as compounds A, B, C, and D (Fig. 1). Two of these metabolites have been identified. Compound C is 12(*R*)-hydroxy-5,8,10,14-eicosatetraenoic acid [12(*R*)-HETE], a potent Na^+,K^+-ATPase inhibitor, and compound D is 12(*R*)-hydroxy-5,8,14-eicosatrienoic acid [12(*R*)DH-HETE], a vasodilator and an angiogenic factor.[7,8]

The assays for cytochrome *P*-450 content, NADPH-cytochrome-*P*-450 (*c*) reductase, and cytochrome *P*-450 arachidonate metabolism including

[1] R. W. Estabrook, N. Chacos, C. Marlin-Wixtrom, and J. Capdevila, *in* "Oxygenases and Oxygen Metabolism" (M. Nozaki, S. Yamamoto, Y. Ishimura, M. J. Coon, L. Ernster, and R. W. Estabrook, eds.), p. 371. Academic Press, New York, 1982.

[2] M. L. Schwartzman, P. Pagano, J. C. McGiff, and N. G. Abraham, *Arch. Biochem. Biophys.* **252,** 635 (1987).

[3] M. Schwartzman, N. G. Abraham, M. A. Carroll, R. D. Levere, and J. C. McGiff, *Biochem. J.* **238,** 283 (1986).

[4] P. Martasek, K. Solangi, A. I. Goodman, R. D. Levere, R. J. Chernick, and N. G. Abraham, *Biochem. Biophys. Res. Commun.* **157,** 480 (1988).

[5] H. Shichi and D. W. Nebert, *in* "Extrahepatic Metabolism of Drugs and Other Foreign Compounds" (T. E. Gram, ed.), p. 333. MTP Press, Lancaster, 1980.

[6] M. L. Schwartzman, J. Masferrer, M. W. Dunn, J. C. McGiff, and N. G. Abraham, *Current Eye Res.* **6,** 623 (1987).

[7] M. L. Schwartzman, M. Balazy, J. Masferrer, N. G. Abraham, J. C. McGiff, and R. C. Murphy, *Proc. Natl. Acad. Sci. U.S.A.* **84,** 8121 (1987).

[8] R. C. Murphy, J. R. Falck, S. Lumin, P. Yadagiri, J. A. Zirrolli, M. Balazy, J. L. Masferrer, N. G. Abraham, and M. L. Schwartzman, *J. Biol. Chem.* **263,** 17197 (1988).

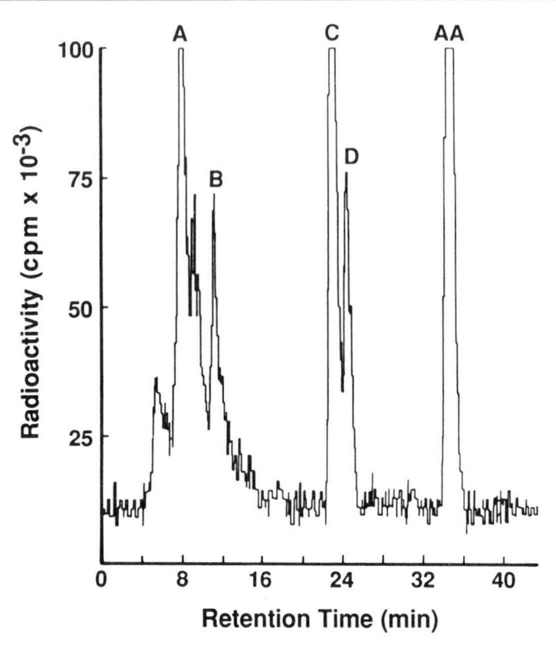

FIG. 1. Reversed-phase HPLC separation of the corneal cytochrome *P*-450 arachidonate metabolites.

isolation and bioassay of the metabolites, 12(*R*)-HETE and 12(*R*)-DH-HETE, are described here. These assays, although described for the corneal epithelium, can be applied to other extrahepatic tissues which have relatively small amounts of spectrally detectable cytochrome *P*-450 hemoprotein. Since arachidonic acid can be metabolized by several enzymatic and nonenzymatic pathways, it is important to determine whether, in a given tissue, the cytochrome *P*-450 system is present by measuring its content and dependent monooxygenase activities.

Preparation of Microsomes

Freshly enaculated calf eyes are obtained from the local abattoir, transported on ice, and dissected within 3 hr after the animals are slaughtered. The corneal epithelium is scraped off with a razor blade and placed in ice-cold 0.1 *M* Tris-HCl buffer, pH 7.6, containing 0.25 *M* sucrose. The epithelium is homogenized in the same buffer using a glass tissue grinder operated at low speed with a Con-Torque power unit (Ederbach, Ann Arbor, MI). Microsomes are obtained by sequential centrifugation steps at 1500 *g* for 10 min, 10,000 *g* for 20 min, and 100,000 *g* for 60 min at 4°. In

TABLE I
CYTOCHROME *P*-450 SYSTEM IN BOVINE CORNEAL EPITHELIAL MICROSOMES

Enzyme	Concentration[a]
Cytochrome *P*-450	16.1 pmol/mg
Aryl hydrocarbon hydroxylase	0.56 pmol/mg/min
7-Ethoxycoumarin *O*-deethylase	5.45 pmol/mg/min
NADPH–cytochrome-*P*-450 (*c*) reductase	164 pmol/mg/min

[a] Results are mean of three determinations: the S.E. was less than 10%.

order to eliminate hemoglobin, which interferes with the spectral measurement of cytochrome *P*-450, the microsomal pellet is washed with 0.1 *M* Tris-HCl buffer, pH 7.6, containing 0.1 *M* pyrophosphate. Protein concentration is measured using the Bio-Rad protein assay (Bio-Rad Chemical Division, Richmond, CA) based on the method of Bradford.[9] Using this procedure, 50 calf eyes yield about 20–30 mg of microsomal protein. Using human tissues, we found that addition of 0.4 m*M* phenylmethylsulfonyl fluoride (PMSF) to the homogenization buffer prevented additional loss of enzyme activity.

Measurement of Cytochrome *P*-450 Content

Cytochrome *P*-450 content is measured from the reduced carbon monoxide difference spectrum. Protein solutions (1–2 mg/ml) are bubbled for 1 min with carbon monoxide. In order to correct for hemoglobin and mitochondrial cytochromes contamination, the solution is divided between sample and reference curvettes to which freshly prepared NADH solution is added at a final concentration of 0.1 m*M*. The sample cuvette is then reduced using a few grains of sodium dithionite. Excess dithionite appears to cause microsomal degradation with consequent loss of spectrally detectable cytochrome. Although all these precautions are taken, a large peak at 424–427 nm still appears (Fig. 2). The spectra are determined with a DW-2C Aminco dual-beam spectrophotometer. The absorbance difference between 450 and 490 nm is used to calculate the cytochrome *P*-450 content with a molar extinction coefficient of 91 mM^{-1} cm^{-1} (Table I).[10]

[9] M. Bradford, *Anal Biochem.* **72,** 248 (1976).
[10] T. Omura and R. Sato, *J. Biol. Chem.* **239,** 2370 (1964).

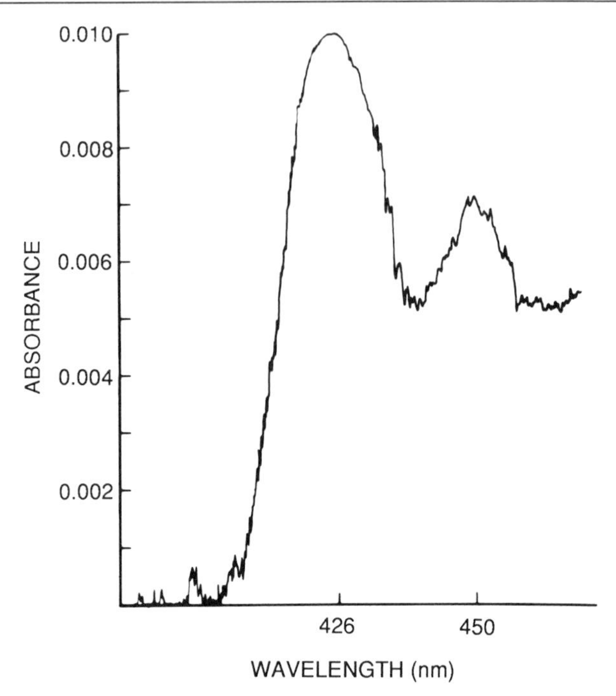

FIG. 2. The reduced carbon monoxide difference spectrum of the corneal epithelial microsomes. [From Schwartzman *et al.*, *Biochem. Biophys. Res. Commun.* **132,** 343 (1985).]

Measurement of NADPH–Cytochrome-*P*-450 (c) Reductase

The reaction mixture for cytochrome *c* reduction contains 0.1 m*M* NADPH, microsomal preparations (0.5–1 mg protein), and the acceptor cytochrome *c* (0.1 m*M*) in a final volume of 1 ml of 0.3 *M* potassium phosphate buffer, pH 7.7. The reaction is initiated by addition of 0.1 ml of a freshly prepared solution of 1 m*M* NADPH in potassium phosphate buffer. The reaction is monitored at 30° by following the reduction of cytochrome *c* at 550 nm during the linear portion of the reaction. The amount of reduced cytochrome *c* is calculated using a molar extinction coefficient of 21 mM^{-1} cm^{-1} (Table I).[11]

Cytochrome *P*-450 Arachidonate Metabolism: Incubation and Extraction Procedures

Determination of cytochrome *P*-450 arachidonate metabolites is based on chromatographic (TLC or reversed-phase HPLC) separation of radio-

[11] Y. Yasukochi and B. S. S. Masters, *J. Biol. Chem.* **251,** 5331 (1976).

TABLE II
EFFECT OF INHIBITORS OF ARACHIDONATE METABOLISM ON NADPH-DEPENDENT CONVERSION IN BOVINE CORNEAL MICROSOMES

Inhibitor	Activity[b] (nmol/mg/30 min)
None	0.33 ± 0.10
NADPH	2.65 ± 0.25
NADPH/carbon monoxide	0.72 ± 0.17
NADPH/SKF 525A (100 μM)	1.05 ± 0.28
NADPH/anti-cytochrome *c* IgG (1 mg)	0.63 ± 0.21
NADPH/indomethacin (10 μM)	2.78 ± 0.30
NADPH/anti-*P*-450-AA IgG (1 mg)[a]	0.52 ± 0.19

[a] Polyclonal antibodies against cytochrome *P*-450-AA epoxygenase.[14]
[b] Results are means ± S.E., n=3.

active products formed in the presence of NADPH. The incubation mixture contains 0.1 *M* potassium phosphate buffer, pH 7.6, microsomal suspension (1–3 mg protein), [^{14}C]arachidonic acid (AA) (0.2 μCi, specific activity of 56 mCi/mmol), and NADPH (1 m*M*) or NADPH-generating system composed of: NADP$^+$ (0.1 m*M*), glucose 6-phosphate (0.4 m*M*), glucose-6-phosphate dehydrogenase (1 unit). The reaction is carried out for 30 min at 37° with continuous shaking. Control incubates of microsomes without NADPH and in the presence of metabolic inhibitors should be included in order to distinguish between cytochrome *P*-450 and cyclooxygenase metabolites (Table II). The reaction is terminated by acidification to pH 4.0–4.5 with 1 *M* citric acid. Radioactive metabolites are extracted twice with 2 volumes of ethyl acetate and the final extract evaporated and reconstituted in 100 μl methanol. Cytochrome *P*-450 metabolite formation can be quantitated by TLC. However, separation of all the products is best achieved by reversed-phase HPLC.

Separation of Arachidonic Acid Metabolites

Separation by TLC

The ethyl acetate extracts are reconstituted in 100 μl methanol and applied onto silica gel G plates (Brinkmann, Westbury, NY) under a continuous stream of nitrogen. Arachidonate metabolites are separated using the A9 solvent system (organic phase of ethyl acetate : isooctane : acetic acid : H_2O, 110 : 50 : 20 : 100, v/v/v/v), visualized by autoradiography on Kodak XAR-5 film and their migration compared to that of known standards (Fig. 3). Radioactive zones are cut and quantitated by a liquid scintillation counter (Beckman Model LS 1801). Product formation is expressed as the percentage of total radioactivity recovered or nanomoles [^{14}C]arachidonate converted.

Separation by HPLC

The ethyl acetate extracts are reconstituted in 100 μl of methanol for separation of the AA metabolites by HPLC. Reversed-phase HPLC is performed on a 5-μm Radial-Pak C_{18} column (Waters, Milford, MA) using a linear gradient solvent system of 1.25%/min from acetonitrile : water : acetic acid (50 : 50 : 1, v/v/v) to acetonitrile : acetic acid (100 : 0.1) at a flow rate of 1 ml/min over 40 min. The elution profile of the products is monitored by radioactivity using an on-line radioactive detector (Radiomatic Instrument & Chemical Co., Inc., Tampa, FL) and by UV absorbance at 237 nm. Using this procedure, four major radioactive metabolites are separated. The unpolar metabolites are designated on Fig. 1 as compounds C and D and are 12(*R*)-HETE and 12(*R*)-DH-HETE, respectively. The polar metabolites, compounds A and B, represent the degradative metabolites of C and D (unpublished data).

Although compounds C and D have a similar chemical structure, they possess quite different biological profiles and can be distinguished by several simple bioassays.

Bioassay for 12(*R*)-HETE and 12(*R*)-DH-HETE

Large quantities of these compounds can be obtained by incubating corneal epithelial microsomes (3 mg/ml) with a mixture of ^{14}C-labeled arachidonic acid (0.05 μCi/ml) and unlabeled arachidonate (10 μg/ml) and NADPH. Extraction and purification are the same as described above; however, rechromatographing the isolated compounds is necessary to achieve purity. The compounds can be aliquoted, sealed dry under nitrogen, and stored at −70°. Under these conditions their biological activities remain unchanged for 2–3 months.

Na^+,K^+-ATPase Assay

Partially purified Na^+,K^+-ATPase enzyme from bovine corneal epithelium is prepared as described by Jorgensen.[12] Briefly, microsomes are

[12] P. L. Jorgensen, this series, Vol. 32, p. 277.

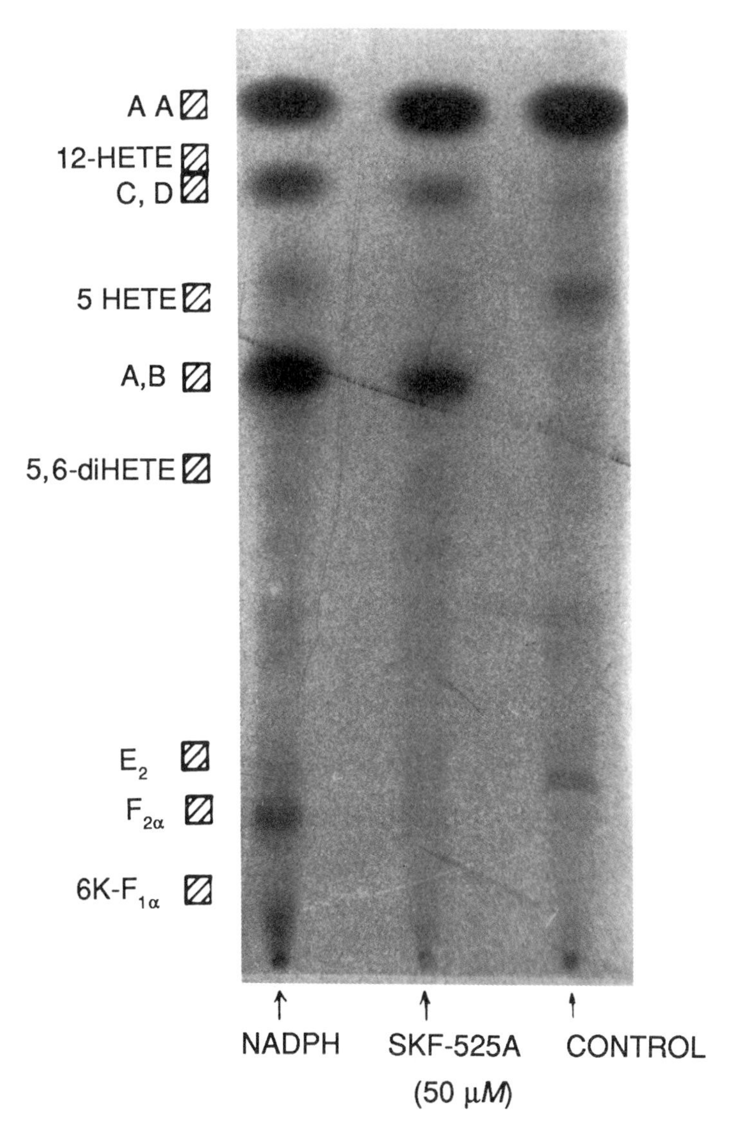

FIG. 3. Autoradiograph of thin-layer chromatography separation of corneal arachidonate metabolites (A, B, C, and D) formed in the presence and absence of NADPH (1 mM) and SKF 525A (100 μM). [From Schwartzman *et al.*, *Biochem. Biophys. Res. Commun.* **132,** 343 (1985).]

solubilized with sodium dodecyl sulfate (SDS) as follows: microsomes (16 mg of protein) are incubated with 6.4 mg SDS, 2 m*M* EDTA, 50 m*M* imidazole, and 3 m*M* ATP in 0.25 *M* sucrose and 30 m*M* histidine buffer, pH 7.2, in a final volume of 7 ml, for 45 min at room temperature with continuous stirring. The solubilized microsomes are then applied on discontinuous density gradients and centrifuged at 110,000 *g* for 90 min at 4°; the gradient consists of three successive layers of sucrose: 29.4, 15, and 10% (w/v). The pellet is resuspended in 25 m*M* imidazole and 1 m*M* EDTA, pH 7.5. The enzyme suspension is divided into aliquots of 100 μg/ml, and immediately placed in liquid nitrogen for storage at −70°. The enzyme is stable and remains active for more than 6 months.

The activity of Na^+,K^+-ATPase assay is measured as the rate of release of inorganic phosphate. The incubation mixture contains 30 m*M* histidine, 20 m*M* KCl, 130 m*M* NaCl, 3 m*M* $MgCl_2$, and 3 m*M* ATP, pH 7.5, in a final volume of 0.5 ml. Enzyme concentrations range between 0.25–5 μg per incubation. Aliquots of the compounds to be tested are dried under nitrogen and resuspended in the reaction mixture without ATP. Ouabain is dissolved in warm water followed by 15-min sonication. Following a 10-min preincubation, ATP is added to start the reaction. The reaction is carried out for 15 min at 30°. During these periods no more than 10% of the substrate is utilized by the enzyme. The reaction is terminated by placing the tubes on ice and adding 0.8 ml of color reagent.[13] The color reagent is prepared as follows: a mixture of 0.045% malachite green and 4.2% ammonium molybdate in 4 *N* HCl (3 : 1) is filtered and 0.1 ml Sterox (nonionic synthetic detergent by Monsanto, St. Louis, MO) added to the 5 ml mixture. One minute after adding the color reagent, 100 μl of 34% (w/v) sodium citrate solution is added and mixed. The final solution is then analyzed at 660 nm in a Beckman DB spectrophotometer using references containing boiled enzyme with all reagents. A standard curve is constructed using inorganic phosphate. Enzyme activity is expressed as micromoles of inorganic phosphate released per hour per milligram of protein. Since the ouabain-sensitive Na^+,K^+-ATPase is defined as the enzyme activity that is inhibited by ouabain, a dose–response curve of ouabain inhibition should be constructed with each preparation. For the corneal epithelial preparation, 66% of the released phosphate is ouabain-sensitive whereas under the same condition in the rat kidney, 77% is ouabain-sensitive Na^+,K^+-ATPase.

Figure 4 shows the dose–response inhibition curve of corneal Na^+,K^+-

[13] P. A. Lanzetta, L. J. Alvarez, P. S. Reinach, and O. A. Candia, *Anal. Biochem.* **100,** 95 (1979).

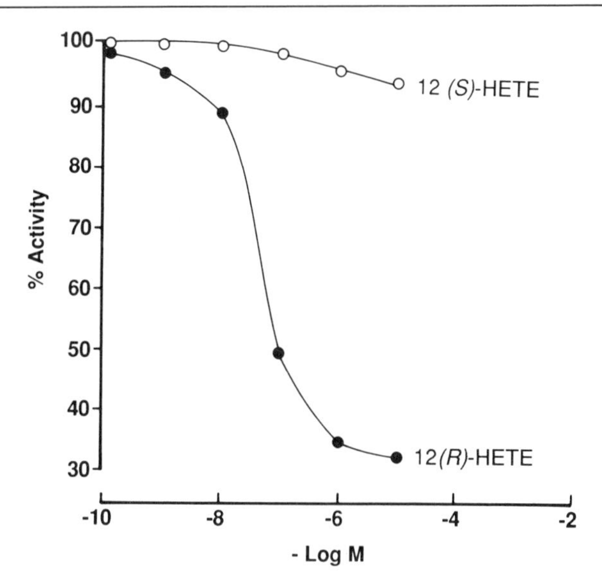

FIG. 4. Effect of 12(*R*)- and 12(*S*)-HETE on the activity of partially purified corneal Na^+,K^+-ATPase.

ATPase by 12(*R*)-HETE. 12(*R*)-HETE also inhibits rat kidney and rat heart Na^+,K^+-ATPases with an IC_{50} of 10^{-6} *M* (unpublished data).

Aqueous Humor Protein Concentration

Protein content in the rabbit aqueous humor is normally low and ranges between 0.3 and 0.7 mg/ml, indicating an intact blood aqueous barrier. Exposure to compounds which compromise the integrity of the blood aqueous barrier results in an increase in protein content of the aqueous humor.

New Zealand White rabbits weighing about 2.5 kg are anesthetized with a mixture of ketamine-HCl (50 mg/kg body weight) and xylazine (10 mg/kg body weight). In addition, 1 drop of proparacaine-HCl 0.5% is applied to each eye. Varying doses of compound dissolved in phosphate-buffered saline, pH 7.4, are injected through the cornea into the anterior chamber of one eye in a final volume of 10 μl using a 27-gauge needle (Hamilton syringe). The other eye receives an equal volume of buffer. Extreme care is taken to ensure that during injections the needle does not touch the iris. After 15 min, samples of aqueous humor are removed from the anterior chamber of both eyes and the protein content measured using

TABLE III
EFFECT OF INTRACAMERAL INJECTION OF CORNEAL CYTOCHROME *P*-450 ARACHIDONATE METABOLITES ON AQUEOUS HUMOR PROTEIN

Intracameral injection[a]	Aqueous humor protein[b] (mg %)
None	80 ± 41
12(*R*)-DH-HETE	1090 ± 166[c]
12(*S*)-DH-HETE	282 ± 125
12(*R*)-HETE	70 ± 20
12(*S*)-HETE	130 ± 40

[a] 10 ng.
[b] Results are mean ± SE, n=4.
[c] $p < 0.01$.

the Bio-Rad protein. Table III summarizes the effects of 12(*R*)-HETE and 12(*R*)-DH-HETE and their isomers on the aqueous humor protein concentration, an index of membrane permeability.

Topical Vasodilation

New Zealand White male rabbits weighing about 2.5 kg are used. Aliquots of compound are resuspended in phosphate-buffered saline, pH 7.4, to a concentration of 2 μg/ml. The animals are treated by topically applying no more than 25 μl containing 5–50 ng of the compound onto the superior rectus muscle in the left eye, while holding the lids open. The same volume of buffer is applied to the right eye and is used as a vehicle control. Photographs of the limbal vessels are obtained 2, 5, and 20 min after topical application of the compound with the use of Zeiss slit-lamp stereomicroscope fitted with a Contax RTS-35 mm camera and an electronic flash (Fig. 5).

Neovascularization Assay: Preparation of Elvax 40P Implants

The polymer Elvax 40P (DuPont Chemical Co., Universal City, CA) is washed in 95% ethanol with continuous stirring for 1 week, after which approximately 100 mg of the polymer is dissolved in 1 ml dichloromethane (10% solution, w/v). Once the compound is dissolved in the Elvax solution the mixture is quickly frozen in an acetone–dry ice bath for 10 min. The

[14] M. Laniado Schwartzman, K. Davis, J. C. McGiff, R. D. Levere, and N. G. Abraham, *J. Biol. Chem.* **263,** 2536 (1988).

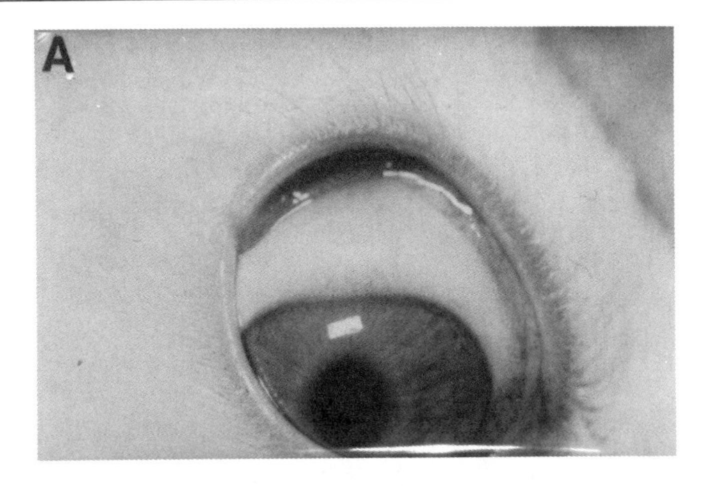

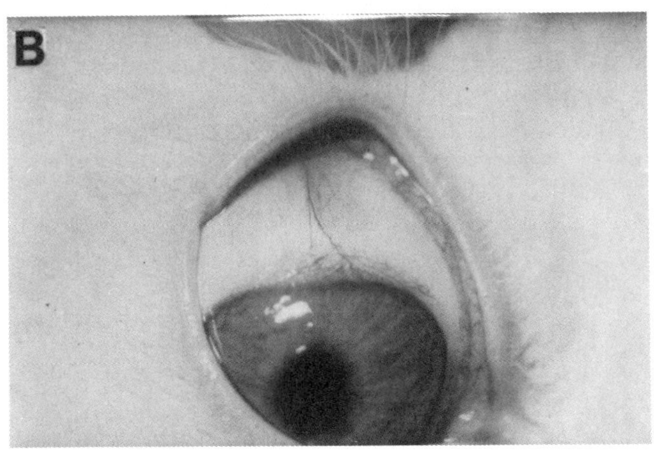

FIG. 5. Effect of 12(R)-DH-HETE on vascular reactivity in the rabbit eye. (A) Vehicle control; (B) 10 ng of 12(R)-DH-HETE. Pictures were taken 5 min after the compound was added.

frozen pellet is then cut into small pieces of 1 to 2 mm. Each piece weighs about 7–10 mg and contains between 0.5 to 1.5 μg of the tested compound. The pellets are transferred to a vial and kept for 2 days at room temperature under a mild vacuum. Rabbits are anesthetized with a mixture of ketamine-HCl (50 mg/kg body weight) and xylazine (10 mg/kg body weight). In addition, a drop of 0.5% of proparacaine-HCl is applied to the eye. The Elvax pellets with or without compound are inserted into pockets made between the corneal epithelium and the stroma. The bottom of the

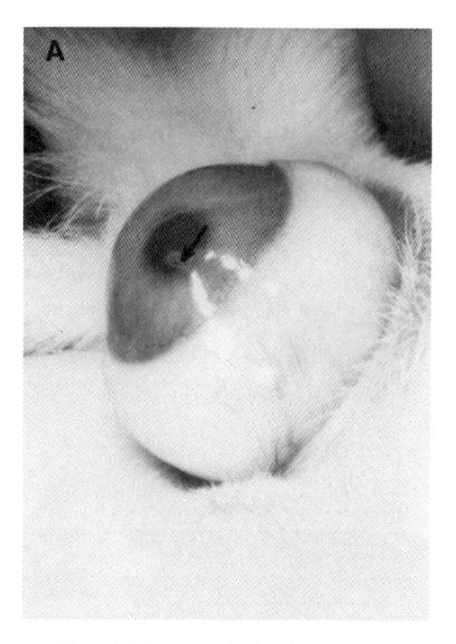

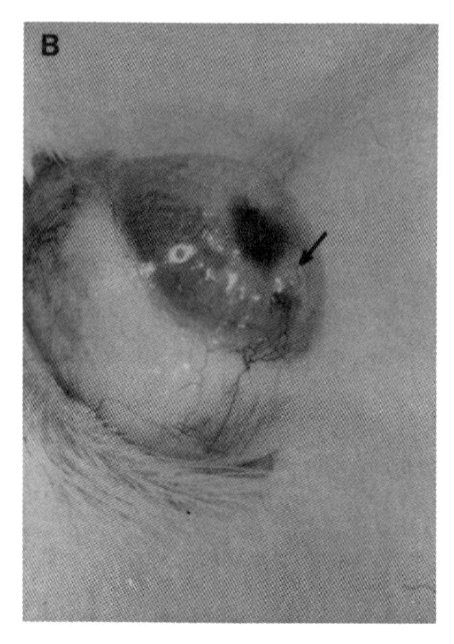

FIG. 6. Neovascularization response to 12(*R*)-DH-HETE. (A) 1 μg; (B) vehicle control. The arrow indicates the inserted Elvax 40P pellet.

pockets are 2 mm from the limbus, and the pellets occupy the interval between 1 and 2 mm. Eyes are examined daily for the presence of hyperemia in the limbal vessels, corneal edema, changes in corneal transparency, and the presence of any new vascular growth. The sustained growth of some vessels from the limbal vasculature toward the implant and ultimately enfolding it constitute a positive neovascular response. Photomicrographs are taken with the use of a Zeiss slit-lamp stereomicroscope, fitted with a Contax RTS-35 mm camera and an electronic flash (Fig. 6).

Acknowledgments

The authors wish to thank P. Blank for typing the manuscript. This work was supported in part by grants EY 06513, AM 29742, and HL 34300 from the National Institutes of Health.

[43] Cytochrome P-450 Arachidonate Oxygenase

By JORGE H. CAPDEVILA, J. R. FALCK, ELIZABETH DISHMAN, and ARMANDO KARARA

Cytochrome P-450, the oxygen-activating component of the microsomal electron transport system, is an efficient catalyst for the oxygenated metabolism of arachidonic acid (AA). The reaction has an absolute requirement for NADPH and proceeds according to the following stoichiometric relationship:

$$AA + NADPH + O_2 + H^+ \rightarrow AAOH + NADP^+ + H_2O$$

Catalytic turnover requires electron transfer from NADPH to the heme iron center of cytochrome P-450. This step is catalyzed by the microsomal membrane-bound flavoprotein, cytochrome P-450 reductase.

The multiple cytochrome P-450 isoenzyme complex, present in microsomal fractions isolated from the livers of phenobarbital-treated rats, oxidizes the fatty acid via three types of reactions: (a) allylic oxidation (lipoxidase-like) to generate six regioisomeric hydroxyeicosatetraenoic acids (HETEs), (b) ω/ω-1 oxidation (ω/ω-1-oxidase) to generate the corresponding 19- and 20-hydroxyeicosatetraenoic acids (ω and ω-1 alcohols), and (c) olefin epoxidation (epoxygenase) to generate four regioisomeric epoxyeicosatrienoic acids (EETs). The relative contribution of these reactions to the overall metabolism of the fatty acid is highly dependent on the tissue source of the microsomal enzymes, the animal hormonal status, or its exposure to several cytochrome P-450 inducers. Work from different laboratories has demonstrated the potent *in vitro* biological activities of several of the cytochrome P-450-derived AA metabolites.[1] Based on the documentation of the EETs as endogenous constituents of several organ tissues, it has been proposed that the cytochrome P-450 epoxygenase may represent an alternate pathway for the *in vivo* metabolism of AA.[1]

We will describe methodology utilized in our laboratory for the biochemical characterization of the hepatic microsomal AA cytochrome P-450 oxidase as well as its reconstitution utilizing solubilized and purified components of the microsomal electron transport system.

[1] F. A. Fitzpatrick and R. C. Murphy, *Pharmacol. Rev.* **40,** 229 (1989).

Microsomal Cytochrome *P*-450 AA Oxygenase

Isolation of Rat Liver Microsomal Fractions

Hepatic microsomal fractions, containing between 0.6–0.9 nmol of cytochrome *P*-450/mg of protein, with little or no cross-contamination by hemoglobin or mitochondrial pigments are isolated as described by Remmer *et al.*[2] Briefly, after animal sacrifice by decapitation, the liver is perfused *in situ* through the portal vein with ice-chilled 0.15 *M* KCl until it decolorizes to a yellowish-brown tone. The organ is then removed free of connective tissue, weighed, minced, and homogenized in 0.25 *M* sucrose, 0.01 *M* Tris-Cl (pH 7.4) (15 ml/g of wet tissue) utilizing a glass–Teflon homogenizer. After 10 gentle passes of the pestle (60–80 rpm), the resulting suspension is successively centrifuged, once at 3000 and twice at 12,000 *g* for 20 min each time. Microsomal fractions are collected from the 12,000 *g* supernatants by centrifugation at 100,000 *g* for 60 min. To decrease contaminant cytosolic enzymes, particularly cytosolic epoxide hydrolases, the 100,000 g pellet is carefully resuspended in 0.15 *M* KCl, using a hand-driven glass–Teflon homogenizer and the microsomal fractions pelleted by spinning 60 min at 100,000 g. The reddish-brown microsomal pellet is finally suspended in 0.25 *M* sucrose, 0.01 *M* Tris-Cl (pH 7.4)(20–30 mg microsomal protein/ml), and maintained at 4°. During isolation, the temperature should be maintained at 2–4°. To eliminate the generation of artifactual, high-polarity AA metabolites, the microsomal suspensions must be discarded after 40 hr of storage. For extended storage (2–3 months), it is recommended to maintain the microsomal fractions as 100,000 *g* pellets at −80°.

The described methodology has also been applied to the isolation of microsomal fractions from rat and/or rabbit kidneys, lungs, brain, pituitaries, ovaries, and canine aorta. When utilizing tissues resistant to homogenization such as the lung or aorta, it is convenient to disrupt the tissue initially with an Ultra-Turrax (Tekman Co., Cincinnati, OH) tissue grinder. The suspension of minced tissue in 0.25 *M* sucrose, 0.01 *M* Tris-Cl (pH 7.4)(15 ml/g wet tissue) is disrupted for 30 sec at 50% full output power, prior to the glass–Teflon homogenization step.

Assay Method

The microsomal metabolism of AA is studied by incubating radiolabeled fatty acid with the microsomal enzymes at pH 7.5 in the presence of NADPH and an NADPH-regenerating system. Aliquots of the incuba-

[2] H. Remmer, H. Greim, J. B. Schenkman, and R. W. Estabrook, this series, Vol. 10, p. 703.

tion mixture are taken at different time points; the reaction products are extracted into an organic solvent, resolved by reversed-phase HPLC, and quantified by liquid scintillation counting. Incubations are performed in open vessels and under constant mixing to assure proper aeration.

Reagents

Preparation of AA Substrate. The presence of contaminant hydroperoxides in commercial samples of radiolabeled and/or nonlabeled AA can result in artifactual, NADPH-independent, formation of oxidized products linked to the peroxidase activity of microsomal cytochrome P-450.[3] Samples of 10 to 20 mg of AA (Nu-Chek Prep, Elysian, MN) can be purified by HPLC on a μBondapak C_{18} column (7.8 × 300 mm, Waters Assoc., Milford, MA). The fatty acid (R_t 32 min) is eluted with a linear solvent gradient from acetonitrile/water/acetic acid (49.9/49.9/0.1, v/v/v) to acetonitrile/acetic acid (99.9/0.1, v/v) over 40 min and at a flow of 3 ml/min. After collection and solvent evaporation, the neat AA can be stored under argon at −80° for 1 month. Radiolabeled samples of [1-^{14}C]AA (55–60 mCi/mmol) are purified, immediately prior to utilization, by chromatography in a silica column (5 × 10 mm) equilibrated with hexane/acetic acid (99.5/0.5, v/v). The AA is loaded and eluted with hexane/acetic acid (99.5/0.5, v/v). Stock solutions of 25 mM sodium arachidonate are prepared by suspending the purified AA in argon-saturated 0.1 M Tris-Cl (pH 8.0). After mixing, small aliquots of 1 N KOH are added until a homogenous, opalescent solution is obtained. The volume is then adjusted to the desired concentration with 0.1 M Tris-Cl (pH 8.0). The AA solution is kept under an argon blanket during the mixing and until use.

Incubation Buffer. The buffer consists of 0.05 M Tris-Cl (pH 7.5), 10 mM $MgCl_2$, 0.15 M KCl, containing 8 mM sodium isocitrate and 0.5 IU of isocitrate dehydrogenase (Type IV from Porcine Heart, Sigma Chemical Co., St. Louis, MO).

NADPH. Stock solutions of 100 mM NADPH in 0.15 M KCl can be maintained at 4° for 1 week. NADPH solutions should not be frozen.

Microsomal Incubations

Incubations are performed under atmospheric air at 30° with a Model 3540 Orbit Shaker Bath set at a mixing speed of 200 (Lab Line Inst., Melrose Park, IL). A suspension of rat liver microsomes (0.5–0.75 mg microsomal protein/ml) in the incubation buffer is preincubated for 2 min

[3] R. H. Weiss, J. L. Arnold, and R. W. Estabrook, *Arch. Biochem. Biophys.* **252,** 334 (1987).

prior to the addition of sodium[1-^{14}C]arachidonate (1.2-0.6 μCi/μmol, 100 μM final concentration). Reactions are started, after 1 min, by the addition of NADPH (1 mM, final concentration). Measured aliquots (corresponding to 1–2 mg of total microsomal protein) are withdrawn at 0 min and after 5, 10, and 15 min of incubation. The reaction is stopped by the addition of an equal volume of ethyl ether containing 0.01% BHT and 0.05% acetic acid, vigorously mixed, and centrifuged until phase separation. After collection of the organic phase, the water phases are extracted once more with an equal volume of ethyl ether. The combined organic phases are evaporated under an argon stream.

HPLC Analysis

The reaction products are separated with a 5 μm Dynamax Microsorb C_{18} column (4.6 × 250 mm, Rainin Inst. Co., Woburn, MA) and a linear solvent gradient from initially acetonitrile/water/acetic acid (49.9/49.9/0.1, v/v/v) to acetonitrile/acetic acid (99.9/0.1, v/v) over 40 min and at ml/min. Rates of product formation are determined by liquid scintillation counting with an on-line radioactivity detector, or with the aid of a fraction collector and a liquid scintillation counter. This HPLC step is intended to be functionality specific and to separate the cytochrome *P*-450 metabolites into three groups: ω/ω-1 alcohols, HETEs, and EETs. Since the first group of NADPH-linked primary metabolites, the ω/ω alcohols, elutes between 19 to 21 min, a single chromatogram can provide information concerning the formation of polar metabolites, such as PGs (3–7 min), trihydroxy derivatives (6–12 min), and well as rates of EET hydration to DHET's (16–19 min).

For a detailed analysis of regioisomeric composition, fractions with retention times corresponding to the ω/ω-1 alcohols (19–21 min), HETEs (22–24 min), and EETs (25–27 min) are pooled batchwise from the C_{18} column eluent and, after solvent evaporation, resolved into the individual positional isomers by normal-phase HPLC in a μPorasil column (4.6 × 250 mm, Waters Assoc., Milford, MA) utilizing the following systems: (a) ω/ω-1 alcohols—A linear solvent gradient from hexane/2-propanol/acetic acid (98.9/1.0/0.1, v/v/v) to hexane/2-propanol/acetic acid (96.9/3.0/0.1, v/v/v) in 30 min at 3 ml/min; (b) HETEs—A linear solvent gradient from hexane/2-propanol/acetic acid (99.4/0.5/0.1, v/v/v) to hexane/2-propanol/acetic acid (98.4/1.5/0.1, v/v/v) over 30 min at 3 ml/min; (c) EETs—isocratically with hexane/2-propanol/acetic acid (99.4/0.5/0.1, v/v/v) at 3 ml/min. Retention times for the different metabolites, obtained under the HPLC conditions described, are shown in Table I.

Comparisons of the HPLC properties of the different cytochrome *P*-450-derived metabolites with those of coinjected samples of synthetic

TABLE I
HPLC RETENTION TIMES FOR PRIMARY METABOLITES OF CYTOCHROME P-450 ARACHIDONATE OXYGENASE

Metabolite	Reversed-phase (min)	Normal-phase (min)
20-Hydroxy	20.2	12.9
19-Hydroxy	20.7	9.2
15-HETE	23.9	8.4
11-HETE	24.5	12.9
12-HETE	25.2	7.4
8-HETE	25.6	21.6
9-HETE	25.8	18.9
5-HETE	26.0	32.3
14,15-EET	28.2	5.4
11,12-EET	29.9	5.8
8,9-EET	30.2	7.8
5,6-EET	30.5	12.6
Arachidonic acid	37.6	—

standards should provide the basis for an initial structural characterization. The GC–MS properties of most of the cytochrome *P*-450 AA metabolites are published (1 and cited references, 4–8).

Comments

Several considerations must be taken into account when performing studies of the NADPH-dependent metabolism of AA by microsomal fractions: (1) A substantial portion of the NADPH-supplied electrons are utilized by the hemeprotein(s) for the reduction of O_2 to H_2O_2. The H_2O_2 is then efficiently metabolized by adventitious catalase present in most microsomal fractions. In the absence of proper mixing and aeration, oxygen

[4] J. Capdevila, Y. R. Kim, C. Wixtrom, J. R. Falck, S. Manna, and R. W. Estabrook, *Arch. Biochem. Biophys.* **243,** 8 (1985).

[5] J. M. Boeynaems, A. R. Brash, J. A. Oates, and W. C. Hubbard, *Anal. Biochem.* **104,** 259 (1980).

[6] N. Chacos, J. R. Falck, C. Wixtrom, and J. Capdevila, *Biochem. Biophys. Res. Commun.* **104,** 916 (1982).

[7] R. Toto, A. K. Siddhanta, S. Manna, B. Pramanik, J. R. Falck, and J. Capdevila, *Biochim. Biophys. Acta* **919,** 132 (1987).

[8] E. Oliw, *J. Chromatogr.* **339,** 175 (1985).

can rapidly become limiting. This is of importance when high protein concentrations of microsomal fractions containing high specific contents of cytochrome *P*-450 are utilized. (2) The $NADP^+$ formed during the course of the reaction is an effective inhibitor of the NADPH–cytochrome-*P*-450 reductase. Therefore, an NADPH-regenerating system is utilized to maintain optimal $NADPH/NADP^+$ ratios during the course of the reaction. (3) Several of the metabolites resulting from the cytochrome *P*-450-catalyzed initial oxygenation step (primary metabolites), are substrates for their further metabolism by the heme protein.[9] Thus, as the ratio of AA to primary metabolites decreases with the incubation time, they compete with the AA substrate for the enzyme active site and can, under certain conditions, i.e., high microsomal protein concentrations and long incubation times, function as inhibitors of the fatty acid metabolism. (4) The reported apparent K_m value for AA metabolism by rat liver microsomal fractions is between 20–40 μM. On the other hand, fatty acid concentrations above 150 μM resulted in a gradual loss of activity, presumably due to the known detergent properties of fatty acids.[10]

AA Oxidation by Purified Cytochrome *P*-450 Enzymes

The role of microsomal cytochrome *P*-450 as the catalyst for the NADPH-dependent metabolism of AA was established by reconstituting the reaction utilizing purified components of the microsomal electron transport system.[11–14] Several methods for the purification and characterization of different cytochrome *P*-450 enzymes, of cytochrome b_5, and of NADPH-cytochrome-*P*-450 reductase have been reported (11–17, and cited references).

Reconstitution studies attempt to mimic those physicochemical parameters of the microsomal system essential for a productive catalytic cycle.

[9] J. H. Capdevila, P. Mosset, P. Yadagiri, S. Lumin, and J. R. Falck, *Arch. Biochem. Biophys.* **261,** 122 (1988).

[10] J. Capdevila, N. Chacos, J. Werringloer, R. Prough, and R. W. Estabrook, *Proc. Natl. Acad. Sci. U.S.A.* **78,** 5362 (1981).

[11] J. Capdevila, L. Parkhill, N. Chacos, R. Okita, B. S. Masters, and R. W. Estabrook, *Biochem. Biophys. Res. Commun.* **101,** 1357 (1981).

[12] J. R. Falck, S. Manna, H. R. Jacobson, R. W. Estabrook, N. Chacos, and J. Capdevila, *J. Am. Chem. Soc.* **106,** 3334 (1984).

[13] E. Oliw, P. Guenguerich, and J. Oates, *J. Biol. Chem.* **257,** 3771 (1982).

[14] M. Laniado-Schwartzman, K. Davis, J. McGiff, R. Levere, and N. Abraham, *J. Biol. Chem.* **263,** 2536 (1988).

[15] F. P. Guengerich, *in* "Hepatic Cytochrome P-450 Monooxygenase System" (J. B. Schenkman and D, Kupfer, eds.), p. 497. Pergamon, New York, 1982.

[16] P. Strittmatter, P. Fleming, M. Connors, and O. Corcoran, this series, Vol. 52, p. 97.

[17] Y. Yasukochi and B. S. S. Masters, *J. Biol. Chem.* **251,** 5337 (1976).

Thus, purified forms of cytochrome *P*-450 and NADPH-cytochrome-*P*-450 reductase are mixed, at a given molar ratio, with a sonicated suspension of dilauroylphosphatidylcholine (DLPC) or, alternatively, incorporated into preparations of unilamellar phospholipid vesicles.[18] Microsomal cytochrome b_5 can participate as an alternate electron donor for the oxygen bound, one electron-reduced form of cytochrome *P*-450[19] and, in many instances, its inclusion in the reconstituted system results in increased rates of product formation.[11]

Assay Method

Studies in which the major phenobarbital or ciprofibrate inducible forms of rat liver cytochrome *P*-450 have been reconstituted,[20] have demonstrated that: (*i*) No important differences in the rates or the regiospecificity of AA oxidation were observed when the enzymes were reconstituted in the presence of dilauroyl- or dimyristoylphosphatidylcholine or incorporated into phospholipid unilamellar vesicles as prepared by Yamakura *et al.*[18] (*ii*) The inclusion of cytochrome b_5, purified from rat liver, resulted in significant increases in the rates of product formation, without detectable changes in regioselectivity.[11,12]

Optimal rates of metabolism are obtained, in most cases, when the cytochrome *P*-450 AA oxygenase is reconstituted by mixing in a 1 : 1 : 1 molar ratio purified cytochrome *P*-450, NADPH-cytochrome-*P*-450 reductase, and cytochrome b_5 in the presence of DLPC (0.05–0.10 mg/ml, final concentration), NADPH, and an NADPH-regenerating system. The time course of the reaction is followed by sampling the incubation mixture at different time points, the reaction products extracted into an organic solvent, resolved, and quantified as described above for the microsomal enzymes.

Reagents

The buffer, NADPH-regenerating system and radiolabeled arachidonic acid (5 m*M*), prepared as described earlier (Microsomal Cytochrome *P*-450 AA Oxygenase).

Dilauroylphosphatidylcholine (DLPC), stock solution, prepared by a 30-sec pulse sonication of a 1 mg/ml suspension of DLPC.

Cytochrome P-450s. The major phenobarbital or ciprofibrate inducible forms of rat liver cytochrome *P*-450[20] are purified, to a specific content of 16 to 17.5 nmol cytochrome *P*-450/mg of protein, from the livers of animals

[18] F. Yamakura, T. Kido, and T. Kimura, *Biochim. Biophys. Acta* **649,** 343 (1981).
[19] A. Hildebrandt and R. W. Estabrook, *Arch. Biochem. Biophys.* **143,** 66 (1971).
[20] F. J. Gonzalez, *Pharmacol. Rev.* **40,** 243 (1989).

treated with phenobarbital or ciprofibrate (80 and 40 mg/kg body weight, respectively) for 5 days prior to sacrifice, according to published procedures.[15,21] Cytochrome *P*-450 concentrations are determined as published.[22] Assuming an average molecular weight of 50,000 for the different cytochrome *P*-450 enzymes, a homogenous preparation of cytochrome *P*-450 holoenzyme should have an specific content of 20 nmol heme protein per milligram of protein.

Cytochrome b_5. Rat liver microsomal cytochrome b_5 is purified to homogeneity and quantified according to Strittmatter.[16]

NADPH-Cytochrome Reductase. Rat liver NADPH-cytochrome-*P*-450 reductase is purified to homogeneity and quantified according to Yasokuchi *et al.*[17]

The purified cytochrome *P*-450 enzymes, cytochrome b_5, and NADPH-cytochrome-*P*-450 reductase are kept at −80° as 20 μM stock solutions in 0.1 *M* Tris-Cl (pH 7.5) containing 20% (v/v) glycerol.

Reconstitution and Metabolism

The AA turnover rate for the different cytochrome *P*-450 enzymes, so far reconstituted in our laboratory, varies between 0.8 to 6.0 nmol of product generated per minute per nanomole of cytochrome *P*-450, at 30°. Therefore, experiments are done in a total volume of 1 ml and with cytochrome *P*-450(s) final concentrations of ≤ 1 μM. Under these conditions, the time course of the reaction can be followed by taking three to four 0.25-ml time aliquots.

The AA oxygenase is reconstituted by adding the purified enzymes to 50–100 μg of DLPC (50–100 μl of the 1 mg/ml sonicated stock) in the following order: 50 μl of 20 μM cytochrome *P*-450, 50 μl of 20 μM NADPH-cytochrome-*P*-450 reductase, and 50 μl of 20 μM cytochrome b_5. After gentle mixing, the concentrated mix of phospholipid, heme proteins, and flavoprotein is allowed to stand, at room temperature, for 5 min and then diluted to 1 ml with 150 m*M* KCl, 0.05 *M* Tris-Cl (pH 7.5) containing 8 m*M* isocitrate, isocitrate dehydrogenase (0.25 IU/ml), and 10% (v/v) glycerol. After mixing for 2 min at 30°, 10 μl of 10 m*M* sodium-[1-^{14}C]arachidonate (2 μCi/mmol, 100 μM, final concentration) is added. The reaction is initiated, 1 min later, by the addition of NADPH (1 m*M*, final concentration). Aliquots of the reaction mixture (0.25 ml) are taken at 0, 5, 10, and 15 min and added to test tubes containing 1.5 ml of 0.2 *M* KCl and 2 ml of ethyl ether/0.05% acetic acid with 0.01% BHT. The reaction

[21] P. Tamburini, H. A. Masson, S. K. Bains, R. J. Makowski, B. Morris, and G. C. Gibson, *Eur. J. Biochem.* **139,** 235 (1984).

[22] T. Omura and R. Sato, *J. Biol. Chem.* **239,** 2370 (1964).

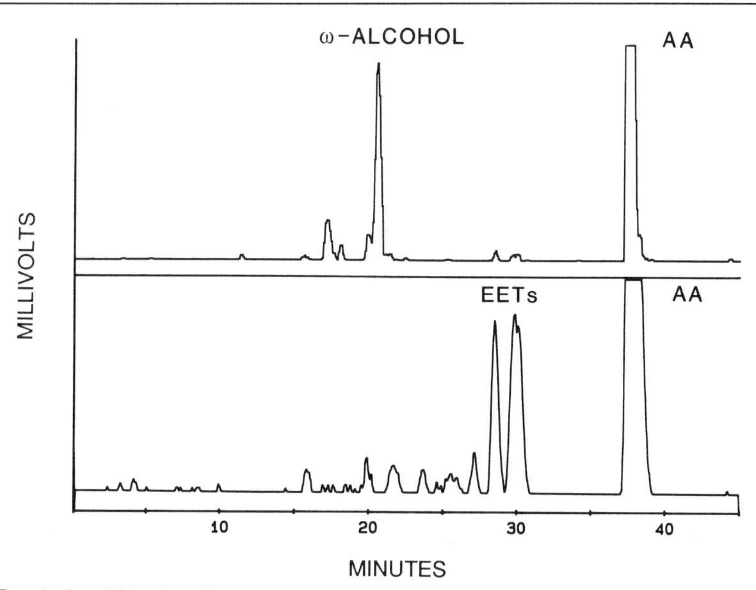

FIG. 1. Arachidonic acid oxidation by purified rat liver cytochrome *P*-450 enzymes. The major ciprofibrate (top)- and phenobarbital (bottom)-inducible forms of rat liver cytochrome *P*-450 are purified to an specific content of 17.6 and 16.0 nmol of cytochrome *P*-450/mg of protein, respectively. The oxygenase activity is reconstituted by mixing cytochrome *P*-450, NADPH-cytochrome-*P*-450 reductase, and cytochrome b_5 (0.5 μM each), in the presence of DLPC (50 μg/ml), sodium[1-^{14}C]arachidonate (0.6 μCi/mmol, 100 μM), NADPH (1 mM), and an NADPH-regenerating system. After 20 min at 30° the reaction products are extracted and analyzed by reversed-phase HPLC. Abscissa: retention time (min). Ordinate: eluent radioactivity expressed as the output (in millivolts) from a Radiomatic Flo-One β-Detector (Radiomatic Instruments, Tampa, FL). To optimize the comparison, the radiochromatograms were normalized to the highest metabolite peak and, therefore, do not correspond to equivalent aliquots of the reaction mixtures.

products are extracted, resolved, and quantified as described above for the microsomal enzymes.

Comments

In addition to the biochemical advantages of working with a defined system, reconstituted systems are excellent tools for the study of the regioselectivity of AA oxidation by different cytochrome *P*-450 enzymes. In the experiment shown in Fig. 1, the AA oxygenase has been reconstituted utilizing either the major phenobarbital-inducible form of rat liver microsomal cytochrome *P*-450 or, alternatively, the major ciprofibrate-inducible form (Fig. 1). As shown by the radiochromatograms, the phenobarbital-inducible enzyme is an active epoxygenase, generating four regioisomeric EETs in an enantioselective manner.[12] On the other hand, the ciprofibrate-inducible form catalyzes, almost exclusively, the ω-oxidation of the fatty acid to form 20-hydroxyeicosatetraenoic acid.

Acknowledgments

This work was supported by NIH grants GM 37922 and DK 38226.

Section III

Pharmacology: Antagonist and Synthesis Inhibitors

A. Prostaglandins
Articles 44 and 45

B. Leukotrienes
Articles 46 and 47

C. Platelet-Activating Factor
Articles 48 through 50

[44] Radioligand Binding Assays for Thromboxane A_2/Prostaglandin H_2 Receptors

By PERRY V. HALUSHKA, THOMAS A. MORINELLI, and DALE E. MAIS

Both thromboxane A_2 (TxA_2) and prostaglandin H_2 (PGH_2) induce platelet aggregation and constriction of vascular smooth muscle via interaction at membrane-bound receptors.[1] Since TxA_2 and PGH_2 share these common pharmacological effects, it is currently thought that they share a common receptor which has been designated TxA_2/PGH_2.[2]

The study of these receptors was impeded by the unstable nature of both TxA_2 and PGH_2. This necessitated the synthesis of stable analogs which allowed for pharmacological and radioligand binding studies to proceed.[3-6] Some of these derivatives have been radiolabeled with either ^3H or ^{125}I and used for the qualitative and quantitative study of TxA_2/PGH_2 receptors. The chemical structures of the ligands that have been utilized are shown in Fig. 1. Both [^3H]U46619 (Refs. 7–9) and [^{125}I]BOP (Ref. 10) are TxA_2/PGH_2 mimetics while the remaining analogs are antagonists.[11-13] Several of these ligands are currently commercially available: [^3H]U46619, [^3H]SQ29548, and [^{125}I]PTA-OH.

[1] S. Moncada and J. R. Vane, *Pharmacol. Rev.* **30,** 293 (1979).

[2] D. L. Saussy, D. E. Mais, D. Knapp, and P. V. Halushka, *Circulation* **72,** 1201 (1985).

[3] G. Bundy, *Tetrahedron Lett.* **24,** 1957 (1975).

[4] E. Corey, K. Nicolaou, Y. Machedra, C. Malmsten, and B. Samuelsson, *Proc. Natl. Acad. Sci. U.S.A.* **72,** 3355 (1975).

[5] K. Nicolaou, R. Magolda, J. Smith, D. Aharony, E. Smith, and A. Lefer, *Proc. Natl. Acad. Sci. U.S.A.* **76,** 2566 (1979).

[6] K. Nicolaou, R. Magolda, and O. Claremon, *J. Am. Chem. Soc.* **102,** 1404 (1980).

[7] N. Liel, D. E. Mais, and P. V. Halushka, *Prostaglandins* **33,** 789 (1987).

[8] E. J. Kattelman, D. L. Venton, and G. C. Le Breton, *Thromb. Res.* **41,** 471 (1986).

[9] T. A. Morinelli, S. Niewiarowski, J. Daniel, and J. B. Smith, *Am. J. Physiol.* **253,** H1035 (1987).

[10] T. Morinelli, J. E. Oatis, A. K. Okwu, D. E. Mais, P. Mayeux, A. Masuda, D. Knapp, and P. V. Halushka, *J. Pharmacol. Exp. Ther.* **251,** 557 (1989).

[11] K. Hanasaki, K. Nakano, H. Kasai, H. Kurihara, and H. Arita, *Biochem. Biophys. Res. Commun.* **151,** 1352 (1988).

[12] A. Hedberg, S. E. Hall, M. L. Ogletree, D. N. Harris, and E. C. K. Liu, *J. Pharmacol. Exp. Ther.* **245,** 786 (1988).

[13] D. E. Mais, R. M. Burch, D. L. Saussy, P. J. Kochel, and P. V. Halushka, *J. Pharmacol. Exp. Ther.* **235,** 729 (1985).

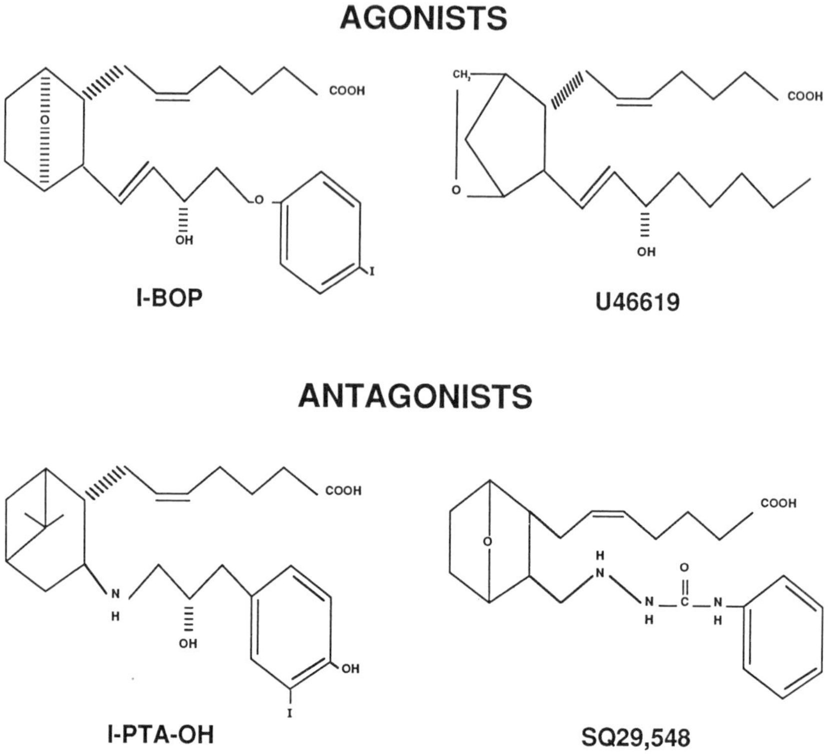

AGONISTS

I-BOP U46619

ANTAGONISTS

I-PTA-OH SQ29,548

FIG. 1. Structures of the radiolabeled agonists and antagonists used for radiolabeled ligand-binding studies. U46619 and SQ29548 are ^{3}H-labeled; I-PTA-OH and I-BOP are labeled with ^{125}I.

Assay Method

Preparation of Washed Human Platelets

Blood is drawn via venipuncture from normal human volunteers, who have not taken any medication for at least 10 days, into syringes containing indomethacin (cyclooxygenase inhibitor) (10 μM) and EDTA (5 mM) (final concentrations).[7,13] The blood is centrifuged at 100 g for 20 min at ambient temperature and the platelet-rich plasma (PRP) is pipetted off and placed into plastic centrifuge tubes. The PRP is centrifuged at 1000 g for 20 min at ambient temperature. The platelet pellet is resuspended in Tris-NaCl buffer (50 mM Tris, 100 mM NaCl, 5 mM dextrose, and 10 μM indomethacin at a final pH of 7.4) to a concentration of 5×10^8 platelets/ml for experiments using [^{125}I]PTA-OH and 1×10^8 platelets/ml for experiments using

[^{125}I]BOP. To prepare washed guinea pig and dog platelets the same procedure may be used.[14,15]

Alternatively, blood may be collected into one-tenth volume of 3.8% trisodium citrate and centrifuged at 180 *g* for 10 min to prepare PRP.[16] One-tenth volume of 77 m*M* sodium EDTA, pH 7.4, is added to PRP and the mixture is centrifuged at 1200 *g* for 10 min. The platelet pellet is resuspended and washed once with the washing buffer (NaCl, 135 m*M*; KCl, 5 m*M*; Na_2HPO_4, 8 m*M*; NaH_2PO_4, 2 m*M*; and EDTA, 10 m*M*, pH 7.2), and recentrifuged. The platelets are finally resuspended in the buffer (NaCl, 138 m*M*; $MgCl_2$, 5 m*M*; EDTA, 1 m*M*; and Tris-HCl, 25 m*M*, pH 7.5) at a density of 10^9 platelets per milliliter.

Preparation of Crude Platelet Membranes

Crude platelet membrane preparations are prepared from the platelet pellet obtained via the above procedure.[17] The pellet is resuspended in ice-cold 5 m*M* Tris-HCl, pH 7.4. The suspension is homogenized with 20 strokes of a tight-fitting Dounce homogenizer at 4°. The suspension is centrifuged at 100,000 *g* for 30 min and the pellet is resuspended using 20 strokes of Dounce homogenizer and recentrifuged at 100,000 *g* for 30 min at 4°. The pellet is resuspended in 25 m*M* Tris-HCl, pH 7.4, at a final protein concentration of 4 to 15 mg/ml, quick frozen in a dry ice–acetone bath, and stored at −70° until used. This procedure is effective for both human and rat platelets.[18]

Preparation of Solubilized Platelet Membranes

Platelet membranes are resuspended in 50 m*M* Tris-HCl (pH 7.4) at 4° containing 20% glycerol (w/v) and 10 μM indomethacin, at a final protein concentration of about 5 mg/ml.[19] The detergent 3-[(3-cholamidopropyl)dimethylammonio]-1-propane sulfonate (CHAPS) is added to the final concentration of 10 m*M* and the solution is stirred in a Vortex mixer at room temperature for 1 min. The solution is placed into polycarbonate tubes and centrifuged at 200,000 *g* for 60 min in a Beckman (San Remos,

[14] P. V. Halushka, D. E. Mais, and M. Garvin, *Eur. J. Pharmacol.* **131,** 49 (1986).

[15] D. E. Mais, P. J. Kochel, D. L. Saussy, Jr., and P. V. Halushka, *Mol. Pharmacol.* **28,** 163 (1985).

[16] S. Narumiya, M. Okuma, and U. Fumitaka, *Br. J. Pharmacol.* **88,** 323 (1986).

[17] D. L. Saussy, Jr., D. E. Mais, R. M. Burch, and P. V. Halushka, *J. Biol. Chem.* **261,** 3025 (1986).

[18] K. Hanasaki and H. Arita, *Biochem. Pharmacol.* **37,** 3923 (1988).

[19] R. M. Burch, D. E. Mais, D. L. Saussy, Jr., and P. V. Halushka, *Proc. Natl. Acad. Sci. U.S.A.* **82,** 7434 (1985).

CA) 75Ti rotor at 4°. The transparent yellow supernatant is removed and diluted in 50 m*M* Tris-HCl (pH 7.4) at 30° to a final concentration of 1 to 2 m*M* CHAPS and 1–2 mg of protein per milliliter and immediately used in the assays. If left undiluted the supernatant can be stored at 4° for several days with no apparent loss of activity. Fifty to one-hundred micrograms of protein are used per assay tube.

Preparation of Pig Aortic Smooth Muscle Membranes

Pig thoracic aorta is separated into the media, intima, and adventia.[20] The media is minced into 1- to 2-mm pieces using a McIlwain tissue chopper and homogenized in 5 volumes of 50 m*M* Tris-HCl (pH 7.4 at 25°) with three bursts of a Brinkmann Polytron homogenizer for 10 sec at speed setting 8. The homogenate is filtered through a double layer of cheesecloth and the filtrate centrifuged at 12,000 *g* for 10 min at 4°. The pellet is discarded and the supernatant centrifuged at 38,000 *g* for 15 min; the resulting pellet is resuspended in Tris buffer and stored at −80° until used.

Preparation of Cultured Rat Endothelial and Vascular Smooth Muscle Cells

Explants from rat thoracic aorta are grown in Dulbecco's modified minimal essential medium (DMEM; Gibco, Grand Island, NY) supplemented with 20% fetal bovine serum.[11,21] Endothelial or vascular smooth muscle cells are subsequently subcultured using 0.125% trypsin/0.01% EDTA to prepare the cell suspensions. Between the 5th to 8th passage the cells are harvested according to Hanasaki *et al.*[11,21] with appropriate modifications from our laboratory for binding assays as follows: The cell monolayer is washed twice with phosphate-buffered saline. Trypsin/EDTA (0.125%/0.01%; Gibco, Grand Island, NY; 1 ml/50 cm^2) is then added to the cells. After a 1-min incubation, the cells are dislodged by repeated washings with a Pasteur pipette. The removed cells are added to a centrifuge tube containing 10 × excess of DMEM and 20% fetal bovine serum. The cells are centrifuged, 132 *g* for 10 min, the pellet is washed with DMEM and the procedure repeated. The final suspension uses Hanks' balanced salt solution containing 0.1% BSA.

^{125}I-Labeled Ligands

The synthesis and characterization of radioiodinated ligands [^{125}I]PTA-OH (Ref. 13) and [^{125}I]BOP (Ref. 10) have been previously described.

[20] S. Mihara, M. Dofeuchi, S. Hara, M. Veda, M. Ide, M. Fjuimoto, and T. Okabayashi, *Eur. J. Pharmacol.* **151,** 59 (1988).

[21] K. Hanasaki, K.. Nakano, H. Kasai, H. Arita, K. Ohtani, and M. Doteuchi, *Biochem. Biophys. Res. Commun.* **150,** 1170 (1988).

^{125}I-Labeled ligands offer the advantage of much greater specific activity (~2000 Ci/mMol) compared to tritium (29 Ci/mMol)-labeled ligands. This represents a major advantage in situations where receptor density may be low.

Binding of ^{125}I-Labeled Ligands to Washed Human Platelets

Incubations (total volume of 200 μl) are conducted in silanized glass tubes (12 × 75 mm) containing either 5 × 10^7 platelets ([^{125}I]PTA-OH)[13] or 1 × 10^7 platelets ([^{125}I]BOP) for 30 min at 37°.[10] The incubation media consists of the Tris-NaCl buffer described above and approximately 0.2 n*M* (~10^5 cpm) [^{125}I]PTA-OH or 0.05 n*M* (~2.5 × 10^4 cpm) [^{125}I]BOP per tube. Various concentrations of the ^{127}I-labeled ligands are also present for equilibrium binding experiments ranging from 10^{-10} to 10^{-6} *M*. For competition binding assays, various concentrations of TxA_2/PGH_2 receptor antagonists or agonists may be included. The incubation is terminated by the addition of 4 ml of ice-cold 50 m*M* Tris, 100 m*M* NaCl buffer at pH 7.4 followed by rapid filtration through Whatman GF/C glass fiber filters. The filters are washed three more times with 4 ml of the ice-cold buffer. The entire filtration process is complete within 10 sec. Nondisplaceable binding is defined as that binding remaining in the presence of 10 μ*M* L657925,[22] a TxA_2/PGH_2 receptor antagonist. For [^{125}I]PTA-OH, displacement binding is typically 60–65% while for [^{125}I]BOP it is 90%. The glass fiber filters are counted in an LKB gamma counter at an efficiency of 80%. Data from equilibrium binding experiments are analyzed according to the method of Scatchard[23] using the LIGAND computer program.[24] The published K_d value for [^{125}I]PTA-OH is 21 ± 5 n*M* and the B_{max} is 42 ± 6.4 fmol/10^7 platelets.[13,16] (See Fig. 2.) For [^{125}I]BOP these values are 2.2 ± 0.3 n*M* and 28 ± 2 fmol/10^7 platelets, respectively.[10] (See Fig. 3.)

For binding to solubilized platelet membranes, the specific binding of [^{125}I]PTA-OH is greater than 90%. Separation of the free from bound [^{125}I]PTA-OH is accomplished using Whatman GF/C filters that have been presoaked for at least 1 hr in a solution of polyethyleneimine (0.3%). Polyethyleneimine solution (0.3%) is prepared by adding 1 ml of 50% polyethyleneimine solution (Sigma, St. Louis, MO) to 165 ml of distilled water and stirring until completely dissolved. Since the polyethyleneimine is very viscous, the tip of a disposable plastic pipette (Pipetman tip) is cut

[22] D. E. Mais, C. Yoakim, Y. Guindon, J. W. Gillard, J. Rokach, and P. V. Halushka, *Biochim. Biophys. Acta* **1012,** 184 (1989).

[23] G. Scatchard, *Ann. N.Y. Acad. Sci.* **51,** 660 (1949).

[24] P. J. Munson and D. Rodbard, *Anal. Biochem.* **107,** 220 (1980).

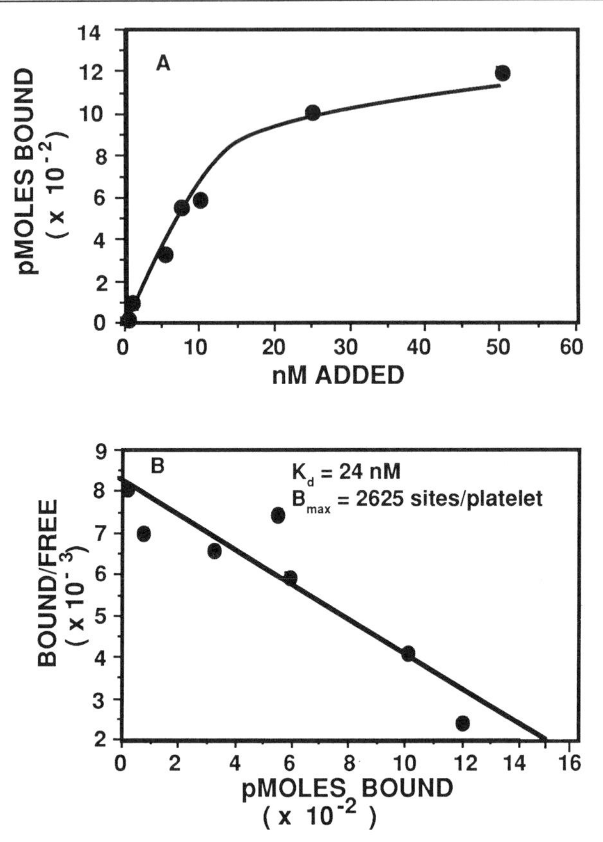

FIG. 2. Representative experiment of [^{125}I]PTA-OH binding to isolated human platelets. [^{125}I]PTA-OH (~0.2 n*M*) is added to isolated human platelets (5×10^7) along with increasing concentrations of [^{127}I]PTA-OH. Specific binding (pmol/5×10^7 platelets) is calculated by subtraction of nonspecific binding (binding that occurred in the presence of 1 μM L657925) from binding that occurred in the absence of any competing ligand (total binding), and is shown as a function of IPTA-OH added. Values from (A) are then analyzed according to Scatchard, using a LIGAND computer program, and plotted as seen in (B). Specific bound/free (pmol/5×10^7 platelets/n*M*) is plotted as a function of picomoles specifically bound/5×10^7 platelets.

off to allow for easier pipetting. The entire pipette is immersed in the distilled water and stirred.

[^{3}H]SQ29548

The TxA_2/PGH_2 receptor antagonist, [^{3}H]SQ29548, has been used to characterize TxA_2/PGH_2 receptors on human platelets,[12] rat vascular

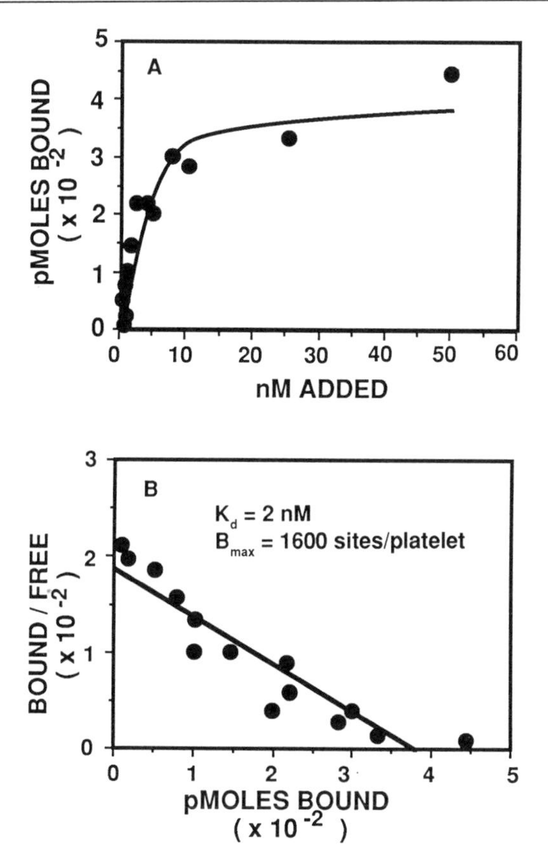

FIG. 3. Representative experiment of [^{125}I]BOP binding to isolated human platelets [^{125}I]BOP (~0.1 nM) is added to isolated human platelets (1×10^7) along with increasing concentrations of [^{127}I]BOP. Specific binding (pmol/10^7 platelets) is calculated by subtraction of nonspecific binding (binding that occurred in the presence of 1 μM L657925) from binding that occurred in the absence of any competing ligand (total binding), and is shown as a function of IBOP added (A). Values from (A) are then analyzed according to Scatchard, using a LIGAND computer program, and plotted as seen in (B). Specific bound/free (pmol/10^7 platelets/nM) is plotted as a function of picomoles specifically bound/10^7 platelets.

smooth muscle cells,[11] and rat platelets and vascular endothelial cells.[21] In all cases, [^{3}H]SQ29548 has high specific (80–97%) binding and high affinity (K_d = 0.87–4.1 nM). Binding assays employing [^{3}H]SQ29548 to characterize TxA_2/PGH_2 receptors on different tissues use [^{3}H]SQ29548 (0.2–30 nM) and excess unlabeled SQ29548 for determination of nonspecific binding.

Specific differences exist in the binding assays between platelets and vascular cells. In particular, platelets should be used at a concentration of 5×10^8/ml; for membranes, 10–50 μg of protein/0.1 ml are used.[12] The optimal concentration for vascular endothelial cells is 1.2×10^6/tube and for smooth muscle cells is 2×10^6/tube. The concentration of [^{3}H]SQ29548 can range from 0.3 to 30 nM in the platelet experiments and from 0.2 to 8.9 nM for the vascular cell experiments. For determination of nonspecific binding, Hedberg *et al.*[12] added 50 μM unlabeled SQ29548, whereas Hanasaki *et al.*[11,21] used 10 μM. The final volume of the incubation mixture was brought to 0.25 ml in the platelet experiments and 0.5 ml in the vascular cell experiments, using Tris/NaCl buffer for platelets and Hanks' BSS for the cells. The optimal incubation temperature appears to be 25°. The time of incubation, however, varies: 20 min for platelets, 30 min for smooth muscle cells, and 40 min for endothelial cells. At the end of the incubation, excess ice cold buffer is added (~3 ml) to the individual assay tubes to prevent further interaction of ligand and receptor, and to decrease nonspecific interaction. The samples are then filtered under vacuum through glass fiber filters (GF/C, Whatman; Schliecher and Schuell #22). The filters are washed from 2 to 4 times with ice-cold buffer (3 ml), dried, soaked in an appropriate scintillation fluid, and counted in a scintillation counter.

[^{3}H]U46619

Platelets. Radioligand binding experiments have been conducted in washed human platelets, rat gel filtered platelets, rat platelet membranes, and pig aortic smooth muscle cell membranes.[20] For washed human platelets, Liel *et al.*[7] used a concentration of [^{3}H]U46619 (22.4 Ci/mmol) of 30 nM (~10^5 cpm) in a final volume of 0.2 ml and Kattleman *et al.*[8] used a concentration of 4 to 5 nM (10 Ci/mmol) in a final volume of 2.2 ml. Morinelli *et al.*[9] varied the concentration of [^{3}H]U46619 from 10 to 1000 nM in a final volume of 0.5 ml. Incubation times were for 30 min at 37°, 5 min at 20°, and 15 min at 37°, respectively. In these assays, the authors claimed that equilibrium was achieved. However, in the method of Kattleman *et al.*, it appeared that if a longer incubation was carried out binding to a nonsaturable pool occurred. The K_d values reported by the three groups varied approximately 5-fold and the B_{max} values varied approximately 4-fold. Johnson *et al.*[25] also used [^{3}H]U46619 for radioligand binding studies in washed human platelets and obtained a K_d and B_{max} value in the range to those reported by Liel, Kattlemann, and Morinelli. [^{3}H]U46619 is a very lipophilic ligand with very high nonspecific binding (~50%) and thus is not a very good ligand for binding studies. Perhaps it is this characteristic that has led to the variable results reported for the ligand in

[25] G. J. Johnson, P. C. Dunlop, L. A. Leis, and A. H. L. From, *Circ. Res.* **62,** 494 (1988).

washed human platelets. Indeed, it has the poorest specific binding of the radiolabeled ligands currently available for TxA_2/PGH_2 receptors.

Binding studies have also been carried out using [^{3}H]U46619 in human platelet membranes.[18] For this assay, 0.45 mg of platelet membranes (see above for preparation of membranes), and 6 n*M* [^{3}H]U46619 (0.054 μCi) in a total volume of 0.4 ml of buffer (50 m*M* Tris-HCl, pH 7.4, 5 m*M* EDTA, 10 μM indomethacin, and 0.3 m*M* PMSF) in silanized glass tubes (12 × 75 mm) were incubated at 24° for 20 min. Magnesium (20 m*M*) is added to increase specific binding.

Binding Assay in Pig Aortic Smooth Muscle Membranes

The binding assay uses 350–430 μg of membrane protein, usually 30 n*M* [^{3}H]U46619, 50 m*M* Tris buffer (pH 7.4 at 25°), and $CaCl_2$ (10 m*M*). The binding assay is terminated by the addition of 2.5 ml of ice cold buffer, which is quickly filtered on Whatman GF/C glass fiber filters, followed by four more washes with 2.5 ml of buffer.

Discussion

Of the ligands commercially available or soon to be made available [^{3}H]SQ29458, [^{125}I]BOP, and [^{125}I]PTA-OH have the highest affinities and specific binding for the platelet and vascular receptors. Presently, these would be the preferred ligands for binding studies because of their favorable binding characteristics. The various methods available for preparation of washed platelets from different species all seem to be adequate and do not appear to differentially influence the radioligand binding. No matter what method is used, a fatty acid cyclooxygenase inhibitor should be present during the collection of blood to ensure that TxA_2/PGH_2 are not generated, which may lead to a desensitization of the receptor.[26,27]

Acknowledgments

This work was supported in part by HL36838 and HL07260. Perry V. Halushka is a Burroughs-Wellcome Scholar in Clinical Pharmacology.

[26] N. Liel, D. E. Mais, and P. V. Halushka, *J. Pharmacol. Exp. Ther.* **247,** 1133 (1988).
[27] R. Murray and G. A. Fitzgerald, *Proc. Natl. Acad. Sci. U.S.A.* **86,** 124 (1989).

[45] Thromboxane A_2/Prostaglandin H_2 Receptor Antagonists

By GUY C. LE BRETON, CHANG T. LIM, CHITRA M. VAIDYA, and DUANE L. VENTON

Human blood platelets and vascular smooth muscle cells have thus far served as the primary tissue types for the study of TxA_2- and PGH_2-mediated cellular activation.[1,2] In this respect, previous studies have provided evidence that TxA_2, and its precursor PGH_2, directly interact with membrane-coupled receptors to initiate the process of platelet stimulation. Whether these molecules interact with the same or distinct platelet receptors remains to be resolved; and because of this consideration, they are collectively referred to as TxA_2/PGH_2 receptors.[3] In addition, studies in vascular smooth muscle have suggested the existence of a tissue heterogenity of TxA_2/PGH_2 receptors.[4–7] Thus, certain agents, e.g., carbocyclic TxA_2,[5] inhibit platelet activation but also stimulate vascular smooth muscle contraction.

Based on these considerations, the decision to apply a particular antagonist to the study of TxA_2/PGH_2 receptors must be tempered by several factors including the pharmacological specificity of the compound in question as well as the tissue being examined. As a general rule, the following characteristics have served as the minimum criteria for establishing the specificity of a putative TxA_2/PGH_2 receptor antagonist in platelets and the vasculature.

First, the compound should not exhibit platelet agonist activity over a wide concentration range, and should be capable of selectively inhibiting

[1] M. Hamberg, J. Svensson, and B. Samuelsson, *Proc. Natl. Acad. Sci. U.S.A.* **72,** 2994 (1975).

[2] M. Hamberg, J. Svensson, T. Wakabayashi, and B. Samuelsson, *Proc. Natl. Acad. Sci. U.S.A.* **71,** 345 (1974).

[3] G. C. Le Breton, D. L. Venton, S. E. Enke, and P. V. Halushka, *Proc. Natl. Acad. Sci. U.S.A.* **76,** 4097 (1979).

[4] P. Needleman, M. Minkes, and A. Raz, *Science* **193,** 163 (1976).

[5] A. M. Lefer, E. F. Smith, H. Araki, J. B. Smith, D. Aharony, D. A. Claremon, R. L. Magolda, and K. C. Nicolaou, *Proc. Natl. Acad. Sci. U.S.A.* **77,** 1706 (1980).

[6] P. T. Horn, J. D. Kohli, Y. Hatano, G. C. Le Breton, and D. L. Venton, *J. Cardiovasc. Pharmacol.* **6,** 609 (1984).

[7] D. E. Mais, D. L. Saussy, Jr., A. Chaikhouni, P. J. Kochel, D. R. Knapp, H. Hamanaka, and P. V. Halushka, *J. Pharmacol. Exp. Ther.* **233**(2), 418 (1985).

TxA_2/PGH_2-mediated platelet activation. This is normally accomplished by using routine platelet aggregometry procedures.[8] Briefly, the compound should act to block platelet aggregation induced by arachidonic acid and the stable TxA_2 mimetic, U46619,[9] but not interfere with primary aggregation induced by adenosine diphosphate, epinephrine, or the divalent cation ionophore A23187. On the other hand, the compound should block ADP- and epinephrine-induced secretion of the platelet-dense bodies, as measured by radiolabeled serotonin release.[10]

Second, it must be established that the enzymatic pathways in the arachidonic acid cascade, i.e., cyclooxygenase, TxA_2 synthase, and prostacyclin synthase are not impaired. Evaluation of activity against cyclooxygenase and TxA_2 synthase is most readily accomplished by using a commercially available radioimmunoassay kit (Amersham Corp., Arlington Heights, IL) to measure platelet production of TxB_2,[11] the stable degradation product of TxA_2. Although these studies can be performed in intact platelets,[3,12] interpretation of the data is less complex if platelet microsomes[13] are employed. This is because platelet TxA_2 production is amplified during the aggregation process and, consequently, specific receptor antagonism can lead to a reduction in measured TxB_2 even though cyclooxygenase and thromboxane synthase activities remain intact.[3] Regarding prostacyclin synthase activity, a commercially available radioimmunoassay kit (Amersham) can be utilized to measure vascular microsome[14] production of 6-keto-$PGF_{1\alpha}$, the stable degradation product of PGI_2.[15]

Third, evidence must be provided that the putative TxA_2 antagonist neither stimulates adenylate cyclase activity nor interferes with the stimulation of such activity by PGI_2, PGE_1, or PGD_2. This can be accomplished by measurement of platelet cAMP production utilizing established procedures.[16] Alternatively, radioligand-binding studies of PGI_2/PGE_1 and

[8] G. V. R. Born, *Nature (London)* **194,** 927 (1962).
[9] G. L. Bundy, *Tetrahedron Lett.* **24,** 1957 (1975).
[10] M. A. Guccione, M. A. Packham, R. L. Kinlough-Rathbone, D. W. Perry, and J. F. Mustard, *Thromb. Haemostasis.* **36,** 360 (1976).
[11] F. A. Fitzpatrick, R. R. Gorman, J. C. McGuire, R. C. Kelly, M. A. Wynalda, and F. F. Sun, *Anal. Biochem.* **82,** 1 (1977).
[12] F. A. Fitzpatrick, G. L. Bundy, R. R. Gorman, and T. Hondhan, *Nature (London)* **275,** 764 (1978).
[13] P. Needleman, S. Moncada, S. Bunting, and J. R. Vane, *Nature (London)* **261,** 558 (1976).
[14] R. N. Gryglewski, S. Bunting, S. Moncada, R. N. Flower, and J. R. Vane, *Prostaglandins* **12,** 685 (1976).
[15] S. Moncada, R. J. Gryglewski, S. Bunting, and J. R. Vane, *Prostaglandins* **12,** 715 (1976).
[16] A. G. Gilman, *Proc. Natl. Acad. Sci. U.S.A.* **67,** 305 (1970).

PGD_2 platelet receptors[17–19] can be performed to demonstrate lack of antagonist competition at these sites.

A fourth criterion involves TxA_2/PGH_2 receptors in smooth muscle. It should be demonstrated that the compound in question antagonizes U46619-induced smooth muscle contraction, and does not itself possess agonist activity. This is normally accomplished by generating cumulative dose–response curves using isolated smooth muscle strips.[5–7,20,21] One important consideration in experiments employing vascular smooth muscle is the endothelial cell layer. Removal of this layer is recommended in order to eliminate the contribution of endothelial-derived relaxing factor to the observed vascular responses.

Last, evidence should be provided using radiolabeled binding studies that the putative antagonist specifically interacts with the TxA_2/PGH_2 receptor site. The specificity of this interaction should be evaluated in competition binding experiments employing agents known to interact at the TxA_2/PGH_2 receptor, e.g., U46619, as well as prostaglandin derivatives which do not interact at this site, e.g., TxB_2, 6-keto-$PGF_{1\alpha}$.

Figure 1 illustrates a number of compounds of diverse structure and potency which have been reported to exhibit TxA_2/PGH_2 receptor antagonist activity in platelets and the vasculature. These compounds include: 13-azaprostanoic acid (13-APA)[22]; pinane TxA_2[23]; derivatives of pinane TxA_2, e.g., EP045,[20] ONO11120,[24] and I-PTA-OH[25]; BM13.177[26]; AH23848[27]; SQ29,548[21]; and S-145.[28] It can be seen that with the exception of BM13.177 (or BM13.505), which are sulfonamides, all of these compounds are either prostaglandin or thromboxane derivatives.

[17] A. M. Siegl, J. R. Smith, and M. J. Silver, *J. Clin. Invest.* **63,** 215 (1979).

[18] B. Cooper and D. Ahern, *J. Clin. Invest.* **64,** 586 (1979).

[19] A. M. Siegl, J. B. Smith, and M. J. Silver, *Biochem. Biophys. Res. Commun.* **90**(1), 291 (1979).

[20] R. L. Jones, V. Peesapati, and N. H. Wilson, *Br. J. Pharmacol.* **76,** 423 (1982).

[21] M. L. Ogletree, D. N. Harris, R. Greenberg, M. F. Haslanger, and M. Nakane, *J. Pharmacol. Exp. Ther.* **234,** 435 (1985).

[22] D. L. Venton, S. E. Enke, and G. C. Le Breton, *J. Med. Chem.* **22,** 824 (1979).

[23] K. C. Nicolaou, R. L. Magolda, J. B. Smith, D. Aharony, D., E. F. Smith, and A. M. Lefer, *Proc. Natl. Acad. Sci. U.S.A.* **76,** 2566 (1979).

[24] M. Katsura, T. Miyamoto, N. Hamanaka, K. Kondo, T. Terada, Y. Ohgaki, A. Kawasaki, and M. Tsuboshima, *Adv. Prostaglandins Thromboxane Leukotriene Res.* **11,** 351 (1983).

[25] D. Mais, D. Knapp, P. Halushka, K. Ballard, and N. Hamanaka; *Tetrahedron Lett.* **25**(38), 4207 (1984).

[26] H. Patscheke and K. Stegmeier, *Thromb. Res.* **33,** 277 (1984).

[27] R. T. Brittain, L. Boutal, M. C. Carter, R. A. Coleman, E. W. Collington, H. P. Geisow, P. Hallett, E. J. Hornby, P. P. A. Humphrey, D. Jack, I. Kennedy, P. Lumley, P. J. McCabe, I. F. Skidmore, M. Thomas, and C. J. Wallis, *Circulation* **72**(6), 1208 (1985).

[28] T. Nakano, K. Hanasaki, and H. Arita, *FEBS Lett.* **234,** 309 (1988).

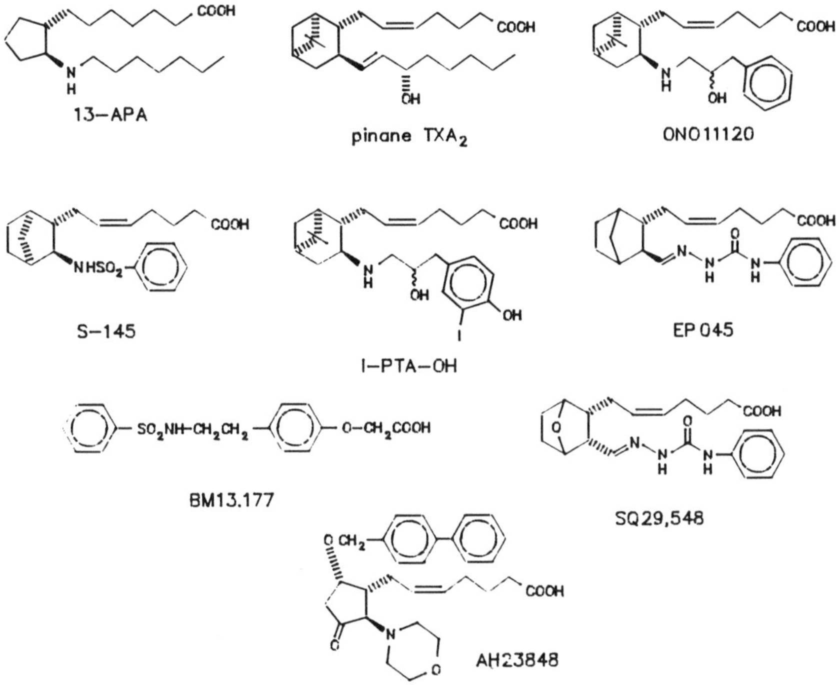

FIG. 1. TxA_2/PGH_2 receptor antagonists of differing structures.

Thus far, published studies have indicated that most of the compounds illustrated in Fig. 1 satisfy the specificity criteria. Exceptions to this include: pinane TxA_2, which exhibits activity against thromboxane synthase in addition to receptor-blocking activity[23]; S-145, which has been reported[28] to have partial agonist activity in platelets; and EP045, AH23848, and ONO1120 for which radiolabeled binding data is not available.

Regarding pharmacological potency, the rank order of the compounds listed in Fig. 1 is difficult to establish accurately since the published experiments characterizing each antagonist employed different conditions. Nevertheless, three general categories emerge: (1) those of relatively low potency, including 13-APA, pinane TxA_2, BM13.177 and EP-045; (2) those of moderate potency, including ONO 11120 and I-PTA-OH; and (3) those of the highest potency including SQ29,548 and S-145.

Consistent with the above criteria regarding pharmacological specificity, 13-APA, I-PTA-OH, BM13.177, and SQ29,548 would all serve as appropriate compounds to evaluate the *in vitro* role of TxA_2/PGH_2 in cellular function. On the other hand, in particular experimental conditions where high biological potency is also of concern, the use of SQ29,548 would presumably be preferable.

Experimental Methods

The following methods can be applied to the characterization of platelet TxA_2/PGH_2 receptors *in vitro*. These procedures provide a means for the measurement of radioligand interaction with the TxA_2/PGH_2 platelet receptor.[29–35] Based on considerations of affinity and specific binding, the most suitable ligand presently available for TxA_2/PGH_2 receptor binding studies is $[^3H]$SQ29,548 (Amersham).

Binding Studies in Intact Human Platelets

Blood is obtained by venipuncture from healthy human donors who have not received medication for 10 days. The blood is collected into a syringe containing one part sodium citrate (3.8% solution in saline) to nine parts whole blood and maintained at room temperature. It should be noted that plastic containers should be employed throughout the handling procedure. In order to prepare platelet-rich plasma (PRP), the blood is transferred to 40-ml centrifuge tubes and centrifuged at 180 *g* for 20 min (23°). The upper fraction, representing PRP, is removed by pipette being careful not to disturb the plasma–red cell interface. PRP prepared by this procedure will yield a platelet concentration of $2–4 \times 10^8$ platelets per milliliter. Frequently, PRP can be directly purchased from local blood banks. However, precautions should be taken to ensure that the preparation of this PRP does not involve high-speed centrifugation or cooling which can lead to partial platelet activation. The PRP is then treated with aspirin or indomethacin (final concentrations 3 m*M* or 20 μM, respectively) to inhibit endogenous TXA_2/PGH_2 biosynthesis, and centrifuged at 160 *g* for 15 min to remove residual contaminating red blood cells.

[29] S. C. Hung, N. I. Ghali, D. L. Venton, and G. C. Le Breton, *Biochim. Biophys. Acta* **728,** 171 (1983).

[30] R. A. Armstrong, R. L. Jones, and N. H. Wilson, *Br. J. Pharmacol.* **79,** 953 (1983).

[31] D. E. Mais, AP.J. Kochel, D. L. Saussy, Jr., and P. V. Halushka, *Mol. Pharmacol.* **28,** 1673 (1985).

[32] E. J. Kattelman, D. L. Venton, and G. C. Le Breton, *Thromb. Res.* **41,** 471 (1986).

[33] D. L. Saussy, Jr., D. E. Mais, R. M. Burch, and P. V. Halushka, *J. Biol. Chem.* **261**(7), 3027 (1986).

[34] E. J. Kattelman, S. K. Arora, C. T. Lim, D. L. Venton, and G. C. Le Breton, *FEBS Lett.* **213**(1), 179 (1987).

[35] A. Heberg, S. E. Hall, M. L. Ogletree, D. N. Harris, and E. C.-K. Liu, *J. Pharmacol. Exp. Ther.* **245,** 786 (1988).

The platelets are then separated from native plasma proteins using the following procedure. PGI_2 (10 ng/ml) is added to aid resuspension of the platelets,[36] and the PRP is centrifuged at 1100 *g* for 15 min to pellet the cells. The upper platelet-free plasma fraction is discarded, and the platelet pellets are gently resuspended in buffer to a final cell count of approximately 12×10^8 platelets per milliliter. The resuspension buffer is composed of NaCl (138 m*M*), KCl (5 m*M*), MgCl (5 m*M*), glucose (5.5 m*M*), and Tris-HCl (25 m*M*), pH 7.4.

Aliquots of platelet suspension (2.2-ml) are incubated with radiolabeled antagonist for 30 min (23°) at a concentration determined by the specific activity and affinity of the ligand. This can range from roughly 0.2 n*M* for [^{3}H]SQ29,548 and [^{125}I]PTA-OH to 30 n*M* for [^{3}H]13-APA. In competition studies, the competing agent (1000-fold molar excess) is mixed with the radioligand prior to addition to the platelets. In displacement studies, the displacing agent (1000-fold molar excess) is added 20–30 min subsequent to radioligand addition. Specific binding is defined as the difference between binding in the presence and absence of excess unlabeled ligand.

At appropriate time points, two 1-ml aliquots of the binding assay mixture are transferred to 1.5-ml Eppendorf tubes, and the incubation is terminated by centrifugation at 7000 *g* for 1 min (23°). The supernatant is aspirated, and the inside of the Eppendorf tube as well as the surface of the pellet is rinsed with 1 ml of ice-cold resuspension buffer, and immediately removed by aspiration. The tip of the tube is cut off, inverted, and placed in a second Eppendorf tube which is centrifuged for 12 sec at 7000 *g*. This results in transfer of the pellet to the second tube which is free of radioligand contamination. The tip of this tube is then cut off, and the pellet is digested overnight in 1 ml of TS-1 tussue solubilizer (RPI, Mt. Prospect, IL). The solubilizer is neutralized with hydrochloric acid before the addition of 10 ml of scintillation fluid (Liquiscint, National Diagnostics, Somerville, NJ). Samples are counted to a 1.5% error in a liquid scintillation spectrometer at an efficiency of 40% and the duplicates from each incubation are averaged.

Alternatively, binding studies in intact washed platelets can be performed by rapid filtration techniques. In this case, two 750-μl aliquots of platelet suspension (5×10^8 platelets per milliliter) are rapidly filtered under suction through glass fiber filters (Whatman, GF/B) which have been presoaked in resuspension buffer. The filters are quickly washed three times with 5 ml of buffer (4°), and placed in scintillation vials containing 5 ml scintillation fluid (Liquiscint). The potential for "specific binding" to the filters themselves is evaluated by performing the above procedure in the absence of platelets.

[36] M. Radomski and S. Moncada, *Thromb. Res.* **30**, 383 (1983).

Binding Studies in Platelet Membranes

Human PRP, obtained as outlined above, is first treated with 3 m*M* aspirin or 20 μM indomethacin aspirin to inhibit endogenous TXA_2/PGH_2 production and then with PGI_2 (10 ng/ml) to prevent platelet activation during the washing procedure. The PRP (240 ml) is added to 40-ml centrifuge tubes and spun (23°) at 4,000 *g* for 15 min. The supernatant is discarded, and the platelet pellet in each tube gently resuspended (23°) in 5 ml of 25 m*M* Tris-HCl and 5 m*M* $MgCl_2$ buffer (pH 7.4) containing PGI_2 (10 ng/ml) using a plastic disposable pipette. This washing procedure is repeated once more, and the platelet suspension in each tube is brought to a total volume of 40 ml using Tris-MgCl buffer containing 5 m*M* EDTA, 2 μg/ml of leupeptin, 2 μg/ml of pepstatin A, 10 μg/ml of trypsin inhibitor, and 44 μg/ml of phenylmethylsulfonyl fluoride. The platelets are then disrupted by nitrogen cavitation procedures. Briefly, the platelet suspension is equilibrated (4°) for 30 min with 50 atm of nitrogen in a pressure homogenizer (Kontes, Vineland, NJ) and then released dropwise. The homogenate is collected (4°) and centrifuged at 1500 *g* for 15 min to remove undisrupted platelets. The resulting supernatant is then centrifuged (4°) at 100,000 *g* for 45 min, the supernatant is discarded, and the pellet is gently rinsed once with 3 ml of homogenizing buffer (4°). The membranes are resuspended in 3 ml of homogenizing buffer (4°) and adjusted to a final protein concentration of 3 mg/ml as determined by Lowry.[37]

Aliquots (0.5 ml) of the membrane suspension are incubated (23°) with radiolabeled antagonist for 30 min at a concentration (0.2 n*M*–30 n*M*) determined by the specific activity and affinity of the ligand. The procedures for the determination of specific binding, competitive binding, and displacable binding are as described above for the whole platelet binding assay (see Fig. 2 for representative data using [^{3}H]SQ29,548).

Binding Studies in Solubilized Platelet Membranes

A recent development which requires further evaluation is the measurement of TxA_2/PGH_2 receptor binding in solubilized membrane preparations.[38] Platelet membranes are prepared according to the pressure homogenization procedure described above. After nitrogen cavitation, the 100,000 *g* membrane pellet is resuspended (4°) in 3 ml buffer containing 50 m*M* Tris-HCl, 5 m*M* EDTA, and 10 m*M* CHAPS [3-[(3-cholami-

[37] O. H. Lowry, N. J. Rosebrough, A. L. Farr, and R. J. Randall, *J. Biol. Chem.* **193,** 265 (1951).

[38] R. M. Burch, D. E. Mais, D. L. Saussy, Jr., and P. V. Halushka, *Proc. Natl. Acad. Sci. U.S.A.* **82,** 7434 (1985).

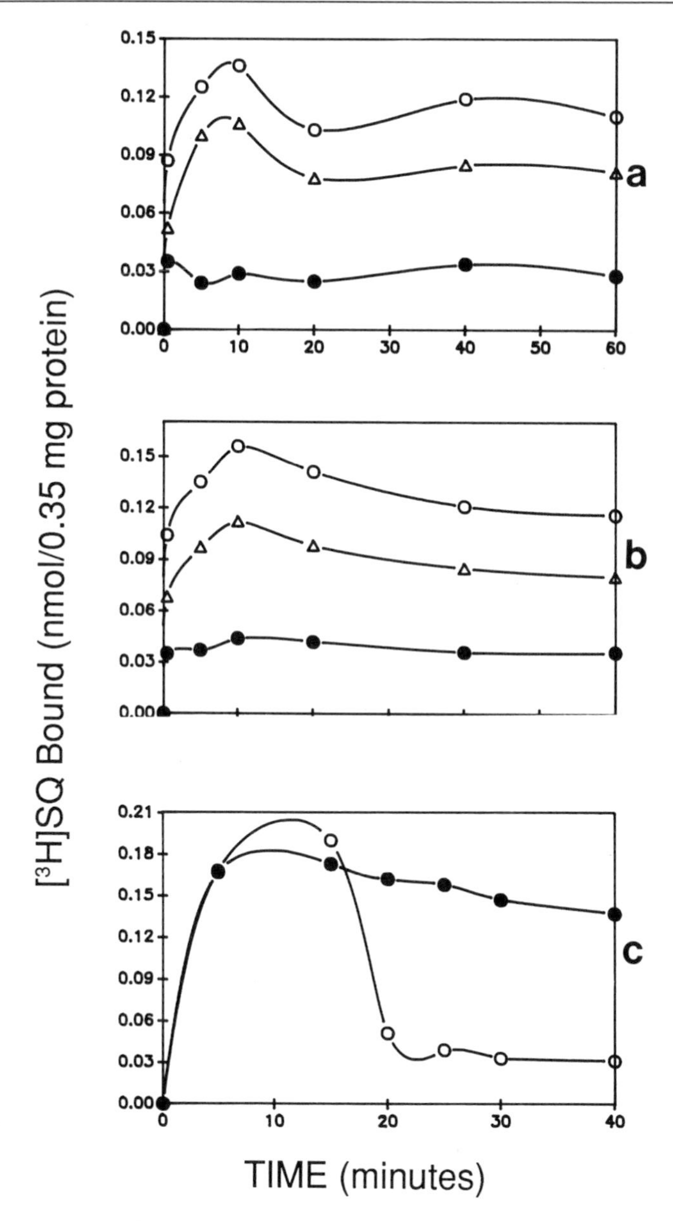

FIG. 2. Binding of [^{3}H]SQ29,548 to human platelet membranes. (a) Total (○), nonspecific (●), and specific (△) binding; 1000-fold molar excess of unlabeled SQ29,548; (b) total (○), nonspecific (●), and specific (△) binding; 25,000-fold molar excess of BM13.505; (c) displacement of [^{3}H]SQ29,548 binding by 1000-fold molar excess of nonlabeled SQ29,548 or TxB_2 (●).

dopropyl)dimethylammonio]-1-propane sulfonate], pH 7.4. The suspension is then homogenized on ice with a Teflon homogenizer, and centrifuged (4°) at 100,000 *g* for 30 min. The clear, slightly yellow supernatant is then diluted with 12 ml of ice-cold buffer (25 m*M* Tris-HCl, 5 m*M* $MgCl_2$, pH 7.4) and contains approximately 0.8 mg/ml of solubilized membrane protein.

Binding of the antagonist [^{3}H]SQ29,548 to solubilized receptor protein is measured by rapid filtration using glass fiber filters (GF/B, Whatman) which have been presoaked for 1 hr in 0.3% polyethyleneimine (in distilled H_2O). Aliquots (200 μl) of the solubilized protein are incubated with [^{3}H]SQ29,548 (1.0 n*M*) at 25° for 30 min in the presence and absence of competing agents. The samples are then filtered according to the methods described above for platelet membranes.

[46] Assessment of Leukotriene D_4 Receptor Antagonists

By DAVID AHARONY

Introduction

Leukotriene C_4 [5(*S*)-hydroxy-6(*R*)-*S*-glutathionyl-7(*E*),9(*E*),11(*Z*),14(*Z*)-eicosatetraenoic acid, LTC_4] is a metabolite of arachidonic acid generated via the 5-lipoxygenase pathway in lung tissues and certain leukocytic cell types.[1] In most tissues or organs, LTC_4 is rapidly metabolized to leukotriene D_4 [5(*S*)-hydroxy-6(*R*)-*S*-cystinylglycyl-7(*E*),9(*E*),11(*Z*),14(*Z*)-eicosatetraenoic acid, LTD_4] which is subsequently converted to leukotriene E_4 [5(*S*)-hydroxy-6(*R*)-*S*-cystinyl-7(*E*),9(*E*),11(*Z*),14(*Z*)-eicosatetraenoic acid, LTE_4]. The facile conversion of LTC_4 suggests that LTD_4 and LTE_4 may actually be the species that exert most of the observed pharmacology. LTD_4, a component of the "slow-reacting substance of anaphylaxis" (SRA), is a potent constrictor of lung and vascular smooth muscles and also induces mucus hypersecretion and vascular permeability.[1] Numerous studies, utilizing functional and radioligand-binding receptor assays have demonstrated that LTD_4 binds with high affinity to specific membrane receptors, which in most tissues examined, are distinct from those activated by its metabolic precursor LTC_4.[2,3] In

[1] B. Samuelsson, *Science* **220,** 568 (1983).

[2] S. T. Crooke, S. Mong, M. A. Clark, G. K. Hogaboom, M. A. Lewis, and J. G. Gleason, *Biochem. Actions Horm.* **14,** 81 (1987).

[3] D. W. Snyder and R. D. Krell, *J. Pharmacol. Exp. Ther.* **231,** 616 (1984).

contrast with LTC_4, LTE_4 binds exclusively to LTD_4 receptors[4,5] and exerts actions similar to LTD_4 via activation of a subset of LTD_4 receptors.

Since the discovery of the first antagonist FPL 55,712,[6] significant progress has been made in the past 5 years in developing several novel, potent, and selective LTD_4 antagonists. These include: (a) structural analogs of FPL 55,712 (generally termed hydroxyacetophenones), such as the orally active, but modestly potent, LY 171,883 (Ref. 7) and YM 16638 (Ref. 8); (b) potent antagonists that resemble the structure of LTD_4, such as SKF 104,353 (Ref. 9) and MK 571 (Ref. 10); (c) highly potent antagonists containing amine rings that bear little apparent resemblance to the above compounds such as ONO-RS-411 (Ref. 11), SR 2640 (Ref. 12), and ICI 198,615 (Fig. 1) (Refs. 13 and 14).

The following describes receptor binding assays utilizing either a radiolabeled agonist (i.e. $[^3H]LTD_4$) or antagonist ($[^3H]$ICI 198,615), which can be used to determine the potency and competitiveness of binding of LTD_4 antagonists.[5,14,15] Some notable differences as well as agreements between data obtained from ligand binding experiments versus that obtained in functional receptor assays will be emphasized.

[4] S. Mong, M. O. Scott, M. A. Lewis, H.-L. Wu, G. K. Hogaboom, M. A. Clark, and S. T. Crooke, *Eur. J. Pharmacol.* **109,** 183 (1985).

[5] D. Aharony, C. A. Catanese, and R. C. Falcone, *J. Pharmacol. Exp. Ther.* **248,** 581 (1988).

[6] J. Augstein, J. B. Farmer, T. B. Lee, P. Sheard, and M. L. Tattersall, *Nature (London)* **245,** 215 (1973).

[7] J. H. Fleisch, L. E. Rinkema, K. D. Haisch, D. Swanson-Bean, T. Goodsen, P. K. P. Ho, and W. S. Marshall, *J. Pharmacol. Exp. Ther.* **233,** 148 (1985).

[8] K. Tomioka, T. Yamada, T. Mase, H. Hara, and K. Murase, *Arz. Forsch. Drug Res.* **38,** 682 (1988).

[9] D. W. P. Hay, R. M. Muccitelli, S. S. Tucker, L. M. Vickery-Clark, K. A. Wilson, J. G. Gleason, R. F. Hall, M. A. Wasserman, and T. J. Torphy, *J. Pharmacol. Exp. Ther.* **243,** 474 (1987).

[10] T. R. Jones, R. Zamboni, M. Balley, E. Champion, L. Charette, A. W. Ford-Hutchinson, F. Frenette, J. Y. Gauthier, S. Leger, P. Masson, C. S. McFarland, H. Piechuta, J. Rokah, H. Williams, R. N. Jones, R. N. DeHaven, and S. S. Pong, *Can. J. Physiol. Pharmacol.* **67,** 17 (1989).

[11] H. Nakai, M. Konno, S. Kosuge, S. Sakuyama, M. Toda, Y. Arai, T. Obata, N. Katsube, T. Miyamoto, T. Okegawa, and A. Kawasaki, *J. Med. Chem.* **31,** 84 (1988).

[12] I. Anfelt-Ronne, D. Kirstein, and C. Kaergaard-Nielsen, *Eur. J. Pharmacol.* **155,** 117 (1988).

[13] D. W. Snyder, R. E. Giles, R. A. Keith, Y. K. Yee, and R. D. Krell, *J. Pharmacol. Exp. Ther.* **243,** 548 (1987).

[14] D. Aharony, R. C. Falcone, and R. D. Krell, *J. Pharmacol. Exp. Ther.* **243,** 921 (1987).

[15] D. Aharony, R. C. Falcone, Y. K. Yee, B. Hesp, R. E. Giles, and R. D. Krell, *Ann. N.Y. Acad. Sci.* **524,** 162 (1988).

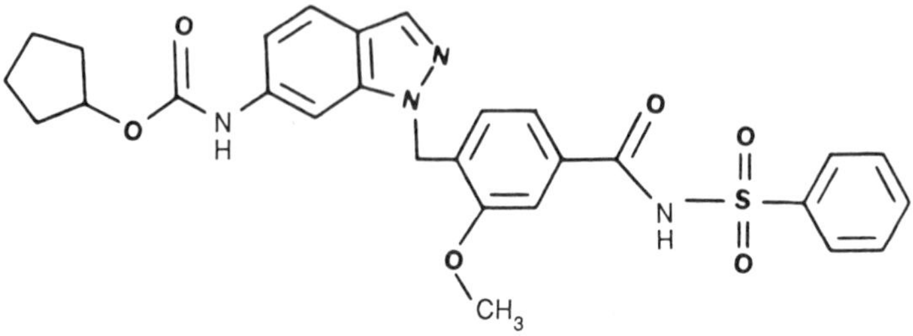

FIG. 1. Structure of ICI 198,615.

Assay Methods

Principle. The potency of a given LTD_4 antagonist is determined by its ability to compete, in a concentration-dependent manner, against binding of $[^3H]LTD_4$ or $[^3H]$ICI 198,615[15] to guinea pig lung membranes (GPLM). Scatchard analysis of saturation of ligand binding, conducted with multiple concentrations of radioligand in the presence of a single concentration of antagonist, is utilized to determine competitiveness.

Reagents

Receptor preparation buffer (A): 20 m*M* Tris-HCl containing 0.25 *M* sucrose and the protease inhibitors: bacitracin, 100 μg/ml; benzamidine, 157 μg/ml; phenymethylsulfonyl fluoride, 87 μg/ml; and soybean trypsin inhibitor, 100 μg/ml, pH 7.4

Assay buffer (B) : 10 m*M* PIPES [piperazine-*N,N'*-bis(2-ethanesulfonic acid)], pH 7.4, containing 10 m*M* $CaCl_2$, 10 m*M* $MgCl_2$, 2 m*M* cysteine, and 2 m*M* glycine

Radioligands: $[^3H]LTD_4$ and $[^3H]$ICI 198,615 with specific activities of ~40 and ~60 Ci/mmol, respectively, purchased from New England Nuclear (Boston, MA).

Preparation of Lung Membranes

Guinea pig lung membranes (GPLM) are prepared as previously described.[5] Male albino, Hartley-strain guinea pigs (300–500 g) are decapitated, lungs perfused with Tyrode's buffer (pH 7.4), and immediately excised. Large blood vessels and all visible necrotic tissue are removed and the remaining tissue is flash-frozen under nitrogen and stored at −70° until processed into membranes. The typical yield from 50 animals is approximately 150 g of lung tissue. A batch of 50 g frozen lung is thawed, chopped with a McIlwain tissue chopper, and washed several times with ice-cold phosphate-buffered saline (PBS) (0.1 *M*, pH 7.4), followed by homogenization with a Brinkman PT-20 Polytron (six pulses of 20 seconds each at setting of 6) in buffer A. The homogenate is centrifuged at 15,000 *g* for 10 min at 4° to remove cell debris. The supernatant is carefully poured through four layers of gauze and is then centrifuged at 40,000 *g* for 30 min at 4°. The pellet is resuspended in 50 ml 20 m*M* Tris-HCl buffer, pH 7.4, utilizing a glass–Teflon motorized homogenizer and centrifuged again as above. This procedure is repeated and the pellets resuspended in 50–100 ml of Tris buffer. The membranes can be stored in convenient 5 to 10-ml aliquots (containing 1–2 mg protein per milliliter) at −70° for periods over a year with no loss of receptor binding activity. This procedure typically yields 150–200 mg membrane protein/50 g lung tissue.

Receptor Binding Assays

The assay composition that allows competition by both agonists and antagonists is identical for the [^{3}H]ICI 198,615 and [^{3}H]LTD_4 binding assays. In competition experiments, 150 μl of buffer B are mixed with 20 μl of either [^{3}H]ICI 198,615 or [^{3}H]LTD_4 (0.5 n*M* final concentration) along with 15 μl of varying concentrations of tested compound. The incubation is initiated by addition of 125 μl of diluted (1 : 4–1 : 6) membranes (150–200 μg protein per milliliter) to a final volume of 310 μl and the incubations carried out at 25° for 30 min. Nonspecific binding is defined by respective unlabeled ligand in 2000-fold excess of labeled ligand. Saturation (Scatchard) experiments are similar to the competition experiments with the exception that the concentrations of the ^{3}H-labeled ligands are varied between 0.006 and 3 n*M* for [^{3}H]ICI 198,615 and 0.05 and 2.5 n*M* for [^{3}H]LTD_4.

The incubation is terminated by dilution with 3 ml of ice-cold 10 m*M* Tris/100 m*M* NaCl followed immediately by vacuum filtration with a total of 16 ml of the Tris/NaCl buffer, utilizing a Brandel Cell Harvester Model M-30 and Whatman GF/C filters. The radioactivity retained on the filters is determined with a scintillation counter. Data from competition or Scatchard experiments are analyzed as published in detail elsewhere.[15]

Both ligands display good specific binding (i.e., ~90%) to lung membranes and bind with high affinity (K_d values are 0.2–0.5 n*M*) and in a saturable manner (B_{max} values are 1000–2000 fmol/mg) to GPLM.[14,15] Figure 2 illustrates that several selective and structurally diverse LTD_4 antagonists inhibit the specific binding of [^{3}H]LTD_4. Antagonists exert even higher potency against [^{3}H]LTE_4, which preferentially binds (under the same conditions as described above) to a subset of high-affinity LTD_4 receptors.[5] Figure 3 illustrates the inhibition of [^{3}H]ICI 198,615 binding by LTD_4 antagonists. K_i values for many antagonists obtained against

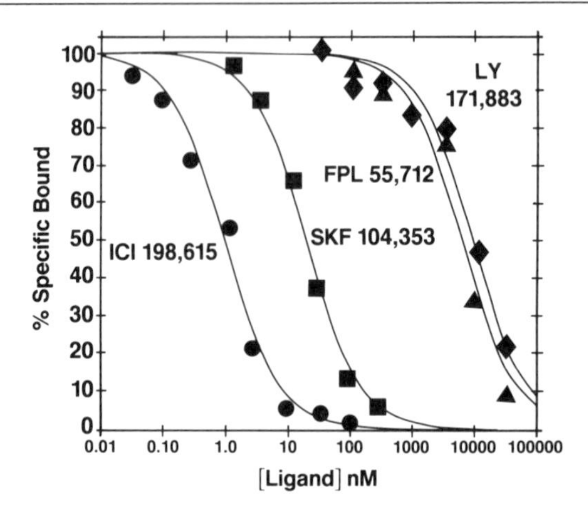

FIG. 2. Inhibition of $[^3H]LTD_4$ (0.5 n*M*) binding to guinea pig lung membranes by structurally diverse antagonists. Nonspecific binding was defined with 1 μM LTD_4. Results are mean of duplicate determinations from a typical experiment.

$[^3H]LTD_4$ binding correlate well (r = 0.962) with those obtained against $[^3H]$ICI 198,615 (Ref. 16).

Moreover, an excellent correlation (r = 0.95) also exists between K_i values from these binding assays and K_B values obtained in functional receptor assays.[14] Similar results with LTD_4 antagonists, have also been demonstrated in membranes from human lung.[17] The results from several such experiments are summarized in Table I which also demonstrates the ability of leukotriene agonists to inhibit binding.

In contrast to antagonists, agonists exert significantly less affinity in competing against the antagonist as compared with the agonist, and they inhibit only 60–70% of the specific binding defined with 2 μM of ICI 198,615 or FPL 55,712. However, when nonspecific binding is defined by 100 n*M* ICI 198,615, 300 n*M* SKF 104,353, or by 3 μM LTD_4, both agonists and antagonists completely displace the specifically bound $[^3H]$ICI 198,615 in a manner compatible with a single class of binding sites. Scatchard analysis demonstrates that selective LTD_4 antagonists, such as ICI 198,615, inhibit $[^3H]LTD_4$ binding in a competitive manner.[14] Similarly, Fig. 4 illustrates that FPL 55,712 is a competitive antagonist of $[^3H]$ICI 198,615 binding to LTD_4 receptors on GPLM. Using similar methods,

[16] D. Aharony, R. C. Falcone, Y. K. Yee, and R. D. Krell, *Biotechnol. Update* **3,** 1 (1988).

[17] D. Aharony and R. C. Falcone, *in* "New Trends in Lipid Mediators Research" (U. Zor, Z. Naor, and A. Danon, eds.), Vol. 3, pp. 67–71. S. Karger, Basel, 1989.

TABLE I
INHIBITION OF ^{3}H-LABELED LIGAND BINDING TO GPLM BY SELECTIVE AGONISTS AND LTD_4 ANTAGONISTS

Compound	K_i [a] (nM) vs.	
	[^{3}H]LTD_4	[^{3}H]ICI 198,615
Agonists		
LTD_4	0.7 ± 0.1	8.7 ± 1.4
LTE_4	2.8 ± 0.6	24 ± 4
LTC_4 [b]	66 ± 12	122 ± 34
YM 17690	5.9 ± 0.7	198 ± 24
Antagonists		
ICI 198,615	0.3 ± 0.1	0.6 ± 0.1
SKF 104,353	4.6 ± 0.8	21 ± 2
FPL 55,712	2552 ± 555	4129 ± 695
LY 171,883	2871 ± 378	5822 ± 1147

[a] K_i values are mean ± SEM of 3 to 7 experiments conducted in duplicate.
[b] LTC_4 was evaluated in the presence of 45 mM serine–borate to prevent its metabolism to LTD_4.

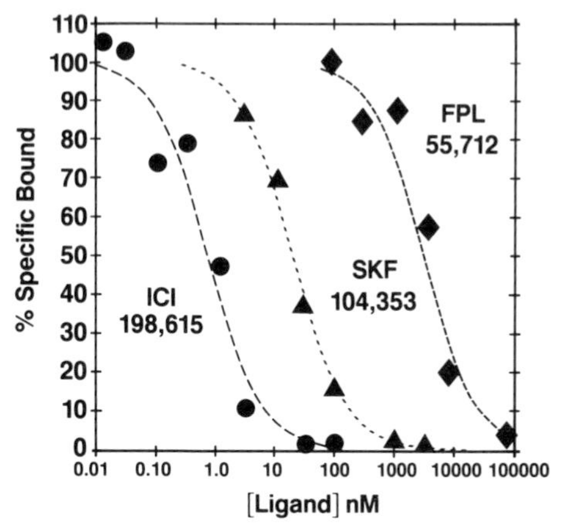

FIG. 3. Inhibition of [^{3}H]ICI 198,615 (0.5 nM) binding to guinea pig lung membranes by structurally diverse antagonists. Nonspecific binding is defined by 100 nM ICI 198,615 or by 300 nM SKF 104,353. Results are mean of duplicate determinations from a typical experiment.

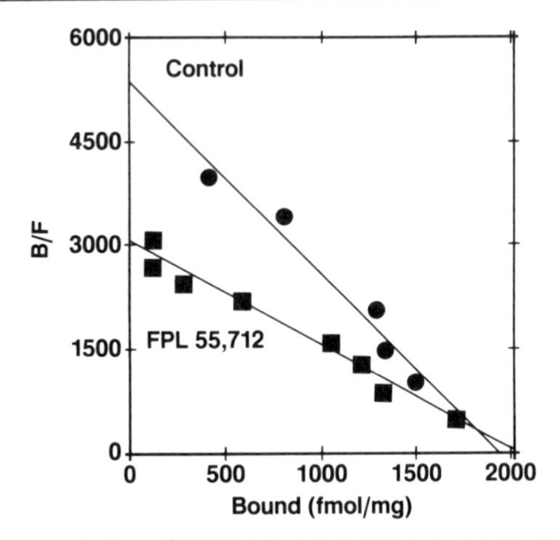

FIG. 4. Scatchard analysis of [^{3}H]ICI 198,615 (0.005–2.0 nM) binding to guinea pig lung membranes in the absence (control) and presence of 1 μM FPL 55,712. K_i = 1.2 μM was calculated from the equation $K_d(\text{app}) = K_d{}^*(1+[\text{I}]/K_i)$, assuming competitive inhibition.

high-affinity (K_d = 0.6–1.0 nM), low-capacity (B_{max} = 80–180 fmol/mg protein) binding sites for both ligands, which are antagonized by LTD_4 antagonists, can also be demonstrated in human lung membranes.[17]

Summary

Both [^{3}H]LTD_4 and [^{3}H]ICI 198,615 bind selectively and with high affinity to LTD_4 receptors on guinea pig and human lung membranes. Binding is inhibited by selective LTD_4 antagonists. However there may be some advantages for preferring one over the other, which is largely due to the specific experimental design. For example, due to the apparent higher affinity of agonists in the [^{3}H]LTD_4 binding assay, one can use this ligand to measure competition by agonists, partial agonists, and to determine the stereoselectivity of LTD_4 analogs.[14]

The disadvantages are susceptibility to oxidation, double-bond isomerization under acidic condition, metabolism by membrane aminopeptidase, and requirement for optimization of "agonist binding conditions"[15] that may limit the use of this ligand in different tissues (i.e., ileum or brain). [^{3}H]ICI 198,615 does not suffer from these disadvantages and allows the direct determination of potency for structurally diverse antagonists without the necessity to optimize the assay for agonist binding.[15] An additional advantage is the ability to distinguish between agonists and antagonists at the receptor binding level, since only agonists inhibition curves (against [^{3}H]ICI 198,615) are shifted to lower affinity by stable GTP analogs.[18]

However, one has to bear in mind that although these binding assays are highly efficient, rapid, and possess high capacity to test antagonist potency and mechanism, they do not provide broad pharmacological information as do functional receptor assays that utilize viable tissues. Some of the more notable examples are the phosphodiesterase inhibitory properties of LY 171,883 (Ref. 7) and the thromboxane A_2 inhibitory activity of FPL 55,712 (Ref. 19) that may enhance antibronchospastic properties in viable tissues or animal models, leading to an apparent overestimation of their potency. An additional important discrepancy is the ability of all LTD_4 antagonists to inhibit LTC_4 and LTD_4 contractile activity on human lung equally well.[20,21] This is in contrast to functional[3] and ligand binding[17] experiments that demonstrate distinct binding sites for LTC_4 and LTD_4 (and ICI 198,615).

Acknowledgments

Mr. R. C. Falcone and Ms. C. A. Catanese are gratefully acknowledged for conducting these experiments and Dr. R. D. Krell for reviewing this manuscript.

[18] C. A. Catanese, R. C. Falcone, and D. Aharony, *J. Pharmacol. Exp. Ther.* **251,** 846 (1989).
[19] A. F. Welton, W. C. Hope, L. D. Tobias, and J. G. Hamilton, *Biochem. Pharmacol.* **30,** 1378 (1981).
[20] C. K. Buckner, R. D. Krell, R. B. Laravuso, D. B. Coursin, P. R. Bernstein, and J. A. Will, *J. Pharmacol. Exp. Ther.* **237,** 558 (1986).
[21] C. K. Buckner, R. Saban, W. L. Castleman, and J. A. Will, *Ann. N.Y. Acad. Sci.* **524,** 181 (1988).

[47] Sulfidopeptide Leukotriene Receptor Binding Assays

By SEYMOUR MONG

Introduction

The sulfidopeptide leukotriene receptor binding assay has been used extensively in antagonist compound screening,[1] in the identification of

[1] C. D. Perchonock, I. Uzinkas, M. E. McCarthy, K. F. Erhard, J. G. Gleason, M. A. Wasserman, R. M. Muccitelli, J. F. Devan, S. S. Tucker, L. M. Vickery, T. Kirchner, B. M. Weichman, S. Mong, M. O. Scott, G. Chi-Rosso, H.-L. Wu, S. T. Crooke, and J. F. Newton, *J. Med. Chem.* **29,** 1442 (1986).

antagonists from fermentation broths and natural products,[2] and also for the quantification of sulfidopeptide leukotrienes in biological samples. Sulfidopeptide leukotrienes are highly lipophilic and tend to interact nonspecifically with plastics, paper, test tubes, glass, and other nonreceptor sites. The radiolabeled agonists are metabolically unstable and sensitive to chemical degradation. These are the major factors to be carefully considered when setting up radioligand binding systems to study the receptors and the mechanisms of receptor regulation. Recent advancement of the synthesis of potent receptor antagonists has alleviated some of the vexing difficulties. When appropriately controlled and validated, the radioligand binding assay has proved to be extremely useful.

Methods

Cell Culture

Any type of media or serum for culturing is sufficient, as long as the cells are in good condition. Cells derived from primary cultures tend to dedifferentiate quickly and lose their leukotriene receptors.[3] Cells are harvested, washed, and resuspended in binding assays. Chemicals used in culturing the cells, such as mercaptoethanol, retinoic acid, vitamins, dimethyl sulfoxide, serum protein, and fetal calf serum should be washed away, unless proved to be necessary. These agents tend to interact with radioligands causing problems. Whole cell binding is only recommended when radiolabeled high-affinity antagonists (e.g., [^{3}H]ICI-198615) are used. Radiolabeled agonist ([^{3}H]LTC_4, [^{3}H]LTD_4, and [^{3}H]LTE_4) binding usually is influenced by processes such as desensitization, i.e., receptor uncoupling or receptor internalization, and hence the data derived from such studies are difficult to evaluate.

Membrane Preparation

Tissues or cells can be preserved by quick freezing, usually by submersion in liquid nitrogen, and stored in an ultra-low temperature (−70° to −80°) freezer for up to 1 year. Primary cultured smooth muscle cells, cell lines, lungs, hearts, and tracheas, from guinea pig, sheep, monkey, or humans have been stored in this manner and have subsequently provided usable membrane preparations. Human lungs, from 18-hr postmortem

[2] Z. Tian, M. N. Chang, M. Sandrino, L. Huang, J. Pan, B. Arison, J. Smith, and Y. K. T. Lam, *Phytochemistry* **26,** 2361 (1987).

[3] S. Mong, J. Miller, H.-L. Wu, and S. T. Crooke, *J. Pharmacol. Exp. Ther.* **244,** 508 (1988).

traffic accident victims, have been preserved this way for up to 8 months and still yielded usable membranes for ligand binding studies.

Tissues (10 g) are minced into 50–500 mm^3 blocks and transferred to 80 ml homogenization buffer [0.25 *M* sucrose in 50 mM PIPES buffer, pH 6.5, containing the following protease inhibitors; phenylmethylsulfonyl fluoride (0.5 m*M*), bacitracin (100 μg/ml), benzamidine (100 μM), and soybean trypsin inhibitor (100 μg/ml)]. Cells are directly resuspended in homogenization buffer at a concentration 10^8–10^9/ml. The pH of the buffer is adjusted with either HCl or KOH. The tissue (or cell) is broken with a Polytron homogenizer at a setting of 6–7, for 40 to 100 sec with 10-sec pulses. The homogenate is then centrifuged (1000 *g*) for 10 min at 4° to remove the tissue clumps and unbroken cells. The supernatant can be centrifuged at 30,000 *g* for 20 min at 4° to yield crude membrane pellets for binding studies. A better way to enrich the receptor-containing plasma membrane preparation[3] is to layer the supernatant (25 ml) from the first centrifugation (1000 *g*) on a 10-ml 40% sucrose cushion in a Beckman SW-27 nitrocellulose centrifuge tube. The centrifuge tubes are spun at 100,000 *g* for 90 min at 4°. The membranes at the 10/40% sucrose interface are collected, pooled, diluted with an equal volume (10 m*M*) of Tris buffer (pH 7.4), and pelleted by centrifugation at 150,000 *g* for 30 min at 4°. The pellets contain plasma membrane-enriched fractions that can be used directly or quickly frozen and stored. It is advisable to use liquid nitrogen to freeze the membrane pellets and avoid repeated cycles of freeze–thawing. Short-term storage (24–48 hr, 0°) of guinea pig lung membranes in protease-containing buffer is associated with gradual loss (10–15% per day) of the receptor binding activity.

Radioligand Binding Conditions

Association and Dissociation of Radiolabeled Agonist and Antagonist. Binding of radiolabeled agonists ([^{3}H]LTD_4 or [^{3}H]LTE_4) is usually regulated by cations and pH. Therefore, each of these factors must be determined independently to minimize nonspecific binding and to avoid metabolic conversion of the radioligands. When these factors and the conditions are well-defined, one can then plan for extensive saturation, competition, and screening experiments. In our laboratory, for kinetic experiments, [^{3}H]LTD_4 (or [^{3}H]LTE_4) is added to an incubation mixture that contains 20 m*M* PIPES buffer (pH 6.5), 10 m*M* cysteine, 10 m*M* glycine, 5 m*M* $CaCl_2$, 5 m*M* $MgCl_2$, 0.5 n*M* [^{3}H]LTD_4 (or [^{3}H]LTE_4) and 1000 μg/ml of membrane protein in a volume of 1 ml. The mixture is incubated at room temperature (22°) from zero to 60 min in Nunc Minisorb tubes (Gibco, Grand Island, NY). Aliquots (100 μl) are retrieved at varying time points and analyzed for membrane-associated radioligand binding. A separate incubation mixture is set up similarly, except that the unlabeled homologous ligand (LTD_4 or LTE_4) is also included in the mixture at 0.5 μM, to determine the nonspecific binding. Bound radioligand is separated from the unbound ligands by filtration and washing. The aliquots are diluted in a reservoir containing ice-cold washing buffer (10 m*M* Tris-HCl, pH 7.5), and filtered through Whatman GF/C filter discs. The filter discs are rapidly washed four times with a total of 20 ml of the washing buffer (0°) in less than 15 sec. It is important to keep the washing buffer cold and the process brief. Under these conditions, approximately 99% of the receptor-associated radioligand remains tightly bound, even with additional washing with 30 ml of ice-cold buffer. Total and nonspecific binding are defined as binding in the absence or presence of the excess cold ligand, respectively. The specific binding component is defined as the difference between total and nonspecific binding of the radioligand. Binding of the radiolabeled high-affinity antagonist ([^{3}H]ICI-198615) can be performed similarly. The nonspecific binding is determined in the presence of the unlabeled homologous ligand (ICI-198615).

To determine if the radioligand binding is reversible, a small volume (2–3 μl) of a high concentration (10 m*M*) of unlabeled agonist or antagonist homologous ligand is added singly, or premixed with other agents (e.g., GTP or NaCl), which could affect the association or dissociation rates of the radioligand, and added to the incubation mixtures at the steady-state of binding. Aliquots of the incubation mixture are retrieved and analyzed for membrane-bound radioactivity.

Equilibrium Saturation Binding of Radioligand. The assumption that radioligand binding to the specific sites is at a equilibrium state is usually determined by the manifestation of a stable, steady-state, reversible binding of the radioligands and the lack of extensive bioconversion or chemical degradation of the radioligand, during the time course of the binding study. Therefore, the exact conditions employed in the saturation binding should be determined from the data derived from the kinetic experiments. Using guinea pig lung membranes, the incubation mixture containing 20 m*M* PIPES buffer (pH 6.5), 5 m*M* cysteine, 5 m*M* glycine, 5 m*M* $CaCl_2$, 5 m*M* $MgCl_2$, membrane protein (10–30 μg/ml), and varying concentrations of radioligand in Nunc Minisorb test tubes are prepared in triplicate and incubated for 60 min at 22°. The concentration range of the radioligand and the volume of the incubation mixture can be adjusted based on the expected density of the receptor and the binding affinity of the radioligand to the receptor. It is also important *not* to add nonspecific binding protein, such as bovine serum albumin, to the incubation mixture. Glass test tubes and exogenous protein tend to interfere with radioligand–receptor interaction, especially in the subnanomolar concentration range of the radioligands used.

Competition of Radiolabeled Ligands. The exact conditions of radioligand competition experiments are dictated by the results obtained from saturation experiments. With guinea pig lung membranes, 0.5 n*M* $[^3H]LTD_4$, 0.8 n*M* $[^3H]LTE_4$ or 0.3 n*M* $[^3H]$ICI-198615 (approximately 3- to 4-fold of the respective K_Ds) are used to label 80–90% of the available sites in the membranes. Ca^{2+}, Mg^{2+}, cysteine, and glycine (5–10 m*M*) are included when $[^3H]LTD_4$ is used to prevent metabolic conversion of $[^3H]LTD_4$. Cysteine and glycine are omitted when $[^3H]LTE_4$ or $[^3H]$ICI-198615 is employed. The membranes are added to the incubation mixtures containing 20 m*M* PIPES buffer (pH 6.5), the radiolabeled ligand, and varying concentrations of the competing ligands in a volume of 0.5 ml in triplicate and incubated at 22° for 60 min.

Data Calculation. The radioactivity (cpm) determined from liquid scintillation spectrometry can be converted to dpm and then to fmol or pmol of the respective radioligand. Quenching of the radioactivity must be carefully determined and corrected. To determine the dissociation constant (K_D) and the maximum binding density (B_{max}), the Scafit program, originally described by Delean *et al.* (1980),[4] can be used to analyze the saturation data. The LIGAND or Superfit programs can also be used and are readily accessible in most research institutions.

Results and Discussion

Association and Dissociation of Radioligand Binding to Membranes

There are two major objectives in studying the kinetics of radioligand binding: (1) to establish the steady-state of radioligand receptor binding, and (2) to optimize the specific binding and minimize the metabolic conversion or degradation of the radioligand. The results in Fig. 1A illustrate the kinetics of $[^3H]LTD_4$ binding to guinea pig lung membranes. Note that binding of $[^3H]LTD_4$ to guinea pig lung membrane reaches a plateau 20 min after the reaction is initiated. Prolonged incubation (greater than 2 hr) or in the absence of cysteine and glycine results in a decrease of the level of specific binding[5] suggesting a loss of the receptors or $[^3H]LTD_4$ due to conversion to $[^3H]LTE_4$. The level of specific binding of $[^3H]LTD_4$ is decreased when EDTA (1–10 m*M*), Na^+ (1–100 m*M*), or GppNHp

[4] A. Delean, J. M. Stadel, and R. J. Lefkowitz, *J. Biol. Chem.* **255,** 7108 (1980).

[5] S. Mong, H.-L. Wu, J. M. Stadel, M. A. Clark, and S. T. Crooke, *Eur. J. Pharmacol.* **29,** 102 (1984).

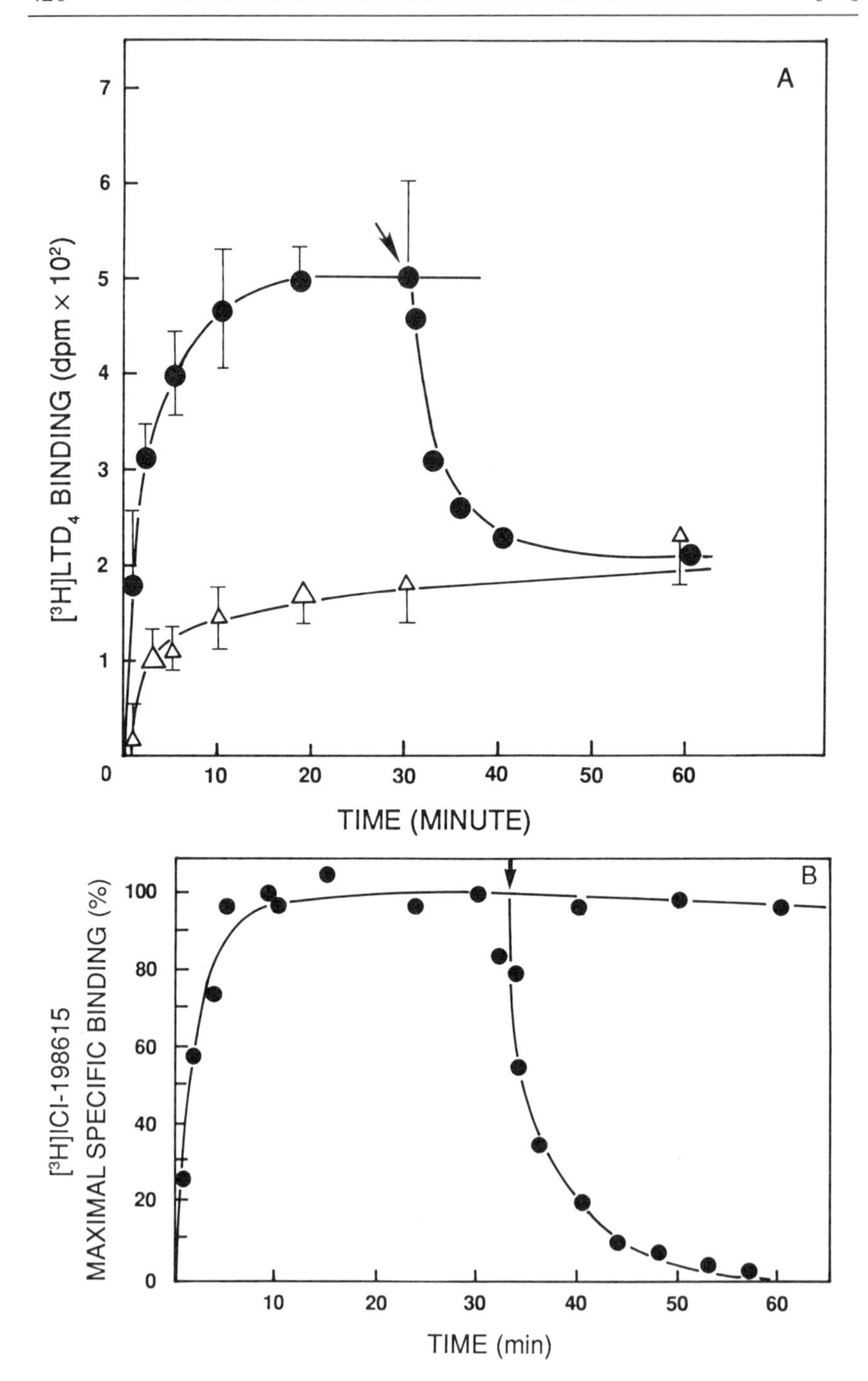
A
[³H]LTD₄ BINDING (dpm × 10²)
TIME (MINUTE)
0
10
20
30
40
50
60
1
2
3
4
5
6
7
B
[³H]ICI-198615
MAXIMAL SPECIFIC BINDING (%)
TIME (min)
0
20
40
60
80
100

(guanosine-5′-[βγ-imido]triphosphate, 1–100 μ*M*) is added to the incubation mixture.[5] The membrane-bound $[^3H]LTD_4$ dissociates slowly when excess LTD_4 is added to the plateau phase (30 min), yet a very rapid dissociation of $[^3H]LTD_4$ is observed when GppNHp is added together with LTD_4 to the incubation mixture[5] (Fig. 1A). These results (not shown) indicate that $[^3H]LTD_4$ binding to the specific sites is time-dependent, reversible, and enhanced by the metabolic conversion enzyme inhibitors. The kinetics of antagonist $[^3H]$ICI-198615 binding to guinea pig lung membranes is shown in Fig. 1B. Specific binding of $[^3H]$ICI-198615 is also time-dependent and reversible but not significantly affected by divalent or monovalent cations and guanine nucleotides (results not shown).

Binding of $[^3H]LTC_4$ to membranes can be determined using methods similar to that described above. In most of the cells or membrane preparations, conversion of $[^3H]LTC_4$ is rapid and not completely inhibitable by serine–borate. The kinetics of $[^3H]LTC_4$ binding must be carefully monitored.

Saturation Binding of Radioligands

The major objectives of the saturation binding experiments are to: (1) demonstrate that the number of the radioligand binding sites is limited and thus saturable, and (2) determine the density of the receptor and the affinity of the receptor for radioligand. The results shown in Fig. 2A illustrate the binding of $[^3H]LTD_4$ to guinea pig lung membranes. Nonspecific binding of $[^3H]LTD_4$ is linearly dependent on the concentration of $[^3H]LTD_4$. Specific binding of $[^3H]LTD_4$ is dependent on the $[^3H]LTD_4$ concentration and reaches near-plateau at higher concentrations. It is generally desirable to have at least 8 to 10 concentrations of $[^3H]LTD_4$ in a single experiment and to have several points near the K_D concentration. The experimental data can be analyzed with the aid of a computer-generated curve-fitting program. $[^3H]LTD_4$ specific binding in guinea pig lung membranes is best described by a model for a single class of specific sites. The K_D and B_{max} are calculated as 0.15 ± 0.05 n*M* and 1080 ± 50 fmol/mg, respectively. The saturation binding data can also be con-

FIG. 1. Association and dissociation of radioligands. (A) Guinea pig lung membranes (1000 μg/ml) are incubated with 0.5 n*M* $[^3H]LTD_4$ as described in the section on Methods. Aliquots of 100 μl are retrieved at varying time points and analyzed. Specific binding (●) and nonspecific binding (△) are calculated as described. LTD_4 (3.5 μ*M*, final concentration) and GppNHp (3 μ*M*, final concentration) are mixed and added to the incubation mixture at the indicated point (by arrow). (B) Guinea pig lung membranes are incubated with 0.3 n*M* $[^3H]$ICI-198615 as described. ICI-198615 (3 μ*M*) is added to the incubation mixture at steady state (indicated by arrow).

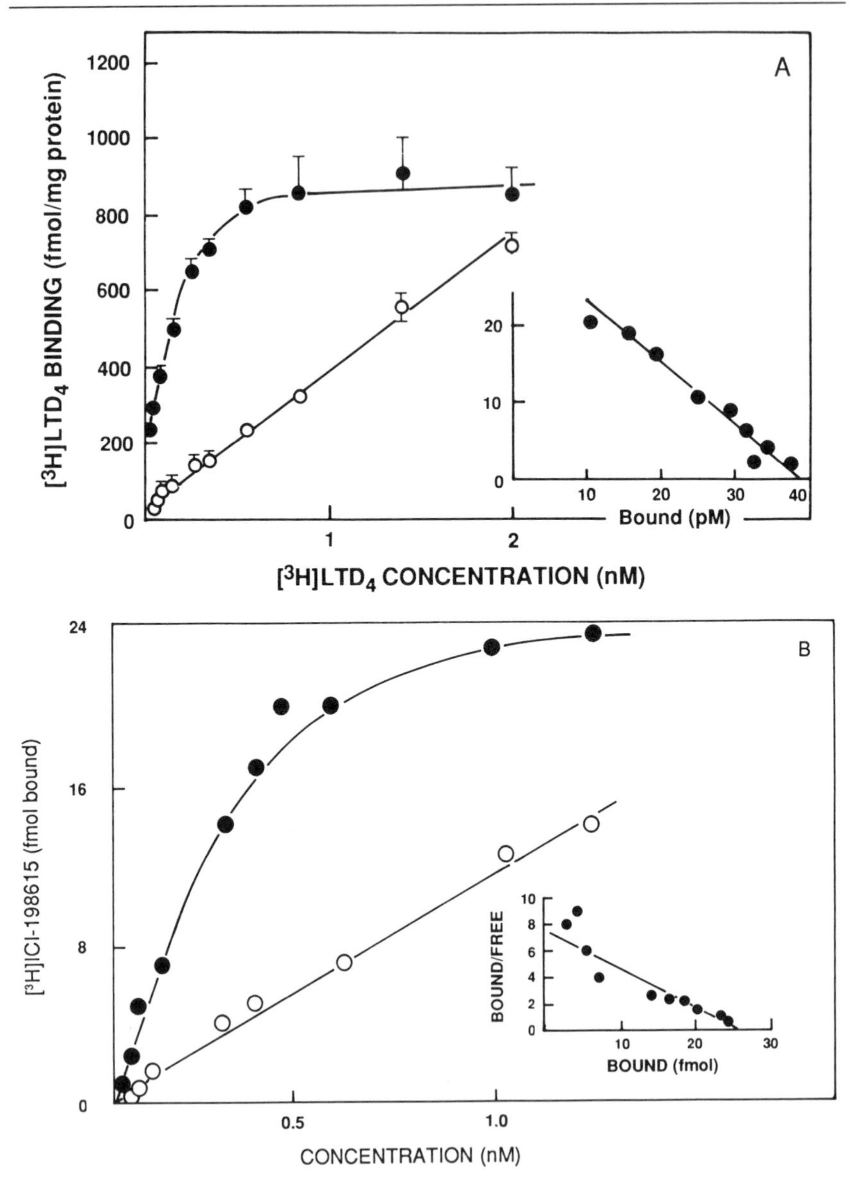

FIG. 2. Saturation binding of radioligands to guinea pig lung membranes. (A) Guinea pig lung membranes are incubated with varying concentrations of [^{3}H]LTD$_4$ as described in the text. Specific binding (●) and nonspecific binding (○) of [^{3}H]LTD$_4$ are calculated and converted by the Scatchard method (see inset). (B) [^{3}H]ICI-198615 was similarly used to label the LTD$_4$/LTE$_4$ receptor as described in the section on Methods.

verted by the Scatchard method to determine the K_D and B_{max} (Fig. 2A, inset). It is very difficult to assess the linearity or lack of linearity of the data points and then to speculate whether the data can be described by either a single- or two-site model of specific binding. The best way to determine this is through the use of the computer-aided best-fit analysis. A statistical significant improvement of the two-site model fit of the experimental data suggests that there are two discernible classes of specific binding sites. However, it does not prove that there are two classes of *receptors*.

These results show that [^{3}H]LTD_4 binds to the guinea pig lung-specific sites in a high affinity and saturable manner—the very first and the most important criterion of establishing the membrane receptor. Radiolabeled [^{3}H]LTE_4 can also be used to label the sulfidopeptide leukotriene receptor sites. [^{3}H]LTE_4 is more hydrophobic than [^{3}H]LTD_4. Nonspecific binding of [^{3}H]LTE_4, to membranes or nonbiological substrates, is higher than that of [^{3}H]LTD_4.[6] However, the advantage of using [^{3}H]LTE_4 is that it is much less susceptible to enzyme conversion or degradation under regular experimental conditions. Many laboratories have reported results using [^{3}H]LTC_4 to label specific binding sites. It was originally thought that [^{3}H]LTC_4 binding was a good assay for the putative LTC_4 receptors. However, so far, most of the [^{3}H]LTC_4 binding studies have yet to demonstrate that the specificity of the [^{3}H]LTC_4 binding site correlates with any physiological or pharmacological function. Invariably, the [^{3}H]LTC_4 specific binding sites turn out to be intracellular or membrane-bound enzymes.[7] It is important to be aware of these pitfalls. A clear demonstration of the physiological and functional correlation of the [^{3}H]LTC_4 binding studies must be established before one plans to conduct a series of experiments to study the "LTC_4 receptors."

Radiolabeled high-affinity receptor antagonist can also be used to study the LTD_4/LTE_4 receptors. Results in Fig. 2B show that specific binding of [^{3}H]ICI-198615 to guinea pig lung membranes is saturable ($B_{max} = 1030 \pm 180$ fmol/mg) and of high affinity ($K_D = 0.08 \pm 0.04$ nM). The nonspecific binding of [^{3}H]ICI-198615 to guinea pig lung membranes is linearly dependent on the radioligand concentrations used. It is important to use the unlabeled homologous ligand (ICI-198615) to define the displacable specific binding component, because other antagonists may bind to adventitious sites, in addition to the LTD_4/LTE_4 receptors (see below).

[6] S. Mong, M. O. Scott, M. A. Lewis, H.-L. Wu, M. A. Clark, G. K. Hogaboom, and S. T. Crooke, *Eur. J. Pharmacol.* **109,** 183 (1985).

[7] F. F. Sun, L. Y. Chau, B. Spur, E. J. Corey, R. A. Lewis, and K. F. Austen, *J. Biol. Chem.* **261,** 8540 (1986).

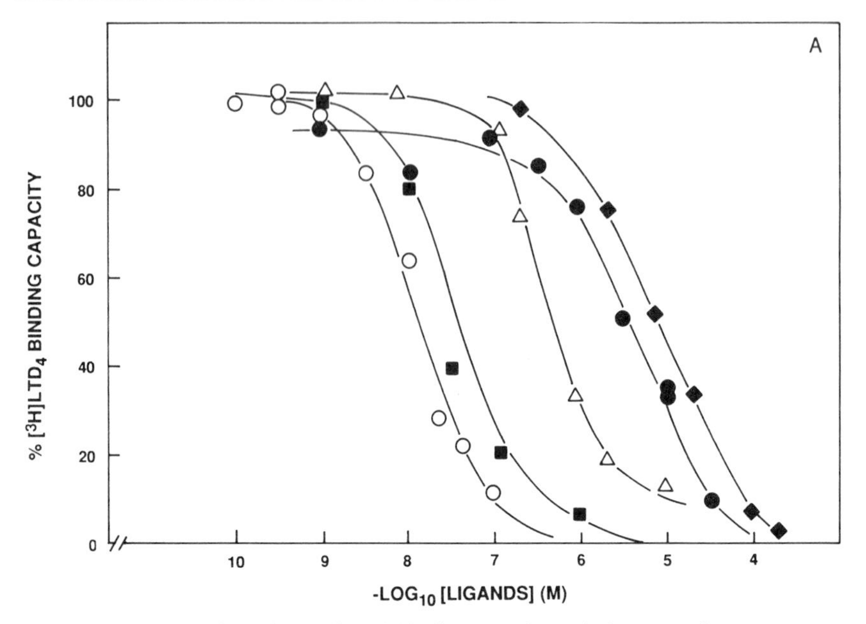

FIG. 3. Competition of radioligand binding to guinea pig lung membrane receptors. Guinea pig lung membranes are incubated with 0.5 n*M* [^{3}H]LTD$_4$ and (A) varying concentrations of LTD$_4$ (○), FPL55712 (●), 5(*R*),6(*S*)-LTD$_4$ (△), LTE$_4$ (■), 5(*R*),6(*S*)-LTE$_4$ (♦), and antagonists (B), SKF 104353 (▲), SKF 104373 (○), ICI-198615 (●), WY48252 (⊠) or R12525 (△) as indicated. (C) Alternatively, the guinea pig lung membranes are labeled with 0.3 n*M* [^{3}H]ICI-198615 and competed with varying concentrations of SKF 104353 (⊗), SKF 104373 (▲), ICI-198615 (●), WY48252 (□), FPL55712 (△), LY171884 (○) or LTD$_4$ (■) as described in the section on Methods.

Specificity of the Receptor Binding Sites

Specificity of the radioligand binding site is usually determined by competition studies. The major objectives of these studies are: (1) to establish a rank-order potency of the competitive ligands and (2) to correlate the rank-order potency to the biological activities, either agonistic or antagonistic activity, in the function assay(s). There are several classes of structurally different receptor antagonists[8] and many agonists for this purpose. The results in Fig. 3A show that LTD$_4$, LTE$_4$, LTC$_4$ and 5(*R*), 6(*S*)-LTD$_4$, in decreasing, rank-order potency, competes with [^{3}H]LTD$_4$ binding to the receptors. Antagonists such as ICI-198615, SKF 104353, R-17525, and WY48252 also compete with [^{3}H]LTD$_4$ binding to the recep-

[8] C. D. Perchonock, T. J. Torphy, and S. Mong, *Drugs Future* **12,** 872 (1987).

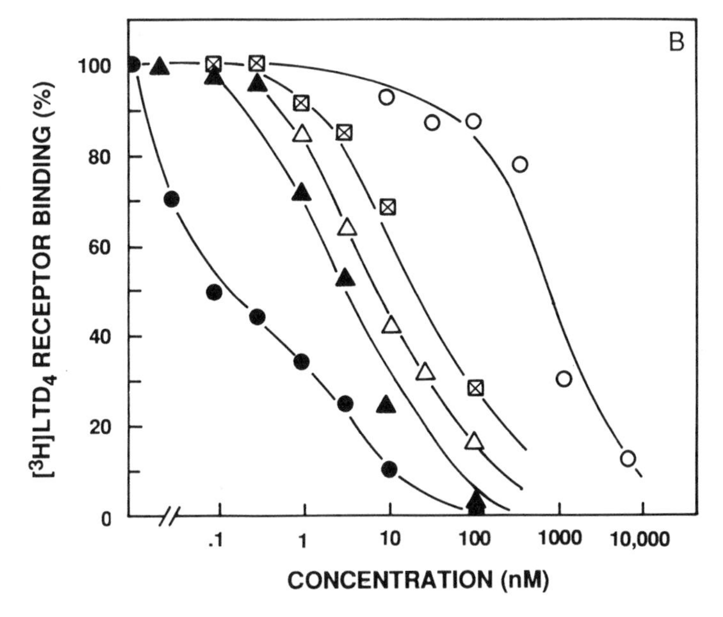

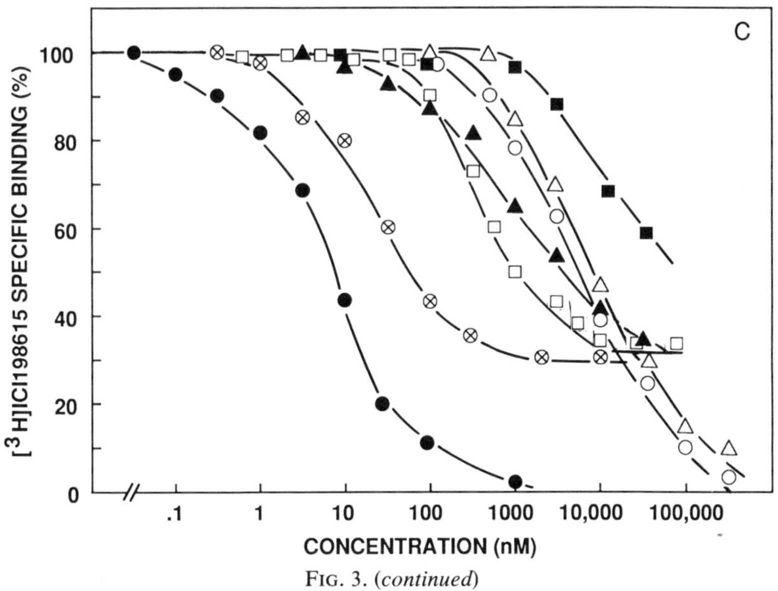

FIG. 3. (*continued*)

tors (Fig. 3B). These results clearly show that [^{3}H]LTD$_4$ binding to the membrane receptors sites is highly stereoselective and specific. The rank-order potency of displacement fully reflect the specificity of agonist activity and the pharmacological antagonist activity in the functional assay systems.[9] The specificity profile can also be studied by using [^{3}H]ICI-198615. The results in Fig. 3C demonstrate that ICI-198615, SKF 104353, WY48252 and SKF 104373, in decrease order of affinity, displaced [^{3}H]ICI-198615-specific binding to the guinea pig lung membranes. However, note that the maximal extent of displacement of [^{3}H]ICI-198615 by SKF 104353 was only 70 ± 5%. The maximal extent of displacement by LTD$_4$ was also 70 ± 10% (result not shown). These results suggest that the functionally defined receptor antagonists compete with the [^{3}H]ICI-198615 binding to the specific sites differently. A major portion of the [^{3}H]ICI-198615 labeled sites represent LTD$_4$/LTE$_4$ receptors in guinea pig lung membranes.

Recent reports demonstrate that, in the sulfidopeptide leukotriene receptor-bearing cells, tissues or organs, a guanine nucleotide-binding protein is critically involved in regulation of agonist binding to the receptor.[3,5-6] The G protein serves as a signal-transducing factor that communicates between the receptor and the intracellular catalyst, a phosphoinositide-specific phospholipase C. LTD$_4$ binding to the plasma membrane receptors promotes a G protein–receptor–ligand ternary complex.[10] However, in the target cells or tissues, formation of the membrane ternary high-affinity complex appears only transiently and leads to phospholipase C activation. Activation of phosphoinositide-specific phospholipase C results in the formation of intracellular messengers and protein kinase C activation,[11] receptor down-regulation[12] and many other biochemical processes. In the study of sulfidopeptide leukotriene receptors, especially involving whole cells, these factors should also be taken into consideration.

A survey of several classes of functionally defined LTD$_4$/LTE$_4$ receptor antagonist compounds and many nonspecific smooth muscle relaxation drugs revealed that the sulfidopeptide leukotriene receptor binding assay is highly selective and specific for the receptor antagonist compounds.[8] When appropriately controlled, this assay has been used for screening of naturally occurring agonists, antagonists or synthetic receptor antago-

[9] S. Mong, H.-L. Wu, M. O. Scott, M. A. Clark, B. M. Weichman, J. G. Gleason, C. Kinzig, and S. T. Crooke, *J. Pharmacol. Exp. Ther.* **234,** 316 (1985).

[10] S. Mong, H.-L. Wu, J. M. Stadel, and S. T. Crooke, *Mol. Pharmacol.* **29,** 235 (1986).

[11] R. V. K. Vagesna, S. Mong, and S. T. Crooke, *Eur. J. Pharmacol.* **147,** 387 (1988).

[12] J. D. Winkler, S. Mong, and S. T. Crooke, *J. Pharmacol. Exp. Ther.* **244,** 449 (1988).

nists. It has been used to study the membrane receptor regulation and desensitization.[12] Most recently, it has also been used, either in conjunction with, or in replacement of, the radioimmunoassay for the detection of sulfidopeptide leukotrienes in biological fluids (Mong *et al.*, results not shown).

[48] Ginkgolides and Platelet-Activating Factor Binding Sites

By D. J. HOSFORD, M. T. DOMINGO, P. E. CHABRIER, and P. BRAQUET

Introduction

Specific membrane receptors are assumed to mediate the various biological responses to platelet-activating factor (PAF),[1] the inflammatory autacoid produced by neutrophils,[2] eosinophils,[3] monocytes,[4] macrophages,[5] platelets,[6] and endothelial cells.[7] However, the failure to isolate and characterize the putative receptor leaves open the question of the nature and heterogenity of the binding site(s) in various cells and tissues. Some progress has been made in the formulation of a model of the PAF receptor by studies with various compounds that can specifically antagonize PAF-induced biological effects both *in vitro* and *in vivo*.[8,9] Among these compounds, the unique 20-carbon cage molecules named ginkgolides[10,11] (Fig. 1) have been shown to specifically inhibit PAF-induced actions, such as platelet aggregation, hypotension, bronchoconstriction,

[1] P. Braquet, L. Touqui, T. S. Shen, and B. B. Vargaftig, *Pharmacol. Rev.* **39,** 97 (1987).
[2] J. M. Lynch, G. Z. Lotner, S. J. Betz, and P. M. Henson, *J. Immunol.* **123,** 1219 (1979).
[3] T. C. Lee, D. J. Lenihan, B. Malone, L. L. Roddy, and S. I. Wasserman, *J. Biol. Chem.* **259,** 5530 (1984).
[4] G. Camussi, M. Aglietta, R. Coda, F. Bussolino, W. Piacibello, and C. Tetta, *Immunology* **42,** 191 (1981).
[5] J. M. Mencia-Huerta and J. Benveniste, *Eur. J. Immunol.* **9,** 409 (1979).
[6] M. Chignard, J. P. Le Couedic, M. Tencé, B. B. Vargaftig, and J. Benveniste, *Nature (London)* **275,** 799 (1979).
[7] S. M. Prescott, G. A. Zimmerman, and T. M. McIntyre, *Med. Sci.* **81,** 3534 (1984).
[8] P. Braquet, P. E. Chabrier, and J. M. Mencia-Huerta, *Adv. Inflammation Res.* **12,** 135 (1987).
[9] D. Hosford, J. M. Mencia-Huerta, C. Page, and P. Braquet, *Phytother. Res.* **2,** 1 (1988).
[10] P. Braquet, B. Spinnewyn, M. Braquet, R. H. Bourgain, J. E. Taylor, A. Etienne, and K. Drieu, *Blood Vessels* **16,** 559 (1985).
[11] P. Braquet, *Drug Future* **12,** 643 (1987).

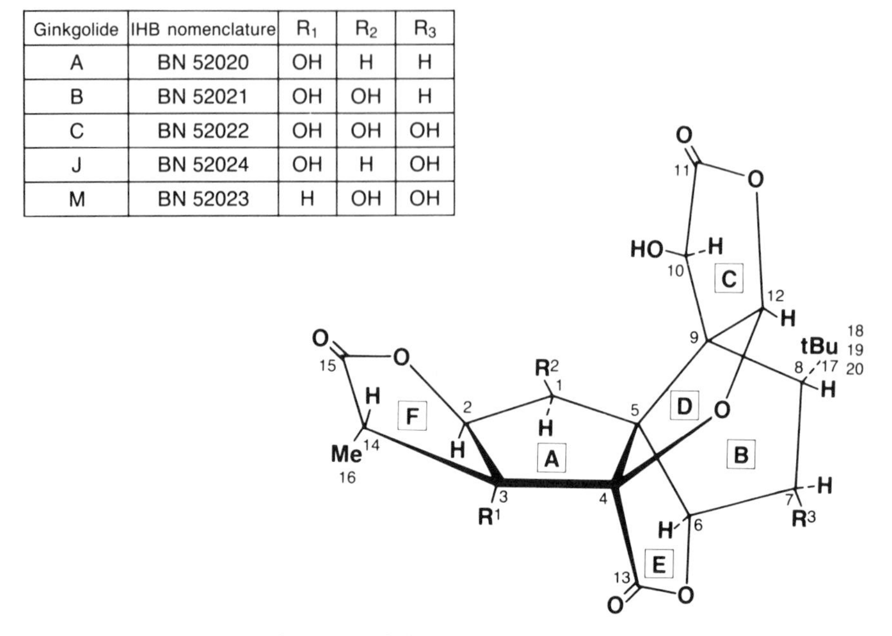

Ginkgolide	IHB nomenclature	R_1	R_2	R_3
A	BN 52020	OH	H	H
B	BN 52021	OH	OH	H
C	BN 52022	OH	OH	OH
J	BN 52024	OH	H	OH
M	BN 52023	H	OH	OH

FIG. 1. Structure of ginkgolides A, B, C, J, and M.

and to exert a protective effect in experimentally induced diseases such as brain ischemia[12,13] and eye inflammation.[14] At the present time, five ginkgolides (A, B, C, M, and J) have been identified. Coded BN 52020, BN 52021, BN 52022, BN 52023, and BN 52024, respectively, by the Institut Henri Beaufour, studies with ginkgolides and other PAF antagonists have implicated PAF in a wide range of pathologies including asthma, shock, ischemia, and graft rejection.[11,15] The development of these compounds into potential therapeutic agents is currently being assessed in clinical trials.[16]

The interaction of ginkgolides with PAF binding sites has been investigated in various tissues by binding studies using tritiated PAF ([^{3}H]PAF),

[12] B. Spinnewyn, N. Blavet, F. Clostre, N. G. Bazan, and P. Braquet, *Prostaglandins* **34,** 333 (1987).

[13] P. Braquet, M. Paubert-Braquet, M. Koltai, R. Bourgain, F. Bussolino, and D. Hosford, *Trends Pharmacol. Sci.* **10,** 23 (1989).

[14] N. L. J. Verbeij and N. J. van Haerigen, *in* "Ginkgolides: Chemistry, Biology, Pharmacology and Clinical Perspectives" (P. Braquet, ed.), Vol. 1, pp. 749-758. Prous Science, Barcelona, 1988.

[15] P. Braquet (ed.), "The Ginkgolides: Chemistry, Biology, Pharmacology and Clinical Perspectives." Vol. 1. Prous Science, Barcelona, 1988.

[16] D. Hosford and P. Braquet, *Prog. Med. Chem.* **27,** 325–379 (1990).

since high specific activity radiolabeled ginkgolides are not currently available. Ginkgolide binding has been measured on membranes from platelet or tissue homogenates, as a function of time and concentration, in order to further elucidate the nature, location, and physiological relevance of PAF receptor sites in these systems. This chapter examines ginkgolide interactions with PAF binding sites in three different preparations: rabbit platelet membranes, gerbil brain tissue, and rabbit iris and ciliary body.

Methods and Materials

Biologically relevant studies on [^{3}H]PAF binding require investigation of the concentration range 10^{-11} to 10^{-6} *M*, in the presence of 0.025% bovine serum albumin (BSA), the latter agent acting as a carrier for the PAF. In our investigations the bound ligand is separated from the free ligand by filtration procedures and characteristics of PAF binding such as dissociation constant (K_D) and number of sites (B_{max}) are obtained from concentration-dependent experiments, subjected to Scatchard analysis, and plotted as bound/free versus bound (B/F vs B).

The equilibrium dissociation constant (K_i) of ginkgolides on PAF binding sites in the various tissues is derived from displacement studies and calculated according to the Cheng–Prusoff equation:

$$K_i = \frac{IC_{50}}{1 + [PAF]/K_D}$$

where [PAF] is the molar concentration of the labeled PAF and K_A is the PAF dissociation constant.

Rabbit Platelet Membranes

In this study, the effect of the ginkgolides on [^{3}H]PAF binding is evaluated using ginkgolide B (BN 52021), ginkgolide A (BN 52020), ginkgolide C (BN 52022), and a mixture (BN 52063) of the three ginkgolides B, A, and C in the molar ratio 2 : 2 : 1.

Chemicals. Synthetic [^{3}H]PAF (specific activity: 59.5 Ci/mmol) in ethanol solution is purchased from New England Nuclear (Boston, MA). Unlabeled PAF and lyso-PAF (Calbiochem, Switzerland), are solubilized in ethanol solution and stored at −80°. BN 52021, BN 52020, BN 52022, and BN 52063 are solubilized in dimethyl sulfoxide (DMSO).

Platelet Membrane Preparation. Rabbit whole blood (6 volumes) is drawn from the central ear artery into 1 volume of ACD solution (citric acid, 1.4 g, sodium citrate, 2.5 g, and dextrose, 2 g per 100 ml of water) and centrifuged at 150 *g* for 15 min at 4°. The platelet-rich plasma (PRP) is carefully removed and centrifuged for an additional 15 min at 1000 *g* at 4°. The platelet pellet is then washed three times, twice in 10 m*M* Tris-HCl buffer, pH 7.4, containing NaCl 150 m*M*, $MgCl_2$ 5 m*M*, and EDTA 2 m*M* and, finally, in the same buffer but without sodium.

The platelet pellet is resuspended in this latter buffer, quickly frozen in liquid nitrogen, and slowly thawed at room temperature. The freeze/thaw cycle is repeated at least three times as described by Shen *et al.*[17] The lysed platelets are then centrifuged at 100,000 *g* for 30 min at 4° in a Beckman L8.55 ultracentrifuge (rotor 50.2 Ti). The platelet membrane homogenate is stored at −80° and used within 2 weeks. Protein content is determined by the Lowry method[18] using BSA as standard.

Binding Assay. Sixty to 100 μg of membrane protein is added to plastic tubes containing 1 n*M* [^{3}H]PAF and 1 ml of 10 m*M* Tris-HCl buffer, pH 7.0, containing 0.025% BSA and incubated with or without unlabeled PAF or ginkgolides. The incubation is carried out for 90 min at 0°. The bound [^{3}H]PAF is separated from the free [^{3}H]PAF by immediate filtration through Whatman GF/C glass fiber filters under a Brandel vacuum system (Brandel Biomedical Research, Gaithersburg, MD). Filters are washed three times with 5 ml of ice-cold buffer. The filters are then placed in polyethylene vials containing 10 ml of liquid scintillation fluid (Instagel, Packard, Paris) and the radioactivity measured by an LKB β-counter with 45% efficiency.

The nonspecific binding is determined in the presence of 10^{-6} *M* of unlabeled PAF. The specific binding is calculated by subtracting nonspecific from total binding. The inhibition by ginkgolides of the specific [^{3}H]PAF binding is expressed as percentage inhibition (% *I*) using the equation:

$$\% I = \frac{[^3\text{H}]\text{PAF (tb)} - [^3\text{H}]\text{PAF bound in presence of ginkgolides}}{[^3\text{H}]\text{PAF specifically bound}} \times 100$$

where [^{3}H]PAF (tb) is [^{3}H]PAF totally bound.

Gerbil Brain Tissue

The involvement of PAF in cerebral ischemia has been suggested by studies showing that PAF antagonists including BN 52021, kadsurenone, and brotizolam improve cerebral metabolism and protect the brain in the postischemic phase induced by bilateral carotid ligature in the gerbil.[12,13] Thus, we investigated the existence of specific [^{3}H]PAF binding sites in gerbil brain membrane homogenate.

[17] T. Y. Shen, S. B. Hwang, M. N. Chang, T. W. Doebber, M. H. T. Lam, M. S. Wu, X. Wang, G. Q. Han, and R. Z. Li, *Proc. Natl. Acad. Sci. U.S.A.* **82,** 672 (1985).

[18] O. H. Lowry, N. J. Rosebrough, A. L. Farr, and J. Randall, *J. Biol. Chem.* **193,** 265 (1951).

Brain Tissue Preparation. Male gerbils *Meriones unguiculatus* (60–70 g) are purchased from Tumblebrook farm (West-Brookfield, UK). Animals are allowed free access to food and water prior to experimentation. Gerbils are killed by decapitation and the brain quickly dissected out and placed into ice-cold 50 mM Tris-HCl buffer, pH 7.4, containing $MgCl_2$ 10 mM, EDTA 2 mM, Trasylol (aprotinin) 50 IU/ml, and PMSF 1×10^{-4} M. In some experiments the brain is perfused using the homogenizing buffer, through the left ventricle and under anesthesia. Brain tissue is then washed and homogenized in the same buffer using a Dounce homogenizer at 4°. The homogenates are centrifuged at 1000 g for 15 min at 4°. The microsome, mitochondria, and synaptosome-rich supernatant is harvested and recentrifuged at 10,000 g for 30 min. The P_2 pellet is then resuspended in the same buffer, assayed for protein content, and used for the binding assay. Protein content is determined by the Bradford method[19] using Bio-Rad protein assay reagent and BSA as the standard. The brain tissue preparation is used within the same day as isolation.

Binding Assay. Protein, 150–350 μg, from the P_2 preparation is added to plastic tubes containing [^{3}H]PAF (3 nM for kinetic and displacement studies) and 1 ml 10 mM Tris-HCl buffer, pH 7.0, containing $MgCl_2$ 5 mM, Trasylol (50 IU/ml), PMSF 1×10^{-4} M, and 0.025% BSA without or with 1000-fold unlabeled PAF (for nonspecific binding). The reaction is carried out at 25° for 30 min. The unbound PAF is separated from the bound by immediate filtration of the reaction mixture through Whatman GF/B glass fiber filters presoaked in the binding buffer. Tubes and filters are washed three times with 5 ml of precooled binding buffer. The filters are then placed into polyethylene vials containing 10 ml of liquid scintillation fluid (Instagel, Packard) and the radioactivity measured in a LKB β-counter with 60% efficiency.

The PAF specific binding is calculated by subtracting the nonspecific binding (binding in presence of 10^{-6} M unlabeled PAF) from the total binding. Binding data are analyzed by the Scatchard method using the computerized program of Vindimian *et al.*[20] The specificity of binding is investigated by competition experiments using lyso-PAF, unlabeled PAF, and BN 52021 to displace 1 nM [^{3}H]PAF.

Rabbit Iris and Ciliary Body

The demonstration that BN 52021 exerts significant protective effects in inflammatory eye diseases,[14] suggests the involvement of PAF in such conditions. In order to investigate the possible presence of PAF receptor

[19] M. Bradford, *Anal. Biochem.* **72,** 248 (1976).

[20] E. Vindimian, C. Robaut, and G. Fillion, *Appl. Biochem.* **5,** 261 (1983).

sites in eye tissue, we carried out binding studies on rabbit iris and ciliary body.

Tissue Preparation. Pigmented rabbits (either sex) are killed by an overdose of pentobarbital 5 min after injection of heparin (1000 IU per rabbit). The eyes are enucleated and placed into precooled 50 m*M* Tris buffer, pH 7.4, containing 10 m*M* $MgCl_2$, 2 m*M* EDTA, 50 IU Trasylol per ml, and 1×10^{-4} *M* PMSF. The eyes are perfused arterially in order to remove the blood from the anterior segment. The iris and ciliary body are carefully dissected out, homogenized in a Dounce homogenizer, the homogenates filtered through two layers of gauze, and centrifugated twice at 40,000 *g* for 20 min at 4°. The pellets are finally resuspended in the homogenization buffer and analyzed for protein content by the Bradford method[19] using Bio-Rad protein assay reagent and BSA as standard.

Binding Assay. Inhibition of [^{3}H]PAF binding to iris and ciliary homogenates is performed in a final volume of 1 ml buffer, using plastic tubes that contain [^{3}H]PAF (10^{-9} *M*), unlabeled PAF (1 μM, for nonspecific binding), lyso PAF or BN 52021, and 40 to 70 μg of protein diluted in the binding buffer. The binding buffer (pH 7.0) contains Tris 10 m*M*, $MgCl_2$ 5 m*M*, Trasylol 50 IU per ml, PMSF 0.1 m*M*, and BSA 0.025%.

After 40-min incubation at 25°, the unbound PAF is separated from the bound by filtration through Whatman GF/B glass fiber filters presoaked for 24 hr in the binding buffer and using a Brandel vacuum system. The assay tubes and filters are washed three times with 3 ml of precooled binding buffer and the filters placed in polyethylene vials containing 10 ml of liquid scintillation fluid (Instagel, Packard). The radioactivity is measured in a LKB β counter with 45% efficiency. The percentage inhibition is determined as previously described. In some experiments, stability of [^{3}H]PAF is analyzed after a 40-min incubation by extraction from the reaction mixture, using the solvent system, chloroform: methanol (2 : 1, v/v) and separation on TLC silica plates using a chloroform : methanol : ammonia (70 : 35 : 7, v/v/v) system. In our binding conditions, intact [^{3}H]PAF represents 93% of the total radioactivity recovered from the plates.

Results

Rabbit Platelet Membranes

The binding of [^{3}H]PAF to rabbit platelet membrane preparations is found to be specific, saturable, time-dependent, reversible, and of high affinity. The specific binding is a linear function of membrane protein concentration from 60 to 120 μg protein. Experiments using increasing concentrations of [^{3}H]PAF from 1 to 8 n*M* demonstrate a specific and

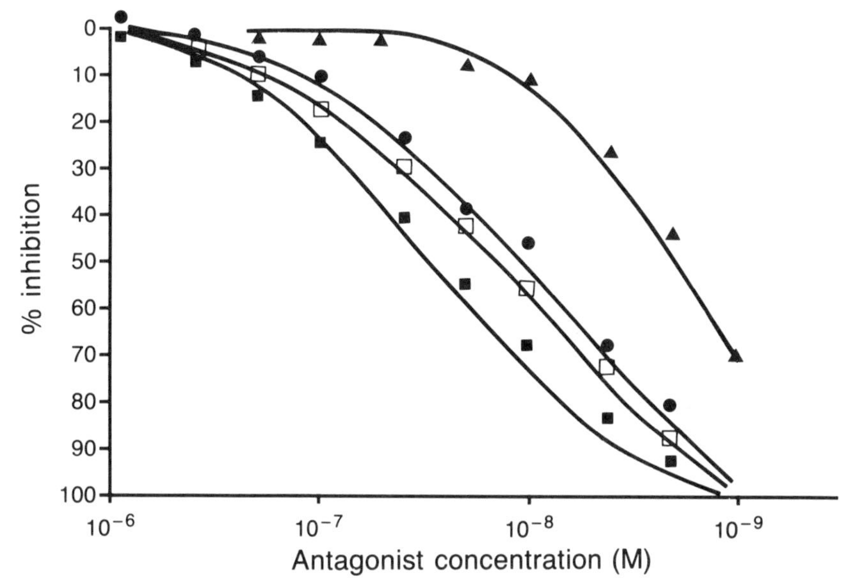

FIG. 2. Inhibition of [^{3}H]PAF (1 nM) specific binding to rabbit platelet membrane by BN 52020 (●), BN 52021 (■), BN 52022 (▲), and BN 52063 (□).

saturable binding, specific [^{3}H]PAF binding accounting for about 45% of the total. Scatchard analysis of the binding data reveals a single class of binding sites with a K_D of 3 nM and a B_{max} of 2.25 ± 0.2 pmol/mg protein.

The displacement of [^{3}H]PAF (1 nM) binding in rabbit platelet membranes by BN 52020, BN 52021, BN 52022, and BN 52063 is shown in Fig. 2. All these compounds inhibit [^{3}H]PAF binding in a dose-dependent manner. Maximum inhibition of [^{3}H]PAF binding by the ginkgolides is observed with BN 52021 with an IC_{50} value of 2.5×10^{-7} M. BN 52063, BN 52020, and BN 52022 exhibit IC_{50} values of 7×10^{-7} M, 8.3×10^{-7} M, and 6×10^{-6} M, respectively. In comparison, the concentration required to inhibit specific [^{3}H]PAF binding by 50% is 2×10^{-9} M for unlabeled PAF, whereas lyso-PAF is unable to displace [^{3}H]PAF binding below concentrations of 10^{-5} M.

The specificity of the ginkgolide compounds for PAF binding sites is also assessed in experiments using other radioligands. BN 52021 and the other ginkgolides (10^{-8}–10^{-5} M) are found to be totally inactive in inhibiting the binding of the appropriate ^{3}H-labeled ligands to α_1-, α_2-, β_1-, and β_2-adrenergic receptors, H_1 and H_2 histamine receptors, $5HT_1$ and $5HT_2$ serotoninergic receptors, μ, δ, and χ opiate receptors, benzodiazepine, and muscarinic and imipraminic receptors.

As reported previously,[21] the binding of [^{3}H]-PAF is modulated by the presence of mono- and divalent cations. A decrease of the binding is observed with the monovalent cations Li^+ and Na^+, whereas K^+ and divalent cations such as Mg^{2+} and Ca^{2+} enhanced the specific [^{3}H]PAF binding (data not shown). In this respect, the inhibition of [^{3}H]PAF binding by BN 52021, the most effective ginkgolide, is investigated in the presence of high concentrations of NaCl or $MgCl_2$. In the presence of 150 m*M* NaCl, the IC_{50} of the displacement curve is shifted to 4.8×10^{-8} *M*, while with 5 m*M* $MgCl_2$ the displacement curve is shifted to the right with an increase of the IC_{50} value to 5×10^{-7} *M*.

Gerbil Brain Tissue

The specific binding is saturable, representing 20–25% of the total binding and constituting about 5% of the total radioactivity added to the incubation medium. Scatchard plot of the binding data from saturation and competition experiments is curvilinear and could be described by a two-component binding model on computerized analysis, indicating the existence of two apparent classes of binding sites (Fig. 3). The first population of PAF recognition sites exhibits a K_{D_1} of 3.66 ± 0.56 n*M* and a B_{max_1} of 0.83 ± 0.23 pmol/mg of membrane protein, while the second population has a K_{D_2} of 20.4 ± 0.56 n*M* and a B_{max_2} of 1.1 ± 0.32 pmol/mg of membrane protein ($n = 5$). The association of [^{3}H]PAF to gerbil brain homogenate at 25° reaches equilibrium within 15 min and remains stable for at least 60 min.

The specificity of binding is established by displacing 1 n*M* [^{3}H]PAF by lyso-PAF, unlabeled PAF, and by BN 52021 in competition experiments. Lyso-PAF is ineffective up to 10^{-5} *M* in competition with [^{3}H]PAF for the binding sites. Unlabeled PAF (10^{-10}–10^{-6} *M*) and BN 52021 (10^{-10}–10^{-5} *M*) dose-dependently inhibits the specific binding of [^{3}H]PAF to brain membrane preparations with IC_{50} values of 2.5×10^{-9} *M* and 3.5×10^{-8} *M*, respectively (Fig. 4). Interestingly, in this tissue, BN 52021 fails to totally inhibit the [^{3}H]PAF binding.

We further evaluated whether the binding of [^{3}H]PAF is associated to discrete brain areas by studying the distribution of [^{3}H]PAF binding in various regions of the gerbil brain. The maximum amount of specific binding is found in the midbrain and the hippocampus, with less binding being apparent in the olfactory bulb, frontal cortex, and cerebellum.

Rabbit Iris and Ciliary Body

Specific [^{3}H]PAF binding is found in cornea, iris, and ciliary body and is characterized in the latter two. The specific binding represents 20% of

[21] S. B. Hwang, M. H. Lam, and S. S. Pong, *J. Biol. Chem.* **261,** 13720 (1986).

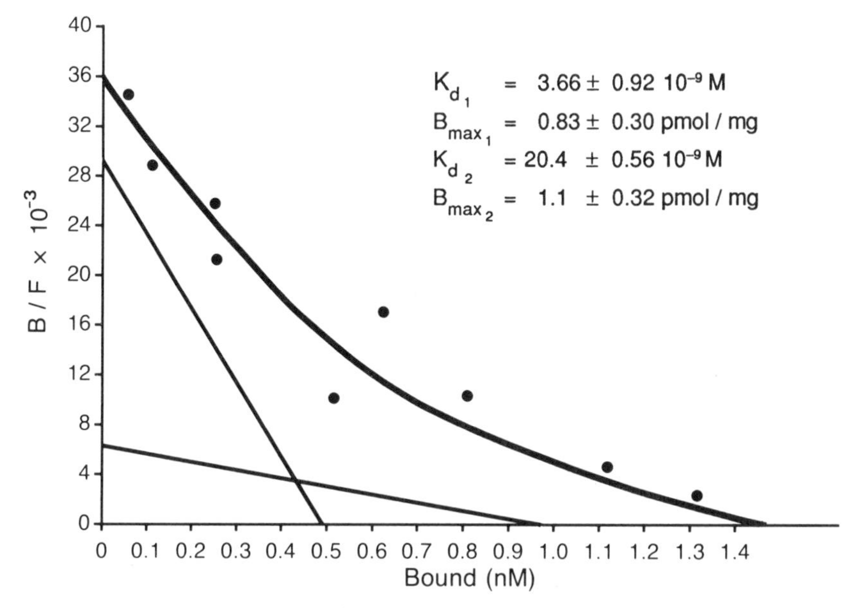

FIG. 3. Scatchard analysis of [^{3}H]PAF binding in gerbil brain homogenate. The experiments are performed at 25° and each point is the average of triplicate determinations ($n = 5$).

the total binding and is linear over the concentration range assayed, up to 90 μg of protein per tube. For kinetic and competition studies, [^{3}H]PAF is used at a concentration of 3 nM. At 25° the association reaches equilibrium within the first 30 min of incubation and remains stable up to 60 min with this latter concentration. An excess of unlabeled PAF added to the reaction mixture at 25 min demonstrates a reversibility of 50% within 15 min. No reversibility is observed at 45 min, even after addition of an excess of unlabeled PAF.

Increasing concentrations of labeled PAF over the range of 1 to 14 nM exhibit saturable binding and show a two-step saturation isotherm. The first maximum is reached at 4–5 nM and Scatchard analysis derived from both saturation and competition studies demonstrates a curvilinear repartition of the points suggesting two subtype components (Fig. 5). For the iris, a high-affinity component with a K_{D_1} of 4.9 ± 0.47 nM and B_{max_1} of 3.17 ± 0.5 pmol/mg protein and a low-affinity component with a K_{d_2} of 11.6 ± 0.33 nM and B_{max_2} of 12.46 ± 2.3 pmol/mg is observed. Similar values are found for ciliary body: $K_{D_1} \simeq 5.7 \pm 0.09$ nM; $B_{max_1} \simeq 3.41 \pm 1$ pmol/mg and $K_{D_2} \simeq 24.4 \pm 0.91$ nM; $B_{max_2} \simeq 16.6 \pm 0.51$ pmol/mg.

In our study, the specificity of PAF binding to iris and ciliary body is confirmed by inhibition of the [^{3}H]PAF binding in the presence of un-

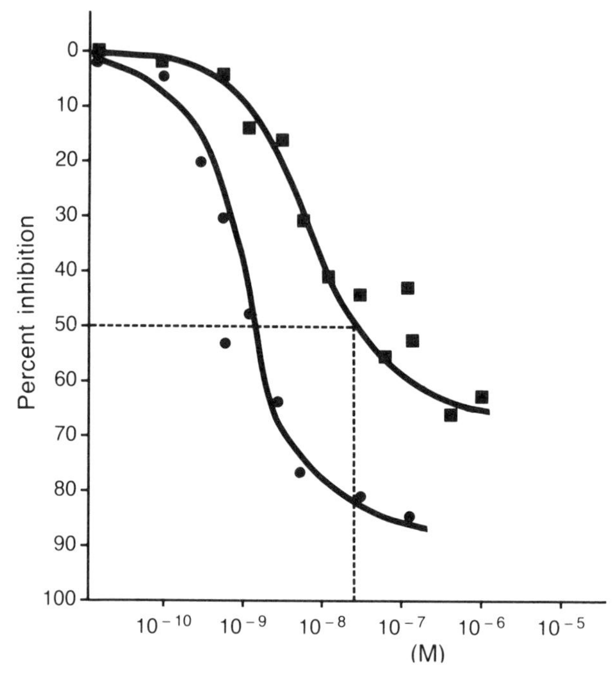

FIG. 4. Inhibition of [^{3}H]PAF (1 n*M*) binding to gerbil brain homogenate by unlabeled PAF (●; 10^{-10}–10^{-6} *M*, $C_{50} = 5 \times 10^{-8}$ *M*) and BN 52021 (■; 10^{-10}–10^{-5} *M*). Each point is the mean of triplicate determinations ($n = 3$).

labeled PAF, lyso-PAF or BN 52021. Displacement of the binding by unlabeled PAF demonstrates a biphasic inhibition curve. Lyso-PAF at a concentration below 10^{-6} *M* fails to inhibit the binding. The inhibition curve obtained with BN 52021 differs from that observed with unlabeled PAF and is only partial (Fig. 6). The maximum inhibition obtained is about 60% and appears to be associated with only one component of PAF displacement curve.

Discussion

Ginkgolides are specific PAF antagonists isolated from the roots and leaves of the fossil tree *Ginkgo biloba*. Extracts of this tree have been used by the Chinese for some 5000 years to alleviate asthma and other inflammatory conditions, the beneficial effects of the extract being largely attributable to the PAF antagonizing effects of its constituents. The chemistry, biology, and pharmacology of these compounds have been extensively

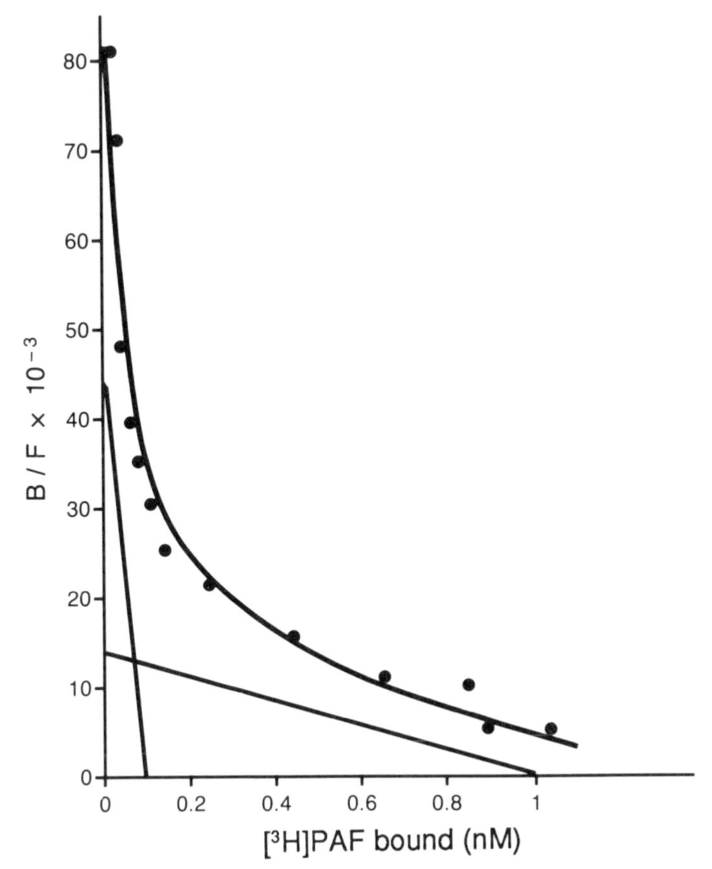

FIG. 5. Scatchard analysis of [^{3}H]PAF binding to iris homogenate in 10 mM Tris buffer, pH 7, containing 5 mM $MgCl_2$, PMSF 1×10^{-4} M, Trasylol 50 IU/ml, and BSA 0.025% at 25°. Each point is the average of triplicate determinations ($n = 4$).

reviewed in various recent publications.[11,15] The results presented here demonstrate the presence of specific binding sites for [^{3}H]PAF in rabbit platelet membranes, gerbil brain tissue, and rabbit iris and ciliary body, the binding of the mediator in these systems being specifically antagonized by BN 52021 and other ginkgolides.

In rabbit platelet membranes, the ginkgolides dose-dependently antagonize specific PAF binding, with IC_{50} values of 2.5×10^{-7} M, 8.3×10^{-7} M, and 6×10^{-6} M for BN 52021, BN 52020, and BN 52022, respectively. The inhibition of PAF recognition sites is highly specific since the ginkgolides do not interact with any other known receptors (see section on Results). Characterized by the presence of two hydroxyl groups on car-

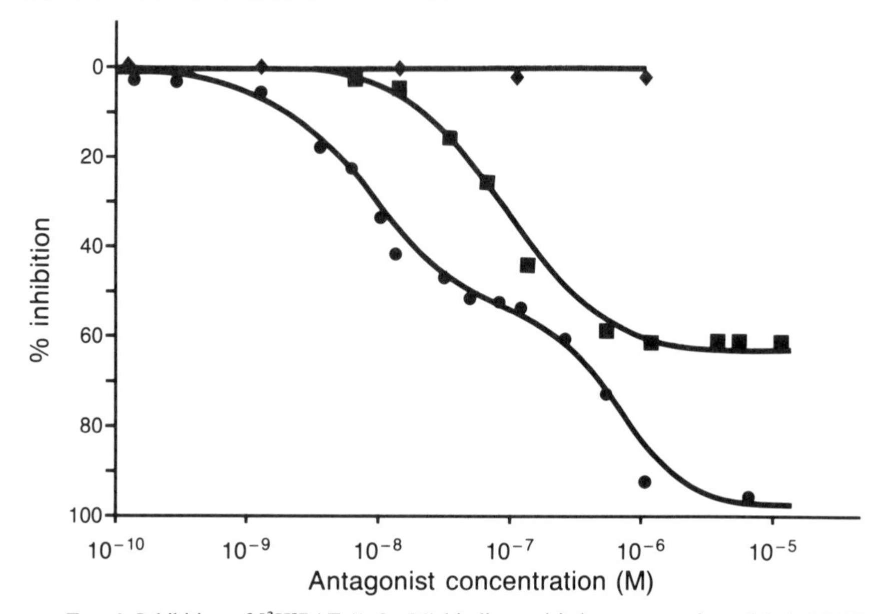

FIG. 6. Inhibition of [^{3}H]PAF (1–3 n*M*) binding to iris homogenate by unlabeled PAF (●—●), BN 52021 (■—■), and lyso-PAF (♦—♦). Each point is the average of triplicate determinations (*n* = 4).

bons 1 and 3, BN 52021 is the most powerful antagonist. The loss of the hydroxyl group on carbon 1 in BN 52020 and the presence of a hydroxyl group on carbon 7, in the α position of the lipophilic *tert*-butyl moiety in BN 52022 render these compounds less active. Studies reported elsewhere[11] have shown BN 52024 to possess an IC_{50} value of 5.4×10^{-5} *M* in this assay, the loss of the hydroxyl group on carbon 1 further decreasing the activity in comparison to BN 52022. These data corroborate physicochemical findings showing that as the ginkgolide becomes less polar, its PAF antagonistic activity increases.[11]

These binding studies corroborate results obtained for the inhibition of PAF-induced platelet aggregation, where BN 52021 is the most effective antagonist followed by BN 52020 and 52022.[11] Moreover, the affinity of BN 52021 for PAF receptor sites is altered by Na^+ and Mg^{2+}. Indeed, the state of affinity of PAF receptor appears to be low in the presence of Na^+ and high in the presence of Mg^{2+}.[21] Accordingly the ability of BN 52021 to displace specific [^{3}H]PAF binding is dependent on the ionic conditions. This phenomenon has also been observed with other PAF antagonists, such as CV-3988, ONO 6240, and L-652,569, but, not with kadsurenone which binds equally well to the PAF receptor at both the high- and low-affinity state.[21]

The finding that BN 52021 inhibits specific [^{3}H]PAF in gerbil brain tissue may account for the protective effect of the compound in various models of cerebral ischemia.[12,13] The direct radioligand binding studies reported here indicate that the PAF recognition sites are most likely located on brain cells and not on blood elements since PAF binding characteristics were similar whether or not the brain was perfused and did not correlate with the level of cerebral vasculature. Scatchard analysis of the binding data indicates the existence of two populations of PAF binding sites in gerbil brain while other tissues such as platelets (see section on Methods and Materials) and lung[22] exhibit only one class of site. Furthermore, the [^{3}H]PAF specific binding to whole brain homogenate is totally displaced by unlabeled PAF but only partially inhibited by BN 52021. Thus, it is possible that BN 52021 and other PAF antagonists interact only with one site, although the present data do not indicate which receptor population is involved.

Similar to the case with brain tissue and that recently reported for rat retina,[23] two populations of PAF binding sites were also detected in iris and ciliary body. The specific binding in these latter tissues (20% of the total binding) is low as compared to that obtained, in the same buffer conditions, for platelets (70–80%), but is similar to the 20–25% found in brain tissue or to the 30–40% in lung tissue.

In rabbit iris and ciliary body, specific PAF binding exhibited a two-step saturation isotherm, suggesting heterogenous binding characteristics. The displacement by unlabeled PAF exhibited a biphasic curve, which could be described by a two-component computerized analysis. Inhibition of [^{3}H]PAF binding by BN 52021 was monophasic, partial (60%), and only involved the first step of the PAF inhibition curve. Despite this lack of total inhibition of PAF binding, BN 52021 significantly inhibits PAF-induced plasma leakage and edema formation in the rat retina[24] and the acute rise in intraocular pressure (IOP) caused by topical administration of the mediator in the rabbit eye.[25] In addition, the ginkgolide also abolishes the deleterious effects of PAF on the electroretinogram of rat eyes *in vitro*,[26] prevents cell infiltration and edema formation provoked by immune kerati-

[22] S. B. Hwang, M. H. Lam, and T. Y. Shen, *Biochem. Biophys. Res. Commun.* **128,** 972 (1985).

[23] M. Doly, B. Bonhomme, P. Braquet, P. E. Chabrier, and G. Meyniel, *Immunopharmacology* **13,** 189 (1987).

[24] M. Doly, P. Braquet, K. Drieu, B. Bonhomme, and M. T. Droy, *in* "Colour Vision Deficiencies" (G. Verrist, ed.), Vol. 8, p. 515. W. Junk, Dordrecht, 1987.

[25] P. Braquet, R. F. Vidal, M. Paubert-Braquet, H. Hamard, and B. B. Vargaftig, *Agents Actions* **15,** 82 (1984).

[26] M. Doly, P. Braquet, B. Bonhomme, and G. Meyniel, *Int. J. Tissue React.* **9,** 33 (1987).

tis in the cornea of the rabbit eye,[14] and inhibits the increased IOP and breakdown of the blood–aqueous barrier after laser irradiation of the rabbit iris.[27]

In conclusion, the identification of specific PAF binding sites in various cells and tissues and their blockade by ginkgolides explains the beneficial effects of ginkgolides in numerous experimental diseases and highlights their potential as valuable therapeutic agents for man.

[27] N. L. Verbeij, J. L. van Delft, N. J. van Haeringen, and P. Braquet, *in* "The Ginkgolides: Chemistry, Biology, Pharmacology and Clinical Perspectives" (P. Braquet, ed.), Vol. 2. Prous Science, Barcelona, 1990, in press.

[49] Kadsurenone and Other Related Lignans as Antagonists of Platelet-Activating Factor Receptor

By T. Y. Shen and I. M. Hussaini

Introduction

Platelet-activating factor (PAF) is a highly potent ether-linked phospholipid (1-*O*-alkyl-2-acetyl-*sn*-glycerol-3-phosphorylcholine) which activates platelets as well as modulates the function of leukocytes and other target cells. Interaction of PAF with a specific membrane recognition site coupled to phosphatidylinositol metabolism produces the biological actions of the phospholipid.[1–3] A number of specific reversible and irreversible PAF receptor antagonists have been developed using both *in vitro* (platelet aggregation; binding studies) and *in vivo* (hypotension; bronchoconstriction) screening methods. Some of these antagonists include PAF analogs, e.g., CV-3989,[4] ONO-6248,[5] SR-63441,[6] natural products,

[1] F. Snyder, *Med. Res. Rev.* **5,** 107 (1985).

[2] D. J. Hanahan, *Annu. Rev. Biochem.* **55,** 483 (1986).

[3] P. Braquet, T. Y. Shen, L. Touqui, and B. B. Vargaftig, *Pharmacol. Rev.* **39,** 97 (1987).

[4] Z. Terrashita, S. Tsuchima, Y. Yoshiota, J. Normura, Y. Inada, and S. Nishikoda, *Life Sci.* **32,** 1975 (1983).

[5] C. M. Winslow, H. U. Gubler, A. K. Delillo, F. J. Davies, G. E. Frisch, J. C. Tomesch, and R. N. Saunders, *Proc. 2nd Intern. Conf. Platelet-Activating Factor and Structurally Related Alkyl Ether Lipids, Gatlinburg, Tennessee,* p. 33 (1986).

[6] T. Mitamoto, H. Ohno, T. Tano, T. Okada, N. Hamanaka, and A. Kawasaki, *Proc. 3rd Intern. Conf. Inflammation, Paris, France,* p. 513 (1984).

METHODS IN ENZYMOLOGY, VOL. 187

e.g., kadsurenone,[7] tetrahydrofuran lignans and their synthetic analogs,[8,9] ginkgolide B (BN 52021),[8] benzodiazepine derivatives,[10] and other synthetic structures. In this chapter the chemistry and pharmacology of kadsurenone and other lignan derivatives will be described.

Methods for Screening Platelet-Activating Factor Receptor Antagonists *in Vitro*

Materials

$PAF_{C\text{-}16}$ (Bachem, Torrance, CA), [^{3}H]PAF, and Aquasol-2 (NEN, Boston, MA). Adenosine diphosphate (ADP), arachidonic acid, collagen, thrombin, prostacyclin (PGI_2), and bovine serum albumin (BSA) (Sigma Chemical Company, St. Louis, MI). Ficoll-Paque (Pharmacia Fine Chemicals, Piscataway, NJ).

Platelet Aggregation Assay

Blood (10 volumes) from volunteers (who had not ingested aspirin-like drugs or steroids in the previous 10 days) is drawn into a plastic syringe containing 3.8% trisodium citrate (1 volume) as an anticoagulant. Heparin interferes with platelet aggregation and, therefore, is not used as an anticoagulant in this assay. Citrated blood is centrifuged at 270 *g* for 10 min at room temperature to obtain platelet-rich plasma (PRP). The platelet count is either determined microscopically following staining with Brilliant Cresol Blue or in a Coulter counter and adjusted to 2.5×10^8 platelets/ml. Platelet-poor plasma (PPP) is prepared by centrifugation of PRP at 1000 *g* for 20 min and is used to calibrate the aggregometer.

Washed platelet suspension (WPS) is prepared from PRP by underlayering the latter with Ficoll-Paque (9 : 2, v/v) and centrifugation at 750 *g* for 15 min at room temperature. The platelets band between the plasma and the separation medium which are carefully collected and suspended in Tyrode's solution (in m*M*: NaCl, 137; KC1, 2.7; $NaHCO_3$, 11.9; NaH_2-PO_4, 0.42; $MgCl_2$, 1.0; $CaCl_2$, 1.0; HEPES, 5.0; and dextrose, 1 g/liter, with 0.25% BSA, w/v, pH 7.4). PGI_2 (1 ng/ml) is usually added to the

[7] T. Y. Shen, S.-B. Hwang, M. N. Chang, T. W. Doebber, M. H. Lam, M. S. Wu, X. Wang, G. Q. Han, and R. Z. Li, *Proc. Natl. Acad. Sci. U.S.A.* **82,** 672 (1985).

[8] T. Y. Shen, S.-B. Hwang, T. W. Doebber, and J. C. Robbins, *in* "Platelet-Activating Factor and Related Lipid Mediators" (P. Snyder, ed.), p. 153. Plenum, New York, 1987.

[9] M. M. Ponpidom, S.-B. Hwang, T. W. Doebber, J. J. Acton, F. J. Leinweber, and L. Klevans *et al., Biochem. Biophys. Res. Commun.* **150,** 1213 (1988).

[10] J. Casals-Stenzel, G. Muacevic, and K.-H. Weber, *J. Pharmacol. Exp. Ther.* **241,** 974 (1987).

suspension to prevent platelet activation. The platelet aggregation inhibitory activity of PGI_2 disappears completely after 10 min of addition at room temperature. The suspension is then spun at 1000 *g* for 10 min and the pellet is finally resuspended in Tyrode's buffer for platelet aggregation. The platelet aggregation produced by PAF in WPS requires calcium.

Platelet aggregation is measured either turbidimetrically (Payton or Chron-log aggregometers) or by using an impedance (Chron-log or Coulter electronic aggregometers) method. Although PRP and WPS can be used in both procedures, whole blood aggregation can only be estimated by the impedance method. In our laboratory, platelet aggregation based on the turbidity of the samples is routinely used to screen potential PAF antagonists.

Platelet samples are maintained at 37° and stirred constantly at 1000 revolutions per min (rpm). PAF (0.05–0.5 μM) and other aggregation agents (ADP, 0.5–10 μM, arachidonic acid, 0.1–5 m*M*, collagen, 1–5 μg/ml, and thrombin, 0.1–0.5 NIH U/ml) are used to induce platelet aggregation and allowed to proceed for 3–5 min. Potential antagonists of PAF (naturally occurring or synthetic) are injected into the cuvettes 2 min before the addition of the aggregation agent. The effect of the antagonist is expressed as percentage inhibition of the control maximum platelet aggregation; the concentration of the antagonist that reduces the maximum response of the platelets by 50% is the IC_{50} value. The aggregation experiment is usually complete within 3 hr of blood collection because after that the platelets tend to lose sensitivity to the aggregation agent.

Radioreceptor Studies

Membrane Preparation. Platelet membranes are prepared using sucrose density gradients following isolation of platelets with Ficoll-Paque. Membranes from other blood cells and smooth muscles have also been used to study [^{3}H]PAF binding.[11] The platelets are resuspended in a solution containing NaCl, 150 m*M*, EDTA, 2 m*M*, and Tris-HCl, 10 m*M* (pH 7.4) and centrifuged twice at 750 *g* for 15 min at room temperature. Finally, the pellet is suspended in a solution of $MgCl_2$, 10 m*M*, EDTA, 2 m*M*, and Tris-HCl, 10 m*M* (pH 7.4). The platelets are then lysed by at least three cycles of repeated freezing with liquid nitrogen and thawing at room temperature. The broken platelets are then layered on top of a discontinuous sucrose density gradient (12 and 27%, w/v) and centrifuged at 63,500 *g* for 3 hr (4°). The platelet membranes sandwiched between 12 and 27% w/v sucrose are carefully collected and used immediately or stored at −80° for

[11] S.-B. Hwang, C.-S. C. Lee, M. J. Cheah, and T. Y. Shen, *Biochemistry* **22,** 4756 (1983).

later use. Protein concentration is determined according to the method of Lowry *et al.*[12]

Binding Studies. The binding of [^{3}H]PAF to the membrane fraction is carried out in a 1-ml reaction mixture containing 100 μg of membrane protein, [^{3}H]PAF (0.5–5 nM) with or without unlabeled $PAF_{C\text{-}16}$ (1000-fold), and a solution of $MgCl_2$, 10 mM, EDTA, 2 mM, and Tris-HCl, 10 mM (pH 7.4). The reaction is incubated at 0° for 1 hr. The free and bound ligand are separated by a filtration technique using Whatman GF/C glass fiber filters. The difference between total amount of bound [^{3}H]PAF in the absence and presence of excess unlabeled $PAF_{C\text{-}16}$ is defined as specific binding of the radiolabeled ligand. In a set of experiments, [^{3}H]PAF (1 nM) is incubated with different concentrations of PAF-receptor antagonists and the effect of the antagonist on the specific binding is expressed as percentage inhibition of the control. The equilibrium dissociation constant (K_i) of the antagonist is calculated from the Chen–Prusoff equation:

$$K_i = \frac{IC_{50}}{1 + [PAF]/K_D]}$$

Chemistry and Pharmacology of Kadsurenone and Related Lignans

Kadsurenone was first isolated from the dichloromethane extract of stems of *Piper futokadsura* Sieb and Zucc (Piperaceae), which is a medicinal plant widely used in Chinese herbal prescriptions to relieve the pain, inflammation, and discomforts associated with arthritis and asthma.[7,13] The discovery of its PAF inhibitory activity was the combined result of an attempt to identify the active ingredients in this plant and the search for novel antagonists of PAF using a newly established platelet membrane-binding assay. Kadsurenone is a member of the lignan family (Fig. 1) derived from biogenetic dimers of alkoxyphenylpropenes. It is a stable and colorless crystalline compound, mp 62.5°, only slightly soluble in the aqueous medium. It is optically active, $[a]_D^{22}$ +3.2, with a λ_{max} 285 nm in the UV spectrum characteristic for its dienone ring. Its total synthesis and NMR spectrum have been published.[14]

In various cellular and membrane binding assays, kadsurenone is a reversible and competitive inhibitor of PAF, active at micromolar or lower

[12] O. H. Lowry, N. J. Rosebrough, A. L. Farr, and R. J. Randall, *J. Biol. Chem.* **193,** 265 (1951).

[13] M. N. Chang, G. Q. Han, B. H. Byron, J. P. Springer, S.-B. Hwang, and T. Y. Shen, *Phytochemistry* **24,** 2079 (1985).

[14] M. M. Ponpipom, R. L. Bugianesi, D. R. Brooker, B. Z. Yue, S.-B. Hwang, and T. Y. Shen, *J. Med. Chem.* **30,** 136 (1987).

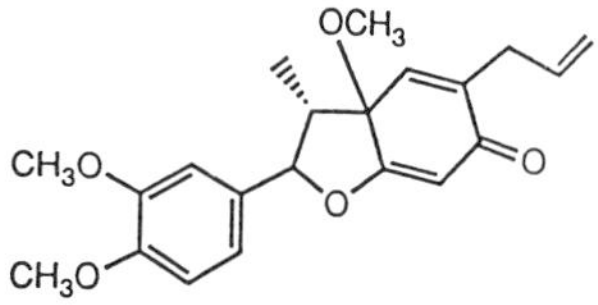

Kadsurenone

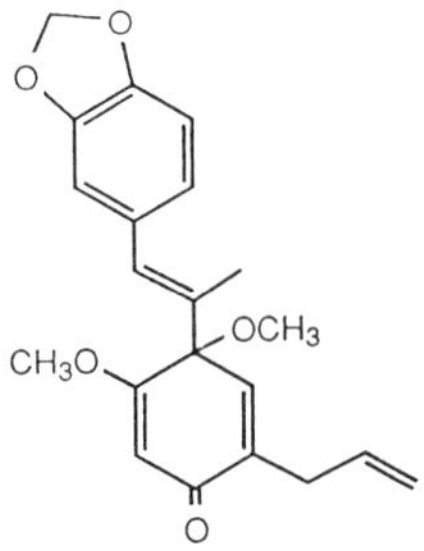

Futoquinol

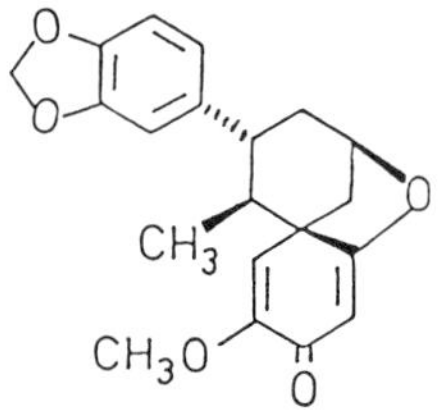

Futoenone

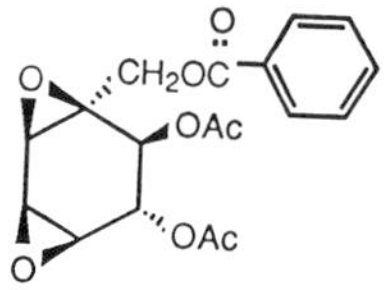

Futoxide

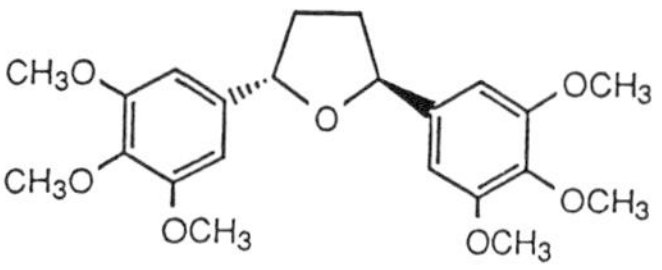

L-652,731

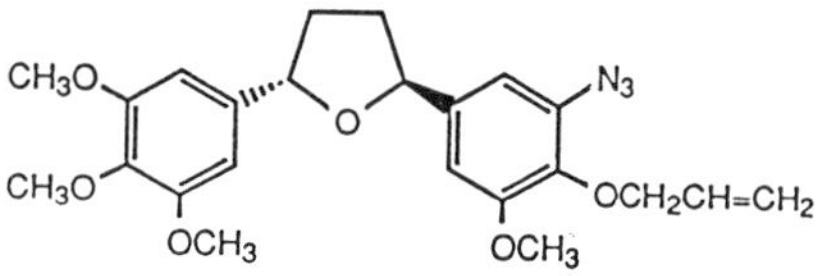

L-662,025

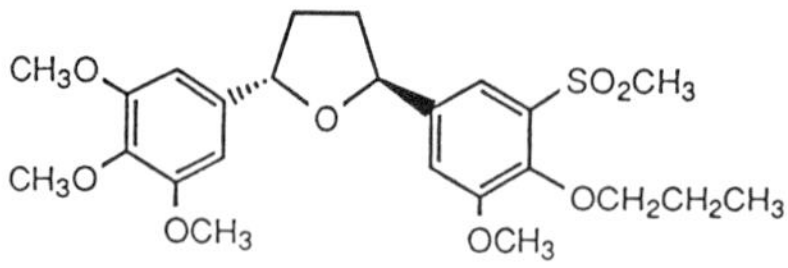

L-659,989

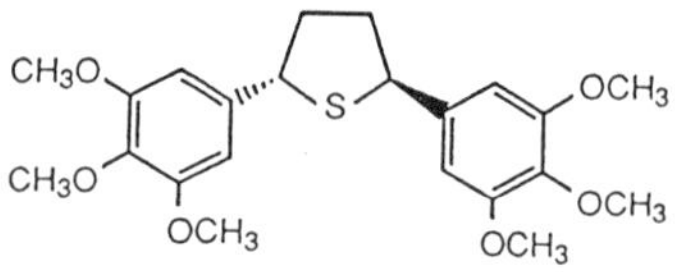

L-653,150

FIG. 1. Chemical structures of kadsurenone and related lignans.

concentrations (see below). Kadsurenone does not inhibit platelet aggregation induced by ADP, collagen, arachidonic acid, or thrombin even at 50 μM. The PAF antagonism of kadsurenone has a high degree of structural and stereochemical specificity. Among many close analogs investigated, only the 9,10-dihydro derivative was found to be equally potent and to interact with the PAF binding site in a similar manner.[15] Several other lignans, i.e., futoquinol, futoenone, and futoxide[16] (Fig. 1), have recently been isolated from samples of the same plant and found to possess comparable PAF antagonistic activities in the platelet aggregation and receptor binding assays. Other types of lignans, especially derivatives of 2,5-diaryltetrahydrofuran, have also shown moderate PAF antagonism.[8,17] Several synthetic analogs of these lignans, e.g., L-652,731, L-653,150, and L-659,989,[9,18] and a photolabile, irreversible PAF-antagonist, L-662,025[16] have been developed.

Upon incubation with platelets for 2 min before addition of PAF, L-662,025 produces a dose-related inhibition of platelet aggregation with parallel shift of the dose–response curve to the right and reproduction of the maximum effect of the control by increasing agonist concentration (Fig. 2A). Following photoactivation (λ = 260–320 nm) of human platelets with L-662,025 (20 μM), the asymptote of the concentration–percentage maximum aggregation curve is dropped below the EC_{50} value of PAF and is not reversed by increasing the ligand concentration (Fig. 2B). This is typical of noncompetitive or irreversible antagonism. Repeated washing (at least three times) by centrifugation at 14,000 *g* for 5 min (4°C) does not reverse the inhibition of [^{3}H]PAF binding to platelet membranes by L-662,025 following photolysis. L-659,989 and a series of 2,5-diarylcyclopentane derivatives reported recently[19] are active *in vitro* at 1–4 nM. The IC_{50} values of a synthetic benzodiazepine derivative (WEB 2086), kadsurenone, and related lignans in human PRP are summarized in Table I. Since PAF binds to a specific recognition site on cell or tissue membranes to elicit its biological action, a direct displacement of [^{3}H]PAF by the antagonists is usually carried out to make certain that the compounds are true PAF receptor antagonists. The equilibrium dissociation

[15] S.-B. Hwang, M. H. Lam, and M. N. Chang, *J. Biol. Chem.* **261,** 13720 (1986).

[16] T. Y. Shen, I. Hussaini, S.-B. Hwang, and M. N. Chang, *Adv. Prostaglandin Thromboxane Leukotriene Res.* **19,** 359 (1989).

[17] S.-B. Hwang, M. H. Lam, T. Biftu, T. R. Beattie, and T. Y. Shen, *J. Biol. Chem.* **260,** 15639 (1985).

[18] S.-B. Hwang, M. H. Lam, and A. H. M. Hsu, *Mol. Pharmacol.* **35,** 48 (1989).

[19] D. W. Grahm, P. Chiang, S. S. Yang, K. L. Thompson, M. N. Chang, T. W. Doebber, S.-B. Hwang, M. H. Lam, M. S. Wu, A. W. Alberts, and J. C. Chabala, *Abstr. 197th Am. Chem. Soc. Meeting, Dallas, TX, MEDI 25, April 9, 1989.*

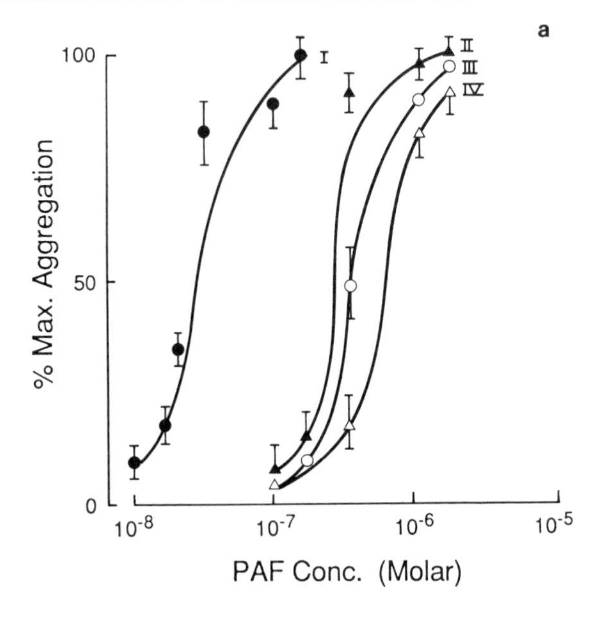

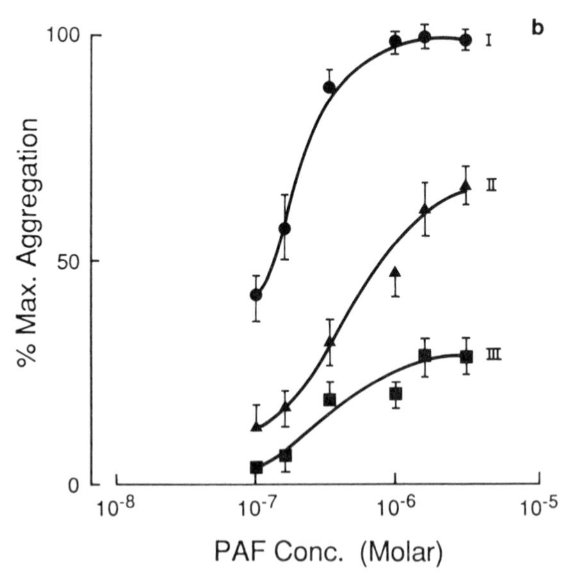

FIG. 2. (a) Dose–percentage maximum aggregations curves for PAF in the absence (I) and presence of L-662,025 6 μM (II), 10 μM (III), and 20 μ (IV) before photoactivation (control). Vertical bars are mean ± SE (n = 5). (b) Effect of L-662,025, 6μM (I), 10 μM (II), and 20 μM (III) on PAF-induced human platelet aggregation after photoactivation (260–320 nM) for 15 min. The control curve (without L-662,025) for PAF was the same as shown in (A).

TABLE I
IC_{50} VALUES FOR PAF RECEPTOR ANTAGONISTS IN INHIBITING PAF-INDUCED PLATELET AGGREGATION[a]

PAF receptor antagonist	IC_{50} (μM)
WEB 2086	0.15 ± 0.05
L-659,989	0.80 ± 0.10
Kadsurenone	3.50 ± 0.30
L-653,150	4.40 ± 0.30
Futoquinol	4.50 ± 0.20
L-662,025	5.60 ± 0.30
L-652,731	6.00 ± 0.50
Futoenone	9.50 ± 0.35

[a] PAF (0.2 μM). In platelet-rich plasma (n = 5).

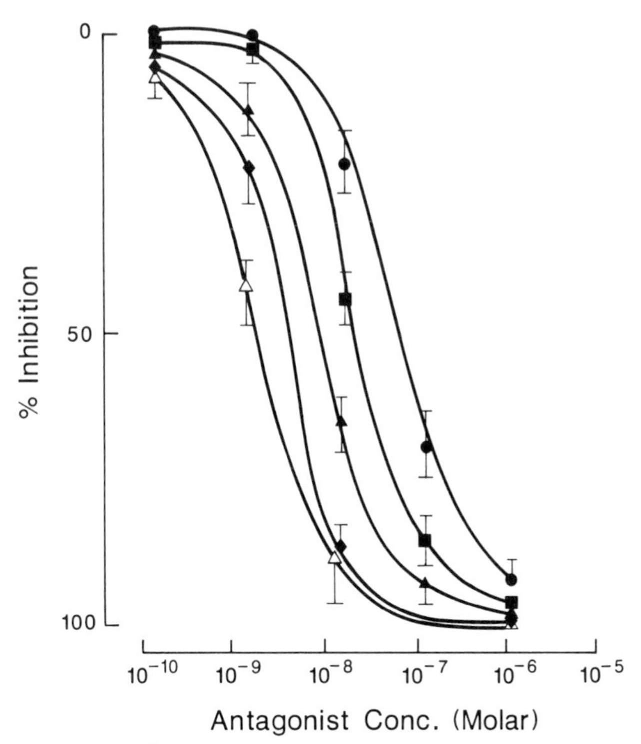

FIG. 3. Inhibition of [^{3}H]PAF binding to human platelet membranes by antagonists of PAF receptor. Vertical bars are mean ± SE (n = 6). △, WEB 2086; ◆, L659,989; ▲, kadsurenone; ■, L652,731; ●, futoenone.

constant (K_D) of PAF in human platelet membranes ranges from 0.5 to 3.8 n*M*. Figure 3 shows the effect of WEB 2086, kadsurenone, and related lignans on the specific binding of [^{3}H]PAF (1 n*M*) to human platelet membranes. The order of relative potency is WEB 2086 > L-659,989 > kadsurenone > L-652,731 > futoenone. This result indicates that the antagonists inhibit PAF-induced platelet aggregation by blocking the receptor with which the agonist interacts to produce its biological activity.

Kadsurenone is well absorbed orally. After oral dosing in the rhesus monkey it is readily hydroxylated at three sites of the allyl side chain and excreted as the corresponding glucuronides in the urine.[20] The major metabolites of lignans with trimethoxyphenyl substituents are mono- or didemethylated derivatives that are usually much less active as PAF-antagonists. In animal experiments, both kadsurenone and L-652,731 have a relatively short serum half-life of less than 2 hr.

Kadsurenone is a specific, potent, and reversible inhibitor of the binding of [^{3}H]PAF to its membrane receptor and PAF-induced aggregation of platelets. It also inhibits PAF-induced aggregation and degranulation of neutrophils. In the rat, kadsurenone inhibits PAF-induced foot edema and the shock syndromes, acute hypotension, neutropenia, extravasation, and secretion of lysosomal enzymes, induced by endotoxin, soluble immune complex or PAF.[8] It also inhibits cutaneous permeability produced by PAF in the guinea pig. The biological profiles of other lignans and more potent synthetic analogs, e.g., L-659,989, are generally similar to that of kadsurenone. Individual studies of some lignan derivatives in special disease models such as antigen-induced bronchospasm in guinea pigs,[21] PAF-induced cardiovascular alterations in the rabbit,[22] and complement-induced pulmonary hypertension in sheep[23] have also been reported.

[20] K. L. Thompson and M. N. Chang, unpublished observation (1986).

[21] C. P. Cox, T. Sata, L. W. Liu, and S. I. Said, *Am. Rev. Respir. Dis.* **133,** A278 (1986).

[22] G. Montrucchio, G. Alloatti, F. Mariano, C. Tetta, G. Emanuelli, and G. Camussi, *Int. J. Tissue React.* **8,** 497 (1985).

[23] W. B. Smallbone, E. N. Taylor, and D. W. J. McDonald, *J. Pharmacol. Exp. Ther.* **242,** 1035 (1987).

[50] Use of WEB 2086 and WEB 2170 as Platelet-Activating Factor Antagonists

By JORGE CASALS-STENZEL and HUBERT O. HEUER

Since the discovery of the platelet-activating factor (PAF, 1-*O*-alkyl-2-acetyl-*sn*-glycerophosphorylcholine) in 1972 by Benveniste *et al.*,[1] the interest in studying its physiological and pharmacological responses as well as its pathophysiological role has created a demand for specific and potent PAF antagonists.[2,3] The search for these antagonists yielded several types of compounds differing in origin and structure.[4] Among these, two synthetic compounds derived from the thienotriazolodiazepine brotizolam,[5] a hypnotic, have recently been described under the code number WEB 2086 and WEB 2170.[6,7] These compounds are specific and very potent PAF antagonists, free from hypnotic and sedative activity. They should facilitate research in this area and may prove to be therapeutically useful. This chapter presents the physicochemical properties and the pharmacological use of these new PAF antagonists.

Physicochemical Properties

WEB 2086 and WEB 2170 belong to the group of thienotriazolo-1,4-diazepines (hetrazepines).[8] Their chemical structures are shown in Fig. 1 and their physical constants are summarized in Table I. The water solubility and the partition coefficient are similar for both compounds. The chemical name for WEB 2086 is 3-{4-(2-chlorophenyl)-9-methyl-6*H*-thieno[3,2-*f*][1,2,4]triazolo[4,3-*a*][1,4]diazepine-2-yl}-1-(4-morpholinyl)-1-propanone and for WEB 2170 8(*R,S*)-6-(2-chlorophenyl)-8,9-dihydro-1-methyl-8-(4-morpholinylcarbonyl)-4*H*,7*H*-cyclopenta(4,5)thieno[3,2-*f*]

[1] J. Benveniste, P. M. Henson, and C. G. Cochrane, *J. Exp. Med.* **136,** 1356 (1972).

[2] B. B. Vargaftig, M. Chignard, J. Benveniste, J. Lefort, and F. Wal, *Ann. N.Y. Acad. Sci.* **370,** 119 (1981).

[3] P. Braquet, L. Touqui, T. Y. Shen, and B. B. Vargaftig, *Pharmacol. Rev.* **39,** 97 (1987).

[4] J. Casals-Stenzel and H. Heuer, *Prog. Biochem. Pharmacol.* **22,** 58 (1988).

[5] J. Casals-Stenzel, *Naunyn-Schmiedeberg's Arch. Pharmacol.* **335,** 351 (1987).

[6] J. Casals-Stenzel, G. Muacevic, and K. H. Weber, *J. Pharmacol. Exp. Ther.* **241,** 974 (1987).

[7] H. Heuer, J. Casals-Stenzel, G. Muacevic, W. Stransky, and K. H. Weber, *Clin. Exp. Pharmacol. Physiol. Suppl.* **13,** 7 (1988).

[8] M. Sanchez-Crespo and K. H. Weber, *Drugs Future* **13,** 242 (1988).

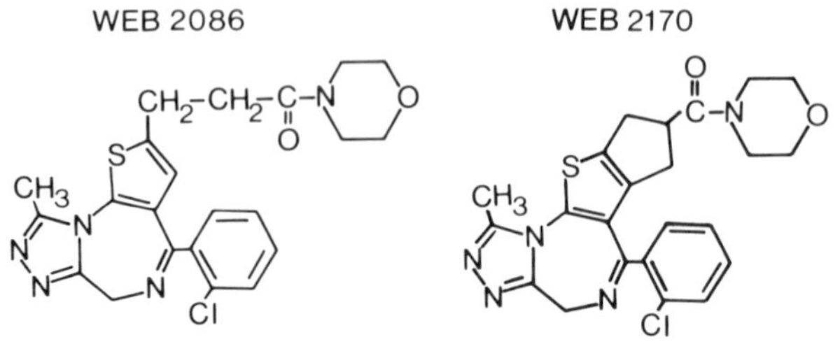

WEB 2086 WEB 2170

FIG. 1. Structure of WEB 2086 and WEB 2170.

[1,2,4]triazolo[4,3-*a*][1,4]diazepine. The synthesis of both compounds has been described elsewhere.[8–10]

Preparation of Solutions

WEB 2086 and WEB 2170 have, to a certain extent, good water solubility. For some pharmacological purposes the following tabulation may prove helpful in preparing solutions for oral, intravenous, or inhaled administration. At concentrations above 0.3% WEB 2086 and 0.5% WEB

Concentration	Procedure
Up to 0.3% for WEB 2086 or 0.5% for WEB 2170	At 20°–30° with the aid of an ultrasonic bath (if necessary)
Up to 1%	Suspend the substance in about 70% of the final volume using distilled water or saline, add 0.1 *N* HCl dropwise to obtain a clear solution (ultrasonic bath), and dilute up to the final concentration; a pH value below 5.5 should be avoided
Up to 3%	To prepare, e.g., a 10 ml solution, dissolve the substance in 1 ml 1 *N* HCl, dilute with distilled water or saline up to 7 ml, adjust to pH 5.5–6.0 with 1 *N* NaOH, and fill up the final volume. The solution should be used the same day as it is prepared as crystallization can sometimes occur after longer storage.

[9] K. H. Weber, G. Walther, A. Harreus, J. Casals-Stenzel, G. Muacevic, and W. Tröger, F. R. G. Patent DE-OS 3502392 (1985).

[10] W. Stransky, K. H. Weber, G. Walther, A. Harreus, J. Casals-Stenzel, G. Muacevic, and W. D. Bechtel, F. R. G. Patent DE 3,701,344 (1987).

TABLE I
PHYSICOCHEMICAL PROPERTIES OF WEB 2086 AND WEB 2170

Compound	WEB 2086	WEB 2170
Empirical formula	$C_{22}H_{22}ClN_5O_2S$	$C_{23}H_{22}Cln_5O_2S$
Molecular weight	455.96	467.9
Melting point (°)	192	189
Color	White	White
Form	Crystalline	Crystalline
Solubility (%) in water	0.3	0.5
log D[a]	0.8	1.0
Stability	Practically unlimited at room temperature; the substance is light stable	

[a] log D = partition coefficient, an index for the lipophilicity and penetrability through the blood–brain barrrier.

2170 the use of 0.1 *N* HCl (about 1 : 20, v/v) is necessary to obtain a clear solution. For *oral administration* either the above described solutions or the same solutions containing 0.5% Tylose (methyl cellulose) can be used. At higher dosages of both compounds 0.5% Tylose should be added to ensure suspension of the substance.

Stability of Solutions

Solutions of WEB 2086 and WEB 2170 are light-stable and can be stored at room temperature. Their activity can be easily monitored by PAF-induced platelet aggregation at a constant concentration of the agonist. Although these solutions are stable, it is recommended that they be used the same day. In addition, pH values below 5.5 of the solutions should be avoided because at a lower pH only partially reversible inactivation accompanied by a yellow coloration can occur.

Concentrations and Doses

The PAF antagonistic activity of this type of compound can be demonstrated and characterized by very simple experiments, such as PAF-induced human platelet or neutrophil aggregation *in vitro*,[6,7] or bronchoconstriction and hypotension *in vivo*.[6,7,11] PAF and the PAF antago-

[11] H. Heuer, J. Casals-Stenzel, G. Muacevic, W. Stransky, and K. H. Weber, *Prostaglandins* **35,** 798 (1988).

nists also act on other systems and organs as well; the concentrations and doses needed in such experiments are mostly in the same range.[12–15] The methodology for these fundamental *in vitro* and *in vivo* experiments and the pharmacology of WEB 2086 have been extensively described elsewhere.[6,12,13] Some recommendations regarding these procedures are given below. For the characterization of WEB 2086 and WEB 2170, the 1-hexadecyl derivative of PAF (racemic PAF, molecular weight 559.73, Bachem Feinchemikalien, Bubendorf, Switzerland) has been used.

In Vitro Experiments

WEB 2086 and WEB 2170 are specific competitive antagonists of PAF because both compounds have little or no effect on the platelet and neutrophil responses to other aggregating agents.[6,7] Their activity is competitive on the basis of binding experiments, as well as other studies. WEB 2086 inhibits the binding of [^{3}H]PAF to human platelets with an equilibrium dissociation constant (K_D) of 15 nM and PAF and WEB 2086 displace [^{3}H]WEB 2086 to the same extent, i.e., with similar K_i values.[16,17]

PAF-Induced Platelet Aggregation

This is a useful tool in PAF antagonist research. When performed in platelet-rich plasma (PRP), it is a quick and simple method for discovering PAF antagonistic activity of new substances (screening) and for assessing their (relative) potency in comparison to other substances *in vitro*. The procedure using PRP is based on the method described by Born and Cross.[18] An appropriate concentration of PAF must be determined from a PAF concentration–response curve. For human PRP, a PAF concentration of 0.05 μM has been found suitable. This concentration induces almost complete aggregation, but is not supramaximal. The platelets of several species (human, rabbit, guinea pig, and dog)[6,7,19,20] can be used for aggregation experiments. It is advantageous to human platelets because of the quantity obtainable and the minimal tendency to spontaneous aggrega-

[12] J. Casals-Stenzel, *Eur. J. Pharmacol.* **135,** 117 (1987).

[13] J. Casals-Stenzel, *Immunopharmacology* **13,** 117 (1987).

[14] J. Casals-Stenzel, J. Franke, T. Friedrich, and J. Lichey, *Br. J. Pharmacol.* **91,** 799 (1987).

[15] A. Brambilla, A. Ghiorzi, and A. Giachetti, *Pharmacol. Res. Commun.* **19,** 147 (1987).

[16] F. W. Birke and K. H. Weber, *Clin. Exp. Pharmacol. Physiol. Suppl.* **13,** 2 (1988).

[17] D. Ukena, G. Dent, F. W. Birke, C. Robaut, G. W. Sybrecht, and P. J. Barnes, *FEBS Lett.* **228,** 285 (1988).

[18] G. V. R. Born and J. J. Cross, *J. Physiol. (London)* **168,** 178 (1963).

[19] S. R. O'Donnell and C. J. K. Barnett, *Br. J. Pharmacol.* **94,** 437 (1988).

[20] A. G. Stewart and G. J. Dusting, *Br. J. Pharmacol.* **94,** 1225 (1988).

TABLE II
CONCENTRATIONS OF WEB 2086 AND WEB 2170 IN PAF-INDUCED (0.05 μM) PLATELET AND NEUTROPHIL AGGREGATION EXPERIMENTS *in Vitro*[a]

Source	Concentration (μM)	
	WEB 2086	WEB 2170
Human PRP[b]	0.17 (0.15–0.19)	0.32 (0.23–0.43)
Rabbit PRP	0.7 (0.66–0.73)	0.55 (0.38–0.72)
Washed rabbit platelets[c]	0.028 (0.023–0.035)	0.057 (0.043–0.074)
Human PMN[d]	0.37 (0.26–0.54)	0.83 (0.57–1.20)

[a] Concentration range: 0.01–1.0 μM.
[b] PRP, Platelet-rich plasma.
[c] PAF: 0.18 nM (=100 pg/ml).
[d] PMN, Polymorphonuclear leukocytes (i.e., neutrophils).

tion in comparison to certain types of animal platelets (e.g., guinea pigs). The platelet aggregation experiments can be performed with platelet-rich plasma (PRP),[6,7] washed platelets, or whole blood.[21] There are differences in sensitivity based on species and method of platelet preparation. The use of washed rabbit platelets may exclude nonspecific binding of PAF and its antagonists to plasma proteins and therefore increases the sensitivity to them. Washed rabbit platelets are highly sensitive to the aggregating effect of PAF requiring about 50 times lower concentrations than in the experiments performed with PRP to induce full aggregation.[22] Similarly, the concentration of WEB 2086 and WEB 2170 are about 10 times lower than those needed in the PRP experiments (Table II). Procedures for preparation of washed platelets have been described and evaluated elsewhere.[23]

PAF-Induced Human Neutrophil Aggregation

This technique is as useful and adequate as platelet aggregation for the search of PAF antagonists, but it is not as simple and quick. The IC_{50}

[21] J. V. Levy and W. Chan, *Prostaglandins* **35,** 841 (1988).
[22] M. Schierenberg, H. Darius, and H. Heuer, *Naunyn-Schmiedeberg's Arch. Pharmacol. (Suppl.)* **338,** R 56 (1988).
[23] J. W. N. Akkerman, M. H. M. Doucet-de-Bruine, G. Gorter, S. de Graaf, S. Hefne, J. P. M. Lips, A. Numeuer, and J. Over, *Thromb. Haemostasis.* **39,** 146 (1978).

values are, in general, similar for both methods (Table II). For preparation of the cells the method of Böyum is used.[24] To facilitate the handling of purified neutrophil leukocytes and to prolong their viability, a mixture of 50% human plasma and 50% HEPES buffer (5 m*M* HEPES, 136.8 m*M* NaCl, 2.61 m*M* KCl, 1.3 m*M* $CaCl_2$, 1 m*M* $MgCl_2$, and 0.1% glucose; pH 7.4) should be used for the final cell suspension.[6] The appropriate PAF concentration for induction of neutrophil aggregation is the same as is used in human PRP experiments, i.e., 0.05 μM.

In Vivo Experiments

For the evaluation of the *in vivo* activity and the relative potency of WEB 2086 and related compounds, PAF-induced bronchoconstriction and hypotension are very simple and suitable experimental models. By continuous and simultaneous measurements of pulmonary airflow (with a pneumotachograph, No. 0000, Fleisch, connected to a tracheal cannula) and mean arterial pressure (carotid artery) in spontaneously breathing, anesthetized guinea pigs (Fig. 2), the anti-PAF activity of a compound can be estimated easily and precisely.[6] For most of the *in vivo* experiments performed in our laboratory, a continuous intravenous PAF infusion [30 ng/(kg × min)] into the jugular vein is used for reliable induction of reproducible PAF effects in guinea pigs and rats. In guinea pigs such PAF infusion causes their death within 15 to 30 min, whereas rats do not die even after infusions lasting for more than 3 hr. This difference is due to the lack of bronchoconstriction in rats (see below).

Bolus injections of PAF are also used. A suitable dose is 100 ng/kg iv. In guinea pigs, this dosage induces an acute and severe bronchoconstriction, accompanied by an initial increase and subsequent decrease in arterial pressure which lasts for 10 to 20 min. The doses of oral, intravenous, or inhaled WEB 2086 or WEB 2170 for inhibition of the effects induced by a PAF bolus are similar to those needed in PAF infusion experiments (Table III). PAF can also be administered by intratracheal instillation. A dose of 300 μg/kg in a volume of 0.1 ml is injected with an insulin syringe and needle and gives a good response. The inhalation of PAF is also possible, although the amounts of PAF needed for this procedure are large and, therefore, expensive. The suitable concentration for this route of administration is 1000 μg/ml. For generation of a PAF aerosol, an ultrasound nebulizer (USV 82 Praxis, Technolab, FRG) can be used.

[24] A. Böyum, *Scand. J. Clin. Lab. Invest.* **21,** 77 (1968).

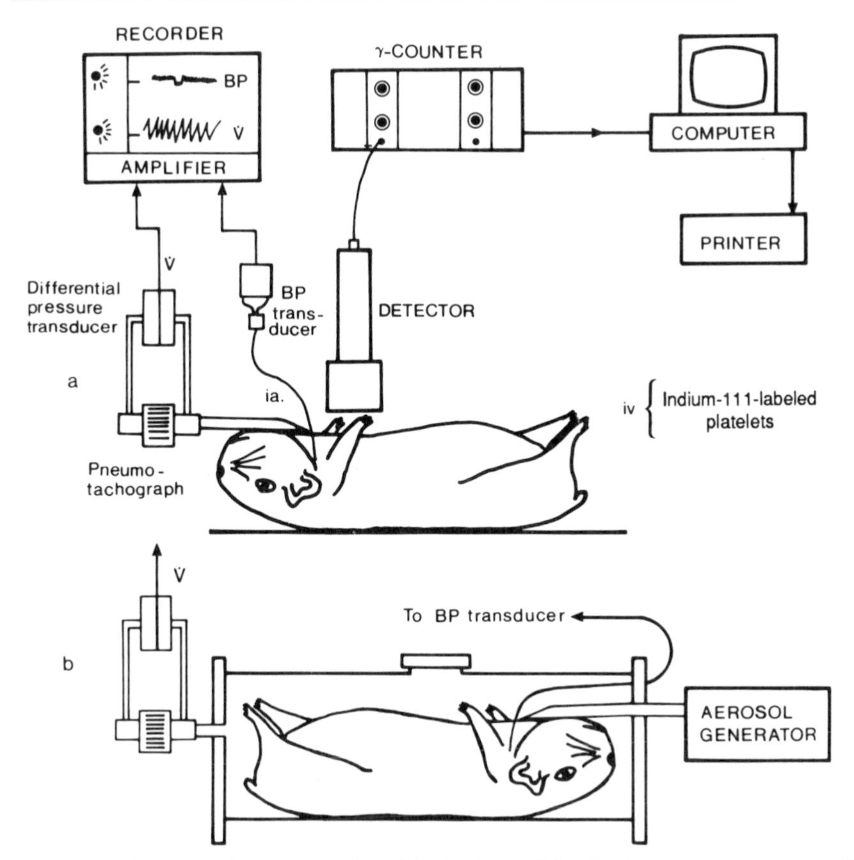

FIG. 2. Diagrammatic representation of the device used for simultaneous measurement of PAF-induced platelet aggregation in the lung, bronchoconstriction ($\dot{V}$; airflow), and blood pressure (BP) depression in anesthetized spontaneously breathing guinea pigs. The animal is (a) connected directly to a pneumotachograph and differential pressure transducer by an intratracheal cannula or (b) placed in a whole-body plethysmograph for inhalation of the PAF antagonist. In this system, the pneumotachigraph is connected to the chamber (indirect airflow measurement) and the intratracheal cannula is attached to the aerosol generator. An intraarterial (ia) catheter is connected to a blood pressure transducer which is linked to an amplifier–recorder system. For measurement of platelet aggregation *in vivo* ^{111}I-labeled platelets are continuously monitored by means of a collimated crystal scintillation probe (detector) connected to a gamma spectrometer (λ-counter).

Choice of Animals

In our laboratory, spontaneously breathing animals are used for all routes of administration of PAF and PAF antagonists. It is important to choose the right species for proper use of the thienotriazolodiazepines.

TABLE III

DOSES,[a] ROUTE OF ADMINISTRATION, AND CHOICE OF ANIMALS FOR *in Vivo* EXPERIMENTS

Parameter	Administration route					
	WEB 2086			WEB 2170		
	po	iv	inh[b]	po	iv	it[c]
Dose range (mg/kg)						
Guinea pigs	0.05– 1.0	0.005–0.1	0.1–2.0	0.01–0.5	0.001–0.1	0.01–0.1
Rats	1.0 –20.0	0.01 –0.5	—	0.05–2.0	0.005–0.1	—
ED_{50} (mg/kg)[d]						
Guinea pigs	0.066 (0.041–0.087)	0.015 (0.009–0.022)	0.205[e]	0.011 (0.001–0.03)	0.006 (0.002–0.010)	0.008 (0.003–0.014)
Rats	9.2 (7.90–10.82)	0.06 (0.009–0.40)	—	0.26 (0.19–0.35)	0.016 (0.010–0.026)	—
$t_{1/2}$ (hr)						
Guinea pigs	5.5	—	1.6	12.1	—	14.8
Rats	3.1	1.2	—	5.4	2.3	—

[a] Inhibition of PAF-induced hypotension (PAF: 30 ng/(kg × min) iv) (see Ref. 6).

[b] inh., Inhalation.

[c] it, Intratracheal administration.

[d] 95% confidence limits.

[e] Graphic determination.

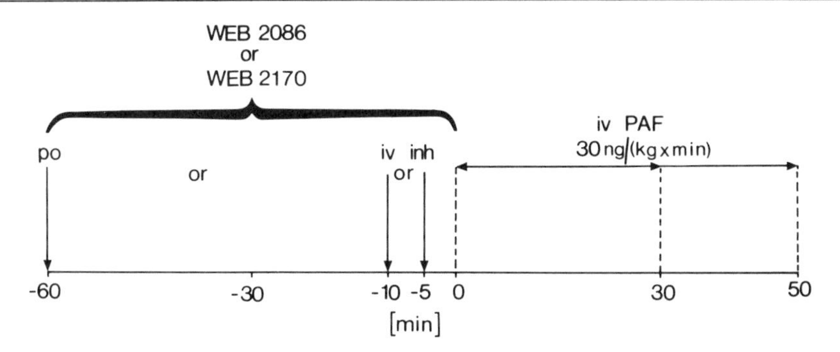

FIG. 3. Diagram of the experimental protocol used for *in vivo* studies with the thienotriazolodiazepines WEB 2086 and WEB 2170.

Guinea pigs are ideal because of their ease of handling, the possibility of registering simultaneously PAF-induced platelet aggregation in the lung,[25] and bronchoconstriction as well as blood pressure depression,[6] i.e., platelet-dependent and -independent reactions to PAF in the same animal (Fig. 2). Moreover, guinea pigs are suitable for studies on lung function and anaphylaxis, thus making possible comparative studies of PAF-induced and anaphylactic bronchoconstriction in the same species.[13] The use of either conscious or anesthetized animals makes some difference. Because the PAF dose required for induction of bronchoconstriction is higher in conscious animals—they are able to compensate actively for the respiratory disturbance—the dose range of the antagonist must also be somewhat higher (0.1–5.0 mg/kg po). In contrast, rats are useful for special purposes like long-lasting PAF-induced depression of arterial pressure. Rat platelets are insensitive to PAF (at the above doses) and, therefore, in this species only a long-lasting, dose-dependent hypotensive effect, but not a bronchoconstrictive one, can be evoked by PAF. At very high PAF doses (6 μg/kg iv) a platelet aggregation with thrombocytopenia can also be achieved in rats.[26]

Choice of Administration Route

Depending on the route of administration of the doses and the timing of administration, the choice of the animal species will differ. The general procedure for administration is shown in Fig. 3.

Oral Administration. When rats and guinea pigs are treated with the same dose of PAF [30 ng/(kg $\times$ min) iv], the oral doses of thieno-

[25] C. P. Page, W. Paul, and J. Morley, *Thromb. Haemostasis.* **47,** 210 (1982).

[26] M. A. Martins, P. M. R. e S. Martins, H. C. Castro Faria Neto, P. T. Bozza, P. M. F. L. Dias, R. S. B. Cordeiro, and B. B. Vargaftig, *Eur. J. Pharmacol.* **149,** 89 (1988).

triazolodiazepines required to produce equivalent inhibition of PAF-induced hypotension are much higher in rats (Table III). Hence, for studies with oral WEB 2086 and WEB 2170, the necessity for higher oral doses when using rats must be balanced against the convenience in using this species and the pharmacological advantages of, for example, the ability to produce reliable and persistent PAF-induced arterial pressure depression (without concomitant bronchospasm) over a long period of time (2–3 hr). In general, PAF antagonists are given orally 1 hr prior to PAF administration (iv infusion or bolus). The duration of action of WEB 2086 and WEB 2170 after oral administration has been determined by following bronchoconstriction and mean arterial pressure changes over a period of 8 to 20 hr. Both substances provided maximal efficacy during the first 3 hr. The biological half-life ($t_{1/2}$) for WEB 2086 in guinea pigs was 5.5 hr, which is similar to the half-life in man; for WEB 2170 it was 12 hr (Table III).

Intravenous Administration. Intravenous thienotriazolodiazepine doses for inhibition of PAF effects are of the same magnitude in guinea pigs and rats, as shown in Table III. As expected, the biological half-life ($t_{1/2}$) is shorter than after oral administration (measured in rats).

Inhalation or Intratracheal Instillation. For administration by inhalation, the anesthetized guinea pig is placed in a whole-body plethysmograph (plethysmograph box 853, ASE, Freiburg, FRG) and attached via an intratracheal cannula to one outlet port, allowing the animal to breath air from outside the chamber. Respiratory airflow is measured indirectly from changes in the thorax volume by a pneumotachograph (Fleisch No. 000) attached to another outlet port of the plethysmograph (Fig. 2). The aerosol of the PAF antagonist solutions is generated by an ultrasound nebulizer (USV 82 Praxis, Technolab, FRG), which is connected to the outlet port of the intratracheal cannula. The spontaneously breathing guinea pigs are exposed to the aerosol for 3, 5, or 10 min prior to the iv PAF infusion or an intratracheal instillation of PAF. The dose administered by *inhalation* can only be quantified by the concentration (mg/ml) used and not by the real inhaled quantity. Hence, the ED_{50} of WEB 2086 in Table III has only a theoretical value and is not comparable with the other doses. Nevertheless, dose–response curves are feasible. The biological half-life ($t_{½}$) is rather short. Intratracheal instillation does not have the problem of dosage, but it does not resemble the actual conditions for the clinical use of an inhaled form of these drugs. As shown in the case of WEB 2170, the instilled dose does not differ from the oral or iv dosage. In guinea pigs, the biological half-lives after oral administration and intratracheal instillation are similar (Table III). The absorption of thienotriazolodiazepines through the mucosa of the airways is excellent.

Use in Disease-Related Experimental Models

There is evidence that PAF plays a major role in some experimental models that resemble human or animal diseases. An increase in (endogenous) PAF synthesis and release is thought to be the cause of the disorders in these models and, therefore, treatment with PAF antagonists like WEB 2086 and WEB 2170 has been attempted. The curative results obtained with these compounds have been described for endotoxin shock in rats and passive and active anaphylaxis in guinea pigs and mice.[12,13] The doses required for these types of experiments cannot be predicted, but, in general, are somewhat higher than those used in experiments with exogenous PAF. This is in agreement with the results obtained with other mediator antagonists.

Conclusion

WEB 2086 and WEB 2170 are potent and selective antagonists of PAF *in vitro* and, more importantly, *in vivo* with a long duration of action. For oral use in rats, WEB 2170 may be preferred to WEB 2086 due to the higher potency and long duration of action. Both compounds are suitable tools for investigating the significance of PAF in the pathophysiology of different diseases.

Section IV

Molecular Biology

A. Prostaglandins
Articles 51 and 52

B. Leukotrienes
Articles 53 and 54

[51] Cloning of Sheep and Mouse Prostaglandin Endoperoxide Synthases

By DAVID L. DEWITT and WILLIAM L. SMITH

Prostaglandin endoperoxide (PGG/H) synthase (E.C. 1.14.99.1) catalyzes two consecutive reactions in prostaglandin biosynthesis[1-3]: (1) a cyclooxygenase reaction involving transformation of one molecule of arachidonate and two molecules of oxygen to prostaglandin G_2 and (2) a hydroperoxidase reaction in which PGG_2 undergoes a two-electron reduction to yield PGH_2. PGG/H synthase activity can be regulated by growth factors, tumor promoters, and steroids in a variety of cell types.[4-10] These observations have prompted the cloning of cDNAs for PGG/H synthase to be used as probes for studying transcriptional regulation of the PGG/H synthase gene.[11-13] The availability of a cDNA containing the entire coding region of PGG/H synthase has also permitted studies of enzyme function using site-directed mutagenesis.[14]

We describe here the protocol used for preparing a λgt10 library and for isolating cDNAs coding for PGG/H synthase from sheep vesicular gland.[11] We have recently used a similar approach to isolate a cDNA coding for a mouse 3T3 cell PGG/H synthase.[14] Radiolabeled oligonucleotide probes used to identify the PGG/H synthase cDNA clone were designed on the

[1] F. J. Van der Ouderaa, M. Buytenhek, D. H. Nugteren, and D. A. van Dorp, *Biochim. Biophys. Acta* **487,** 315 (1977).
[2] S. Ohki, N. Ogino, S. Yamamoto, and O. Hayaishi, *J. Biol. Chem.* **254,** (1979).
[3] W. R. Pagels, R. J. Sachs, L. J. Marnett, D. L. DeWitt, J. A. Day, and W. L. Smith, *J. Biol. Chem.* **258,** 6517 (1983).
[4] A. J. Habenicht, M. Goerig, J. Grulich, D. Rother, R. Gronwald, V. Loth, G. Scheith, G. Krommerell, and R. Ross, *J. Clin. Invest.* **75,** 1381 (1985).
[5] R. L. Huslig, R. L. Fogwell, and W. L. Smith, *Biol. Reprod.* **21,** 589 (1979).
[6] J. M. Bailey, B. Muza, T. Hla, and K. Salata, *J. Lipid Res.* **26,** 54 (1985).
[7] L. Hedin, D. Gaddy-Kurten, D. L. DeWitt, W. L. Smith, and J. S. Richards, *Endocrinology* **121,** 722 (1987).
[8] B. B. Weksler, *Adv. Prostaglandin Thromboxane Leukotriene Res.* **17A,** 238 (1987).
[9] P. S. Whitely and P. Needleman, *J. Clin. Invest.* **74,** 2249 (1984).
[10] M. Goerig, A. J. R. Habenicht, R. Heitz, W. Zeh, H. Katus, B. Kommerell, R. Ziegler, and J. A. Glomset, *J. Clin. Invest.* **79,** 903 (1987).
[11] D. L. DeWitt and W. L. Smith, *Proc. Natl. Acad. Sci. U.S.A.* **85,** 1412 (1988).
[12] J. P. Merlie, D. Fagan, J. Mudd, and P. Needleman, *J. Biol. Chem.* **263,** 3550 (1988).
[13] C. Yokoyama, T. Takai, and T. Tanabe, *FEBS Lett.* **231,** 347 (1988).
[14] D. L. DeWitt, E. A. El-Harith, S. A. Kraemer, E. F. Yao, R. L. Armstrong, and W. L. Smith, *J. Biol. Chem.,* in press (1990).

basis of amino acid sequence information obtained from the purified sheep vesicular gland enzyme.

Materials. Guanidine thiocyanate is from Fluka BioChemika; CsCl from Bar Lac Oid Chemical Company. Oligo(dT)$_{12-18}$, dATP, dCTP, dGTP, dTTP, ATP, and *Escherichia coli* RNase H are from Pharmacia Molecular Biology; Moloney murine leukemia virus reverse transcriptase is from Bethesda Research Laboratories (Gaithersburg, MD). [α-^{32}P]dCTP (800 Ci/mmol, 10 mCi/ml), [γ-^{32}P]ATP (3000 Ci/mmol, 10 mCi/ml), and [γ-^{32}P]ATP (NEG 035C, 6000 Ci/mmol, $>$ 100 mCi/ml) are from New England Nuclear (Boston, MA). *Escherichia coli* DNA polymerase, T4 polynucleotide kinase, T4 DNA ligase, and *Eco*RI endonuclease are from Boehringer Mannheim Biochemicals (Indianapolis, IN). *Eco*RI methylase and *Eco*RI linkers (#1004) are from New England Biolabs. Affi-Gel 10 and BioGel A-50m are from Bio-Rad Laboratories (Richmond, CA). Nitrocellulose filters (BA-85, 0.45 μm, 132 mm) are from Schleicher & Schuell. *N*-Tosyl-L-phenylalanine chloromethyl ketone (TPCK)–trypsin and agarose (type II: medium EEO) are from Sigma (St. Louis, MO). The bacterial strains C600 and C600 (*hfl*$^-$), packaging extracts, and λgt10 DNA are available from Stratagene. All other reagents are from standard sources.

Methods

Preparation of Oligonucleotide Probes

PGG/H synthase is purified from sheep vesicular glands. Briefly, the glands (100 g wet weight) are homogenized at 4° in 400 ml of 0.1 *M* Tris-Cl, pH 8.0, using a Polytron homogenizer. The sample is centrifuged at 10,000 *g* for 10 min, and the resulting supernatant is centrifuged at 100,000 *g* for 60 min to yield the microsomal membrane fraction. Microsomes (500 mg of protein) are solubilized by homogenization in 50 ml of 0.1 *M* Tris-Cl, pH 8.0, containing 1% Tween 20 and incubated for 1 hr at 4°[11]; the sample is then centrifuged at 100,000 *g* for 60 min at 4°, and the resulting supernatant containing the solubilized PGG/H synthase passed over a 1 × 5 cm immunoaffinity column containing an anti-PGG/H synthase monoclonal antibody [IgG(*cyo*-7)][11] attached to Affi-Gel 10.

The IgG (*cyo*-7)–Affi-Gel 10 column is prepared as follows. IgG (*cyo*-7) (10 mg) is dissolved in 10 ml of 0.1 *M* 4-(2-hydroxyethyl)-piperazine-1-ethanesulfonic acid (HEPES), pH 7.5, and incubated with 2.0 ml Affi-Gel-10 at 4° for 12 hr. The gel is collected by centrifugation at 1000 *g* for 5 min and 20 ml of 0.1 *M* ethanolamine is added. After 1 hr at 24°, the gel is poured into a 5-ml plastic syringe equipped with a Luer-lok tip and washed with 20 volumes of equilibration buffer (0.1 *M* Tris-Cl, pH 8.0, containing 0.1% Tween 20).

Solubilized microsomes (ca. 50 ml) are added to the IgG (*cyo*-7)–Affi-Gel-10 column running at a flow rate of 2.5 ml/min. The column is washed with 20 volumes of 0.1 *M* Tris-Cl, pH 8.0, containing 1% Tween 20. The pure PGG/H synthase is then eluted with 10 ml of 0.2 *M* acetic acid. The protein is extensively dialyzed against water and then lyophilized. It is next reduced, carboxymethylated, and trypsinized with TPCK-trypsin using a procedure identical to that detailed by Gracy.[15] Following lyophilization, the sample is suspended in 0.1 ml of 0.1% trifluoroacetic acid and peptides are isolated by reversed-phase HPLC on a Varian protein C_{18} column in 0.1% aqueous trifluoroacetic acid with a 0–60% acetonitrile gradient programmed at 1%/min. Fractions containing peptides, as determined by absorbance at 214 nm, are collected by hand into 1.5-ml Eppendorf tubes. Selected peak fractions are subjected to a second round of HPLC using the same system. Purified peptide samples are sequenced on a Beckman 890M protein sequencer. A sample of the reduced and carboxymethylated PGG/H synthase is also subjected to protein sequencing to determine the sequence of the N terminus of the protein.

Oligonucleotide probes modeled from the amino acid sequences of the N terminus and one of the tryptic peptides are synthesized: (a) DD-5; a 128-fold-redundant mixed oligomer 5′-d(GGRTARTARCARCANGGRTT)-3′, complementary to nucleotides coding for the peptide sequence NPCCYYP near the amino terminus; and (b) DD-6; a 192-fold-redundant mixed oligomer 5′-d(AGRTTRTCNCCRTADAARTG)-3′, complementary to nucleotides coding for the tryptic peptide HIYGDNL. (Single-letter abbreviations for nucleotides: R is A or G; Y is C or T; D is A, T, or G; and N is A, C, G, or T.)

Preparation of Sheep Vesicular Gland λgt10 cDNA Library

Sheep vesicular glands are removed immediately after slaughter and homogenized with a Polytron homogenizer in 10 volumes of 6 *M* guanidine thiocyanate containing 5 m*M* sodium citrate, pH 7.0, 0.1 *M* 2-mercaptoethanol, and 0.5% sodium *N*-laurylsarcosine. RNA is then isolated by centrifugation through a 5.7 *M* CsCl gradient containing 0.1 *M* EDTA, pH 8.0, as described by Maniatis *et al.*[16] Because of the high nuclease content of vesicular glands, high molecular weight mRNA is obtained only from freshly isolated tissue. RNA isolated from glands more

[15] R. L. Gracy, this series, Vol. 47, p. 195.

[16] T. Maniatis, E. F. Fritsch, and J. Sambrook, "Molecular Cloning: A Laboratory Manual," p. 196. Cold Spring Harbor Laboratory, Cold Spring Harbor, New York, 1982.

than 10–15 min after slaughter or from glands initially frozen with dry ice or liquid nitrogen is always degraded. Since mRNA of high quality is the single most important factor in the preparation of a library containing full-length recombinants, care should be taken in its isolation. Our experience is that the mRNA obtained using the guanidine thiocyanate procedure is the most suitable template for synthesizing full-length complementary DNAs (cDNAs).

After centrifugation, mRNA is separated from total RNA by oligo(dT)-cellulose chromatography.[17] cDNA is then synthesized by the Gubler and Hoffman procedure[18] as modified by D'Alessio *et al.*[19] For first-strand synthesis, 10 μg of vesicular gland poly$(A)^+$ mRNA is combined into a reaction mix (50 μl final volume) containing 50 m*M* Tris-Cl, pH 8.3, 75 m*M* KCl, 10 m*M* dithiothreitol, 3 m*M* $MgCl_2$, 500 μM each of dATP, dCTP, dGTP, and dGTP, 25 μg of oligo$(dT)_{12-18}$, and 500 units of Moloney murine leukemia virus reverse transcriptase. This reaction is kept on ice and a 10-μl aliquot is transferred to a separate vial containing 1 μl of $[\alpha\text{-}^{32}P]$dCTP (10 μCi; 800 Ci/mmol). The two samples are incubated at 37° for 1 hr, and a portion of the reaction mixture containing the radiolabel is processed to determine the yield of first-strand cDNA.[20] We synthesize 2.5 μg of first-strand cDNA from 10 μg of vesicular mRNA, but yields range from 10 to 50% of the mass of the starting mRNA.

Synthesis of the second strand is initiated by dilution of the 40 μl first-strand reaction mix to a final volume of 320 μl containing 25 m*M* Tris-Cl, pH 8.3, 100 m*M* KCl, 5 m*M* $MgCl_2$, 250 μM each of dATP, dCTP, dGTP, dTTP, and $[\alpha\text{-}^{32}P]$dCTP (10 μCi; 800 Ci/mmol), 5 m*M* dithiothreitol, 250 units of *E. coli* DNA polymerase I per ml, and 8.5 units of *E. coli* RNase H per ml. This reaction mixture is incubated for 2 hr at 16°. The reactions are stopped by adding 25 μl of 0.2 *M* EDTA. An aliquot (10 μl) of the second-strand reaction is then removed, and the yield is determined as described above for the first strand.[20] A typical second-strand yield is 80–100% of the mass of the first strand. The double-stranded cDNA is precipitated by adding 0.5 volumes of 7.5 *M* ammonium acetate and three volumes of ethanol and then immediately centrifuging for 15 min in a microcentrifuge at room temperature. Additional aliquots of the ^{32}P-labeled first- and second-strand reactions are electrophoresed on a 0.8%

[17] T. Maniatis, E. F. Fritsch, and J. Sambrook, "Molecular Cloning: A Laboratory Manual," p. 197. Cold Spring Harbor Laboratory, Cold Spring Harbor, New York, 1982.

[18] U. Gubler, and B. J. Hoffman, *Gene* **25,** 263 (1985).

[19] J. M. D'Alessio, M. C. Noon, H. L. Ley, and G. F. Gerard, *Focus* **9,** 1 (1987).

[20] T. Maniatis, E. F. Fritsch, and J. Sambrook, "Molecular Cloning: A Laboratory Manual," p. 473. Cold Spring Harbor Laboratory, Cold Spring Harbor, New York, 1982.

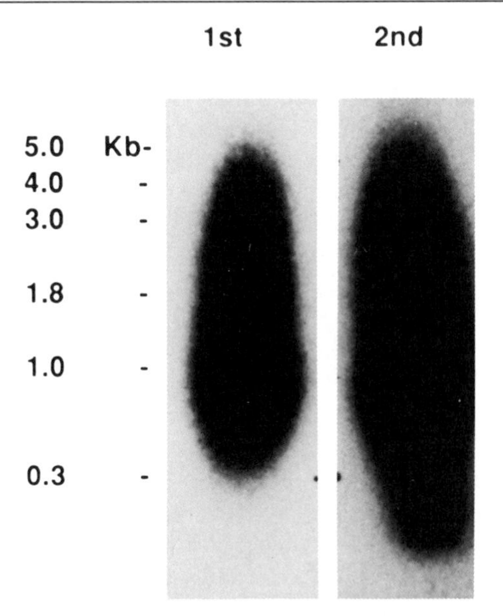

FIG. 1. Size distribution of newly synthesized first- and second-strand sheep vesicular gland cDNA. Aliquots of the ^{32}P-labeled first- and second-strand reaction (~10^5 cpm), prepared as described in the text, are electrophoresed on a 0.8% TBE agarose gel which is subjected to autoradiography.

TBE agarose[21] minigel to determine the size distribution of the cDNAs (Fig. 1). After electrophoresis, the intact gel is wrapped in Saran wrap and exposed to Kodak XAR-5 X-ray film using an intensifying screen at −80°; upon development, a smear of radioactivity is visible beginning near 5 kilobases (kb) and continuing to 100 base pairs (bp) for both reactions (Fig. 1).

The cDNA is next resuspended into 50 μl of a reaction mix containing 100 m*M* NaCl, 100 m*M* Tris-Cl, pH 8.0, 1 m*M* EDTA, 80 μM *S*-adenosylmethionine, and 400 units of *Eco*RI methylase per ml and incubated at 37° for 1 hr. The entire reaction mixture is extracted with an equal volume of phenol/chloroform/isoamyl alcohol (25 : 24 : 1; v/v/v). The cDNA is then precipitated by adding 0.1 volumes of 3.0 *M* sodium acetate, pH 5.2, and 2.5 volumes of ethanol. This sample is kept at −20° overnight before centrifugation.

[21] T. Maniatis, E. F. Fritsch, and J. Sambrook, "Molecular Cloning: A Laboratory Manual," p. 150. Cold Spring Harbor Laboratory, Cold Spring Harbor, New York, 1982.

*Eco*RI linkers (10-mer) are kinased as follows: 0.02 A_{260} units of linkers are added to a reaction mixture (10 μl final volume) containing 100 m*M* Tris-Cl, pH 7.5, 20 m*M* dithiothreitol, 10 m*M* $MgCl_2$, 5 μCi [γ-^{32}P]ATP (3000 Ci/mmol), and 500 units of T4 polynucleotide kinase per ml. After incubation at 37° for 15 min, 0.5 μl of 100 m*M* ATP is added, and the reaction allowed to proceed for an additional 30 min. This reaction mixture is then added to the precipitated cDNA together with 1 μl of T4 DNA ligase (1000 units/ml), and the sample incubated at 14° overnight. The ligated linker–cDNA mixture is heated to 70° for 10 min to inactivate the ligase, and 3 μl of 10 × *Eco*RI buffer (500 m*M* NaCl, 1.0 *M* Tris-Cl, pH 7.5, 50 m*M* $MgCl_2$, 1 mg bovine serum albumin/ml), 14 μl of water, and 3 μl of *Eco*RI restriction endonuclease (20,000 units/ml) are added. The digestion is allowed to proceed at 37° for 2 hr. Aliquots of (a) the linker after kinasing, (b) the cDNA and linker after ligation, and (c) the cDNA and linker after ligation and *Eco*RI digestion are electrophoresed on a 10% polyacrylamide gel.[22] Autoradiography of this gel reveals the presence of concatamerized (i.e., polymerized) linkers in the ligation reaction, and the subsequent cleavage of these linker–multimers to monomers upon *Eco*RI digestion. The cDNA, which has a much lower specific activity than the kinased linkers, is not clearly visible in the autoradiograph.

The sample containing linker-ligated, *Eco*RI-treated cDNA is applied to a 0.2 mm × 32 cm BioGel A-50m column and size-fractionated as described by Huynh *et al.*[23] The column is eluted with 100 m*M* Tris-Cl, pH 7.5, containing 100 m*M* NaCl and 1 m*M* EDTA. Fractions (20–40 μl) are collected and 5-μl aliquots of each fraction are subjected to electrophoresis on a 0.8% TBE agarose gel which is then subjected to autoradiography (Fig. 2). Those fractions containing cDNA with a minimum size greater than about 1 kb are pooled (e.g., fractions 7–9 of Fig. 2) and precipitated with ethanol.

Bacteriophage λgt10 are grown, and phage DNA is isolated as described by Huynh *et al.*[24] Purified λgt10 DNA (10 μg) is digested in a total of 20 μl of *Eco*RI restriction digest buffer containing 2000 units of *Eco*RI endonuclease/ml. The digested DNA is heated to 70° for 5 min and then cooled to 45° for 30 min to anneal the lambda cohesive ends. Finally, the DNA is put on ice and 1 μl of 200 m*M* dithiothreitol and 0.5 μl of 100 m*M* ATP are added. This *Eco*RI-digested stock vector DNA is stored at −20°.

[22] T. Maniatis, E. F. Fritsch, and J. Sambrook, "Molecular Cloning: A Laboratory Manual," p. 173. Cold Spring Harbor Laboratory, Cold Spring Harbor, New York, 1982.

[23] T. B. Huynh, R. A. Young, and R. W. Davis, *in* "DNA Cloning: A Practical Approach" (D. M. Glover, ed.), p. 64. IRL, Washington, DC, 1985.

[24] T. B. Huynh, R. A. Young, and R. W. Davis, *in* "DNA Cloning: A Practical Approach" (D. M. Glover, ed.), p. 57. IRL, Washington, DC, 1985.

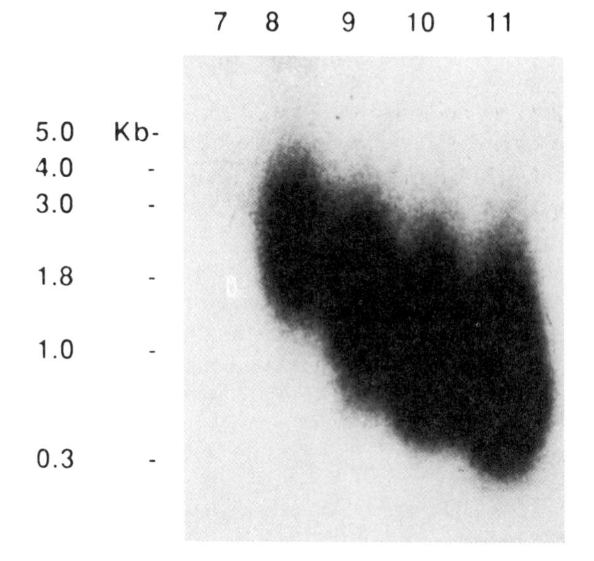

FIG. 2. Size distribution of the linker-ligated, *Eco*RI-treated, double-stranded sheep vesicular gland cDNA after fractionation on a 0.2 mm × 32 mm BioGel A-50m column. Twenty 40-μl fractions are collected and 5-μl aliquots of these fractions electrophoresed on a 0.8% TBE agarose gel which is subjected to autoradiography. Fractions 7–11 are shown; fractions 7–9 are pooled, ethanol-precipitated, and then ligated with *Eco*RI-digested λgt10 to construct the library.

To confirm that this DNA could be religated, 2 μl of *Eco*RI-digested λgt10 DNA is mixed with 1 μl of 50 m*M* Tris-Cl, pH 7.5, containing 10 m*M* $MgCl_2$, 1 μl of TE (10 m*M* Tris-Cl, pH 7.5, 1 m*M* EDTA), and 0.4 μl of T4 DNA ligase (1000 units/ml). After 2 hr at 14°, the ligated DNA, as well as equal masses of unligated *Eco*RI-digested λgt10 DNA and uncut λgt10 DNA, are packaged with *in vitro* packaging extracts (see Ref. 25, or these are commercially available). The resulting phage are then plated on C600 and C600 (hfl^-) bacteria.[26] *Eco*RI digestion should reduce the infectivity of the lambda DNA by at least three orders of magnitude when compared to uncut λgt10 DNA (from about 10^8 to $< 10^5$ pfu/μg DNA), and the *Eco*RI-digested, ligated DNA should have about 30% of the infectivity of the uncut DNA. The number of cI^- plaques [those that form on the C600 (hfl^-) lawn] in the uncut DNA and the *Eco*RI-digested, ligated DNA should be $< 0.1\%$ of the total ($cI^+ + cI^-$) plaques that form on the C600 lawn.

[25] T. Maniatis, E. F. Fritsch, and J. Sambrook, "Molecular Cloning: A Laboratory Manual," p. 264. Cold Spring Harbor Laboratory, Cold Spring Harbor, New York, 1982.

[26] N. E. Murray, W. J. Brammar, and K. Murray, *Mol. Gen. Genet.* **150,** 53 (1977).

Size-selected cDNA (Fig. 2) is resuspended in 10 μl of TE, and 1 μl of this sample is mixed with 2 μl of the *Eco*RI-digested stock vector DNA together with 1 μl of 50 m*M* Tris, pH 7.5, containing 10 m*M* $MgCl_2$, and 0.4 μl of T4 DNA ligase (1000 units/ml). After 2 hr at 14°, the mixture is packaged using *in vitro* packaging extracts and aliquots are plated on both C600 and C600 (*hfl*$^-$) bacteria. Insertion of cDNA into the *Eco*RI site of λgt10 DNA interrupts the coding region of the lambda phage *cI* repressor which results in the production of *cI*$^-$ phage. After ligation, the percentage of *cI*$^-$ phage should increase to 0.5–2% of the total phage. If *cI*$^-$ levels greater than 2% are observed, the cDNA/vector ratio should be reduced to decrease the possibility of obtaining phage-containing multiple unrelated inserts. We obtain about 10^6 recombinant phage/μg of vesicular gland mRNA; typical yields range from $10^5-5 \times 10^6$ phage/μg mRNA.

Screening of Sheep Vesicular Gland cDNA Library

Oligonucleotides DD5 and DD6 are radiolabeled as follows. Each oligonucleotide (50 pmol) is added to separate 10-μl reaction mixtures containing 50 m*M* Tris-Cl, pH 7.6, 10 m*M* $MgCl_2$, 5 m*M* dithiothreitol, approximately 60 pmol of [γ-^{32}P]ATP (6000 Ci/mmol, >100 mCi/ml), and 500 units of T4 polynucleotide kinase per ml. Labeled oligonucleotides are separated from unincorporated [γ-^{32}P]ATP by chromatography on a 2-ml column of Sephadex G-25 prepared in a Pasteur pipette and eluted with TE. Typically, the probes have a specific activity of $>10^7$ cpm/pmol.

The vesicular gland library is plated with C600 (*hfl*$^-$) bacteria on ten 150-mm NZCYM agar plates at a density of 20,000 plaques/plate and, after 6–7 hr, phage are transferred to duplicate nitrocellulose filters by standard procedures.[27] The filters are baked at 80° for 2 hr in a vacuum oven, then prehybridized using for each filter 2 ml of 5x SSPE (20×SSPE is: 3.6 *M* NaCl, 0.2 *M* $NaPO_4$, 20 m*M* EDTA) containing 1× Denhardt's solution (50× Denhardt's is: 1% Ficoll, 1% polyvinylpyrrolidone, 1% bovine serum albumin), 0.1% SDS, and 100 μg of fragmented calf thymus DNA per ml at 42°. The hybridization temperature, T_H, is determined by the formula[28]:

$$T_H = 2° \text{ (No. A + T bases)} + 4° \text{ (No. G + C bases)} - 5°$$

The base composition of the oligonucleotide in each mixture that has the highest A/T content (e.g., lowest melting temperature) is used for calcula-

[27] T. Maniatis, E. F. Fritsch, and J. Sambrook, "Molecular Cloning: A Laboratory Manual," p. 320. Cold Spring Harbor Laboratory, Cold Spring Harbor, New York, 1982.

[28] S. V. Suggs, T. Hirose, T. Miyake, E. H. Dawashime, J. M. Johnson, K. Itakure, and R. B. Wallace, *in* "Developmental Biology Using Purified Genes" (D. D. Brown, ed.), p. 683. Academic Press, New York, 1981.

tion of the hybridization temperature. After 2 hr, the hybridization buffer is replaced with new hybridization buffer, 2×10^6 cpm/ml of labeled oligonucleotide probe is added, and the filters hybridized at 42°; one set of filters are used for each oligonucleotide probe. After 48 hr, the filters are washed three times (30 min per wash) in 2× SSC (20× SSC is: 3.0 *M* NaCl, 0.30 *M* sodium citrate, pH 7.0), containing 0.1% SDS at room temperature and one time in the same buffer at T_H (42°). The filters are then dried and subjected to autoradiography at −80° with intensifying screens.

When screening with mixed probes at temperatures that allow the hybridization of the oligonucleotides with the highest A/T content (lowest duplex melting temperature), false positives may be obtained; these result from oligonucleotides with a higher G/C content (higher duplex melting temperature) hybridizing with incompletely homologous sequences. To circumvent this problem, we use two mixed probes and consider only those plaques that hybridize with both probes to be positive. It is unlikely that two different mixed probes will hybridize with the same false-positive phage plaques.

Approximately 0.5% of the phage plaques hybridize with both DD5 and DD6, consistent with the observation that PGG/H synthase comprises about 0.5% of the sheep vesicular gland protein. Cores containing representative positive phage plaques are removed from the petri dishes with the wide end of a sterile Pasteur pipette and transferred to 1 ml of 10 m*M* Tris-Cl, pH 8.0, containing 10 m*M* $MgCl_2$. These phage are then rescreened at lower concentrations (100–500 pfu/100 mm plate) and individual positive phage plaques isolated. These phage are amplified, and their DNA isolated and digested with *Eco*RI endonuclease. Electrophoresis of *Eco*RI-treated DNAs on a 0.8% TBE agarose gel reveals a similar restriction pattern for the inserts from each clone, indicating that they are derived from related poly(A)$^+$ RNAs. The clone with the largest total insert size (2.7 kb) is sequenced.[11]

Primary Structure of Sheep and Mouse PGG/H Synthases

The nucleotide sequence and deduced amino acid sequences of the sheep PGG/H synthase cDNA is shown in Fig. 3. Comparison of the amino acid sequences of the sheep and mouse proteins reveals that there is 88% amino acid sequence identity and 98% similarity when conservative substitutions are included.[14] The sheep PGG/H synthase contains a signal peptide encompassing amino acids 1–26; the putative mouse signal peptide contains an additional two amino acids. Sheep PGG/H synthase contains four potential sites for *N*-glycosylation including Asn-68, -104, -144, and -410[11]; there are corresponding sites in the mouse protein (Asn-70, -106,

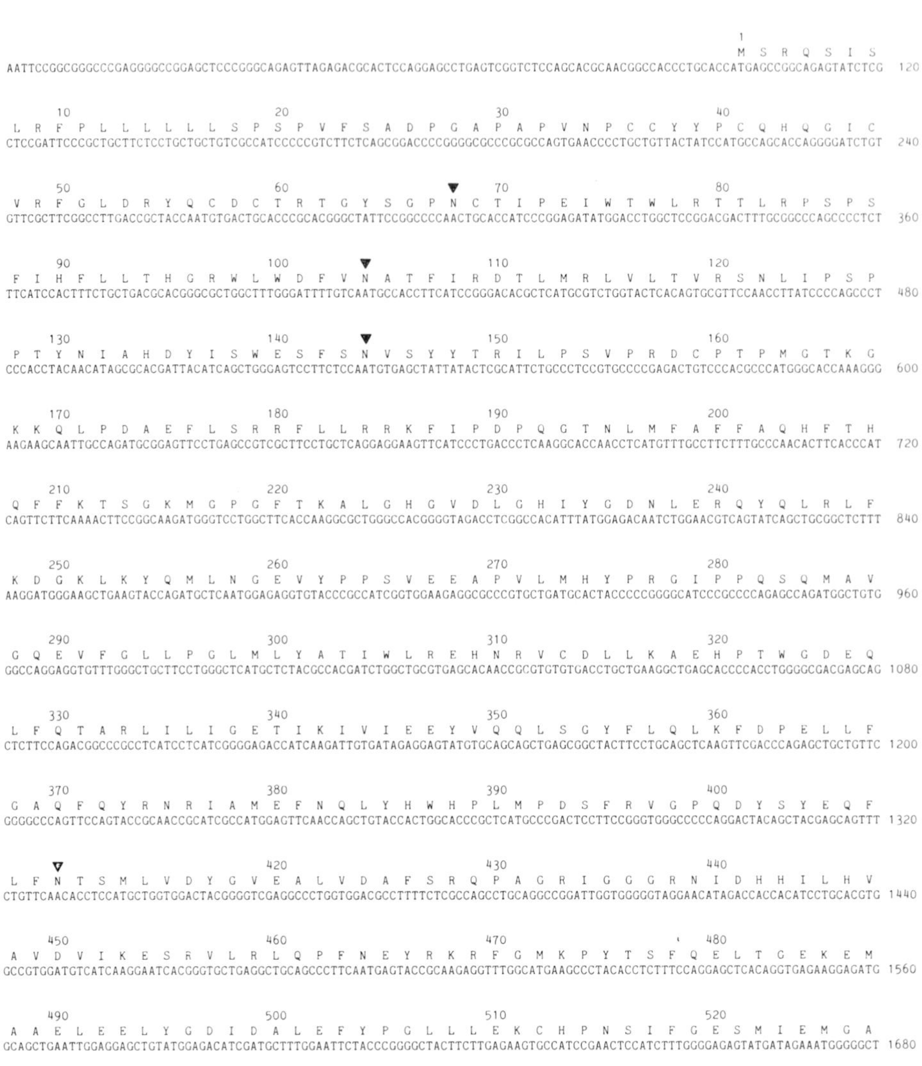

```
◊          540                 550                 560
P F S L K G L L G N P I C S P E Y W K A S T F G G E V G F N L V K T A T L K K L
CCTTTTTCCCTTAAGGGCCTCTTAGGAAACCCCATCTGTTCTCCAGAGTACTGGAAGGCGAGCACATTTGGCGGTGAGGTGGGCTTCAACCTTGTCAAGACGGCCACGCTAAAGAAGCTG 1880

570        580                 590                 600
V C L N T K T C P Y V S F H V P D P R Q E D R P G V E R P P T E L *
GTTTGCCTCAACACCAAGACTTGTCCCTATGTCTCCTTCCACGTGCCAGACCCCCGTCAGGAGGACAGGCCTGGGGTGGAGCGGCCACCCACAGAGCTCTGAAGGGGCTGGGCAGCAGCA 1920

TTCTGGATGGTAGAGCTTCCTGCTTGCCATTCCAGAATGCCACGGGGTGGATTGTCTTTGATGTTGGGTTTCTGATTTGGTGTCGAGAGCATCAGTGTGGACGTTTAGAACTCTAGGTCT 2040
CTCACCCCATGGTCTGGAATACTGTGTTCCTTGTTTGTTGTTCTAGAATGCTGAATTCCTGGTAAACCATTGAGAATGTTAGGAGTGGTTATCCCTTCAGCATTGCCAGAACACTGGGTT 2160
CCTGGGTGACCACCTAGAATGTCAGATTTCTAGTTGATCCGGAATTTAGGCACTCTGAAATATGGACTCCTGATGGAATCATCTGGAAAGTGAGGGGGTTTTTATTTTGCATTCTAGAAT 2280
TCTGGGTGGCCCTCCAGAATGTCGACTTTCTGACTGGTTATCCGGAATGTTGTGCTCCGAGTTGCTGATCCAGAACAGTGGCTGGCATTCTAGATCAGTCCTGATCCGAATGTCTAGAGT 2400
GTTGAGAATTCATTTTCCTGTTCAGTGAGACAGCCACGGAGCAGGAGGATCTCGTGTCCTACAAGAACGCATTGCCTGGATCTGTGCCTGCATGGAGAGGGCAAGGAAGTGGGGTGTTCG 2520
TCTTCTCAGTGGGACCCCTGATGAGACCCTGGATATGGAGAGAACAGGTGGCTTTCTCCAGGCCATTGGTTGGAAGCCACCAGAGCTCTGTCCTCATCCAGGTCTCAACTCACGGCAGCT 2640
GTTTTTCATGAAGTTAATAAAATGCTTTTTCCGAAAAAAAAAAACGGAATT                                                                      2690
```

FIG. 3. Nucleotide and deduced amino acid sequence of PGG/H synthase from sheep vesicular gland. Potential sites of *N*-glycosylation are indicated by closed triangles. Ser-530 is indicated with an open diamond.

-146, and -412).[14] Ser-530 in the sheep enzyme is the site of acetylation by aspirin[11]; the corresponding residue in the mouse is Ser-532.[14] Residues 303–312 in the sheep protein[11] and 305–314 in the mouse protein[14] are identical and are closely related to a sequence TLW(L)LREHNRL which may form the axial heme-binding site in thyroid peroxidase and myeloperoxidase, respectively.[29–32]

Acknowledgments

This work was supported in part by U.S.P.H.S. Grants DK22042 and GM40713.

[29] K. R. Johnson, W. M. Nauseef, A. Care, J. J. Weelock, and G. Rovera, *Nucleic Acids Res.* **15,** 2013 (1987).

[30] K. Morishita, N. Kubota, S. Asano, Y. Kaziro, and S. Nagata, *J. Biol. Chem.* **262,** 3844 (1987).

[31] S. Kimura, T. Kotani, O. W. McBride, K. Umeki, K. Hirai, T. Nakayama, and S. Ohtaki, *Proc. Natl. Acad. Sci. U.S.A.* **84,** 5555 (1987).

[32] S. Kimura and M. Ikeda-Saito, *Proteins Struct. Func. Gene.* **3,** 113 (1988).

[52] Preparation and Proteolytic Cleavage of Apoprostaglandin Endoperoxide Synthase

By REBECCA ODENWALLER, YING-NAN PAN CHEN, and LAWRENCE J. MARNETT

Partial or complete proteolytic digestion of prostaglandin endoperoxide (PGH) synthase is fundamental to the characterization of structural domains of the protein or labeling of its active site.[1,2] Studies of limited tryptic cleavage have led to a hypothesis about the location of the heme-binding site associated with peroxidase activity.[3] Exhaustive digestion by pepsin has been used to identify the serine residue selectively acetylated by aspirin.[2] Treatment of apoPGH synthase with trypsin under nondenaturing conditions appears to effect a single cleavage between Arg-253 and Gly-254.[1] Binding of heme to the apoprotein protects this bond from trypsin

[1] Y. -N. P. Chen, M. J. Bienkowski, and L. J. Marnett, *J. Biol. Chem.* **252,** 16892 (1987).

[2] G. J. Roth, E. T. Machuga, and J. Ozols, *Biochemistry* **22,** 4672 (1983).

[3] L. J. Marnett, Y. -N. P. Chen, K. R. Maddipati, P. Plé, and R. Labeque, *J. Biol. Chem.* **263,** 16532 (1988).

cleavage rendering holoPGH synthase refractory to trypsin treatment.[1,4] The procedure given in the literature for preparation of apoPGH synthase requires conditions that cause oxidative inactivation of a significant fraction of the protein.[5,6] Oxidatively inactivated PGH synthase is much more susceptible to tryptic cleavage than holoenzyme and cannot be reconstituted by addition of heme.[1] This can lead to artifacts in quantitative studies of the differential sensitivity of apo- and holoenzyme to proteolytic cleavage. We describe below a procedure for preparation of apoPGH synthase under conditions that do not cause its oxidative inactivation and procedures for limited or complete proteolysis of the apoprotein by trypsin.

Assay Method (Cyclooxygenase Activity)

Reagents

Buffer A is 100 m*M* Tris-HCl containing 500 μM phenol. It is titrated at 37° to pH 8.

A hematin shock solution is prepared to a concentration of 500 μM in dimethyl sulfoxide (DMSO). It can be stored at room temperature for up to 1 month. A 50 μM heme solution is made by diluting an aliquot of the stock in 100 m*M* phosphate (pH 7.8) containing 20% glycerol. An aliquot of the 50 μM solution is added to the assay mixture to a final concentration of 1 μM.

Arachidonic acid is stored in methanol containing 10 μM butylated hydroxyanisole under nitrogen at −20°. From this stock solution, a 10 m*M* arachidonic acid solution is made which contains 10% methanol and 10 m*M* NaOH. An aliquot of the 10 m*M* arachidonic acid stock solution is added to the assay mixture to a final concentration of 100 μM.

Both the heme and arachidonic acid solutions are kept on ice during the course of a day's assays.

Procedure. Oxygen uptake is measured using a Gilson model 5/6H oxygraph with a YSI 5331 oxygen probe (Yellow Springs Instrument Co., Yellow Springs, OH) and a water-jacketed cell. Cyclooxygenase activity is determined at 37° by measuring the rate of oxygen uptake following addition of arachidonic acid to a solution of buffer A containing 1 μM hematin and an aliquot of the enzyme preparation. Determination of the percentage of apoenzyme is accomplished by assaying the cyclooxygenase activity

[4] R. J. Kulmacz and W. E. M. Lands, *Biochem. Biophys. Res. Commun.* **104,** 758 (1982).
[5] B. G. Titus and W. E. M. Lands, this series, Vol. 86, p. 69.
[6] B. G. Titus, R. J. Kulmacz, and W. E. M. Lands, *Arch. Biochem. Biophys.* **214,** 824 (1982).

with and without the addition of exogenous heme to the cuvette. The cuvette must be entirely void of heme when assaying for apoenzyme. To minimize heme contamination of the apoenzyme, the cyclooxygenase assay is carried out several times with low percentage holoPGH synthase and no exogenous heme.

Preparation of ApoPGH Synthase

Reagents. There are several published procedures for the purification of PGH synthase.[7–9] The enzyme is also commercially available from Oxford Biomedical Research (Oxford, MI), Cayman Chemicals (Ann Arbor, MI), and BioMol (Philadelphia, PA). The protein used for the experiments below was purified from ram seminal vesicles and was 30–60% holoenzyme.[10] It is stored in 50 m*M* Tris (pH 8) containing 300 μM diethyl dithiocarbamate and 0.1% Tween 20. Buffer B is 100 m*M* Tris-HCl, 300 μM diethyl dithiocarbamate, 0.1% deoxycholate, 5 m*M* glutathione (reduced form, free acid), and 5 m*M* EDTA; the pH is adjusted to 8 at 4°. Buffer C is 80 m*M* Tris-HCl (pH 8), 300 μM diethyl dithiocarbamate, and 0.1% Tween 20.

Procedure. The following procedure is performed at 4°. PGH synthase (8 mg in a total volume of 3 to 7 ml) is concentrated to less than 1 ml using an Amicon concentrator (YM30 membrane); the concentrator is then rinsed twice with buffer B. The rinses are combined with the concentrated enzyme to obtain 8 mg of PGH synthase in 1 ml. This is loaded on a 37 × 2.5 cm Sephadex G-200 column and eluted with buffer B. ApoPGH synthase, occasionally followed by any remaining holoenzyme, comes off of the column shortly after the void volume. The flow rate may determine whether the eluting enzyme is 99% apo or 100% apoPGH synthase. One percent holoPGH synthase may remain if the flow rate is 0.2 ml/min, whereas a flow rate of 0.07–0.1 ml/min yields 100% apoenzyme. The cyclooxygenase activity of each 1.35-ml column fraction should be assayed in the area where cyclooxygenase activity elutes to determine an accurate elution profile. Any remaining native PGH synthase may elute either as a small hump trailing the apoPGH synthase peak, or it may only

[7] S. Yamamoto, this series, Vol. 86, p. 55.

[8] F. J. G. van der Ouderaa and M. Buytenhek, this series, Vol. 86, p. 60.

[9] R. J. Kulmacz and W. E. M. Lands, *in* "Prostaglandins and Related Substances: A Practical Approach (C. Benedetto, R. G. McDonald-Gibson, S. Nigam, and T. F. Slater, eds.), p. 209. IRL Press, Washington, D.C., 1987.

[10] L. J. Marnett, P. H. Siedlik, R. C. Ochs, W. D. Pagels, M. Das, K. V. Honn, R. H. Warnock, B. E. Tainer, and T. E. Eling., *Mol. Pharmacol.* **26,** 328 (1984).

broaden the base of the activity peak on the trailing side. Care must be taken to pool only those fractions containing apoprotein. The apo fractions are concentrated (usually to 1–2 ml), immediately diluted with 6 volumes buffer C, then dialyzed against buffer C for 20 to 24 hr. During dialysis, the buffer is changed at least five times. (Dialysis of the apoenzyme from buffer B into C must be done immediately to avoid the formation of a precipitate.) After concentrating the apoenzyme a final time, it is aliquoted into Eppendorf tubes and stored at −80°. If the Sephadex column is to be reused, it should be rinsed with 100 m*M* Tris-HCl (pH 8) to prevent formation of a precipitate either in the column or in the capillary tubing leading out of the column. Alternatively, gel filtration with a 13 × 1 cm Sephadex G-25 column may be used to transfer the apoPGH synthase from buffer B to buffer C.

Milligram quantities of 100% apoPGH synthase are routinely prepared in 75% yield with an increase in specific activity. The use of glutathione in the buffer to make apoenzyme presumably aids in removal of the heme group from PGH synthase since glutathione is known to form a reversible complex with hematin.[11] Use of deoxycholate below its critical micellar concentration in buffer B may allow hematin to be more available to glutathione. The gel filtration/buffer B combination provides a relatively gentle method of apoPGH synthase preparation that retains both cyclooxygenase and peroxidase activities.

Trypsin Digestion of PGH Synthase

Reagents. Trypsin is either from porcine pancreas (type IX, Sigma, St. Louis, MO) or from bovine pancreas (Sigma, type VIII, L-tosylamido-2-phenylethyl chloromethyl ketone-treated). Trypsin inhibitor is from chicken egg white (Sigma). Soybean trypsin inhibitor is insolubilized on agarose (Sigma). Buffer D is 80 m*M* Tris-HCl, pH 8.

Procedure. Trypsin and trypsin inhibitor are dissolved at concentrations of 1 and 10 mg/ml in buffer D, respectively. ApoPGH synthase in buffer C is incubated with 1–3% (w/w) trypsin for 1 hr at room temperature. The reaction may be quenched with a 10-fold excess of trypsin inhibitor, or it can be passed through a trypsin inhibitor agarose column. HPLC or SDS–PAGE analysis determines the percentage of PGH synthase (70 kDa) digested to 33 and 38 kDa fragments. Laser densitometry of the 10% polyacrylamide gel usually reveals 85–90% cleavage to 33 and 38 kDa fragments after a 2–3% (w/w) tryptic treatment of apoPGH synthase.

[11] Y. Shviro and N. Shaklai, *Biochem. Pharmacol.* **36,** 3801 (1987).

Reversed-Phase HPLC of Trypsin-Cleaved PGH Synthase

Solvents

Solvent A is 0.1% (v/v) trifluoroacetic acid (TFA) in 10% acetonitrile
Solvent B is 0.1% TFA in 90% acetonitrile

Procedure. HPLC is performed with a Varian Model 5020 pump and a 2050 variable wavelength absorbance monitor. Separation of PGH synthase (70 kDa) from its two tryptic cleavage products is accomplished by reversed-phase (RP) chromatography on a Vydac C-4 column (0.46 × 25 cm). Samples of the reaction mixture are injected onto the column using a 50-μl injection loop. The column is initially equilibrated in 40% solvent B

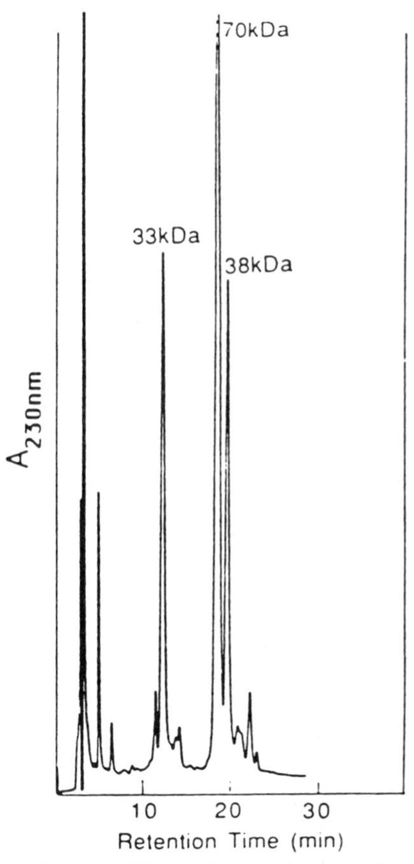

FIG. 1. HPLC separation of PGH synthase and the products of its limited tryptic cleavage.

at a flow rate of 1 ml/min. A linear gradient of 40 to 60% B runs from 0 to 30 min, holds at 60% B from 30 to 35 min, and returns to 40% B from 35 to 40 min. Eluting protein is detected by absorbance at 230 nm and recorded on a 10 m V Soltec chart recorder or on a Hewlett-Packard integrator (3380A). Individual fractions are collected and dried with a Speed Vac for SDS–PAGE and amino acid analysis. A typical chromatographic profile is shown in Fig. 1.

Peptide Mapping of PGH Synthase Proteolytic Fragments

Reagents. Chymotrypsin is from bovine pancreas (Sigma, type VII); elastase from porcine pancreas (Sigma, type IV); pepsin, from porcine stomach mucosa, thermolysin (type X), and protease from *Staphylococcus aureus* strain V8 (type XVII) are all also from Sigma.

Solvents

Solvent A is 0.1% TFA in water
Solvent B is 0.1% TFA in 80% acetonitrile

Procedure. Samples of PGH synthase are dissolved in 1.0 ml of 0.5 *M* Tris-HCl (pH 8.5) containing 6 *M* guanidine hydrochloride, 10 m*M* dithiothreitol, and 0.2% EDTA. The incubation proceeds for 2 hr at 37° or 5 hr at room temperature. The reaction mixtures are cooled to room temperature and 20 μl iodoacetic acid or iodoacetamide is added (final concentration 10 mM). The solutions are maintained at pH 8.0 for 30 min in the dark with dropwise additions of 1 *N* NaOH to maintain a constant pH, and then dialyzed against 0.1 *M* NH_4HCO_3 to remove salt. The carboxymethylated proteins are digested with trypsin, chymotrypsin, elastase, pepsin, thermolysin, or protease V8 at 37° for 24 hr. A ratio of protein to protease of 100 : 2 (w/w) is used. The digestion is terminated by freezing and lyophilization. One to three nanomoles of protein digests are redissolved in 5% acetic acid and injected onto a Beckman ODS or a Vydac C_{18} reversed-phase column (0.46 × 25 cm). The protein is eluted at a flow rate of 1 ml/min with a 115 min gradient of 0–50% B in 75 min, then 50–75% B in 25 min, followed by 75–100% B in 15 min. To detect most peptides, absorbance of the effluent is monitored at 230 nm. To detect those peptides that contain aromatic amino acids, fluorescence due primarily to tryptophan are detected using a Kratos Spectroflow 980 fluorescence detector with 280 nm excitation wavelength and a 370 nm emission filter. Data are collected using a Soltec chart recorder or a Varian DS-65 data system. When desired, peaks are collected and amino acid sequence analysis of the

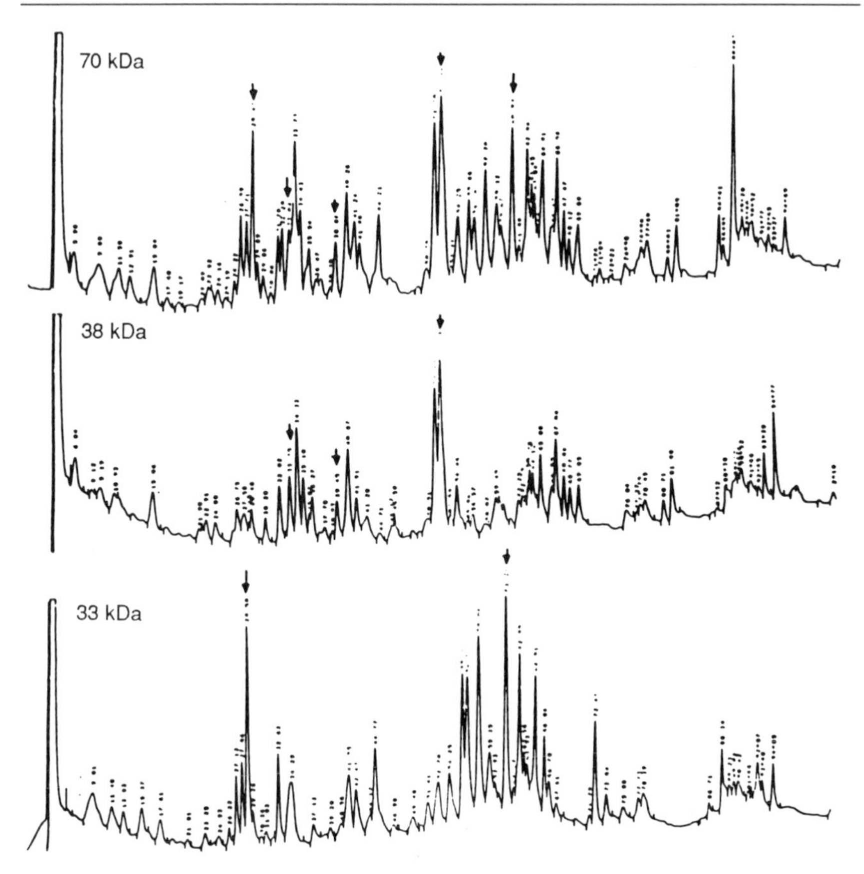

FIG. 2. HPLC separation of peptide fragments resulting from exhaustive tryptic proteolysis of carboxymethylated PGH synthase and its 33 and 38 kDa fragments. Retention times above individual peaks are illegible. Arrows indicate fluorescent peaks resulting from presence of tryptophan in the peptide.

eluted peptides is determined. Figure 2 displays the tryptic maps of apoPGH synthase and its 33 and 38 kDa fragments.

Acknowledgments

We are grateful to Coleen Young O'Gara for helpful discussions. This work was supported by research grants from the National Institutes of Health (GM23642 and CA47479).

[53] Cloning of Leukotriene A_4 Hydrolase cDNA

By OLOF RÅDMARK, COLIN FUNK, JI YI FU, TAKASHI MATSUMOTO, HANS JÖRNVALL, BENGT SAMUELSSON, MICHIKO MINAMI, SHIGEO OHNO, HIROSHI KAWASAKI, YOUSUKE SEYAMA, KOICHI SUZUKI, and TAKAO SHIMIZU

Leukotriene A_4 hydrolase (EC 3.3.2.6) is a soluble epoxide hydrolase which catalyzes the enzymatic hydrolysis of leukotriene A_4 (LTA_4) to the chemotactic agent leukotriene B_4.[1] After initial purification and characterization of the enzyme,[2] it was determined to be different from other epoxide hydrolases, and further studies were initiated.

The cloning of LTA_4 hydrolase was carried out successfully using two different strategies, i.e., screening of λgt11 cDNA libraries (which express recombinant protein) with polyclonal antiserum, and screening of a λgt10 cDNA library with a single isomer oligonucleotide probe (48-mer) whose structure was based on mammalian codon usage frequencies.[3,4] Prior to this, screening of plasmid (pBR322) cDNA libraries with mixed shorter oligonucleotide probes (16-mers) was unsuccessful. Thus, the use of better quality cDNA synthesized according to the RNase H method,[5] in λgt 10/11 vectors,[6] as well as the use of more efficient probes, were beneficial.

Identification of Clones Using Antiserum

A commercially available λgt11 cDNA library (Clontech, Palo Alto, CA) constructed from human lung poly(A) RNA was screened with polyclonal antiserum. The serum was obtained from rabbits injected with LTA_4 hydrolase purified from human leukocytes.[2] In addition to the regular purification procedure, the enzyme used as antigen was further refined by SDS–polyacrylamide gel electrophoresis according to Laemmli.[7] For each

[1] P. Borgeat, and B. Samuelsson, *Proc. Natl. Acad. Sci. U.S.A.* **76,** 3213 (1979).

[2] O. Rådmark, T. Shimizu, H. Jörnvall, and B. Samuelsson, *J. Biol. Chem.* **259,** 12339 (1984).

[3] C. D. Funk, O. Rådmark, J. Y. Fu, T. Matsumoto, H. Jörnvall, T. Shimizu, and B. Samuelsson, *Proc. Natl. Acad. Sci. U.S.A.* **84,** 6677 (1987).

[4] M. Minami, S. Ohno, H. Kawasaki, O. Rådmark, B. Samuelsson, H. Jörnvall, T. Shimizu, Y. Seyama, and K. Suzuki, *J. Biol. Chem.* **262,** 13873 (1987).

[5] U. Gubler and B. J. Hoffman, *Gene* **25,** 263. (1983).

[6] T. V. Huynh, R. A. Young, and R. W. Davis, *in* "DNA Cloning" (D. M. Glover, ed.), Vol. 1, pp. 49–78. IRL, Oxford, 1983.

[7] U. K. Laemmli, *Nature (London)* **227,** 680 (1970).

booster, a gel slice containing approx 40 μg of protein is homogenized in PBS (1 ml) plus Freund's complete adjuvant (1 ml) and injected.

The specificity of the antiserum was estimated from Western blots. The proteins in the 10,000 *g* supernatant (50 μl) obtained after sonication of a human leukocyte suspension (10^8 cells/ml) are separated by electrophoresis in SDS–polyacrylamide (8% gel) and electroblotted[8] to nitrocellulose (Hybond C, Amersham, Buckinghamshire, England). When probed with the antiserum (1 : 100) one major band appeared, with the same retention as standard LTA_4 hydrolase. In some samples, bands also appeared at MW 50,000–55,000, probably representing breakdown products of LTA_4 hydrolase.

Before use in the screening of λgt11 libraries, antibodies reactive to *Escherichia coli* epitopes were removed as described.[6] Briefly, a suspension of nontransformed *E. coli* is sonicated and nitrocellulose filters are soaked with this lysate. After washing the filters with TBS (50 m*M* Tris-HCl, pH 8; 100 m*M* NaCl), they are soaked with the antiserum (1 : 10, v/v in TBS), thus binding antibodies that recognize *E. coli* proteins. The purified antiserum is diluted further to 1 : 100, v/v and used in screening as described.[6]

Phage (10^4 per plate) are plated on *E. coli* Y1090 and incubated at 42° for 3.5 hr. The plates are then overlaid with nitrocellulose filters (Millipore HATF) presoaked with 10 m*M* isopropyl-β-D-galactoside (IPTG), in order to induce synthesis of recombinant protein. After further incubation at 37° for 3.5 to 4 hr, or preferably overnight, filters are removed, briefly air-dried, and washed with TBS. Filters are treated with 20% fetal calf serum in TBS for 30 min (blocking) and incubated with the antiserum for 2 hr at room temperature. Next, filters are washed with TBS, TBS plus 0.1% Nonidet (NP-40) (Sigma, St. Louis, MO), and again with TBS (5 min each), before incubation in a solution of ^{125}I-labeled protein A in TBS plus 20% fetal calf serum for 1 hr at room temperature. The specific activity of ^{125}I-labeled protein A is 1110 MBq/mg (Amersham IM 144). Typically 10^6 cpm per filter (82 mm) is used and the volume is 2–5 ml/filter. After repeated washing, filters are air-dried and autoradiographed at −70° for 1 to 2 days (Fuji RX film) with an intensifying screen. Putative positive clones are carried through two or three additional rounds of screening at reduced density until homogeneous clones are obtained. In the original procedure,[6] incubation of IPTG-impregnated filters on the plates for 3.5 to 4 hr at 37°, prior to probing with antiserum, was recommended. In our hands, incubation overnight gives much clearer results.

Thus, a clone λluH6-1 was isolated from a human lung λgt11 cDNA

[8] H. Towbin, T. Staehelin, and J. Gordon, *Proc. Natl. Acad. Sci. U.S.A.* **76,** 4350 (1979).

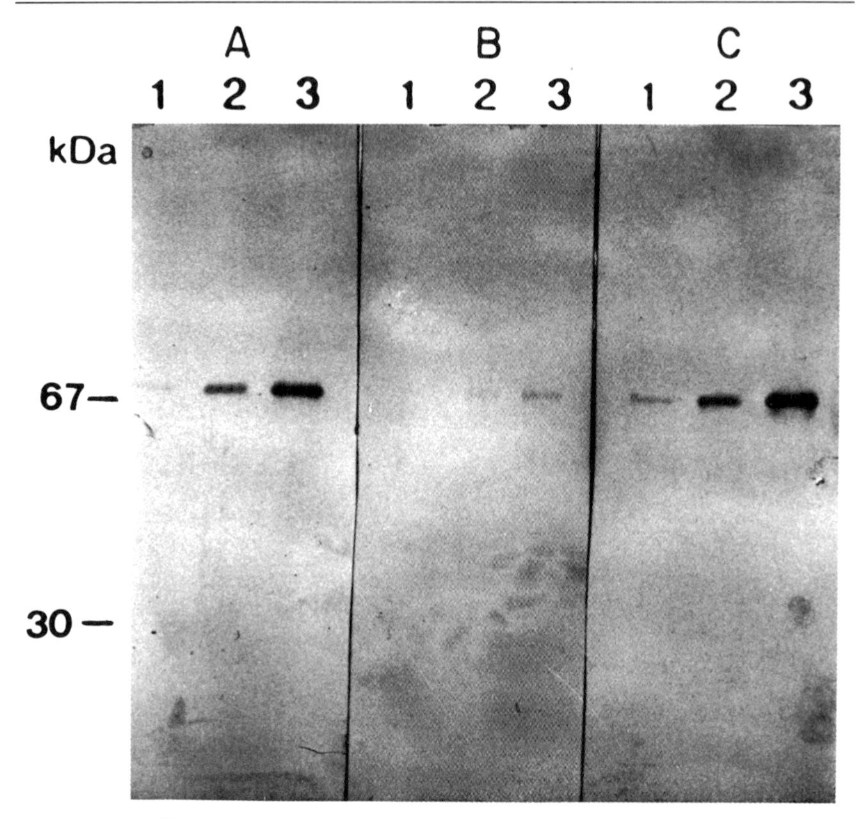

FIG. 1. Antibody selection assay of clone λluH6-1. The clone is plated lytically and synthesis of *lacZ* fusion protein induced by IPTG on overlaid nitrocellulose filters. Proteins transferred to the filters are incubated with LTA_4 hydrolase antiserum (1 : 100 dilution, 10 ml). Antibodies selectively bound are eluted at pH 2.6 and immediately neutralized to pH 8.0. Human leukocyte LTA_4 hydrolase (200 ng, lanes 1; 400 ng, lanes 2; 800 ng, lanes 3) is subjected to electrophoresis in an 8% SDS-polyacrylamide gel, transferred to nitrocellulose, and allowed to react with the selected antibodies. (A) Antibodies selected from clone λluH6-1; (B) antibodies selected from the parent vector λgt11, lacking a cDNA insert; (C) non-selected LTA_4 hydrolase antiserum (1 : 100 dilution, 10 ml). Markers indicate retentions of bovine serum albumin (67 kDa) and carbonate dehydratase (30 kDa).

library. Antibody selection assay (Fig. 1) confirmed the identity of this clone. When digested with *Eco*RI two inserts were found, around 200 and 570 base pairs (bp), respectively. These were separated and religated to λ arms by standard procedures[9] and the ligation mixture added to a phage protein extract (Promega, WI, according to the manufacturer) to give

[9] T. Maniatis, E. F. Fritsch, and J. Sambrook, "Molecular Cloning: A Laboratory Manual." Cold Spring Harbor Laboratory, Cold Spring Harbor, NY, 1982.

infectious phages. When screened with the antiserum, only some phages containing the shorter (200 bp) insert (λluH6-1a) gave positive signals, thus indicating that nonrelated cDNAs were ligated together in one clone in the cDNA synthesis. The insert of λluH6-1a was subcloned in M13 for production of single-strand DNA which was sequenced by the Sanger dideoxy chain-termination method. A coding sequence was found, corresponding to 29 amino acids of a CNBr fragment of LTA_4 hydrolase. This identified λluH6-1a as a partial LTA_4 hydrolase cDNA clone. The sequence of the 570 bp fragment of λluH6-1 was also determined; it was subsequently confirmed that it had no relation to the LTA_4 hydrolase cDNA clones.

The artifact of one clone carrying multiple nonrelated inserts was encountered once more in the further screening for longer clones. Thus the insert of λluH6-1a was labeled by nick translation and the same lung λgt11 cDNA library screened by plaque hybridization. In this procedure, phages are plated on *E. coli* Y1090 and grown for 6 hr at 37°. Nitrocellulose filters (Millipore HATF) are overlaid for 1 min and processed as described.[9] 5×10^5 clones were screened; the longest clone hybridizing with λluH6-1a gave two inserts upon *Eco*RI digestion (770 and 405 bp). Sequencing showed that the 405 bp fragment overlaps with λluH6-1a. The 770 bp fragment, however, had multiple termination codons in all reading frames, and was judged as nonrelated.

Full-length cDNA was isolated from a human placenta λgt11 library, also from Clontech. Here, 4×10^5 clones were screened, and 19 clones that hybridized strongly exhibited a characteristic 1.1 kb/0.8 kb double insert pattern after *Eco*RI digestion, indicating the presence of an internal *Eco*RI site. The original λluH6-1a hybridized to the 0.8 kb insert. Appropriate restriction fragments of a clone λpl16A were subcloned in M13 and sequenced. λpl16A contained a 1910 bp insert (excluding the *Eco*RI linkers) with a continuous reading frame of 1830 bp terminated by a stop codon (TAA), encoding a protein of 610 amino acids. The noncoding regions were short, especially at the 3′ end (47 bp) and a polyadenylation signal as well as poly(A) tail were lacking. This is explained by the presence of another internal *Eco*RI site (see below: cloning with 48-mer).

Screening with 48-mer

A 48-mer oligonucleotide was designed from the amino acid sequence of a cyanogen bromide fragment of LTA hydrolase, according to the mammalian codon usage frequencies (see Fig. 2)[4].[10] An 8-mer primer, complementary to the 5′ end of the 48-mer is also prepared. Two complementary probes are labeled with ^{32}P as follows. First, the 48-mer is phos-

[10] R. Lathe, *J. Mol. Biol.* **183,** 1 (1985).

```
Primer E8                                                        3'-TGTTCAGG-5'

Probe E48          5'-AAGTTCACCCGGCCCCTGTTCAAGGACCTGGCCGCCTTTGACAAGTCC-3'
                      ::::: ::::::::: : :::::::: :: :: ::::::::::: :::
LTA85 (1696-1743)     AAGTTTACCCGGCCCTTATTCAAGGATCTTGCTGCCTTTGACAAATCC

Peptide #11           LysPheThrArgProLeuPheLysAspLeuAlaAlaPheAspLysSer
```

FIG. 2. Sequence of oligonucleotide probe. The 48-mer oligonucleotide and antistrand 8-mer oligonucleotide primer are synthesized, based on the sequence of a LTA_4 hydrolase CNBr fragment [peptide 11 (ref. 4)]. The corresponding sequence found in clone LTA85 was homologous (about 85%) to the probe E48.

phorylated at the 5′ end, using [λ-^{32}P]ATP and T_4 polynucleotide kinase. The 8-mer primer dissolved in annealing buffer (50 mM Tris-HCl, pH 7.5) is added to give a ratio of 50 pmol of 8-mer per 10 pmol of 48-mer. After incubation, first at 85° for 5 min, and then at 65° for 10 min, the mixture is gradually cooled to room temperature during more than 1 hr. Additions are now made to the annealed mixture to give a final volume of 50 μl containing 1 m*M* dGTP, 1 m*M* dATP, 1 m*M* dTTP, 10 μCi [α-^{32}P]dCTP (3000 Ci/mmol, Du Pont), 10 m*M* $MgCl_2$, and 10 m*M* dithiothreitol. The Klenow fragment of DNA polymerase I is added and the mixture incubated for 1 hr at 12°. After addition of unlabeled dCTP (1 m*M*), the incubation is continued for another 20 min to complete the synthesis of antistrand oligonucleotide. Unincorporated radioactivity is removed by gel filtration on Sephadex G-50.

A human spleen λgt10 cDNA library was constructed as described.[11] Double-stranded cDNA longer than 2 kb is purified by electrophoresis on low-gelling temperature agarose, before ligation to the λgt10 arms.[6] About 5×10^4 plaques were transferred to nylon membranes (Biodyne A, Pall) using standard methods.[9] The 48-mer probe is hybridized at 50° in a solution of 6× SSC, 5× Denhardt's, 0.5% sodium dodecyl sulfate (SDS), 400 μg/ml heat-denatured salmon sperm DNA, 20 μg/ml *E. coli* DNA, and the radiolabeled probe (2×10^6 cpm/ml). After washing with 6 × SSC, 0.1% SDS at 50° the filters are exposed to Fuji RX films at −80° using Du Pont Lightning Plus intensifying screens.

The screening of 5×10^4 plaques of the spleen library gave 5 positive clones, whose inserts were practically identical, as determined by restriction mapping. A clone with an insert of 2.1 kb (LTA85) was further analyzed. Digestion of LTA85 with *Eco*RI gave three fragments of sizes 1.0, 0.85, and 0.1 kb showing the presence of two internal *Eco*RI sites.

[11] S. Ohno, Y. Emori, H. Sugihara, S. Imajoh, and K. Suzuki, this series, Vol. 139, pp. 363–379.

Various fragments of LTA85 were subcloned in vectors pUC8 or pUC18, and sequenced by the dideoxy-sequencing method, as modified.[12] Of the complete sequence (2060 bp), 1833 bp encodes an open reading frame corresponding to a protein of 610 amino acids (excluding the initial methionine). A polyadenylation signal (AATAAA) appears at base 1971.

When the sequence of the clones λpl16A and LTA85 are compared, the open reading frames are identical, encoding a protein with MW 69,158. Longer 5′ as well as 3′ noncoding sequences are present in LTA85. Regarding the 5′ noncoding sequences, these are identical up to position −27, with in-frame stop codons at −15. Computer-aided analysis of the deduced amino acid sequence shows no significant homology with other proteins, including rabbit and rat liver microsomal epoxide hydrolases. Regarding the properties of LTA_4 hydrolase, a segment from amino acids 170–185 is the most hydrophobic part of the enzyme, possibly involved in binding of the hydrophobic substrate LTA_4. A plasmid for expression of LTA_4 hydrolase in *E. coli* is constructed, using the cDNA insert of LTA85.[13]

Acknowledgments

This study was supported by grants from the Swedish Medical Research Council (03X-217, 03X-3532, 03X-7467), and from the Ministry of Education, Science and Culture of Japan.

[12] M. Hattori and Y. Sakaki, *Anal. Biochem.* **152,** 232 (1986).

[13] M. Minami, Y. Minami, Y. Emori, H. Kawasaki, S. Ohno, K. Suzuki, N. Ohishi, T. Shimizu, and Y. Seyama *FEBS Lett.* **229,** 279 (1988).

[54] Molecular Biology and Cloning of Archidonate 5-Lipoxygenase

By COLIN D. FUNK, TAKASHI MATSUMOTO, and BENGT SAMUELSSON

Arachidonate 5-lipoxygenase (Ec1.13.11.34) is a complex enzyme requiring calcium, ATP, and various unknown cellular stimulatory factors for maximal conversion of arachidonic acid (20 : 4) to 5-hydroperoxy-6,8,10,14-eicosatetraenoic acid (5-HPETE) and its subsequent metabolite 5,6-oxido-7,9,11,14-eicosatetraenoic acid [leukotriene A_4 (LTA_4)].[1–4] The

[1] C. A. Rouzer, T. Matsumoto, and B. Samuelsson, *Proc. Natl. Acad. Sci. U.S.A.* **83,** 857 (1986).

[2] T. Shimizu, T. Izumi, Y. Seyama, K. Tadokoro, O. Rådmark, and B. Samuelsson, *Proc. Natl. Acad. Sci. U.S.A.* **83,** 4175 (1986).

enzyme has been isolated from human leukocytes and has been extensively characterized (see this volume [34]). To further enhance our knowledge of this enzyme and the regulation of its synthesis, this chapter describes the isolation and characterization of 5-lipoxygenase cDNA clones and a method for overexpressing the enzyme in a baculovirus–insect cell system.

Isolation of 5-Lipoxygenase cDNA Clones

Our strategy for obtaining 5-lipoxygenase cDNA clones is outlined in Fig. 1 and can essentially be used for cloning the cDNA for any protein that lends itself to relatively straightforward purification. 5-Lipoxygenase was purified from human leukocytes as described (this volume [34]).

Peptide Sequence Analysis

Purified 5-lipoxygenase (400 μg) is reduced with dithiothreitol and carboxymethylated with neutralized iodo[2-^{14}C]acetic acid in 6 *M* guanidine hydrochloride/0.4 *M* Tris/2 m*M* EDTA, pH 8.1. Reagents are removed by extensive dialysis against distilled water, and the carboxymethylated protein is tested for sensitivity to acid cleavage of Asp–Pro bonds by treatment with 70% formic acid at room temperature for 3 days. SDS–PAGE reveals that specific cleavage products are largely absent. The acid is removed, and the protein dissolved in 9 *M* freshly deionized urea and ≈0.1 *M* ammonium bicarbonate (pH 8.0). Lysine-specific protease (50 μg) from *Achromobacter lyticus* (Wako Chemicals, Neuss, F. R. G.) is added immediately, and proteolysis is allowed to proceed for 4 hr at 37°. The peptides generated are separated by reversed-phase HPLC on an Ultropac (LKB) TSK ODS-12OT (5 μm particle size) column, utilizing a gradient of acetonitrile (0–80%) in 0.1% trifluoroacetic acid. All peptides purified are analyzed for amino acid sequence with an Applied Biosystems (Foster City, CA) 470A gas-phase sequencer. Phenylthiohydantoin derivatives are identified by reversed-phase HPLC with a Hewlett-Packard 1090 system or an Applied Biosystems 120A on-line system.

Sequence information is obtained for about one-third of the protein chain (≈200 residues) at internal segments. In addition, the intact protein is directly analyzed without cleavage, yielding the amino acid sequence of the first 17 residues, since the amino terminus is not blocked.

---

[3] N. Ueda, S. Kaneko, T. Yoshimoto, and S. Yamamoto, *J. Biol. Chem.* **261,** 7982 (1986).

[4] G. K. Hogaboom, M. Cook, J. F. Newton, A. Varrichio, R. G. L. Shorr, H. M. Sarall, and S. T. Crooke, *Mol. Pharmacol.* **30,** 510 (1986).

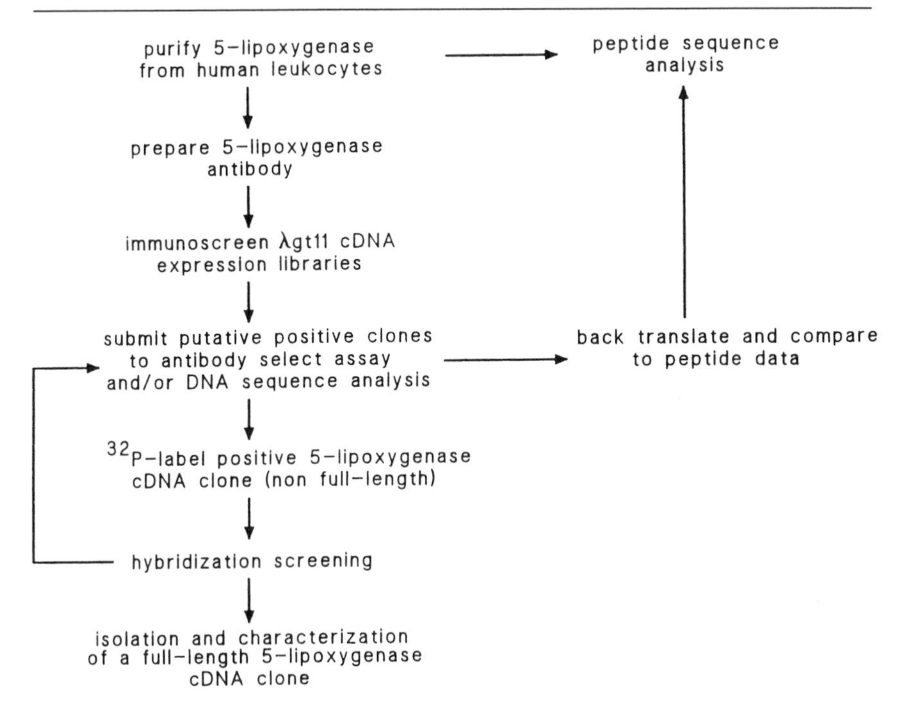

FIG. 1. Strategy for isolation of 5-lipoxygenase clones.

Preparation of Polyclonal 5-Lipoxygenase Antibody

Protein (≈100 μg) containing purified 5-lipoxygenase (≈85–90% pure) is electrophoresed in a SDS–polyacrylamide gel. The region at ≈80 kDa containing 5-lipoxygenase is cut out, minced, and mixed with Freund's complete adjuvant, and subsequently used to immunize rabbits. The IgG fraction is further purified by protein A–Sepharose CL-4B chromatography and then is absorbed with *E. coli* strain Y1090 lysate to remove antibodies that recognize *E. coli* antigens (see this volume [53]). Antibody is diluted 1 : 100 in TBS (50 m*M* Tris-HCl, pH 8.0, 100 m*M* NaCl) containing 0.05% sodium azide.

Immunoscreening of λgt11 cDNA Expression Libraries

The technique for isolating clones in λgt11 libraries with antibodies is described in detail elsewhere.[5] We follow a protocol identical to that used

[5] R. C. Mierendorf, C. Percy, and R. A. Young, this series, Vol. 152, p. 458.

for isolation of LTA_4 hydrolase cDNA clones (this volume [53]). Fourteen positive clones are isolated.

Clone Characterization: Antibody Selection Assay and DNA Sequence Analysis

Phages (from the 14 clones), adsorbed to *E. coli* Y1090, are plated lytically at $\approx 10^5$ plaque-forming units (pfu) per 100 mm diameter LB-agar plate. The plates are overlaid with isopropyl-β-D-thiogalactoside-soaked (then dried) nitrocellulose filters for at least 4 hr at 37°. The filters are washed briefly in TBS and incubated with the 5-lipoxygenase antiserum for ≈6 hr at room temperature followed by further washing. Filter-bound selected antibodies are eluted by 100 m*M* glycine/150 m*M* NaCl (pH 2.6) treatment followed by immediate neutralization to pH 8.0. Purified leukocyte 5-lipoxygenase (50–500 ng) is dot-blotted onto small pieces of nitrocellulose membrane and these membranes are reacted with the selected antibodies in a procedure analogous to the immunoscreening. Selected antibodies from 9 of the 14 clones recognized purified 5-lipoxygenase. Although not a foolproof method for ascertaining clone specificity, this step, therefore, helped to eliminate five false-positive clones. The cDNA inserts of the 9 positive phages are isolated and found to range in size from 0.3 to 1.5 kb. To check the relatedness of the 9 clones, the clone with the longest insert (λluI01) is nick-translated with [^{32}P]dCTP using a kit from Amersham (Arlington Heights, IL) and used as a probe for a Southern blot containing the 9 cDNAs. Of the 9 clones, 8 cross-hybridized, whereas 1 clone (λluSl) did not.

The cDNA inserts of λluSl (0.4 kb) and λluI01 (1.5 kb) are subcloned by standard procedures[6] into phage vectors M13mp18 and M13mp19 and are sequenced by the dideoxy chain termination method[7] using a kit from Pharmacia (Bromma, Sweden). The sequences are examined for open reading frames and are back-translated to obtain peptide sequences. This data is compared with the directly sequenced 5-lipoxygenase peptide data. The λluSl insert is found to contain a coding sequence for a 17-amino acid segment of a carboxyl-terminal 5-lipoxygenase peptide fragment.[8] On the other hand, the λluI01 insert (and related clones) did not match any of the 13 sequenced 5-lipoxygenase peptide fragments. When the cDNA sequence of the λluI01 insert is compared to sequences in the Genebank database it is found to completely match the sequence of human α-enolase.

[6] J. R. Greene and L. Guarente, this series, Vol. 152, p. 512.

[7] F. Sanger, S. Nicklen, and A. R. Coulson, *Proc. Natl. Acad. Sci. U.S.A.* **74,** 5463 (1977).

[8] T. Matsumoto, C. D. Funk, O. Rådmark, J. -O. Höög, H. Jörnvall, and B. Samuelsson, *Proc. Natl. Acad. Sci. U.S.A.* **85,** 26 (1988) [and correction, **85,** 3406 (1988)].

We cannot explain exactly why we obtained so many α-enolase clones during immunoscreening. 5-Lipoxygenase and α-enolase do not share any appreciable homologies so they would not be expected to share common epitopes. In addition, virtually all of the sequenced peptides from the purified protein correspond to 5-lipoxygenase. Apparently, somehow the 48 kDa α-enolase protein is present in the sample used for injection of the rabbits for preparation of antiserum and α-enolase antibodies are produced in addition to 5-lipoxygenase antibodies.

Hybridization Screening

The 397-bp insert of clone λluSl is ^{32}P-labeled by nick-translation to $\approx 10^7$ dpm/μg and used to screen a human placenta λgt11 library (Clontech) in order to obtain longer inserts. Phages are grown for 6 to 15 hr at 37°, after which duplicate nitrocellulose filters are overlaid for 1 min each. The DNA bound to these filters is subsequently prehybridized in a solution containing 50% (v/v) formamide, 5× Denhardt's solution (1x is 0.02% Ficoll/0.02% polyvinylpyrrolidone/0.02% bovine serum albumin), 5× SSPE (1× is 0.14 *M* NaCl/10 m*M* sodium phosphate, pH 7.4/1 m*M* EDTA), 0.1% SDS, and 100 μg of denatured salmon sperm DNA per ml at 42° overnight. Nick-translated probe is added and hybridization continued for 24 hr. Filters are washed according to standard procedures[9] and are autoradiographed. Several positive clones are isolated.

Characterization of a Full-Length 5-Lipoxygenase Clone

The complete DNA sequence of the insert of clone λp15BS (2.5 kb) is obtained.[8] Within the sequence are two exact 51-bp adjacent repeats, one of which is deemed to be a cloning artifact on the basis of sequence analysis of two additional 5-lipoxygenase clones. The cDNA sequence,[8,10] thus encodes for a mature protein of 673 amino acids (the initiator methionine is not present in the mature protein: direct sequence analysis revealed the subsequent proline residue to constitute the amino terminus) with a calculated molecular weight of 77,856. The sequence shares homologies with all sequenced lipoxygenases[11–16] (rat basophilic leukemia cell 5-

[9] T. Maniatis, E. F. Fritsch, and J. Sambrook, "Molecular Cloning: A Laboratory Manual," Cold Spring Harbor Laboratory, Cold Spring Harbor, NY, 1982.

[10] R. A. F. Dixon, R. E. Jones, R. E. Diehl, C. D. Bennett, S. Kargman, and C. A. Rouzer, *Proc. Natl. Acad. Sci. U.S.A.* **85,** 416 (1988).

[11] J. M. Balcarek, T. W. Theisen, M. N. Cook, A. Varrichio, S. -M. Hwang, M. W. Strohsacker, and S. T. Crooke, *J. Biol. Chem.* **263,** 13937 (1988).

[12] D. Shibata, J. Steczko, J. E. Dixon, M. Hermodson, R. Yazdanparast, and B. Axelrod, *J. Biol. Chem.* **262,** 10080 (1987).

lipoxygenase, 93%; soybean and pea seed lipoxygenases, ≈40%; human 15-lipoxygenase, 61%) and a short segment related to the interface binding domain of hepatic lipase and lipoprotein lipase.[10] The protein based on its predicted structure is hydrophilic overall but contains several short hydrophobic portions in the carboxyl-terminal half which are likely related to the structure and activity of the enzyme.

RNA Isolation and Northern Blots

Total RNA is extracted from human peripheral leukocytes, placenta, and lung (obtained at the time of surgery) by the guanidinium isothiocyanate method.[17] Poly(A) RNA is obtained by two steps of oligo(dT)-cellulose chromatography.[18] RNA is size-fractionated by electrophoresis in 1% agarose gels containing 0.22 *M* formaldehyde and then is transferred to nitrocellulose. Blots are prehybridized for 3 to 4 hr in a solution containing 4× SSC (1× is 0.15 *M* NaCl/0.015 *M* trisodium citrate, pH 7), 40% (v/v) formamide, 10% (w/v) dextran sulfate, 1× Denhardt's solution, and 15 m*M* Tris-HCl (pH 7.5) at 43°. Nick-translated probes (≈10^8 dpm/μg of DNA, ≈10^5 cpm/ml) are added to the prehybridization solution and incubated for a further 15–20 hr. Hybridized blots are washed four times (20 min per wash) with 0.1× SSC containing 0.1% SDS and are autoradiographed for 6 to 15 hr at −70° with two intensifying screens. Hybridization to a discrete mRNA of ≈2700 nucleotides occurs in all three tissues. A faint band at ≈3200 nucleotides is also present in the leukocyte sample. The hybridization intensity is greatest for leukocyte poly(A) RNA, followed by the preparations from placenta and lung tissue.

Expression of 5-Lipoxygenase in Insect Cells Using Baculovirus Vector

High-level expression of many foreign genes has been achieved using a baculovirus/insect cell system.[19] For a detailed consideration of the prin-

[13] D. Shibata, J. Steczko, J. E. Dixon, P. C. Andrews, M. Hermodson, and B. Axelrod, *J. Biol. Chem.* **263,** 6816 (1988).
[14] R. L. Yenofsky, M. Fine, and C. Liu, *Mol. Gen. Genet.* **211,** 215 (1988).
[15] P. M. Ealing and R. Casey, *Biochem. J.* **253,** 915 (1988).
[16] E. Sigall, C. S. Craik, E. Highland, D. Grunberger, L. L. Costello, R. A. F. Dixon, and J. A. Nadel, *Biochem. Biophys. Res. Commun.* **157,** 457 (1988).
[17] R. J. MacDonald, G. H. Swift, A. E. Przybyla, and J. M. Chirgwin, this series, Vol. 152, p. 219.
[18] A. Jacobson, this series, Vol. 152, p. 254.
[19] V. A. Luckow and M. D. Summers, *Biotechnology* **6,** 47 (1988).

ciples, methods, and expression vectors involved in this system, the reader is referred to other references.[20,21]

Maintenance of Insect Cells

Spodoptera frugiperda insect cells (Sf9) can be obtained from the American Type Culture Collection. The cells are maintained in TNM-FH medium (Nord Vac, Stockholm, Sweden) containing 10% heat-inactivated fetal bovine serum, 100 units/ml penicillin, and 100 μg/ml streptomycin (complete medium) at 27°.[20] Carbon dioxide is not required. The source of fetal serum is very important and cells must be prescreened for long-term viability. We usually maintain our cells as monolayers in 75-cm^2 tissue culture flasks, although cells can be grown as spinner cultures. Cells are seeded at a density of $\approx 5-10 \times 10^6$ cells/flask in Grace's medium (Nord Vac, Stockholm) in the absence of fetal bovine serum.[22] Under these conditions the cells rapidly attach to the flasks. After 20 min, the medium is gently removed and fresh complete medium is added. Cell doubling time should be ≈24 hr otherwise cells are not completely healthy. Medium is changed every 2 days and cells should be subcultured when confluency is reached (2–3 days). Cells are detached by gently scraping with a piece of Tygon tubing attached to a curved glass rod.

Construction of Recombinant 5-Lipoxygenase Transfer Vector

Since the baculovirus genome is very large (128 kb) one can not directly subclone the gene of interest but must proceed via a transfer vector (Fig. 2). The *Eco*RI–*Bcl*I fragment of the clone λp15BS,[8] which contains the cDNA sequence of human 5-lipoxygenase, is first inserted into the *Eco*RI and *Bam*HI sites of the plasmid pERAT 308.[23] The 51-bp cloning artifact within a *Xma*I–*Xma*I fragment is removed, followed by insertion of the corresponding fragment of clone λp19AS,[8] a partial-length 5-lipoxygenase cDNA clone without the artifact. The orientation of the inserted fragment is checked by digestion with *Pvu*II. A 2.2 kb *Eco*RI–*Sal*I fragment from this construct is isolated, treated with *Bam*HI methylase (New England Biolabs, Beverly, MA) to protect the internal *Bam*HI site, and then ligated with excess *Eco*RI/*Bam*HI and *Sal*I/*Bam*HI converters. After *Bam*HI

[20] M. D. Summers and G. E. Smith, "A Manual of Methods for Baculovirus Vectors and Insect Cell Culture Procedures," Texas Agricultural Expt. Stn., College Stn., Bulletin 1555, 1987.

[21] G. E. Smith, M. D. Summers, and M. J. Fraser, *Mol. Cell. Biol.* **3,** 2156 (1983).

[22] T. D. C. Grace, *Nature (London)* **195,** 788 (1962).

[23] B. Nilsson and L. Abrahmsen, this series, Vol. 185, p. 144.

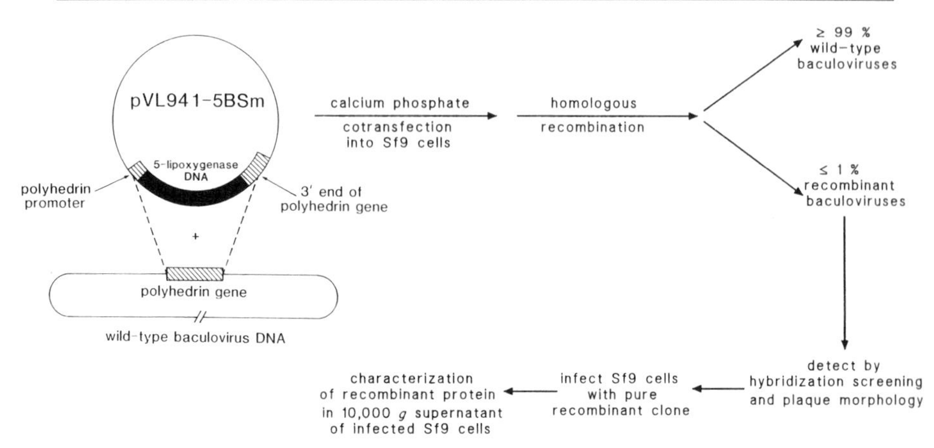

FIG. 2. Expression of recombinant human 5-lipoxygenase in insect cells using a baculovirus vector.

digestion and agarose gel purification, the fragment is ligated into *Bam*HI-digested pVL941[24] transfer vector. The orientation of the insert with respect to the polyhedrin promoter is checked by *Xho*I and *Pst*I digestion. A 12 kb construct, pVL941-5BSm, is purified by CsCl gradient centrifugation and used for cotransfection.

Cotransfection and Selection of Recombinant Virus

Wild-type (1 μg) AcNPV (*Autographa californica* nuclear polyhedrosis virus; gift from Dr. Max D. Summers, Texas A & M University) and 2 μg transfer vector pVL941-5BSm are mixed in a 1.5 ml Eppendorf tube. To this are added 950 μl of IX HEBS/CT (0.137 *M* NaCl/6 m*M* glucose/5 m*M* KCl/0.7 m*M* sodium phosphate/20 m*M* HEPES, pH 7.05–7.10, plus 15 μg/ml sonicated calf thymus DNA) and 50 μl 2.5 M $CaCl_2$. The mixture is vortexed and allowed to stand at room temperature for 30 min. Sf9 cells 2–3 $\times 10^6$, in a 25-cm^2 culture flask in 2 ml Grace's medium[22] containing 10% fetal bovine serum are gently overlaid with the finely precipitated calcium phosphate/DNA mixture and allowed to incubate for 4 hr at 27°. The cells are washed twice with fresh medium and allowed to incubate in complete medium for 4 days, until all cells become infected. Signs of infection include: vast increases in cell size, changes in nuclear morphology, and cell detachment, and lysis. The medium (containing the extracellular form of the virus) is removed and stored at 4° under sterile conditions.

[24] V. A. Luckow and M. D. Summers, *Virology* **170,** 31 (1989).

To select for recombinant 5-lipoxygenase baculoviruses, a plaque assay is performed.[20] Briefly, $\approx 3 \times 10^6$ Sf9 cells are seeded in 60-mm diameter culture dishes and transfection medium (serial dilutions, 10^{-3}–10^{-6}; 1.5 ml) is added. After a 1-hr incubation (infection period), the medium is removed and 4 ml of a molten agarose (SeaPlaque FMC BioProducts, Rockland, ME), low-melting point; 1.5% final concentration) complete medium mixture is gently overlaid on the cells starting from the edge of the plate. The soft agarose is allowed to harden for 1 hr at room temperature and the plates then incubated 3–5 days at 27° in a humidified chamber. Recombinant baculoviruses can be distinguished from wild-type by visualization under low power in an inverted phase-contrast microscope for distinctive nuclear plaque morphologies and by hybridization screening.[20] Positive clones are picked and the selection process is repeated at least three times until only pure recombinant plaques (complete absence of wild-type baculoviruses) can be visualized.

5-Lipoxygenase Assays of Recombinant Virus-Infected Cells

Monolayer Sf9 cells ($\approx 5 \times 10^6$) in 25-cm^2 flasks are inoculated with virus stock for 1 hr in complete medium. The medium is replaced with 5 ml of fresh medium and the cells incubated at 27°. At various times after infection, cells are removed for counting, centrifuged at 400 *g* at 4°, and resuspended in 0.5 ml of KPBI buffer (50 m*M* potassium phosphate/0.1 *M* NaCl/1 m*M* EDTA/1 m*M* DTT/1 m*M* phenylmethylsulfonyl fluoride, pH 7.1, containing 60 μg of soybean trypsin inhibitor per milliliter). The cells are disrupted by sonication (Branson S125 Sonifier, setting 4) for two pulsatile 1-sec bursts in an ice/water bath. The sonicates are centrifuged at 10,000 *g* for 10 min. The supernatant is removed and aliquots taken for enzyme assay, SDS/PAGE-immunoblot analysis, and protein analysis by the method of Bradford.[25]

The standard enzyme assay mixture consists of 90 μl of KPBI buffer containing various amounts of protein and 10 μl of 20 m*M* $CaCl_2$/20 m*M* ATP. The reaction is initiated by the addition of a mixture of 20 : 4 (16 nmol) and 13-hydroperoxy-9,11-octadecadienoic acid (0.5 nmol) in 3 μl of ethanol and is allowed to proceed for 10 min at 37°. Two hundred microliters of stop solution (acetonitrile/methanol/acetic acid, 350 : 150 : 3, v/v/v) containing the internal standard, 16-hydroxy-9,12,14-heneicosatrienoic acid (0.5 nmol), is added and the proteins precipitated on ice for 10 min. After centrifugation, the supernatants are analyzed directly by HPLC (LDC/Milton Roy system with variable-wavelength UV detector coupled with Waters 712 WISP automatic injector and Wa-

[25] M. M. Bradford, *Anal. Biochem.* **72,** 248 (1976).

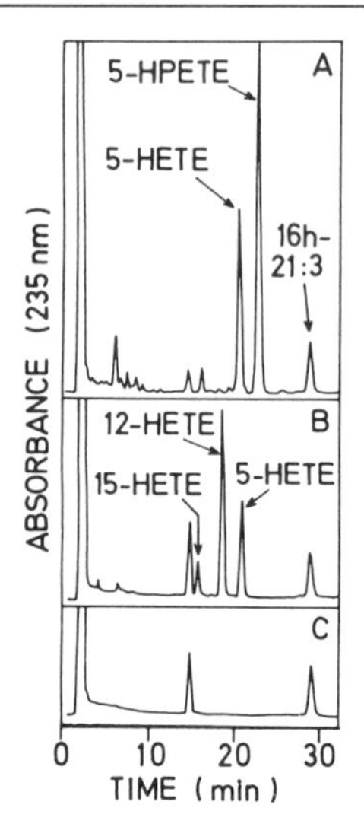

FIG. 3. HPLC profiles (UV detection at 235 nm) of products formed from arachidonic acid by 10,000 *g* supernatants. (A) Recombinant baculovirus-infected cells [5.2×10^5; 129 μg of protein; 4.1 nmol of 5-H(P)ETE (5-HPETE + 5-HETE)]. (B) Human leukocytes (9.2×10^5; 48 μg protein; 0.75 nmol 5-HETE) plus some platelet contamination. (C) Wild-type baculovirus-infected cells (5.8×10^5; 0 nmol of 5-H(P)ETE). 5-H(P)ETE is quantitated by comparison of peak areas to the internal standard, 16-hydroxy-9, 12, 14-heneicosatrienoic acid (16h-21 : 3; 0.5 nmol; t_R 29 min). The peaks at 15.5 and 14.5 min are 13-hydroperoxy-9,11-octadecadienoic acid, the hydroperoxide activator added to the enzyme assay, and its metabolite, 13-hydroxy-9,11-octadecadienoic acid, respectively.

ters 745 data module; 5 μm Nucleosil C_{18}, 250 × 4.6 mm column) with UV detection at 235 nm, an acetonitrile/methanol/water/acetic acid (350 : 150 : 250 : 1, v/v/v/v) mobile phase, and a flow rate of 1.5 ml/min.

Characterization of Recombinant 5-Lipoxygenase

Recombinant baculovirus-infected Sf9 cell 10,000 *g* supernatants contain abundant 5-lipoxygenase enzyme activity (Fig. 3) and protein (Fig. 4) 2 days postinfection. The recombinant enzyme is indistinguishable from the human leukocyte enzyme in immunoblot analysis and is present in amounts 50–200 times that of leukocytes on a per cell basis. Although the enzyme activity of 5-lipoxygenase in insect cells is always higher than in corresponding incubations of leukocytes, the relative specific activity is 80–90% lower than expected in the crude preparations. This difference is likely due to the lack of 5-lipoxygenase stimulatory factors in insect cells which are normally present in leukocytes. The enzyme activity in the 10,000 *g* supernatants is moderately stable if kept in an ice/water bath for

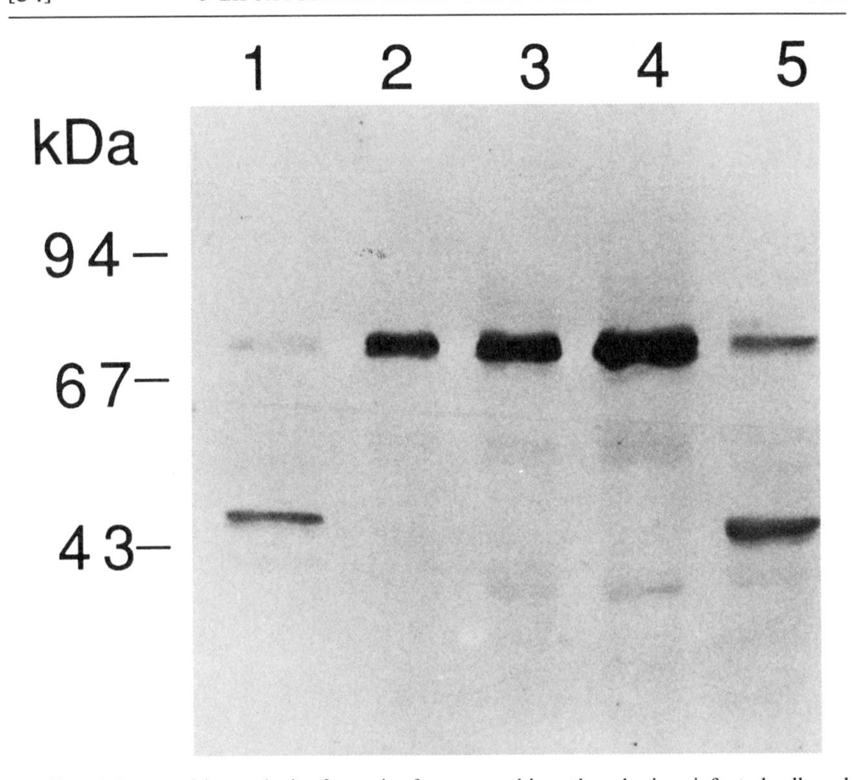

FIG. 4. Immunoblot analysis of proteins from recombinant baculovirus-infected cells and human leukocytes. The 10,000 *g* supernatant proteins are electrophoresed in a SDS–polyacrylamide (8%) gel, transferred to nitrocellulose, and incubated first with a polyclonal 5-lipoxygenase antiserum and then with goat anti-rabbit IgG horseradish peroxidase conjugate. Detection of immunoreactive bands is carried out by incubation with 3,3′-diaminobenzidine and hydrogen peroxide. Lanes: 1 and 5, human leukocyte proteins (21 μg from 4.1×10^5 cells and 50 μg from 10^6 cells, respectively; two different leukocyte preparations); 2–4, recombinant baculovirus-infected cell proteins (6 μg from 2.1×10^4 cells, 12 μg from 4.2×10^4 cells, and 24 μg from 8.4×10^4 cells, respectively; same sample). Markers at left indicate positions of phosphorylase *b* (94 kDa), bovine serum albumin (67 kDa), and ovalbumin (43 kDa) run in parallel.

36 hr or less (20% loss of activity) or at $-70°$ for 10 days (40% loss of activity). In a scaled-up purification scheme, 1–2 mg of purified enzyme should be isolated from 10^9 cells per liter of culture medium. The recombinant 5-lipoxygenase baculovirus–insect cell system described here offers a distinct advantage over human leukocytes, not only in terms of increased enzyme levels, but also in the relative ease of obtaining large numbers of cells, and the absence of hepatitis and AIDS (autoimmune deficiency syndrome) risk factors associated with the handling of human blood.

Section V

Cell Models of Lipid Mediator Production

A. Isolated Cell Preparations
Articles 55 through 59

B. Cell–Cell Interactions
Articles 60 through 64

C. *In Vitro* Tissue Preparations
Articles 65 through 67

[55] Parietal Cell Preparation and Arachidonate Metabolism

By N. ANN PAYNE and JOHN G. GERBER

Introduction

The eicosanoids, prostaglandins and leukotrienes are involved in various aspects of gastric function, including gastric acid secretion. The gastric mucosa is inhabited by numerous cell types, but it is the parietal cell that is responsible for the hydrogen ion production and secretion. There is voluminous literature on the effect of prostaglandins on gastric acid secretion in both human and animal *in vivo* models, but none of those studies could elucidate whether the effect of prostaglandins is directly on the parietal cells, what the underlying molecular mechanisms of prostaglandins are in affecting acid production, and whether the parietal cell is capable of synthesizing prostaglandins to affect its own function. Isolated gastric cell preparations of widely varying parietal cell content have been a valuable tool in exploring the answers to some of these questions. Prostaglandins (PG) E_2 and I_2, and arachidonic acid have been found to be potent inhibitors of histamine-stimulated acid production in isolated canine parietal cell-enriched preparations.[1,2] This acid inhibitory effect could not be observed if the cells were stimulated with either gastrin or the muscarinic receptor agonist, carbachol. Prostaglandins E_2 and I_2 were found to inhibit histamine-stimulated adenylate cyclase in the isolated cells.[1] In addition, the PGE_2 inhibition of both acid production and cyclic AMP generation required an intact GTP regulatory protein, G_i, since ADP-ribosylation of G_i with pertussis toxin blocked the effect of prostaglandin E_2.[3] The ability of parietal cells to synthesize PGE_2 has been reported, but it is still unclear which cell type from the gastric mucosa is the major source of the mucosal PGE_2 production.[4,5] Nonetheless, prostaglandin inhibition *in vivo* results in enhanced acid secretory response to histamine and pentagastrin suggesting that mucosal prostaglandins have a modulatory role in acid production.[6,7]

[1] A. H. Soll, *J. Clin. Invest.* **65,** 1222 (1980).
[2] M. L. Skoglund, A. S. Nies, and J. G. Gerber, *J. Pharmacol. Exp. Ther.* **220,** 371 (1982).
[3] M. C. Y. Chen, D. A. Amirian, M. Toomey, M. J. Sanders, and A. H. Soll, *Gastroenterology* **94,** 1121 (1988).
[4] S. Ota, M. Razandi, W. Krause, A. Terano, H. Hiraishi, and K. J. Ivey, *Prostaglandins* **36,** 589 (1988).
[5] J. G. Gerber, *Prostaglandins* **36,** 581 (1988).
[6] J. F. Gerkens, D. G. Shand, C. Flexner, A. S. Nies, J. A. Oates, and J. L. Data, *J. Pharmacol. Exp. Ther.* **203,** 646 (1977).
[7] I. H. M. Main and B. J. R. Whittle, *Br. J. Pharmacol.* **53,** 217 (1975).

There are essentially three techniques that have been used to enrich the parietal cells from the crude gastric cells. All three methods utilize collagenase digestion to generate single cells, but differ in their cell separation techniques. The cells can be separated on a discontinuous Ficoll gradient under unit gravity,[2] a procedure that requires 2–3 hr. Another technique to separate the mucosal cells utilizes sedimentation velocity in a Beckman elutriator rotor. The parietal cell enriched fraction contains a large percentage of chief cells, which can then be separated from the chief cells using a discontinuous density gradient of bovine serum albumin.[8] The procedure we utilize and describe in detail below separates the gastric cells using two consecutive Percoll gradients, the first discontinuous and the second continuous. In addition, the technique utilizes sieving through a series of 100% polyamide nylon fibers of different pore size to remove cell clumps.[9]

Materials

Hanks' Balanced Salt Solution

Hanks' balanced salt solution (HBSS)[10] is prepared from a stock of HBSS that is three times the physiological concentration (HBSS×3). The following salts are dissolved in a total of 3 liters of glass-distilled water: NaCl, 73.5 g; KCl, 3.60 g; $MgCl_2 \cdot 6H_2O$, 0.90 g; $MgSO_4 \cdot 7H_2O$, 0.90 g; KH_2PO_4, 0.54 g; and $NaHCO_3$, 3.15 g. In a separate beaker $NaHPO_4 \cdot 7H_2O$, 0.81 g, is dissolved by heating and added to the above salts. Only certified reagent-grade chemicals should be used.

This HBSS×3 stock is stored at 4° and diluted 1:3 with glass-distilled water on the day of the experiment. Dextrose, 1 g/liter, penicillin G, potassium 125 units (U)/ml, streptomycin 50 μg/ml (Pfizer Co., New York), $CaCl_2$, 0.14 g/liter, and bovine serum albumin (BSA), either 0.1 or 0.2% (Sigma, St. Louis, MO) are added and the medium filtered through a 0.22 μm Millex-GS filter (Millipore Corp., Bedford, MA). In some of the steps (where indicated), calcium and/or BSA is omitted. The osmolality of HBSS×3 is measured and adjusted with NaCl to 870 (3 × 290) mOsm/liter. All glassware that is used in the preparation of the solutions and in the cell isolation procedure is ethanol-rinsed to remove traces of detergent.

[8] A. H. Soll, D. A. Amirian, L. P. Thomas, T. J. Reedy, and J. D. Elashoff, *J. Clin. Invest.* **73,** 1434 (1984).

[9] N. A. Payne and J. G. Gerber, *J. Pharmacol. Exp. Ther.* **243,** 511 (1987).

[10] M. M. Bashor, this series, Vol. 58, p. 119.

Digestion Media

The collagenase medium is prepared in basal medium Eagle (BME), containing Earle's balanced salt solution without L-glutamine (Whittaker/M. A. Bioproducts, Walkersville, MD). To the BME is added the following: 25 m*M* *N*-2-hydroxyethylpiperazine-*N'*-2-ethanesulfonic acid (HEPES), 0.1% BSA, 0.38 mg/ml crude collagenase, Type 1, all from Sigma, and 2 m*M* L-glutamine (Whittaker/M. A. Bioproducts). After thorough mixing, the medium is warmed to 37° gassed with 95% O_2/5% CO_2 for 5 min and the pH adjusted to 7.4 with 1 *N* HCl to ensure that under the incubation conditions during the digestion the pH will be 7.4.

There is a large variability among the lots of the Type 1 crude collagenase in their ability to digest the mucosal tissue. Therefore, we request samples from numerous lots and when a lot is found that results in satisfactory digestion, we purchase a large quantity of Type 1 crude collagenase from that specific lot number.

The medium containing EDTA is prepared in Earle's balanced salt solution that is calcium and magnesium free (Gibco Laboratories). To the medium, HEPES, 25 m*M*, BSA, 0.1%, and EDTA, 2 m*M*, are added, and the medium then warmed to 37°, gassed with 95% O_2/5% CO_2, and the pH adjusted to 7.4.

Percoll Gradients

Cell separation is performed on two consecutive Percoll (Pharmacia Fine Chemicals, Piscataway, NJ) gradients. The first is a discontinuous gradient of densities 1.04, 1.065, and 1.09 g/ml. Each layer is prepared by combining various volumes of Percoll, HBSS×3, and glass-distilled water. The volume of Percoll required to give a particular density is first calculated using the following formula[11]

$$V_0 = V \frac{\rho - 0.1\rho_{10} - 0.9}{\rho_0 - 1}$$

where V_0 is volume of Percoll (ml); V, volume of final solution (ml); ρ, desired density of final solution; ρ_0, density of Percoll (from bottle); and ρ_{10}, density of 1.5 *M* NaCl or 1.058 g/ml (minor differences for other salts like HBSS).

To a given volume of HBSS×3, dextrose (3 g/liter) and BSA (0.6%) are added. Calcium is omitted. The medium is filtered through a 0.22 μm

[11] Percoll Methodology and Applications. Density Marker Beads for calibration of gradients of Percoll, Pharmacia Fine Chemicals AB, Uppsala, Sweden. Publication of Pharmacia for Peroll application.

Millex-GS filter. The volume of HBSS×3 in each gradient layer is approximately one-third the final volume, but the volume may be varied to result in a final gradient osmolality of 260 mOsm/liter. The osmolality of the Percoll gradients is slightly below physiological to enhance the buoyancy of parietal cells. The calculated volumes of Percoll, HBSS×3, and filtered, glass-distilled water are combined, dithiothreitol (0.1 mg/ml) is added, and the pH adjusted to 7.4. The three solutions are kept at 22° until use.

The second gradient is a continuous gradient prepared from 44% Percoll in HBSS. Upon centrifugation of a Percoll solution, a nonlinear gradient is formed whose shape depends on several factors: the concentration of Percoll, the volume of the solution, the speed and time of centrifugation, and the angle of the rotor. A concentration of Percoll is chosen that will generate a gradient whose center has a density corresponding to that of the cells of interest. It is desirable these cells to band around the center of the gradient since this region displays the greatest discrimination between the differences in the cell densities. It was found using Density Marker Beads (Pharmacia) that 30 ml of 44% Percoll (v/v) spun in a Sorvall SS-34 rotor at 15000 *g* for 1 hr at 22° generated a gradient whose center had a density of 1.06 g/ml, which corresponds to the buoyant density of parietal cells.

The Percoll medium is prepared by combining Percoll with HBSS×3 and glass-distilled water. After the dextrose (3 g/liter) and BSA (0.6%) are added to HBSS×3, the solution is filtered. The volume of HBSS×3 that is combined with Percoll is approximately one-third the final volume, but there can be some variability so that a final osmolality of the solution is 260 mOsm/liter. The remainder of the volume is filtered glass-distilled water. After the addition of dithiothreitol (0.1 mg/ml), the medium is adjusted to pH 7.4, portioned into centrifuge tubes, and spun as described above.

Preparation of Parietal Cells

After an overnight fast, mongrel dogs of either sex are anesthetized with sodium pentobarbital (30 mg/kg) and the corpus of the stomach, the region most dense in parietal cells, is removed through a wide flank incision. After washing with ice-cold saline, the tissue is placed in HBSS containing 0.2% BSA for subsequent handling (see Fig. 1).

The muscularis is dissected away and after scraping with a glass microscope slide to remove surface mucus, the mucosa is separated from the submucosa using a dull razor blade. A superficial cut is first made and the mucosa then peeled off using a lateral movement of the razor blade. This procedure is performed while the mucosal strip is on a Styrofoam board immersed in ice-cold HBSS with 0.2% BSA.

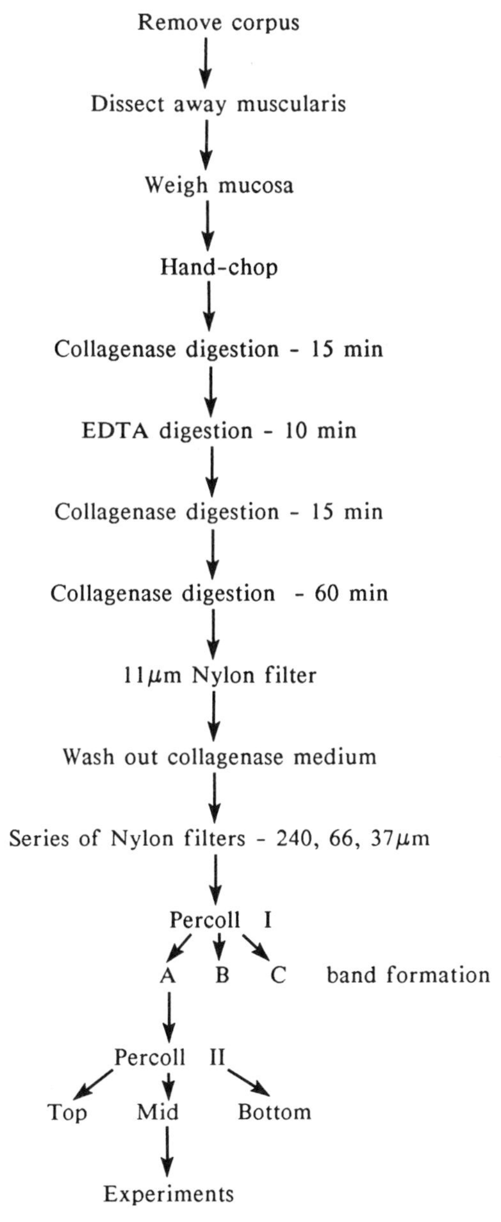

FIG. 1. Schematic representation of gastric cell generation and parietal cell enrichment in the dog mucosa.

Digestion

The mucosa is weighed in 10-g aliquots and chopped by hand for 2 min using two scalpels in a crossing motion. The mucosal fragments are digested with collagenase medium alternated with EDTA medium as follows. The tissue is first incubated in a 250-ml Nalgene flask containing 10 g of tissue with 50 ml collagenase medium for 15 min at 37° under an atmosphere of 95% O_2/5% CO_2 in a Dubnoff metabolic shaking incubator. The flasks are shaken at 120 cpm. Thirty grams of mucosal tissue are digested for each cell preparation.

After 15 min, the contents from each flask are poured into a beaker, the tissue is allowed to settle, the medium is decanted, and the tissue is rinsed with EDTA medium. The tissue is then incubated with fresh EDTA medium for 10 min at 37° under an atmosphere of 95% O_2/5% CO_2. The tissue is then rinsed with the collagenase medium and incubated in fresh collagenase medium for 15 min. After a final rinse, the tissue is digested for 1 hr in the collagenase medium. In the final 20 min, dithiothreitol (0.1 mg/ml) is added to the incubation mixture to decrease mucus content.

Sieving and Washing of Cells

The cell suspension is filtered through an 11 μm nylon mesh (Nitex nylon, Titko, Inc., Elmsford, NY), which filters out small cells that tend to persist through the cell separation procedure. Parietal cells, whose diameters range from 15 to 35 μm, remain on top of the nylon mesh, and after washing with HBSS containing 0.2% BSA, are reclaimed and counted. In ten cell preparations, parietal cell content increased from 20 $\pm$ 1% (mean $\pm$ SEM) in the crude mucosal cell preparation to 32 $\pm$ 2% after the 11 μm sieving.

The cell suspension is washed twice with HBSS containing 0.2% BSA. The cells are portioned into 40-ml centrifuge tubes (20 ml per tube containing 80 million cells), spun at approximately 50 *g* for 6 min, and the supernatant aspirated. Twenty milliliters of HBSS with 0.2% BSA is added to each tube, and the cells resuspended first by mixing with a plastic pipette and then by transferring the suspension into a 30 ml plastic syringe and using three strokes of the plunger. This force in the syringe breaks up cell clumps but does not lyse the cells. The cells are spun as described above and the wash repeated. The cell suspension is next passed through a series of nylons of decreasing pore size, 240, 66, and 37 μm, to remove residual clumps.

Clumping

Clumping has been a major problem in the purification of parietal cells. It is crucial that the cells are as individual cell suspensions in order to obtain effective separation on the Percoll gradients. Clumping is caused by several factors, including mucus formation, generation of cell debris caused by cell lysis, and the presence of calcium. Various measures are followed throughout the procedure to prevent clumping or to remove clumps. Scraping the mucosa not only removes mucus but also some of the surface mucous cells that produce it. This procedure and the use of dithiothreitol, an agent which degrades mucus, minimize clumping. Cell lysis is avoided by spinning the cells at low speed, 50 *g*, and by keeping the cells at room temperature. At 4°, the cell membranes are more rigid, so that the cells tend to lyse during resuspension. Initially the cells were kept in calcium-free medium after the digestion, but this resulted in decreased cell viability. Consequently, the cells are only in calcium-free medium during the gradient separation. Sieving of cells through the series of nylons of different pore size separates out large clumps from the single cells prior to the Percoll gradients.

Separation on Percoll Gradients

After the sieving and washing of cells, the cells are separated on two consecutive Percoll gradients. The first is a discontinuous gradient consisting of three layers of densities 1.04 (10 ml), 1.065 (25 ml), and 1.09 (5 ml) g/ml. The cell suspension is divided into several 40-ml centrifuge tubes, 75 $\times 10^6$ cells/tube, and spun. The cell pellet is resuspended in 25 ml of 1.065 g/ml Percoll. The suspension is then filtered through 66 μm nylon and drawn into a 30 ml syringe and thoroughly mixed with 12 strokes of the plunger. The gradient is formed by upward displacement. The cell suspension is drawn into a 30 ml syringe which is fitted with a 3.5-inch 18-gauge spinal needle, and then layered under the 1.04 Percoll in clear polycarbonate tubes. The 1.09 Percoll is subsequently layered under the 1.065 Percoll. The gradients are spun at 1000 *g* for 40 min at 22° in a Beckman model TJ-6 centrifuge. Three bands are formed: A, at the 1.04/1.065 interface, contains approximately 60% parietal cells; B, in the mid-region of density 1.065, is about 35% parietal cells; and C, at the 1.065/1.09 interface, is primarily chief cells. The bands are harvested and the cells diluted 1 : 1 with HBSS containing 0.2% BSA to dilute the Percoll before spinning. The cells are spun at 50 *g* for 6 min. The cells are rinsed twice with calcium-free HBSS with 0.1% BSA.

The second gradient is a continuous gradient prepared by spinning 30 ml of 44% Percoll at 15,000 *g* for 1 hr at 22° in a Sorvall SS-34 rotor. The cells from band A (Percoll 1) are resuspended in 10% Percoll (containing the same ingredients as the 44% Percoll), passed through a 66 μm nylon to remove clumps, and resuspended with 12 strokes of the plunger in a 30 ml syringe. The cells are layered, 20×10^6 cells/gradient, on the 30-ml preformed continuous Percoll gradient, and spun at 800 *g* for 12 min at 22°. Three bands form. The top band corresponds to density 1.048 and is primarily nonviable cells. The mid-band between densities 1.048 and 1.062 is approximately 85% parietal cells. The bottom band corresponding to density 1.062 is 40–50% parietal cells. The mid-band is used for the experiments where a highly enriched parietal cell preparation is desired. In eight recent cell preparations, the mid-band parietal cell enrichment was $84 \pm 2\%$. Cell viability was $90 \pm 2\%$.

Parietal cell viability and purity are initially determined using trypan blue exclusion in which the parietal cells are identified by size. The purity is later corroborated by hematoxylin and eosin staining of fixed cells. Parietal cells can easily be identified because of their pink stain due to the acidic canaliculi in the cytoplasm. In addition, parietal cells are the largest cells recovered from the gastric mucosa averaging a cell diameter of 25 μm.

Parietal Cell Experiments

Effect of Prostaglandins on Parietal Cell Acid Production

Numerous laboratories have demonstrated that PGE_2 inhibits histamine-stimulated acid production in isolated gastric cells. Parietal cell enrichment is not critical in these experiments because cellular uptake of aminopyrine is used to evaluate acid production, and only parietal cells actively take up aminopyrine in an energy-dependent manner. Aminopyrine is a weak base with a $pK_a \sim 5$, and thus in the acidic canaliculi of parietal cells the drug is ionized and trapped. The correlation between cellular oxygen utilization and cellular aminopyrine uptake in secretagogue-stimulated cells suggests that there is a close correlation between hydrogen ion production and active aminopyrine accumulation in the gastric cells.[12]

Arachidonic acid is a potent inhibitor of acid production, an effect that can be blocked by indomethacin.[2] These data suggest that parietal cells may well be capable of producing prostaglandins that inhibit acid production. However, the physiological stimulus for the synthesis of prostaglandins is unknown.

[12] T. Berglindh, H. F. Helander, and K. J. Obrink, *Acta Physiol. Scand.* **97,** 401 (1976).

Effect of Gastric Secretagogues on Parietal Cell PGE_2 Production

For these experiments, a highly enriched canine parietal cell preparation is necessary to minimize the contamination of PGE_2 production by other cells. Measuring PGE_2 concentration in the incubation media is an appropriate technique to evaluate cellular synthesis of PGE_2 since the gastric cells do not metabolize PGE_2 (unpublished results). Increasing concentrations of the three gastric acid secretagogues, carbachol, histamine, and pentagastrin with the parietal cells result in the release of PGE_2 only with carbachol. Maximum PGE_2 release occurs by 5 min in a pulsed manner.[9]

Similar results are obtained when the effect of secretagogues on cellular arachidonic acid release is examined in [^{14}C]arachidonic acid-enriched parietal cells. Again, only carbachol stimulated the release of [^{14}C]arachidonic acid.[9] These data suggest that activation of parietal cell muscarinic receptors modulates parietal cell prostaglandin synthesis. Thus, acid secretion per se is not a stimulus for prostaglandin production by the parietal cell.

Miscellaneous

A highly enriched parietal cell preparation would be necessary to study the regulation of PGE_2 receptors responsible for gastric acid inhibition. This has not always been performed.[13,14] Thus the possibility that PGE_2 receptors on other gastric cells confound the results cannot be ruled out. A highly enriched parietal cell preparation is also important when measuring the effect of PGE_2 on intracellular second messengers like cyclic AMP. PGE_2 can stimulate adenylate cyclase in numerous cells, but it appears to inhibit parietal cell adenylate cyclase.[1] Finally, a highly enriched parietal cell preparation is necessary when examining the parietal cell synthesis of other eicosanoids such as leukotrienes in order to minimize eicosanoid contamination by other cells.

In conclusion, the parietal cell-enriched preparation has allowed the examination of the cellular and intracellular mechanism of action of prostaglandins, as well as the regulation of their production.

Acknowledgments

This work was supported in part by Public Health Service Grants HL 21308 and AM 31607. Dr. Gerber is a Burroughs Wellcome Scholar in Clinical Pharmacology.

[13] D. B. Barr, J. A. Duncan, J. A. Kiernan, B. D. Soper, and B. L. Tepperman, *J. Physiol. (London)* **405,** 39 (1988).

[14] U. Seidler, M. Beinborn, and K.-F. Sewing, *Gastroenterology* **96,** 314 (1989).

[56] Culture of Bone Marrow-Derived Mast Cells: A Model for Studying Oxidative Metabolism of Arachidonic Acid and Synthesis of Other Molecules Derived from Membrane Phospholipids

By EHUD RAZIN

Introduction

For many years, research in mast cell biology and chemistry has been directed at the ability of these cells to respond to activation with the release not only of histamine, but also of other equally or more important biological substances, such as oxidative products of arachidonic acid and platelet activating factor (PAF).

Interleukin-3 (IL-3), a product of activated helper T lymphocytes, has been shown to stimulate the development of mast cells, defined by various criteria, from rodent hematopoietic tissues[1] and human bone marrow cells.[2,3] The IL-3-dependent bone marrow-derived cultured mouse and human mast cells (BMMC) synthesize chondroitin sulfate E proteoglycan rather than heparin proteoglycan.[3,4]

Upon immunological or calcium ionophore stimulation, these cultured cells metabolize arachidonic acid to leukotriene C_4 (LTC_4) and leukotriene B_4 (LTB_4) and generate only a minimal amount of prostaglandin D_2 (PGD_2).[1,3] Upon immunological challenge these cultured cells also synthesize and release PAF.[1,3] The concerted actions of these highly potent lipid mediators could modulate components of the inflammatory response in the microenvironment, such as, microvasculature, platelets, and neutrophils. Therefore, these culture systems can be used as a model of homogeneous cell population responding to a biologically relevant activating secretion stimulus, with the release of lipid mediators originating from membrane phospholipids.

This chapter provides information regarding the techniques of growing BMMC, their activation by various stimuli, and ways of activating separately either arachidonic acid metabolism or exocytosis in these cells.

[1] E. Razin, J. N. Ihle, D. Seldin, J. -M. Mencia-Huerta, R. H. Katz, P. A. Leblanc, A. Hein, J. P. Caulfield, K. F. Austen, and R. L. Stevens, *J. Immunol.* **132,** 1479 (1984).

[2] L. Gilead and E. Razin, *Abstr. FASEB J.* **2,** A687 (1988).

[3] L. Gilead, E. Rahamim, I. Ziv, R. Or, and E. Razin, *Immunology* **63,** 669 (1988).

[4] E. Razin, R. L. Stevens, F. Akiyama, K. Schmid, and K. F. Austen, *J. Biol. Chem.* **257,** 7229 (1982).

Procedures

Cell Culture

Mouse BMMC. Bone marrow cells obtained from femurs of 2-month-old male BALB/c mice are cultured at 37° in a humidified atmosphere containing 5% CO_2 at a starting density of 0.1×10^6 cells/ml in RPMI-1640 supplemented with 10% fetal calf serum, 2 m*M* L-glutamine, 0.1 m*M* nonessential amino acids, 100 U/ml penicillin, 100 μg/ml streptomycin, and 50 μM 2-mercaptoethanol, pH 7.2 (enriched medium). The enriched medium is supplemented with either 20 U/ml (4 ng/ml or $\simeq 2 \times 10^{-10}$ *M*) of purified IL-3 or with 50% WEHI-3 conditioned medium (WEHI-3-CM). WEHI-3-CM is produced by seeding WEHI-3 cells at 1×10^6/ml into enriched medium and incubating them for 4 days at 37° in 5% CO_2. Every 7 days the nonadherent cells from the bone marrow cultures are transferred into fresh enriched medium containing either 20 U/ml IL-3 or 50% WEHI-3-CM; at the same time, a sample of the nonadherent cells is stained with toluidine blue, pH 3.5, to determine the percentage of metachromatic cells. The cultures are maintained for 14 days in IL-3 or for 14 days in WEHI-3-CM to obtain a homogeneous population of metachromatic cells for chemical analysis.[1]

Human BMMC.[2,3] Human bone marrow mononuclear cells are isolated by Ficoll-Hypaque density sedimentation (1.072 g/ml) from bone marrow aspirated from normal donors (16–40 years). These cells are washed twice with saline and resuspended at a concentration of 2.5×10^6 cells/ml in 50% RPMI-1640 medium containing 10% fetal calf serum, 2 m*M* L-glutamine, 0.1 m*M* nonessential amino acids, 100 μ/ml penicillin, 100 μg/ml streptomycin, and 50 μM 2-mercaptoethanol (enriched medium) supplemented with 50% CM derived from peripheral blood mononuclear cells (PBMC) as will be described below. Twice a week the nonadherent population of the human bone marrow-derived cells is harvested, centrifuged, and resuspended in fresh medium. It should be emphasized that this culture technique is highly reproducible if the human bone marrow cells are obtained from a healthy, relatively young, adult. All the characterization of the cultured cells is performed on cells derived from 12- to 14-day cultures after the nonadherent cells are stained by toluidine blue to determine the percentage of metachromatic cells.

For the CM preparation, PBMC are isolated from a pool of buffy coats derived from the blood of two healthy adult donors by centrifugation over Ficoll-Hypaque for 10 min at 800 *g*. CM is prepared by culturing 5×10^6 PBMC/ml in enriched medium containing 10 μg/ml concanavalin A (Con A) for 48 hr at 37° in 5% CO_2 humidified atmosphere.

Exocytosis of Preformed and Newly Generated Mediators

Cells are washed and suspended at a concentration of 1×10^6 cells/0.5 ml in Tyrode's buffer containing 0.32 mM calcium, 0.2 mM magnesium, and 0.05% gelatin (TG) and are incubated for 30 min at 37° with the calcium ionophore A23187. Alternatively, 1×10^6 of BMMC are sensitized for 60 min either with 0.5 μg mouse monoclonal IgE against DNP–BSA (mouse cells) or with 5 μg human myeloma IgE (human cells). The cells are then washed twice and triggered with DNP-BSA (mouse) or with 3 μg rabbit anti-human IgE.[1,3] Only the mouse BMMC are also activated by appropriate doses of either thrombin (3000 NIH U/mg protein) or peanut agglutinin for specified time in 37°.[5,6]

The cells are sedimented, the supernatants collected, and the cell pellets resuspended in TG and sonicated. The exocytosis of secretory granule constituents is assessed by the measurement of histamine or β-hexosaminidase. The percentage release is calculated by the amount of histamine or β-hexosaminidase in the supernatant divided by the total in the supernatant and the pellet; the data are expressed as net percentage release, which is the difference between percentage release with and without activation.[1]

The generation and release of LTC_4 and LTB_4 is determined by radioimmunoassays both before and after product resolution by reversed-phase high-performance liquid chromatography (RP-HPLC). The eluate from the reversed-phase HPLC is monitored for absorbance at 280 nm; the columns are standardized with synthetic LTC_4 and LTB_4. The release of PGD_2 is assessed by measurement of immunoreactive PGD_2 in the cell supernatants by radioimmunoassay.[7]

For measuring PAF activity released by stimulated cells, IgE-sensitized cells are stimulated with DNP-BSA (mouse cells) or anti-human IgE (human cells) or unsensitized cells are stimulated with calcium ionophore A23187 in TG containing 0.25% fatty acid-free BSA instead of gelatin. PAF activity is measured by aggregation of rabbit platelets that had been prepared by differential centrifugation, treatment with 0.1 mM aspirin for 15 min at room temperature, and washing.[8,9] Two to five times 10^8 platelets in 400 μl of TG containing the adenosine diphosphate scav-

[5] E. Razin and G. Marx, *J. Immunol.* **133,** 3282 (1984).

[6] E. Razin, *J. Immunol.* **134,** 1142 (1985).

[7] L. Levine, R. A. Morgan, R. A. Lewis, K. F. Austen, D. A. Clark, A. Marfat, and E. J. Corey, *Proc. Natl. Acad. Sci. U.S.A.* **78,** 7692 (1981).

[8] J. Benveniste, P. M. Henson, and C. G. Cochrance, *J. Exp. Med.* **136,** 1356 (1972).

[9] J. Benveniste, J. P. Le Couedic, J. Polansky, and M. Tence, *Nature* (*London*) **269,** 170 (1977).

enger complex, creatine phosphate (1 m*M*) and creatine phosphokinase (10 U/ml), are aggregated in an aggregometer and variation in light transmission recorded.[8] PAF activity in the supernatants, with or without dilution in saline containing 0.25% fatty acid-free BSA, is measured over the linear portion of the calibration curve of 5 to 25 pg obtained with synthetic PAF. Results are expressed in picogram equivalents of PAF released by 10^6-activated BMMC. PAF is extracted from the supernatants with 4 volumes of absolute ethanol and is characterized by its mobility relative to sphingomyelin, lysophosphatidylcholine, and PAF standard on thin-layer chromatography and its inactivation by phospholipases A_1, A_2, and C.

Phosphatidylinositol (PI) Turnover

myo-[2-^{3}H]Inositol (40 μCi; 16.5 Ci/mmol) is added to each set of culture containing 20×10^6 cells in 15 ml culture medium. After 17 hr at 37° the cells are washed with TG. Duplicate samples of 4×10^6 cells are suspended in 500 μl TG with or without 1.5 m*M* LiCl and activated as described in the previous section.

The assay of [^{3}H]IPs is performed as described by Beaven *et al.*[10] The methanol/water solutions are applied to 3.5×0.5 cm columns containing 1 ml anion-exchange resin (Dowex 1-X8, 20–50 mesh), converted to the formate form by extensive washing with 1 *N* formic acid. Inositol, inositol 1-phosphate (IP_1), IP_2, and IP_3 are sequentially eluted with increasing concentrations of ammonium formate.[10] The eluates are collected and their ^{3}H content determined by liquid scintillation counting. Unlabeled myo-inositol, IP_1, IP_2, and IP_3 are used as markers for the column. Of the unlabeled IPs, 95% are recovered from the column after elution. Total [^{3}H]inositol lipids are estimated by mixing the chloroform phase of the cell extracts with 3 ml chloroform–methanol (2 : 1). The solution is washed twice with 2.5 ml methanol containing 1 m*M* KCl and 10 m*M* *myo*-inositol and after removal of the solvent by evaporation at 22° its ^{3}H content is determined by liquid scintillation counting.

Production of Lipid Metabolites in Activated BMMC

Both mouse and human cultured BMMC after activation produce comparable amounts of arachidonic acid metabolites and PAF. Both cultures contain approximately 500 ng histamine/10^6 cells and the net percentage release of preformed mediators and the release of LTs and PAF had the

[10] M. A. Beaven, J. P. Moore, G. A. Smith, T. R. Hesketh, and J. C. Metcalf, *J. Biol. Chem.* **259,** 7137 (1984).

same time courses, reaching maximal values approximately 5 min after immunological challenge. After activation with DNP-BSA 1×10^6 IgE-sensitized mouse BMMC typically released 24 ng LTC_4, 3.5 ng LTB_4, and 4000 pg PAF, whereas upon calcium ionophore A23187 stimulation these cells released 78 ng LTC_4, 8 ng LTB_4, and 8000 pg PAF. IgE-sensitized human BMMC (1×10^6) triggered with anti-human IgE released 20 ng LTC_4 whereas upon calcium ionophore A23187 stimulation these cells released 160 ng LTC_4, 10 ng LTB_4, and 500 pg PAF.

Differentiation of 5-Lipoxygenase Activation and Exocytosis in Stimulated Cultured Mouse BMMC

We were able to distinguish the processes of degranulation and 5-lipoxygenase activation in mouse BMMC using three methods: (1) drug intervention; (2) selective stimuli, and (3) oxyen depletion.

Drug Intervention. The inhibitory effect of the drug BW755C on the 5-lipoxygenase pathway was analyzed for murine BMMC.[11] The drug prevented the formation of 5-HETE from exogenous [^{14}C]arachidonic acid when IgE-sensitized cells were challenged by the antigen. BW755C also prevented formation of LTC_4 in a dose-dependent fashion when IgE-sensitized BMMC, preincubated with the drug, were activated with either the specific antigen or the calcium ionophore A23187. LTC_4 inhibition occurred with a minimal drug preincubation period of 1 min before the cells were subjected to antigen-dependent activation. BW755C-did not affect the degranulation response of these cells. Thus, the transmembrane activation of BMMC through their IgE Fc receptors, which lead to granule secretion, is not dependent upon corresponding metabolites from the 5-lipoxygenase pathway.

Selective Stimuli. We have shown that thrombin, a highly specific procoagulant enzyme that has numerous affects on platelets, endothelial, and smooth muscle cells could activate mouse BMMC to degranulate.[5] However, exposure of BMMC to thrombin did not lead to the generation and release of LTC_4 as observed when IgE-sensitized cells were stimulated with antigen or when cells were challenged by calcium ionophore A23187 (Table I). To exclude the possibility that thrombin itself interfered with the lipoxygenase system, 10^6 BMMC were first exposed to 0.5 U of thrombin for 5 min at 37° followed by the addition of 0.2 μM calcium ionophore A23187. Table I shows that cells stimulated by thrombin did not generate LTC_4 whereas those activated by calcium ionophore A23187 generated similar quantities of LTC_4, whether or not thrombin was added.

[11] H. Shoam and E. Razin, *Biochim. Biophys. Acta* **837,** 1 (1985).

TABLE I
EFFECT OF THROMBIN, IgE ANTIGEN, AND CALCIUM IONOPHORE A23187 ON BMMC[a]

Stimulator	Net release %	
	β-Hexosaminidase	LTC_4 (ng/10^6 cells)
IgE antigen	20 ± 6	15 ± 2
Thrombin	20 ± 4	<1
A23187	30 ± 4	72 ± 5
A23187 + thrombin	32 ± 5	73 ± 6

[a] Results represent the mean ± SE of three separate experiments.

Similar results were obtained in human neutrophil exposed to thrombin.[12] In human neutrophils, thrombin rapidly triggers lysozyme release without concomitant activation of the 5-lipoxygenase pathway.

Stimulation of murine BMMC either by thrombin or immunologically resulted in a rapid formation of inositol phosphates.[13] An increase in all three IPs (IP_1, IP_2, and IP_3) could be detected 20 sec after stimulation. The depletion of Ca^{2+} from the medium resulted in more than 80% reduction in preform mediators release from either thrombin- or IgE antigen-stimulated cells. However, both thrombin and IgE antigen increased the formation of IP_3 under these conditions independent of the presence of extracellular Ca^{2+}. The role of calcium in the mechanism of thrombin activation of BMMC was explored by measuring the changes in the uptake of $^{45}Ca^{2+}$ into quiescent BMMC and into cells stimulated by thrombin or by IgE antigen. The results indicate that activation of BMMC by either thrombin or IgE antigen is Ca^{2+}-dependent.[14] BMMC, 1×10^6, activated by 0.05–5 U thrombin, accumulated $^{45}Ca^{2+}$ in a concentration-dependent manner, which leveled off at around 1 U thrombin. Extracellular $^{45}Ca^{2+}$ uptake of thrombin-stimulated cells is saturable within 90 sec and corresponds to the kinetics of histamine release, whereas that of IgE antigen-exposed cells continues unabated for over 5 min. The pattern of $^{45}Ca^{2+}$ uptake of IgE-sensitized BMMC exposed to thrombin suggests that the prostimulatory

[12] D. Baranes, J. Matzner, and E. Razin, *Inflammation* **10,** 455 (1986).
[13] D. Baranes, F. -T. Liu, and E. Razin, *FEBS*, **206,** 64 (1986).
[14] R. Pervin, B. I. Kanner, G. Marx, and E. Razin, *Immunology* **56,** 667 (1985).

locus of thrombin action on the surface membrane is distinct from that of IgE.

Oxygen Depletion. The exocytosis of performed mediators from either IgE antigen or calcium ionophore A23187-stimulated murine BMMC was not affected by oxygen-depleted conditions regardless of the absence of glucose from the medium.[15] No detectable changes in the content of ATP were observed when the cells were triggered immunologically under anaerobic conditions in absence of glucose in the medium. However depletion of oxygen from the cells activated by both stimuli almost completely inhibited the specific release of arachidonic acid indicating that arachidonate does not play a significant role in the exocytosis.

[15] M. Lerner, A. Sammuni, and E. Razin, *Immunol. Lett.* **16,** 121 (1987).

[57] Endothelial Cells for Studies of Platelet-Activating Factor and Arachidonate Metabolites

By GUY A. ZIMMERMAN, RALPH E. WHATLEY, THOMAS M. MCINTYRE, DONELLE M. BENSON, and STEPHEN M. PRESCOTT

The development of methods to study endothelium in *ex vivo* systems has contributed greatly to our understanding of the biochemical bases for the complex interactions between the cells of the vascular wall and the cellular and humoral components of blood. Studies of cultured endothelial cells (EC) have been particularly useful since they represent a homogeneous population of cells that maintain many features of endothelium *in vivo*.[1-3] These include the production of eicosanoids and platelet-activating factor (PAF; 1-*O*-alkyl-2-acetyl-*sn*-glycero-3-phosphocholine), which are of particular interest because they contribute to the regulation of the interaction of blood cells with the intima, and to vascular resistance and permeability.[4,5] Furthermore, cultured EC produce eicosanoids and PAF in response to receptor-mediated agonists, which makes them an

[1] E. A. Jaffe, "Biology of Endothelial Cells." Martinus Nijhof, Boston, MA, 1984.
[2] M. A. Gimbrone, *Prog. Hemostasis Thromb.* **3,** 1 (1976).
[3] I. Hüttner and G. Gabbiani, *Lab. Invest.* **47,** 409 (1982).
[4] P. W. Majerus, *J. Clin. Invest.* **72,** 1521 (1983).
[5] G. Feuerstein and R. E. Goldstein, *in* "Platelet-Activating Factor and Related Lipid Mediators" (F. Snyder, ed.), p. 403. Plenum, New York, 1987.

attractive model system for studies of signal transduction and other regulatory systems.[6] The ability to maintain them in culture allows manipulations that cannot be applied in systems that must be studied acutely, e.g., isolated cells. We have studied the synthesis, distribution, and degradation of eicosanoids and PAF, and the role of these lipids in intercellular interactions, using *ex vivo* preparations of EC. Several experimental systems have been described (Table I); this chapter emphasizes the use of primary and early passage cultures of human and bovine macrovascular EC, and *in situ* EC on freshly isolated vascular segments.

Ex Vivo Models for Studies of Endothelial Cells

Primary Cultures of Human Endothelial Cells (as Modified[1,7]*)*

Solutions

Cord buffer[7]: add 49.09 g NaCl, 1.79 g KCl, and 11.89 g glucose to 6 liters of deionized water. To this, add 4.8 ml of Na_2HPO_4 (*M*) and 1.2 ml of NaH_2PO_4 (*M*). The final composition will be: 140 m*M* NaCl, 4 m*M* KCl, 1 m*M* sodium phosphate, and 11 m*M* glucose. Adjust the pH to 7.35 with *M* NaOH. Filter through a 0.22 μm filter (Nalgene, Rochester, NY) into a sterile container. Store at 4°.

Collagenase: Type I is obtained from Worthington Biochemical, (Freehold, NJ; the specific activity varies from lot to lot). Dilute the collagenase to 0.2% in cord buffer. Store at −20°. Immediately prior to use, thaw in a water bath at 37°, filter through a 0.2 μm filter (Acrodisc, Gelman), and keep at 37° until used.

Culture medium: medium 199 in Earle's balanced salt solution (BSS) containing 25 m*M* HEPES buffer and supplemental L-glutamine (2.0 m*M*) is obtained from Whittaker Bioproducts (Walkersville, MD). Add pooled human serum to 20%, penicillin (196 U/ml), and streptomycin (196 μg/ml), and adjust the pH to 7.35–7.45. We prepare pooled serum from 10 to 15 normal donors,[8] store it at −80°, and pass it twice through 0.2 μm serum filters (Nalgene, Ann Arbor, MI) prior to addition to the medium. Penicillin and streptomycin are purchased as a premixed solution from Whittaker Bioproducts.

[6] R. E. Whatley, G. A. Zimmerman, T. M. McIntyre, R. Taylor, and S. M. Prescott, *Semin. Thromb. Hemostasis* **13,** 445 (1987).

[7] E. A. Jaffe, R. L. Nachman, C. G. Becker, and C. R. Minick, *J. Clin. Invest.* **52,** 2745 (1973).

[8] G. A. Zimmerman, T. M. McIntyre, and S. M. Prescott, *J. Clin. Invest.* **76,** 2235 (1985).

TABLE I
Ex Vivo ENDOTHELIAL CELL PREPARATIONS USED TO EVALUATE SYNTHESIS AND ACTIVITIES OF LIPID MEDIATORS[a]

	References to biochemical and functional responses[c]			
Ex vivo EC preparation	Eicosanoid synthesis	PAF synthesis	Signal transduction	Cell–cell interaction
Cultured EC[b]				
Macrovascular EC monolayers	*1–4*	*4–7*	*8,9*	*5,10–13*
Macrovascular EC microcarrier culture	*14*	—	*15*	*15*
Macrovascular EC suspension	*8,16*	—	*8*	*16*
Microvascular EC monolayers	*17–19*	*20*	—	—
In situ, ex vivo EC				
Perfused vessels	*21–23*	—	—	*21,23*
Vascular fragments	*19,24–26*	—	—	*4,13*
Exposed, *in situ* endothelium	*7,27,28*	*7*	—	—

[a] *Ex vivo* preparations that have been used for the study of biologically active lipid mediators. The table is not comprehensive, but rather lists representative examples of a particular approach.

[b] A continuous cell line, derived by fusion of primary human umbilical vein EC with a human carcinoma, is reported to synthesize PGI_2 in response to thrombin [J. E. Suggs, M. C. Madden M. Friedman, C.-J. S. Edgell, *Blood* **68**, 825, 1986]. Early passage HUVEC are available from the American Type Culture Collection, but their synthesis of eicosanoids and PAF has not been studied in detail. The references in this table describe studies with primary or passed EC isolated and maintained in the reporting laboratories, rather than experiments with cell lines.

[c] Key to references: (*1*) B. B. Weksler, C. W. Ley, and E. A. Jaffe, *J. Clin. Invest.* **62,** 923 (1978); (*2*) F. Alhenc-Gelas, S. J. Tsai, K. S. Callahan, W. B. Campbell, and A. R. Johnson, *Prostaglandins* **24,** 723 (1982); (*3*) G. Fry, T. Parsons, J. Hoak, H. Sage, R. D. Gingrich, L. Ercolani, D. Ngheim, and R. Czervionke, *Arteriosclerosis* **4,** 4 (1984); (*4*) T. M. McIntyre, G. A. Zimmerman, K. Satoh, and S. M. Prescott, *J. Clin. Invest.* **76,** 271 (1985); (*5*) S. M. Prescott, G. A. Zimmerman, and T. M. McIntyre, *Proc. Natl. Acad. Sci. U.S.A.* **81,** 3534 (1984); (*6*) F. Bussolino, F. Brevario, C. Tetta, M. Aglietta, A. Mantovani, and E. Dejana, *J. Clin. Invest.* **77,** 2027 (1986); (*7*) R. W. Whatley, G. A. Zimmerman, T.M. McIntyre, and S. M. Prescott, *Arteriosclerosis* **8,** 321 (1988); (*8*) E. A. Jaffe, J. Grulich, B. B. Weksler, G. Hampel, and K. Watanabe, *J. Biol. Chem.* **262,** 8557 (1987); (*9*) R. W. Whatley, P. Nelson, G. A. Zimmerman, D. L. Stevens, C. J. Parker, T.M. McIntyre, and S. M. Prescott, *J. Biol. Chem.* **264,** 6325 (1989); (*10*) S. J. Feinmark and P. J. Cannon, *J. Biol. Chem.* **261,** 16466 (1986); (*11*) G. A. Zimmerman and D. Klein-Knoeckel, *J. Immunol.* **136,** 3839 (1986); (*12*) G. A. Zimmerman, T. M. McIntyre, and S. M. Prescott, *J. Clin. Invest.* **76,** 2235 (1985); (*13*) P. G. Milner, N. J. Izzo, J. Saye, A. L. Loeb, R. A. Johns, and M. J. Peach, *J. Clin. Invest.* **81,** 1795 (1988); (*14*) A. Ager and J. L. Gordon, *J. Exp. Med.* **159,** 592 (1984); (*15*) P. Ganz, P. F. Davies, J. A. Leopold, M. A. Gimbrone, and R. W. Alexander, *Proc. Natl. Acad. Sci. U.S.A.* **83,** 3552 (1986); (*16*) A. A. Marcus, B. B. Weksler, E. A. Jaffe, and M. J. Broekman, *J. Clin. Invest.* **66,** 979 (1980); (*17*) M. E. Gerritsen and C. D. Cheli, *J. Clin. Invest.* **72,** 1658 (1983); (*18*) I. F. Charo, S. Shak, M. A. Karasek, P. M. Davison, and I. M. Goldstein, *J. Clin.*

Establishing Cultures

1. Obtain umbilical cords. It is ideal to harvest cells by 4–6 hr after delivery, but specimens up to 36 hr are acceptable. Store at 4° in a sterile container until processed. Cords can be covered with a physiological buffer, but it is not necessary.

2. Cleanse the cord of blood and other debris with physiological buffer and then transfer to a sterile tray in a clean hood. *Use sterile technique for all following steps in culture.*

3. Excise and discard a few millimeters of each end of the cord, and any clamped areas. Identify and gently cannulate the umbilical vein at each end of the cord with a 14-gauge, 2-inch, blunt metal needle (Popper & Sons, New Hyde Park, NY); secure the needles (we use two umbilical cord clamps on each end with a ligature between them). If a clamped segment has been excised, the cut ends can be rejoined with a hubless cannula (clamp it in place) prior to the addition of collagenase.

4. Gently perfuse the vein with sterile, prewarmed cord buffer until all blood has been removed (we usually use 240 ml of buffer delivered by a hand-driven syringe in 60-ml portions via a three-way stopcock).

5. Replace the cord buffer with approximately 10 ml of 0.2% collagenase per 15 cm cord length, and incubate for 10 to 15 min at 37° (usually optimal, but can vary depending on the lot of collagenase). Midway through the incubation gently agitate the cord and then turn it over to ensure that the protease solution contacts the intima evenly.

6. Gently agitate the cord and then add 25–30 ml of cord buffer to rinse the collagenase solution and detached EC from the vein. Collect the cells in a 50-ml conical centrifuge tube containing culture medium *with serum.* (The serum antiproteases will neutralize the collagenase.)

7. Mix the contents by gentle inversion and then centrifuge at 250 *g* at 20° for 10 min. The EC will form a small, translucent pellet. Gently resuspend EC in 6 ml of complete medium. Repeat the centrifugation step and resuspend in a known volume (usually 18–36 ml).

Invest. **74,** 914 (1984); (*19*) S. A. Moore, A. A. Spector, and M. N. Hart, *Am. J. Physiol.* **254** (*Cell Physiol.* **23**) C37 (1988); (*20*) J. E. Lynch and P. M. Henson, *J. Immunol.* **137,** 2653 (1986); (*21*) S. R. Saba and R. G. Mason, *Thromb. Res.* **5,** 747 (1974); (*22*) R. S. Kent, S. L. Diedrich, and A. R. Whorton, *J. Clin. Invest.* **72,** 455 (1983); (*23*) S. Moncada, E. A. Higgs, and J. R. Vane, *Lancet* **1,** 18 (1977); (*24*) B. B. Weksler, S. B. Pett, D. Alonso, R. C. Richter, P. Stelzer, V. Subramanian, K. Tack-Goldman, and W. A. Gay, *N. Engl. J. Med.* **308,** 800 (1983); (*25*) G. G. Neri Sereni, R. Abbate, G. F. Gensini, A. Panetta, G. C. Casolo, and M. Carini, *Prostaglandins* **25,** 753 (1983); (*26*) D. S. Rush, M. D. Kerstein, J. A. Bellan, S. M. Knoop, P. R. Mayeux, A. L. Hyman, P. J. Kadowitz, and D. B. McNamara, *Arteriosclerosis* **8,** 73 (1988); (*27*) C. Goldsmith, C. T. Jafvert, P. Lollar, W. G. Owen, and J. C. Hoak, *Lab. Invest.* **45,** 191 (1981); (*28*) J. C. Goldsmith and S. W. Needleman, *Prostaglandins* **24,** 173 (1982).

8. Plate the cells. In most cases, the EC from one vein are placed in 12–24 35-mm dishes (1.5 ml of cell suspension/dish), or on an equivalent surface area. It is critical that the surface of the culture dish/flask be coated with a matrix that favors EC attachment and spreading. A variety of proteins have this property,[9] but we routinely use gelatin to minimize expense. To coat with gelatin, add a sterile solution of gelatin (0.2%) to culture dishes for 1 hr at 37° or overnight at 4°. Wash once with medium 199 immediately before plating.

Maintaining the Cultures

1. Keep cultures in an incubator at 37° in a humidified atmosphere of 5% CO_2/95% air.
2. Wash the monolayers at 2–4 hr after plating (medium 199 at 37°; use a volume equal to the volume in the culture vessel). This removes contaminating cells, which do not adhere as rapidly as EC. If the vessels contain many nonadherent cells and/or much acellular debris, several washes may be required.
3. Change the medium every 3 days.
4. The cultures can be passed using the method described by Jaffe.[1] For passed cells, we supplement the medium with endothelial cell growth supplement (ECGS; purchased from Bionetics Research Institute, Rockville, MD, or prepared as in Ref. 10).
5. For routine experiments, we use the cultures 24–96 hours after they become tightly confluent, which occurs within 3–7 days after plating. At this point, the cells form a continuous monolayer of uniform, flattened, tightly apposed polygonal cells ("postconfluent morphology"), with few or none of the EC having the elongate, "juvenile" morphology that is present in preconfluent cultures, as determined by inspection under phase-contrast microscopy (Fig. 1). In addition to characteristic morphology, the EC have characteristic biochemical markers.[1,11] In the postconfluent state, the number of cells in each monolayer will be relatively constant at approximately 10^5 cells/cm^2. The cell number can be determined by removal of the EC from the plate with trypsin/EDTA, suspension in divalent cation-free buffer, and counting in a hemocytometer or electronic particle counter. We have found that counting with a hemocytometer is more reliable than particle counting because there are frequently aggregates of EC as well as single cells in the suspension. An alternative method in-

[9] D. Cheresh, *Proc. Natl. Acad. Sci. U.S.A.* **84,** 6471 (1987).

[10] T. Maciag, J. Cerundolo, S. Ilsley, P. R. Kelley, and R. Forand, *Proc. Natl. Acad. Sci. U.S.A.* **76,** 5674 (1979).

[11] R. E. Whatley, G. A. Zimmerman, T. M. McIntyre, and S. M. Prescott, *Arteriosclerosis* **8,** 321 (1988).

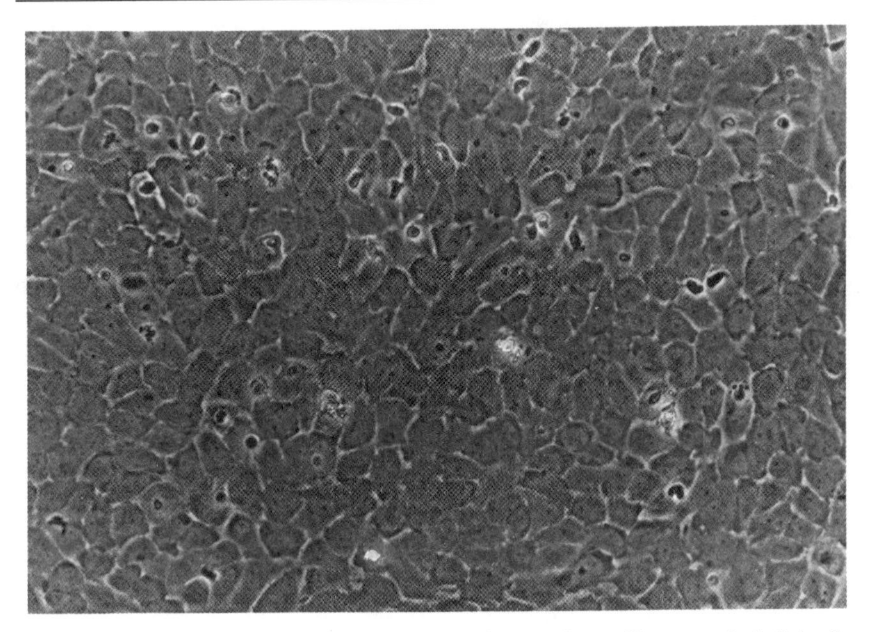

FIG. 1. Phase-contrast photomicrograph of a primary culture of human endothelial cells. Endothelial cells derived from an umbilical vein were cultured as described. After 5 days, the EC reached confluence. At 7 days they had a uniform, flattened, polygonal shape with tight apposition to adjacent cells and indistinct cell borders (postconfluent morphology), as shown.

volves counting crystal violet-stained nuclei from lysed EC (see this series, Vol. 58, pp. 143–144, for a general method). Living EC can be counted without removing them from the monolayer using a micrometer eyepiece and phase-contrast microscopy, followed by calculation of the number of cells in the monolayer based on the number per unit surface area. Because human umbilical vascular endothelial cells (HUVEC) have indistinct cell borders (Fig. 1; Ref. 7), parallel fixation and staining of the monolayers may be required for validation.

COMMENTS. To minimize contamination with other cells use collagenase rather than trypsin, standardize the type of collagenase and the duration of incubation, and avoid trauma to the cord. Do not use cords that have been clamped in multiple places. If the collagenase digestion and other variables are carefully controlled, the EC monolayers will contain few, if any, contaminating cells (smooth muscle cells, fibroblasts, nucleated blood cells). The "purity" of the cultures can be determined by a number of methods, of which immunofluorescent staining for von Willebrand factor (vWF) is the most commonly used because of the fact that only EC and megakaryocytes are known to synthesize vWF and commercial antisera are readily available.[1] On a day-to-day basis, we depend heavily on the uniform appearance of EC in confluent culture to exclude isolates with a significant number of contaminating smooth muscle cells or fibroblasts. The morphological and other features that differentiate the latter cells from EC have been described.[1,2,7]

Optimal growth may depend on the use of medium 199 and human serum.[1] Fetal bovine serum is used by many laboratories, but will alter the growth characteristics and may change phenotypical features (see p. 7 in Ref. 1 for a discussion of this issue). We do not use amphotericin B as an antifungal agent because of its effects on plasma membranes. We do not routinely use ECGS or heparin[12] for primary cultures. The plasticware supplied by several manufacturers is satisfactory. However, it is crucial that the sterilization procedure be one that does not leave toxic residues; we use plates sterilized with radiation.

We have used minor variations of this method to culture EC from human umbilical artery, pulmonary artery, and aorta.

Culture of Bovine Macrovascular Endothelial Cells[11]

1. Obtain vascular segments (pulmonary artery, aorta, inferior vena cava) immediately after slaughter. Ligate both the ends of the segments (we use umbilical tape), dissect them free, and transport to the laboratory in ice-cold saline. Process within 2–6 hr. *Use sterile technique in steps that follow.*
2. Insert 16-gauge catheters into the vessel near each ligature. Rinse the lumen with sterile Hanks' balanced salt solution (HBSS) until free of blood, and then infuse approximately 25 ml of 0.2% collagenase in HBSS. These volumes assume a segment of about 12 $\times$ 6 cm.
3. Incubate in a water bath at 37° for 20 min with rotation and gentle agitation every 5 min.
4. Withdraw the collagenase solution with detached EC and add to 20 ml of complete medium at 4° (see below). Centrifuge (500 *g* for 10 min at 20°) and then resuspend in warm (37°) complete medium consisting of medium 199 containing 20% fetal bovine serum (Hyclone Laboratories, Logan, UT), 100 U/ml penicillin, and 100 μg/ml streptomycin.
5. Plate onto gelatin-coated 35-mm dishes (approximately 20 for a segment of this size) or other vessels.
6. After 4 hr, wash the cells (see above). Maintain as above.
7. The cultures will usually grow to confluence within 5 to 7 days, at which time there will be approximately 10^5 cells/cm^2.

[12] S. C. Thornton, S. N. Mueller, and E. M. Levine, *Science* **222,** 623 (1983).

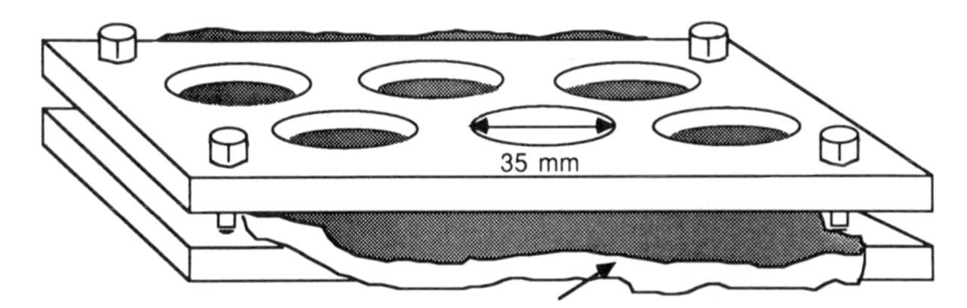

FIG. 2. Template used for *ex vivo* experiments of endothelium *in situ*. This custom-made device allows incubations of the intact vessel in which the experimental reagents are in contact only with the endothelium.

Ex Vivo, in Situ Endothelial Cells[11]

1. Obtain a segment of bovine pulmonary artery or aorta as above.

2. Incise the vessel lengthwise, and place it in a plastic template with the luminal surface upward (Fig. 2). In our custom-made template, the top portion has holes that are 35 mm in diameter, so that we form a well of the same size as a culture dish with an *in situ* layer of endothelium covering the bottom.

3. Wash the wells gently several times with HBSS to remove any remaining blood, and cover the cells with buffer or medium.

4. Perform incubations just as in culture dishes. The endothelial cells can be harvested by gentle scraping with a scalpel blade.

Production of Platelet-Activating Factor by Endothelial Cells

Endothelial cells from different species and from different vascular beds rapidly synthesize PAF when they are appropriately stimulated.[11,13–17] They make little, or no, PAF in the absence of an agonist. One important feature of this response by EC is that virtually all of the PAF is retained by the cells—it is not secreted as a fluid-phase mediator.[11,14] At least a

[13] S. M. Prescott, G. A. Zimmerman, and T. M. McIntyre, *Proc. Natl. Acad. Sci. U.S.A.* **81,** 3534 (1984).
[14] T. M. McIntyre, G. A. Zimmerman, K. Satoh, and S. M. Prescott, *J. Clin. Invest.* **76,** 271 (1985).
[15] T. M. McIntyre, G. A. Zimmerman, and S. M. Prescott, *Proc. Natl. Acad Sci. U.S.A.* **83,** 2204 (1986).
[16] G. Camussi, M. Aglietta, F. Malavasi, C. Tetta, W. Piacibello, F. Sanavio, and F. Bussolino, *J. Immunol.* **131,** 2397 (1983).
[17] J. M. Lynch and P. M. Henson, *J. Immunol.* **137,** 2653 (1986).

portion of the PAF is expressed transiently on the surface of the EC, where it mediates the interaction of endothelium with polymorphonuclear leukocytes (PMNs).[6,8,15] This would be advantageous *in vivo* since it would provide a mechanism for adhesion of PMNs to the vascular wall only when there was an appropriate, e.g., inflammation, stimulus to the endothelium. Also, the response is both initiated and concluded rapidly, which could be advantageous in physiological responses.

Principle. Two routes for the synthesis of PAF have been described,[18] but the "remodeling pathway" appears to be the important one in EC. This pathway begins with the removal of a fatty acid from the *sn*-2 position of an appropriate precursor phospholipid. The resultant lysophospholipid is then acetylated to yield PAF. Our usual assay measures the incorporation of [^{3}H]acetate into a phospholipid with the chromatographic characteristics of PAF. These systems do not separate choline phosphoglycerides that have a 1-acyl linkage from PAF. However, we have shown that under many circumstances this is not an important consideration since less than 10% of the labeled product in EC has a 1-acyl group.[11,14] If EC are studied under markedly different conditions using this method, the contribution of this lipid should be determined.

Solutions

1. Hanks' balanced salt solution or an equivalent is satisfactory. Add HEPES buffer to 10 m*M* (pH 7.4). The synthesis of PAF is dependent on the extracellular concentration of calcium,[19] which should be 1–10 m*M* for an optimal response. Thus, calcium should be added to the buffer just prior to use to prevent precipitation or complex formation. In experiments with human EC, we often use a buffer that is similar to HBSS except that it does not contain bicarbonate, which can complex with calcium. Albumin should be included for incubations longer than 1 hr. However, albumin increases the production of PAF in response to agonists,[6] so appropriate controls should be included.

2. The agonist should be made up as a concentrated solution, i.e., suitable for 100-fold dilution in the final incubation (we usually add it in 10 μl to an incubation mixture with a final volume of 1 ml).

3. Add [^{3}H]acetate (3.4 Ci/mmol) to the amount of incubation buffer needed for the experiment to a final concentration of 25 μCi/ml. CAUTION: some ready-made buffers and media (e.g., medium 199) contain unlabeled acetate, which will lower the specific radioactivity of the tracer.

[18] F. Snyder, *in* "Platelet-Activating Factor and Related Lipid Mediators" (F. Snyder, ed.), pp. 89–113, Plenum, New York, 1987.

[19] R. E. Whatley, P. Nelson, G. A. Zimmerman, D. L. Stevens, C. J. Parker, T. M. McIntyre, and S. M. Prescott, *J. Biol. Chem.* **264,** 6325 (1989).

4. Extraction solvents: chloroform, methanol, methanol containing 50 m*M* acetic acid, and 0.1 M sodium acetate. Also, prepare and save the upper phase from a mock Bligh–Dyer[20] extraction of the incubation buffer. This solution, called preequilibrated upper phase (PEU), is used to wash the lower phase of the experimental extractions. The solvents for the mock extraction are: buffer/methanol/chloroform/0.1 *M* sodium acetate, 1.0/2.5/3.75/1.0 (v/v/v/v).

5. Chromatography solvent: mix 50 ml of chloroform, 25 ml of methanol, 8 ml of glacial acetic acid, and 4 ml of water. This is an adequate volume for a usual thin-layer chromatography tank, but can be scaled up as needed.

Stimulation and Extraction of the Endothelial Cells

1. Remove the culture medium from EC and wash twice with buffer (use twice the volume of the medium for each wash). This is important as it removes any residual plasma PAF acetylhydrolase,[21] which can degrade the EC-produced PAF and alter the results.

2. Cover the cells with incubation buffer that contains [^{3}H]acetate (for a 35-mm dish we use 1 ml; this protocol will assume that 1 ml = 1 volume). Add the agonist to start the reaction (tilt the dish gently to ensure mixing). The subsequent incubation can be at room temperature or 37°, since there is little difference in the response.[22]

3. At the appropriate time (in the case of stimulation with thrombin the peak accumulation is at 10 min), add 0.5 volume of the acidified methanol. This stops the synthesis and degradation of PAF.

4. Scrape the cells, in the buffer and acidified methanol, from the dish with a rubber policeman, and add all of these to 1.25 volumes of chloroform in a screw-top tube. (We include 7 nmol of unlabeled, "carrier" PAF to the chloroform to improve recovery, but it can be omitted.) Wash the dishes twice with 1 volume of methanol (scrape the plate each time) and add both washes to the extraction mixture. Place caps on the tubes and shake them vigorously. The mixture should be a monophase: if it is not, add a small amount of methanol. Then, add 1.25 volumes of chloroform and 1 volume of 0.1 *M* sodium acetate, shake vigorously, and separate into phases by brief centrifugation in a table top centrifuge. Discard the upper phase, which will contain most of the radioactivity (primarily unincorporated [^{3}H]acetate). Wash the lower phase twice by mixing with 2 volumes PEU and centrifuge as above to remove the remaining water-soluble radiolabel.

[20] E. G. Bligh and W. J. Dyer, *Can. J. Biochem. Physiol.* **37,** 911 (1959).
[21] D. M. Stafforini, T. M. McIntyre, and S. M. Prescott, this volume [39].
[22] G. A. Zimmerman, T. M. McIntyre, and S. M. Prescott, *Circulation* **72,** 718 (1985).

5. Dry the lower phase under nitrogen. Redissolve in a small (0.1 volume), precisely measured amount of chloroform/methanol (9/1). A positive displacement micropipette (e.g., Rainin) is useful for this. Remove 25% of this solution, place it directly in a scintillation vial, allow the chloroform to evaporate, and determine the radioactivity in a scintillation spectrometer. Multiply this value by four to obtain the total radioactivity in lipids.

6. Apply the remaining 75% of the lipid extract to a TLC plate (it is useful to first reduce the volume under nitrogen). The plate should be silica gel 60 (Merck & Co., Darmstadt) that has been dried for 60 min at 100° prior to use. Develop the plate in a TLC tank that has been equilibrated with the solvent described above. Dry the plate and locate the fraction that contains PAF. We mist the plate lightly with water until the carrier PAF appears as a white spot on a gray background, at an R_f of 0.23. Alternative methods include running standards of PAF in parallel lanes and identifying them by radiochemical or staining procedures. Scrape the fraction (about 1 cm) that contains PAF into a scintillation vial. Then scrape the remainder of the lane into one or more vials, and determine the radioactivity. Determine the percentage of the radioactivity in the TLC lane that is accounted for by PAF (PAF dpm/total dpm in TLC lane × 100), and multiply this value by the total lipid radioactivity (above). This gives the [^{3}H]acetate incorporation into PAF. This approach is superior to simply scraping and counting the PAF fraction from TLC because it corrects for losses and/or inefficient counting at the TLC step, which can be substantial.

Alternatives. PAF can be measured by bioassay,[11,13,17] which is sensitive but not always specific. Also the extract can be assayed by one of several mass spectrometric techniques.[23,24] If either of these methods is to be used, do not include the unlabeled PAF in the extraction as it will be detected in the assay. PAF can be isolated by a variety of HPLC and TLC systems[14,15] in addition to the TLC system described here.

Comments. This assay has the virtue of measuring only PAF synthesized by the cells during the incubation, but it is not a precise measure of synthetic rate since degradation occurs simultaneously. However, we have shown that the degradation rate of PAF in EC does not change under a variety of conditions. Thus, the accumulation of PAF reflects changes in the activity of the synthetic pathway. If it is important to measure the synthetic rate, this can be achieved by briefly pulsing the cells with labeled acetate at different times after the addition of the agonist.[14,15] The metabolic labeling assay should be used only for relatively short periods (<1 hr)

[23] K. L. Clay, this volume [16].

[24] W. C. Pickett and C. S. Ramesha, this volume [17].

since at longer times there is substantial incorporation of radioactive acetate into other lipids and macromolecules, which decreases the specificity of the assay.

Arachidonic Acid Metabolism in Endothelial Cells

Endothelial cells take up arachidonic acid, incorporate it into phospholipids and, upon appropriate stimulation, release it for conversion to eicosanoids.[25] As in other cell types, the rate-limiting step for the production of eicosanoids appears to be the release of arachidonate from the *sn*-2 position of phospholipids. The subsequent metabolism is mostly via cyclooxygenase. In endothelial cells from the macrovasculature the predominant product is prostaglandin I_2.[26] Other metabolites produced by macrovascular EC in lesser amounts, or as major products in microvascular EC, include prostaglandins E_2, $F_{2\alpha}$, D_2, and HETEs.[27-31]

Assays of Eicosanoid Production

Endothelial cells should be prepared as described above. Handle them gently during the removal of the culture medium and washing of the monolayer because even mild mechanical stimuli can induce eicosanoid synthesis. Stimulate the cells with an agonist, or control, solution. The synthesis of PGI_2 is rapid—it will be complete within 5 to 10 min after stimulation.[14] Remove the supernatant for assay.

Because of its relative ease and sensitivity, radioimmunoassays (RIA) of arachidonate metabolites, such as prostaglandin I_2 and E_2, are widely used for the study of arachidonate metabolism in endothelial cells. The disadvantages are that measurement of a single metabolite may not reflect all of the arachidonate metabolism under all conditions, and if cross-reactivity occurs, the values will be quantitatively imprecise. For many studies of eicosanoids in macrovascular EC these are not major considerations since PGI_2 is the predominant product. One advantage of RIA is that the eicosanoids can usually be measured directly in the supernatant without extraction or derivatization.

[25] A. A. Spector, J. C. Hoak, G. L. Fry, G. M. Denning, L. M. Stoll, and J. B. Smith, *J. Clin. Invest.* **65,** 1003 (1980).

[26] B. B. Weksler, A. J. Marcus, and E. A. Jaffe, *Proc. Natl. Acad. Sci. U.S.A.* **74,** 3922 (1977).

[27] A. R. Johnson, *J. Clin. Invest.* **65,** 841 (1980).

[28] G. E. Revtyak, A. R. Johnson, and W. B. Campell, *Am. J. Physiol.* **254,** C8 (1988).

[29] F. Alhenc-Gelas, S. J. Tsai, K. S. Callahan, W. B. Campbell, and A. R. Johnson, *Prostaglandins* **24,** 723 (1982).

[30] S. A. Moore, A. A. Spector, and M. N. Hart, *Am. J. Physiol.* **254,** C37 (1988).

[31] I. A. Blair, this series, Vol. 86, pp. 467–477.

Various mass spectrometric assays have been described for the common eicosanoids from endothelial cells,[31] and should be employed under conditions in which the precise structure of the products has not been defined. Mass spectrometry is the most specific assay and can be performed quantitatively. Methods to extract[32] and separate[33] eicosanoids have been described.

Metabolic labeling of endothelial cells using radioactive arachidonate is a useful technique for studies of incorporation of arachidonate, its stimulated release from phospholipids, and the spectrum of oxygenated metabolites. It is not a quantitatively accurate method since the tracer does not come to isotopic equilibrium with all of the endogenous arachidonate unless long labeling periods (several doublings) are used, and these protocols present confounding variables such as metabolism of the arachidonate by additional pathways (e.g., elongation).

Labeling of Endothelial Cells with [^{3}H]Arachidonate

1. Dry an amount of [^{3}H]arachidonate (several commercial sources) sufficient to provide 0.5 μCi per 10^6 cells under nitrogen in a polypropylene tube. Add serum-free M199 (or other standard medium) containing 1 mg/ml of fatty acid-free bovine serum albumin (FAF-BSA) and sonicate for 5 min in a bath sonicator. Pass the solution through a 0.2 μm filter.

2. Remove the culture medium and wash the monolayer once with a sterile balanced salt solution (e.g., HBSS). Replace the wash buffer with the labeling medium and return the cells to the incubator (37°; 5% CO_2).

3. Sufficient label will be incorporated within 2 hr for most purposes. At the end of the labeling period, remove the medium and wash the monolayer twice with label-free medium (containing FAF-BSA). Replace the standard culture medium and return the cells to the incubator.

Comments

1. More label can be incorporated by increasing the amount of tracer or with longer incubations; however, essentially all of the label is incorporated within 6 to 8 hr.

2. Radiolabeled arachidonate undergoes characteristic shifts between phospholipid classes in the 12–24 hr after labeling.[34,35] Label shifts out of phosphatidylcholine into the diacyl- and plasmalogen species of phospha-

[32] W. S. Powell, this series, Vol. 86, pp. 467–476.

[33] P. Borgeat, S. Picard, P. Vallerand, S. Bourgoin, A. Odeimat, P. Sirois, and P. E. Poubelle, this volume [12].

[34] H. E. Wey, J. A. Jakubowski, and D. Deykin, *Biochim. Biophys. Acta* **878,** 380 (1986).

[35] H. Takayama, M. H. Kroll, M. A. Gimbrone, and A. I. Schafer, *Biochem. J.* **258,** 427 (1989).

tidylethanolamine. Such a shift may involve up to 20% of incorporated arachidonate and must be considered in studies performed in the immediate postlabeling period.

3. The radiolabeled arachidonate should be of high specific radioactivity so that labeling does not substantially alter the total amount of arachidonate in the cells.

4. The purity of the arachidonate should be monitored regularly (e.g., every few weeks), and it should be repurified when more than a few percentage impurities are present. Store under an inert atmosphere.

5. If the goal is to study the incorporation, distribution, and release of arachidonate, the complex lipids can be recovered by Bligh–Dyer[20] extraction (it would not need the carrier PAF or sodium acetate described above), and then separated by TLC or HPLC.[6] If secreted metabolites are the focus, remove the supernatant and extract[32] and analyze as described.[33]

Factors Influencing Production of Eicosanoids and PAF by Endothelial Cells *ex Vivo*

The species of origin influences receptor-mediated responses (Table II). For example, thrombin is a potent agonist for arachidonate metabolism and PAF synthesis in human, but not bovine, EC.[11,13–15,36,37] Conversely, bradykinin and ATP are potent agonists in bovine endothelium, but are weak in the human.[11,14]

Culture conditions affect PGI_2 synthesis; significant variables include the concentration and types of fatty acids and lipoproteins (HDL, LDL) in the serum, other serum factors, the presence of ECGF and heparin, the duration of time that the cells have been in culture, and whether the EC are in the replication phase or are confluent (see Chapter 24 in Ref. 1 for a discussion of some of these variables). In addition, it was recently reported that the number of cumulative population doublings affects the synthesis of PGI_2.[38] These factors indicate that the density at which the cells are plated, the degree of confluence of the monolayers at the time of study, the schedule of changes of the medium, and the number of times the culture has been passed will affect eicosanoid synthesis. The influence of culture conditions on the synthesis of PAF by activated EC has not been studied as thoroughly. Stimulated production of PAF in response to receptor-

[36] S. L. Hong, *Thromb. Res.* **18,** 787 (1980).

[37] C. Goldsmith, C. T. Jafvert, P. Lollar, W. G. Owen, and J. C. Hoak, *Lab. Invest.* **45,** 191 (1981).

[38] C. M. Ingerman-Wojenski, M. J. Silver, S. N. Mueller, and E. M. Levine, *Prostaglandins* **36,** 127 (1988).

TABLE II
RECEPTOR-MEDIATED AGONISTS THAT CAUSE PRODUCTION OF PAF AND PGI_2 IN ENDOTHELIAL CELLS[a]

Human umbilical vein EC (HUVEC)	Bovine macrovascular EC
Thrombin	Bradykinin
LTC_4	ATP
LTD_4	Angiotensin II
Histamine	Histamine
Bradykinin	
ATP	

[a] The agonists are listed in order of relative potency.

mediated agonists persists at least through the third passage in human EC[6] but declines by the tenth passage.[39] These various effects of culture conditions on EC metabolism are largely unexplained at a mechanistic level. However, the important practical point to be drawn from them is that until the mechanisms are well understood it is essential to rigorously standardize the procedures for establishing and maintaining the cultures and, particularly, the conditions under which the EC are studied. For most of our experiments we have chosen to study tightly confluent EC with postconfluent morphology (as defined previously) because of the phenotypic similarities of these cells and *in situ* EC[1–3,7,11,40] and because of evidence that the cells are relatively homogeneous in terms of biochemical activities under these conditions.[7,40] An added advantage is that the number of cells in tightly confluent monolayers is relatively constant from dish to dish, resulting in close agreement in replicate determinations and usually eliminating the need to normalize PAF or PGI_2 concentrations. If preconfluent, or "wounded" EC (monlayers that have been altered by scraping or another means that removes some of the EC[40]) are studied, normalization to cell number, protein content, or lipid phosphate content[41] will be required. Normalization to DNA levels may not be useful under the latter conditions since there may be a severalfold difference in DNA synthetic rate in different areas of a "wounded" EC monolayer.[40]

[39] F. Breviario, F. Bertocchi, E. Dejana, and F. Bussolino, *J. Immunol.* **141,** 3391 (1988).
[40] M. A. Gimbrone, R. S. Cotran, and J. Folkman, *J. Cell Biol.* **60,** 673 (1974).
[41] R. E. Whatley *et al.*, manuscript in preparation.

Acknowledgments

This work was supported by the Nora Eccles Treadwell Foundation and by Grants from the National Institutes of Health (HL 34127 and HL 35828) and the American Heart Association (Grant-in-Aid 871147). Drs. Prescott and Zimmerman are Established Investigators of the American Heart Association. Dr. Whatley was supported by a National Research Service Award (HL 07529). We thank Dan Fennell for photographs of EC monolayers and Julie Wald for preparing the manuscript.

[58] Use of Cultured Cells to Study Arachidonic Acid Metabolism

By ROBERT R. GORMAN, MICHAEL J. BIENKOWSKI, and CHRISTOPHER W. BENJAMIN

Introduction

As is often the case in modern biology, there is no one cell system that is ideal for the study of a particular metabolic pathway and arachidonic acid metabolism is no exception. For example, the human platelet is an excellent cell in which to study thromboxane A_2 (TxA_2) synthesis, and the endothelial cell is useful for studies of prostacyclin (PGI_2) synthesis, but, neither cell type makes both PGI_2 and TxA_2. However, there are enough cultured cells available to study the synthesis of a majority of the various eicosanoids. In this chapter we describe the culture of the human lymphoma cell line U937 and the human lung fibroblast WI-38 as useful cell lines in which to study thromboxane A_2 synthesis, human umbilical vein endothelial cells to investigate PGI_2 biosynthesis, and the murine fibroblast line NIH-3T3 to study PGE_2 synthesis and the influence of cellular transfection and transformation on arachidonate metabolism. These same cell culture systems and techniques can also be used to study the regulation of the various enzymes of the arachidonate cascade as well as the phospholipase(s) that regulate arachidonate concentration in cells.

Thromboxane A_2 Synthesis by U937 Cells

U937 cells were originally isolated from the pleural fluid of a patient with diffuse histiocytic lymphoma.[1] These cells exist as immature myelomonocytic cells that can be terminally differentiated by treatment with

[1] C. Sundstrom and K. Nilsson, *Int. J. Cancer* **17,** 565 (1976).

phorbol 12-myristate 13-acetate (PMA) into cells with the phenotypical characteristics of monocytes/macrophages.[2] These differentiated cells can produce TxA_2 as the major product in response to a variety of agonists,[3] and are an excellent system in which to study the influence of glucocorticoids on eicosanoid biosynthesis.

Reagents. U937 cells (American Type Culture Collection, Rockville, MD); RPMI-1640, Fungi Bact, and fetal calf serum (FCS) (Irvine Scientific, Santa Ana, CA); phorbol 12-myristate 13-acetate (PMA), Ca^{2+} ionophore A23187, lipopolysaccharide (*E. coli* serotype 055 : B5), zymosan A, and dexamethasone (Sigma Chemical, St. Louis, MO); arachidonic acid, NuCheck Preps (Elysian, MN); EIA Kit (Cayman Chemical Co., Ann Arbor, MI).

Cell Culture Procedure. U937 cells are maintained in suspension culture in RPMI-1640 supplemented with 1× FungiBact, 10% heat-inactivated FCS. For differentiation, cells are plated at 2.0×10^5 cells/ml in medium containing 100 n*M* PMA and allowed to attach for 48 hr. The cells are then fed with PMA-free medium and cultured overnight prior to use.

Measurement of Synthesis

1. Differentiated U937 cell monolayers are washed twice with serum-free RPMI-1640 followed by addition of fresh medium containing 1% FCS and either *E. coli* LPS (10 μg/ml), heat-activated zymosan A (500 μg/ml), A23187 (10 μg/ml), or arachidonic acid (10 μM) both initially dissolved in dimethyl sulfoxide.
2. After 4 hr (LPS or zymosan) or 15 min (A23187 or arachidonic acid), the medium is removed and TxB_2 levels measured using a TxB_2-specific enzyme immunoassay (EIA) (see [3], this volume).
3. Steroid treatments (i.e., dexamethasone) are performed by adding the appropriate steroid to RPMI-1640 medium containing 1% FCS and pretreating the cells for varying lengths of time. The medium is removed, replaced with steroid-free medium containing the appropriate agonist, and TxA_2 synthesis quantified as described above.

Table I shows a typical result following exposure of differentiated U937 cells to LPS, zymosan A, arachidonic acid, or A23187: 10 μg/ml LPS stimulates 2.7 ng $TxA_2/10^6$ cells, 500 μg/ml zymosan A stimulates 1.5 ng $TxA_2/10^6$ cells in 4 hr, 1 μg/ml A23187 stimulates 1.0 ng $TxA_2/10^6$ cells, and 10 μM arachidonate stimulates 0.8 ng $TxA_2/10^6$ cells in 15 min.

[2] P. Harris and R. Peter, *J. Leukocyte Biol.* **37,** 407 (1985).

[3] M. J. Bienkowski, M. A. Petro, and L. J. Robinson, *J. Biol. Chem.* **264,** 6536 (1989).

TABLE I
THROMBOXANE SYNTHESIS IN U937 CELLS[a]

Agonist	ng $TxA_2/1 \times 10^6$ cells
None, basal	0.54 ± 0.04
LPS, 10 μg/ml	2.7 ± 0.15
Zymosan A, 500 μg/ml	1.5 ± 0.09
Arachidonic acid 10 μ*M*	0.8 ± 0.06
A23187, 1 μg/ml	1.0 ± 0.10

[a] U937 cells are differentiated and cultured as described. TxA_2 synthesis is initiated by washing the monolayer with serum-free RPMI-1640 and replacing the culture medium with fresh medium containing 10 μg/ml LPS, 500 μg/ml zymosan A, 10 μ*M* arachidonic acid, 1 μg/ml A23187. After 4 hr (LPS or zymosan) or 15 min (A23187, arachidonate) at 37°, the culture medium is removed and the TxB_2 content determined by EIA.

Figure 1 shows a similar experiment when cells are pretreated with dexamethasone (10^{-5}–10^{-11} *M*) for 2 hr prior to stimulation with LPS (10 μg/ml) or zymosan A (500 μg/ml) for 4 hr, or arachidonic acid (10 μ*M*) or A23187 (1 μg/ml) for 15 min. Dexamethasone has no effect on A23187 or

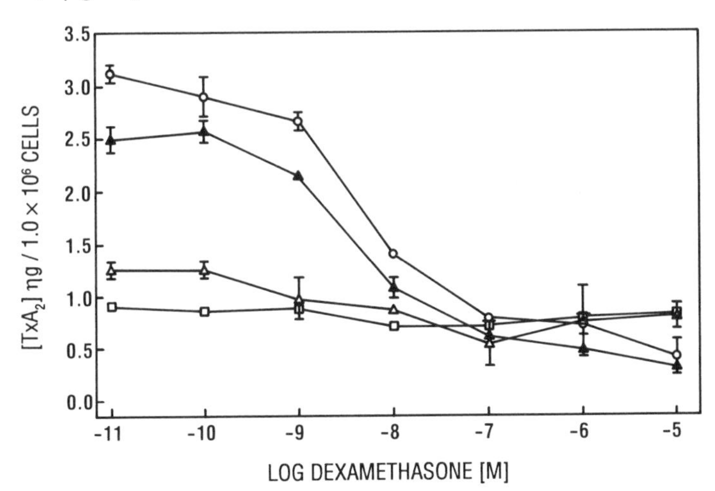

FIG. 1. Dexamethasone inhibition of TxA_2 synthesis initiated by various agonists. Cells are pretreated with various concentrations of dexamethasone for 2 hr. They are then stimulated with either 10 μg/ml LPS (○—○) or 500 μg/ml zymosan (▲—▲) for 4 hr or with 10 μ*M* arachidonic acid (□—□) or 1 μg/ml A23187 (△—△) for 15 min in steroid-free medium, and the TxB_2 content of the medium determined by EIA. Results are expressed as mean ± SE.

arachidonate stimulation of TxA_2, however, it dose-dependently inhibits both LPS- and zymosan A-stimulated TxA_2 release.

Thromboxane A_2 Synthesis by WI-38 Cells

WI-38 cells are unique because they are the only fibroblast cell line that has been shown to produce thromboxane A_2.[4] The following outlines step by step procedures for measuring thromboxane synthesis in these cells.

Reagents. Human lung fibroblast WI-38 (Flow Laboratories, Rockville, MD); 75 mm tissue culture flasks (Falcon Plastics, Oxnard, CA); 35-mm tissue culture wells (Costar Plastics, Cambridge, MA); Eagle's minimum essential medium (Earle's base), FCS (inactivated 56°, 30 min), and EDTA-trypsin (Irvine Scientific, Santa Ana, CA); precoated silica gel GF thin-layer plates (Analtech, Inc., Newark, DE); 9,11-azoprosta-5,13-dienoic acid (azo analog I) and prostaglandin standards (Upjohn Co., Kalamazoo, MI); arachidonic acid (NuCheck Preps., Elysian, MN). [1-^{14}C]PGH_2 is synthesized according to Gorman *et al.*[5]

Cell Culture Procedure. Cells are split 1 : 5 using EDTA–trypsin, into culture bottles (75 cm^2) and grown in Eagle's minimum essential medium (Earle's base) supplemented with 10% FCS. This medium with serum will be referred to as MEM-10. Confluent cells are fed with 25 ml of MEM-10 every 5 days, and passaged every 13–14 days. For individual biochemical experiments, the cells are seeded into 35-mm Costar wells, and grown under a humidified atmosphere of 95% air–5% CO_2 at 37° until confluent (3–4 days). Final density is approximately 1×10^6 cells/35-mm well.

Experimental Procedures

1. Confluent WI-38 cells in 35-mm wells are washed twice with ice-cold MEM and incubated for 15 min at 37° with fresh MEM.
2. After 15 min, the cells are exposed to 0.1–1.0 μM [1-^{14}C]PGH_2 (dissolved in MEM immediately before use) or 0.1–10 μM [1-^{14}C]arachidonic acid (initially dissolved in ethanol) and incubated for 5–30 min at 37°.
3. The reactions are stopped by acidification with 100 μl of 1 N HCl, followed by three extractions with ethyl acetate. All tubes then receive 1 μg of unlabeled TxB_2, $PGF_{2\alpha}$, PGE_2, and PGD_2 to assist recoveries.
4. The ethyl acetate extracts are pooled, taken to dryness, and resuspended in 100 μl of benzene.

[4] N. K. Hopkins, F. F. Sun, and R. R. Gorman, *Biochem. Biophys. Res. Commun.* **85,** 827 (1978).

[5] R. R. Gorman, F. F. Sun, O. V. Miller, and R. A. Johnson, *Prostaglandins* **13,** 1043 (1977).

5. The resuspended samples are then spotted onto a silica gel GF plate, and chromatographed in a solvent system of 1% acetic acid, 99% ethyl acetate.

6. The appropriate regions are then scraped from the thin-layer plate and quantitated by liquid scintillation spectroscopy.

Table II shows a typical experiment in which 0.1 μM [1-^{14}C]PGH_2 is added to WI-38 cells in culture, and TxB_2 production quantitated by thin-layer chromatography. The data show that WI-38 cells produce nearly equivalent amounts of TxB_2 and PGE_2 from PGH_2, with lesser amounts of $PGF_{2\alpha}$ and PGD_2 being formed. There is also a considerable amount of HHT formed from nonenzymatic breakdown of the endoperoxide in aqueous medium. If radioactive PGH_2 is not readily available, either [1-^{14}C]- or [^{3}H]arachidonic acid can be substituted for [1-^{14}C]PGH_2. Alternatively, unlabeled PGH_2 or arachidonic acid can be used, and thromboxane quantitated by radioimmunoassay or EIA.

Human Umbilical Vein Endothelial Cells

Human umbilical vein endothelial cells are an ideal system in which to study PGI_2 (prostacyclin) biosynthesis. The PGI_2 synthase can be studied directly by adding exogenous cold or [1-^{14}C]PGH_2, and the cyclooxygenase and PGI_2 synthase can be studied simultaneously by adding exogenous unlabeled or radiolabeled arachidonic acid, or the entire activation sequence, including phospholipase activity, can be studied by stimulating the release of endogenous arachidonic acid by exposing the cells to thrombin. The procedures for using exogenous PGH_2 or arachidonate are essentially identical to those outlined for WI-38 cells, except that the thin-layer chromatography system is the organic phase of ethyl acetate–acetic acid–isooctane–water (110 : 20 : 50 : 100). The following outlines the step-by-step procedures for using thrombin to initiate PGI_2 synthesis in endothelial cells.

TABLE II
[1-^{14}C]PGH_2 METABOLISM IN WI-38 CELLS[a]

Additions	cpm/radioactive zone				
	PGF_{2a}	PGE_2	TxB_2	PGD_2	HHT
[1-^{14}C]PGH_2, 0.6 μM	852 ± 76	4126 ± 312	4419 ± 511	977 ± 35	7154 ± 762

[a] WI-38 cells are incubated with 0.60 μM PGH_2 for 15 min at 22°. Zones of radioactivity that corresponded to standard prostaglandins are scrapped and quantitated by liquid scintillation counting. Data reported as mean ± SEM of triplicate determinations.

Reagents. Human umbilical cords; T-75 plastic tissue culture flasks (Costar Plastics, Cambridge, MA); bovine thrombin, collagenase, and trypsin (Sigma Chemical, St. Louis, MO); medium 199, Earle's base, HEPES buffer, and L-glutamine (Microbiological Associates, Walkersville, MD); sodium penicillin G (Upjohn Co., Kalamazoo, MI); streptomycin sulfate (Eli Lilly, Indianapolis, IN). Human serum is prepared from fresh whole blood. Cord buffer consists of 137 m*M* NaCl, 4 m*M* KCl, 10 m*M* HEPES, pH 7.5, and 11 m*M* glucose.

Cell Culture Procedure. Endothelial cells are derived from human umbilical cord veins as described by Jaffe *et al.*[6] Cords are stored in sterile containers with 20 ml of cord buffer. A disposable syringe is attached to a blunt 18-gauge needle and the vein flushed with 50 ml of cord buffer to remove any blood. The vein is then perfused with 10 ml of 0.1% collagenase in cord buffer, and incubated at 37° in a bath of cord buffer for 10 min. The collagenase–cell mixture is flushed from the vein with 30 ml of cord buffer into a plastic centrifuge tube that contains 10 ml of medium 199 and centrifuged at 200 *g* for 5 min at 22°. The cell pellet is resuspended in 10 ml of fresh medium 199 + 20% serum (v/v) and added to a T-75 flask. Cells are exposed to an atmosphere of 95% air–5% CO_2 and are fed twice a week until subculturing. Confluency is usually reached in 4–5 days.

For subculturing, the medium is removed and the cells rinsed once with cord buffer. A 5-min incubation with a 1 : 1 mixture of 0.02% EDTA, 0.2% collagenase followed by centrifugation is used to harvest the cells. Cells are grown in either T-150 culture flasks or distributed into 35-mm wells. Suspended cells are counted using a Coulter Model ZB-I cell counter.

Experimental Procedures

1. Endothelial cells are grown to confluency in 35-mm dishes, and washed twice with ice-cold medium 199–HEPES (50 m*M*, pH 7.5) and allowed to equilibrate with an additional 1.0 ml of medium 199–HEPES for 10 min at 37°. Cells grown in this manner are characterized by (a) presence of Weibel-Palade bodies as observed by electron microscopy[7]; (b) the ability of the medium from cultured cells to support ristocetin-induced agglutination of washed human platelets[8]; and (c) immunofluorescence studies of Factor VIII antigens.[9]

[6] E. A. Jaffe, R. L. Nachman, C. G. Becker, and C. R. Minick, *J. Clin. Invest.* **52,** 2745 (1973).

[7] E. R. Weibel and G. E. Palade, *J. Cell Biol.* **23,** 101 (1964).

[8] E. A. Jaffe, L. W. Hoyer, and R. L. Nachman, *Proc. Natl. Acad. Sci. U.S.A.* **71,** 1906 (1974).

[9] E. A. Jaffe, L. W. Hoyer, and R. L. Nachman, *J. Clin. Invest.* **52,** 2757 (1973).

2. After the equilibration period, the cells are challenged with 0.1–3.0 U/ml thrombin, and allowed to incubate for the appropriate time interval at 37°.

3. The media is then immediately withdrawn and quickly frozen using liquid nitrogen.

4. Aliquots of the media are subsequently assayed for 6-keto-$PGF_{1\alpha}$ by radioimmunoassay or enzyme-linked immunoassay (EIA) (see [3], this volume).

Table III shows a typical profile of PGI_2 (measured as 6-keto-$PGF_{1\alpha}$) synthesis in thrombin-stimulated endothelial cells. Synthesis of PGI_2 increases 10-fold with 2 U/ml thrombin and this synthesis is completely blocked by cyclooxygenase inhibitors such as aspirin or indomethacin. In addition, the addition of exogenous PGH_2 or arachidonic acid can markedly stimulate PGI_2 production.

NIH-3T3 Murine Fibroblasts

NIH-3T3 cells are very useful because they produce prostaglandin E_2 in response to growth factors, and can be easily transfected and transformed by a number of oncogenes. These characteristics make it possible to study the influence of various transfected gene products or drugs on growth factor-regulated eicosanoid metabolism.

Reagents. NIH-3T3 cells (American Type Culture Collection (Rockville, MD); Dulbecco's modified Eagle's medium (DMEM), penicillin–streptomycin, glutamine, sodium pyruvate, EDTA, fetal calf serum (Irvine Scientific, Santa Ana, CA); sterile disposable plasticware (Costar Plastics, Cambridge, MA); platelet-derived growth factor (purified according to

TABLE III
THROMBIN-STIMULATED PGI_2 BIOSYNTHESIS[a]

Treatment	6-keto-$PGF_{1\alpha}$/10^6 cells (ng)
None, basal	1.7 ± 0.2
Thrombin, 2 U	18.8 ± 1.0
Thrombin + 1 m*M* aspirin	1.9 ± 0.3
Thrombin + 10 μ*M* indomethacin	1.6 ± 0.3

[a] Confluent monolayers of endothelial cells in 35-mm wells are washed once with 2.0 ml of warm MEM containing 25 m*M* HEPES. The cells are allowed to equilibrate with 1.0 ml of MEM–HEPES with or without the cyclooxygenase inhibitors for 10 min at 37° and then stimulated with 2.0 U/ml thrombin. Data are reported as mean ± SEM of triplicate samples.

Raines and Ross[10]) or recombinant c-*sis* BB-PDGF (Amgen Biologicals, Thousand Oaks, CA); plasma-derived serum (PDS) from freshly drawn human blood (prepared as described by Habenicht *et al.*[11]); PGE_2 EIA Kit (Cayman Chemical, Ann Arbor, MI).

Cell Culture Procedure. NIH-3T3 fibroblasts are maintained in Dulbecco's modified Eagle's medium (DMEM) containing 10% FCS (EC-10 media). The cells are incubated at 37° in 5% CO_2/95% air, and 90% relative humidity. Semiconfluent cultures are passed twice weekly at a 1/10 dilution, by washing the monolayer with EDTA, followed by trypsinization. The suspended cells are centrifuged at 300 *g* for 5 min at 22°, resuspended in EC-10, and plated in 10-cm^2 tissue culture dishes.

Experimental Procedures. NIH-3T3 cells can be stimulated to produce prostaglandin E_2 by the addition of 5–20 ng/ml of either purified human platelet-derived growth factor (PDGF) or the recombinant β-chain homodimer of PDGF, c-*sis* PDGF.[12] The addition of nonexternal receptor agonists, such as phorbol myristate acetate and/or the calcium ionophore, A23187, will also elicit a response from these cells.[13]

1. Cells are grown in EC-10 media in 35-mm culture dishes, or 35-mm "6-pack" culture dishes, to about 75% confluency. The media is replaced with DMEM containing either 1.25% PDS, or 0.2% FCS to remove endogenous growth factor and thus maximize the response 18 hr prior to an experiment.

2. To begin the experiment, the media is replaced with DMEM with no serum, and 5–20 ng/ml of PDGF (or c-*sis* PDGF) or vehicle (0.1 *M* acetic acid) is added.

3. Media are collected at various times and frozen at −70° until analysis of PGE_2 by radioimmunoassay or EIA (see [3], this volume).

4. Exogenous arachidonate metabolism may be studied by adding 0.33–33 μM arachidonate to cultures treated as above, without addition of PDGF. Again, media is collected and frozen at −70°, and PGE_2 levels determined by RIA or EIA.

Figure 2 shows a typical experiment in which purified PDGF is added to four stably transfected NIH-3T3 cell lines and PGE_2 measured by EIA over time. The four cell lines are: (1) control, cotransfected with calf thymus DNA; (2) v-*src*, transformed by pSRA-2, a permuted clone of Rous

[10] E. W. Raines, and R. Ross, this series, Vol. 109, p. 733.

[11] A. J. R. Habenicht, J. A. Glomset, and R. Ross, *J. Biol. Chem.* **255,** 5134 (1980).

[12] C. W. Benjamin, W. G. Tarpley, and R. R. Gorman, *Proc. Natl. Acad. Sci. U.S.A.* **84,** 546 (1987).

[13] C. W. Benjamin, W. G. Tarpley, and R. R. Gorman, *Biochem. Biophys. Res. Commun.* **145,** 1254 (1987).

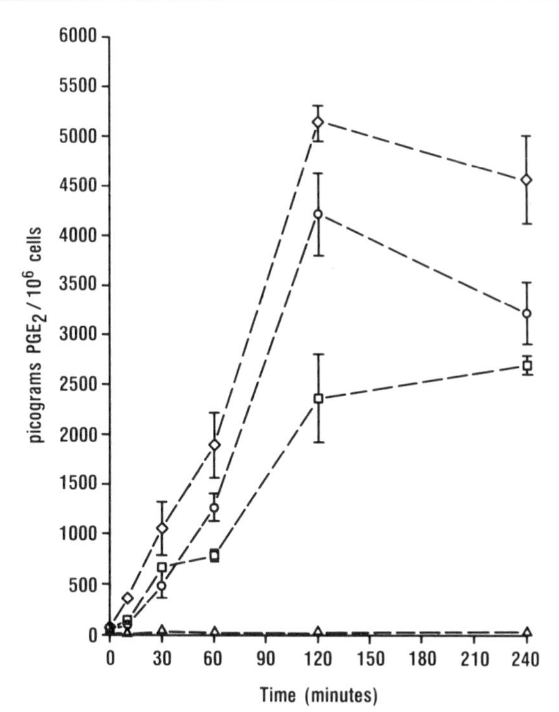

FIG. 2. PDGF-stimulated PGE_2 release from control, EJ-*ras*-, v-*src*-, and c-*ras*-transfected cells. Cells (0.6×10^6 cells per 35-mm well) are grown for 24 hr in 1.25% PDS and then stimulated by PDGF at 20 ng/ml. PGE_2 levels are quantitated by EIA at the indicated times. Data are presented as mean ± SEM of triplicate determinations. ○—○, Control + PDGF; ◇—◇, v-*src* transfected + PDGF; □—□, c-*ras* transfected + PDGF; △—△, EJ-*ras* transfected + PDGF.

sarcoma virus DNA; (3) EJ-*ras,* transformed by the EJ human bladder carcinoma oncogene; and (4) *c*-ras, a cell line that overexpresses the normal human c-Harvey-*ras* allele, but is not transformed. PDGF markedly stimulates PGE_2 release in control cells, with maximal stimulation at 2 hr. v-*src*-Transformed cells show a similar pattern of PDGF-stimulated PGE_2 synthesis, but EJ-*ras*-transformed cells show no PDGF-stimulated E_2 synthesis. Finally, cells overexpressing c-*ras* display slightly reduced PDGF-stimulated PGE_2 levels. This type of approach led to the discovery that the *ras* gene product inhibits phospholipase C activity in NIH-3T3 cells.[14]

[14] C. W. Benjamin, J. A. Connor, W. G. Tarpley, and R. R. Gorman, *Proc. Natl. Acad. Sci. U.S.A.* **85,** 4345 (1988).

[59] Eicosanoid Biochemistry in Cultured Glomerular Mesangial Cells

By MICHAEL S. SIMONSON and MICHAEL J. DUNN

Eicosanoids play an important role in the autocrine and paracrine regulation of glomerular hemodynamics.[1-3] Vasodilatory prostaglandins (PGE_2, PGI_2) function as negative feedback signals to attenuate vasoconstruction in the glomerular arterioles and mesangium. In contrast, $PGF_{2\alpha}$, TxA_2, and the peptidoleukotrienes (LTC_4, LTD_4) are potent vasoconstrictors in the renal vasculature, and these eicosanoids probably mediate inflammatory reactions in glomerular disease.[4] Glomerular mesangial cells play a central role in the kidney's synthesis of and response to eicosanoids. Techniques are available for isolating and culturing mesangial cells *in vitro*, and cultured mesangial cells provide a useful model for investigating eicosanoid biochemistry and cell biology (see Refs. 1–9 for review). This chapter focuses on (*i*) the use of mesangial cells in eicosanoid research; and (*ii*) basic techniques for culturing, characterizing, assessing viability, and measuring contraction of mesangial cells.

Structure and Function of Glomerular Mesangium

In addition to endothelial and epithelial cells, the glomerulus contains a perivascular, smooth musclelike mesangial cell.[10] The mesangial cell and its surrounding matrix comprise the central portion of the glomerular tuft

[1] M. J. Dunn, *in* "Contemporary Nephrology" (S. Klahr and S. Massry, eds.), pp. 133. Plenum, New York, 1987.
[2] B. Brenner, L. D. Dworkin, and I. Ichikawa, *in* "The Kidney" (B. M. Brenner and F. C. Rector, eds.), pp. 124. Saunders, Philadelphia, PA, 1986.
[3] M. J. Dunn, *in* "Renal Endocrinology" (M. J. Dunn, ed.), p. 1. Williams and Wilkins, Baltimore, 1983.
[4] L. Scharschmidt, M. S. Simonson, and M. J. Dunn, *Am. J. Med.* **81**(S2B), 30 (1986).
[5] P. Mene', M. S. Simonson, and M. J. Dunn, *Physiol. Rev.* **69,** 1347 (1989).
[6] J. I. Kreisberg and A. Hassid, *Miner. Electrolyte Metab.* **12,** 25 (1986).
[7] D. Schlondorff, *FASEB J.* **1,** 272 (1987).
[8] G. E. Striker and L. J. Striker, *Lab. Invest.* **53,** 122 (1985).
[9] P. Mene', M. S. Simonson, and M. J. Dunn, *Am. J. Physiol.* **256,** F375 (1989).
[10] H. Latta, *in* "Handbook of Physiology" (J. Orloff, R. W. Berliner, and S. R. Geiger, eds.). pp. 1. Williams and Wilkins, Baltimore, MD, 1973.

between the capillary loops. Mesangial cells serve several specialized functions including synthesis and assembly of the mesangial matrix, endocytosis and processing of plasma macromolecules, control of glomerular hemodynamics via mesangial contraction, and mesangial synthesis and release of vasoactive autacoids, especially prostaglandins and thromboxane.[1,4,5,7,8] Current evidence suggests that eicosanoids have diverse effects in the mesangium, but the best-studied role of mesangial eicosanoids is the regulation of glomerular hemodynamics.[1,3,4]

Cultured Mesangial Cells as Model Systems for Eicosanoid Research

Selected examples illustrating the use of mesangial cells in eicosanoid research are highlighted below. A detailed account is available in several reviews.[1,4–9] Glomerular epithelial cells synthesize smaller amounts of eicosanoids[1,2,6,8] and are not considered here.

Eicosanoid Biosynthesis

Rat mesangial cells *in vitro* express abundant cyclooxygenase activity, yielding PGE_2 and $PGF_{2\alpha}$, and lesser amounts of poorly characterized lipoxygenase and cytochrome *P*-450 enzymatic products.[5,9] Of cyclooxygenase products, rat cells in culture synthesize mainly $PGE_2 > PGF_{2\alpha} \gg TxA_2 \gg PGI_2$. By contrast, human mesangial cells in culture display modest cyclooxygenase activity with PGI_2 or PGE_2 representing the major metabolite.[11,12] Mesangial prostaglandin synthesis is enhanced by both physiological and pathophysiological stimuli. Most agents that activate the phosphoinositide cascade, which, in turn, elevates cytosolic-free Ca^{2+} concentration and activates protein kinase C, also stimulate prostaglandin synthesis. These agents include angiotensin II (ANG II), bradykinin, arginine vasopressin, serotonin, platelet-activating factor, TxA_2 mimetics, platelet-derived growth factor, and endothelin.[5,7,9] Peptidoleukotrienes are a notable exception as they are compounds that evoke the phosphoinositide cascade but fail to increase prostaglandin synthesis in human or rat cells.[12,13] Other stimuli of prostaglandin synthesis include Ca^{2+} ionophores, melittin, *Escherichia coli* lipopolysaccharide, and endocytosis of opsinized zymosan as well as immune complexes.[5,7,9] In addition, the

[11] N. Ardaillou, J. Hagege, M. P. Nivez, R. Ardaillou, and D. Schlondorf, *Am. J. Physiol.* **248,** F240 (1985).

[12] M. S. Simonson, P. Mené, G. R. Dubyak, and M. J. Dunn, *Am. J. Physiol.* **255,** C771 (1988).

[13] R. Barnett, P. Goldwasser, L. A. Scharschmidt, and D. Schlondorff, *Am. J. Physiol.* **250,** F838 (1986).

cytokines interleukin 1 and tumor necrosis factor cause a delayed (≥6 hr) increase in prostaglandin synthesis that requires RNA and protein synthesis.[5,14] Mesangial cells are also useful for studying the inhibition of prostaglandin synthesis by nonsteroidal antiinflammatory drugs (NSAIDs) and cyclooxygenase inhibitors. For example, blockade of PGE_2 synthesis by NSAIDs greatly amplifies basal- and vasoconstrictor-induced mesangial cell contraction.[4,5,15] Eicosanoids attenuate mesangial cell contraction by a cAMP-dependent mechanism.[15]

Eicosanoid Receptors

Functional evidence suggests that mesangial cells contain receptors for PGE_2, $PGF_{2\alpha}$, TxA_2, and PGI_2. PGE_2, $PGF_{2\alpha}$, and TxA_2 activate phospholipase C, whereas PGE_2 and PGI_2 stimulate adenylate cyclase.[5,9] However, radioligand binding studies in mesangial cells have not been performed to directly evaluate prostaglandin receptors and their coupling to phospholipase C or adenylate cyclase. We recently demonstrated that human mesangial cells express a single class of saturable, specific binding sites for $[^3H]LTD_4$ (K_D = 12.0 nM, B_{max} = 987 fmol/mg protein) coupled to phospholipase C but not to adenylate cyclase or phospholipase A_2.[12] The regulation of eicosanoid receptor affinity and number in mesangial cells, as well as GTP-binding protein coupling, remains unknown.

Transmembrane Signaling

In mesangial cells, eicosanoids evoke several pathways of transmembrane signaling to produce diverse second messengers.[5,9] Eicosanoids activate the phosphoinositide and adenylate cyclase cascades, and electroneutral, amiloride-inhibitable Na^+/H^+ exchange. It remains unknown whether eicosanoids affect membrane potential or tyrosine kinase activity in the mesangium. Ion-exchange chromatography and HPLC techniques have been employed in mesangial cells to study radiolabeled phosphoinositide turnover.[16,17] Furthermore, mesangial cell monolayers on plastic coverslips are well-suited for fluorescent probes (i.e., fura-2, and BCECF) to measure receptor-triggered Ca^{2+} signaling and changes in intracellular pH.[12,17]

[14] L. Baud, J. Perez, G. Friedlander, and R. Ardaillou, *FEBS Lett.* **239,** 50 (1988).

[15] P. Mené and M. J. Dunn, *Circ. Res.* **62,** 916 (1988).

[16] J. V. Bonventre, K. Skorecki, J. I. Kreisberg, and J. Y. Cheung, *Am. J. Physiol.* **251,** F94 (1986).

[17] P. Mené, G. R. Dubyak, H. E. Abboud, A. Scarpa, and M. J. Dunn, *Am. J. Physiol.* **255,** F1059 (1988).

Mesangial Cellular Functions

Eicosanoids help control diverse mesangial cell functions including contraction,[3,5] mitogenesis,[5,9] and matrix accumulation.[18] Techniques to study cell contraction by computer-aided microscopy,[4,5] mitogenesis by [^{3}H]thymidine incorporation and measurements of cell number in 96-well plates,[19] and fibronectin accumulation by immunoprecipitation and immunocytochemical staining[20] have been published using cultured mesangial cells.

Protocols for Mesangial Cell Culture

Isolation of glomeruli by selective sieving is the first step in mesangial cell culture. Mesangial cell strains are then cultured from the outgrowth of explanted glomerular "cores" (Figs. 1 and 2) formed when isolated glomeruli are partially digested with bacterial collagenase.[5,8] The "cores" consist mostly of mesangium and capillary loops. Described below is a protocol for culturing mesangial cells from isolated rat glomeruli; with small changes this protocol applies to other species including dog,[21] mouse,[22] guinea pig,[23] and humans.[8] Isolation of glomeruli is performed on ice using standard aseptic techniques.

Primary Mesangial Cell Culture

Procedure. Kidneys are removed from four ether-anesthetized, exsanguinated Sprague-Dawley rats (150–200 g) and placed in Hanks' balanced salt solution (HBSS) containing 100 U/ml penicillin and 100 μg/ml streptomycin at 4°. The kidneys are decapsulated and halved sagitally, the medulla dissected out with iridectomy scissors, and the cortex minced with a sharp razor blade in a petri dish on ice. If the kidney is well-blanched, the medulla is easily identified by its dark red color in contrast to the light-colored cortex. The cortical paste is pressed through a 106 μm

[18] P. Klotman, L. Bruggeman, J. Hassell, E. Horigan, G. Martin, and Y. Yamada, *Kidney Int.* **35,** 294A (1989).

[19] M. S. Simonson, S. Wann, P. Mené, G. R. Dubyak, M. Kester, Y. Nakazato, J. R. Sedor, and M. J. Dunn, *J. Clin. Invest.* **83,** 708 (1988).

[20] M. S. Simonson, L. A. Culp, and M. J. Dunn, *Exp. Cell Res.* **184,** 484 (1989).

[21] S. R. Holdsworth, E. F. Glasgow, R. C. Atkins, and N. H. Thomson, *Nephron* **22,** 454 (1978).

[22] J. S. Hunt, A. E. Jackson, W. A. Day, and A. R. McGiven, *Br. J. Exp. Pathol.* **62,** 52 (1981).

[23] T. D. Oberley, J. E. Murphy-Ullrich, B. W. Steinert, and J. VicMuth, *Am. J. Pathol.* **104,** 181 (1981).

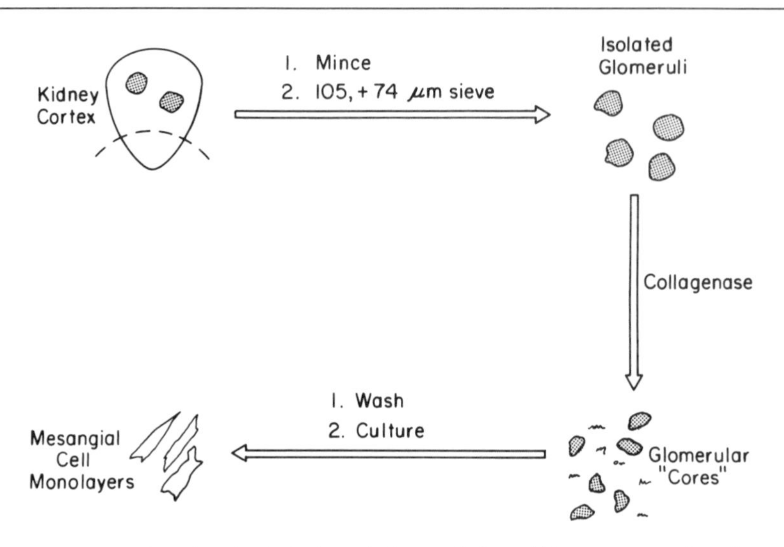

FIG. 1. Procedure for the isolation of glomeruli from kidney cortex, digestion with collagenase to form "glomerular cores," and the subsequent propagation of mesangial cells. Sieve sizes are specific for isolation of rat glomeruli.

metal sieve (Fisher), using the curved end of a stainless-steel spatula, and brei from the underside is suspended in 20 ml fresh HBSS. To remove Bowman's capsule, this suspension is passed twice through a 21-gauge needle, and aliquots are poured through a 74 μm nylon sieve (Nitex, Elmsford, NY) to capture the isolated glomeruli. Glomeruli are collected from the top surface (by aspirating into a pipette) and suspended in 10 ml ice-cold HBSS in a conical 50-ml polycarbonate centrifuge tube. The suspension is centrifuged at 80 g for 5 min, and the resulting pellet is suspended in 4 ml Ca^{2+}- and Mg^{2+}-free HBSS (no antibiotics) with 400–600 U/ml of bacterial collagenase (Worthington Type I). Incubate at 37° until most glomeruli are approximately one-third of their original size. Examine often, using a phase-contrast inverted microscope, to determine when the glomeruli are sufficiently digested. The incubation is usually complete by 30 min. Incubations greater than 45 min cause cellular damage. We avoid shaking the digestion mixture as this promotes clumping; instead, the mixture is mildly agitated every 5 min. Terminate the incubation by adding 10 ml of complete RPMI-1640 medium containing 17% fetal bovine serum (FBS), penicillin, and streptomycin as above, 5 μg/ml each of insulin and transferrin, and 5 ng/ml selenium (ITS Premix, Collaborative Research) and wash the glomerular explants by centrifuging three times at 200 g. Resuspend the final pellet in 6 ml fresh medium. Plate 1 ml of the glomerular core suspension into 6-well plates (9.6 cm^2) and incubate

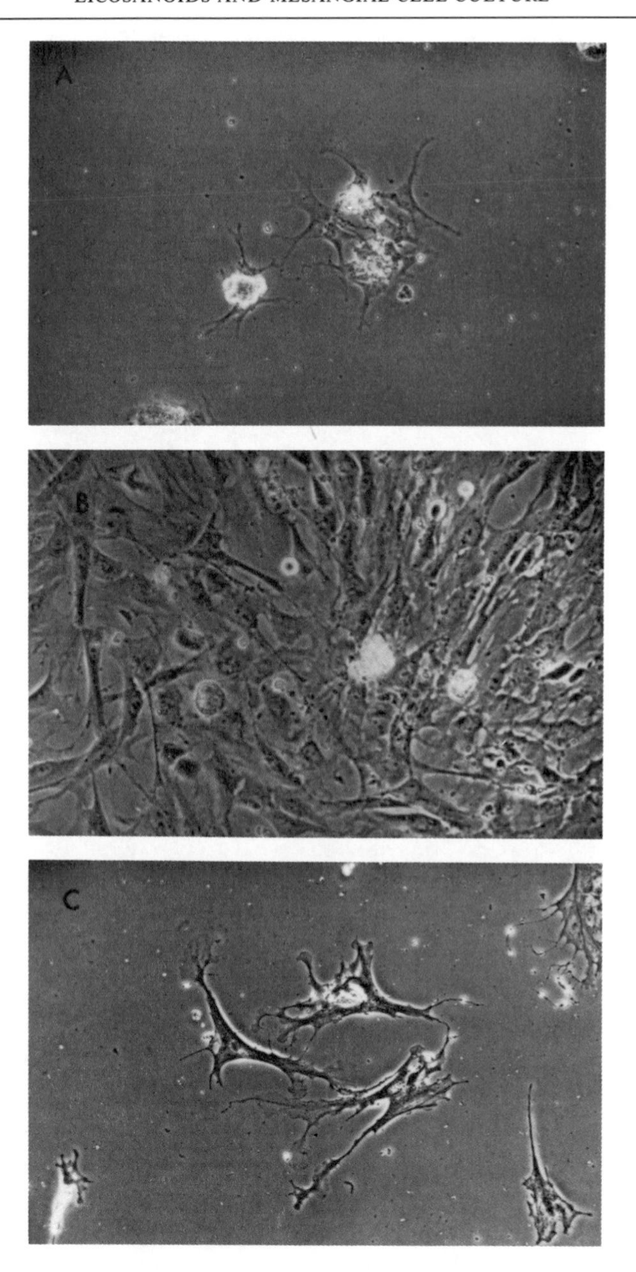

FIG. 2. Phase-contrast photomicrographs of (A) mesangial cells emerging from "glomerular cores" on day 2 of primary culture; (B) a confluent monolayer; and (C) mesangial cells stained with Giemsa to illustrate typical morphology with numerous cytoplasmic extensions.

at 37° in 5% CO_2 and 95% air for 2 days without disturbing the plates. On day 3 *gently add* 1 ml fresh medium (do not aspirate spent medium), and on day 5 aspirate and replace with 2 ml fresh medium. Replace medium every 2 days thereafter. By days 8–10 the primary cultures are usually ready for subculture (Fig. 2). Expose the monolayers to trypsin (0.25% in 0.1% EDTA) only long enough to release the cells but leave behind the more adherent glomerular explants. Subcultured cells are split 1 : 3 or 1 : 4 and plated into flasks (75 cm^2) for maintenance or into multiwell plates for experiments. Complete RPMI medium is replenished every 2 days.

Comments. Freshly plated glomerular "cores" must be allowed to attach undisturbed for 2 days (Fig. 2); unattached "cores" will not support mesangial outgrowth. Glomeruli from young rats (i.e., 150–200 g) yield better mesangial outgrowth, however, viable cultures can be established from older rats and from diseased human kidneys and experimental animal models.[8] Mesangial cells retain a stable phenotype for 15 to 20 passages in culture, but later passages appear to selectively adapt for culture and exhibit a loss of differentiated phenotype (especially ANG II-induced contraction), an increase in growth fraction, and alterations of other metabolic functions. Mesangial cells derived from clonal selection have been propagated for 50 to 60 passages.[5,6,8] We routinely freeze stock cultures in their fourth or fifth passage and thaw them as required, discarding cells beyond the 20th passage or earlier if the growth fraction jumps sharply. Mesangial cells are difficult to culture long-term in serum-free medium, however, we propagate rat cells up to 2 weeks in Dulbecco's minimum essential medium (DMEM) Hams' F12 medium (1 : 1), 10 μg/ml each of insulin and transferrin, 1 mg/ml fatty acid-free bovine serum albumin, 100 μg protein/ml of high-density lipoprotein (HDL, Sigma), and low concentrations of FBS (0.5–2.0%). Mesangial cells in serum-free culture grow best when plated on a matrix-coated substratum such as polystyrene adsorbed with fibronectin, collagen IV, or laminin (20 μg/ml for 1 hr).[20] For most studies of eicosanoid biosynthesis, mesangial cells are best studied at 80 to 90% confluence.

Characterization. The characterization of cultured mesangial cells is well-documented.[5–8,10] Light microscopy reveals large, stellate or spindle-shaped cells with many irregular cytoplasmic extensions (Fig. 2C). Electron microscopy reveals features typical of mesangial cells *in vivo*[10] including an oval, uniformly elongated nucleus with one or more nucleoli, well-developed endoplasmic reticulum and Golgi apparatus, and prominent bundles of peripheral microfilaments and dense bodies which constitute the contractile apparatus.[3,10] Immunocytochemical analysis is negative for cytokeratins, factor VIII, and glial fibrillary proteins, but is positive for fibronectin, thrombospondin, actin, myosin, desmin, vimentin, and Thy

1.1. The presence of Thy 1.1 appears to be a selective marker for mesangial cells; in thin sections of rat glomeruli, Thy 1.1 localizes specifically over mesangial cells[24] and antisera to Thy 1.1 lyse mesangial cells *in vivo*.[25] Mesangial cells also lack angiotensin-converting enzyme activity, nonspecific esterase stain, common leukocyte antigen, and contain <2% of Ia-positive cells in early passages and no Ia positivity after passage 5. Mesangial cells also exhibit receptor-dependent contraction in response to vasoconstrictor peptides.[3–7] Taken together, these findings rule out contamination by other glomerular or blood-borne cells. Protocols for analysis of mesangial antigens by immunofluorescene are given below.

Immunofluorescent Staining. Mesangial cells grown in four-chamber Lab-Tek culture slides (Nunc) are convenient for immunocytochemical studies. For identifying surface antigens, wash mesangial cells twice with 1.0 ml Dulbecco's PBS (DPBS) and fix with 3.7% formaldehyde in DPBS for 10 min at 23°. Wash five times with DPBS and stain with 10 μl/well of the appropriate monoclonal or polyclonal antibody diluted in 1.0% fatty acid-free BSA. Incubate for 30 min at 23° in an enclosed, humidified petri dish on a rotary shaker. Wash five times, for 2 min each wash, in DPBS and incubate with the appropriate isothiocyanate derivatives of fluorescein isothiocyanate (FITC)- or rhodamine-labeled second antibody for 30 min; wash as above. Mount under glycerol/DPBS (4 : 1) containing 1 mg/ml *p*-phenylenediamine to reduce fluorochrome quenching. Seal the edges with nail polish and examine under epifluorescence. For staining cytoplasmic antigens, mesangial cells are permeabilized after fixation in formaldehyde by incubating with 0.1% Triton X-100 for 3 min followed by two washes in DPBS. The use of $F(ab')_2$ fragments, especially for the second antibody, significantly reduces background staining in mesangial cells. Negative controls are run by omitting the first antibody.

Freezing Cells for Storage. Rat and human mesangial cells are stable for at least 1 year when stored frozen in liquid nitrogen. Mesangial cells in two 75-cm^2 flasks, passages 4–5, are grown to 80% confluence, trypsinized, centrifuged at 200 *g* for 3 min, and the pellet thoroughly resuspended in 2 ml complete RPMI-1640 medium containing 10% (v : v) dimethyl sulfoxide (DMSO) before aliquoting into two 1.8-ml Costar Biofreeze vials (Costar Plastics, Cambridge, MA). After 1 hr each at −20° and −70°, transfer to liquid nitrogen for long-term storage. To thaw frozen cells, transfer 1 vial to a 37° water bath and thaw rapidly (<2 min) while mixing. Dilute the 1-ml suspension with 12 ml complete RPMI-1640 medium and

[24] E. Yaoita, T. Kazama, K. Kawasaki, S. Miyazaki, T. Yamamoto, and I. Kihara, *Virchows Arch. Cell Pathol.* **49,** 285 (1985).

[25] W. Baghus, PhJ. Hoedemaeker, J. Rozing, and W. W. Bakker, *Lab. Invest* **55,** 680 (1986).

seed into one 75-cm^2 flask. Thawed mesangial cells grow poorly if sparsely plated.

Viability Assays. We routinely use three techniques to assess viability of mesangial cells after experimental manipulations or exposure to pharmacological agents: (*i*) uptake and hydrolysis of carboxyfluorescein diacetate, (*ii*) mitochondrial reduction of the tetrazolium salt, MTT, and (*iii*) measurements of membrane permeability using ethidium bromide. All cells take up 5(6)-carboxyfluorescein diacetate, but only in viable cells do cytosolic esterases convert this nonfluorescent compound into the fluorescent probe carboxyfluorescein.[26] Mesangial cells in Lab-Tek chambers are washed once with RPMI-1640 medium without serum, and 5(6)-carboxyfluorescein diacetate (Molecular Probes) is added to 35 μM from a 10 mM stock solution in DMSO (stored at −20°). After incubating for 15 min in a 5% CO_2 incubator at 37°, the monolayer is washed once with RPMI-1640 and immediately examined separately under phase contrast and epifluorescence (FITC excitation filter) to determine the percentage of cells displaying green fluorescence. In viable cells, [3-(4,5-dimethylthiazoyl-2-yl)-2,5-diphenyltetrazolium bromide (MTT, Sigma, St. Louis, MO) is reduced by mitochondrial dehydrogenases to a purple formazan, the absorbance of which is quantified at 570 nm after solubilization.[27] Mesangial cells in Costar 96-well plates (100 μl medium/well) are treated with 20 μl of a fresh, sterile solution (5 mg/ml in DPBS) of MTT. After incubation for 3 hr at 37°, discard the supernate by sharply inverting the plate onto absorbant paper. Solubilize each monolayer in 0.1 ml of 0.04 N HCl in 2-propanol. [If the formazan is difficult to solubilize, add 20 μl of 3.0 % sodium dodecyl sulfate (SDS) to each well and mix.] Read each plate on a microplate spectrophotometer at 570 nm using as a blank a well without cells but treated with MTT. Percentage cytotoxicity curves can be calculated as described.[27] Because ethidium bromide (MW = 394) is normally impermeant and fluorescent only in the presence of nucleic acids, it is a sensitive probe of plasma membrane permeability.[28] Mesangial cells in four-chamber Lab-Tek slides are washed once with BSS (137 mM NaCl, 2.7 mM KCl, 20 mM HEPES, pH 7.4, 5.6 mM glucose, and 1 mg/ml BSA). Add 0.25 ml of 10 μM ethidium bromide in BSS and incubate for 1 min at 23°. Mix in 0.25 ml of 2.5 μM acridine orange and incubate for 30 sec to counterstain viable cells. Immediately remove the supernate, mount the monolayer under a coverslip, and examine using epifluorescence with fluorescein excitation filters. Ethidium-permeable cells appear bright orange, whereas viable cells display a dull green color. The time course of

[26] B. Rotman and B. Papermaster, *Proc. Natl. Acad. Sci. U.S.A.* **55,** 134 (1966).

[27] L. M. Green, J. L. Reade, and C. F. Ware, *J. Immunol. Methods* **70,** 257 (1984).

ethidium permeability can be monitored in monolayers grown on cover-slips as described by Gomperts.[28]

Contraction of Mesangial Cells. Low-density monolayers of mesangial cells on three-dimensional, type I collagen gels are incubated with contractile agonists and the decrease in cross-sectional area of individual cells is measured using computer-aided microscopy or photomicroscopy.[29,30] Type I collagen gels are prepared by mixing 8.0 ml of 2.9 mg/ml collagen in 0.012 *N* HCl, 1.0 ml of 1 *N* NaOH, and 1.0 ml of 10x minimal essential medium, all from sterile solutions at 4°, and adjusting to pH 7.4. Aliquots (1 ml) are plated into two-chamber Lab-Tek slides and gelation initiated by warming to 37° for 1 hr. Contraction is tested 24 hr after subculture onto the collagen gel by incubating cells plus or minus agonists in HBSS with 10 m*M* HEPES (pH 7.4), and 2 mg/ml fatty acid-free bovine serum albumin. The change in cross-sectional for individual cells area over time is measured using computer-aided microscopy as previously described.[29,30] A cell is scored positive for contraction with a decrease in cross-sectional area ≥7%.

Acknowledgments

This work was supported by National Institutes of Health Grants HL-22563 and HL-37117.

[28] B. D. Gomperts, *Nature (London)* **306,** 64 (1983).
[29] M. S. Simonson and M. J. Dunn, *Kidney Int.* **30,** 524 (1986).
[30] M. S. Simonson, *BioTechniques* **3,** 484 (1985).

[60] Leukotriene B_4 Biosynthesis by Erythrocyte–Neutrophil Interactions

By DAVID A. JONES and FRANK A. FITZPATRICK

Enzymatic cooperation between different cell types has been observed for biosynthesis of several eicosanoids.[1-12] This process, termed transcellular biosynthesis, represents a novel mechanism for the production of

[1] S. Bunting, R. Gryglewski, S. Moncada, and J. R. Vane, *Prostaglandins* **12,** 897 (1976).
[2] A. Marcus, B. Weksler, E. Jaffe, and J. Broekman, *J. Clin. Invest.* **66,** 979 (1980).
[3] A. Marcus, M. J. Broekman, B. Weksler, E. Jaffe, L. Safier, H. Ullman, and K. Tack-Goldman, *Philos. Trans. R. Soc. London* **B294,** 343 (1981).

prostaglandins and leukotrienes. Transcellular biosynthesis is important from several perspectives. For instance, the quantitative and qualitative features of eicosanoid production via cell–cell cooperation differ from the sum of the separate cellular biosynthetic capabilities. Second, cell combinations encountered in diseases often differ from those encountered in normal circumstances. Thus, better characterization of transcellular eicosanoid biosynthesis should improve our comprehension of the role of eicosanoids in the cellular biology of inflammation and thrombosis. Experiments on transcellular biosynthesis should conform to certain criteria: (*i*) are the cells under investigation likely to occur together *in vivo*; (*ii*) are the ratios of the respective cells likely to occur together *in vivo*; (*iii*) are the cells sufficiently pure to provide unambiguous conclusions regarding the origin of the eicosanoid under study; and (*iv*) do transformations of synthetic substrate and cellularly "donated" substrate correspond.

We outline below the experimental procedures to assess these criteria for investigations on erythrocyte–neutrophil interactions. The original investigations demonstrated that combinations of human neutrophils plus erythrocytes, with appropriate stimulation, produce more leukotriene B_4 (LTB_4) than neutrophil suspensions alone.[12] This was significant because erythrocytes do not contain detectable phospholipase A_2 or 5-lipoxygenase activity and have therefore been regarded physiologically inert in terms of leukotriene formation. This opinion is no longer warranted.

Experimental Procedure

Reagents

Sodium citrate (Sigma, St. Louis, MO): 3.8% (w/v) in sterile saline
NaCl (Sigma): 0.9% (w/v) in sterile, distilled water
Cellulose mixture: microcrystalline cellulose (Sigma cat. No. C-8002), α-cellulose (Sigma cat. No. S-5504) (1 : 1, w/w) suspended in 0.9% NaCl

[4] J. Maclouf and R. C. Murphy, *J. Biol. Chem.* **263,** 174 (1988).
[5] C. Dahinden, T. Clancy, M. Gross, J. Chiller, and T. Hugh, *Proc. Natl. Acad. Sci. U.S.A.* **82,** 6632 (1985).
[6] A. Marcus, L. Safier, H. Ullman, M. Broekman, N. Islam, T. Oglesby, and R. Gorman, *Proc. Natl. Acad. Sci. U.S.A.* **81,** 907 (1984).
[7] S. Feinmark and P. Cannon, *J. Biol. Chem.* **261,** 16466 (1986).
[8] S. Feinmark and P. Cannon, *Biochim. Biophys. Acta* **922,** 125 (1987).
[9] J. A. Claesson and J. Haeggstrom, *Adv. Prostaglandin. Thromboxane Leukotriene* **17,** 115 (1987).
[10] P. Needleman, A. Wyche, and A. Raz, *J. Clin. Invest.* **63,** 345 (1979).
[11] G. Hornstra, E. Haddeman, and J. Don, *Nature (London)* **279,** 66 (1979).
[12] J. E. McGee and F. A. Fitzpatrick, *Proc. Natl. Acad. Sci. U.S.A.* **83,** 1349 (1986).

Hanks' balanced salt solution (Gibco Labs, Gaithersburg, MD)
A23187 divalent calcium ionophore (Calbiochem, San Diego, CA)
Ethyl acetate (Burdick & Jackson, Muskegon, MI)
Mobile phase: methanol/water/glacial acetic acid (70:30:0.1)
Prostaglandin B_1 (Sigma)
LTA_4 methyl ester (The Upjohn Co., Kalamazoo, MI)
Human serum albumin (Sigma)
LiOH (Sigma)
Tetrahydrofuran (Burdick & Jackson)

Isolation of Human Erythrocytes

This isolation procedure is a modification of that presented by Beutler *et al.*[13] The coagulation step is omitted to avoid activation of different cell types.

Human venous blood (from drug-free volunteers) is collected using sterile sodium citrate as an anticoagulant (9 : 1, v/v). The whole blood is then centrifuged at 200 *g* for 30 min at 25° and the platelet-rich plasma pipetted off, removing as much of the buffy coat as possible. The sedimented cells are resuspended to the original volume in 0.9% NaCl and again centrifuged at 200 *g* for 20 min at 25°. The supernatant is discarded and the wash repeated.

Suspended erythrocytes are further purified by filtering through a microcrystalline cellulose-α-cellulose column to remove residual neutrophils and platelets. Columns can be prepared in 5 cm^3 syringes, sedimenting the cellulose mixture to a depth of 1.5 cm. Suspensions are filtered at a flow rate of 2 ml/min using 0.9% NaCl as an elution buffer. The erythrocytes are resuspended to the original whole blood volume in saline. Finally, erythrocytes, neutrophils, and platelet concentrations are determined by differential cell counting with a Coulter Diff. Typically, at each stage of sample purification, 20 μl of the cell suspension are diluted with 20.0 ml of Isoton II counting buffer prior to particle counting. Values for six representative preparations (Table I) illustrate that purified erythrocytes typically contain $\leq 5 \times 10^6$ platelets/ml.[12]

When these suspensions are incubated for 10 min at 37° with 20 μ*M* synthetic LTA_4 they produce 1.18 ± 0.12 (mean ±SE) nmol LTB_4/ml. This production could not be attributed to residual leukocytes: suspensions containing only neutrophils (2×10^7/ml) produce only 0.41 ± 0.17 nmol LTB_4/ml. Suspensions containing only platelets (1×10^6/ml) produce no detectable LTB_4. In erythrocyte suspensions containing 20–200 times

[13] E. Beutler, C. West, and K. G. Blume, *J. Lab. Clin. Med.* **88**, 328 (1976).

TABLE I
PURITY OF ERYTHROCYTE SUSPENSIONS FOLLOWING FILTRATION THROUGH CELLULOSE COLUMNS

Component	Preparation					
	1	2	3	4	5	6
Erythrocytes (10^9/ml)	5.56	3.24	4.35	3.58	3.07	2.97
Leukocytes (10^6/ml)	0.50	0.2	<0.05	<0.05	<0.05	<0.05
Platelets (10^6/ml)	9	3	2	8	9	4
LTB_4 (nmol/ml)	1.3	1.2	2.0	1.7	1.4	1.0

fewer neutrophils (5×10^3 to 9×10^4 neutrophils/ml) formation of LTB_4 is appreciable and corresponded to the erythrocyte content.[12]

Isolation of Human Polymorphonuclear Neutrophils

Human whole blood is collected as for erythrocyte preparation. Suspensions of neutrophils are prepared by sedimenting human whole blood through dextran T-500 as described originally by Boyum[14] and elsewhere in this volume ([61] and [62]).

Neutrophil–Erythrocyte Leukotriene Biosynthesis

The transcellular metabolism of neutrophil-derived LTA_4 is initiated by stimulating various neutrophil/erythrocyte mixtures with the calcium ionophore A23187. Typically, 1.5 ml of a neutrophil/erythrocyte mixture (2×10^7 neutrophils/ml, 1×10^9 erythrocytes/ml) are preincubated at 37° for 5 min. These cells are then challenged with A23187 (5 μM, 37°) to stimulate leukotriene biosynthesis. After a 2.5-min incubation, prostaglandin B_1 (0.5 μg) is added to the cell mixture as a quantitative internal extraction standard. The reaction is terminated by adding 5 ml of ethyl acetate. The enzymatic and nonenzymatic LTA_4 products are extracted from the cell suspensions by vortexing for 3 min and removing the organic phase. The extraction procedure is carried out three times.

In addition to the neutrophil/erythrocyte mixtures, neutrophils alone (2×10^7 per ml) and erythrocytes alone (2×10^9 per ml) are also incubated with A23187 and treated identically to the above combinations. This establishes the capacity of each cell type to synthesize LTB_4 following calcium ionophore challenge.[12] In this system only the neutrophil can produce the substrate LTA_4. Typically, neutrophils alone have been shown to produce

[14] A. Boyum, *Scand. J. Clin. Lab. Invest. Suppl.* **97,** 77 (1968).

0.60 ± 0.04 nmol of LTB_4 (mean ± SEM, $n = 8$) while neutrophil–erythrocyte combinations produce 1.29 ± 0.06 nmol of LTB_4 showing a relative increase of 112%.[12]

LTB_4 Quantitation

The ethyl acetate from the above extraction is evaporated and the residue reconstituted in 0.5 ml of mobile phase. A portion (250 μl) of this solution is then used to quantitate leukotrienes by reversed-phase high-performance liquid chromatography (HPLC).[15] Components are separated with a C_{18} (250 × 6 × 4 mm) reversed-phase column using methanol/water/acetic acid (70/30/0.1, v/v/v) as mobile phase. Adequate separation and identification can be achieved using a flow rate of 1 ml/min with a UV detector set at 270 nm. In addition, LTB_4 can be identified quantitatively and qualitatively by radioimmunoassay,[16] bioassay,[17] and UV spectrophotometry. It is always advisable to verify the identity of a chromatographic peak by alternate methods.

Verifying LTB_4 Origin

Verifying that the increase in LTB_4 production in neutrophil/erythrocyte combinations is due to LTA_4 conversion by erythrocytes is an important consideration in this system. This can be accomplished by exploiting the suicide inactivation of LTA_4 hydrolase by its substrate.

In this protocol, 1.5 ml (2×10^9 cells/ml) of isolated erythrocytes are incubated with synthetic LTA_4 (30 μ*M*) for 5 min at 37° to inactivate the epoxide hydrolase. The cells are then washed twice with 0.9% NaCl containing human serum albumin (25 mg/ml) to remove LTA_4 metabolites. These inactivated erythrocytes are then resuspended with PMNs and challenged with A23187 as described previously. LTA_4 metabolites are extracted and quantitated as before. LTB_4 production is reproducibly diminished under these conditions indicating that the erythrocytes must play a role in LTB_4 biosynthesis. Control groups of inactivated and normal cells can be incubated with synthetic LTA_4 to verify inactivation.[12]

Saponification of LTA_4 Methyl Ester

LTA_4 for the inactivation experiments can be prepared from the methyl ester as follows: add 0.6 μmol of the methyl ester to a small vial and evaporate the solvent under nitrogen. Reconstitute the residue in 500 μl of

[15] P. Borgeat and B. Samuelsson, *J. Biol. Chem.* **254,** 7865 (1979).
[16] J. Salmon, P. Simmons, and R. Palmer, *Prostaglandins* **24,** 225 (1982).
[17] P. Sirois, P. Borgeat, and A. Jeanson, *J. Pharm. Pharmacol.* **33,** 466 (1981).

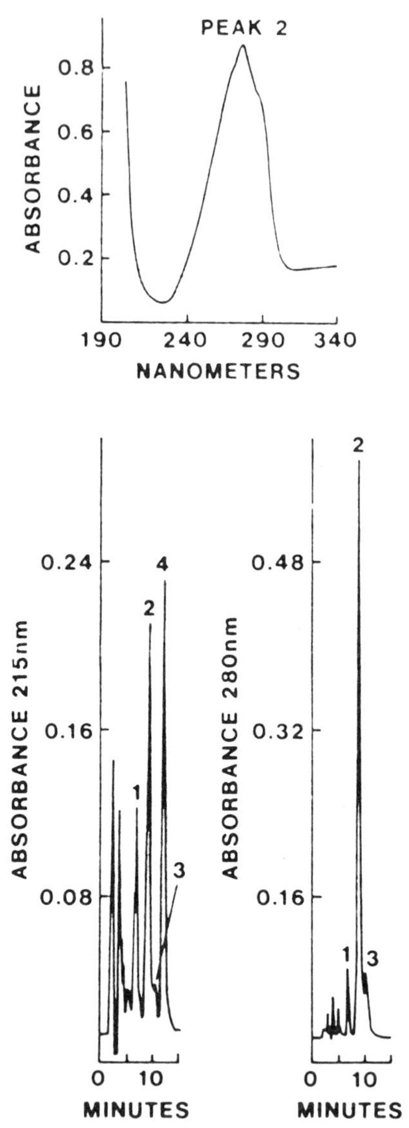

FIG. 1. Reversed-phase high-performance liquid chromatographic analysis of intact leukotriene A_4. Column: Lichrosorb RP8. Mobile phase: acetonitrile/0.01 *M* borate, pH 10. Flow: 1.0 ml/min. Peak 1, 12-*O*-ethyl-5-hydroxyeicosatetraenoic acid; Peak 2, leukotriene A_4; Peak 3, 11-*trans*-leukotriene A_4; Peak 4, arachidonic acid. Left profile is monitored at 215 nm; right profile is monitored at 280 nm. The ultraviolet absorbance spectrum of peak 2 (upper trace) was obtained under stopped-flow conditions in a Tracor model 970 detector. The UV spectrum corresponds to the UV spectrum of leukotriene A_4.

THF. Add 30 μl of 2 N LiOH followed by 15 μl water. Stir this mixture at 4° for 24 hr. Saponification can be verified by reversed-phase HPLC using a Hamilton PRP-1 column eluted with acetonitrile/0.01 M borate, pH 10 (Fig. 1).[18] The concentration of LTA_4 can be determined by UV spectroscopy at 280 nm and an extinction coefficient of 52,000.

[18] M. Wynalda, D. Morton, R. Kelly, and F. A. Fitzpatrick, *Anal. Chem.* **54,** 1079 (1982).

[61] Leukotriene C_4 Biosynthesis during Polymorphonuclear Leukocyte–Vascular Cell Interactions

By STEVEN J. FEINMARK

Polymorphonuclear leukocytes (PMNL) must adhere to and cross the endothelial cell (EC) lining in order to emigrate from the circulation to sites of developing inflammation. The sustained contact between EC and PMNL that ensues enhances the potential for novel biochemical interactions between these cells. One such interaction is the augmented production of leukotriene (LT)C_4 as a consequence of PMNL–EC coincubation *in vitro*.[1] A similar interaction occurs between aortic smooth muscle (SMC) and PMNL.[2] Several approaches have been taken to demonstrate the transcellular nature of LTC_4 production during these incubations. Each cell type was evaluated individually to determine which metabolites could be produced either from endogenous arachidonate or from exogenous biosynthetic intermediates. Then, cultured cells and PMNL were coincubated and the products quantified. Finally, the glutathione (GSH) pools of each cell type were labeled individually and incorporation of this label into leukotriene was used to determine the site of the final synthetic step, the conversion of LTA_4 to LTC_4.

Cell Preparations

Human Peripheral Blood Leukocytes

Blood is collected from normal drug-free volunteers who had granted informed consent. Typically, blood is drawn through a 19-gauge butterfly

[1] S. J. Feinmark and P. J. Cannon, *J. Biol. Chem.* **261,** 16466 (1986).
[2] S. J. Feinmark and P. J. Cannon, *Biochim. Biophys. Acta* **922,** 125 (1987).

into 60-ml syringes and immediately transferred to EDTA-containing vacutainers (Becton Dickinson, Rutherford, NJ). The anticoagulated blood is pooled into 50-ml polycarbonate tubes and subjected to centrifugation at 200 *g* for 15 min at room temperature. The platelet-rich plasma (upper layer) is removed and the residual blood (approx. one-half of the original volume) is transferred to a Nalgene cylinder and combined with an equal volume of dextran [prepared by dissolving 6 g of dextran T-500 (Pharmacia, Piscataway, NJ) plus 2.7 g of NaCl in 300 ml of water]. After several inversions to ensure adequate mixing, the blood is left at room temperature to permit the aggregated erythrocytes to sediment. When the blood/dextran solution separates into two roughly equal phases (approx. 45–60 min), the PMNL-rich upper phase is collected and concentrated by centrifugation at 200 *g* for 15 min at room temperature. The cell pellets are gently resuspended in 5–10 ml of lysis buffer [prepared by dissolving 2.06 g of Tris(hydroxymethyl)aminoethane and 7.47 g of ammonium chloride/liter of water and finally adjusting the pH to 7.4) and kept at 37° for 7 min. The tubes are gently agitated and a sodium metrizoate/Ficoll solution (density 1.077 g/ml; Lymphoprep, Nycomed) is gently layered under the cell suspension to form a discontinuous gradient (5 ml is slowly dispensed through an 18-gauge needle placed at the bottom of the centrifuge tube). These tubes are spun at 400 *g* for 40 min at room temperature. The cell pellets are resuspended in a physiological salt solution (for example, Dulbecco's phosphate-buffered saline, pH 7.4; PBS; GIBCO) with bovine serum albumin (BSA; essentially fatty acid-free, Sigma, St. Louis, MO), if desired. The resulting suspension is counted in a hemocytometer and diluted as necessary. This preparation contains primarily neutrophils with some eosinophils (mononuclear cells are generally less than 1%).[3,4]

Porcine Aortic Endothelial Cells

Porcine aortas are collected within minutes of slaughter at a local abatoir under a license from the United States Department of Agriculture (USDA). The vessels are packed in sterile bags on ice and transported to the laboratory within 2–3 hr. Subsequent steps are carried out in a sterile hood using standard sterile technique. The vessels are washed free of blood with Dulbecco's modified Eagle medium (DMEM; GIBCO), opened longitudinally, and laid on a piece of sterile foil. The exposed intima is gently scraped with a scalpel taking care to avoid collateral vessels, any residual blood clots, and the cut ends of the tissue. Cells are collected by gently swirling the scalpel blade in DMEM (10 ml in a sterile tube) followed by centrifugation (300 *g*, 15 min). The cell pellet is resuspended in DMEM

[3] A. Böyum, *Scand. J. Immunol.* **5** (Suppl. 5), 9 (1976).

[4] H.-E. Claesson and S. J. Feinmark, *Biochim. Biophys. Acta* **804,** 52 (1984).

containing fetal bovine serum (FBS, 10%), penicillin (400 U/ml), streptomycin (400 μg/ml), L-glutamine (2 mM), chlortetracycline (50 μg/ml), and Fungizone (2.5 μg/ml), and washed twice more with this medium before being plated into 25-cm^2 tissue culture flasks and grown overnight. The following day, the primary cultures are washed vigorously to remove any nonadherent cells or debris and refed with the same medium. On the second day, the cultures are rewashed and refed with normal growth medium consisting of DMEM plus FBS (10%), penicillin (100 U/ml), streptomycin (100 μg/ml), and L-glutamine (2 mM). Cultures are routinely fed every 3–4 days and subcultured as needed. Confluent cultures are washed with calcium, magnesium-free PBS (GIBCO) to remove the growth medium, and subdivided with 1–2 ml of trypsin–EDTA solution (containing 0.5 mg/ml trypsin and 0.2 mg/ml Na_4EDTA in calcium, magnesium-free Hanks' balanced salt solution; GIBCO). The cells round up and can be detached within minutes by gentle tapping of the tissue culture flask. The cell suspensions are then diluted in growth medium and subdivided as needed. From time to time, cultures are grown in the absence of antibiotics and shown to be free of mycoplasma (MycoTect, BRL, Gaithersburg, MD). Cultures are verified to be endothelium by their typical cobblestone morphology and by the organization of actin-based cytoskeleton (by rhodamine–phalloidin) and factor VIII staining, and the absence of detectable smooth muscle actin.[5]

Human umbilical vein endothelial cells (HUVEC) are prepared from term umbilical cords by collagenase treatment essentially as reported by others.[6,7] Cells are maintained in HEPES-buffered M-199 supplemented with human serum (15%), penicillin (50 U/ml), streptomycin (50 μg/ml), L-glutamine (2 mM), endothelial cell growth factor (ECGF) (15 μg/ml), and heparin (90 μg/ml)[8] and subcultured as necessary.

Vascular smooth muscle cells are cultured by explant from porcine aortas after removal of the endothelium by intimal scraping. The aortic media is isolated by blunt dissection and chunks of vascular smooth muscle (approximately 1 mm^3) are placed in a culture dish. The tissue is fed with DMEM plus FBS (10%), penicillin (100 U/ml), streptomycin (100 μg/ml), and L-glutamine (2 mM) and covered with a sterile glass coverslip. The dishes are left undisturbed until cells are observed to be growing out of the explants. The tissue pieces and coverslip are then removed and

[5] A. I. Gotlieb, W. Spector, M. K. K. Wong, and C. Lacey, *Arteriosclerosis* **4,** 91 (1984).

[6] E. A. Jaffe, R. L. Nachman, C. G. Becker, and C. R. Minick, *J. Clin. Invest.* **52,** 2745 (1973).

[7] G. A. Zimmerman, R. E. Whatley, T. M. McIntyre, D. M. Benson, and S. M. Prescott, this volume [57].

[8] S. C. Thornton, S. N. Mueller, and E. M. Levine, *Science* **222,** 623 (1983).

the primary cultures refed and grown to confluence. These cells are subcultured as described above. The cultures are determined to be smooth muscle by morphological criteria and the presence of smooth muscle actin.

Incubation and Purification Procedures

Exogenous LTA_4. Synthetic LTA_4 methyl ester is obtained from Upjohn Co. (Kalamazoo, MI, through the courtesy of Dr. F. Sun) or from BIOMOL Research Laboratories (Plymouth Meeting, PA). The ester is hydrolyzed by dissolution in methanol: 50% NaOH (9/1, v/v; usually at a concentration of 5 m*M*) and kept at $-20°$ overnight. In many cases, the LTA_4 is tested to confirm that the hydrolysis is complete and that the labile epoxide is intact. An aliquot of the hydrolyzate is diluted in acidic methanol which rapidly converts intact LTA_4 into two isomers of 5(*S*)-hydroxy-12-methoxyeicosatetraenoic acid.[9] Samples are then dried, redissolved in mobile phase, and analyzed by HPLC on a Nucleosil C_{18} column (4.6×250 mm, Alltech, Deerfield, IL) eluted with methanol : water : acetic acid (75 : 25 : 0.01, v/v/v) at 1 ml/min. The products are detected by UV absorbance at 270 nm. The methanolysis products of LTA_4 are readily separated from the isomeric 5(*S*),12-dihydroxyeicosatetraenoic acids which result from prior decomposition of LTA_4. In addition, the free acids resolve from their corresponding methyl esters thus confirming that the ester hydrolysis is complete.

LTA_4 is added directly to cultured cells bathed in PBS buffered with HEPES (15 m*M*, pH 7.4). Media from these experiments are directly injected on to the HPLC which is fitted with a Nucleosil C_{18} column and eluted with methanol : water : acetic acid (67 : 33 : 0.08, v/v/v) buffered to pH 5.8 with ammonium hydroxide and flowing at 1 ml/min. Columns are routinely treated with EDTA.[10] The eluent is monitored at 280 nm or with a photodiode array spectrophotometer. Fractions (1-ml) are collected for subsequent assay.

PMNL/Vascular Cell Coincubations. Cultured cells are used at confluence (typically in 25-cm^2 flasks). PMNL are purified as described above and resuspended in PBS plus BSA (0.5–1%). PMNL, $1–2 \times 10^7$, in 1 ml are added to a washed culture of vascular cells or incubated alone. A23187 (final concentration, 5 μM) is added to the incubation buffer in a concentrated ethanolic solution (final solvent concentrations never exceeded 0.2%). Simultaneous addition of arachidonic acid is achieved by the dissolution of a measured amount of fatty acid in the A23187 solution, followed

[9] P. Borgeat and B. Samuelsson, *Proc. Natl. Acad. Sci. U.S.A.* **76,** 3213 (1979).
[10] S. A. Metz, M. E. Hall, T. W. Harper, and R. C. Murphy, *J. Chromatogr.* **233,** 193 (1982).

by brief sonication. fMLP (formylmethionylleucylphenylalanine) stimulation is carried out by the initial addition of arachidonic acid (10 μM) in ethanol followed 1 min later by fMLP (1 μM) in dimethyl sulfoxide (DMSO). In some experiments, EC monolayers are treated with aspirin (ASA; 0.2 mM) added in ethanol for 30 min. The cultures are then washed twice more and the PMNL suspension added.

Incubations are continued for 30 min at 37° in a humidified CO_2 incubator (5% in air). After the completion of the incubation, the media are mixed with ice-cold methanol (1–2 volumes) containing tracer amounts of [^{3}H]LTC_4 as internal standard and kept at −20° for several hours; any protein precipitate is removed by centrifugation. Aliquots of the samples are evaporated to dryness, reconstituted in mobile phase, and applied to the reversed-phase HPLC as above.

LTC_4 Quantification

LTC_4 is measured by radioimmunoassay after HPLC purification. Aliquots of the HPLC fractions are tested for the presence of [^{3}H]LTC_4 internal standard to unequivocally identify the LTC_4-containing fractions and to correct for purification losses. The remainder of each HPLC fraction is evaporated to dryness and reconstituted in PBS. Each fraction is assayed in duplicate for the presence of immunoreactive LTC_4. Details of the radioimmunoassay (DuPont-NEN, Wilmington, DE) have been described previously.[11] The anti-LTC_4 antiserum has a cross-reactivity of 55.3% with LTD_4 and 8.6% with LTE_4.[12] The intraassay variability is 8.5%, the interassay variability is 19–22%.

Measurement of Cellular Integrity

PMNL or cultured vascular cells are labeled with ^{111}In-oxine by a modification of published methods.[1,13,14] Monolayer cultures (grown in 24-well plates) and PMNL are washed twice with Hanks' balanced salt solution (Hanks' BSS; GIBCO). PMNL are resuspended in Hanks' (2×10^7/ml) and the cultures covered with the same medium (0.5 ml). ^{111}In-oxine (2 μCi) is added to each well or the suspended PMNL and the cells are left at room temperature for 15 min.

[11] D. Piomelli, S. J. Feinmark, and P. J. Cannon, *J. Pharmacol. Exp. Ther.* **241,** 763 (1987).

[12] E. C. Hayes, D. L. Lombardo, Y. Girard, A. L. Maycock, J. Rokach, A. S. Rosenthal, R. N. Young, R. W. Egan, and H. J. Zweerink, *J. Immunol.* **131,** 429 (1983).

[13] B. Zakhireh, M. L. Thakur, H. L. Malech, M. S. Cohen, A. Gottschalk, and R. K. Root, *J. Nucl. Med.* **20,** 741 (1979).

[14] T. Collins, A. M. Krensky, C. Clayberger, W. Fiers, M. A. Gimbrone, Jr., S. J. Burakoff, and J. S. Pober, *J. Immunol.* **133,** 1878 (1984).

After washing the monolayers with Tyrode's buffer containing 4% BSA (Tyrode's–BSA), the cells are covered with the same buffer and incubated at 37° for 90 min. Each well is rewashed with Tyrode's–BSA and then incubated as described with unlabeled PMNL. At the end of the incubation, the buffer is recovered, centrifuged (200 *g*, room temperature) to remove any cells, and aliquots are assayed for the presence of ^{111}In by gamma counting. The cells remaining in the wells are lysed by the addition of 2% Nonidet P-40 (NP-40). Aliquots of the detergent lysate are assayed for ^{111}In and the percentage of total ^{111}In released is calculated.

PMNL are washed twice in Hanks' BSS after the labeling period and then resuspended either in phosphate-buffered saline (PBS) or PBS–BSA as appropriate and incubated as described above. At the end of the incubation, the buffer is recovered and the PMNL are concentrated by centrifugation at 200 *g*, at room temperature. The pellet is lysed with 2% NP-40 and aliquots of the supernatant and of the lysed pellet are assayed for the presence of ^{111}In.

Incorporation of Label into Cellular Glutathione Pools

PMNL (2×10^7/ml) or monolayer cultures are washed twice with PBS plus 1% BSA and then incubated for 1 hr at 37° with [^{35}S]cysteine (5 μCi; >600 Ci/mmol) in PBS–BSA. Unincorporated cysteine is removed by two washes with PBS-BSA. Prelabeled cells are incubated as described above and the incubation media purified by HPLC. The fractions are analyzed for the presence of ^{35}S-labeled material by liquid scintillation counting.

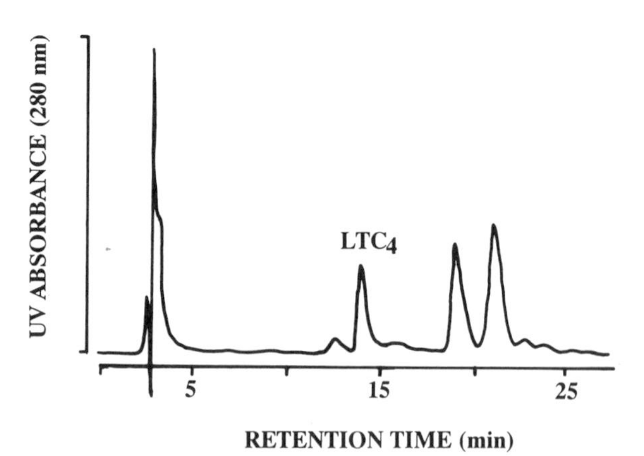

FIG. 1. Monolayer cultures of human umbilical vein endothelial cells are incubated with LTA_4 (10 μM) in HEPES-buffered PBS plus 1% BSA. The culture medium is recovered, diluted with methanol (2 volumes), and directly injected on the HPLC as described in the text.

TABLE I
LTC_4 PRODUCTION BY MIXED PMNL AND VASCULAR CELLS [a]

Stimulus	LTC_4 production (pmol/10^7 PMNL-flask)			
	PMNL	PMNL+EC	PMNL+SMC	PMNL+EC (ASA)
A23187 (5 μM)	3.8 ± 1.8 (n = 8)	15.9 ± 6.1 (n = 3)	11.0 ± 3.5 (n = 5)	
A23187 (5 μM) +20:4 (150 μM)	10.5 ± 2.1 (n = 11)	24.3 ± 6.7 (n = 3)	39.8 ± 8.1 (n = 5)	
fMLP (1 μM) +20:4 (10 μM)	0.75 ± 0.39 (n = 5)	0.23 ± 0.08 (n = 5)		1.7 ± 0.4 (n = 5)

[a] PMNL and vascular cells are coincubated as described in the text with the stimuli noted above. LTC_4 is purified by HPLC and quantified by radioimmunoassay. Portions of these data have been reported elsewhere in a different form. [1,2,19]

Results

LTA_4 Metabolism

HUVEC are incubated with exogenous synthetic LTA_4 and the presence of metabolites in the medium measured by HPLC. Several peaks of UV-absorbing material are detected at 280 nm (Fig. 1). The two largest peaks are also found in control incubations and correspond to the nonenzymatic decomposition products of LTA_4. In the cell incubations, a third peak appears that elutes at the retention time of synthetic LTC_4. The production of LTB_4 is not observed in this experiment. The absence of LTB_4 cannot be explained by its metabolic removal since neither EC[1,15] nor SMC[2] metabolize this lipid. Claesson and Haeggström,[15] however, have reported that HUVEC produce LTB_4 under these conditions. Earlier studies of HUVEC,[16] porcine EC,[1] or porcine SMC[2] did not detect the conversion of LTA_4 to LTB_4.

Quantitative Aspects of PMNL–Vascular Cell Interactions

A23187 Stimulation. A23187-stimulated PMNL preparations generate LTC_4 and the levels are enhanced by the addition of exogenous arachidonic acid (Table I). The precise cellular source of this leukocyte product is not clear since portions of the LTC_4 production may be due to eosino-

[15] H.-E. Claesson and J. Haeggström, *Eur. J. Biochem.* (1988).
[16] B. O. Ibe and W. B. Campbell, *Biochim. Biophys. Acta* **960,** 309 (1988).

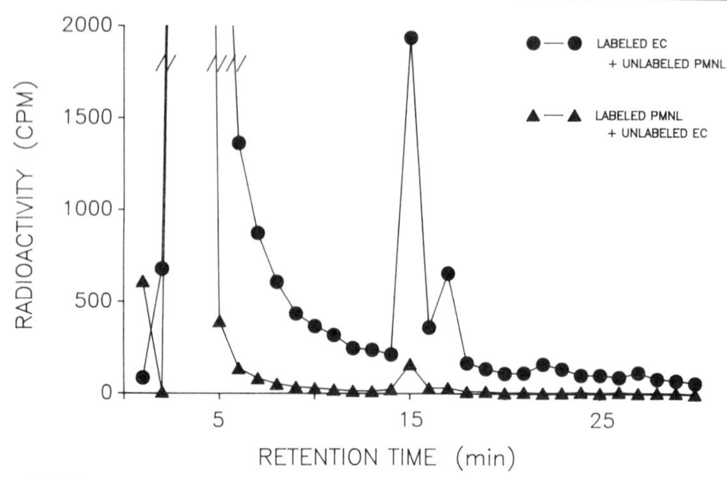

FIG. 2. Either PMNL or EC are prelabeled with [^{35}S]cysteine and then washed. PMNL and EC are coincubated, stimulated with A23187 (5 μM) and arachidonic acid (150 μM), and the incubation buffer recovered. The LTC_4 is purified by HPLC as described in the text and fractions analyzed for the presence of ^{35}S-labeled products. Data have been normalized to equal LTC_4 mass production from a paired incubation.

phils[17] or to transcellular metabolism of neutrophil-derived LTA_4 by contaminating platelets.[18] Nevertheless, A23187 stimulation of PMNL in the presence of vascular cells invariably leads to increased LTC_4 production compared to matched control incubations of PMNL alone (Table I). Exogenous arachidonate also enhances the leukotriene production during these experiments.

fMLP Stimulation/EC Feedback Regulation. The bacterial tripeptide fMLP is a very weak inducer of PMNL LTC_4 synthesis, generating less than 20% of the product released after A23187 treatment (Table I). Unlike A23187, fMLP stimulation of mixed PMNL plus EC fails to produce a quantitative increase in LTC_4 production but rather, PMNL LTC_4 synthesis is significantly inhibited (Table I).[19] EC-dependent inhibition of fMLP-induced LTC_4 synthesis is abolished by treatment of the EC cultures with the cyclooxygenase inhibitor, aspirin (0.2 mM). fMLP stimulation of PMNL in the presence of aspirin-treated EC generates more LTC_4 than either PMNL alone or PMNL plus untreated EC. Thus, transcellular

[17] P. F. Weller, C. W. Lee, D. W. Foster, E. J. Corey, K. F. Austen, and R. A. Lewis, *Proc. Natl. Acad. Sci. U.S.A.* **80,** 7626 (1983).

[18] J. A. Maclouf and R. C. Murphy, *J. Biol. Chem.* **263,** 174 (1986).

[19] S. J. Feinmark, J. Edasery, and P. J. Cannon, *Adv. Prostaglandin Thromboxane Leukotriene Res.* **19,** 255 (1989).

metabolism of LTA_4 to LTC_4 can be demonstrated even after relatively weak stimulation.

[^{35}S]LTC_4 Production during PMNL/Vascular Cell Coincubation. To demonstrate that the conversion of PMNL-derived LTA_4 to LTC_4 occurred within the vascular cell cultures, [^{35}S]cysteine is used to label the cellular GSH pools. Prelabeled PMNL incubated alone or with unlabeled vascular cells releases low levels of [^{35}S]LTC_4 after A23187 stimulation. However, [^{35}S]cysteine-prelabeled EC[1] or SMC[2] incubated with A23187-stimulated unlabeled PMNL produces significantly more [^{35}S]LTC_4 than in the reverse-labeling experiment (Fig. 2).

Conclusions

Using the methods described in this chapter, it has been possible to demonstrate that activated PMNL release LTA_4 which can be converted to LTC_4 by adjacent cultured vascular cells. The level of LTC_4 produced depends on the activating stimulus. Weak PMNL activators, such as fMLP, may permit the observation of a feedback regulation loop in which PMNL leukotriene synthesis is inhibited by a vascular cell product which is probably a prostaglandin.

Acknowledgments

Some of the experiments described in this chapter were supported by AI-26702, HL-38312, and HL-21006. Portions of this work were done during the tenure of an Established Investigatorship award from the American Heart Association and Boehringer Ingelheim, Inc. The author would like to thank Dr. J. Brett for the morphological examination of the cultured cells and Dr. R. R. Sciacca for statistical analyses and critical reading of this manuscript.

[62] Release and Metabolism of Leukotriene A_4 in Neutrophil–Mast Cell Interactions

By CLEMENS A. DAHINDEN and URS WIRTHMUELLER

Introduction

The cellular sources, biosynthesis, structure, and different biological activities of the 5-lipoxygenase (5-LOX) metabolites leukotriene (LT) B_4, LTC_4, LTD_4, and LTE_4 are well established. However, there is an interesting particularity of the 5-LOX pathway in that at each metabolic step the precursor molecules are formed in amounts largely exceeding the capacity of the next enzyme. When released outside the cell, the different precursor molecules can be taken up and further metabolized by different cell types, resulting in the generation of lipid mediators that are not produced by an individual cell type alone. Indeed, there is increasing evidence from work of different laboratories, that such cellular cooperations and intercellular transfer of precursor lipids exist among a variety of leukocytes and tissue cells.

In contrast to the cyclooxygenase and other lipoxygenases, the 5-LOX is a calcium-dependent enzyme, and requires agonist-induced increases in intracellular calcium for activity.[1,2] Since the activity of the phospholipase(s) liberating arachidonic acid, and of the 5-LOX are regulated by different second messengers,[1] cell cooperation by exchange and metabolism of free arachidonic acid may already be present at this first step of the cascade. 5-Hydroxy-1-eicosatetraenoic acid (5-HETE) formed by reduction of excess 5-HPETE, can be further metabolized by the 12-LOX of platelets or the 15-LOX, i.e., of eosinophils. Alternatively, 5-HETE may be reincorporated into phospholipids, possibly altering membrane properties and cellular functions. Finally, in neutrophils (PMN) LTA_4 is formed in excess of what is enzymatically hydrolyzed to LTB_4. This is clearly indicated by the fact that whenever LTB_4 is formed *in vitro*, the all-trans stereoisomers of LTB_4 are also present. Here, we review our findings that LTA_4 formed by Ca^{2+} ionophore-stimulated PMN does not appreciably decay within the cell, but is released into the extracellular medium as the intact epoxide in relatively large amounts. Low concentrations of albumin-bound LTA_4 are efficiently and rapidly taken up by mast cells and metabolized into LTC_4. It is remarkable that more recent studies showed that neutrophil-derived LTA_4 can be metabolized into biologically active leukotrienes even by cells (like erythrocytes, platelets, and endothelial cells), which are lacking the 5-LOX and thus are unable to form LTA_4. Therefore, interactions between various cell types that release or utilize LTA_4 may provide an important metabolic pathway for the production of leukotrienes (Fig. 1).

Release of LTA_4 by Neutrophils

Materials. LTA_4 methyl ester, LTB_4, LTC_4, LTD_4, and LTE_4 are a gift of Dr. Rockach (Merck-Frosst, Pointe Claire, PQ, Canada). Organic solvents are HPLC-grade and chemicals of the highest purity available are used as supplied. Synthetic all-*trans*-LTB_4 derivatives are prepared by

[1] C. A. Dahinden, J. Zingg, F. E. Maly, and A. L. De Weck, *J. Exp. Med.* **167,** 1281 (1988).
[2] T. Puustinen, M. M. Scheffer, and B. Samuelsson, *Biochim. Biophys. Acta* **960,** 261 (1988).

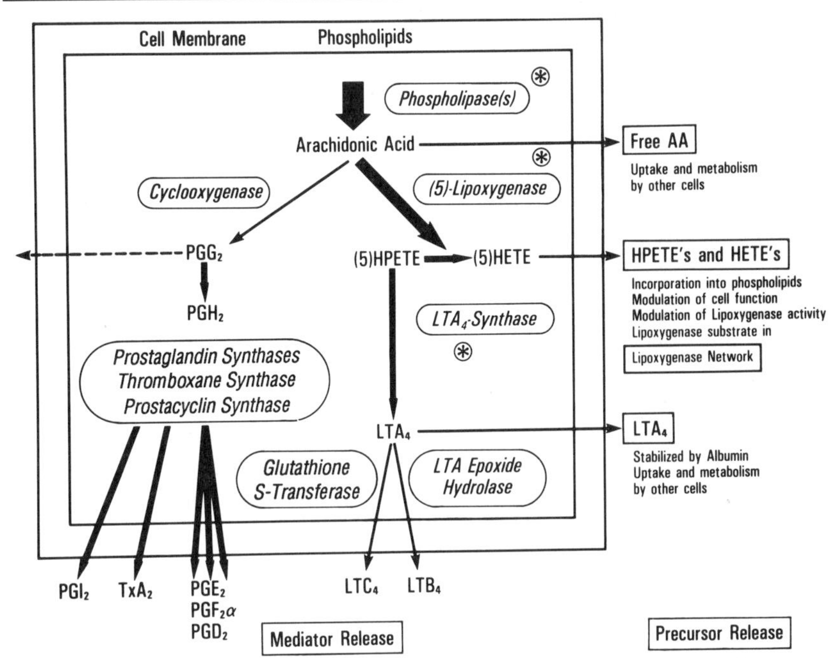

FIG. 1. Hypothetical scheme representing the proposed consequences of precursor release in the 5-LOX pathway. Asterisks indicate that the enzyme is not activated in resting cells.

reacting synthetic LTA_4 methyl ester with water, methanol, or ethanol acidified with HCl, respectively, and purification by reversed-phase high-performance liquid chromatography (RP-HPLC) before and after base hydrolysis of the ester.

Preparation of Neutrophils

Portions of blood, 20 ml, anticoagulated with 20 U/ml preservative-free heparin (Novo Industries, Copenhagen, Denmark) are carefully layered over 15 ml Methocel metrizoate [10 g Methocel MC 25cP (Fluka AG, Buchs, Switzerland) is dissolved in 500 ml distilled water, degased, and mixed with Isopaque (Nyegaard & Co., Oslo, Norway) to obtain a density of 1.080 at 20°]. The erythrocytes which agglutinate at the interface are allowed to sediment at room temperature. The leukocytes are then aspirated leaving 0.5 cm of plasma at the interface to avoid contaminating the leukocyte suspension with Methocel, and further purified by means of Ficoll-Hypaque (Pharmacia, Uppsala, Sweden) density separation (30 min, 400 *g*, room temperature). After hypotonic lysis of residual erythrocytes (65 mOsmol, 20 sec, at 0°) PMN are washed twice in Gey's solution at 4° and suspended in Dulbecco's PBS (both from GIBCO, Glasgow, Scottland) supplemented with 10 mg/ml fatty acid-free BSA (Pentex) and 2 m*M* glucose (DPBS/A) at a cell density of 10^7 to 10^8 ml.

This PMN preparation is essentially free of platelets as judged by the abundant production of 20-COOH-LTB_4, 20-OH-LTB_4, LTB_4, all-*trans*-LTB_4, and 5-HETE in the almost complete absence of 20-OH-5(*S*),12(*S*)-DiHETE, 5(*S*),12(*S*)-DiHETE, HHT, and 12-HETE (determined by HPLC as described below), after stimulation with 10 μM Ca^{2+} ionophore for 5 min. We believe this to be a better estimation of platelet contamination from morphological methods, due to their high capacity of 12-HETE and HHT formation and the difficulty inherent in microscopically counting the platelets, particularly when they are adherent to PMN. Platelet contamination may be considerable in isolation procedures that involve mixing of the blood with erythrocyte-agglutinating agents. Also cooling below room temperature has to be avoided before PMN are separated from the platelets. Platelets may affect LTA_4 recoveries since they are able to transform LTA_4 into LTC_4[3,4] Another unsolved problem is the influence of eosinophils in the PMN preparation. Eosinophils are the major (or possibly even the sole) source of LTC_4 in stimulated polymorphonuclear cells, but it is not yet known if they are able to metabolize low concentrations of albumin-bound LTA_4.

HPLC Analysis of Lipoxygenase Products

The HPLC system is composed of a U6K injector, two model 510 pumps, gradient controller, a column oven, and a variable-wavelength detector (Waters, Milford, MA). In more recent studies, we also used a diode array detector (System 990-plus, from Waters), which permits on-line spectral analysis of individual peaks of the chromatograms and even allows the distinction of the discrete spectral differences of LTB_4 isomers and derivatives at the picomole level.[1] The neutrophil products are separated on a Nucleosil C_{18} column (250 × 4.6 mm; Alltech, Deerfield, IL) with a mobile phase of methanol/water/acetic acid (75 : 25 : 0.01, v/v/v) at a flow rate of 1 ml/min at 35° (Fig. 2).[5] Retention times (in minutes) of synthetic trienes are: ω-oxidized leukotrienes, 4.55; PGB_2 (internal standard), 7.81; 6-*trans*-LTB_4, 9.5; 12-epi-6-*trans*-LTB_4, 10.0; LTB_4 and 5(*S*),12(*S*)-DiHETE, 10.8; LTB_4 lactone, 15.0; 5,6-DiHETEs, 16.6 and

[3] J. A. Maclouf and R. C. Murphy, *J. Biol. Chem.* **263,** 174 (1988).
[4] C. Edenius, K. Heidvall, and J. A. Lindgren, *Eur. J. Biochem.* **178,** 81 (1988).
[5] C. A. Dahinden, R. M. Clancy, M. Gross, J. M. Chiller, and T. E. Hugli, *Proc. Natl. Acad. Sci. U.S.A.* **82,** 6632 (1985).

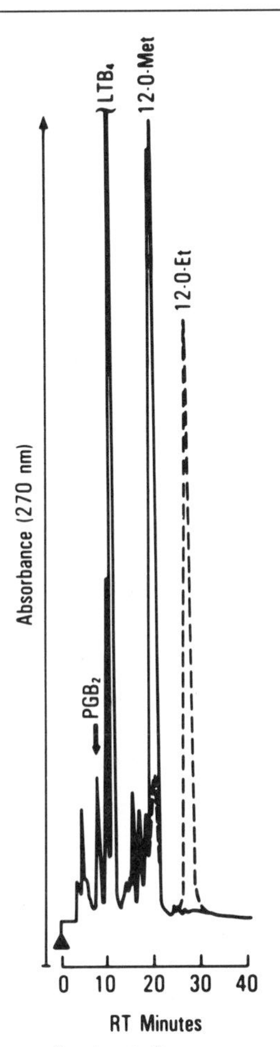

FIG. 2. HPLC chromatograms of extracts from supernatants trapped with either methanol (solid line) or ethanol (interrupted line) obtained after stimulation of 10^8 PMN with ionophore for 5 min.

18.0; 12-*O*-methyl-6-*trans*-LTB4 isomers, 19.2 and 19.7; and 12-*O*-ethyl-6-*trans*-LTB_4 isomers, 27.2. More recently, we have established the following HPLC procedure for the analysis of LOX metabolites of PMN and platelets[1]: Stationary phase: Nucleosil C_{18}-100-5 μm in a 25 $\times$ 0.4 cm column and a 5 $\times$ 0.4 cm precolumn, thermostatted at 35°. Mobile phase: Buffer A, water/acetonitrile/tetrahydrofuran/acetic acid (75 : 25 : 0.15 : 0.01, v/v/v/v); Buffer B, acetonitrile/methanol/acetic acid (57 : 43 : 0.01, v/v/v). Flow rate 1 ml/min. Gradient program: Initial, 100% A, 0% B; 0.01 min, 80% A, 20% B; isocratic until 7.5 min; 7.5–8.0 min, linear to 45% A, 55%

B; isocratic until 17.5 min; 17.5–18 min, linear to 25% A, 75% B; isocratic until 27.7 min; 27.7–28.2 min, back to initial. Retention times with baseline separation for each compound are as follows (in min): 20-COOH-LTB_4, 12.39; 20-OH-LTB_4, 13.17; 20-OH-5,12-DiHETE, 14.06; PGB_2 (internal standard), 18.03; 6-*trans*-LTB_4, 20.36; 12-epi-6-*trans*-LTB_4, 20.66; LTB_4, 21.05; 5(*S*),12(*S*)-DiHETE, 21.55; HHT (cyclooxygenase product), 23.71; 5,6-DiHETEs, 25.42 and 25.76; 12-*O*-methyl-6-*trans*-LTB_4, 26.54; 15-HETE, 27.09; 12-HETE, 27.71; 5-HETE, 28.43.

Measurements of LTA_4 Release by Alcohol Trapping

Trapping is performed according to the procedure of Borgeat and Samuelsson.[5,6] Five minutes after exposure of PMN to 10 μm ionophore A23187 (Calbiochem, La Jolla, CA) the cells are sedimented in microfuge tubes (15,000 *g*, 30 sec, room temperature) and the supernatant immediately mixed with 9 volumes of methanol or ethanol previously acidified with sufficient HCl to lower the pH of the aqueous buffer to 3. After 3 min, the samples are neutralized with NaOH and the alcohol evaporated until approx. 2 volumes alcohol is left, PGB_2 is added as an internal standard, and precipitated proteins are removed by centrifugation. (It should be noted that prolonged exposure of the leukotrienes to acidified alcohol can result in the formation of lactones). The sample is mixed with 2.5 volumes of diethyl ether and after a second acidification to pH 3 with HCl, a biphasic mixture is formed by the addition of 4 volumes water. The organic phase is dried under nitrogen and further purified by silicic acid chromatography [Silicar CC-4, Mallinckrodt; the sample is applied and washed in diethyl ether/hexane, 20 : 80 (v/v) and eluted with ethyl acetate]. Recoveries of LTB_4 and the LTA_4 hydrolysis products vary between 75–90% and are identical to that of the PGB_2 internal standard. Clean up of the samples by silicic acid chromatography is not always necessary, and other extraction methods may also be useful, since any method that allows equally good recoveries of LTB_4 and 5-HETE will also be suitable for alcohol-trapped LTA_4, these products being intermediate in hydrophobicity.

The LTA_4 derivatives can be identified by (1) coelution with synthetic standards in HPLC using different solvents; (2) the typical UV absorption spectra with λ_{max} at 268.0 nm and shoulders at 258.5 and 279.5 nm indicative of an all-*trans*-triene configuration (Fig. 3); (3) gas chromatography–mass spectrometry (GC–MS) of the purified compounds as described.[5,6] However, we feel that GC–MS is not always necessary for identification of trapped LTA_4, if the products are only formed after alcohol trapping and

[6] P. Borgeat and B. Samuelsson, *Proc. Natl. Acad. Sci. U.S.A.* **76,** 3213 (1979).

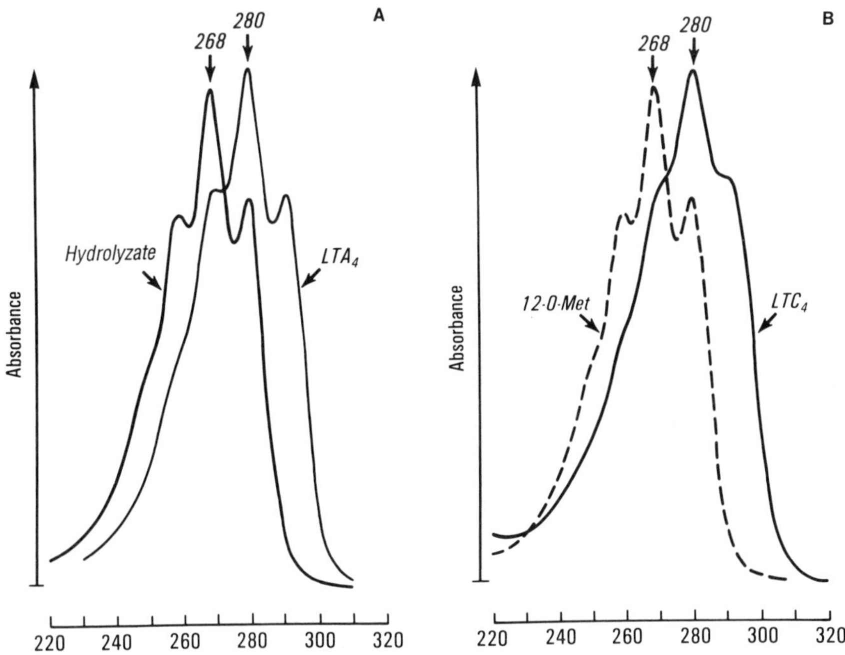

FIG. 3. UV spectra of trienes in methanol. (A) UV spectra of synthetic LTA_4 and of the products (12-*O*-methyl derivatives) which are immediately formed after addition of HCl. (B) UV spectra of 12-*O*-methyl-6-*trans*-LTB_4 purified from methanol-trapped supernatants of stimulated PMN (interrupted line), and of LTC_4 purified from MC after exposure to albumin-bound LTA_4 (solid line).

are absent under otherwise identical conditions. Also, 12-*O*-methyl-6-*trans*-LTB_4 epimers always elute after the 5,6-DiHETEs (λ_{max} at 272 nm) in all solvent systems examined so far, and we never detect a leukocyte-derived triene with a λ_{max} at 268 nm other than trapped LTA_4 in this region of the chromatogram.

Figure 2 demonstrates that the epoxide LTA_4 is released by ionophore-stimulated PMN without considerable intracellular nonenzymatic hydrolysis. In seven experiments performed in duplicates we found that a mean of 136 pmol $LTA_4/10^7$ PMN (range 40–300) could be trapped by alcohols in the supernatants of PMN 5 min after ionophore stimulation.[5] These data demonstrate that LTA_4 is a major LOX product released by stimulated PMN. However, these values are only minimal estimates, since (1) some LTA_4 may be released but remain cell-associated; (2) some LTA_4 may decay until trapping is performed even in the presence of albumin[7]; (3) a

[7] A. Fitzpatrick, D. R. Morton, and M. A. Wynalda, *J. Biol. Chem.* **257,** 4680 (1982).

fraction of LTA_4 will react with residual water after acidification. Our findings have been confirmed and extended by several groups which showed that PMN-derived LTA_4 can be metabolized into LTB_4 and LTC_4 by cells which are by themselves unable to produce leukotrienes.[3,4,8–10] The enhancement of LTB_4 or LTC_4 formation through intercellular transfer of LTA_4 is (values expressed per 10^7 PMN): 250–300 pmol LTB_4 (PMN–erythrocytes)[8]; 350 pmol LTC_4 by coincubation 8 min after PMN activation, or 105 pmol from PMN supernatants (PMN-platelets)[3]; 72 pmol LTC_4 from coincubations (PMN-platelets)[4]; 90 pmol $LTC_4/D_4/E_4$ and 75 pmol LTB_4 by coincubation of PMN with endothelial cells.[9] Although other types of cellular interactions may exist in coincubation experiments,[4,9] these values are in good agreement with our trapping experiments.

It is interesting that a recent report demonstrated the formation of lipoxins in coincubation experiments of PMN and platelets stimulated with ionophore, probably through the intermediate of LTA_4 released by PMN.[11] Up to now, lipoxins were only formed under very artificial conditions. Therefore, intercellular transfer of LTA_4 may also be the major pathway of lipoxin formation.

There are no other established methods for the measurement of cell-derived LTA_4. Gut *et al.*[12] recently described a procedure for the measurement of low amounts of LTA_4 which involves reversed-phase HPLC in a volatile buffer followed by direct chemical ionization mass spectrometry, but methods for efficient LTA_4 extraction from complex biological fluids have not been worked out. It is quite possible, however, that intracellular decay of LTA_4 does not occur at all. Indeed, no all-*trans*-LTB_4 isomers were detected and large amounts of LTC_4 were formed when ionophore-stimulated PMN were infused in lung preparations,[13] indicating that all the PMN-derived LTA_4 was enzymatically metabolized by the tissue cells. Thus, an estimate of LTA_4 release may be simply made by measuring the sum of the nonenzymatic products. However, estimates of LTA_4 release or consumption based upon the measurement of all-*trans*-LTB_4 isomers should be interpreted with caution if alcohols are added to a sample that has not been previously acidified (see, i.e., Refs. 4 and 9; in

[8] J. E. McGee and F. A. Fitzpatrick, *Proc. Natl. Acad. Sci. U.S.A.* **83,** 1349 (1986).

[9] H. E. Claesson and J. Haeggstrom, *Eur. J. Biochem.* **173,** 93 (1988).

[10] S. J. Feinmark and P. J. Cannon, *J. Biol. Chem.* **261,** 16466 (1986).

[11] C. Edenius, J. Haeggstrom, and J. A. Lindgren, *Biochem. Biophys. Res. Commun.* **157,** 801 (1988).

[12] J. Gut, J. R. Trudell, and G. C. Jamieson, *Biomed. Environ. Mass. Spectrom.* **15,** 509 (1988).

[13] F. Grimminger, M. Menger, G. Becker, and W. Seeger, *Blood* **72,** 1687 (1988).

these papers the possible formation of 12-*O*-ethyl-LTA_4 derivatives has not been considered).

Metabolism of Albumin-Bound LTA_4 into LTC_4 by Mast Cells

Analysis of Sulfidoleukotrienes

Cellular reactions are stopped by adding 0.6 volumes of 2-propanol and allowing the mixture to stand at room temperature (T) for at least 5 min (leukotrienes are stable in this mixture for at least several hours at 4°). Immediately after acidifying with 0.03 volumes of 5 M formic acid, 1.5 volumes of ether is added, resulting in the development of two phases. The organic phase is harvested and 0.015 volumes of 10 N NH_4OH is added. After evaporation to dryness under nitrogen the residue is dissolved in 0.5 ml of $HCCl_3$/methanol (1 : 1) and 0.22 ml of 10 mM NH_4OH is added. The upper water/methanol phase containing the sulfidoleukotrienes.[14] is collected, dryed, and dissolved in HPLC solvent. The sulfidoleukotrienes are separated on a Nucleosil C_{18}-100 5-μm column (250 × 4.6 cm) with a mobile phase of methanol/3.5 mM ammonium acetate in water, pH 5.7 (apparent pH) (65 : 35, v/v), at a flow rate of 1 ml/min.[5,15]

LTC_4 Formation by Mast Cells

Mast cells (MC) are obtained from cultures of bone marrow cells of BDF_1 mice in Dulbecco's modified Eagle's medium supplemented with 10% heat-inactivated horse serum and 10% WEHI 3 supernatant (containing interleukin 3).[15] After 8 weeks of culture and selection of the nonadherent cells every 3–4 days, the cells are >95% MC as judged by staining with toluidine blue, the presence of FcϵI receptors, and the production of LTC_4 upon activation with ionophore or IgE receptor cross-linking.[5] MC are washed and suspended in DPBS/A at 10^7 cells/ml.

After evaporation to dryness under nitrogen, LTA_4 methyl ester (1 mM) is hydrolyzed in ethanol/50% NaOH in water (9 : 1), at 4° for 3 hr. Alternatively, the lithium salt is obtained as described.[8] The concentration, integrity, and purity of LTA_4 can be estimated by UV spectroscopy (Fig. 2), HPLC,[8,11] and trapping with alcohols followed by HPLC. LTA_4, which is prepared fresh for each experiment, is rapidly mixed with ice-cold DPBS/A at the desired concentration, warmed to 37° just before use, and added to MC in DPBS/A in a 1 : 1 volume ratio (the final concen-

[14] R. M. Clancy and T. E. Hugli, *Anal. Biochem.* **133,** 30 (1983).

[15] E. Razin, L. C. Romeo, S. Krilis, F.-T. Liu, R. A. Lewis, E. J. Corey, and K. F. Austen, *J. Immunol.* **133,** 983 (1984).

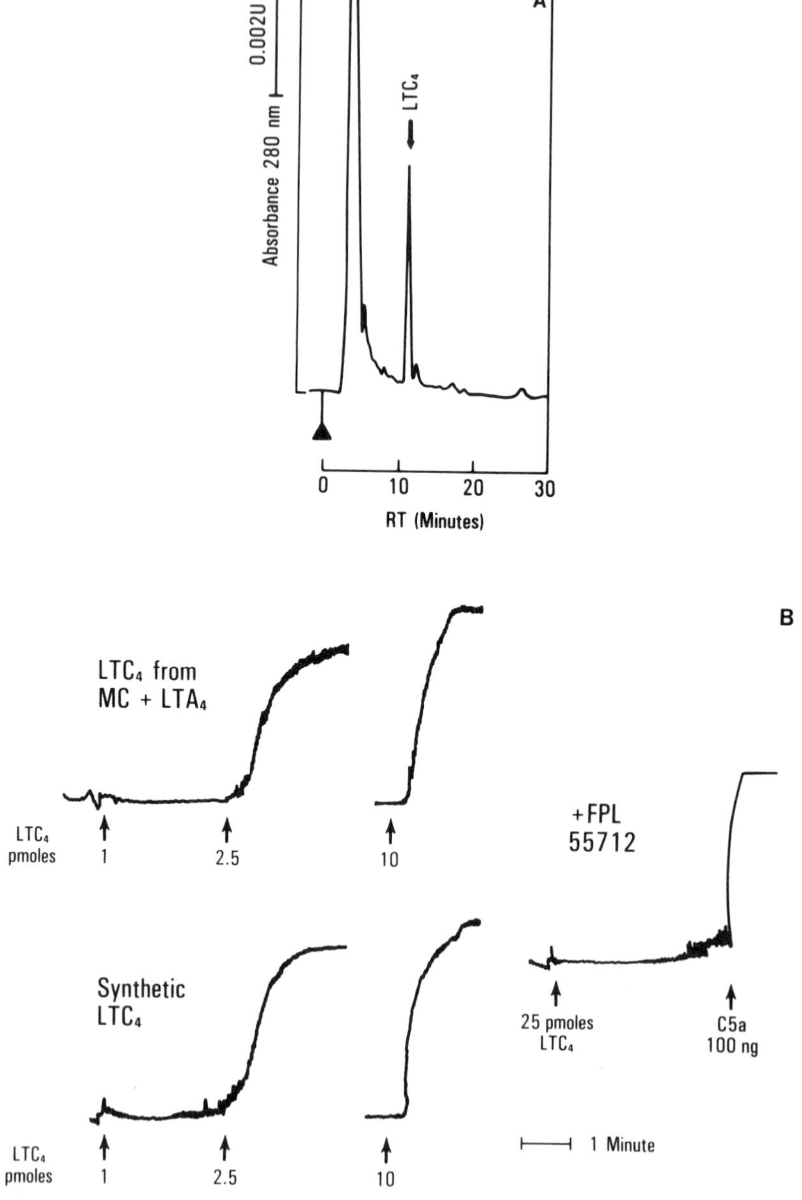

FIG. 4. LTC_4 produced by mast cells from albumin-bound LTA_4. (A) HPLC chromatogram of the extract from 5×10^6 MC after incubation with 100 pmol LTA_4 for 10 min. (B) Ileum contractions induced by synthetic LTA_4 and MC-derived LTC_4 purified by HPLC. The response to LTC_4 but not C5a is abolished by 5 μM FPL 55712.

tration of ethanol does not exceed 0.2%). After the indicated time, the reaction is stopped by adding 0.6 volumes of 2-propanol.

The metabolism of albumin-bound LTA_4 by MC is very efficient and rapid.[5] More than 50% of 2 nmol LTA_4 is metabolized into LTC_4 by 10^7 MC within 10 to 15 min and no conversion to LTD_4 and LTE_4 is observed during up to 30 min incubation at 37°.[5] The production of LTC_4 is linearly correlated with the concentration of LTA_4 added from 0.1 to 2 μM LTA_4. Most important, when 5×10^6 MC are exposed to 100 pmol albumin-bound LTA_4 for 10 min, 80–87 pmol LTC_4 (mean 83, $N = 6$) could be measured in the cell cultures (Fig. 4). Thus, MC are able to transform TLA_4 even if bound to albumin and in amounts shown to be released by stimulated PMN.

The pronounced stabilizing effect of albumin[7] together with the exceedingly short half-life of LTA_4 in neutral buffers indicates that LTA_4 is bound to albumin with high affinity. The efficient and rapid metabolism of low amounts of LTA_4 by MC is, therefore, surprising and may suggest the presence of an LTA_4 binding site on MC membranes.

Labeling experiments demonstrate that LTC_4 produced by MC derived exclusively from exogenous LTA_4.[5] When tritiated LTA_4 at variable concentrations is added to MC, the LTC_4 produced has the same specific activity. Furthermore, MC are labeled with 0.5 μCi tritiated arachidonic acid per milliliter (86 Ci/mmol, Amersham) overnight, washed, and cultured for an additional 3 hr in label-free medium. After two washes, the labeled MC are then exposed to either 5 μM ionophore or 1 nmol LTA_4 for 10 min. After purification by HPLC, radioactivity is found only in the LTC_4 fraction from ionophore-stimulated MC (95,000 cpm/nmol).

LTC_4 produced by MC after exposure to albumin-bound LTA_4 is identified after purification by HPLC by the following criteria: coelution with synthetic LTC_4 on reversed-phase HPLC (Fig. 4A); UV spectroscopy (Fig. 3B), bioactivity on guinea pig ileum (Fig. 4B); and hydrolysis of 4 nmol of LTC_4 in 6 M HCl at 110° for 24 hr and amino acid analysis with a Beckman model 121M analyzer, which demonstrates the presence of glutamic acid and glycine in equimolar concentrations.[5]

Acknowledgments

This work was supported in part by the Swiss National Science Foundation grant #3.058-0.87. We are grateful for the contributions of R. M. Clancy, M. Gross, J. M. Chiller, and T. E. Hugli.

[63] Interaction between Platelets and Lymphocytes in Biosynthesis of Prostacyclin

By KENNETH K. WU and AUDREY C. PAPP

Prostacyclin is a potent arachidonate metabolite with diversified biological activities. It is synthesized by endothelial cells via the cyclooxygenase pathway catalyzed by three major enzymes, i.e., phospholipase, which liberates arachidonic acid (AA) from phospholipids, prostaglandin endoperoxide (G/H) synthase converting arachidonic acid into PGG_2 and then PGH_2, and prostacyclin synthase which generates PGI_2 from PGH_2. PGI_2 can also be generated through the metabolic cooperation of two different types of cells. For example, interaction between platelets and lymphocytes leads to synthesis of PGI_2.[1] Human lymphocytes are capable of taking up arachidonic acid and incorporating it into the membrane phospholipids, yet they possess minimal ability to convert AA into metabolites.[2] Human blood platelets, on the other hand, are capable of synthesizing and releasing a large quantity of prostaglandin endoperoxides, i.e., PGH_2, but cannot convert PGH_2 into PGI_2. When platelets mixed with lymphocytes are stimulated with physiological agonists, a significant quantity of PGI_2 is generated. In this chapter we describe the strategies and methods used to unravel this important mechanism of eicosanoid synthesis.

Preparation of Platelets and Lymphocytes

Platelets and lymphocytes are isolated from the same donor's blood. We usually draw 64 ml of venous blood from healthy subjects into polypropylene tubes containing 16 ml acid–citrate–dextrose (ACD) anticoagulant [blood to ACD (v/v) = 4/1]. After mixing, the samples are centrifuged at 200 *g* for 10 min at ambient temperature. The platelet-rich supernatant is collected. The remaining pellet is diluted with calcium-free Tyrode's buffer, pH adjusted to 6.8 with ACD, to a final volume of 80 ml. After mixing, the sample is centrifuged under similar conditions and the supernatant is again collected and added to the initial supernatant. This fraction is further

[1] K. K. Wu, A. C. Papp, C. E. Manner, and E. R. Hall, *J. Clin. Invest.* **79**, 1601 (1978).
[2] M. E. Goldyne and J. D. Stobo, *Prostaglandins* **24**, 623 (1982).

processed for platelet isolation while the pellet is used for lymphocyte preparation.

For platelet preparation, the platelet-rich fraction is diluted with 40 ml of calcium-free ACD–Tyrode's buffer, pH 6.8, laid over 5 ml of 40% bovine serum albumin, and centrifuged at 1000 *g* for 15 min at room temperature. The upper plasma layer is discarded and the platelet layer is gently transferred into a polypropylene tube containing 6 ml of ACD–Tyrode's buffer. This washing procedure is repeated once and the platelet suspension is laid over 2 ml of Ficoll-Hypaque and centrifuged at 1000 *g* for 15 min at room temperature. The platelet layer is collected and resuspended in Tyrode's buffer, pH 7.35, containing 5 m*M* calcium and 0.5 mg/ml fibrinogen.

For lymphocyte preparation, the pellet is diluted with calcium-free ACD-Tyrode's buffer, pH 6.8, to a final volume of 140 ml, laid over Ficoll-Hypaque, and centrifuged at 800 *g* for 30 min at room temperature. The lymphocyte layer is collected, diluted with HBSS, and centrifuged at 600 *g* for 10 min to remove contaminated platelets present in the upper layer. The lymphocyte layer is collected. At times, the lymphocyte layer is contaminated by red blood cells which are lysed by adding 2 ml of distilled water to the lymphocyte preparation. The preparation is vortexed at room temperature for 10 sec and centrifuged at 1000 *g* at room temperature for 10 min to remove the red cell debris. The lymphocytes are then suspended in calcium-free Tyrode's buffer, pH 7.35.

The platelet concentration is determined in a Coulter counter and the lymphocyte concentration by hemocytometer. As the lymphocyte preparation often contains other cells as well, we stain the lymphocyte preparation on a smear and perform differential counts. In a typical experiment, the lymphocyte preparation is contaminated with 13.7% platelets, 0.6% monocytes, 0.4% erythrocytes, and 0.2% neutrophils. Although monocytes may be a potential source of 6-keto-$PGF_{1\alpha}$, the monocyte concentration in the lymphocyte preparation, i.e., 6×10^4 monocytes per 10^7 lymphocytes, is too small to contribute to 6-keto-$PGF_{1\alpha}$ synthesis. The platelet contamination generally is between 10 and 15%. The platelet preparation has minimal contamination of other cells. The cell contamination is confirmed by scanning electron microscopy.

General Strategies for Cell–Cell Interaction Experiments

The first step is to establish the arachidonate metabolic profile of the individual lymphocyte and platelet preparation and compare the individual metabolic profile with that of a mixture of platelets and lymphocytes. This is best achieved by treating the cells with [1-^{14}C]arachidonic acid and analyzing the metabolites by reversed-phase HPLC. The 12-HETE generation in the lymphocyte preparation can serve as a useful marker of significant platelet and monocyte contaminations. In establishing the platelet metabolic profile, inhibitors of cyclooxygenase activity (indomethacin) and thromboxane synthase [1-benzylimidazole (1-BI)] are used. The thromboxane synthase inhibitor is particularly important since it should cause an increased amount of PGH_2. The metabolic profile of lymphocyte–platelet mixture should have a discernible 6-keto-$PGF_{1\alpha}$ peak

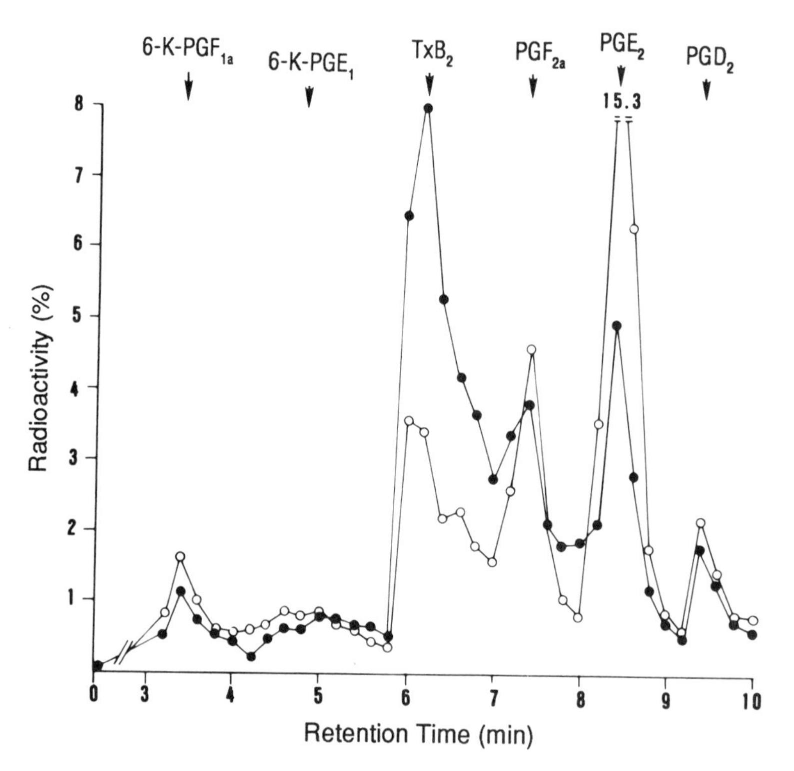

FIG. 1. Analysis of 6-keto-$PGF_{1\alpha}$ and other prostanoids by reversed-phase HPLC using the initial isocratic solvent system which excluded HHT, 12-HETE, and AA. One milliliter of lymphocyte preparation (1×10^7/ml), platelet preparation (3×10^8/ml) or a mixture of lymphocyte–platelet preparation is incubated with 10 μM of [1-^{14}C]AA at 37° for 3 min. The reaction is terminated and the eicosanoids extracted and analyzed by reversed-phase HPLC. Each arrow refers to the relative retention time of each eicosanoid determined with tritiated standards. Lymphocytes alone generated minimal 6-keto-$PGF_{1\alpha}$ (data not shown). The mixed lymphocyte–platelet preparation (●—●) produced discernible 6-keto-$PGF_{1\alpha}$ and other eicosanoids. Addition of 1-BI (○—○) led to enhanced 6-keto-$PGF_{1\alpha}$ peak. By contrast, addition of aspirin or indomethacin eliminated the 6-keto-$PGF_{1\alpha}$ peak (not shown). Taken from Wu *et al.*[3]

if, indeed, the two cells cooperate in synthesizing a significant quantity of PGI_2. Addition of thromboxane synthase inhibitors to the cell mixture should enhance this peak while cyclooxygenase inhibitors should suppress it. The 6-keto-$PGF_{1\alpha}$ so generated should be further confirmed quantitatively by RIA. Figure 1 illustrates the validity of this approach.

The second step is to determine the direction of metabolic cooperation. This is done by prelabeling platelets and lymphocytes individually with [1-^{14}C]AA. Labeled platelets are mixed with unlabeled lymphocytes and physiological agonists of platelets such as thrombin, collagen, and ionophore A23187 are added and the metabolites analyzed by reversed-phase HPLC. The unlabeled lymphocyte preparation is treated with aspirin or indomethacin to block the cyclooxygenase activity. The labeled platelet preparation is treated with a thromboxane synthase inhibitor for further proving the shunting of PGH_2 from activated platelets to lymphocytes. Experimental data that validate this approach are shown in Table I.

TABLE I
EICOSANOID GENERATION BY PLATELET, LYMPHOCYTE, AND MIXED PLATELET–LYMPHOCYTE PREPARATIONS[a]

	Eicosanoids (%)						
Agonists	6-KP	Tx	$F_{2\alpha}$	E_2	D_2	12-HETE	HHT
Ionophore (10 μM)							
*L	0	7.1	0	1.3	0	13.3	6.3
*P	0	13.7	0	1.9	0	14.0	6.1
*P + BI	0	2.1	2.4	11.5	13.2	13.0	3.0
*P + L	1.2	10.0	0	0	0	13.7	7.9
*P + L + BI	2.0	2.5	3.5	5.9	2.7	7.8	2.0
Thrombin (0.5 U/ml)							
*P	0	8.3	0	1.8	1.2	12.8	5.9
*P + BI	0	2.7	3.3	4.7	5.2	7.5	1.6
*P + L	1.0	9.3	0	1.3	1.2	10.3	7.4
*P + L + BI	1.5	2.4	4.2	6.5	5.0	8.0	3.1
Collagen (10 μg/ml)							
*P	0	4.3	0	2.1	1.6	3.2	2.6
*P + BI	0	1.8	1.9	2.1	2.3	4.9	3.0
*P + L	1.3	4.6	0	1.4	0	3.2	2.6
*P + L + BI	2.0	4.0	4.2	6.1	4.2	4.5	2.6

[a] Experimental details are described in the text. Following the incubation of cell preparations with each agonist, the reaction is terminated, and the eicosanoids are extracted and analyzed by reversed-phase HPLC. The percentage of each eicosanoid refers to the percentage at the peak point. Zero denotes no discernible peak. *L and *P denote lymphocytes and platelets prelabeled with [^{14}C]arachidonic acid, respectively. L denotes aspirin-treated unlabeled lymphocytes. BI, 1-Benzylimidazole; 6-KP, 6-keto-$PGE_{1\alpha}$; Tx, TxB_2, $F_{2\alpha}$, E_2, and D_2, $PGF_{2\alpha}$, PGE_2, and PGD_2, respectively. Taken from Wu *et al.*[3]

The third step is to test directly that lymphocytes possess prostacyclin synthase activity. We treat lymphocytes with ultrapurified PGH_2 and meaure the production of 6-keto-$PGF_{1\alpha}$ by RIA.

The final step is to evaluate the influence of lymphocytes on platelet aggregation and release reaction in response to physiological agonists. In these experiments, the influence of lymphocytes on light transmission in a platelet aggregometer should be corrected by adding lymphocytes of comparable concentration to platelet-poor plasma which serves as an arbitrary 100% of light transmission.

Methods for carrying out each step of experiments are given below.

Arachidonic Acid Metabolism in Individual and Mixed Platelet and Lymphocyte Preparations

One milliliter of lymphocyte preparation (10^7/ml), platelet preparation (3×10^8/ml), or a mixture of lymphocyte–platelet preparation containing equal cell concentrations as in the individual cell preparation is placed into a platelet aggregometer cuvette under constant stirring at 1000 rpm at 37°. [1-^{14}C]AA, 10 μM (56 mCi/mmol) is added and at the end of 3 min the reaction is terminated by acidification to pH 3.0. The eicosanoids are extracted using a Sep-Pak cartridge and analyzed by reversed-phase HPLC, using an isocratic solvent system of 35% acetonitrile in acidified water (0.1% acetic acid in water, pH 3.2) to separate the prostanoid. After the initial isocratic period, the solvent composition is changed to 50% acetonitrile in acidified water and runs for 10 min to elute HHT, leukotrienes, and HETE. Next, a slight convex gradient is run from 50 to 75% acetonitrile for 8 min and then to 100% acetonitrile. The fractions are collected and the radioactive counts determined in a scintillation counter. Neither platelets nor lymphocytes alone have a noticeable 6-keto-$PGF_{1\alpha}$ peak. The lymphocyte preparation generates small TxB_2, PGE_2, and $PGF_{2\alpha}$ peaks due to platelet contamination. The mixture of lymphocyte and platelet preparation yields a discernible 6-keto-$PGF_{1\alpha}$ peak and large TxB_2 and PGE_2 peaks (Fig. 1). Treatment of the mixture with 1-BI enhances the 6-keto-$PGF_{1\alpha}$ peak with a concurrent reduction in the TxB_2 and PGE_2 peaks (Fig. 1). To quantify 6-keto-$PGF_{1\alpha}$, the platelet (3×10^8/ml) and lymphocyte (1×10^7/ml) preparation individually or in mixture is incubated with unlabeled AA at 37° for 3 min, the 6-keto-$PGF_{1\alpha}$ separated by HPLC, and its content is measured by RIA.[3] Only the mixture produces a significant quantity of PGI_2 (1.39 ng/ml/3 min) and pretreating the mixture with 1-BI increases the 6-keto-$PGF_{1\alpha}$ content (2.56 ng/ml/3 min). The

[3] K. K. Wu, E. R. Hall, Rossi, E. C., and A. C. Papp, *J. Clin. Invest.* **75,** 168 (1985).

mixture of lymphocytes and platelets also produces a significant quantity of 6-keto-$PGF_{1\alpha}$ in response to ionophore stimulation. The level is comparable to that produced by the cell mixture treated with AA (1.21 ng/10^7 lymphocytes/3 min vs 1.39 ng). To further demonstrate cell cooperation, the buffy coat cells from citrated venous blood are washed and incubated with 10 μM AA, 10 μg/ml collagen, or 0.1 U/ml thrombin at 37° for 3 min. The reaction is terminated by rapid centrifugation and the 6-keto-$PGF_{1\alpha}$ content in the supernatant measured by RIA. The 6-keto-$PGF_{1\alpha}$ content is typically 3.42, 1.42, and 1.6 ng, respectively, while the vehicle-treated controls have no measurable 6-keto-$PGF_{1\alpha}$.[3] Taken together, these studies clearly show that a significant quantity of PGI^2 is generated through the cooperation of platelets and lymphocytes.

Evidence that Platelet-Derived PGH_2 Is Used by Lymphocytes in Biosynthesis of PGI_2

The lymphocyte and platelet preparations are incubated with 2 μM of [1-^{14}C]AA at 37°6 for 2 hr. The cells are washed twice to remove free [1-^{14}C]AA. Lymphocytes labeled with [1-^{14}C]AA are stimulated with phytohemagglutinin (PHA) and the eicosanoids analyzed by HPLC. PHA fails to induce any eicosanoid synthesis. Labeled lymphocytes also do not generate 6-keto-$PGF_{1\alpha}$ when treated with ionophore, but TxB_2, HHT, and 12-HETE peaks are apparent. When prelabeled platelets are suspended together with aspirin-treated lymphocytes and stimulated with ionophore, thrombin or collagen, a 6-keto-$PGF_{1\alpha}$ peak is discernible with each of the three agonists and the peak is enhanced by 1-BI (Table I). A mixture of boiled lymphocytes (100°, 3 min) and labeled platelets no longer synthesize PGI_2 in response to these agonists. These results indicate that the stimulated platelets synthesize and release PGH_2 which is taken up by lymphocytes for the biosynthesis of PGI_2. Blocking of platelet thromboxane synthase with 1-BI leads to the an increment in PGH_2 and, hence, increased PGI_2 synthesis.

Prostacyclin Synthase Activity in Lymphocytes

To provide direct evidence that lymphocytes possess prostacyclin synthase, we incubate the lymphocyte suspension with 5 μM ultrapurified PGH_2 which was prepared according to Zulak *et al.*[4] The supernatant is applied to HPLC and the 6-keto-$PGF_{1\alpha}$ content in the HPLC fractions corresponding to the 6-keto-$PGF_{1\alpha}$ is measured by RIA. The conversion of

[4] I. M. Zulak, M. L. Puttermans, A. B. Schilling, E. R. Hall, and D. L. Venton, *Anal. Biochem.* **154,** 152 (1986).

PGH_2 to 6-keto-$PGF_{1\alpha}$ is 4.1 ng/10^7 lymphocytes/3 min. PGH_2 incubated in buffer alone under similar experimental conditions is not spontaneously hydrolyzed to 6-keto-$PGF_{1\alpha}$, although it is converted to PGE_2 and $PGF_{2\alpha}$.

Influence of Lymphocytes on Platelet Function

The premise is that PGI_2 produced through the cooperation of activated platelets and lymphocytes should by physiologically significant in inhibiting platelet function. We evaluate platelet aggregation by the procedure of Born[5] and platelet release reaction using [1-^{14}C]serotonin as the marker.[6] In the aggregation experiments, the light transmission is standardized by adding lymphocytes to the buffer. The platelet–lymphocyte mixture is placed in the aggregometer under constant stirring at 37°. Agonists (thrombin, collagen, AA) or vehicles are added and the change in light transmission is monitored. Addition of vehicle has no effect on platelet aggregation while each specific agonist causes a significant inhibition in platelet aggregation. Similar results are obtained with the release of [1-^{14}C]serotonin. Inhibition of platelet aggregation by lymphocytes is related to lymphocyte concentration present in the platelet preparation. The 6-keto-$PGF_{1\alpha}$ content in these mixture experiments corresponds to the degree of inhibition by concurrent *in vitro* experiments. These experiments clearly show that lymphocytes even at 1×10^6/ml inhibit platelet aggregation and secretion induced by physiological stimuli.

Conclusion

By several different assays coupled with stringent controls, human blood lymphocytes can be shown to possess prostacyclin synthase activity and that they can readily take up PGH_2 released from activated platelets and convert it into PGI_2.[3] The quantity of PGI_2 generated by physiological concentrations of platelets and lymphocytes under the stimulation of physiological agonists (collagen and thrombin) exhibits inhibitory effects on platelet aggregation and release. This represents an important mechanism whereby PGI_2 acts as an effective autacoid in modulating inflammation, immune responses, and hemostasis and thrombosis.

Acknowledgments

This work was supported by a grant from the National Institutes of Health (P01 NS-18494 and P50 NS-23327). We wish to thank Nancy Fernandez for excellent secretarial assistance.

[5] G. V. R. Born, *Nature (London)* **194,** 927 (1962).

[6] H. J. Weiss, T. Tschopp, H. Brand, and J. Rogers, *J. Clin. Invest.* **54,** 421 (1974).

[64] Eicosanoid Interactions between Platelets, Endothelial Cells, and Neutrophils

By AARON J. MARCUS

Introduction

Eicosanoid research initially focused on metabolites produced by individual cells upon activation.[1,2] Given that eicosanoids are autacoids which exert their biological effects in the microenvironment, a multitude of cell types in proximity may be capable of metabolizing precursors, intermediates, and presumed end products of this pathway.[3] We initially noted this phenomenon[4] in 1980 when stimulated suspensions of platelets and endothelial cells were found to produce more prostacyclin (PGI_2) than activated endothelial cells alone. Also, we demonstrated biochemically that aspirin-treated endothelial cells utilized released platelet endoperoxides to synthesize PGI_2. This concept was later extended to platelet–neutrophil interactions,[5] wherein under conditions of ionophore A23187 stimulation, leukotriene B_4 (5*S*,12*R*-DiHETE), 5*S*,12*S*-dihydroxyeicosatetraenoic acid (5*S*,12*S*-DiHETE), and 5*S*-hydroxyeicosatetraenoic acid (5-HETE) were produced by neutrophils from platelet precursors. These interactions were stimulus-specific in the sense that ionophore had activated *both* the platelets and neutrophils. Addition of [^{3}H]5-HETE to stimulated platelets resulted in 5*S*,12*S*-DiHETE production. Conversely, when ^{3}H-labeled 12*S*-hydroxyeicosatetraenoic acid (12-HETE), a platelet product, was added to stimulated neutrophils, 5*S*,12*S*-DiHETE also formed.

When thrombin or collagen was added to a suspension of platelets and neutrophils, only traces of 5*S*,12*S*-DiHETE formed, because only the platelets were stimulated by these agonists. Instead, a new metabolite,

[1] A. J. Marcus, *J. Lipid Res.* **19,** 793 (1978).

[2] A. J. Marcus, *J. Lipid Res.* **25,** 1511 (1984).

[3] A. J. Marcus, *in* "Inflammation: Basic Principles and Clinical Correlates" (J. I. Gallin, I. M. Goldstein, and R. Snyderman, eds.), pp. 129–137. Raven, New York, 1988.

[4] A. J. Marcus, B. B. Weksler, E. A. Jaffe, and M. J. Broekman, *J. Clin. Invest.* **66,** 979–986 (1980).

[5] A. J. Marcus, M. J. Broekman, L. B. Safier, H. L. Ullman, N. Islam, C. N. Serhan, L. E. Rutherford, H. M. Korchak, and G. Weissmann, *Biochem. Biophys. Res. Commun.* **109,** 130 (1982).

12*S*-20-dihydroxyeicosatetraenoic acid (12,20-DiHETE) was synthesized by unstimulated neutrophils from platelet 12-HETE.[6] Subsequently, it was found that 12,20-DiHETE had formed via an ω-hydroxylation of 12-HETE by a cytochrome *P*-450 system in the neutrophil.[7] We recently found that the neutrophil continues to metabolize 12,20-DiHETE to 12-HETE-1,20-dioic acid via a dehydrogenase mechanism.[8] These reactions involving lipoxygenase products from the neutrophil and platelet are of potential significance for the inflammatory process, hemostasis, and thrombosis, because they are *aspirin-insensitive*. They also provide evidence for the occurrence of multicellular metabolic events (transcellular metabolism) which are an integral component of host-defense mechanisms.

It is now possible to construct a working classification of cell–cell interactions in the eicosanoid pathway.[3,9] In cell–cell interaction, Type I, different cells are capable of sharing a common precursor. In the subgroup Type IA, a given cell is capable of producing its own precursor, but if it obtains the same precursor from a neighboring cell, more endogenous eicosanoids can be synthesized. Thus, platelet-derived endoperoxides were utilized by endothelial cells for the formation of PGI_2. In Type IB, a given cell is unable to produce a precursor endogenously, but it possesses mechanisms for further processing a precursor which it has obtained from another stimulated cell. Examples of this interaction are the transformation of LTA_4 to LTB_4 by the human erythrocyte[10] and to LTC_4 by the human platelet.[11] Several other examples of Type IB have been elucidated.[9]

In Type II cell–cell interactions, one cell is capable of transforming an eicosanoid of another into a new compound which neither cell can synthesize alone. The Type II interactions have two subdivisions, depending upon whether the cells have been activated. Thus, platelet 12-HETE can be utilized by activated neutrophils in the production of 5*S*,12*S*-DiHETE when both cells are exposed to a common stimulus in the form of the model compound ionophore A23187 (Type IIA). The reverse of this

[6] A. J. Marcus, L. B. Safier, H. L. Ullman, M. J. Broekman, N. Islam, T. D. Oglesby, and R. R. Gorman, *Proc. Natl. Acad. Sci. U.S.A.* **81,** 903 (1984).

[7] A. J. Marcus, L. B. Safier, H. L. Ullman, N. Islam, M. J. Broekman, and C. von Schacky, *J. Clin. Invest.* **79,** 179 (1987).

[8] A. J. Marcus, L. B. Safier, H. L. Ullman, N. Islam, M. J. Broekman, J. R. Falck, S. Fischer, and C. von Schacky, *J. Biol. Chem.* **263,** 2223 (1988).

[9] A. J. Marcus, *in* "Progress in Hemostasis and Thrombosis" (B. S. Coller, ed.), pp. 127–142. Grune & Stratton, Orlando, FL, 1986.

[10] F. Fitzpatrick, W. Liggett, J. McGee, S. Bunting, D. Morton, and B. Samuelsson, *J. Biol. Chem.* **259,** 11403 (1984).

[11] J. A. Maclouf and R. C. Murphy, *J. Biol. Chem.* **263,** 174 (1988).

reaction is also demonstrable, in that addition of radiolabeled 5-HETE to activated platelets results in formation of 5*S*,12*S*-DiHETE as well.[5] In cell interaction Type IIB, only one of the two cells under study is activated. Thus, if thrombin or collagen are added to a combined suspension of resting neutrophils and platelets, the neutrophils will metabolize 12-HETE to 12,20-DiHETE. Thrombin and collagen do not activate eicosanoid metabolism in neutrophils. This was verified in control studies involving addition of thrombin or collagen to neutrophils, wherein no eicosanoids were formed.[6]

In Type III cell–cell interactions, an eicosanoid or intermediate synthesized by one cell can act as an agonist or even inhibit production of eicosanoids in another cell. Thus, leukotrienes (LT) C_4, D_4, and E_4 can induce release of thromboxane and possibly other eicosanoids in perfused lung preparations.[12]

In studying eicosanoid biochemistry and function, the importance of cell isolation and ascertainment of a complete profile of eicosanoids synthesized by the cells of interest (under varying conditions of stimulation) cannot be overemphasized. Techniques developed in our laboratory for evaluation of synthetic mechanisms, metabolism, and catabolism of eicosanoids produced by platelets, neutrophils, and endothelial cells will be summarized.

Collection and Processing of Platelets and Neutrophils

Venipuncture Procedure

Prospective blood donors should be questioned in detail concerning medication ingestion during the past 10 days. A complete blood count should be performed as a screening procedure. Donors should be in the fasting state and well hydrated by ingestion of at least 12 ounces of water prior to blood collection. Venipuncture is carried out with a Plasma Transfer Set (4C2242, Fenwal Laboratories, Deerfield, IL). This apparatus contains 76 cm of plastic tubing with a 15-gauge needle at one end. The coupler at the other end can be cut off. Contrary to popular belief, a clean venipuncture with a 15-gauge needle is less painful than a comparable venipuncture with a smaller needle. Free flow through the tubing is controlled with a hemostat following application of the tourniquet. Venipuncture is initiated and the first 5 ml of blood are discarded. Collection is accomplished by this free flow technique into four to six conical calibrated polypropylene tubes containing 4.6 ml, 3.2% (3.2 g/100 ml) sodium citrate.

[12] P. J. Piper, *Physiol. Rev.* **64,** 744 (1984).

Tubes are filled to the 46-ml mark, rapidly capped, and inverted. Centrifugation is performed at 200 *g* for 15 min at 20°–25°. Samples of platelet-rich plasma (PRP) are taken for platelet counting and morphological examination. The platelet count in PRP obtained by this method averages 460,000/μl. We find this technique preferable to collection with syringes because of the possibility of cell activation and hemolysis during exertion of negative pressure. This procedure is used for preparation of PRP for platelet aggregation studies.

Preparation of Washed Platelet Suspensions

Whole blood is collected as described above, but the procedure outlined here is designed to process two tubes at a time. In this instance, 6 ml acid-citrate-dextrose (ACD, USP Formula A) is used instead of citrate for every 40 ml of whole blood [ACD: citric acid (monohydrate), 38 m*M*; sodium citrate (trisodium, dihydrate), 75 m*M*; glucose, 135 m*M*]. PRP is initially prepared by centrifugation of whole blood at 200 *g* for 15 min (25°). The PRP is then removed, and the contents of two tubes combined and centrifuged a second time at 90 *g* for 10 min. This eliminates most of the residual erythrocytes and leukocytes. The resulting PRP is transferred to another tube and acidified with 0.1 volume of "citrate solution" (38 m*M* citric acid, 75 m*M* sodium citrate) and a pellet obtained by centrifugation at 2000 *g* for 15 min (22°). The resulting platelet-poor plasma (PPP) is discarded, and the pellet initially suspended in 2 ml of buffer by repeated gentle aspiration with a plastic transfer pipette (Falcon 7524). Buffer A used for this purpose consists of the following: Tris, 63 m*M*; NaCl, 95 m*M*; KCl, 5 m*M*; citric acid, 12 m*M* (pH 6.5). During this transfer, any residual erythrocytes which may have been trapped in the cone of the tube can be avoided. After pipetting of this suspension into a clean conical tube, buffer is added to a total volume of 45 ml. The suspension is inverted two to three times, and centrifuged at 1450 *g* for 15 min (4°), and the resulting pellet resuspended in 3 ml of 0.15 *M* NaCl (isotonic). This serves as a stock solution for further experiments. Usually, the platelets are enumerated, and the concentration adjusted with saline to suit the needs of the particular experiment. For example, we use $1 \times 10^6/\mu$l. The suspension is kept in a closed container at 4° and periodically flushed with 5% CO_2–95% air.

Aspirin (ASA) Treatment of Platelets or Endothelial Cells

A stock solution of aspirin for platelets is prepared as follows: To 18.02 mg acetylsalicylic acid (ASA) (Sigma, St. Louis, MO), add 10 ml of 0.15 *M* NaCl. Dissolve by warming (37°) and mixing on a magnetic stirrer for 10 to 15 min. Stock equals 10 m*M*. Add 1/10th volume of stock solution to PRP (for example, 3 ml stock solution to 30 ml PRP). Incubate for 10 min (37°) with gentle shaking. Continue platelet washing procedure as required. Platelets prepared as above should not aggregate to added sodium arachidonate.

The procedure for endothelial cells is as follows: To 10 ml of buffer (as used by the investigator) over the monolayer in the T-75 flask, add 10 μl of 1000 m*M* ASA in ethanol to equal 1 m*M* final concentration of ASA. The flask is gently rotated and incubated for 30 min at 37°. Cells are then removed with collagenase-EDTA solution, washed, and finally resuspended in ASA-free buffer.

As a precautionary measure for possible recovery of endothelial cell cyclooxygenase from ASA acetylation in the course of the experiment, we add indomethacin to the final suspension of cells at a concentration of 10 μM. Indomethacin stock solution (1 μg/μl) is prepared as follows: 5 mg indomethacin (Sigma) is brought to 0.5 ml in 0.1 *M* Na_2CO_3, and then to 5 ml with 0.15 *M* NaCl. We adjust to pH 7.4 (which is the pH of buffer used in our experiments, with approximately 3 μl 6 *N* HCl). For the endothelial cell suspensions, 3.58 μl indomethacin stock solution is used per milliliter.

Methods for Studying Arachidonic Acid Metabolism in Human Platelets

Use of radiolabeled arachidonate to study eicosanoid metabolism in cells and tissues offers high sensitivity and specificity. A profile of all metabolic products from stimulated cells and tissues can be obtained without requiring a specific assay for each compound. The radioisotope procedure might not provide direct information on the relative importance of endogenous precursors and intermediates. This is because the distribution and fate of incorporated label may not correspond to that of endogenous arachidonate in the cell. However, the approach is particularly valuable for time-course studies designed to examine stimulus–response coupling phenomena. For example, it can be seen from a time-course study that thromboxane production in washed platelets occurs very early and is complete in a few seconds.[13] In contrast, production of 12-HETE commences more gradually, and the quantities increase with time. Therefore, early sampling of a stimulated platelet supernatant for 12-HETE production may be misinterpreted as being quantitatively low. In contrast, a sample taken at a later time point will show a higher quantity of 12-HETE production. Most importantly, metabolic studies of interactions of platelet

[13] A. J. Marcus, *in* "Hemostasis and Thrombosis" (R. W. Colman, J. Hirsh, V. J. Marder, and E. W. Salzman, eds.), 2nd Ed., pp. 676–688. J. B. Lippincott, Philadelphia, Pennsylvania, 1987.

precursors and intermediates, as well as end products, with other cells, can readily be carried out with the isotope approach. We demonstrated metabolism of radiolabeled PGH_2 produced by platelets to PGI_2 by unlabeled endothelial cells in which the cyclooxygenase had been inactivated by aspirin. The radiolabeled PGI_2 detected could only have originated from the radiolabeled platelet precursor.

Preparation of Sodium[^3H]Arachidonate

[^3H]Arachidonic acid (50 μCi, specific activity 60-100 Ci/mmol; New England Nuclear, Boston, MA) is converted to its sodium salt in the following manner: The compound, which is provided as an ethanol solution, is taken to dryness under 99.998% nitrogen and the sides of the vial are subsequently washed down at least three times with hexane, which has been previously bubbled with nitrogen. The hexane is then carefully evaporated while maintaining the arachidonate at the bottom of the vial. This is followed by addition of 50 μl of 0.01 *M* Na_2CO_3 plus 150 μl deionized water, both of which have been bubbled with nitrogen. The sodium carbonate solution is manually agitated in the container for 15 min at room temperature, followed by the addition of 3 ml of buffer B (Tris, 15 m*M*; NaCl, 134 m*M*; glucose, 5 m*M*; defatted bovine serum albumin, 0.01%; pH 7.4). To evaluate exactly how much [^3H]arachidonate has actually been added to the platelets, a 3-μl sample is removed for scintillation counting (Aquasol-2, New England Nuclear, Boston, MA). This procedure is also used for other fatty acids.

Procedure for Labeling Platelets with Radioactive Arachidonate

Venipuncture is performed as previously described, but in this instance, 6 ml of acid–citrate–dextrose (ACD) is used as anticoagulant instead of citrate. Collection is carried out in calibrated conical polypropylene tubes to the 46-ml mark. PRP is prepared by centrifugation of the tubes at 200 *g* for 15 min at 20°. A second centrifugation is carried out at 90 *g* for 10 min for removal of residual erythrocytes. The average donor will yield 20–30 ml PRP, containing $10–15.5 \times 10^9$ platelets from the original two tubes of whole blood. The PRP is then acidified with 0.1 volume of "citrate solution" (38 m*M* citric acid, 75 m*M* sodium citrate) and a pellet prepared by centrifugation at 2000 *g* for 15 min (20°–25°). The pellet is initially suspended in 2 ml of buffer A by repeated gentle aspiration with a plastic transfer pipette (Falcon 7524), and the suspension transferred to another calibrated polypropylene tube. Residual erythrocytes which may still be trapped in the cone of the tube are avoided during this resuspension procedure. Final volume of the suspension is 3 ml in buffer C (Tris, 63 m*M*; NaCl, 95 m*M*; KCl, 5 m*M*; citric acid, 12 m*M*; glucose, 5.5 m*M*; defatted bovine serum albumin, 0.01%; pH 6.5). For the labeling procedure, a total of 7 to 9 × 10^9 platelets is the optimal concentration. The platelet suspension is diluted with buffer such that the final incubation mixture for labeling will contain 0.2 × 10^9 platelets per milliliter. The platelet suspension is transferred to a disposable screw-capped plastic specimen container and 3 ml sodium[^{3}H]arachidonate added. The suspension is incubated at 37° for 45 min with gentle shaking. The incubation container is then cooled on ice for 10 min, and the radiolabeled suspension transferred to a calibrated conical plastic tube and centrifuged at 1450 *g* for 15 min (4°). The supernatant, which contains unincorporated arachidonate is removed and the platelets washed once with cold buffer A. The final suspension is prepared by adding cold 0.15 *M* NaCl to the 3-ml mark. A platelet count is again performed on 3.3 μl of a 4-fold dilution of the suspension. The average range of such counts is 1.47–2.16 × 10^6/μl. The radiolabeled suspension is then further diluted to a final concentration of 1 × 10^6/μl and placed at 4° until used. A 20-μl aliquot is immediately taken for scintillation counting. This labeling procedure should yield an average recovery of added radioactivity in the final suspension of 56%.

Experimental Use of Radiolabeled Platelet Suspensions

Platelets or combined suspensions of platelets, endothelial cells, neutrophils, or erythrocytes can be studied in aggregometer cuvettes. Stirred cell suspensions in aggregometer tubes also provide maximal cell contact and serve as *in vitro* models of hemostatic and thrombotic phenomena, as well as inflammatory responses.

The aggregometry tubes must be siliconized in order to prevent adsorption of radioactivity to the glass surface. A Teflon-coated stir bar is added to each tube. Buffer B, minus albumin (0.4 ml) is added, followed by 0.1 ml of labeled platelet suspension (1 × 10^8 platelets). The tube is preincubated in the aggregometer for 4 min at 37° prior to addition of the agonist. The aggregation response is recorded over a 5-min period and the reaction stopped by addition of 5 μl 1 *M* citric acid (this will bring the pH to 3–3.5). The Teflon bar is removed with a magnet and the tubes placed on ice. Using a siliconized Pasteur pipette, the contents of the cuvettes are transferred to 12 × 75 mm polypropylene tubes (Falcon 2063). Chloroform/methanol 2 : 5 (v/v) is added in a volume of 0.88 ml to the emptied cuvette for the purpose of rinsing, and this material is transferred to the polypropylene tube. The rinsing process is repeated with a second aliquot of 0.88 ml chloroform/methanol 2 : 5 (v/v). The latter is the starting material for lipid extraction, utilizing the method of Bligh and Dyer.[14] The polypropylene tubes are vortexed, and at this point in the procedure can be stored at −70° for processing at a later time.

If aggregation recordings are not desired and the investigator is solely interested in metabolic effects of agonists on cell suspensions, the entire procedure can be carried out in Falcon 2063 tubes with stirring in the presence of 0.5 m*M* EGTA. Stimuli, such as ionophore, thrombin, arachidonate, bradykinin, or other agonists, can still be used. In some instances, such as when collagen is employed, external calcium is required and EGTA cannot be used. This must be determined in preliminary experiments. If the major interest is in products released, a centrifuge step can be introduced prior to addition of chloroform/methanol 2 : 5. The supernatant is then removed and extracted. Caution is required in studies of hydroxy acids because they tend to sediment with the cells at low pH and will be partially lost from the supernatant. This can be prevented by stopping the reaction with immediate rapid centrifugation (15,000 *g*, Eppendorf) at 5° for 3 min prior to acidification.

Procedure for Lipid Extraction

If samples have been frozen, they are brought to room temperature with occasional vortexing. Deionized water (1 volume) and chloroform (1 volume) are added to the samples and the mixture vortexed several times. The tubes are then centrifuged at 1450 *g*, 15 min (22°), which should result in a sharp demarcation boundary between the two phases that have formed. The tube is marked at the meniscus of the upper phase and the lower phase is transferred to a conical 4-ml polypropylene tube (57.512, 12 × 75 mm, Sarstedt, Inc., Princeton, NJ). Transfer is carried out with a Drummond Pipet-Aid (Drummond Scientific Co., Broomall, PA), containing a siliconized Pasteur pipette. The remaining upper phase in the Falcon tube is carefully washed by addition of chloroform to the level of the mark previously indicated on the tube. The Falcon tube is vortexed, centrifuged as above, and the new lower phase that formed combined with the first phase already present in the Sarstedt tube. Caps for these tubes must be ordered separately from Sarstedt (65.722).

The combined lower phases are then evaporated under nitrogen at 37°, in an apparatus which can accommodate multiple samples, such as the N-Evap, 11155 (Organomation Associates, Inc., South Berlin, MA). It is important to concentrate all the material in the cone of the tube. This is accomplished more efficiently by washing the sides of the tube twice with

[14] E. G. Bligh and W. J. Dyer, *Can. J. Biochem. Physiol.* **37,** 911 (1959).

chloroform/methanol (2 : 1, v/v), and the solvent evaporated. In the final step, 0.1 ml of chloroform/methanol (2 : 1, v/v) is added, the tube gently vortexed, and a 4-μl aliquot removed for scintillation counting. At this point, the sample can be stored at $-70°$ for further evaluation by high-performance liquid chromatography (HPLC), thin-layer chromatography (TLC),[15] or other procedures.

Preparation of Neutrophil Suspensions for Studies of Eicosanoid Metabolism

A common cause of error and misinterpretation of results in experiments involving eicosanoid metabolism, is the presence of other cell types in the preparation under study. When radiolabeled precursors or intermediates are involved, the contaminating cell can take up the label and may or may not respond to the stimulus delivered to the cell type presumed to be under study. For example, if a putative pure neutrophil suspension produces 12-HETE when stimulated with an agonist to which a platelet will also respond, it means that the neutrophils are likely contaminated with platelets. Similarly, if thromboxane is identified in a "neutrophil preparation," the presence of platelets can be assumed. Human monocyte preparations invariably contain platelets and, therefore, a comparable problem arises. Neutrophil preparations, free of platelets and mononuclear cells, can be obtained in the following manner[7,8]: Starting with 276 ml of ACD anticoagulated blood, PRP is initially removed by centrifugation at 200 *g* (15 min, 25°). This is much more preferable than processing the whole blood immediately, because it eliminates most of the platelets, which might contaminate the subsequent cell preparations. After removal of PRP, the remaining leukocytes and erythrocytes in each tube are diluted to the 46 ml starting volume by addition of saline. The mixture is then divided between four 100-ml clear plastic graduated cylinders (69 ml in each). Dextran T-500 (Pharmacia Inc., Piscataway, NJ) is added to a final concentration of 1.22% (37 ml of 3.5% Dextran in saline added to each cylinder). The cylinders are covered with Parafilm and gently mixed by about seven inversions. The erythrocytes are allowed to sediment at room temperature for approximately 30–45 min, at which time the supernatant leukocyte-rich plasma (LRP) is ideally straw-colored or pink, and the demarcation between sedimented red cells and LRP is sharp. LRP is then aspirated with a plastic pipette into six conical polypropylene tubes

[15] A. J. Marcus, *in* "Methods in Hematology, Vol. 8: Measurements of Platelet Function" (L. A. Harker and T. S. Zimmerman, eds.), pp. 126–143. Churchill Livingstone, New York, 1983.

and centrifuged at 280 *g* (10 min, 4°). Resulting supernatants are discarded and each pellet, which is still red due to remaining erythrocytes, gently suspended in 7.5 ml cold saline by sucking up and down with plastic Falcon transfer pipettes (No. 7524). Resuspended cells from every two tubes are combined into three 15-ml pools. Fifteen milliliters of suspended cells are then carefully layered on each of three 10-ml aliquots of Ficoll-Hypaque (*d* 1.077) (Pharmacia) and the tubes centrifuged at 350 *g* (30 min, 4°). Before removal of the total resulting supernatant from above the neutrophil-containing pellet, the cloudy band at the Ficoll interface (platelets, monocytes, lymphocytes) is initially discarded by penetrating the surface with a siliconized Pasteur pipette directly to the band and aspirating while slowly rotating the tube. The remaining supernatant is then discarded and the inner wall of the tube, above the pellet, wiped with cotton swabs. These maneuvers are helpful in preventing contamination of the final neutrophil preparation by the above-mentioned cells.

Contaminating erythrocytes are next removed from each neutrophil pellet by lysis with 6 ml of distilled water, sucking up and down with a Falcon transfer pipette for a period of exactly 25 sec. Each suspension is rapidly returned to isotonicity with 2 ml of 3.5% NaCl and 10 ml of phosphate-buffered saline [0.9 m*M*, $CaCl_2$; 2.7 m*M*, KCl; 1.5 m*M*, KH_2PO_4; 0.5 m*M*, $MgCl_2$; 136.9 m*M*, NaCl; 8.1 m*M*, Na_2HPO_4 (pH 7.4)] added. The suspension is then centrifuged at 280 *g* (5 min, 4°), and the neutrophils suspended in HEPES buffer [5 m*M*, HEPES; 140 m*M*, NaCl; 5 m*M*, KCl; 1.29 m*M*, $CaCl_2$; 1.20 m*M*, $MgCl_2$ (pH 7.45)].[5]

If the neutrophils are to be subsequently fractionated into subcellular particles, the HEPES buffer is modified to contain no calcium, but 2.5 m*M* magnesium. Trypan blue exclusion by these neutrophil preparations averages 95% and, on stained smears, no platelets are seen. Purity of these neutrophil preparations is verified by adding thrombin or collagen, stimuli which do not induce arachidonic acid metabolism in neutrophils. When this preparation is examined by HPLC, no 12-HETE can be identified–thus, confirming the absence of platelet contamination.[6]

Assay of 12,20-DiHETE Production by Neutrophils Using HPLC

Neutrophils (3×10^7) or aliquots of subcellular fractions are adjusted to 1 ml in HEPES buffer. Preincubation (1 min, 37°) is carried out in polypropylene tubes with stirring. Inhibitors or NADPH (when used) are added during the preincubation period. 12-HETE is prepared as the sodium salt as described above for arachidonate. The reaction is started with 4.4 μM 12-HETE and is terminated at 5 min by adding 1.5 ml acetone and placement on ice. When subcellular fractions are under study, reaction time is increased to 15 or 30 min. In the latter case, calcium is added back to the buffer at a final concentration of 1.29 m*M* at the start of preincubation. Intact cells or cell fractions can be assayed without loss of activity if they are stored at 4° overnight.

Acetone-precipitated protein is removed by centrifugation at 1400 *g* (15 min, 25°) and the acetone evaporated from supernatants under nitrogen at 37°. The material is then acidified to pH 4.0 with 1 *N* H_3PO_4. This is followed by extraction three times with 1-ml aliquots of ethyl acetate and evaporation to dryness. The dried residue is dissolved in 50 μl methanol/water (3 : 1, v/v). Tubes are vortexed and centrifuged for 5 min at 225 *g*. The supernatant is now ready for injection (5 μl) into an HPLC column.[7] Separation is accomplished by reversed-phase HPLC on an 8-mm id Radial-PAK C_{18} (10 μm column) at room temperature (Millipore, Waters Chromatography Division, Milford, MA). The eluting solvent is methanol/water/acetic acid (75 : 25 : 0.01, v/v/v), pH 6.1. Flow rate is 0.6 ml/min and absorbance properties of the effluent are monitored at 237 nm (Fig. 1).[7]

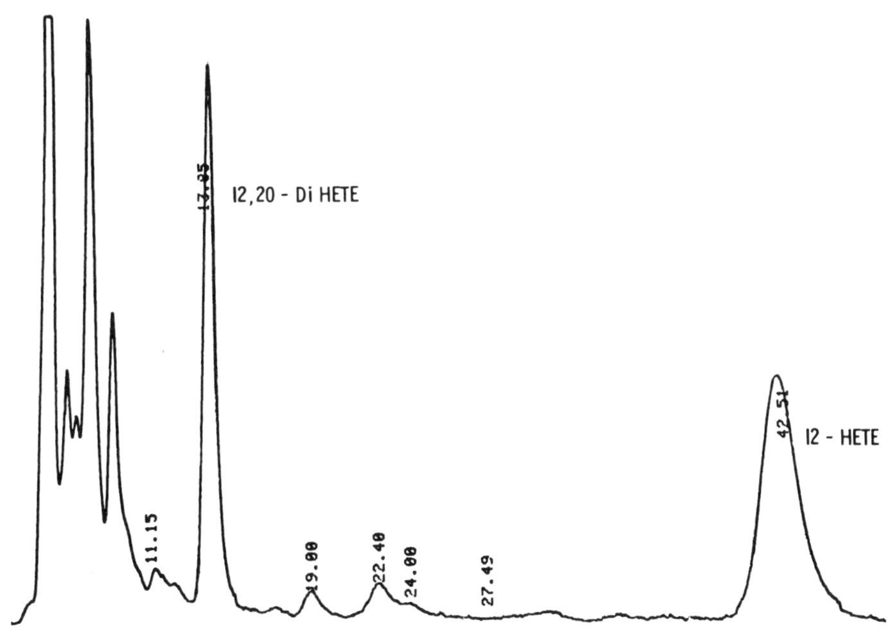

FIG. 1. HPLC assay for 12,20-DiHETE production by unstimulated neutrophils from platelet-derived 12-HETE. Unstimulated neutrophils (3×10^7) are incubated with 4.4 μM of the sodium salt of 12-HETE (5 min). Reactions are terminated with acetone and precipitated protein removed. Samples are then acidified and lipids extracted with ethyl acetate. Finally, the lipids are redissolved in methanol/water (3 : 1, v/v) and injected into a reversed-phase HPLC column as described herein. Absorbance of effluents is monitored at 237 nm. The 12-HETE peak at 42.5 min represents unconverted substrate remaining after the 5-min incubation period.[7]

12,20-DiHETE and 12-HETE are quantified by comparison of peak areas with those of purified external standards. The 12-HETE standard is synthesized from ionophore-stimulated platelets. 12,20-DiHETE is prepared by incubating the purified 12-HETE with neutrophils or from thrombin-stimulated mixtures of platelets and neutrophils.[6] Both standards are purified by reversed-phase HPLC and quantitated spectrophotometrically using an extinction coefficient at 237 nm of 30,500.

When 12,20-DiHETE production is assayed in broken cell preparations or subcellular fractions, optimal results are obtained if an NADPH-generating system is used. This consists of 0.02 *M* DL-isocitric acid (trisodium salt), 0.1 mg of isocitrate dehydrogenase/ml (grade 1, Boehringer Mannheim, Indianapolis, IN), and 1 m*M* NADPH. Results of a typical HPLC assay for 12,20-DiHETE production are depicted in Fig. 1.

Diisopropyl Fluorophosphate (DFP) Treatment of Neutrophil Suspensions

In contrast to cytochrome *P*-450 enzyme systems in other tissues, such as kidney or liver, mechanical disruption of neutrophils results in loss of 12-HETE ω-hydroxylase activity. This is due to proteolytic enzyme release from the neutrophil upon cell disruption. DFP is the only proteolytic enzyme inhibitor that can block this loss of enzyme activity. DFP stock solutions (~2.5 *M*) are prepared in a fume hood by injection of 1 ml of anhydrous 2-propanol into a vial containing 1 g of DFP (Sigma Chemical Co., St. Louis, MO). Prior to injection, the bottle is vented by puncturing with a 3/8 inch 26-gauge needle. After the 1 ml of anhydrous 2-propanol is injected with a disposable plastic tuberculin syringe and a 1 inch 22-gauge needle, the bottle is gently mixed by hand. The seal is then opened with pliers and contents transferred to a small Teflon-lined screw-cap vial with a Pasteur pipette for further storage.

During preparation of the DFP stock solution, a full-face respirator mask with organic vapor cartridges is worn (Willson Safety Products, Reading, PA). Aliquots of the DFP stock solution in microliter quantities are added to neutrophil suspensions (average 7×10^7 neutrophils/ml, total volume, 5–7 ml) bringing the final DFP concentration to 2 m*M*. The tubes are then vortexed and incubated in ice for 5 min. The excess DFP is washed out twice with 25 ml of cold, modified (no calcium, 2.5 mM magnesium) HEPES buffer, and centrifuging at 400 *g* (5 min, 25°). The aspirated supernatants are discarded into a large plastic beaker containing 1 *N* NaOH (which destroys DFP). Final suspension is made in buffer to the original volume with a plastic pipette.[16] Samples are then taken for cell counting.

[16] P. C. Amrein and T. P. Stossel, *Blood* **56,** 442 (1980).

Additional precautions include: wearing of gloves throughout the procedure and all work carried out in a fume hood until final resuspension. All pipettes, tubes, and other apparatus touched by DFP should be soaked in 1 *N* NaOH prior to washing. Disposal material should be soaked in the 1 *N* NaOH prior to discarding.

HPLC Assay for 12-HETE-1,20-dioic Acid Production from 12,20-DiHETE by Neutrophils

If a time-course study is carried out on the metabolism of 12,20-DiHETE by unstimulated neutrophils, it can be discerned that a dehydrogenation reaction occurs in which a new product, more polar than 12,20-DiHETE, is formed. The dehydrogenase reaction may also occur during conversion of 20-hydroxy-LTB_4 to 20-carboxy-LTB_4. The 12,20-DiHETE dehydrogenase system requires NAD as cofactor and has subcellular components in both cytosolic and microsomal fractions. These components are synergistic in their activity.[8]

In the actual assay, 3×10^7 neutrophils or appropriate quantities of subcellular fractions are adjusted to 1 ml with HEPES buffer. Test samples are preincubated in polypropylene tubes for 1 min at 37° with stirring. Cofactors, such as NAD or potential inhibitors, are added during this preincubation period. Reactions are initiated by adding 2.8 μM 12,20-DiHETE (sodium salt, prepared as described above), and termination is accomplished after 20 to 40 min with 1.5 ml acetone and placement at 4°. Also, as previously mentioned, in subcellular fractionation work, calcium is added back to the buffer at a final concentration of 1.29 m*M* at the beginning of the preincubation period. Precipitated protein is removed by centrifugation at 1400 *g* (15 min, 25°). Acetone is evaporated from the supernatants with nitrogen (37°). The pH of the samples is brought to 4.0 with 1 *N* H_3PO_4 and extraction carried out three times with 1- ml aliquots of ethyl acetate. After evaporation to dryness, residues are dissolved in 50 μl of methanol/water (70 : 30, v/v). Tubes are vortexed and centrifuged at 225 *g*, 25°, for 5 min. The supernatant (5 μl) is injected into an HPLC column. Reversed-phase HPLC is carried out on a 3.9 mm id × 30 cm μBondapak C_{18} (10 μm) column (Millipore, Waters Chromatography Division, Milford, MA). The eluting solvent is methanol/water/acetic acid (70 : 30 : 0.01, v/v/v), pH 6.1, and flow rate 0.46 ml/min. Column temperature is maintained at 25° by immersion in a circulating water bath. Absorbance is monitored at 237 nm (see Fig. 2).

12-HETE-1,20-dioic acid is quantified using 12,20-DiHETE as external standard and an extinction coefficient at 237 nm for both compounds of 30,500 (comparable to that of 12-HETE). 12,20-DiHETE is prepared by incubating purified 12-HETE with neutrophils, followed by isolation by

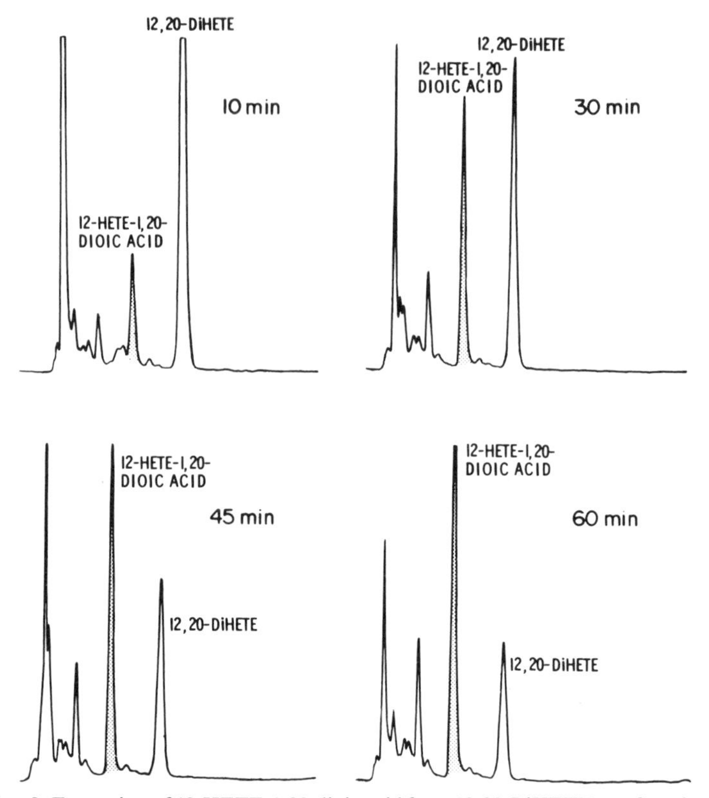

FIG. 2. Formation of 12-HETE-1,20-dioic acid from 12,20-DiHETE as a function of time. Neutrophils (3×10^7) are incubated with 3 μM 12,20-DiHETE for 10 to 45 min. Lipids are extracted and chromatographed on a μBondapak C_{18} column as described. The eluting solvent is methanol/water/acetic acid (70 : 30 : 0.01, v/v/v), pH 6.1. Flow rate is 0.46 ml/min. Absorbance is monitored at 237 nm. 12,20-DiHETE decreases with time and is accompanied by an increase in production of 12-HETE-1,20-dioic acid.[8]

reversed-phase HPLC. In addition, 12,20-DiHETE can now be prepared by chemical synthesis.[8]

When 12-HETE-1,20-dioic acid production is under study in cell homogenates or subcellular fractions, NAD at a final concentration of 1 mM is added. NAD stock solutions in HEPES buffer are adjusted to pH 7 with 0.5 N NaOH.

When 12-HETE-1,20-dioic acid is to be biosynthesized and isolated for further characterization, the following methodology is recommended: 10–20 ml of neutrophil suspensions at a concentration of 3 to 4×10^7/ml are incubated with 20 μM purified 12,20-DiHETE or 12-HETE (sodium salts) for 30 to 80 min (37° with stirring). Tubes are centrifuged at 280 g for

5 min (4°) and the pellets discarded. Acetone (1.5 volumes) is added to the supernatants which contain 12-HETE-1,20-dioic acid. Protein precipitation, acidification, and lipid extraction are carried out as described above. In this instance, 12-HETE-1,20-dioic acid is purified by HPLC on the μBondapak C_{18} column at 25° with a solvent system consisting of methanol/water/acetic acid (65 : 35 : 0.01, v/v/v), pH 6, at 0.42 ml/min.

It is of critical importance to realize that 12-HETE-1,20-dioic acid undergoes spectral alterations if maintained and/or dried from the above acidic eluting solvent. Spectral integrity is best preserved by adding NH_4OH to an approximate concentration of 0.4 m*M* after collection. Drying is then accomplished by rapid, direct flow of argon into the sample tube. The compound is then stored at −70° in the dry state under argon.[8]

Acknowledgments

I would like to acknowledge the research and intellectual collaboration of the following colleagues: Lenore B. Safier, Harris L. Ullman, Naziba Islam, M. Johan Broekman, M. Teresa Santos, and Juana Valles. I would also like to acknowledge the expertise provided by Ms. Evelyn M. Ludwig with regard to preparation of this manuscript. Research work mentioned in this chapter was supported by grants from the Veterans Administration, National Institutes of Health Grant HL-18828-14 SCOR, the Edward Gruenstein Fund, the Sallie Wichman Fund, and the S.M. Louis Fund.

[65] Isolated Perfused Rat Lung in Arachidonate Studies

By SHIH-WEN CHANG and NORBERT F. VOELKEL

The isolated perfused rat lung preparation has been used for the past 20 years by both physiologists and pharmacologists interested in the circulatory, biochemical, and metabolic aspects of this complex organ. In particular, studies of arachidonate metabolism in isolated perfused rat lungs have yielded important information regarding the significant *in vivo* metabolic pathways of arachidonic acid metabolites, the action of various eicosanoids in the physiological regulation of the pulmonary circulation. In this section, we will detail the method of isolated lung perfusion currently used in our laboratory. We will also highlight some important results related to arachidonate metabolites obtained using this preparation.

General Methodology

Lung Isolation and Perfusion

Pathogen-free, young adult Sprague-Dawley rats with body weight from 250 to 350 g are anesthetized with an ip injection of pentobarbital sodium (70 mg/kg). After removal of the overlying skin, a median incision is made in the center of the neck and the trachea exposed by blunt dissection. The trachea is partially transected and a 15-gauge blunt metal cannula is inserted and secured with a 0-silk suture. Following tracheal cannulation, ventilation is initiated with a small animal ventilator using a humidified gas mixture containing 21% O_2, 5% CO_2, and 74% nitrogen. The ventilator is set at a rate of 55 breaths per minute, using either a pressure-limited ventilator with a maximal inspiratory pressure of 8 cm water, or a volume-limited ventilator with a tidal volume of 2 to 4 ml/breath (approximately 1 ml/100 g body weight). We do not impose a positive end-expiratory pressure (PEEP) initially to allow partial lung collapse and to minimize the chance of nicking the lung during chest opening and lung removal.

The sternum is split at the midline using a blunt-tipped scissors, taking care to avoid the ventilated lungs. The chest is then opened with a rib spreader and 100 units of heparin (in 0.5 ml saline) are immediately injected into the right ventricle to prevent intravascular clotting. Using a small, curved-tip hemostatic forceps, a suture is placed around the outflow tracts of both aorta and pulmonary artery and secured with a loose knot. At this time, it is prudent to be sure that the perfusion circuit is filled with the perfusate solution and that no air bubbles remain in the tubing. A small nick is made in the right ventricular-free wall and the pulmonary artery cannula (mounted at the end of the perfusion tubing) is introduced into the pulmonary artery through the preexisting ligature. The aorta and pulmonary artery are then tightly ligated. Another incision is made at the cardiac apex to allow insertion of a second cannula into the left ventricule to collect effluent perfusate coming from the lungs. The trachea, lungs, and the heart are removed *en bloc* from the chest cavity and suspended by the tracheal tubing in a constant-temperature, humidified chamber (Fig. 1).

Perfusion is begun with a peristaltic Holter pump, starting at a slow rate, and gradually increasing to a final rate of 0.03 ml/g body weight/min. If blood-free perfusion is desired, the initial 50 ml of the perfusate containing residual plasma and blood cells are discarded prior to initiation of recirculation. When blood is used to perfuse the lungs, recirculation of the perfusate can be established from the beginning. Pulmonary artery perfusion pressure is measured (via a side arm of the inflow perfusion tubing) with a Statham P23AA transducer, the signal is amplified, and continu-

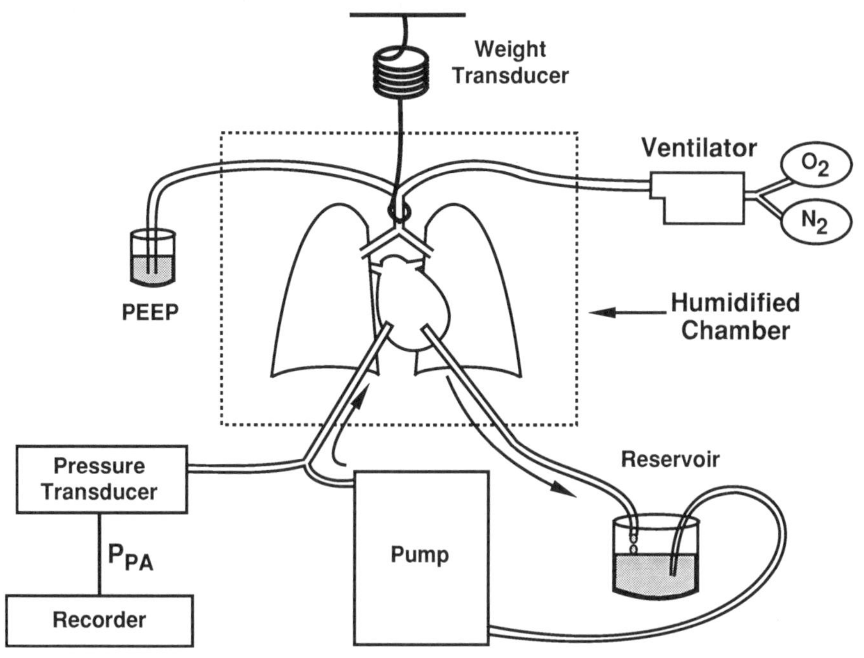

FIG. 1. Isolated perfused rat lung. The lungs are suspended on a weight transducer within a humidified chamber, and ventilated with a small animal ventilator. The perfusate is recirculated using a Holter peristaltic pump. PEEP, Positive end-expiratory pressure; P_{PA}, pulmonary arterial pressure.

ously recorded. Since perfusion rate is constant, changes in perfusion pressure reflect changes in pulmonary vascular resistance. At this time, the lungs are hyperinflated to remove any residual atelectasis and PEEP is initiated at 2 to 3 cm water. In a long perfusion protocol, periodic (every 30–40 min) hyperinflation of the lungs may be necessary to prevent atelectasis. Throughout the experiment, the lungs and the perfusate are maintained at 37° to 38° by recirculation of warmed water through water-jacketed chambers around the lungs, the humidification unit, and the perfusate reservoir.

Choice of Perfusate

Depending on the experimental design, the lungs can be perfused with whole blood, plasma, or a blood- and plasma-free physiological salt solution (PSS). Initial perfusion pressure averages around 12–16 mmHg with blood perfusate and 5–7 mmHg with PSS, using the ventilation and perfusion parameters outlined above. Blood-perfused lungs closely resemble the *in vivo* state, maintain excellent vascular reactivity, and are often used in the studies of hypoxic pulmonary vasoconstriction. Ether-anesthetized, retired breeder rats are used as blood donors. Approximately 8–10 ml of whole blood may be removed from each donor rat via percutaneous cardiac aspiration. In general, 30 ml of whole blood are sufficient for each perfused lung experiment. However, in studies assessing the vascular action of lipid mediators or measuring lung metabolism of eicosanoids, binding of the mediator or metabolites to plasma proteins and metabolism by blood cells may complicate the interpretation of results. For these studies, perfusion with PSS containing either albumin or Ficoll is recommended.

We use a modified Krebs–Henseleit PSS containing: 119 m*M* NaCl, 4.7 m*M* KCl, 1.17 m*M* $MgSO_4$, 19.1 m*M* $HaHCO_3$, 1.18 m*M* KH_2PO_4, 1.6 m*M* $CaCl_2$, and 5.5 m*M* glucose. Immediately prior to use, the solution is equilibrated with a gas mixture containing 95% O_2 and 5% CO_2 and Ficoll 70 (a synthetic copolymer of sucrose and epichlorohydrin with a molecular weight of 70,000) is dissolved to 40 g/liter. In experiments where bovine serum albumin is used in place of Ficoll, the PSS is modified to 22.6 m*M* $NaHCO_3$ and 3.2 m*M* $CaCl_2$. The use of PSS-Ficoll as perfusate is appropriate for short-term experiments in which mediator binding to albumin is a concern. Lungs erfused with albumin-free perfusate (i.e., PSS–Ficoll) are more susceptible to the development of edema, especially with perfusion periods greater than 2–3 hr.

Stability of Preparation

The major factors that affect the performance and stability of isolated perfused lung preparations are: (1) health status of the experimental animal, (2) technical expertise and care during lung isolation and removal, (3) perfusate composition, and (4) length of perfusion time. Viral infections are not infrequent in rat colonies and superimposed bacterial pneumonitis can follow and complicate the viral respiratory tract infection. Lungs with macroscopic consolidation or scarring should be discarded as they develop edema easily and may have elevated baseline levels of eicosanoids.

Extreme care must be taken not to traumatize the lung tissue during the isolation procedure. Some degree of flow interruption is unavoidable while the lung is removed from the chest cavity but, with practice, this ischemic period can be as short as 3–4 minutes; continued ventilation during this period of time should prevent any ischemia reperfusion-related problems. Some laboratories have reported rapid development of spontaneous edema in isolated lungs and the question is raised whether isolated lungs are unavoidably injured. In our experience, however, spontaneous edema is rare when healthy lung donor rats are used and stable and reproducible lung preparations are consistantly obtained. Moreover, we have recently found comparable values for lung permeability–surface area product between isolated rat lungs perfused with either blood or PSS–albumin and intact rats.[1] Blood-perfused lungs often maintain excellent vascular reactivity and remain edema free for up to 4 hr of perfusion. Even in lungs perfused with PSS-albumin, significant edema is not observed up to 4 hr of perfusion.[2] Although lungs perfused with PSS-Ficoll are somewhat more susceptible to edema than those perfused with PSS-albumin or blood, they should remain stable for up to 2 hr of perfusion.

An important complication is the occasional development of ventricular spasm, which is observed most commonly between 5 and 20 min after the initiation of lung perfusion and account for most cases of "spontaneous" edema that we have observed. Presumably, postmortem spasm in the heart muscle partially or completely interrupts the flow of effluent perfusate out of the left atrium, resulting in increased pulmonary venous and microvascular pressures. If uncorrected, the lungs rapidly become edematous. When ventricular spasm is suspected, the ventricle should be firmly squeezed with a tweezer and the left ventricular cannula repositioned. This should be followed by a rapid decrease in the pulmonary arterial pressure and a marked decrease in the size of the congested left atrial appendage. This complication can often be prevented by vigorously disrupting the mitral valve apparatus during the insertion of the left ventricular cannula and by a close monitoring of the lung preparation.

Physiological Parameters

Mean pulmonary artery pressure is continuously measured and recorded on a strip chart recorder. Left atrial pressure is typically set at 0 cm water but can be varied by simply raising the platform supporting the perfusate reservoir. In some experiments, setting the left atrial pressure at 5 cm water may be desirable to fully recruit the rat lung vasculature. Since perfusion rate and left atrial pressure are fixed, changes in the pulmonary artery pressure reflect changes in total pulmonary vascular resistance. The total vascular resistance can be further partitioned to pre- and post-capillary sites by measuring the pulmonary microvascular pressure (Pmv). This is done by using the double occlusion technique initially described by Dawson and colleagues[3] and recently validated against the traditional

[1] J. Czartolmna, N. F. Voelkel, and S. Chang, *FASEB J.* **3,** A1140.

[2] A. B. Fisher, C. Dodia, and J. Linask, *Exp. Lung Res.* **1,** 13 (1980).

[3] C. A. Dawson, J. H. Linehan, and D. A. Rickaby, *Ann. N.Y. Acad. Sci.* **384,** 90 (1982).

gravimetric technique by Townsley *et al.*[4] The inflow and outflow tubings are simultaneously clamped for a period of 20 sec and the difference between this equilibration pressure and a stop-flow pressure is taken to be Pmv. This pressure averages around 2–2.5 mmHg in PSS-Ficoll perfused lungs and increases to 4.6 ± 0.4 mm Hg after a bolus injection of LTC_4 (2 μg) into the pulmonary artery.[5]

In experiments investigating mediator effects on lung edema, the isolated lungs can be suspended from a force-displacement transducer for continuous measurement of lung weight. Alternatively, the perfusate reservoir can be placed on a scale and changes in lung weight can be inferred by the loss of perfusate in a recirculating system. At the end of the experimental protocol, the lungs are dissected from the mediastinal tissues, weighed, and the wet lung-to-body weight ratio can be taken as a rough estimate of fluid accumulation. If lung tissues are not used for biochemical analysis, the lungs can be dried to constant weight and the lung wet-to-dry ratio calculated as an index of lung edema. Assessment of lung mechanics and airway reactivity using a heated pneumotachograph and airway pressure transducer in isolated perfused rat lungs have also been reported.[6]

Assessment of lung vascular permeability can be done by measuring either capillary filtration coefficient[7] or the extravascular leakage of radiolabeled protein. We prefer the later method and use a technique modified from Kern *et al.*[8] After an equilibration period of 20 min, during which time the lungs are in an isogravimetric state, approximately 1 μCi of ^{125}I-labeled albumin (specific activity 1.06 mCi/mg; ICN radiochemicals, Irvine, CA; greater than 99.9% binding to albumin) is added to the perfusate and allowed to recirculate for exactly 3 min. After 1 ml of the perfusate is collected, the lungs are perfused with a fresh, nonradioactive perfusate for the additional 3 min without recirculation to remove intravascular ^{125}I-labeled albumin. The lungs are dissected, weighed, and their radioactivity determined in an auto-γ-scintillation spectrometer. The lung permeability–surface area product (PS) can be calculated as follows:

$$\text{Lung PS} = \text{total lung } ^{125}\text{I activity}/(^{125}\text{I activity in 1.0 g of initial perfusate} \times 3 \text{ min})$$

[4] M. I. Townsley, R. J. Korthius, B. Rippe, J. C. Parker, and A. E. Taylor, *J. Appl. Physiol.* **61,** 127 (1986).

[5] A. Sakai, S. Chang, and N. F. Voelkel, *J. Appl. Physiol.*, in press (1989).

[6] K. B. Nolop, J. H. Newman, J. R. Sheller, and K. C. Brigham, *Am. Rev. Respir. Dis.* **129,** A231 (1984).

[7] K. A. Gaar, Jr., A. E. Taylor, L. J. Owens, and A. C. Guyton, *Am. J. Physiol.* **213,** 910 (1967).

[8] D. F. Kern, D. Levitt, and D. Wangensteen, *Am. J. Physiol.* **245,** H229 (1983).

While lung PS can be affected by capillary recruitment (by altering perfused surface area available for macromolecular exchange), this effect is relatively small and changes in PS of greater than 50% are likely due to changes in vascular permeability.

Advantages and Disadvantages

The isolated perfused rat lung allows the studies of an intact, physiologically stable organ preparation in which influences from plasma-borne mediators, hormones, and circulating blood cells can be excluded. The intact pulmonary circulation is under investigation and is capable of vasoreactivity and local regulation, yet the investigator has control over blood flow and vascular pressures. In addition, the effects of varying gas tensions on metabolism and vascular responses are easily studied. All relevant cell types intrinsic to the lung are present and the cell–cell relationships are preserved in their normal state. With care, this preparation remains stable and metabolically active for up to 4 hr.

Concerns which limit the use of the isolated perfused rat lung are: (1) long-term studies are not feasible; (2) the lung is composed of a mixture of many cell types, and, therefore, it is difficult to attribute mediator synthesis or metabolism to a specific cell type; finally (3), one needs to take into account the possibility of activation of certain synthetic and metabolic pathways by the lung isolation and perfusion procedures.

Use of Isolated Perfused Rat Lung in Arachidonate Studies

Isolated Perfused Rat Lung as Bioassay Organ for Assessment of Eicosanoid Action

Infusion of arachidonic acid into the pulmonary artery of isolated perfused rat lungs results in vascular effects attributable to both cyclooxygenase and lipoxygenase metabolites.[9] The acute pulmonary vasoconstrictive effect of arachidonic acid is inhibited by indomethacin and probably due to thromboxane A_2. In the presence of indomethacin, arachidonic acid causes a delayed pressure rise and lung edema that can be attributed to the actions of leukotrienes since both can be inhibited by 5-lipoxygenase inhibitors. Furthermore, injections of individual purified lipoxygenase metabolites showed that LTC_4 is the most potent vasoconstrictor in the rat lung, while both LTC_4 and LTB_4 cause lung edema.[9]

[9] N. F. Voelkel, K. R. Stenmark, J. T. Reeves, M. M. Mathias, and R. C. Murphy, *J. Appl. Physiol.* **57,** 860 (1984).

Subsequently, LTE_4 was also found to be a pulmonary vasoconstrictor in the rat lung.[10] An important concept from this and other studies[9,11,12] is the marked attenuation of the vascular activity of eicosanoids in the presence of albumin perfusate, presumably due to binding of the mediators to albumin. Avid binding of these lipid mediators by plasma proteins may serve to localize their vasoactive effects to near sites of biosynthesis.

Injection of another lipid mediator, platelet-activating factor (PAF), into isolated rat lung causes pulmonary vasoconstriction and lung edema. This effect is, in a large part, mediated by the production of LTC_4 by the lung tissue.[13] We have recently found that injection of both PAF and LTC_4 cause significant increase in Pmv, indicating a significant degree of pulmonary venoconstriction. Furthermore, inhibition of the vascular pressure changes using dibutyryl-cAMP completely blocked LTC_4-induced lung edema and albumin leak in perfused lungs.[5] Thus, pulmonary venoconstriction is an important mechanism in lipid mediator-induced lung edema.

Another important concept in regards to the vasoactivity of lipid mediators in the pulmonary circulation is that the observed response is often dependent on both the dose of the mediator as well as the tone of the pulmonary circulation. Under basal conditions, the pulmonary circulation is nearly maximally vasodilated. Thus, it is difficult to demonstrate vasodilatory responses unless the tone is first increased with vasoconstrictors or exposure to alveolar hypoxia (hypoxic pulmonary vasoconstriction). For example, a low dose (10 ng) of PAF has minimal effect during basal condition but causes vasodilation during hypoxic vasoconstriction; a higher dose of PAF (1 μg) causes vasoconstriction during basal condition but vasodilation followed by vasoconstriction during hypoxic vasoconstriction. Both the vasodilatory and vasoconstrictive actions of PAF are receptor-mediated[14] and can be inhibited by receptor blockers such as CV3988 or WEB 2086. The vasodilatory effect is likely due to the release of endothelium-dependent relaxing factor, while the vasoconstrictive effect may be due to the release of leukotriene C_4.

[10] C. O. Feddersen, M. Mathias, R. C. Murphy, J. T. Reeves, and N. F. Voelkel, *Prostaglandins* **26,** 869 (1983).

[11] V. J. Iacopino, S. Compton, T. Fitzpatrick, P. Ramwell, J. Rose, and P. Kott, *J. Pharmacol. Exp. Ther.* **229,** 654 (1984).

[12] T. C. Noonan, W. M. Selig, K. E. Burhop, C. A. Burgess, and A. B. Malik, *J. Appl. Physiol.* **64,** 1989 (1988).

[13] N. F. Voelkel, S. Worthen, J. T. Reeves, P. M. Henson, and R. C. Murphy, *Science* **218,** 286 (1982).

[14] N. F. Voelkel, S. Chang, K. Pfeffer, S. G. Worthen, I. F. McMurtry, and P. M. Henson, *Prostaglandins* **32,** 359 (1986).

Injection of arachidonic acid during hypoxic vasoconstriction in PSS-Ficoll perfused rat lungs also causes a biphasic response with acute, transient vasoconstriction followed by a prolonged vasodilatory effect. In the presence of cyclooxygenase inhibition, only the acute vasodilation is evident.[15] Recent results from our laboratory suggest that this cyclooxygenase-independent pulmonary vasodilatory effect of arachidonic acid is, in part, mediated by cytochrome *P*-450-dependent metabolites.

Role of Eicosanoids in Pulmonary Vasoregulation

Cyclooxygenase metabolites are generated by respiratory movement[16] as well as during blood flow alterations.[17] Inhibition of the cyclooxygenase pathway enhances and preserves vascular reactivity in isolated rat lungs, suggesting that endogenous vasodilatory prostaglandins such as PGI_2 modulate vascular responses in this preparation. This activation of vasodilatory prostaglandins is markedly amplified during lung infection or vascular injury and studies using diverse experimental preparations have suggested an important role for PGI_2 in the depression of hypoxic pulmonary vasoconstriction observed in these settings.[18]

Thromboxane A_2, a potent vasoconstrictor, is also released following activation of the cyclooxygenase pathway. In species such as guinea pig and sheep, large amounts of thromboxane A_2 are released in response to injury and play an important role in the observed pulmonary hypertension. In the rat lung, however, the amount of thromboxane A_2 synthesized is modest compared with PGI_2 and thromboxane A_2 plays a minor role in mediating the vascular response to injury.

5-Lipoxygenase metabolites are clearly activated during acute lung injury. LTB_4 is a potent chemoattractant for neutrophils, but whether it has a direct effect on vascular permeability is unclear.[12] LTC_4 and other peptidoleukotrienes are markedly increased in lung tissue after endotoxin-, α-staphylotoxin-, and protamine-induced lung injury. At least in the latter two injury models, the production of LTC_4 appears to promote lung edema by enhancing pulmonary venoconstriction.[19] Whether LTC_4 plays a role in the physiological regulation of the normal pulmonary circulation is less clear. Although earlier studies from this laboratory suggested

[15] C. O. Feddersen, S. Chang, J. Czartoloma, and N. F. Voelkel, *J. Appl. Physiol.*, submitted.
[16] R. Korbut, J. Boyd, and T. Eling, *Prostaglandins* **21,** 491 (1981).
[17] A. van Grondelle, G. S. Worthen, D. Ellis, M. M. Mathias, R. C. Murphy, R. J. Strife, J. T. Reeves, and N. F. Voelkel, *J. Appl. Physiol.* **57,** 388 (1984).
[18] S. Chang and N. F. Voelkel, *in* "Eicosanoids in Cardiovascular and Renal Systems" (P. V. Halushka and D. E. Mais, eds.), p. 62. MTP Press, Lancaster, 1988.
[19] S. Chang, J. Czartolomna, and N. F. Voelkel, unpublished observations (1989).

a potential role for leukotrienes as mediator of hypoxic pulmonary vasoconstriction,[20] recent results have not supported this hypothesis.[21]

There are currently no experimental data to support or refute a vasoregulatory role for other arachidonic acid metabolites, such as lipoxins and epoxyeicosatrienoic acids, in the normal pulmonary circulation.

Eicosanoid Metabolism in Isolated Perfused Rat Lung

The lung is a metabolic organ and is capable of synthesis as well as metabolism of a large number of eicosanoids. Using the isolated perfused lung technique, Westcott *et al.*[22] observed marked species dependence in lung eicosanoid synthesis. When the perfused rat lung is treated with the calcium ionophore A23187, a large amount of PGI_2 (as 6-keto-$PGF_{1\alpha}$), LTB_4, and LTC_4 and lesser amounts of thromboxane A_2 (as TxB_2), PGE_2, and LTD_4 are synthesized.[22] Significant amounts of 6-keto-$PGF_{1\alpha}$ TxB_2 and LTB_4 are released into the lung perfusate. In guinea pig lungs, the cyclooxygenase metabolites, especially thromboxane A_2, is the predominant eicosanoid product after A23187 stimulation, and very little 5-lipoxygenase metabolites can be measured. In the ferret lung, the lipoxygenase products, especially LTB_4, are prominent. In rabbit and hamster lungs, roughly equal amounts of cyclooxygenase and lipoxygenase products are synthesized, after A23187 stimulation.[22]

Physiological factors that are important in stimulation of lung eicosanoid synthesis include respiratory motion,[16] and alterations in the hydrodynamic forces in the circulation.[17] Indeed, we find that the "baseline" levels of eicosanoids in isolated perfused lungs are generally severalfolds higher than those in lungs removed from anesthetized rats and immediately homogenized in methanol. This is particularly true for the PGI_2 metabolite 6-keto-$PGF_{1\alpha}$ which averages around 15 ng/g wet lung in nonperfused lungs and may reach 340 ng/g in lungs perfused with PSS-Ficoll for 1 hr (Table I). Thus, PGI_2 synthesis appears to be stimulated in the perfused lungs and likely contributes to the impaired vasoreactivity observed after prolonged perfusion. It is not surprising that cyclooxygenase inhibitors enhance and prolong vascular reactivity in isolated perfused rat lung.

[20] M. L. Morganroth, J. T. Reeves, R. C. Murphy, and N. F. Voelkel, *J. Appl. Physiol.* **56,** 1340 (1984).

[21] A. J. Lorigro, R. S. Sprague, A. H. Stephenson, and T. E. Dahnms, *J. Appl. Physiol.* **64,** 2538 (1988).

[22] J. Y. Westcott, T. J. McDonnell, P. Bostwick, and N. F. Voelkel, *Am. Rev. Respir. Dis.* **138,** 895 (1988).

TABLE I
COMPARISON OF LUNG TISSUE EICOSANOID LEVELS IN CONTROL RATS BEFORE AND AFTER LUNG PERFUSION

Lung tissue	Thromboxane B_2[a] (ng/g)	6-keto-PGF_1 (ng/g)	LTC_4 (ng/g)
Unperfused[b] (11)[d]	6 ± 1	15 ± 4	1.5 ± 0.3
Perfused[c] (6)[d]	15 ± 3	338 ± 55	5.0 ± 1.2

[a] Data are mean + SEM and expressed as nanogram per gram wet lung weight.
[b] Lung are removed from anesthetized rats and immediately homogenized in methanol. Thromboxane B_2 and 6-keto-PGF_1 are measured by enzyme-linked immunoassay. LTC_4 is measured by enzyme-linked immunoassay after HPLC purification. Methods are as described by J. Y. Westcott *et al.* [*Prostaglandins* **32,** 857 (1986)].
[c] Lungs are isolated and perfused in a physiological salt solution containing Ficoll for 60 min prior to lung homogenization and eicosanoid measurements.
[d] Number of experiments.

Harper *et al.*[23] have also used isolated perfused rat lung to study the metabolism of LTB_4 and LTC_4 and found that LTC_4 is rapidly metabolized to LTC_4 sulfoxide, LTD_4, and LTE_4. In contrast, LTB_4 appears not to be significantly metabolized by the rat lung, either when added to the lung perfusate or when instilled into the airspace. In a subsequent paper, the time course of metabolism and transfer of eicosanoids instilled into the alveolar space was further characterized by Westcott *et al.*[24] They confirmed the rapid metabolism of instilled LTC_4 to LTD_4 to LTE_4, but found that LTE_4 was further metabolized by the lung to *N*-acetyl-LTE_4. Significantly, a large portion of peptidoleukotrienes instilled into the alveolar space was retained in the air space or lung tissue 2–15 minutes later, while much of instilled PGD_2, PGE_2, thromboxane B_2, LTB_4, and 5-HETE were quickly removed from the lung.[24]

An important concept derived from these studies is that in the isolated lung three distinct compartments can be studied: the air space, the interstitium, and the vascular space. The synthetic and metabolic capabilities for eicosanoids are markedly different for each of these compartments—probably because of the differences in the metabolic pathways of the predominant cell types. Moreover, the ability to transfer from one compartment to another may differ between different eicosanoids.[24] It is therefore not surprising that lung lavage levels of 6-keto-$PGF_{1\alpha}$ and PAF do not reflect lung tissue levels in rats injured by systemic bacterial endotoxin.[18]

[23] T. W. Harper, J. Y. Westcott, N. F. Voelkel, and R. C. Murphy, *J. Biol. Chem.* **259,** 14437 (1984).
[24] J. Y. Westcott, T. J. McDonnell, and N. F. Voelkel, *Am. Rev. Respir. Dis.* **139,** 80 (1989).

Localization of Synthesis/Metabolism

While it is difficult to be certain of the exact cell types involved in a certain metabolic process defined for the intact lung, Haroldsen *et al.*[25] have used autoradiographic analysis in isolated perfused rat lung to infer the cell types involved in mediator metabolism. They found rapid metabolism of [^{3}H]PAF instilled into the airspace of isolated perfused rat lung. With autoradiography, they localized PAF uptake to the type II cell and Clara cell in the lung. This technique can be useful for localization of eicosanoid metabolism in the intact lung.

Concluding Remarks

We have described a method for isolated perfused rat lung studies and have given examples in which useful information on eicosanoid action and metabolism have been obtained in this preparation. Similar techniques can be used for lungs from different species of animals. We believe that studies in intact lungs can yield information relevant to the *in vivo* state and complement results from studies in isolated cells in culture.

Acknowledgments

This work was supported by NIH Program Project Grants HL-14985 and HL-07171 and Clinical Investigator Award HL-01966 and by the American Lung Association. We thank Marcia Brassor for her excellent secretarial assistance.

[25] P. E. Haroldsen, N. F. Voelkel, J. E. Henson, P. M. Henson, and R. C. Murphy, *J. Clin. Invest.* **79,** 1860 (1987).

[66] Isolated Coronary-Perfused Mammalian Heart: Assessment of Eicosanoid and Platelet-Activating Factor Release and Effects

By ROBERTO LEVI and KEVIN M. MULLANE

Langendorff Heart

The isolated coronary-perfused whole-heart preparation was originally devised by Langendorff.[1] In the "Langendorff heart" the coronary vascular bed is perfused by retrograde flow from the aorta in the absence of

[1] O. Langendorff, *Pflügers Arch. Ges. Physiol.* **61,** 291 (1895).

TABLE I
PHYSIOLOGICAL SALT SOLUTIONS

Salt/Sugar	Concentration (g/liter)		
	Krebs–Henseleit	Tyrode	Ringer
NaCl	6.87	8.00	9.00
KCl	0.40	0.20	0.42
$MgCl_2 \cdot 6\ H_2O$	—	0.21	—
$MgSO_4 \cdot 7\ H_2O$	0.30	—	—
$CaCl_2 \cdot 2\ H_2O$	0.36	0.26	0.32
$NaH_2PO_4 \cdot H_2O$	0.16	0.06	—
$NaHCO_3$	2.10	1.00	0.50
Glucose	2.00	1.00	1.00

pulmonary or systemic circulation. Either blood or a variety of oxygenated physiological solutions is infused at constant flow or constant pressure into a cannula inserted in the ascending aorta. As the aortic valves remain closed, the perfusing fluid enters the coronary arteries and exits through the coronary sinus into the right atrium and from there to the exterior through the superior and inferior venae cavae; depending on the particular experimental protocol, the effluent can either be discarded or recirculated through the heart. Thus, in the "Langendorff heart," the ventricular and atrial chambers remain practically empty and, although they contract rhythmically, they do not pump fluid. In contrast, in the "working heart" version, the left ventricle performs pressure–volume work using a second or "systemic" circulation (*vide infra*).

Most commonly, guinea pigs are chosen for this preparation, although rabbits and rats are also used. The animals are deprived of food, but not water, overnight; heparin (500 USP/100 g body weight, iv or ip) is administered to maintain patency of the coronary vasculature. Thirty minutes after heparinization, the animals are anesthetized with CO_2 vapors or with sodium pentobarbital (30 mg/kg ip), killed by cervical dislocation, and exsanguinated. The aorta is cut as far cranially as possible and the heart is rapidly excised from the surrounding tissue (the procedure should take not more than 1 minute). The aorta is cannulated ensuring that the tip of the cannula is not placed beyond the semilunar valves and that it does not obstruct the ostia of the coronary arteries.[2,3] The composition of the most frequently used physiological solutions is shown in Table I.

[2] G. Allan and R. Levi, *J. Pharmacol. Exp. Ther.* **214,** 45 (1980).

[3] R. Levi, J. A. Burke, Z.-G. Guo, Y. Hattori, C. M. Hoppens, L. M. McManus, D. J. Hanahan, and R. N. Pinckard, *Circ. Res.* **54,** 117 (1984).

In the constant-pressure version, the heart is perfused at a pressure closely resembling diastolic pressure *in vivo*, corrected to account for the lower viscosity of buffered solutions as compared with blood. In this model, increases or decreases in coronary vascular resistance translate into decreases or increases in coronary flow, respectively. The latter are assessed by measuring the volume of coronary venous effluent in a given time or by the use of a flowmeter. In the constant-flow version, increases or decreases in coronary vascular resistance translate into increases or decreases in perfusion pressure, respectively. These changes are monitored with a pressure transducer.[3]

Left ventricular contractile force is measured most times isometrically by connecting the tip of the left ventricle to a force transducer via a thread or by monitoring left ventricular pressure (*LVP*) with a small balloon connected to a pressure transducer via a catheter inserted into the left ventricle through the left atrium and the mitral valve. For a more complete assessment of the changes in contractility, the derivative of *LVP* can be obtained (i.e., $LV\ dp/dt$).

Spontaneous heart rate is best determined by recording the bipolar electrogram from the surface of the right atrium and left ventricle and then measuring the interval between two consecutive *P* or *R* waves. Further, the state of atrioventricular conduction can be assessed by measuring the duration of the *P–R* interval and by determining the number of *P* and *R* waves in a given time frame. Aside from atrioventricular conduction blocks of various degrees, the electrogram facilitates the recognition of idioventricular arrhythmias due to enhanced automaticity or reentry. Inasmuch as changes in rate will affect contractility, at times it may be necessary to pace the heart at constant rate with rectangular pulses generated by a stimulator and delivered by a pair of electrodes placed on the right atrium with the interposition of a stimulus isolation unit. In a more sophisticated version, the right atrium is opened to expose the atrioventricular node and additional recording electrodes are placed in the region of the bundle of His; stimulus-to-His bundle interval measurements are continuously monitored with the aid of a microcomputer.[4] This system allows for the most reliable assessment of atrioventricular conduction on a beat-to-beat basis.

Hearts can also be isolated with an intact sympathetic innervation[5]; right and left sympathetic ganglia are stimulated with square pulses of appropriate intensity and duration. Overflow of endogenous norepinephrine is assayed in the coronary effluent in response to increasing frequencies of sympathetic nerve stimulation.[5] With this model, one can de-

[4] J. R. Jenkins and L. Belardinelli, *Circ. Res.* **63,** 97 (1988).

[5] S. S. Gross, Z.-G. Guo, R. Levi, W. H. Bailey, and A. A. Chenouda, *Circ. Res.* **54,** 516 (1984).

termine the modulatory effect of a variety of substances, including eicosanoids, on norepinephrine release from the heart.[6]

Cardiac Effects of Prostaglandins

The cardiac effects of prostaglandins (PG) have been studied in the isolated guinea pig heart perfused at constant pressure.[2] Arachidonic acid (AA; 0.1–10 μg), PGE_2 (0.01–1 μg), and prostacyclin (PGI_2; 0.01–10 μg) administered by intraaortic bolus injection elicit dose-dependent increases in coronary flow rate. Dose-dependent coronary vasodilatation also occurs when PGI_2 or PGE_2 (0.3 nM–0.3 μM) is infused in isolated guinea pig hearts perfused at constant flow.[7] In contrast, PGD_2 (0.01–10 μg), $PGF_{2\alpha}$ (1–100 μg), and the stable thromboxane (TxA_2) analog compound U46,619 (0.01–1 μg) cause dose-dependent decreases in coronary flow rate.[2] Indeed, TxA_2 generated by conversion of PGH_2 with platelet microsomes elicits a dose-dependent coronary-vasoconstricting effect.[8] Over the dose range studied, PGE_2, AA, PGI_2, and $PGF_{2\alpha}$ increase, and PGD_2 decreases, the spontaneous sinus rate of the isolated guinea pig heart, whereas U46,619 has no consistent chronotropic effects.[2] AA, $PGF_{2\alpha}$, and U46,619 cause a dose-dependent decrease in contractile force; however, the negative inotropic effect of U46,619 is probably secondary to its coronary-vasoconstricting action.[8] PGE_2 has no effect on the contractility of the isolated guinea pig heart, whereas PGI_2 elicits a modest positive inotropic response.[2] PGD_2 has a primary positive inotropic action; on the other hand, because PGD_2 is also a potent coronary vasoconstrictor, a secondary negative inotropic response predominates.[9]

Cardiac Effects of Leukotrienes

Coronary vasoconstriction and contractile failure are the hallmark of partially purified slow-reacting substance of anaphylaxis (SRS-A), HPLC-purified SRS, and synthetic leukotrienes (LT)C_4 and D_4 injected into isolated guinea pig hearts perfused at either constant flow or pressure.[9] Although the dose-dependent decrease in contractility parallels the decrease in coronary flow, the two phenomena are likely to be independent

[6] Å. Wennmalm, *Acta Physiol. Scand.* **105,** 254 (1979).

[7] K. Schrör and S. Moncada, *Prostaglandins* **17,** 367 (1979).

[8] Z.-I. Terashita, H. Fukui, K. Nichikawa, M. Hirata, and S. Kikuchi, *Eur. J. Pharmacol.* **53,** 49 (1978).

[9] R. Levi, Y. Hattori, J. A. Burke, Z.-G. Guo, U. Hachfeld del Balzo, W. A. Scott, and C. A. Rouzer, *in* "Prostaglandins, Leukotrienes, and Lipoxins" (J. M. Bailey, ed.), p. 275. Plenum, New York, 1985.

both in the Langendorff and working heart models.[9,10] Nevertheless, the possibility remains that the negative inotropic effect of leukotrienes may be amplified by local ischemia resulting from their concomitant coronary-constricting effect. The rank order of potency for the negative inotropic and coronary-vasoconstricting effects of leukotrienes is $LTD_4 > LTC_4 > LTE_4$ (LTD_4, 0.01–1 ng; LTC_4, 0.3–30 ng; LTE_4, 100 ng–3 μg).[9] Both the negative inotropic and coronary-vasoconstricting effects of leukotrienes are antagonized by specific leukotriene receptor antagonists, but are unaffected by cyclooxygenase inhibitors.[9]

Cardiac Effects of Platelet-Activating Factor

In the isolated guinea pig heart perfused at constant pressure, the intraaortic administration of platelet-activating factor (PAF) by bolus injection (0.01 pmol to 3 nmol) elicits dose-related decreases in left ventricular contractile force and coronary flow (−5 to −85%).[3] The negative inotropic effect of PAF persists also in hearts perfused at constant flow.[3] The deacetylated derivative of PAF has only minimal cardiac effects in doses as high as 10 nmol.[3] Neither the coronary vasoconstriction nor the decrease in contractility caused by PAF are affected by indomethacin or FPL 55172, indicating that although cyclooxygenase and lipoxygenase products may be released by PAF,[11] they do not play a functional role in the cardiac response to PAF.[3,12] Indeed, the cardiac effects of PAF are specifically antagonized by pharmacological agents known to block the platelet-activating effects of PAF.[13]

Platelet-Perfused Hearts

For the investigation of platelet–vessel wall interactions in eicosanoid research, hearts of guinea pigs,[14] rats,[15] or rabbits[16] have been perfused with buffer containing either human[15] or rabbit[15,16] platelets. The

[10] O. G. Björnsson, K. Kobayashi, and J. R. Williamson, *Eur. J. Clin. Invest.* **17,** 146 (1987).

[11] P. J. Piper and A. G. Stewart, *Br. J. Pharmacol.* **88,** 595 (1986).

[12] G. L. Stahl, D. J. Lefer, and A. M. Lefer, *Arch. Pharmacol.* **336,** 459 (1987).

[13] R. Levi, T. Y. Shen, S. J. Yee, D. A. Robertson, A. Genovese, O. W. Isom, and K. H. Krieger, *in* "New Horizons in Platelet Activating Factor Research" (C. M. Winslow and M. L. Lee, eds.), p. 255. John Wiley & Sons, Chichester, England, 1987.

[14] K. Schrör, P. Köhler, M. Müller, B. A. Peskar, and P. Rösen, *Am. J. Physiol.* **241,** H18 (1981).

[15] M. Purchase, G. J. Dusting, D. M. F. Li, and M. A. Read, *Circ. Res.* **58,** 172 (1986).

[16] G. Montrucchio, G. Alloatti, C. Tetta, R. DeLuca, R. N. Saunders, G. Emanuelli, and G. Camussi, *Am. J. Physiol.* **256,** H1236 (1989).

"Langendorff heart" is prepared as described above. Washed platelets are prepared from platelet-rich plasma. The details of platelet preparation vary, with some investigators using PGE_1 (Ref. 4) or PGI_2 (Ref. 15) to limit platelet activation during the steps of centrifugation and suspension of the platelet pellet. The platelets are finally resuspended in an albumin-containing buffer lacking any PGE_1 or PGI_2. The platelets are infused into the aortic inflow tract just above the heart at a flow rate of 0.1 ml/min for guinea pig or rat hearts, or 0.1 to 1 ml/min in the rabbit heart, representing 1–3% of the total perfusate flow. The concentration of platelets used varies from $\sim 1 \times 10^6$ to 2×10^8 platelets per milliliter perfusate.

The predominant use of this preparation has been in the study of interactions between PGI_2 released from the heart, and TxA_2 produced mainly by the platelets. The formation of these two mediators has been determined by bioassay where the coronary effluent continuously superfuses strips of vascular tissue (e.g., rabbit aorta, porcine or bovine coronary arteries; all three are contracted by TxA_2, but only bovine coronary artery is relaxed by PGI_2) or by radioimmunoassay of the stable hydrolysis products 6-keto-$PGF_{1\alpha}$ and TxB_2. Perfusion of hearts with a platelet-containing buffer enhances the production of PGI_2 and, to a lesser extent, TxA_2, while the formation of both mediators can be increased substantially by the addition of arachidonic acid.[14,15] Despite the formation of approximately three times more TxA_2 than PGI_2 in the presence of arachidonic acid, coronary vasodilatation predominates.

Infusions of low concentrations of epinephrine (0.6–6 pmol/ml), that are insufficient to provoke platelet aggregation, potentiate platelet TxA_2 production from arachidonic acid within the heart.[15] It is proposed that perfused hearts produce a stable nonprostanoid factor that potentiates TxA_2 release when incubated with platelets.[15]

Neutrophil-Perfused Hearts

Rabbit hearts perfused in the Langendorff mode with rabbit neutrophils added to the perfusate have been used to study interactions between leukocytes and the heart.[17,18] Neutrophils are prepared from citrated whole blood (3.8% citrate at ratio 1 : 10 citrate : blood), mixed with 6% dextran (molecular weight = 500,000; ratio 1 : 5) for sedimentation of erythrocytes. The supernatant is centrifuged (400 *g* for 5 min), contaminating erythrocytes are eliminated by hypotonic lysis, and the cells separated

[17] M. N. Gillespie, S. Kojima, J. O. Owasoyo, H. H. Tai, and M. Jay, *J. Pharmacol. Exp. Therap.* **241,** 812 (1987).

[18] R. Kraemer and K. M. Mullane, *J. Pharmacol. Exp. Therap.* **251,** 620 (1989).

by Ficoll/Hypaque density gradient centrifugation (600 *g* for 20 min). The pellet is resuspended in Hanks' balanced salt solution and is generally comprised of 90–95% neutrophils with cell viability close to 100%, as assessed by trypan blue exclusion. Neutrophils can also be radiolabeled with either indium-111[17] or chromium-51,[18] and radioactivity determined using a γ-counter.

The perfused heart preparation is essentially similar to that described above, except that the superior and inferior venae cavae and pulmonary veins are ligated, and the pulmonary artery cannulated to capture the coronary outflow and to perfuse the hearts in a recirculating mode. The reservoir volume is 400 ml, containing 5×10^3 neutrophils per milliliter in Krebs–Henseleit buffer with 1% BSA; it is gassed with $95\%O_2 + 5\%CO_2$ by either blowing the gas across the surface of the buffer in a revolving reservoir or by using a membrane lung system, to minimize "frothing" of the protein-rich buffer. The hearts are perfused at a constant flow of 25 to 30 ml/min at a left ventricular end-diastolic pressure (LVEDP) of 10 to 15 mm Hg and paced at a constant rate of 200 to 400 beats/min. Myocardial hypoxia is induced by gassing the perfusion fluid with either $12\%O_2 + 83\%N_2 + 5\%CO_2$ or $100\%N_2$ for 30 min and then reoxygenating with the original $95\%O_2 + 5\%CO_2$. Hypoxia and reoxygenation provoke the accumulation of radiolabeled neutrophils in the heart.[17,18] Hypoxia-induced neutrophil accumulation is attributed to the generation of LTB_4 by the myocardium, as determined by increases in the tissue content of LTB_4-like immunoreactivity and suppression of cell accumulation by diethylcarbamazine or nordihydroguaiaretic acid, purported lipoxygenase inhibitors.[17] Neutrophil sequestration is associated with a reduced recovery of function during reoxygenation.[18]

Ischemia–Reperfusion in Langendorff Hearts

Studies on the release and role of eicosanoids in myocardial ischemia and reperfusion have benefitted enormously from the isolated perfused heart technique. Hearts of rats,[19,20] guinea pigs,[21] rabbits,[16,22] and cats[23] have been used in these studies. Generally, global ischemia is induced by decreasing the flow rate to <1 ml/min for 30 to 120 min before restoring preischemic flow for a 15- to 30-min reperfusion period.[16,19,20,23] The degree of ischemia is more severe if the heart is paced at a fixed rate

[19] M. Karmazyn, *Am. J. Physiol.* **251,** H133 (1986).
[20] W. G. Nayler, M. Purchase, and G. J. Dusting, *Basic Res. Cardiol.* **79,** 125 (1984).
[21] M. P. Moffat, *J. Pharmacol. Exp. Ther.* **242,** 292 (1987).
[22] A. Edlund, K. Sahlin, and Å Wennmalm, *J. Mol. Cell. Cardiol.* **18,** 1067 (1986).
[23] H. Araki and A. M. Lefer, *Circ. Res.* **47,** 757 (1980).

throughout the procedure; this enables the ischemic period to be abbreviated and thus to minimize time-related changes in cardiac function. An alternative procedure that has been used in the rabbit heart is to ligate the left anterior descending branch of the main coronary artery for 2 hr to study the effect of PGI_2 on regional ischemia.[22] Ischemia does not provoke prostaglandin production; however, reperfusion of the globally ischemic rat[19] or guinea pig[21] heart is a potent stimulus for PGI_2 synthesis: PGI_2 release is increased from 0.01 to 0.3–0.4 ng/ml/min or from 0.2 to 1.5 ng/ml/min (measured as 6-keto-$PGF_{1\alpha}$ by radioimmunoassay) for the rat and guinea pig, respectively. This release is sustained in the rat heart, but rapidly declines in the guinea pig preparation. The magnitude and duration of PGI_2 release at reperfusion is independent of the duration of the preceding ischemic period when varied from 5 to 60 min.[19]

The significance of the PGI_2 production at reperfusion on the postischemic recovery of cardiac function is a subject of controversy. PGI_2 is reported to be either beneficial,[23] detrimental,[19,21,24] or without effect.[22] While this discrepancy may relate in part to PGI_2 improving the rate, but not the magnitude, of functional recovery,[20] it appears that the discrepancy may also result from different concentration-dependent effects of PGI_2. Small PGI_2 concentrations that mimic the endogenously released material (~0.5–1 ng/ml) impair recovery of contractile function, whereas larger concentrations of PGI_2 may be cardioprotective.[19,20,23,24] However, the true effects of endogenous PGI_2 produced by ischemia and reperfusion remain to be elucidated.

Ischemia and reperfusion of the Langendorff rabbit heart is also a stimulus for transient PAF release. PAF is undetectable during control perfusion and ischemia (<1 ng/min), but increases to $\sim 11 \pm 13$ (mean $\pm$ SD) ng/min during the first minute of reperfusion, with a range of 1.2 to 37 ng/min.[16] Reperfusion with a platelet-containing buffer exacerbates the postischemic cardiac dysfunction that is abrogated by the PAF antagonist, SDZ 63675. In the absence of platelets, the formation of PAF at reperfusion has no functional significance.[16]

Ex Vivo Perfused Hearts

There is increasing evidence that platelets and leukocytes play important roles in the evolution of myocardial injury associated with ischemia and reperfusion.[25] Because these blood elements are also major sources of eicosanoids, experiments designed to address the role of eicosanoids in

[24] M. Karmazyn and J. R. Neely, *J. Mol. Cell. Cardiol.* **21,** 335 (1989).

[25] K. Mullane, *in* "Prostaglandins in Clinical Research: Cardiovascular Systems" (K. Schrör and H. Sinzinger, eds), p. 39. Liss, New York, 1989.

myocardial ischemia and reperfusion in buffer-perfused hearts *in vitro* would underestimate their contribution. One alternative has been to add neutrophils or platelets to the perfusate as described above. Another variation has been the production of myocardial ischemia and reperfusion *in vivo*, prior to removal of the heart for perfusion in the Langendorff mode *ex vivo*.[26–29] Histological examination reveals platelet and leukocyte accumulation in the injured hearts[26,28,29] that resolves as the infarct repairs.[26]

Only rabbits have been used in these experiments. The animals are anesthetized with either pentobarbital (20–35 mg/kg iv) or a mixture of ketamine (50 mg/kg), acepromazine (0.1 mg/kg), and xylazine (20 mg/kg), administered intramuscularly, intubated, and ventilated with room air or 100% oxygen. Following thoracotomy, a major branch of the coronary artery is occluded using a snare ligature (5–0) for 60 min; the snare is then released to allow flow to return. The chest is closed in layers and the animals allowed to recover. Reperfusion periods of 30 min to 30 days have been studied.

Subsequently, the heart is removed from the anesthetized, heparinized rabbit, and retrogradely perfused at 20 to 35 ml/min in a constant flow system with Krebs–Henseleit buffer gassed with 95% O_2 + 5% CO_2 at 37°. The effluent superfuses a series of bioassay tissues, including a segment of guinea pig ileum, a rat stomach strip, the chick rectum, and a spiral strip of rabbit thoracic aorta, to continuously monitor leukotriene, prostaglandin, and TxA_2 release, respectively.[26,27,29] Alternatively, selected samples are collected to measure eicosanoid production by specific radioimmunoassay.[26,28,29] The hearts are then challenged with a leukocyte agonist (e.g., *N*-formylmethionylleucylphenylalanine; fMLP) or with platelet-activating factor (PAF), an agonist of both platelets and leukocytes, and the release of eicosanoids is monitored and related to changes in cardiac function.

Administration of fMLP (10–100 ng) to the postinfarcted heart augments the release of various eicosanoids, including LTB_4, LTC_4, TxB_2, 6-keto-$PGF_{1\alpha}$, and PGE_2, when compared to control noninfarcted hearts.[26,27,29] The peak increase in LTB_4 production occurs at 24 hr when neutrophils predominate,[29] while maximal levels of LTC_4 and TxB_2 are observed after 48 to 96 hr and are associated with an invasion of mononuclear cells.[26,29] Prevention of leukocyte infiltration by treatment of the

[26] A. S. Evers, S. Murphree, J. E. Saffitz, B. A. Jakschik, and P. Needleman, *J. Clin. Invest.* **75,** 992 (1985).

[27] S. Barst and K. Mullane, *Eur. J. Pharmacol.* **114,** 383 (1985).

[28] J. K. Mickelson, P. J. Simpson, and B. R. Lucchesi, *J. Mol. Cell. Cardiol.* **20,** 547 (1988).

[29] M. S. Freed, P. Needleman, C. G. Dunkel, J. E. Saffitz, and A. S. Evers, *J. Clin. Invest.* **83,** 205 (1989).

rabbits with intravenous BW 755C (25 mg/kg tid), or pretreatment with nitrogen mustard to induce leukopenia, attenuates the production of eicosanoids in response to fMLP in the postinfarcted heart.[29]

Mickelson and co-workers[28] have measured PAF-induced release of immunoreactive TxB_2 and LTC_4 into the small "lymphatic" drainage which passes through the interstitium and drips from the apex of the heart. The apparent advantage of this modification is that the eicosanoids become concentrated, and enhanced PAF effectiveness is evident within only 4 hr of reperfusion, while by 6 hr TxB_2 and LTC_4 synthesis is increased ~8-fold over noninfarcted control hearts. Elevations in TxB_2 and LTC_4 are associated with coronary vasoconstriction, elevations in LVEDP, and contractile failure.[28] Selective inhibition of thromboxane synthase with CGS 13080 prevents the PAF-induced release of TxB_2 and LTC_4, and attenuates the associated cardiac dysfunction.[30]

Cardiac Anaphylaxis: Useful Model for Study of Release and Effects of Eicosanoids

Cardiac anaphylaxis *in vitro* is characterized by tachycardia, arrhythmias, contractile failure, decrease in coronary flow, and mediator release.[31] These changes occur in response to specific antigen challenge, most usually 1 mg protein, administered by intraaortic bolus injection into isolated hearts obtained from presensitized guinea pigs. Sensitization can be either active or passive. In the former, guinea pigs are immunized with a variety of foreign proteins or hapten–protein conjugates generally without adjuvants, such as ovalbumin, hemocyanin, or benzylpenicilloyl conjugates.[32–34]

Passive sensitization can be achieved either *in vivo* or *in vitro*. In the former, heterocytotropic IgG_2 antibodies made in rabbits against penicilloyl–protein conjugates are transferred to nonsensitized guinea pigs by injection into the foot vein 18–24 hr prior to isolation of the heart for Langendorff perfusion.[31] Alternatively, homocytotropic IgG_1 or IgE antibodies made in guinea pigs against dinitrophyl–protein conjugates are

[30] J. K. Mickelson, P. J. Simpson, and B. R. Lucchesi, *Circulation* **76,** IV-202 (1987).

[31] R. Levi, *in* "Human Inflammatory Disease" (G. Marone, L. M. Lichtenstein, M. Condorelli, and A. S. Fauci, eds.), p. 93. B. C. Decker, Inc., Toronto, 1988.

[32] G. A. Feigen, E. M. Vaughan Williams, J. K. Peterson, and C. B. Nielsen, *Circ. Res.* **8,** 713 (1960).

[33] H. Anhut, B. A. Peskar, and W. Bernauer, *Arch. Pharmacol.* **305,** 247 (1978).

[34] U. Aehringhaus, A. Dembinska-Kiéc, and B. A. Peskar, *Arch. Pharmacol.* **326,** 368 (1984).

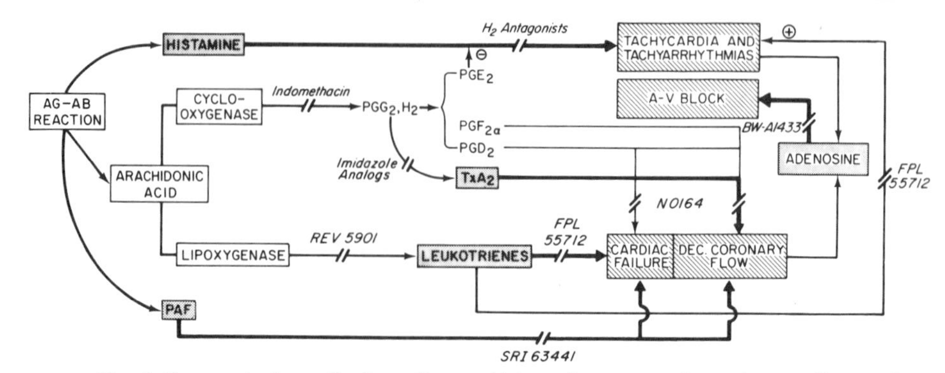

FIG. 1. Proposed scheme for the actions and interactions among the various mediators of cardiac anaphylaxis. [Reprinted with permission from R. Levi, *in* "Human Inflammatory Disease, Vol 1: Clinical Immunology" (G. Marone, L. M. Lichtenstein, M. Condorelli, and A. S. Fauci, eds.). B. C. Decker, Inc., Toronto, 1988.]

transferred to nonsensitized guinea pigs by iv administration.[3,9,35] A convenient and fast procedure of passive sensitization is to perfuse hearts from naïve guinea pigs with rabbit IgG_2 anti-ovalbumin in a Langendorff apparatus.[36] The advantage of passive sensitization is that a more predictable level of sensitization can be achieved; graded anaphylactic responses are obtained by increasing the amount of antibody protein used for sensitization, while maintaining constant the quantity of challenging antigen.[31,36]

Following challenge of sensitized hearts with the specific antigen, several mediators are released into the coronary venous effluent; some are preformed, such as histamine, while others are synthesized *de novo*, such as cyclooxygenase and 5-lipoxygenase metabolites of arachidonic acid, and PAF.[31] Prostaglandin, leukotrienes, and PAF cause a marked decrease in coronary flow and a protracted negative inotropic effect.[3,9,35] TxA_2 is another major contributor to anaphylactic coronary vasoconstriction.[35] The metabolites of arachidonic acid interact with histamine by modulating its release and its effects.[31] Leukotrienes potentiate and prolong the tachyarrhythmic effects of histamine, whereas PGE_2 attenuates these effects.[31] In addition, PGE_2 could increase, and PGD_2 decrease, the quantity of histamine released from the anaphylactic heart.[31] In all likelihood, PAF is a major participant in this interplay of mediators, and either modulates or is modulated by the other autacoids. Thus, an intricate interrelationship exists among the various mediators of cardiac anaphylaxis[31] (see Fig. 1).

[35] G. Allan and R. Levi, *J. Pharmacol. Exp. Ther.* **217,** 157 (1981).
[36] R. Levi, *J. Pharmacol. Exp. Ther.* **182,** 227 (1972).

Acknowledgment

Supported by NIH grants HL 34215 and HL 18828 to RL and HL 31591 to KM.

[67] Preparation of Human and Animal Lung Tissue for Eicosanoid Research

By T. VIGANÒ, M. T. CRIVELLARI, M. MEZZETTI, and G. C. FOLCO

Eicosanoids are fatty acids of great biological significance because of their ubiquity, high potency, and multitude of other effects. Among all the organs that have been investigated, the lungs are the most active in forming and metabolizing eicosanoids; human and animal lung tissue can oxidize arachidonic acid (AA), generating cyclooxygenase as well as lipoxygenase-derived products that appear to play a role in the development of asthma, allergic reactions, and inflammation. In fact, leukotrienes (LT) have potent spasmogenic actions on human lung tissue *in vitro*, affect mucus production and clearance, and may be important factors in the pathogenesis of airway hyperresponsiveness.

A sufficient number of *in vitro* methods has been developed utilizing human and animal lung tissue for eicosanoid research; this chapter will describe techniques that are useful in understanding the physiopathological role of eicosanoids in the pulmonary system as well as their functional effects.

Preparation of Human Lung Parenchyma to Produce Eicosanoids

Normal human lung parenchyma is a preferential tissue to use and probably represents the richest source of sulfidopeptide LT and of PGD_2. The tissue is usually available following surgery for pulmonary carcinoma or bronchiectasis; the lung parenchyma should be immediately collected in a buffered solution such as Tyrode's [composition (in g/liter): NaCl, 8.0, KCl, 0.25, CaCl, 0.15 $MgCl_2 \cdot 6\ H_2O$, 0.01, $NaH_2PO_4 \cdot H_2O$, 0.066, glucose, 1.0, $NaHCO_3$, 1.0] or Krebs–Henseleit [composition (in g/liter): NaCl, 6.9, KCl, 0.35, KH_2PO_4, 0.16, $MgSO_4 \cdot 7\ H_2O$, 0.29, $CaCl_2$, 0.28, glucose, 1.0, $NaHCO_3$, 2.1], pH 7.4, kept at 4° and gassed with a mixture of 95% O_2–5% CO_2. It is imperative that the tissue not be frozen as it loses its capacity to respond to different stimuli.

Small biopsies of approximately 100 mg each are sampled at random from the excised lung portion and used for histological examination to exclude the possibility of neoplastic infiltrations. Usually the available peripheral tissue is free of macroscopically evident bronchi or vessels and is chopped manually to fragments of approximately 100 mg each. The fragments are washed again using the same buffer and can now be utilized for release experiments. It is possible to follow at least two different experimental protocols in order to mimic a release of lipid mediators that is likely to occur during immediate hypersensitivity reactions.

The fragments can be passively sensitized using diluted reaginic serum from hyperimmune patients (1500–5000 ng/ml of specific IgE). The tissue is incubated overnight (1 g tissue/5 ml diluted serum) at room temperature in a Dubnoff shaking incubator without need for direct oxygenation; alternatively, the passive sensitization can be carried out using a shorter incubation step (3 hr) at 37°. Lung parenchyma is then washed extensively using the same buffer and divided into portions of about 1 g resuspended in 10 ml of fresh Tyrode's or Krebs buffer. The parenchymal fragments are preincubated for 10 min under gentle agitation and challenged with the appropriate antigen (usually 500–1000 ng/ml), agonists or vehicle. At the end of the incubation, an aliquot (max. 0.5 ml) is collected for prostaglandin assay using RIA[1] or EIA[2] while another is immediately tested for the presence of eicosanoid-like activity using appropriate bioassay systems.[3] The remaining volume (approximately 8 ml) can be added to the required internal standards (PGB_2 500 ng or tracer amounts of radiolabeled LT, 20,000 dpm) and frozen or further processed for quantitative assays of prostanoids using physicochemical methods.

It is also possible to trigger the immunological release of different autacoids from pulmonary tissue using specific anti-human IgE antibodies which are usually raised in sheep or goat. Our personal experience, accumulated in about 40 experiments with an equal number of human specimens, indicates that the mediator release obtained in this experimental condition does not require the time-consuming step of passive sensitization. Usually after mincing and careful washing of the lung fragments, the tissue is preconditioned in Tyrode's buffer for approximately 30–60 min at 37° (1 g/5 ml), rewashed, and subjected to final incubation in Tyrode's buffer (1 g/10 ml). The challenge is performed using a final anti-human IgE

[1] E. Granstrom and H. Kindahl, *Adv. Prostaglandin Thromboxane Res.* **5,** 119 (1978).

[2] P. Pradelles, J. Grassi, and J. Maclouf, *Anal. Chem.* **57,** 1170 (1985).

[3] S. Moncada, S. H. Ferreira, and J. R. Vane, *Adv. Prostaglandin Thromboxane Res.* **5,** 211 (1978).

antibody dilution varying between 1 : 300 to 1 : 1000. Preparation and storage of aliquots is carried out as described above.

As the availability of good hyperimmune sera as well as that of anti-human IgE antibodies can present a problem, a preliminary check of the tissue releasability should be performed with the calcium ionophore A23187 (1–5 μM), stopping the reaction 20–30 min after challenge. By the concomitant use of a bioassay setup which utilizes a strip of longitudinal smooth muscle of guinea pig ileum, it is possible to follow rather carefully the development of the experiment. It is advisable to carry on the immunological challenge only in tissues that have responded successfully to A23187. Experience with approximately 200 human lung parenchymal specimens, indicates that while good releasability after A23187 treatment occurs in about 80–90% of the cases, a successful immunological challenge takes place in about 50% of the tissues.

Preparation of Human Bronchial Tissue

Both large and small human airways are suitable models to assess the pharmacological activity of eicosanoids as well as their role in pathophysiology. Bronchi of different caliber, from primary bronchi to terminal bronchioles, are utilized for such studies, however the latter, with an internal caliber of 1 to 2 mm, are better suited because they have no cartilage plates. These can be identified and differentiated from pulmonary arteries and veins of similar caliber. The vascular lumen can be pinched with small forceps and with gentle pulling can be detached from the surrounding parenchyma, whereas, the bronchioles cannot be freed. In addition, particularly following surgery for bronchiectasis, the bronchial lumen shows presence of mucous plugs. A fine polyethylene catheter is gently slipped through the full length of the bronchus in order to facilitate its localization. Further confirmation that one is dealing with a bronchiole comes from the fact that a gentle administration of air into the catheter leads to tissue inflation. The cannula is left in place and the bronchioles are carefully dissected free of parenchyma and blood vessels; it is advisable to leave a minimum layer of parenchyma to avoid inadvertent damage of the bronchiole smooth muscle. During this, as well as the following preparation steps, the tissue is kept moist with lukewarm saline. The polyethylene tubing is now fixed in a vertical position and the bronchiole spiralized to obtain a strip 2–3 mm wide and of varying length (usually 20–30 mm). Great care is taken to avoid the stretching of the preparation as this may damage its contractile function; a silk surgical thread is tied to both ends of the strip.

The smooth muscle strips are suspended in a 10-ml organ bath, under an isometric resting tension of 200 to 500 mg, at 37°, in a Krebs–Henseleit solution oxygenated with a mixture of 95% O_2–5% CO_2. The time required to attain a steady baseline tone may vary considerably. Whenever possible the tissues should be used the day of surgery or stored overnight at 4° under a slow constant flow of oxygenated buffer and then set up the following day, without appreciable loss of sensitivity or contractility. Transducers are coupled to a multichannel pen recorder and during the stabilization period (1–2 hr) it is advisable to change the buffer every 10–15 min.

Pharmacological studies can be performed by adding agonist eicosanoids as well as their potential antagonists in small volumes (0.05–0.1 ml). In this way, cumulative dose–response curves (for sulfidopeptide LT, 0.1–100 n*M*) can be constructed as tachyphylaxis for LT or bronchoactive prostaglandins (e.g., PGD_2) has not been described in human bronchi. By using a similar technique, Dahlen *et al.*[4] were able to demonstrate that bronchi from allergic asthmatics respond to antigen exposure with a contraction that correlates with the release of sulfidopeptide LT.

Human bronchiolar strips represent a relatively homogeneous structure and their functional changes are not affected by vascular smooth muscles that lie in close proximity, as is the case of the parenchyma. They can be used as detector organs for eicosanoids and, in general, for bronchoactive substances, such as histamine or PAF. The classical nonflow organ bath technique and also the perifusion technique (i.e., the tissue is set up in a tissue chamber where the buffer is constantly changed by means of a pump) require that the active compound be diluted in a volume that rarely is smaller than 5 ml. On the other hand, superfusion techniques involve only a very transient contact of the tissue with the substances added as bolus, but also require a rather high flow, 5–10 ml/min, once again leading to high dilution of the samples being tested.

A technique that we have used successfully is laminar flow superfusion, first suggested by Ferreira[5] in 1976. The isolated assay tissue is immersed in mineral oil thermostatted at 37° and superfused with a drip of Tyrode's solution at a very low rate of flow (0.1–0.2 ml/min). Each drop of aqueous buffer floats slowly downward at the interface between the tissue and the oil, allowing a contact time that is approx. 1 sec/1 cm of strip length. It is important to ensure a close contact between the catheter delivering the buffer and the thread which connects the tissue with the

[4] S. E. Dahlen, G. Hansson, P. Hedqvist, T. Bjorck, E. Granstrom, and B. Dahlen, *Proc. Natl. Acad. Sci. U.S.A.* **80,** 1712 (1983).

[5] S. H. Ferreira and F. De Souza Costa, *Eur. J. Pharmacol.* **39,** 379 (1976).

transducer, because lateral drops may be formed. This particular assay method should be used only if the partition coefficient between the water and oil phases of the compound under study guarantees full retention in the buffer.

Spiral strips of human bronchioles are easily adjusted to the laminar flow technique and amounts as small as 0.3–1 pmol of sulfidopeptide leukotrienes can be detected. This method is preferred when only very minute amounts of different eicosanoids or drugs are available.

Preparation of Guinea Pig Isolated Trachea, Parenchyma, and Lung

The airway tree of the guinea pig has been extensively studied, particularly for eicosanoid research. The first evidence of the existence of thromboxane (Tx) A_2 and sulfidopeptide LT came from experiments using this particular organ.

The tracheopulmonary system of the guinea pig can be easily excised and the trachea and peripheral parenchymal strips can be isolated. These isolated preparations represent suitable models of the central and peripheral airways, respectively. Guinea pig lungs may also be used whole, perfused through the pulmonary artery, and occasionally even ventilated with a respiratory pump. It is, therefore, possible to perform functional studies at different airway levels as well as to investigate the capacity of the organ to respond to different stimuli with a secondary formation of eicosanoids.

The tracheal smooth muscle of the guinea pig is well known for its outstanding responsiveness to multiple agonists; a contractile dose–response curve to sulfidopeptide LT ranges from 0.03 to 100 n*M* and no tachyphylaxis occurs. Moreover, this response is likely to be modulated by the concomitant formation of other dilator prostanoids (perhaps PGE_2) as the maximal response is potentiated by pretreatment of the tissue with indomethacin or meclofenamic acid (1 μM).

Several preparations have been described using the isolated guinea pig trachea and with careful dissection the muscle fibers can be arranged to run longitudinally. A chain of tracheal sections can be made by cutting individual rings of cartilage and tying them together. At least three rings are required to obtain contractions that are adequately recorded: resting tension 1–2 *g* and isotonic tone recording. Tyrode's or Krebs physiological solutions are normally used. Alternatively, the trachea can be set up as a spiral or a zigzag section and this technique brings the muscle sections close to each other; the smooth muscle fibers are located dorsally and usually the cartilaginous rings are cut ventrally along the middle, from the larynx to the first bifurcation. The organ is opened, fixed on a cork disc with fine pins at its corners and zigzag sections cut with a sharp scalpel. Fine silk threads are then bound to the extremities for connection to the transducer; resting load is usually adjusted to 2.5–3.0 g.

The isolated tracheal tube of the guinea pig can also be set up intact, by sealing the lower end (linked to a glass hook) and by connecting the upper end to a pressure transducer. The system is filled with buffer to obtain a resting intraluminal pressure of 1 mm Hg and the preparation can be stimulated intra- and extraluminally via platinum wire electrodes with square wave pulses.[6] All the preparations of the guinea pig airways described above can be easily set up using standardized isolated organ baths, thermostatted at 37°, and oxygenated with a mixture of 95% O_2–5% CO_2. Tracheal rings and zigzag preparations, as well as the lung parenchymal strips, can also be superfused at constant flow of 8 to 10 ml/min or arranged according to the laminar flow technique.

Guinea pig lung parenchymal strips are also widely used preparations for the study of eicosanoids and related drugs. After sacrifice by bleeding, the lung are excised and care should be taken to save approximately 1 cm of trachea. After a gentle wash with Krebs–Henseleit buffer, the lungs are slightly inflated with air and strips about 3 mm wide, 3 mm thick, and 2–3 cm long are cut along the outer curvature of the lobe rim. The extremities are then connected using a silk thread to an isotonic force transducer under a resting tension of 1 to 1.5 *g*, at 37° and gassed with a mixture of 95% O_2–5% CO_2; the tissues are equilibrated in approximately 30–60 min. Usually a contractile dose response curve to LTC_4 and LTD_4 ranges from 0.1 to 1 μM and involves, at least in part, cyclooxygenase-derived constrictor products of arachidonic acid, namely TxA_2. LTB_4 is equiactive with sulfidopeptide LT and significant induction of tachyphylaxis might take place. It is important to keep in mind that both guinea pig trachea and lung parenchyma are capable of metabolizing the peptide LT and therefore their functional response might change according to the tissue content of transpeptidase or aminopeptidase.[7] Guinea pig lungs can also be set up intact, isolated, perfused, and ventilated. For this study guinea pigs are anesthetized with pentobarbital (30 mg/kg, ip) and the trachea cannulated for artificial ventilation using a respiratory pump (50–60 cpm, 1.5 ml tidal volume/100 g body weight and approx. 2 cm water end-expiratory pressure). In order to stop spontaneous breathing, the animals are paralyzed with pancuronium bromide (4 mg/kg, iv). Before getting ready to transfer the lungs from the chest cavity to a thermostatted glass chamber, a thoracotomy is performed and the pulmonary artery dissected; the right ventri-

[6] R. A. Coleman and G. P. Levy, *Br. J. Pharmacol.* **52,** 167 (1974).

[7] D. W. Snyder and R. D. Krell, *J. Pharmacol. Exp. Ther.* **231,** 616 (1984).

cle is cut open and the pulmonary artery cannulated with the cannula used to deliver the perfusing buffer (Krebs–Henseleit) by means of a peristaltic pump. Once cannulation is completed, the lungs are freed from the left atria and other mediastinal structures and suspended in an appropriate chamber; perfusion is carried out at 8 to 12 ml/min. The trachea is then connected via a T piece to the respiratory pump and to a pressure transducer for continuous recording of ventilation pressure. A second pressure transducer inserted between the peristaltic pump and the pulmonary artery allows recording of pulmonary arterial pressure.

The perfusion buffer is collected from the bottom of the glass chamber and recirculated using the same peristaltic pump for several hours (3–4), if necessary. Alternatively, the lung effluent can superfuse a number of selective bioassay tissues for detection of prostaglandin or SRS-A-like activity. Challenge can be performed by bolus injections of different agonists into the cannula inserted in the left atrium.

By using this experimental approach the lungs can be challenged repeatedly and maintain their capacity to release (and metabolize, e.g., PGE_2) eicosanoids depending on the type and intensity of the stimulus. Usually no tachyphylaxis occurs.

It is also possible to study the profile of eicosanoid release that takes place after an anaphylactic shock *in vitro*. For this purpose guinea pigs are sensitized to allergens; the one most widely used is ovalbumin. A variety of protocols for inducing sensitization can be followed and all seem to induce antibody production successfully. Two major sensitization procedures[8] have been proposed for (a) production of high titer specific IgE- and moderate quantities of specific IgG-like antibodies or (b) production of IgG-like antibodies. (1) Guinea pigs of either sex are given one intraperitoneal injection of 0.5 ml saline containing 1 μg ovalbumin and 100 mg $Al(OH)_3$. One milliliter of adjuvant is added to the antigen solution 1 hr before injection. The animals can be challenged *in vivo* with 5 μg/kg ovalbumin 2 months later. (2) Guinea pigs are injected ip with 5 mg ovalbumin on day 0 and 10 mg on day 2 (injection volume 0.1 ml). The animals are ready for *in vivo* challenge with 120 μg/kg ovalbumin 2 months later. A modification of procedure (1) has been reported[9] as successful in sensitizing guinea pigs: two subcutaneous injections of 0.5 ml saline containing 10 μg of ovalbumin dispersed in 1 mg of $Al(OH)_3$ are given with an interval of 2 weeks. The animals are then used 10 days after the second injection. Another protocol is based on intraperitoneal (100 mg) and subcutaneous (100 mg) injections of ovalbumin; the animals are killed 3 weeks

[8] P. Andersson and H. Bergstrand, *Br. J. Pharmacol.* **74,** 601 (1981).

[9] J. Randon, J. Lefort, and B. B. Vargaftig, *Br. J. Pharmacol.* **92,** 683 (1987).

later. When the anaphylactic shock is to be carried out *in vitro*, it is possible to modulate its severity by using different amounts of antigen. For a massive release of eicosanoids, a bolus injection of 2–5 mg ovalbumin into the pulmonary artery (or in case of the isolated tracheal spiral or parenchymal strip a concentration of 0.1–0.5 mg/ml ovalbumin in the organ bath) is advised. However, amounts as small as 0.1–0.2 μg have been reported to induce TxB_2 release from sensitized lung strips. These sensitization procedures are successful in about 90% of the treated animals.

Author Index

Numbers in parentheses are footnote reference numbers and indicate that an author's work is referred to although the name is not cited in the text.

A

Abboud, H. E., 546
Abraham, N. G., 175, 365, 366, 373, 382
Abraham, N., 390
Abrahmsen, L., 497
Acton, J. J., 447, 451(9)
Adams, J., 169
Adesnik, M., 253
Aehringhaus, U., 117, 118, 619
Aehringhaus, W., 82
Ager, A., 522
Aglietta, M., 134, 433, 522, 527
Aharony, D., 312, 313(4), 318(4), 319(4), 397, 406, 408, 409(23), 415, 416(5, 15), 417(5, 14, 15), 418, 420(14, 15, 17), 421
Ahern, D., 408
Akiyama, F., 514
Akkerman, J. W. N., 459
Albarez, L. J., 380
Albert, D. H., 159
Alberts, A. W., 451
Albright, J. O., 302, 359
Alder, L., 167, 168(4), 191
Alessandrini, P., 44
Alexander, R. W., 522
Alhenc-Gelas, F., 522, 531
Ålin, P., 306, 309(3)
Allan, G., 611, 613(2), 620
Allen, M. G., 367
Alloatti, G., 454, 614, 616(16), 617(16)
Alonso, D., 523
Alonso, F., 216
Amirian, D. A., 505, 506
Amrein, P. C., 596
Anderson, F. S., 77, 281
Andersson, P., 627
Andre, J. C., 193
Andrews, P. C., 496
Andriadis, N. A., 63, 70(13)
Anfelt-Ronne, I., 415
Anggard, E. E., 226, 232(4)
Anggard, E., 3
Anhut, H., 619
Aoshima, H., 300, 301(10)
Arai, Y., 415
Araki, H., 406, 408(5), 616, 617(23)
Ardaillou, N., 545
Ardaillou, R., 545
Ardailou, R., 546
Ardlie, N. G., 126, 131, 132(2)
Arison, B., 422
Arita, H., 397, 399, 400, 403(11, 21), 404(11, 21), 405(18), 408, 409(28)
Armstrong, R. A., 410
Armstrong, R. L., 469, 470(14), 477(14)
Arnold, J. L., 387
Arnout, J., 115
Arora, S. K., 410
Asano, S., 479
Aster, R. H., 131, 233
Atherton, R. S., 231
Atkins, R. C., 547
Atkinson, J. G., 83, 87(12)
Aubert, F., 73, 77
Augstein, J., 415
Austen, K. F., 51, 82, 89, 117, 120, 121, 284, 307, 320, 321(9), 322(8, 9), 335, 337(3), 429, 514, 515(1), 516, 566, 575
Axelrod, B., 4, 495, 496

B

Baba, M., 367, 371(11)
Bach, M. K., 76, 82, 117, 118, 307, 335

Baenziger, N. L., 239
Baertschi, S. W., 193, 194(14, 15)
Baghus, W., 551
Bailey, J. M., 4, 469
Bailey, W. H., 612
Bains, S. K., 392
Baird, S., 221
Baker, R. H., 18, 48, 63, 67(2)
Bakker, W. W., 551
Balakrishnan, K., 233, 234(19)
Balazy, M., 16, 34, 46, 82, 90, 91, 357, 373
Balcarek, J. M., 269, 318, 495
Ballan, J. A., 523
Ballard, K., 408
Balley, M., 415
Barnes, D., 519
Barnes, P. J., 458
Barnett, C. J. K., 458
Barnett, R., 545
Barr, D. B., 513
Barrow, S. E., 13, 14, 17(9), 18, 176
Barst, S., 618
Bartfai, T., 231, 235(13)
Barton, A. E., 302
Bashor, M. M., 506
Baud, L., 546
Baudisch, H., 63
Bauer, R. F., 75
Bazan, A. C., 69
Bazan, N. G., 434, 436(12), 445(12)
Baze, M. E., 359
Baze, M., 358
Beattie, T. R., 451
Beaubien, B. C., 117, 120
Beaucourt, J. P., 71, 73, 77, 286
Beaurain, G., 128
Beaven, M. A., 517
Beaver, T. H., 113, 115
Bechtel, W. D., 456
Becker, C. G., 521, 525(7), 526(7), 534(7), 540
Becker, G., 574
Behl, B., 233
Beinborn, M., 513
Belardinelli, L., 612, 615(4)
Bend, J. R., 257
Benigni, A., 43, 48
Benjamin, C. W., 542, 543
Bennett, C. D., 269, 318, 495, 496(10)
Benveniste, J., 125, 126(1, 5), 127, 128, 129, 134, 433, 455, 516, 517(8)
Bergholte, J. M., 260, 264(19)
Berglindh, T., 512
Bergman, T., 287, 291(6), 331, 333(15), 334(15)
Bergstrand, H., 627
Bergström, S., 173, 174(15)
Bernauer, W., 619
Bernstein, P. R., 421
Bertocchi, F., 534
Beskar, B. A., 82
Betz, S. J., 132, 134, 433
Beutler, E., 555
Bianki, R. G., 75
Bidault, J., 125
Bienkowski, M. J., 479, 480(1), 536
Biftu, T., 451
Billah, M. M., 157, 344
Bilman, H., 71
Binkley, F., 4
Birke, F. W., 458
Bito, H., 287, 291(7)
Bjorck, T., 168, 624
Bjork, J., 344
Bjornsson, O. G., 614
Blair, I. A., 13, 14, 15, 17(9), 18, 19, 44, 48, 63, 67(2), 79, 176, 190, 191(4), 531, 532(31)
Blank, M. L., 125, 148, 158, 159, 164, 345, 347(8)
Blavet, N., 434, 436(12), 445(12)
Bleasdale, J. E., 230, 231, 232(6)
Bligh, E. C., 127, 138, 143, 146(6), 153, 161, 197, 218, 348, 529, 533(20), 592
Blume, K. G., 555
Boeynaems, J. M., 389
Bon, S., 25, 83
Bone, R. C., 63, 70(13)
Bonhomme, B., 445
Bonventre, J. V., 546
Borgeat, P., 1, 99, 102(1), 105, 109(1), 110, 111, 112, 113, 115, 268, 301, 304, 331, 486, 532, 557, 562, 572
Born, G. V. R., 407, 458, 584
Bornemeier, D., 314
Bossant, M. J., 127
Bostwick, P., 608
Bothwell, W. M., 63, 68(6), 70(6)

Boulikas, T., 260
Boullais, C., 71
Boullet, C., 125
Bourgain, R., 433, 434, 436(13), 445(13)
Bourgion, S., 110, 532
Bouska, J., 314
Boutal, L., 408
Boyd, J., 607, 608(16)
Boyd, M. R., 257
Boyer, J. L., 233, 237(20)
Bøyum, A., 233, 460
Boyum, A., 556, 560
Bozza, P. T., 463
Bradford, M. M., 499
Bradford, M., 375, 437
Braithwaite, S. S., 3
Brambilla, A., 458
Brammar, W. J., 475
Brand, H., 584
Braquet, P., 99, 105, 112, 113, 433, 434, 436(12), 443(11, 15), 444(11), 445, 446, 455
Brash, A. R., 15, 18, 47, 167, 168(4), 175, 178(3), 191, 192(9), 193, 194(14, 15), 347, 389
Brashler, J. R., 82, 307, 335
Brass, E. P., 46, 278, 279
Bratton, D. L., 133
Brenner, B., 544, 545(2)
Brevario, F., 522
Breviario, F., 534
Bridson, W., 30
Brigham, K. C., 604
Brike, F. W., 458
Brion, F., 122
Brittain, R. T., 408
Broekman, J., 553
Broekman, M. J., 522, 553, 585, 586, 587(6), 593(7, 8), 594(6), 595(7), 596(6), 597(8), 598(8)
Bronsey, B. G., 63
Brooker, D. R., 449
Brooks, B., 125
Brooks, P. M., 16, 77
Brosher, J. R., 117, 118
Bross, T. E., 237
Brown, V. R., 4
Bruggeman, L., 547
Brunengraber, H., 213
Bryan, D. B., 287
Bryant, R. W., 4, 195
Buckner, C. K., 421
Bugianesi, R. L., 449
Bugliese, G., 52
Bull, H. G., 312, 313(5), 314(5), 318(5), 319(5)
Bundy, G. L., 76, 77(3), 81(3), 407
Bundy, G., 397
Bunting, S., 407, 553, 586
Burakoff, S. J., 563
Burch, R. M., 397, 398(13), 399, 401(13), 410, 412
Burg, M. B., 367
Burgess, C. A., 606, 607(12)
Burhop, K. E., 606, 607(12)
Burke, J. A., 611, 612(3), 613, 614(3, 9), 620(3, 9)
Burlingame, A. L., 69
Burnatowska-Hledin, M., 367
Burnstein, M., 352
Bussolino, F., 134, 433, 434, 436(13), 445(13), 522, 527, 534
Buytenhek, M., 469, 481
Byers, L. W., 125
Byron, B. H., 449

C

Callahan, K. S., 522, 531
Campbell, W. B., 522, 531, 565
Camussi, G., 134, 433, 454, 527, 614, 616(16), 617(16)
Camussi, J., 129
Candia, O. A., 380
Cann, S., 213
Cannon, P. J., 522, 559, 563, 565(1, 2), 566, 574
Cannon, P., 554
Capdevila, J., 19, 175, 178, 183(4), 186, 190, 191(4), 193, 264, 265(22), 357, 358, 359, 363(16), 364(16), 365, 366, 373, 389, 390, 391(11, 12), 394 (12)
Care, A., 479
Carey, F., 117, 123
Carini, M., 523
Carminati, C., 48, 63
Carroll, M. A., 365, 366, 367, 371(11), 373

Carter, G. W., 314
Carter, M. C., 408
Carter, M. E., 350, 351(13)
Cartledge, F. K., 71
Carty, T. J., 115
Casals-Stenzel, J., 447, 455, 456, 457, 458, 459(6, 7), 460(6), 462(6), 463(6, 13), 465(12, 13)
Casey, R., 496
Castle, L., 3
Castleman, W. L., 421
Castro Faria Neto, H. C., 463
Catanese, C. A., 415, 416(5), 417(5), 421
Catella, F., 17, 35, 39(2), 41, 42, 43, 44, 45(3, 7)
Caulfield, J. P., 514, 515(1), 516(1)
Cazenave, J. P., 125, 126(5), 128(5)
Cepa, S. R., 246
Cerundolo, J., 524
Chabala, J. C., 451
Chabardes, D., 27
Chabrier, P. E., 433
Chacos, N., 175, 178, 183(4), 186, 193, 357, 358, 365, 373, 389, 390, 391(11, 12), 394(12)
Chaikhouni, A., 406, 408(7)
Chambrier, P. E., 445
Champion, E., 415
Chan, H. W.-S., 268
Chan, W., 459
Chance, C., 264
Chander, P. N., 365, 366(2), 367(2)
Chang, D. B. B., 3
Chang, M. N., 422, 436, 447, 449, 451, 454
Chang, S., 34, 90, 603, 604, 606, 607
Channon, J. Y., 216, 223, 224(1)
Charette, L., 35, 41(4), 43, 415
Charo, I. F., 522
Chau, L. Y., 429
Cheah, M. J., 448
Cheli, C. D., 522
Chen, M. C. Y., 505
Chen, Y.-N. P., 479, 480(1)
Chenouda, A. A., 612
Cheresh, D., 524
Chernick, R. J., 373
Cheung, J. Y., 546
Chi-Rosso, G., 421
Chiabrando, C., 43, 48
Chiang, P., 451
Chignard, M., 433, 455
Chiller, J. M., 554, 570, 572(5), 573(5), 575(5), 577(5)
Chilton, F. H., 158, 159, 166
Chirgwin, J. M., 496
Chottard, J.-C., 301
Christ-Hazelhof, E., 3
Christman, B. W., 19
Chung, S.-K., 359
Ciabattoni, G., 34, 35, 36, 39(2, 3), 41(2), 44, 45, 89
Cirino, M., 35, 41(4), 43
Claesson, H. E., 565, 574
Claesson, J. A., 554
Claeys, M., 194
Clancy, R. M., 570, 572(5), 573(5), 575, 577(5)
Clancy, T., 554
Claremon, D. A., 406, 408(5)
Claremon, O., 397
Claret, M., 226
Clark, D. A., 2, 75, 82, 83, 117, 120, 284, 516
Clark, M. A., 414, 415, 425, 427(5), 429, 432
Clay, K. L., 46, 78, 93, 134, 143, 148(1), 149(2), 184, 530
Clayberger, C., 563
Clostre, F., 434, 436(12), 445(12)
Cochrane, C., 125, 126(1), 134, 455, 516, 517(8)
Cockcroft, S., 226
Coda, R., 433
Coene, M. C., 115
Coene, M.-C., 194
Cohen, M. S., 563
Colca, J., 182, 186(17)
Cole, P. J., 15
Coleman, R. A., 408, 626
Collington, E. W., 408
Collins, P. W., 75
Collins, P., 334
Collins, T., 563
Comens, P. G., 182, 186(17)
Compton, S., 606
Concoran, D., 254, 261(9)
Connel, T. R., 158, 159(3)
Connor, J. A., 543
Connors, M., 254, 261(9), 390
Cook, M. N., 269, 495
Cook, M., 296, 312, 313(3), 314(3), 318, 492
Coon, M. J., 253, 267

Cooper, B., 408
Corcoran, O., 390
Cordeiro, R. S. B., 463
Corey, E. J., 2, 75, 82, 83, 117, 120, 121, 122, 176, 194, 284, 302, 359, 397, 429, 516, 566, 575
Costello, L. L., 318, 496
Cotran, R. S., 534
Coulson, A. R., 494
Coursin, D. B., 421
Cowen, D. S., 236
Cox, C. P., 344, 347(7), 454
Cox, J. W., 63, 68(4, 6), 69(4), 70(4, 5, 6), 78
Cozzi, E., 48, 63
Cragoe, E. J., 18
Craik, C. S., 318
Crawford, C. G., 248
Crooke, S. T., 269, 296, 312, 313(3), 314(3), 318, 414, 415, 421, 422, 425, 427(5), 429, 432, 433(12), 492
Cross, J. J., 458
Culp, L. A., 547
Czartolomna, J., 603, 607
Czervionke, R., 522

D

D'Alessio, J. M., 472
Dahinden, C. A., 568, 570, 572(5), 573(5), 575(5), 577(5)
Dahinden, C., 554
Dahlen, B., 624
Dahlén, S.-E., 167, 168, 169, 170(3, 12), 173(3, 12), 268, 320, 624
Dahnms, T. E., 608
Daniel, J., 397
Danielson, U. H., 307
Darius, H., 459
Das, B. C., 125
Das, M., 481
Dassa, E., 127
Data, J. L., 505
Daugherty, J., 15
Davi, G., 35, 39(3)
Davies, F. J., 446
Davies, P. F., 522
Davis, K. L., 175
Davis, K., 382, 390
Davis, R. W., 474, 486, 487(6)
Davison, P. M., 522
Dawashime, E. H., 476
Dawson, C. A., 603
Dawson, M., 16, 77
Day, J. A., 469
Day, W. A., 547
de Graaf, S., 459
de Silva, J. A. F., 13
De Souza Costa, F., 624
De Weck, A. L., 568, 570(1)
DeChatelet, L. R., 160
Deckmyn, H., 115
DeHaven, R. N., 415
Dejana, E., 522, 534
Delean, A., 425
Delillo, A. K., 446
DeLuca, R., 614, 616(16), 617(16)
Dembinska-Kiéc, A., 619
Demopoulos, C. A., 125
Denizot, Y., 127, 129
Denning, G. M., 531
Dennis, E. A., 351
Dent, G., 458
DePierre, J. W., 325, 327, 328, 331(6), 334(6, 13)
Deschamps, J., 71
DeSousa, D. M., 77
Devan, J. F., 421
DeWitt, D. L., 3, 469, 470(11, 14), 477(11, 14)
Deykin, D., 159, 532
Di Renzo, G. C., 230, 231, 232(6)
Dias, P. M. F. L., 463
Diedrich, S. L., 523
Diehl, R. E., 269, 318, 495, 496(10)
Dishman, E., 19
Dixon, F. E., 4
Dixon, J. E., 495, 496
Dixon, M., 332, 333(20)
Dixon, R. A. F., 269, 318, 495, 496
Dobia, C., 603
Doebber, T. W., 436, 447, 449(7), 451, 454(8)
Dofeuchi, M., 400, 404(20)
Doherty, N., 113, 115
Dole, V. P., 217
Dollery, C. T., 13, 14, 15, 17(9), 18, 176
Doly, M., 445
Domazet, Z., 16
Don, J., 554

Doran, J., 18, 47
Doteuchi, M., 400, 403(21), 404(21)
Doucet-de-Bruine, M. H. M., 459
Douglas, M. B., 102, 108(3)
Downes, C. P., 226, 232(4), 233, 237(20)
Dray, F., 31
Drazen, J. M., 120, 284
Drieu, K., 433, 445
Drugge, E. D., 367
Dubertret, L., 129
Dubyak, G. R., 236, 545, 546, 547
Ducep, J. B., 313, 318(8)
Dulery, B., 313, 318(8)
Duncan, J. A., 513
Dunkel, C. G., 618, 619(29)
Dunn, M. J., 544, 545, 546, 547, 550(5), 551(4, 5), 553
Dunn, M. W., 373
Dupuis, P., 287, 291(2), 333
Dusting, G. J., 458, 614, 615(15), 616, 617(20)
Dworkin, L. F., 544, 545(2)
Dyer, W. F., 348
Dyer, W. J., 127, 138, 143, 146(6), 153, 161, 197, 218, 529, 533(20), 592

E

Eadie, G. S., 331
Ealing, P. M., 496
Eastabrook, R., 175, 183(4), 264, 265(22), 391
Edasery, J., 566
Edenius, C., 167, 168(9), 570, 574, 575(11)
Edlund, M. P., 616, 617(22)
Egan, R. W., 82, 118, 563
Eisenstadt, T. C., 344
El-Harith, E. A., 469, 470(14), 477(14)
Elashoff, J. D., 506
Eling, T., 481, 607, 608(16)
Elleman, G. L., 119
Eller, T. D., 18, 48, 63, 67(2, 3), 69(3), 78
Ellis, D., 607, 608(17)
Ellis, J. M., 159
Emanuelli, G., 454, 614, 616(16), 617(16)
Emori, Y., 295, 490, 491
Enke, S. E., 406, 407(3), 408
Ennis, M., 13, 358, 359
Ercolani, L., 522
Erhard, K. F., 421
Ernest, I., 74
Ernest, M. J., 314
Esko, J. C., 242
Esko, J. D., 241
Eskra, J. D., 115, 314
Estabrook, R. W., 175, 193, 253, 254, 357, 358, 365, 373, 386, 387, 389, 390, 391(11, 12), 394(12)
Esterbauer, H., 46
Etienne, A., 433
Evans, E. A., 73
Evans, J. F., 169, 287, 291(2), 295, 328, 331(11), 333
Evans, R. W., 238
Evers, A. S., 618, 619(26, 28)

F

Fagan, D., 469, 470(12)
Falck, J. R., 19, 175, 176, 178, 183(4), 186, 190, 191(4), 193, 264, 265(22), 357, 358, 359, 363(16), 366, 373, 389, 390, 391(12), 394(12), 586, 593(8), 597(8), 598(8)
Falcone, R. C., 415, 416(5, 15), 417(5, 14, 15), 418, 420(17), 421
Falgueyret, J.-P., 340, 343(4)
Falkenhein, S. F., 90
Falqueyret, J.-P., 191
Fanelli, R., 43, 48, 63
Farmer, J. B., 415
Farr, A. L., 182, 254, 297, 412, 436, 449
Farr, R. S., 344, 347(7)
Fasulo, J. M., 161, 165(21), 218
Fayer, L., 314
Feddersen, C. O., 606, 607
Feigen, G. A., 619
Feinmark, S. J., 522, 554, 559, 560, 563, 565(1, 2), 566, 574
Feinstein, M. B., 231
Fenselau, C., 90
Ferber, E., 216
Ferreira, S. H., 622, 624
Ferreri, N. R., 365, 366(2), 367(2, 3)
Feuerstein, G., 520
Fiers, W., 563
Fillion, G., 437
Fine, M., 496

Fischer, S., 586, 593(8), 597(8), 598(8)
Fisher, A. B., 603
FitzGerald, D. J., 16
FitzGerald, G. A., 15, 16, 17, 18, 35, 36, 39(2), 41, 42, 43, 44, 45, 47, 405
Fitzgerald, V., 164, 345, 347(8)
Fitzpatrick, A., 573, 577(7)
Fitzpatrick, F. A., 59, 63, 68(6), 70(6), 76, 77(3), 81(3), 82, 90, 117, 118, 287, 291(3), 292, 295(3), 331, 333, 357, 358, 385, 407, 554, 555(12), 556(12), 559, 574, 575(8), 586, 606
Fitzsimmons, B. J., 167, 168(4, 5), 169, 295, 328
Fitzsimmons, B., 320, 321(9), 322(9), 343
Fjuimoto, M., 400, 404(20)
Fleisch, J. H., 415, 421(7)
Fleming, P., 254, 261(9), 390
Flexner, C., 505
Flojurnier, M., 112
Fogwell, R. L., 469
Folch, J., 197
Folkman, J., 534
Forand, R., 524
Ford-Hutchinson, A. W., 287, 291(2), 295, 328, 331(11), 333, 415
Forder, R. A., 117, 123
Forest, M. J., 77
Foster, D. W., 566
Fradin, A., 14
Fragetta, J., 16
Franchi, A., 241
Franke, J., 458
Franzen, L., 168
Fraser, M. J., 497
Freed, M. S., 618, 619(29)
Frenette, F., 415
Frenette, R., 295
Friedlander, G., 546
Friedrich, T., 458
Frisch, G. E., 446
Fritsch, E. F., 471, 472, 473, 474, 475, 476, 488, 490(9), 495
Frolich, J. C., 245, 247(2)
Fruteau de Laclos, B., 99, 105
Fry, G., 522, 531
Fu, J. Y., 295, 334, 486
Fujii-Kuriyama, Y., 267
Fujimura, K., 157
Fujita, V. S., 267
Fukui, J., 159
Fukui, Y., 613
Fumitaka, U., 399, 401(16)
Funk, C. D., 4, 268, 295, 318, 486, 494, 495(8), 497(8)
Funk, M. O., 193
Furst, O., 18, 49

G

Gaar, K. A., Jr., 604
Gabbiani, G., 520, 534(3)
Gaddy-Kurten, D., 469
Gaget, C., 313, 318(8)
Galliard, T., 268
Ganz, P., 522
Garland, W. A., 13, 46
Garrity, M. J., 107, 278, 285
Garvin, M., 399
Gasiecki, A., 75
Gaskell, S. J., 63
Gauthier, J. Y., 415
Gay, W. A., 523
Gee, A., 260
Geisow, H. P., 408
Genovese, A., 614
Gerard, G. F., 472
Gerber, J. G., 46, 505, 506, 512(2), 513(9)
Gerritsen, M. E., 522
Ghali, N. I., 410
Ghiorzi, A., 458
Giachetti, A., 458
Gibian, M. J., 340
Gibson, G. C., 392
Gilead, L., 514, 515(2, 3), 516(3)
Giles, H., 51
Giles, R. E., 415, 416(15), 417(15), 420(15)
Gill, L., 366
Gill, S., 175, 183(4), 324, 357
Gillard, J., 35, 41(4), 43, 401
Gillespie, M. N., 615, 616(17)
Gilman, A. G., 407
Gimbrone, M. A., 520, 522, 526(2), 532, 534, 563
Gingrich, R. D., 522
Girard, Y., 82, 83, 87(12), 118, 121, 169, 563
Girerd, J.-J., 301
Glasgow, E. F., 547

Gleason, J. G., 287, 414, 421, 432
Gleispach, H., 16, 46
Glesch, I., 216
Glomset, J. A., 469, 542
Goding, J. W., 341
Goerig, M., 469
Goetze, A. M., 314
Goetzl, E. J., 82, 117, 118
Goldsmith, C., 533
Goldsmith, J. C., 523
Goldstein, A. L., 160
Goldstein, I. M., 320, 322(7), 522
Goldstein, R. E., 520
Goldwasser, P., 545
Goldyne, M. E., 578
Gomperts, B. D., 553
Gonzalez, F. J., 253, 267, 391
Goodsen, T., 415, 421(7)
Goore, M. Y., 4
Gordon, J. L., 522
Gordon, J., 487
Gorman, R. R., 407, 538, 542, 543, 554, 586, 587(6), 594(6), 596(6)
Gorter, G., 459
Goto, G., 2, 83, 120
Gotoh, O., 267
Gottschalk, A., 563
Grace, T. D. C., 497, 498(22)
Gracy, R. L., 471
Graff, G., 245
Grahm, D. W., 451
Grandel, K. E., 344
Granström, E., 35, 42, 36(6), 622, 624
Grassi, J., 24, 25, 27, 82, 83, 90, 622
Gree, R., 359
Green, D. G., 158
Green, K., 14, 245, 247(2)
Green, L. M., 552
Green, N., 367
Greenberg, R., 408
Greene, J. R., 494
Greim, H., 254, 386
Gresele, P., 115
Grimminger, F., 574
Gronwald, R., 469
Gross, M., 554, 570, 572(5), 573(5), 575(5), 577(5)
Gross, S. S., 612
Grulich, J., 469, 522
Grunberger, D., 318, 496
Gryglewski, R. J., 407, 533
Grynkiewiz, G., 220
Guarente, L., 494
Gubler, H. U., 446
Gubler, U., 472, 486
Guccione, M. A., 407
Guengerich, F. P., 175, 186(1), 253, 254, 256(7), 365, 390, 392(15)
Guenguerich, P., 390
Guido, D. M., 76, 77(3), 81(3)
Guindon, Y., 71, 83, 87(12), 401
Guiso, N., 27
Gullikson, G. W., 75
Gunsalus, I. C., 253
Guo, Z.-G., 611, 612, 613, 614(3, 9), 620(3, 9)
Gut, J., 574
Guthenberg, C., 306
Guyton, A. C., 604

H

Habenicht, A. J., 469, 542
Hachfeld del Balzo, U., 613, 614(9), 620(9)
Haddeman, E., 554
Hadley, J. S., 14, 159
Haeggström, J., 167, 168, 287, 291(6), 324, 327, 331, 333(15), 334, 554, 565, 574, 575(11)
Hagege, J., 545
Haisch, K. D., 415, 421(7)
Hale, S. E., 253, 254(3), 264(3)
Halech, H. L., 563
Hall, E. R., 245, 246, 249(3), 250(3), 251(3), 252(5), 578, 582, 583, 584(3)
Hall, M. E., 562
Hall, S. E., 397, 402(12), 404(12), 410
Hallett, P., 408
Halonen, M., 344
Halushka, P. V., 18, 48, 63, 67(2), 397, 398(7, 13), 399, 401, 404(7), 405, 406, 407(3), 408, 410, 412
Hamanaka, H., 406, 408(7)
Hamanaka, N., 408, 446
Hamard, H., 445
Hamberg, M., 9, 52, 167, 170(1), 245, 247(2), 268, 302, 304, 324, 327, 331(2), 406
Hamel, P., 295
Hamilton, J. G., 421

Hammarström, S., 1, 2, 4, 75, 82, 83, 90, 117, 118, 120, 306, 308(12), 309(3), 311
Hammock, B. D., 175, 183(4), 324, 357
Hammonds, T. D., 190, 191(4)
Hampel, G., 522
Han, G. Q., 436, 447, 449
Hanahan, D. J., 125, 154, 446, 611, 612(3), 614(3), 620(3)
Hanasaki, K., 397, 399, 400, 403(21), 404(11, 21), 405(18), 408, 409(28)
Hancock, R., 260
Handler, J., 367
Haniu, M., 267
Hansen, H. S., 3
Hansson, G., 320, 624
Hara, H., 415
Hara, S., 400, 404(20)
Harden, T. K., 233, 237(20)
Harding, K. E., 75
Hardwick, J. P., 267
Haroldsen, P. E., 145, 164, 610
Harper, T. W., 107, 285, 320, 322(8), 562, 609
Harreus, A., 456
Harris, D. N., 397, 402(12), 404(12), 408, 410
Harris, P., 536
Harris, T. M., 54, 193, 194(14, 15), 347
Hart, M. N., 523, 531
Hashimoto, S., 194, 302
Haslanger, M. F., 408
Hassell, J., 547
Hassid, A., 544, 545(6), 550(6), 551(6)
Hatanaka, A., 300, 301(10)
Hatano, Y., 406, 408(6)
Hatayama, I., 306
Hattori, M., 491
Hattori, Y., 611, 612(3), 613, 614(3, 9), 620(3, 9)
Haurand, M., 3
Hawkins, D. J., 191, 192(9), 193(9)
Hawthorne, J. N., 231
Hay, D. W. P., 415
Hayaishi, O., 3, 469
Hayashi, M., 3
Hayashi, Y., 24
Hayes, E. C., 82, 118, 121, 124, 563
Hayes, R., 260, 264(19)
Healy, D., 17, 41, 42, 45(3)
Heavey, D. J., 89
Heberg, A., 410
Hébert, J., 105, 112
Hedberg, A., 397, 402(12), 404(12)
Hedin, L., 469
Hedqvist, P., 320, 624
Hefne, S., 459
Heidvall, K., 570, 574(4)
Hein, A., 514, 515(1), 516(1)
Heitz, R., 469
Helander, H. F., 512
Hemler, M. E., 248
Henson, J. E., 610
Henson, P. M., 125, 126(1), 132, 133, 134, 216, 433, 455, 516, 517(8), 523, 527, 530(17), 606, 610
Herman, A. G., 194
Hermodson, M., 4, 495, 496
Hesketh, T. R., 517
Hesp, B., 415, 416(15), 417(15), 420(15)
Heuer, H., 457, 459
Higgs, E. A., 523
Highland, E., 318, 496
Hildebrandt, A., 391
Hirai, K., 479
Hiraishi, H., 505
Hirata, M., 613
Hirose, T., 63, 476
Hirsch, U., 167, 168(7)
Hjerten, S., 254
Hla, T., 469
Ho, P. K. P., 415, 421(7)
Hoak, J. C., 522, 523, 531, 533
Hoedemaeker, PhJ., 551
Hoffman, B. J., 472, 486
Hofmann, S. L., 232
Hofstee, B. H. J., 331
Hogaboom, G. K., 296, 312, 313(3), 314(3), 414, 415, 429, 492
Holdsworth, S. R., 547
Holland, P. T., 69
Hollingsworth, D. R., 75
Holme, G., 83, 87(12)
Holub, B. J., 230
Honda, Z., 307, 311(10), 335
Hondhan, T., 407
Hong, S. L., 533
Honn, K. V., 481
Hoober, D., 117
Höög, J.-O., 4, 268, 318, 494, 495(8), 497(8)
Hoover, D., 121
Hope, W. C., 421

Hopkins, N. K., 538
Hopkins, P. B., 122
Hoppens, C. M., 611, 612(3), 614(3), 620(3)
Horigan, E., 547
Horn, P. T., 406, 408(6)
Hornby, E. J., 408
Hornstra, G., 554
Horrocks, L. A., 207, 209(11)
Hosaka, K., 237
Hosford, D., 433, 434, 436(13), 445(13)
Hoshimaru, M., 237
House, H. O., 93
Hoyer, L. W., 540
Hsu, A. H. M., 451
Huang, L., 422
Hubbard, H. L., 18, 48, 63, 67(2)
Hubbard, W. A., 14
Hubbard, W. C., 52, 176, 389
Huber, M. M., 90
Huberman, E., 267
Hugh, T., 554
Hughes, K. T., 233, 235(21)
Hughes, R., 2
Hugli, T. E., 570, 572(5), 573(5), 575, 577(5)
Humphrey, P. P. A., 408
Humple, M., 63
Hung, S. C., 410
Hunt, J. S., 547
Huslig, R. L., 469
Hussaini, I., 451
Hüttner, I., 520, 534(3)
Huynh, T. B., 474
Huynh, T. V., 486, 487(6)
Hwang, A., 318
Hwang, S.-B., 436, 440, 444(21), 445, 447, 448, 449, 451, 454(8)
Hwang, S.-M., 269, 495
Hyman, A. L., 523

I

Iacopino, V. J., 606
Ibe, B. O., 565
Ichihara, K., 253, 264(6)
Ichikawa, I., 544, 545(2)
Ide, M., 400, 404(20)
Iguchi, S., 3
Ihle, J. N., 514, 515(1), 516(1)
Ikeda-Saito, M., 479
Ikezawa, H., 206
Ilsley, S., 524
Imai, Y., 254
Imajoh, S., 490
Inada, Y., 446
Ingebretsen, W. R., Jr., 279
Ingerman-Wojenski, C. M., 533
Ingram, C. D., 193, 194(15), 347
Inoue, K., 216
Irvine, R. F., 157, 226, 232(4)
Ishikawa, T., 306
Islam, N., 554, 585, 586, 587(6), 593(7, 8), 594(6), 595(7), 596(6), 597(8), 598(8)
Isom, O. W., 614
Itakure, K., 476
Ivey, K. J., 505
Izumi, T., 4, 189, 287, 296, 302(6), 303, 305, 306, 307, 311(10), 324, 328, 331(10), 334(3), 335, 491
Izzo, N. J., 522

J

Jack, D., 408
Jackson, A. E., 547
Jacobson, A., 496
Jacobson, H. R., 175, 193, 357, 358, 390, 391(12), 394(12)
Jacoby, M., 358
Jaffe, E. A., 520, 521, 522, 524(1), 525(7), 526(1), 531, 534(1, 7), 540, 553, 585
Jafvert, C. T., 523, 533
Jakobs, U., 63
Jakschik, B. A., 182, 186(17), 618, 619(26)
Jakubowski, J. A., 532
Jamieson, G. C., 574
Jandl, J. H., 131, 233
Jarabak, J., 3
Jay, M., 615, 616(17)
Jean-Louis, F., 129
Jeanson, A., 557
Jenkins, J. R., 612, 615(4)
Jensson, H., 306, 309(3)
Johns, R. A., 522
Johnson, A. R., 522, 531
Johnson, E. F., 253, 267
Johnson, J. M., 476
Johnson, K. R., 479
Johnson, R. A., 538

Johnston, J. M., 157, 230, 231(6), 232(6), 344
Jonces, P. H., 75
Jones, R. E., 269, 318, 495, 496(10)
Jones, R. L., 408, 410
Jones, R. N., 415
Jones, T. R., 415
Jorgensen, P. L., 365, 378
Jorgensen, R., 344, 347(7)
Jörnvall, H., 4, 268, 287, 290(1), 291(1, 6), 295, 306, 318, 331, 332, 333(15, 19), 334(15), 486, 448(4), 490(4), 494, 495(8), 497(8)
Jouvenaz, G. H., 194
Jouvin-Marche, E., 128
Jubiz, W., 320

K

Kadowitz, P. J., 523
Kaduce, T. L., 158
Kaergaard-Nielsen, C., 415
Kajiwara, T., 300, 301(10)
Kaku, M., 253, 264(6)
Kakushima, M., 117, 118, 119(3), 121(3)
Kaneko, S., 268, 296, 312, 313(2), 314(2), 338, 341, 342(1), 343(1), 492
Kanner, B. I., 519
Karasek, M. A., 522
Kargman, S., 269, 314, 318, 495, 496(10)
Karmazyn, M., 616, 617
Kasai, H., 397, 400, 403(11, 21), 404(11, 21)
Kasama, T., 189, 303
Katayama, O., 159
Katsube, N., 415
Katsura, M., 408
Kattelman, E. J., 397, 404(8), 410
Katus, H., 469
Katz, R. H., 514, 515(1), 516(1)
Katzen, D., 3
Kawahara, Y., 63
Kawasaki, A., 3, 408, 415, 446
Kawasaki, H., 4, 295, 486, 488(4), 490(4), 491
Kawasaki, K., 551
Kazama, T., 551
Kaziro, Y., 479
Keith, R. A., 415
Kelley, P. R., 524
Kelly, R. C., 407, 559
Kelsey, C. R., 15
Kemper, B., 253
Kennedy, I., 408
Kennett, R. H., 341
Kent, R. S., 523
Kern, D. F., 604
Kerstein, M. D., 523
Kester, M., 547
Kido, T., 391
Kierman, J. A., 513
Kihara, I., 551
Kikuchi, S., 613
Kim, D. K., 216
Kim, H.-Y., 127, 143
Kim, K. S., 122
Kim, Y. R., 264, 265(22), 389
Kimura, S., 479
Kimura, T., 391
Kindahl, H., 622
Kinlough-Rathbone, R. L., 407
Kinne, R., 367
Kinzig, C. M., 287
Kinzig, C., 432
Kirchner, T., 421
Kirstein, D., 415
Kishore, V., 19
Kitamura, S., 4, 189, 303, 307, 311(10), 328, 331(10), 335
Kito, 145
Klein-Knoeckel, D., 522
Klerks, J. P. M., 167, 168(6)
Klevans, L., 447, 451(9)
Klotman, P., 547
Knapp, D. R., 18, 48, 63, 67(2, 3), 69, 78, 397, 401(10), 406, 408
Knapp, H. R., 44
Knight, R. K., 15
Knoop, S. M., 523
Kobayashi, K., 614
Kochel, P. J., 397, 398(13), 399, 401(13), 406, 408(7), 410
Köhler, P., 614, 615(14)
Kohli, J. D., 406, 408(6)
Kojima, S., 615, 616(17)
Koltai, M., 434, 436(13), 445(13)
Kommerell, B., 469
Kondo, K., 408
Konig, W., 82, 117, 118
Konno, M., 415
Korbut, R., 607, 608(16)

Korchak, H. M., 585
Korff, J. M., 3
Kornberg, A., 237
Kortius, R. J., 604
Kostner, G. M., 16, 46
Kosuge, S., 415
Kotani, T., 479
Kott, P., 606
Kraemer, R., 615, 616(18)
Kraemer, S. A., 469, 470(14), 477(14)
Kramer, R. M., 159
Kramer, S. W., 75
Krause, W., 63, 505
Kreisberg, J. I., 544, 545(6), 546, 550(6), 551(6)
Krell, R. D., 414, 415, 416(15), 417(14, 15), 418, 420(14, 15), 421, 626
Krensky, A. M., 563
Krieger, K. H., 614
Krilis, S., 575
Kroll, M. H., 532
Krommerell, G., 469
Kubota, N., 479
Kudo, I., 216
Kuehl, F. A., Jr., 77
Kuhl, P. G., 18, 49
Kühn, H., 167, 168(4), 187, 191, 192(9), 193(9), 268
Kulmacz, R. J., 480, 481
Kumlin, M., 35, 36(6), 42, 324
Kurihara, H., 397, 400(11), 403(11), 404(11)
Kusunose, E., 253, 264(6), 267
Kusunose, M., 253, 264(6), 267

L

Laemmli, U. K., 257, 260(17), 261(17), 486
Laesson, H.-E., 560
Lam, B. K., 2, 167, 168(8)
Lam, M. H., 436, 440, 444(21), 445, 447, 449(7), 451
Lam, Y. K. T., 422
Landon, D. N., 15
Lands, W. E. M., 62, 64(1), 248, 480, 481
Langendorff, O., 610
Laniado-Schwartzman, M., 175, 382, 390
Lankin, V. Z., 191
Lanzetta, P. A., 380
Laposata, M., 238, 240(7)
Lapuerta, L., 175
Laravuso, R. B., 421
Larue, M., 83, 87(12)
Lathe, R., 489
Latta, H., 544, 550(10)
Lau, C. K., 71
Laviolette, M., 105
Lawson, J. A., 17, 18, 36, 41, 47
Lawson, J., 42, 45, 175, 178(3)
Lazarus, H. M., 236
Le Breton, G. C., 397, 404(8), 406, 407(3), 408, 410
Le Couedic, J. P., 433, 516
Leahy, K. M., 25, 83
Lebeque, R., 479
Leblanc, J.-P., 301
Leblanc, P. A., 514, 515(1), 516(1)
Leblanc, Y., 169, 191, 295, 328
LeCouedic, J. P., 125
Lee, C. W., 566
Lee, C.-S. C., 448
Lee, D. Y., 233, 235(21)
Lee, S. Y., 226
Lee, S.-C., 3
Lee, T-c., 345, 347(8)
Lee, T. B., 415
Lee, T. C., 433
Lees, M., 197
Lefer, A. M., 397, 406, 408, 409(23), 614, 616, 617(23)
Lefer, D. J., 614
Leff, P., 51
Lefkowitz, R. J., 425
Lefort, J., 455, 627
Leger, S., 295, 415
Leibach, F. H., 4
Leinweber, F. J., 447, 451(9)
Leis, H. H., 16
Leis, H. J., 46
Leithauser, M. T., 260
Lellouche, J. P., 71, 73, 77, 286
Lemm, U., 63
Lenihan, D. J., 433
Leopold, J. A., 522
Lerner, M., 520
Leslie, C. C., 216, 223, 224(1)
Letcher, A. J., 226, 232(4)
Léveillé, C., 295, 328
Levere, R. D., 175, 366, 373, 382, 390
Levi, R., 611, 612, 613, 614, 619, 620

Levin, O., 254
Levin, W., 253
Levine, E. M., 526, 533
Levine, L., 3, 82, 117, 120, 516
Levine, S. D., 358
Levitt, D., 604
Levy, G. P., 626
Levy, J. V., 459
Lewis, M. A., 414, 415, 429
Lewis, P. J., 14, 17(9), 176
Lewis, R. A., 51, 82, 89, 117, 120, 121, 284, 307, 335, 429, 516, 566, 575
Ley, C. W., 522
Ley, H. L., 472
Li, D. M. F., 614, 615(15)
Li, R. Z., 436, 447, 449(7)
Lichey, J., 458
Lichtenstein, L. M., 124
Liel, N., 397, 398(7), 404(7), 405
Liggett, W., 358, 586
Lim, C. T., 410
Limbrid, L. E., 18
Limjuco, J. A., 124
Lin, Y. M., 3
Linask, J., 603
Lincoln, F. H., 46
Lindgren, J. Å., 82, 117, 118, 167, 168(9), 268, 320, 570, 574, 575(11)
Linehan, J. H., 603
Linne-Saffran, E., 367
Lips, J. P. M., 459
Liston, T. E., 51, 52
Liu, C., 496
Liu, E. C. K., 397, 402(12), 404(12), 410
Liu, F.-T., 519, 575
Liu, L. W., 454
Liu, M. C., 124
Loeb, A. L., 522
Lohman, I. C., 344
Lollar, P., 523, 533
Lombardo, D. L., 82, 118, 121, 563
Lord, A., 35, 41(4), 43
Lorigro, A. J., 608
Loth, V., 469
Lotner, G. Z., 132, 134, 433
Lowenstein, J. M., 213
Lowry, O. H., 182, 254, 297, 412, 436, 449
Lucchesi, B. R., 618, 619
Luckow, V. A., 496, 498
Lumin, S., 357, 359, 373
Lumley, P., 408
Lundqvist, G., 327
Lunel, J., 125
Lynch, J. E., 523
Lynch, J. M., 132, 134, 433, 527, 530(17)

M

Maas, R. L., 193
Mc Lauchlin, L., 25, 83
McBride, O. W., 479
McCabe, P. J., 408
McCarthy, M. E., 421
McCluer, R. H., 207
McDaniel, M. L., 14, 182, 186(17)
MacDermot, J., 15
McDonald, D. W. J., 454
MacDonald, P. C., 230, 231(6), 232(6)
MacDonald, R. J., 496
McDonnell, T. J., 608, 609
McGee, C. M., 16, 77
McGee, J. E., 287, 291(3), 295(3), 331, 358, 554, 555(12), 556(12), 574, 575(8), 586
McGiff, J. C., 365, 366, 367, 371(11), 373, 382, 390
McGill, J. C., 175
McGiven, A. R., 547
McGuire, J. C., 407
Machedra, Y., 397
Machuga, E. T., 479
Maciag, T., 524
McIntyre, T. M., 157, 241, 344, 345, 348(9), 350, 351(13), 354(9), 355(9), 356(9, 15), 433, 521, 522, 524, 526(11), 527, 528, 529, 530(11, 13, 14, 15), 531(14), 533(11, 14), 534(11)
Maclouf, J., 24, 27(3), 31, 34, 35, 39(2), 41(2), 44, 82, 90, 99, 195, 554, 566, 570, 574(3), 586, 622
McManus, L. M., 344, 611, 612(3), 614(3), 620(3)
McMenamin, M. G., 257
McMurtry, I. F., 606
McNamara, D. B., 523
Maddipati, K. R., 268, 479
Maeda, T., 233, 234(19)
Magolda, R. L., 406, 408, 409(23)
Magolda, R., 397
Mahadevappa, V. G., 230

Main, A. J., 74
Main, I. H. M., 505
Mais, D. E., 18, 48, 63, 67(2), 397, 398(7, 13), 399, 401, 404(7), 405, 406, 408, 408(6), 410, 412
Majerus, P. W., 232, 237, 238, 239, 240(7), 520
Makane, M., 408
Makowski, R. J., 392
Malavasi, F., 134, 527
Malik, A. B., 606, 607(12)
Malle, E., 16, 46
Mallet, A. I., 19
Malmsten, C., 320, 397
Malone, B., 433
Malous, J., 82
Maltby, D., 69
Maly, F. E., 568, 570(1)
Maniatis, T., 471, 472, 473, 474, 475, 476, 488, 490(9), 495
Manna, S., 19, 175, 178, 183(4), 186, 193, 264, 265(22), 357, 358, 359, 363(16), 364(16), 389, 390, 391(12), 394(12)
Manner, C. E., 578
Mannervik, B., 306, 307, 309(3), 311
Mantovani, A., 522
Marcinkiewicz, E., 52
Marcus, A. H., 586, 587(6), 589, 594(6), 596(6)
Marcus, A. J., 531, 553, 554, 585, 593
Marfat, A., 2, 75, 82, 83, 117, 120, 122, 359, 516
Mariano, F., 454
Marlin-Wixtrom, C., 373
Marnett, L. J., 366, 469, 479, 480(1), 481
Marron, B. E., 169, 170(12), 173(12)
Marshall, W. S., 415, 421(7)
Martin, G., 547
Martin-Wixtrom, C., 175, 183(4), 264, 265(22), 357
Martins, M. A., 463
Martins, P. M. R. eS., 463
Marus, A. A., 522
Maruyama, T., 341
Marx, G., 519
Mase, T., 415
Masferrer, J. L., 357, 373
Mason, J. I., 366
Mason, R. G., 523
Masson, H. A., 392
Masson, P., 83, 87(12), 415
Massoulié, J., 25, 83
Masters, B. S. S., 253, 254, 260, 261(8), 264(3, 19), 267, 376, 390, 392(17)
Masters, B. S., 390, 391(11)
Masuda, A., 397, 401(10)
Mathews, W. R., 76, 77(3), 81(3), 90, 105
Mathias, M. M., 605, 607, 608(17)
Mathias, M., 606
Matsubara, S., 267
Matsuda, H., 168
Matsumoto, T., 4, 268, 295, 312, 313(1), 318, 486, 491, 494, 495(8), 497(8)
Matsushita, Y., 63
Matzner, J., 519
Maunder, R. J., 63, 70(13)
Maycock, A. L., 77, 82, 118, 121, 563
Mayeux, P. R., 523
Mayeux, P., 397, 401(10)
Medina, J. F., 324
Meese, C. O., 14, 18, 49
Mehdi, S. Q., 233, 234(19)
Meier, R. M., 75
Meijer, J., 324, 325, 327, 328, 331(2, 6), 334
Meijer, L., 195
Meinertz, H., 217
Menasse, R., 74
Mencia-Huerta, J.-M., 117, 121, 433, 514, 515(1), 516(1)
Mené, P., 544, 545(5, 9), 546, 547, 550(5), 551(5)
Menger, M., 574
Merlie, J. P., 469, 470(12)
Metcalf, J. C., 517
Metz, S. A., 562
Meuller, H. W., 347, 348(10)
Meyniel, G., 445
Micallef, S., 18
Michel, L., 129
Mickelson, J. K., 618, 619
Mierendorf, R. C., 493
Mihara, S., 400, 404(20)
Miki, I., 189, 287, 291(7), 303
Miller, J., 422
Miller, M. J. S., 367, 371(11)
Miller, O. V., 538
Milner, P. G., 522
Min, B. H., 13
Minami, M., 4, 287, 291(7), 295, 328, 331(10), 486, 488(4), 490(4)

Minami, Y., 295, 491
Minick, C. R., 521, 525(7), 526(7), 534(7), 540
Minkes, M., 406
Mintyre, T. M., 522
Mioskowski, C., 75, 83, 120
Mishina, M., 237
Mitamoto, T., 446
Miwa, M., 152, 157
Miyake, T., 476
Miyamoto, T., 3, 408, 415
Miyamoto, Y., 195, 340, 341(6)
Miyazaki, S., 551
Mizuno, T., 157
Moffat, M. P., 616, 617(21)
Moldeus, P., 175, 176(5), 178(5)
Moltz, J., 186
Moncada, S., 397, 407, 411, 523, 553, 613, 622
Mong, S., 414, 415, 421, 422, 425, 427(5), 429, 432, 433(12)
Monroe, J. E., 122
Montrucchio, G., 454, 614, 616(16), 617(16)
Moore, J. P., 517
Moore, S. A., 523, 531
Morel, F., 371
Morgan, R. A., 82, 117, 120, 516
Morganroth, M. L., 608
Mori, M., 206
Morinelli, T. A., 397, 401(10)
Morishita, K., 479
Morley, J., 463
Morris, B., 392
Morris, H. R., 117, 120
Morrow, J. D., 54
Morton, D. R., 82, 117, 118, 292, 307, 333, 335, 559, 573, 577(7), 586
Morton, H., 35, 41(4), 43
Morton, R. A., 284
Moser, R., 46
Mosset, P., 359, 390
Moustakis, C. A., 359, 363(16), 364(16)
Muacevic, G., 447, 455, 456, 457, 459(6, 7), 460(6), 462(6), 463(6)
Muccitelli, R. M., 415, 421
Mudd, J., 469, 470(12)
Mueller, H. W., 158
Mueller, S. N., 526, 533
Muirhead, E. E., 125
Mullane, K. M., 615, 616(18), 617, 618
Müller, M., 614, 615(14)
Mulliez, E., 301
Munson, P. J., 401
Muramatsu, S., 63
Murase, K., 415
Murphree, S., 618, 619(26)
Murphy, R. C., 1, 13, 14, 16, 34, 46, 76, 78, 82, 90, 91, 93, 105, 107, 134, 159, 164, 166, 184, 260, 264(19), 278, 281, 283(1), 285, 286, 307, 308(12), 320, 322(8), 357, 373, 385, 554, 562, 566, 570, 574(3), 586, 605, 606, 607, 608, 609, 610
Murphy-Ullrich, J. E., 547
Murray, J. J., 18
Murray, K., 475
Murray, N. E., 475
Murray, R., 405
Mustard, J. F., 125, 126, 128(5), 131, 132(2), 407
Muza, B., 469

N

Naccache, P. H., 110
Nachman, R. L., 521, 525(7), 526(7), 534(7), 540
Nadel, J. A., 318, 496
Nagata, S., 479
Nakagawa, A., 63
Nakagawa, Y., 158, 159, 207, 209(11)
Nakai, H., 415
Nakajima, M., 158
Nakano, K., 397, 400, 403(11, 21), 404(11, 21)
Nakano, T., 408, 409(28)
Nakao, A., 367
Nakayama, T., 479
Nakazato, Y., 547
Nakoniz, I., 221
Napoli, J. L., 175, 357
Narumiya, S., 9, 399, 401(16)
Nathaniel, D. J., 295, 328, 331(11), 340, 343(4)
Nauseef, W. M., 479
Navé, J. F., 313, 318(8)
Nayler, W. G., 616, 617(20)
NcFarland, C. S., 415
Nebert, D. W., 253, 373

Needleman, P., 3, 25, 83, 406, 407, 469, 470(12), 554, 618, 619(26, 29)
Needleman, S. W., 523
Neely, J. R., 617
Negro-Vilar, A., 175
Nekrasov, A., 191
Nelson, D. R., 253
Nelson, P., 522, 528
Neufeld, E. J., 237, 238
Neukom, C., 74
Newman, J. H., 604
Newton, J. F., 296, 312, 313(3), 314(3), 421, 492
Ng, K., 195
Ngheim, D., 522
Niaudet, P., 128
Nichikawa, K., 613
Nicklen, S., 494
Nicolaou, K. C., 167, 168, 169, 170(3, 12), 173(3, 12), 397, 406, 408, 409(23)
Nielsen, C. B., 619
Nies, A. S., 46, 505, 506(2), 512(2)
Nieuweboer, B., 63
Niewiarowski, S., 397
Nilsson, B., 497
Nilsson, K., 535
Nimpf, J., 16, 46
Ninio, E., 127, 128
Nishikoda, S., 446
Nivez, M. P., 545
Niwa, H., 176, 359
Nolop, K. B., 604
Noon, M. C., 472
Noonan, T. C., 606, 607(12)
Nordenstrom, A., 42
Normura, J., 446
Nowak, J., 18
Nugteren, D. H., 3, 4, 194, 469
Numa, S., 237
Numeuer, A., 459

O

O'Donnell, S. R., 458
O'Flaherty, J. T., 158, 159, 347, 348(10)
Oates, J. A., 18, 35, 43, 51, 175, 178(3), 186(1), 358, 365, 389, 390, 505
Oatis, J. E., 397, 401(10)
Obata, T., 415
Oberdisse, E., 233
Oberley, T. D., 547
Obrink, K. J., 512
Ochi, K., 195, 340, 341(6)
Ochs, R. C., 481
Oda, M., 134, 149
Odeimat, A., 532
Oehme, M., 81
Ogino, N., 3, 469
Ogita, K., 253, 264(6)
Oglesby, T. D., 554, 586, 587(6), 594(6), 596(6)
Ogletree, M. L., 397, 402(12), 404(12), 408, 410
Ohgaki, Y., 408
Ohishi, N., 4, 287, 291(7), 295, 305, 307, 311(10), 324, 328, 331(10), 334(3), 335, 491
Ohkawa, S., 4, 287, 328, 331(10)
Ohki, S., 3, 469
Ohno, H., 446
Ohno, S., 4, 295, 486, 488(4), 490, 491
Ohtaki, S., 479
Ohtani, K., 400, 403(21), 404(21)
Ohyabu, T., 206
Ojeda, S., 175
Okabayashi, T., 400, 404(20)
Okada, T., 446
Okazaki, T., 230, 231(6), 232(6)
Okegawa, T., 415
Okita, J. R., 230, 231(6), 232(6)
Okita, R. T., 253, 254(3), 260, 264, 320, 321, 322(9), 390, 391(11)
Okuma, M., 399, 401(16)
Okwu, A. K., 397, 401(10)
Oliw, E., 175, 176(5), 178(3, 5), 186(1), 191, 365, 389, 390
Olson, S. C., 159
Omura, T., 254, 375, 392
Or, R., 514, 515(3), 516(3)
Orchard, M. A., 18
Orellana, M., 366
Örning, L., 4, 306, 309(3)
Ota, S., 505
Over, J., 459
Owasoyo, J. O., 615, 616(17)
Owen, W. G., 523, 533
Owens, L. J., 604
Ozols, J., 479

P

Pace-Asciak, C. R., 13, 16, 18, 61
Packham, M. A., 126, 131, 132(2), 407
Pagano, P., 373
Page, C. P., 433, 463
Pagels, W. D., 481
Pagels, W. R., 469
Paggioli, J., 226
Palade, G. E., 540
Pålmblad, J., 167, 168(7), 320
Palmer, J. D., 344
Palmer, R. M. J., 117, 122
Palmer, R., 557
Palmer, S., 233, 235(21)
Pan, J., 422
Pao, J., 13
Papermaster, B., 552
Papp, A. C., 578, 582, 583(3), 584(3)
Parandoosh, Z., 267
Parker, C. J., 522, 528
Parker, C. W., 90
Parker, J. C., 604
Parkhill, L., 390, 391(11)
Parsonnet, M., 13
Parsons, P., 237
Parsons, To., 522
Parsons, W. G., II, 58
Patrignani, P., 34, 35, 41(4), 43
Patrono, C., 34, 35, 36, 39(2, 3), 41(2, 4), 42, 43, 44, 45, 82, 117, 118
Patscheke, H., 408
Patton, G. M., 159, 201, 213, 218
Patton, G. W., 161, 165(21)
Paubert-Braquet, M., 434, 436(13), 445
Paul, W., 463
Payne, N. A., 506, 513(9)
Peach, M. J., 522
Pedersen, A. K., 42
Peesapati, V., 408
Pelletier, G., 112
Perchonock, C. D., 421, 430
Percy, C., 493
Pereira, M. J., 314
Perez, J., 546
Perico, N., 43
Perrin, P., 286
Perry, D. W., 407
Pervin, R., 519
Peskar, B. A., 117, 118, 614, 615(14), 619
Peskar, B. M., 82, 117, 118
Peter, R., 536
Peters, S. P., 124
Peterson, G. L., 309
Peterson, J. K., 619
Petro, M. A., 536
Pett, S. B., 523
Petty, E. H., 191, 192(9), 193(9), 358
Pfeffer, K., 606
Philipp, D. P., 237
Phillips, I. R., 253
Phillips, M. A., 14, 52, 176
Philpot, R. M., 253, 257, 267
Piacibello, W., 134, 527
Picard, S., 99, 102(1), 105, 109(1), 111, 532
Piccinelli, A., 63
Piccinelli, C., 48
Pickett, W. C., 13, 102, 108(3), 134, 143, 149, 150(14, 15), 151(15), 152, 530
Piechuta, H., 415
Pierucci, A., 35, 39(3)
Pinckard, R. N., 125, 344, 611, 612(3), 614(3), 620(3)
Piomelli, D., 563
Piper, P. J., 587, 614
Pitt, G. A., 284
Pitton, C., 129
Plé, P., 479
Pluznik, D. H., 160
Pober, J. S., 563
Poenie, M., 220
Polansky, J., 125, 129, 516
Pong, S. S., 3, 415, 440, 444(21)
Ponpidom, M. M., 447, 449, 451(9)
Porter, A. T., 193
Posner, G. H., 364
Poubelle, P. E., 99, 110, 113, 115, 532
Pough, R. A., 365
Pouyssegur, J., 241
Powell, M. L., 46
Powell, W. S., 94, 99, 102, 104, 105(4), 108(3, 4), 169, 170(13), 173(13), 253, 325, 532
Pradel, M., 31
Pradelles, P., 24, 25, 27, 31, 34, 82, 83, 90, 622
Prakash, C., 15
Pramanik, B., 19, 175, 357, 389
Prasad, G., 277

Prescott, S. M., 157, 237, 344, 345, 348(9), 350, 351(13), 354(9), 355(9), 356(9, 15), 433, 521, 522, 524, 526(11), 527, 528(6, 8, 11, 13, 14), 529, 530(11, 13, 14, 15), 531(14), 533(11, 14), 534(11)
Pricer, W. E., 237
Pritzker, C. R., 159
Proctor, K. G., 175
Prough, R., 390
Przybyla, A. E., 496
Pugliese, F., 35, 39(3), 44
Pullen, R. H., 63, 68(4, 6), 69(4), 70(4, 5, 6, 12, 13, 14)
Purchase, M., 614, 615(15), 616, 617(20)
Purdon, A. D., 158, 159
Puttemans, M. L., 245, 246, 249(3), 250(3), 251(3), 252(5)
Puttermans, M. L., 583
Puustinen, T., 167, 168, 169(3), 170(3), 173(3), 317, 318(13), 568

R

Raaflaub, J., 231, 235(12)
Rabbinovitch, H., 105
Radin, N. S., 197
Rådmark, O., 4, 268, 287, 290(1), 291(1, 6), 295, 296, 297(3), 301(3), 302, 318, 320, 324, 327(1), 331, 332, 334, 486, 488(4), 490(4), 491, 494, 495(8), 497(8)
Radomski, M., 411
Raetz, C. R. H., 241, 242
Rahamim, E., 514, 515(3), 516(3)
Raines, E. W., 542
Ralph, P., 221
Ramesha, C. S., 13, 134, 143, 148, 149, 150(14, 15, 16), 151(15), 530
Ramwell, P., 606
Randall, J., 436
Randall, R. J., 182, 254, 297, 412, 449
Random, J., 627
Rao, M. K., 277
Rapoport, S. M., 4, 187, 268
Raschke, W. C., 221
Raud, J., 168
Rayford, P., 30
Raz, A., 406, 554
Razandi, M., 505
Razin, E., 514, 515(1, 2, 3), 516, 518, 519, 520, 575
Read, M. A., 614, 615(15)
Reade, J. L., 552
Reddanna, P., 268, 277
Reddy, C. C., 268, 277
Reddy, P. S., 268
Reedy, T. J., 506
Reeves, J. T., 606, 607, 608
Reeves, M. M., 605
Reich, E. L., 238, 240(7)
Reilly, I. A. G., 44
Reinach, P. S., 380
Reingold, D. F., 3
Remazzi, G., 63
Remmer, H., 254, 386
Remuzzi, G., 43, 48
Renard, D., 226
Resch, K., 233
Revtyak, G. E., 531
Rhee, S. G., 226
Richards, J. S., 469
Richmond, R., 15
Richter, R. C., 523
Rickaby, D. A., 603
Riendeau, D., 191, 340, 343(4)
Riguad, M., 301
Rinkema, L. E., 415, 421(7)
Rippe, B., 604
Rittenhouse, S. E., 226, 230(1), 236(1)
Rittenhouse-Simmons, S., 232
Robaut, C., 437, 458
Robbins, J. B., 85, 89(13)
Robbins, J. C., 447, 451(8), 454(8)
Roberts, L. J., II, 16, 35, 43, 51, 52, 54, 58, 320, 321(10), 322(10)
Robertson, D. A., 614
Robertson, E. P., 278
Robertson, G. C., 257
Robertson, I. G. C., 253
Robertson, R. M., 52
Robins, S. J., 161, 165(21), 201, 218
Robinson, C., 18
Robinson, L. J., 536
Robinson, M., 159, 164
Rodbard, D. W., 30, 401
Roddy, L. L., 433
Roerig, D. L., 260
Rogers, D. Z., 364
Rogers, J., 584

Rokach, J., 35, 41(4), 43, 71, 82, 83, 87(12), 90, 105, 117, 118, 119(3), 121, 167, 168(4, 5), 169, 191, 320, 321(9), 322(9), 340, 343, 401, 415, 563
Romeo, L. C., 575
Root, R. K., 563
Rose, J., 606
Rosebrough, N. J., 182, 254, 297, 412, 436, 449
Rösen, P., 614, 615(14)
Rosenberger, M., 74
Rosenthal, A. L., 82
Rosenthal, A. S., 118, 121, 563
Ross, R., 469, 542
Ross, T., 85, 89(13)
Rossi, E. C., 582, 583(3), 584(3)
Roth, G. J., 479
Rother, D., 469
Rotman, B., 552
Rouzer, C. A., 167, 268, 269, 296, 312, 313, 314, 316(6), 317(6), 318, 319, 491, 495, 496(10), 613, 614(9), 620(9)
Rovera, G., 479
Royer, M. E., 63, 68(4), 69(4), 70(4)
Rozing, J., 551
Ruane, R. J., 105
Rush, D. S., 523
Russell, D. H., 69
Rutherford, L. E., 585
Ryhage, R., 173, 174(15)
Ryu, S.-H., 226

S

Saba, S. R., 523
Saban, R., 421
Sachs, R. J., 469
Saffitz, J. E., 618, 619(26, 28)
Safier, L. B., 585, 586, 587(6), 593(7, 8), 594(6), 595(7), 596(6), 597(8), 598(8)
Safier, L., 553, 554
Sagawa, N., 231
Sage, H., 522
Sahlin, K., 616, 617(22)
Said, S. I., 454
Saito, K., 145, 157
Saito, L., 134
Sakai, A., 604
Sakaki, Y., 491
Sakuyama, S., 415
Salari, H., 78, 99, 112, 113
Salata, K., 469
Saleh, S., 15
Salem, N., Jr., 127, 143
Salmon, J. A., 4, 117, 122
Salmon, J., 557
Salomon, J. C., 241
Sambrook, J., 471, 472, 473, 474, 475, 476, 488, 490(9), 495
Sameshina, Y., 157
Sammuni, A., 520
Samuel, M. P., 158
Samuelsson, B., 1, 2, 3, 4, 52, 75, 83, 90, 120, 167, 168, 169, 170(1, 3, 12), 173, 174(15), 245, 247(2), 268, 287, 290(1), 291(1), 295, 296, 297(3), 301, 302, 304, 307, 308(12), 312, 313, 314, 316(6), 317, 318, 319, 320, 331, 332, 333(19), 397, 406, 414, 486, 488(4), 490(4), 491, 494, 495(8), 497(8), 557, 562, 568, 572, 586
Sanavio, F., 134, 527
Sanchez-Crespo, M., 455, 456(8)
Sanders, M. J., 505
Sandrino, M., 422
Sanger, F., 494
Sano, H., 189, 303
Sarau, H. M., 296, 312, 313(3), 314(3), 492
Sata, T., 454
Sato, A., 189, 303
Sato, K., 306, 307, 311(10), 335
Sato, R., 253, 254, 375, 392
Satoh, K., 306, 527, 528(14), 530(14), 531(14), 533(14)
Satouchi, K., 134, 145, 148
Saunders, R. N., 446, 614, 616(16), 617(16)
Saussy, D. L., Jr., 397, 398(13), 399, 401(13), 406, 408(7), 410, 412
Saye, J., 522
Scarpa, A., 546
Scatchard, G., 401
Schacht, J., 197
Schafer, A. I., 532
Scharschmidt, L., 544, 545, 545(4), 546(4), 547(4), 551(4)
Scheffer, M. M., 317, 318(13), 568
Scheith, G., 469
Schenkman, J. B., 254, 386
Schewe, T., 4, 187, 191, 268
Schierenberg, M., 459

Schilling, A. B., 245, 246, 249(3), 250(3), 251(3), 252(5), 583
Schlondorff, D., 358, 544, 545, 550(7), 551(7)
Schmid, K., 514
Schmidt, B., 216
Schneider, W. P., 76, 77(3), 81(3)
Scholnick, H. R., 352
Schraby, S., 367
Schrader, N. L., 113, 115
Schrör, K., 613, 614, 615(14)
Schueler, V. J., 357
Schulze, P. E., 63
Schwartzman, M. L., 365, 366, 367, 373
Schweer, H., 18, 49
Scott, A. I., 359
Scott, D. M., 367
Scott, M. O., 415, 421, 429, 432
Scott, W. A., 613, 614(9), 620(9)
Sebek, O. K., 46
Sedor, J. R., 547
Seeger, W., 574
Seglen, P. O., 280
Seibert, K., 16, 51
Seibert, T. E., 52
Seidler, U., 513
Sekiguichi, N., 158
Sekiya, J., 300, 301(10)
Seldin, D., 514, 515(1), 516(1)
Selig, W. M., 606, 607(12)
Serabjit-Singh, C. J., 253
Serhan, C. N., 2, 167, 168, 169, 170(1, 3, 12), 173(1, 3, 12), 268, 585
Setlzer, P., 523
Sewing, K.-F., 513
Seyama, Y., 4, 189, 287, 291(7), 295, 296, 302(6), 303, 305, 307, 311(10), 324, 328, 331(10), 334(3), 335, 486, 488(4), 490(4), 491
Seyberth, H. W., 18, 49
Shabata, D., 4
Shak, S., 320, 322(7), 522
Shaklai, N., 482
Shand, D. G., 505
Sheard, P., 415
Sheller, J. R., 51, 52(3), 604
Shen, R.-F., 3
Shen, T. S., 433
Shen, T. Y., 436, 445, 446, 447, 448, 449, 451, 454(8), 455, 614
Sherman, W. R., 14
Shibata, D., 495, 496
Shichi, H., 373
Shimizu, T., 3, 9, 189, 268, 287, 290(1), 291(1, 7), 295, 296, 297(3), 301(3), 302, 303, 305, 306, 307, 311(10), 313, 324, 328, 331(10), 332, 333(19), 334(3), 339, 486, 488(4), 490(4), 491
Shimuzu, T., 335
Shirley, P. S., 160
Shively, J. E., 267
Shoam, H., 518
Shorr, R. G. L., 296, 312, 313(3), 314(3), 492
Shulman, E. S., 124
Shurin, S. B., 236
Shviro, Y., 482
Siddhanta, A. K., 19, 357, 359, 363(16), 364(16), 389
Siedlik, P. H., 481
Siegl, A. M., 408
Sigal, E., 318
Silver, M. J., 408, 533
Silvestre, P., 241
Simmons, P. A., 117, 122
Simmons, P., 557
Simonetti, B. M., 35, 39(3)
Simonson, M. S., 544, 545, 546(4, 5, 9, 12), 547, 550(5), 551(4, 5), 553
Simpson, P. J., 618, 619
Sirois, P., 99, 112, 113, 532, 557
Sisson, J. H., 157
Sjövall, J., 173, 174(15)
Skidmore, I. F., 408
Skoglund, M. L., 505, 506(2), 512(2)
Skorecki, K., 546
Sloane Stanley, G. H., 197
Smallbone, W. B., 454
Smedegard, G., 344
Smigel, M., 245, 247(2)
Smith, C. D., 233, 236(17)
Smith, D. E. F., 408, 409(23)
Smith, D. L., 2
Smith, E. F., 406
Smith, E., 397
Smith, G. A., 517
Smith, G. E., 497, 499(20)
Smith, J. B., 158, 159, 397, 406, 408, 409(23), 531
Smith, J. R., 408
Smith, J., 397, 422
Smith, L. L., 202

Smith, W. L., 3, 62, 64(1), 367, 469, 470(11, 14), 477(11, 14)
Snyder, D. W., 414, 415, 421(3), 626
Snyder, F., 125, 127, 158, 159, 164, 345, 347(8), 446, 528
Snyder, G., 175, 357
Snyderman, R., 233, 236(17)
Soberman, R. J., 89, 117, 121, 260, 264(19), 307, 320, 321, 322(8, 9, 10), 335, 337(3)
Söderström, M., 306, 311
Sogawa, K., 267
Soll, A. H., 505, 506, 513(1)
Solomon, S., 253
Sommermeyer, H., 233
Song, B. J., 267
Songu-Mize, E., 365, 367(3)
Sonnenburg, W. K., 367
Soper, B. D., 513
Spector, A. A., 158, 523, 531
Spielman, W. S., 367
Spinnewyn, B., 433, 434, 436(12, 13), 445(12, 13)
Spokas, E. G., 52
Sprague, R. S., 608
Sprecher, H., 191, 195, 238
Springer, J. P., 449
Spur, B., 89, 307, 320, 321(9), 322(9), 335, 337(3)
Stadel, J. M., 425, 427(5), 432
Staehelin, T., 487
Stafforini, D. M., 345, 348(9), 350, 351(13), 354(9), 355(9), 356(9, 15), 529
Stahl, G. L., 614
Statham, C. N., 257
Steczko, J., 4, 495, 496
Steele, R., 367
Steffenrud, S., 78
Stegmeier, K., 408
Stehle, R. G., 28
Stein, R. L., 312, 313(4), 318(4), 319(4)
Steinert, B. W., 547
Stene, D. O., 34, 90, 278, 283(1), 286
Stenmark, K. R., 605
Stephenson, A. H., 608
Stevens, D. L., 522, 528
Stevens, R. L., 514, 515(1), 516(1)
Stewart, A. G., 458, 614
Stobo, J. D., 578
Stockman, P. T., 83
Stockmann, P. T., 25
Stoll, L. M., 531
Stoll, S. E., 236
Stossel, T. P., 596
Stransky, W., 455, 456, 457, 459(7)
Stremler, K. E., 350, 356(15)
Strife, R. J., 13, 16, 607, 608(17)
Striker, G. E., 544, 545(8), 547(8), 550(8)
Striker, L. J., 544, 545(8), 547(8), 550(8)
Strittmatter, P., 254, 261(9), 390
Strohsacker, M. W., 269, 495
Stroschaker, S. M., 318
Subramanian, V., 523
Sugatani, J., 157
Suggs, S. V., 476
Sugihara, H., 490
Sugiura, T., 158, 159
Suh, P.-G., 226
Summers, M. D., 496, 497, 498, 499(20)
Sun Lumin, 390
Sun, F. F., 46, 407, 429, 538
Sundstrom, C., 535
Surles, J. R., 152
Sutyak, J. P., 320, 321(10), 322(10)
Suzuki, K., 4, 295, 486, 488(4), 490, 491
Svensson, J., 406
Swanson-Bean, D., 415, 421(7)
Sweatt, J. D., 18
Sweeney, F. J., 115
Sweetman, B. J., 15, 35, 43, 51
Swendsen, C. L., 159
Swift, G. H., 496
Sybrecht, G. W., 458

T

Taber, D. F., 14, 15, 52, 176
Tack-Goldman, K., 523, 553
Tadokoro, K., 296, 302(6), 491
Taguchi, R., 206
Tahir, M. K., 306
Tai, H. H., 3, 615, 616(17)
Tainer, B. E., 481
Takahagi, H., 63
Takahashi, Y., 189
Takai, T., 469, 470(13)
Takaku, F., 4, 287, 305, 324, 328, 331(10), 334(3)
Takasaki, W., 63
Takayama, H., 532

Tamburini, P., 392
Tanabe, T., 287, 291(7), 469, 470(13)
Tanaka, T., 237
Tanaka, Y., 3
Tanczer, J., 63
Tanishima, Y., 63
Tano, T., 446
Tantengco, M. V., 52
Tare, N. S., 160
Tarpley, W. G., 542, 543
Tattersall, M. L., 415
Taylor, A. E., 604
Taylor, B. M., 46
Taylor, E. N., 454
Taylor, J. E., 433
Taylor, R., 521, 528(6)
Ténce, M., 125, 128, 433, 516
Teng, J. I., 202
Tepperman, B. L., 513
Terada, T., 408
Terano, A., 505
Terao, S., 4, 287, 328, 331(10)
Terashita, Z.-I., 446, 613
Tetta, C., 134, 454, 522, 527, 614, 616(16), 617(16)
Thakur, M. L., 563
Theisen, T. W., 269, 318, 495
Thomas, L. P., 506
Thomas, M., 408
Thomas, Y., 127, 129
Thompson, J. F., 9
Thompson, K. L., 451, 454
Thomson, N. H., 547
Thornberry, N. A., 312, 313(5), 314(5), 318(5), 319(5)
Thorton, S. C., 526
Tian, Z., 422
Tippins, J. R., 117, 120
Tiselius, A., 254
Titus, B. G., 480
Tobias, L. D., 421
Toda, M., 415
Togo, T., 300, 301(10)
Tomesch, J. C., 446
Tomioka, K., 415
Tonai, T., 341
Toomey, M., 505
Torphy, T. J., 415
Toto, R., 19, 357, 389
Touqui, L., 433, 446, 455
Towbin, H., 487
Townsley, M. I., 604
Tröger, W., 456
Trophy, T. J., 430
Trudell, J. R., 574
Tsai, S. J., 522, 531
Tschopp, T., 584
Tsien, R. Y., 220
Tsubshima, M., 408
Tsuchida, S., 306, 307, 311(10), 335
Tsuhima, S., 446
Tucker, S., 421
Turk, J., 14, 182, 186(17)

U

Udén, A.-M., 320
Ueda, N., 167, 168(5), 189, 268, 296, 299, 312, 313(2), 314(2), 338, 340, 341, 342(1), 343, 492
Ullman, H. L., 585, 586, 587(6), 593(7,8), 594(6), 595(7), 596(6), 597(8), 598(8)
Ullman, H., 553, 554
Ullman, M. D., 207
Ullrich, V., 3
Umeki, K., 479
Ursprung, J. J., 63, 70(13)
Ukena, D., 458
Uzinkas, I., 421

V

Vagesna, R. V. K., 432
Vaïtukaitis, J., 85, 89(13)
Vallerand, P., 532
van Delft, J. L., 446
van der Ouderaa, F. J. G., 481
van Dorp, D. A., 469
van Grondelle, A., 607, 608(17)
van Haerigen, N. J., 434, 446
Vandenberg, P., 340
Vanderhoek, J. Y., 231
Vanderlugt, J. T., 63, 68(6), 70(6)
Vane, J. R., 397, 407, 523, 553, 622
Varenne, P., 125
Vargaftig, B. B., 433, 445, 446, 455, 463, 627

Varrichio, A., 269, 296, 312, 313(3), 314(3), 318, 492, 495
Vassar, M. J., 63, 70(13)
Vaughan Williams, E. M., 619
Veale, C. A., 167, 168, 169, 170(3, 12), 173(3, 12)
Veda, M., 400, 404(20)
Veldink, G. A., 167, 168(6)
Venton, D. L., 245, 246, 249(3), 250(3), 251(3), 252(5), 397, 404(8), 406, 407(3), 408, 410, 583
Vergeij, N. L. J., 434, 446(14)
Verhagen, J., 167, 168(6)
Vermeer, M. A., 167, 168(6)
Vermylen, J., 115
Viala, J., 359, 363(16), 364(16)
Vickery, L. M., 421
VicMuth, J., 547
Vidal, R. F., 445
Vindimian, E., 437
Vine, J. H., 16, 77
Vliegenthart, J. F. G., 167, 168(6)
Voelkel, N. F., 90, 34, 603, 604, 605, 606, 607, 608, 609, 610
Voelker, D. R., 216, 223(1), 224(1)
von Schacky, C., 586, 593(7,8), 595(7), 597(8), 598(8)
Vrbanac, J. J., 18, 48, 63, 67(2, 3), 69(3), 78

W

Waddell, K. A., 13, 14, 15, 17(9), 18, 176
Wadkins, J. D., 3
Wagle, S. R., 279
Wakabayashi, T., 406
Wakatsuka, H., 3
Wakelam, J. O., 233, 235(21)
Waku, K., 158, 159
Wal, F., 455
Walenga, R., 231
Wall, M. W., 216, 223(1), 224(1)
Wallace, R. B., 476
Wallach, D. P., 4
Wallis, C. J., 408
Walstra, P., 167, 168(6)
Walther, G., 456
Wanderer, A. A., 344
Wang, X., 436, 447, 449(7)
Wangensteen, D., 604
Wann, S., 547
Ward, S. L., 3
Wardlow, M. L., 344, 347(7)
Ware, C. F., 552
Warholm, M., 306
Warnock, R. H., 481
Wasserman, M. A., 421
Wasserman, S. I., 344, 433
Watanabe, K., 3, 522
Waterman, M. R., 253
Watson, T. R., 16, 77
Webb, E. C., 332, 333(20)
Webber, S. E., 167, 168, 169, 170(3, 12), 173(3, 12)
Weber, K. H., 455, 456, 457, 458, 459(6, 7), 460(6), 462(6), 463(6)
Weber, K.-H., 447
Weber, P., 3
Weelock, J. J., 479
Wei, Y., 25, 83
Weibel, E. R., 540
Weichman, B. M., 421, 432
Weintraub, S. T., 143, 154
Weiss, H. J., 584
Weiss, R. H., 387
Weissmann, G., 585
Weksler, B. B., 469, 522, 523, 531, 553, 585
Wellby, J., 13
Weller, P. F., 566
Welton, A. F., 421
Wendelborn, D. F., 16, 51, 320, 321(10), 322(10)
Wennmalm, Å., 613, 616, 617(22)
Wermuth, M. M., 241
Werringloer, J., 365, 390
West, C., 555
Westcott, J. Y., 34, 90, 184, 281, 608, 609
Westlund, P., 42, 168
Westrich, G. L., 113, 115
Wetterholm, A., 324, 327, 331(2)
Wey, H. E., 532
Whatley, R. E., 521, 524, 526(11), 527(11), 528, 530(11), 533(11), 534
Whatley, R. W., 522
Whelan, J., 268, 277
White, D. A., 231
Whitely, P. S., 469
Whittle, B. J. R., 505
Whorton, A. R., 523
Wiesner, R., 167, 168(4), 191

Wilkstrom, E., 168
Will, J. A., 421
Williams, D. E., 253, 254(3), 264(3)
Williams, H., 415
Williamson, J. R., 614
Willis, A. L., 2
Wilson, D. B., 237
Wilson, I. D., 105
Wilson, N. H., 408, 410
Wing, D. C., 241
Winkler, J. D., 432, 433(12)
Winquist, S. M., 260
Winslow, C. M., 446
Wixtrom, C., 389
Wolbling, R. H., 117, 118
Wolbling, R., 82
Wolf, B. A., 14, 182, 186(17)
Wolf, C. R., 253, 257
Wolfe, L. S., 61
Wollard, P. M., 193
Wong, P. Y.-K., 2, 52, 167, 168(8), 169
Woollard, P. M., 357
Worthen, G. S., 606, 607, 608(17)
Wray, V. P., 260
Wray, W., 260
Wu, H.-L., 415, 421, 422, 425, 427(5), 429, 432
Wu, K. K., 578, 582, 583(3), 584(3)
Wu, K.-Y., 3
Wu, M. S., 436, 447, 449(7), 451
Wurm, M., 16
Wyche, A., 554
Wykle, R. L., 158, 159, 347, 348(10)
Wynalda, M. A., 59, 63, 68(6), 70(6), 76, 77(3), 81(3), 82, 117, 118, 292, 333, 358, 407, 559, 573, 577(7)

Y

Yadagiri, P., 357, 359, 366, 373, 390
Yamada, T., 415
Yamada, Y., 547
Yamakura, F., 391
Yamamoto, S., 3, 24, 167, 168(5), 189, 195, 253, 264(6), 267, 268, 296, 299, 312, 313(2), 314(2), 338, 339, 340, 341, 342(1), 343, 469, 481, 492
Yamamoto, T., 551
Yamashita, K., 19
Yang, S. S., 451
Yano, T., 24
Yao, E. F., 469, 470(14), 477(14)
Yaoita, E., 551
Yasdanparast, R., 4
Yasukochi, Y., 254, 261(8), 376, 390, 392(17)
Yasunaga, K., 134
Yazdanparast, R., 495
Yee, S. J., 614
Yee, Y. K., 415, 416(15), 417(15), 418, 420(15)
Yenofsky, R. L., 496
Yoakim, C., 401
Yokotani, N., 267
Yokoyama, C., 469, 470(13)
Yoshida, R., 3
Yoshimoto, T., 195, 268, 296, 307, 312, 313(2), 314(2), 335, 337(3), 338, 340, 341, 342(1), 343(1), 492
Yoshiota, Y., 446
Yotsumoto, H., 4, 328, 331(10)
Young, R. A., 474, 486, 487(6), 493
Young, R. N., 82, 83, 87(12), 117, 118, 119(3), 121, 563
Yuan, B., 3
Yue, B. Z., 449

Z

Zakhireh, B., 563
Zamboni, R., 71, 117, 121, 295, 328, 331(11), 415
Zatz, M. M., 160
Zeh, W., 469
Zeiger, E., 257
Zelarney, P. T., 216, 223(1), 224(1)
Ziegler, R., 469
Zierold, K., 367
Zimmerman, G. A., 157, 344, 350, 351(13), 356(15), 433, 521, 522, 524, 526(11), 527(11), 528, 529, 530(11, 13, 14, 15), 531(14), 533(11, 14), 534(11)
Zingg, J., 568, 570(1)
Zipkin, R., 324, 331(2), 331(2)
Zirrolli, J. A., 281, 286, 357, 373
Ziv, I., 514, 515(3), 516(3)
Zulak, I. M., 245, 246, 249(3), 250(3), 251(3), 252(5), 583
Zweerink, H. J., 82, 118, 121, 124, 563

Subject Index

A

A23187. *See* Calcium ionophore, A23187
AA. *See* Arachidonic acid
Acetylcholinesterase, from *Electrophorus electricus*
  asymmetric, 25
  enzyme immunoassay for eicosanoids using, 24–34, 82–89
  globular, 25
  as label for eicosanoids, 82
  preparation of, 24–25
Acetylsalicylic acid. *See* Aspirin
Acyl-CoA synthetases
  separation of, by hydroxylapatite chromatography, 240–241
  specific for arachidonate and other eicosanoid precursor fatty acids, 237–238
Adenylate cyclase, histamine-stimulated, in parietal cells, 505
Aggregometer cuvettes, cell preparations studied in, 591–592
AHA23848, 409
  TxA_2/PGH_2 receptor antagonist activity in platelets and vasculature, 408–409
Antibodies, immobilized, preparation of, 63–66
Antioxidants, in analysis of phospholipid molecular species, 196–197
Aortic smooth muscle membranes, pig, preparation of, 400
Aqueous humor, protein content in, effect of corneal cytochrome *P*-450 AA metabolites on, 381–382
Arachidonate
  in inflammatory cells, 157–158
  radioactive, labeling platelets with, procedure for, 590–591
  radiolabeled, for studying arachidonic acid metabolism in platelets, 589
  released on stimulation of human platelets with epinephrine, quantification of, 18–19
  at *sn*-1 position, 157–158
  at *sn*-2 position, 157, 166
    phospholipase A_2 with specificity for, 216
  tritiated sodium salt, preparation of, 590
Arachidonate cascade, epoxygenase branch of, 357
Arachidonate-CoA ligase, mutant cell line defective in, 237
Arachidonate-containing phosphoglycerides, 157–167
  HPLC, 159–160
  in inflammatory cells, 157–159
  molecular species
    $[^3H]$arachidonate/arachidonate ratio of, 166
    HPLC separation of, 164–167
  separation of
    methods, 160–167
    reagents for, 160
  subclasses, 157–158
    choline- and ethanolamine-linked, TLC separation of, 162–164
  TLC, 159–160
Arachidonate 5-lipoxygenase
  activity, 296, 338, 491
    cellular components that stimulate, 313
  Ca^{2+}-requiring, 296
  cDNA clones
    hybridization screening, 495
    isolation of, 492–496
  cofactor requirements, 491
  comments on, 9
  expression of, in insect cells, using baculovirus vector, 496–502

full-length clone, characterization of, 495–496
HPLC assay, 313–314
kinetics of, 312, 318–319
λt11 cDNA expression libraries, immunoscreening of, 493–494
leukocyte, 296, 312–319
  activity, 312
  assay, 312–314
  Ca^{2+} stimulation of, 317–318
  cellular fractions that stimulate, 316–317, 319
  molecular cloning of cDNA for, 317–318
  porcine
    immunoaffinity purification of, 338–343
    properties of, 342–343
    purification, 340–342
  properties of, 317–319
  purification of, 314–317
  yield of, 317–318
metabolites, in pulmonary vasoregulation, 607
from murine mast cells, 296
pathway, 567–568
  precursor release in, 567–569
  products of, nomenclature for, 1–9
peptide sequence analysis, 492
polyclonal antibody, preparation of, 493
potato, 296–306
  activity, 301–302
  arachidonic acid oxidation products, 273–274
  assay, 269–270, 297–299
  catalytic properties, 275–277
  HPLC analysis, 298–299
  isoelectric point, 301
  kinetics, 301–302
  molecular weight, 301
  oxygen monitor assay, 297–299
  peroxidase activity, 277
  physical properties, 275
  properties of, 273–277, 301–302
  purification of, 268–277, 300–301
  purity, 275
  reaction products, 277
  spectrophotometric assay, 297–298
  stability, 275
  storage, 275
  substrate specificity, 301
  thin-layer chromatographic assay, 298–299
radiolabeled arachidonate transformation by, assay of, 338–339
from rat basophilic leukemia cells, 296, 314
reaction, 9
recombinant, characterization of, 500–501
sequence homologies, 318
from soybeans, 318–319
spectrophotometric assay, 339–340
stability, 318–319
substrate inhibition, 318–319
systematic name, 9
transfer vector, construction of, 497–498
Arachidonate 12-lipoxygenase
  comments on, 9
  reaction, 9
  systematic name, 9
Arachidonate 15-lipoxygenase
  type I
    comments on, 9
    reaction, 9
    systematic name, 9
  type II
    comments on, 9
    reaction, 9
    systematic name, 9
Arachidonic acid, 14
  action of, in isolated perfused rat lung, 605–607
  C-20 trideuterated, NCI mass spectrum, 15
  deuterated, 52
  eicosanoids derived from, GC–MS of, 44
  epoxidation of, 360–361
  and gastric acid secretion, 505
  incorporation or release of, into or from cellular phosphoglycerides, 159
  lipoxygenase products of, HPLC of, 98–116
  molecular species of, 196
  NADPH-dependent oxygenation of, demonstration of, by stable-isotope dilution GC–NCI–MS in liver microsomes, 183

NCI mass spectrum of, 14–15
oxidation of
noncyclooxygenase, 60
by rabbit lung cytochrome *P*-450 prostaglandin ω-hydrolase, 264–265
platelet aggregation and release reaction in response to, effect of lymphocytes on, 582, 584, 587
sources of, 176
for eicosanoid biosynthesis, 159–160
Arachidonic acid metabolism, 157
in cultured mast cells, 514
cytochrome *P*-450 monooxygenase pathway, 365
in isolated perfused rat lung, studies of, 599–610
NADPH-dependent, by microsomal fractions, 385, 389–390
oxygenated, 385
in platelets, procedure for studying, 589–599
third pathway of, 365, 372
use of cultured cells to study, 535–543
Arachidonic acid metabolites, 76
antiserum for
cross-reactivity with 2,3-dinor metabolites, 64
preparation of, 64
in bovine corneal microsomes
HPLC, 378
TLC, 377–378
cytochrome *P*-450-dependent, 365–366
analysis of trace quantities of, 19
biological activities of, 385
production of
influence of pharmacological inhibitors on, 185
by liver microsomes in presence of NADPH, 184
regional isomeric composition of, 185
enzyme immunoassays, 90
extraction–purification procedures for, 62
gas chromatography–mass spectrometry of, 62
high-performance liquid chromatography of, 62
immunoaffinity purification–chromatographic quantitative analysis of, 62–70
immunoassay techniques, 90
mTALH cell, stimulated release of, 371–372
polar, quantitative on-line extraction of, 102
in pulmonary vasoregulation, 607
quantitative analysis of, 62
radioimmunoassay, 62, 90
reversed-phase HPLC of, 370–371
Arachidonoyl-CoA synthetase
assays, 238
chromatographic separation of, 238
deficiency of, 238
mutant cell line defective in, production of, 241–242
solubilization of, 238–240
substrate specificity, 238
Arginine vasopressin, effect on mTALH cell release of AA metabolites, 371–372
ASA. *See* Aspirin
Aspirin
selective acetylation of prostaglandin endoperoxide synthase serine residue by, 479
in syndromes associated with platelet activation, 42
treatment of endothelial cells
and fMLP-induced leukotriene C_4 synthesis, 566
method, 588–589
and prostacyclin production, 585
treatment of platelets, method, 588–589
13-Azaprostanoic acid
in evaluation of *in vitro* role of TxA_2/PGH_2 in cellular function, 409–410
TxA_2/PGH_2 receptor antagonist activity in platelets and vasculature, 408–409

B

Baculovirus/insect cell system, expression of 5-lipoxygenase in, 496–502
Benzodiazepine derivatives, as PAF antagonists, 447, 451–454

Bicyclo-PGE, enzyme immunoassays
binding parameters, using acetylcholinesterase as label, 29
interassay variation of, at low, medium, and high concentration, 33
BM13.177
in evaluation of *in vitro* role of TxA_2/PGH_2 in cellular function, 409–410
TxA_2/PGH_2 receptor antagonist activity in platelets and vasculature, 408–409
BN 52020. *See* Ginkgolides
BN 52021. *See also* Ginkgolides
protective effect of, in cerebral ischemia, 445
BN 52022. *See* Ginkgolides
BN 52023. *See* Ginkgolides
BN 52024. *See* Ginkgolides
BOP, radiolabeled, 397–398
affinities and specific binding for platelet and vascular receptors, 405
binding assays for thromboxane A_2 receptors, 401
for qualitative and quantitative study of TxA_2/PGH_2 receptors, 397–405
synthesis and characterization of, 400
Bovine serum albumin
as carrier for leukotriene C_4 and leukotriene E_4, 84–85
polyamino, preparation, 118
Brain, gerbil
binding of [^{3}H]PAF, 440, 445
[^{3}H]PAF binding sites in, 436–437
preparation, 437
Bronchial tissue, human, preparation, for eicosanoid studies, 623–625
Bronchoconstriction, 433–434
PAF-induced, 460–464
Brotizolam, as PAF antagonist, 436
BW755C, effect of
on 5-lipoxygenase pathway, in mast cells, 518
on production of DHET by rat liver microsomes, 185

C

Cahn–Ingold–Prelog system, 187
Calcitonin, effect on mTALH cell release of AA metabolites, 371–372
Calcium ionophore, A23187
effect on mast cells, 518–520
stimulation of human blood with, reversed-phase HPLC analysis of lipoxygenase and cyclooxygenase products generated by, 114
stimulation of PMNL preparations, effect on leukotriene metabolism, 565–566
Carbons, number of, nomenclature for, 2
20-Carboxyl-leukotriene B_4. *See* Leukotriene, B_4, 20-carboxyl
Cardiac anaphylaxis, for study of release and effects of eicosanoids, 619–620
Cardiolipin, chromatography, 198–201
Cell–cell interactions
in eicosanoid pathway, classification of, 586
general experimental strategies for, 579–582
Cerebral ischemia, ginkgolides as PAF antagonists in, 445
Chemotaxis, 319–320
Chen–Prusoff equation, 435, 449
Chief cells, 506
Chiralcel columns, 187
advantages of, 188
chiral stationary phases of, 189
comparison between DNBPG and, 192–193
resolution on, 188–192
separation of enantiomers of 5,6-EET methyl ester on, 191
separation of racemic HETE methyl esters on, 189–190
stereofidelity of, 189–190
Chiralty
assignment of, 187
nomenclature, R/S and D/L systems, 187–188
Collagen, platelet aggregation and release reaction in response to, effect of lymphocytes on, 582, 584, 587
Corneal epithelium, microsomes, preparation of, 374–375
Cutaneous bullous mastocytosis, 57
CV-3988, 444
as PAF antagonist, in isolated perfused rat lung, 606
CV-3989, 446

Cyclic adenosine monophosphate, effect of PGE_2 on, 505, 513
Cyclooxygenase pathway, 407
products, HPLC analysis of, 114–115
in pulmonary vasoregulation, 607–608
Cysteinyl-glycine dipeptidase
comments on, 9
reaction, 9
systematic name, 9
Cytochrome b_5
characterization, 390
purification, 390
in reconstituted cytochrome *P*-450 isoenzyme system, 391–393
storage, 392
Cytochrome *P*-450, 357
arachidonate metabolism, in corneal epithelium, 376–377
reversed-phase HPLC separation of products, 373–374
in bovine corneal epithelial microsomes, 375
measurement of, 375–376
gene expression, modulation of, 373
hemoprotein, 373
isoenzyme complex, activity, 385
isoenzymes
AA oxidation by, 390–393
regioselectivity, 393–394
proteolytic degradation of, 373
microsomal, purified components of, reconstitution studies, 390–393
rabbit lung, N-terminal sequence analysis of, 265–267
rat liver
ciprofibrate-inducible forms of, 390–393
phenobarbital-inducible forms of, 390–393
storage, 392
Cytochrome *P*-450 arachidonate oxygenase, hepatic microsomal, 385–394
assay method, 386–387
metabolites
GC–MS properties of, 389
HPLC analysis, 388–389
reconstituted, assay, 391
Cytochrome *P*-450 epoxygenase, 385
Cytochrome P-450_{LTB}
activity, 320
ω-oxidation catalyzed by, 320, 322–324
assay, 322–324
substrate specificity, 320
Cytochrome *P*-450 monooxygenases, microsomal, 372–373
Cytochrome *P*-450 pathway, conversion of AA through, 365–366
Cytochrome *P*-450 prostaglandin ω-hydroxylase
aminooctyl-Sepharose 4B chromatography, 255–256
conversion of radiolabeled prostaglandins and other eicosanoids to ω-hydroxy derivatives, 260–264
detergent removal from, 258–259
hydrophobic absorption chromatography, for detergent removal, 259
hydroxylapatite chromatography, for detergent removal, 259
ω-hydroxylase activities of, 264–265
ion-exchange chromatography on DEAE-Sepharose, 256–258
measurement of activity, 260–264
molecular weight, 260
N-terminal sequence analysis of, 265–267
purity, 260
rabbit lung, 253–267
pregnancy-inducible, 253
purification of, 253–261
reactions catalyzed, 253
reaction products, reversed-phase HPLC of, 262–264
SDS–PAGE of, 260–261
Cytochrome *P*-450 reductase, 385

D

DCI-MS, of prostaglandins, 246
Dehydroabietylamine, isocyanate of, 193
11-Dehydrothromboxane B_2. *See also* Thromboxane, metabolites
antigenic determinant, 37
antiserum L4, immunological cross-reactivity of, 39–40
capillary gas chromatography–negative-ion chemical ionization mass spectrometry, 45
chemical properties of, 35–36

coupling, to human serum albumin, 37
covalent coupling of, to protein carriers, 36–37
derivatization of, 49–50
deuterated, production of, 46
enzyme immunoassay, 44
binding parameters, using acetylcholinesterase as label, 29
equilibrium between closed, δ-lactone form and open, acyclic form, 35–36
formation of, 35
gas chromatography–mass spectrometry, vs. immunoassay, 44–45
δ-lactone form, 35
half-life of, in human circulation, 41
HSA conjugate, immunization of rabbits with, 37
hydrolysis, 35–37
as index of TxB_2 metabolism, 43–45
^{18}O-labeled, formation, by enzymatic exchange, 46
lactonization, 35–36
measurement of, in plasma and urine, 39–41
open ring structure, 35
in plasma
analysis, 43
measurement, 45
purification of, 49
radioimmunoassay of, 34–42, 44
antibody production for, 36–39
with antiserum L4, standard curve of, 39–40
materials for, 36
sensitivity, 39–40
specificity, 39
titer and affinity of antisera, 37–39
validation of, 41
radioiodinated, binding to antiserum R1, 39
species distribution measurement of, as reflection of changes in TxA2 production, 34, 41
tritiated, binding to antiserum L4, pH-dependence of, 38–39
urinary
analysis of, 43
excretion of, 43
reversed-phase high-performance liquid chromatography of, 41
DFP. *See* Diisopropyl fluorophosphate
12(*R*)DH-HETE. *See* 12(*R*)-Hydroxyeicosatetraenoic acid
DHT. *See* Dihydroxyeicosatrienoic acid
Diacylglycerides
benzoylated, subclasses
molecular species, separation of, 210–213
separation of, 209–210
benzoylation of, 207–209
purification of, 207–208
Diacylphospholipids, separation of, 205
DiHETE. *See* Dihydroxyeicosatetraenoic acid
(5*S*,12*R*)-Dihydroxy-(6*Z*,8*E*,10*E*,14*Z*)-eicosatetraenoic acid. *See* Leukotriene, B_4
(5,6)-Dihydroxyeicosatetraenoic acid, synthesis, 324
5*S*,12*R*-Dihydroxyeicosatetraenoic acid, production, by neutrophils, from platelet precursors, 585
5(*S*),12(*S*)-Dihydroxyeicosatetraenoic acid
preparation of, 305–306
production, by neutrophils, from platelet precursors, 585–587
stability, 306
storage, 306
synthesis, 296
5(*S*),12(*S*)-Dihydroxy-6,10-*trans*-8,14-*cis*-eicosatetraenoic acid. *See* 5(*S*),12(*S*)-Dihydroxyeicosatetraenoic acid
12,20-Dihydroxyeicosatetraenoic acid
12-hydroxyeicosatetraenoic acid-1,20-dioic acid production from, by neutrophils, HPLC assay for, 597–599
production, by neutrophils, HPLC assay of, 594–596
12*S*,20-Dihydroxyeicosatetraenoic acid, production, by neutrophils, from platelet precursors, 586
Dihydroxyeicosatrienoic acid
formation, 175, 365, 373
$^{18}O_2$-labeled, preparation of, 184–186
pentafluorobenzyl ester (PFBE) derivatives
analysis, 179
formation, 179

GC–NCI–MS, 179–181
possible mediator role for, 175–176
production of
by intact islets, 186
by islet microsomes, 186
by rat liver, stable isotope-dilution GC–NCI–MS, 179–184
quantitation of, 175–186
5,6-Dihydroxyeicosatrienoic acid, $^{2}H_8/^{3}H_8$-labeled standard, preparation of, 177
8,9-Dihydroxyeicosatrienoic acid, $^{2}H_8/^{3}H_8$-labeled standard, preparation of, 178–179
11,12-Dihydroxyeicosatrienoic acid, $^{2}H_8/^{3}H_8$-labeled standard, preparation of, 178–179
14,15-Dihydroxyeicosatrienoic acid, $^{2}H_8/^{3}H_8$-labeled standard, preparation of, 178–179
Diisopropyl fluorophosphate, treatment of neutrophil suspensions, 596–597
(*R*)-(–)-*N*-3,5-Dinitrobenzoylphenylglycine, HPLC columns prepared with, 191–193
chiral stationary phases of, 189
2,3-Dinorthromboxane B_2. *See also* Thromboxane, metabolites
derivatization of, 49–50
as methoxime, TMS ether, PFB ester, 49–50
deuterated, production of, 46
enzyme immunoassay, binding parameters, using acetylcholinesterase as label, 29
enzyme immunoassays, 44
extraction of, 46–48
gas chromatography–mass spectrometry, vs. immunoassay, 44–45
as index of TxB_2 metabolism, 43–45
^{18}O-labeled, formation, by enzymatic exchange, 46
purification of, 48–49
radioimmunoassays, 44
stable complex with bonded-phase phenylboronic acid, formation of, 46–47
urinary excretion of, 43
DNBPG. *See* (*R*)-(–)-*N*-3,5-Dinitrobenzoylphenylglycine
Double-antibody radioimmunoassays, of leukotriene, 124
Double bonds
nomenclature for, 3
number of, nomenclature for, 2

E

EET. *See* Epoxyeicosatrienoic acid
Eicosanoid metabolism
in isolated perfused rat lung, 608–609
neutrophil suspensions for studies of, preparation of, 593–594
Eicosanoids
action of, in isolated perfused rat lung, 605–607
activated, preparation of, 25
analysis of, 76
biosynthesis, by erythrocyte-neutrophil interactions, 553–559
in cardiac anaphylaxis, release and effects of, 619–620
chiral analysis of, HPLC for, 187–195
in control of mesangial cell functions, 547
enzyme immunoassays of, using acetylcholinesterase, 24–34
formation, *in vivo*, investigation of time course of, 44–45
and gastric acid secretion, 505
GC-MS analysis, 69
high-performance liquid chromatography–fluorescence detection, 69–70
immunoaffinity chromatography, 18
methoxime derivatives, 14
methoxime-pentafluorobenzyl ester-trimethylsilyl ether (MO-PFB-TMS) derivatives, 13, 14, 69
nomenclature, R/S and D/L systems, 187–188
pentafluorabenzoyl (PFBO) ester, 13
pentafluorobenzyl (PFB) ester, 13
on phenyl boronate minicolumn chromatography, 18
in plasma, analysis of, 44
production of, in endothelial cells, 520–521

in pulmonary vasoregulation, 607–608
quantitative analysis, 69–70
racemic
conversion to diastereomer, 192–195
reaction with resolving agents, 192–194
reaction with lipoxygenase, 194–195
in regulation of glomerular hemodynamics, 544
reversed-phase chromatography on C_{18} minicolumns, 17
tandem mass spectrometry, 18
transcellular metabolism of, 585–599
urinary
gas chromatography–negative-ion chemical ionization mass spectrometry, 62
measurement of, 67
Eicosapentaenoic acid, eicosanoids derived from, GC–MS of, 44
Eicosa-5,8,11,14-tetrynoic acid, effect on production of EET and DHET in rat liver microsomes, 182–183, 185–186
Electron-capture negative-ion chemical ionization mass spectrometry. *See also* Negative-ion chemical ionization mass spectrometry
of lipid mediators, 13–23
sensitivity of, 13–14
Elvax 40P implants, ocular, neovascularization response to, effect of 12(*R*)-DH-HETE, 382–384
Endothelial cells
arachidonate in, 158
arachidonic acid metabolism in, 531–533
aspirin-treated
and fMLP-induced leukotriene C_4 synthesis, 566
prostacyclin production by, 585
aspirin treatment of, 588–589
bovine macrovascular, culture of, 526
cell–cell interaction, *ex vivo* cell preparation used to evaluate, 522
cultured, studies of, 520
eicosanoid production
assays of, 531–532
ex vivo cell preparation used to evaluate, 522
ex vivo, production of eicosanoids and PAF by, factors influencing, 533–534
ex vivo, in situ, 527
human, primary cultures of, 521–526
human umbilical vein
leukotriene A_4 metabolism, 564–565
prostacyclin biosynthesis, 539–541
labeling of, with [^{3}H]arachidonate, 532–533
and PMNL, biochemical interactions between, 559
porcine aortic, preparation, 560–562
production of PAF, 527–531
and culture conditions, 533–534
ex vivo cell preparation used to evaluate, 522
receptor-mediated agonists that cause, 533–534
prostacyclin synthesis
and culture conditions, 533–534
receptor-mediated agonists that cause, 533–534
rat, binding assays to characterize TxA_2/PGH_2 receptors, 403–404
signal transduction, *ex vivo* cell preparation used to evaluate, 522
studies of, *ex vivo* models for, 521–527
for studies of PAF and arachidonate metabolites, 520–535
Endothelial-derived relaxing factor, 408
α-Enolase, 494–495
Enzyme immunoassay
of arachidonic acid metabolites, 90
of 11-dehydrothromboxane B_2, 29, 44–45
of 2,3-dinorthromboxane B_2, 44–45
of eicosanoids, 24–34, 82–89
advantage over radioimmunoassay, 34
antiserum titer/assay dilution
determination of, 27–28
preparation of standards, 27–28
assessment of, 31–34
automation, 34
binding parameters of, 29
competitive binding step, 27–31
incubating standards and samples, 29–30
inter- and intraassay reproducibility, 32–33
materials for, 25
measurement of solid-phase bound AChE, 30–31
preparation of plates, 26–27
procedure, 28–29

reagents for, 26
sensitivity, 31, 34
specificity of, validation with gas chromatography–mass spectrometry, 33–34
titer and sensitivity of different antisera using enzyme tracer, 31
validation, 34
for leukotrienes C_4 and E_4, generation of, 86–89
of thromboxane B_2, 29–32
Enzymes
involved in eicosanoid metabolism, nomenclature for, 1–9
involved in hydroxy acid and leukotriene formation, 9
involved in prostaglandin and thromboxane biosynthesis, 5
involved in prostaglandin catabolism, 7–8
Eosinophils, 11-ketoreductase activity, 58
EP045, 409
TxA_2/PGH_2 receptor antagonist activity in platelets and vasculature, 408–409
Epoxide hydrolase, 286–287, 324
cytosolic, 324
kinetics, 331–332
from mouse liver cytosol, purification of, 328
properties, 334
in liver tissue, subcellular distribution of, 327
microsomal, 324
occurrence in mammalian tissues, 324
substrate specificity, 327
Epoxyeicosatrienoic acid
asymmetric syntheses of, 359
cis-14,15-aziridine heteroatom analog, preparation of, 364
conversion of, to *cis*-thiiranes and *cis*-aziridines, 359
14,15-EET PFB ester, NCI mass spectrum of, 19–20
enantiomers, chiral syntheses of, 358–359
exogenous, biological activity of, 175
formation, 175, 365, 385
$^2H_6$-14,15-EET PFB ester, NCI mass spectrum of, 19–20
heteroatom analogs, structure of, 358
5,6-[$^2H_8$/$^3H_8$], preparation, 176
insulin and glucagon secretagogue properties of, 186
$^{18}O_2$-labeled, preparation of, 184–186
methyl 5,6, saponification of, 363
methyl 8,9,11,12, or 14,15, saponification of, 363
NCI mass spectrum, 15
pentafluorobenzyl ester (PFBE) derivatives
analysis, 179
formation, 179
GC–NCI–MS, 179–181
PFB esters, NCI–MS methods for, 19
PFB regioisomers, NCI mass spectra, 19–20
production of
by intact islets, 186
by islet microsomes, 186
by rat liver, stable isotope-dilution GC–NCI–MS, 179–184
quantitation of, 175–186
racemic, regioisomers of, resolved on Chiralcel OB or OD columns, 190–191
structure of, 358
synthesis, 373
by nonselective peracid epoxidation of arachidonic acid, 359–364
cis-14,15-thiirane (episulfide) heteroatom analog, preparation of, 363–364
unlabeled, source, 176
cis-Epoxyeicosatrienoic acid, stability, 357–358
8,9-Epoxyeicosatrienoic acid, $^2H_8$/$^3H_8$-labeled standard, preparation of, 178–179
11,12-Epoxyeicosatrienoic acid, $^2H_8$/$^3H_8$-labeled standard, preparation of, 178–179
14,15-Epoxyeicosatrienoic acid, $^2H_8$/$^3H_8$-labeled standard, preparation of, 178–179
Epoxy fatty acids, nomenclature for, 1–8
Epoxygenase metabolites, 357
Epoxygenase pathway, 357
Erythrocyte–neutrophil interactions
eicosanoid production via, 554–559
leukotriene biosynthesis via, 556–557
Erythrocytes, human, isolation of, 555–556
Eye. *See also* Rabbit, iris and ciliary body

experimental disease of, effects of ginkgolides in, 445–446

F

FA. *See* Fatty acids
Fast atom bombardment–mass spectrometry, of arachidonate-containing phosphoglyceride molecular species, 165–166
Fatty acids, 1
Fibroblasts, human lung. *See* WI-38 cells
Fischer D/L designation, 187–188
Fish oils, antithrombotic properties, 44
Formylmethionylleucylphenylalanine, and PMNL leukotriene C_4 synthesis, 566–567
FPL 55,712, 415
thromboxane A_2 inhibitory activity of, 421
Futoenone
inhibition of PAF-induced platelet aggregation, IC_{50} values for, 451–454
as PAF antagonist, 451–454
structure of, 450
Futoquinol
inhibition of PAF-induced platelet aggregation, IC_{50} values for, 451–454
as PAF antagonist, 451–454
structure of, 450
Futoxide
as PAF antagonist, 451–454
structure of, 450

G

Gas chromatography–mass spectrometry
of EET and DHET, 176–186
preparation of standards, 176–179
procedure, 176–186
of eicosanoids, 13–14
correlation with enzyme immunoassay, 33–34
of hepatic microsomal cytochrome *P*-450 arachidonate oxygenase metabolites, 389
of leukotriene B_4, 76–77
internal standard, 77–78
method, 77–79
principle, 77
reagents, 77
sample preparation, 78
sensitivity, optimization of, 81
standard conditions, 80–81
standard curve, 79–80
of 6-oxo-$PGF_{1\alpha}$, 22–23
of PAF, 134–142
advantage over bioassay procedures, 134–135
in biological fluids, 137–142
internal standards, 136–137
deuterated, 137–138
principle of, 137
procedure, 135–137
reagents, 135
use of stable isotopically labeled variants, 135
quantitation of thromboxane metabolites by, 50
selected-ion monitoring (SIM) mode, 50
of silyl ethers, 14
of sulfidopeptide leukotrienes, 90–98
internal standards, 92–93
principle of, 91–92
reagents, 91
standard curve, 92
preparation of, 97
of thromboxane metabolites, 42–50
with capillary columns, 50
sensitivity, 50
specificity, 50
Gas chromatography–mass spectrometry coupling, for elucidation of complex arachidonic acid metabolism, 74–75
Gas chromatography–negative-ion chemical ionization–mass spectrometry
assay for leukotriene B_4, 76–81
of dihydroxyeicosatrienoic acid pentafluorobenzyl ester derivatives, 179–181
of EET and DHET, 179–181
of epoxyeicosatrienoic acid pentafluorobenzyl ester derivatives, 179–181
of PAF, 19, 143–152
gas chromatography, 147
interfering compounds, 149
mass spectrometry, 146–147

procedure, 143–152
reagents, 143
sample preparation, 143–145
sensitivity, 147–148, 152
of production of arachidonic acid metabolites, by rat liver, 179–184
quantitative analysis of 6-oxo-PGF$_{1\alpha}$ by, 20–23
Gastric cells
digestion media, 507
Percoll gradients for, 507–508
separation techniques, 506
materials, 506–508
Gerbil. *See* Brain, gerbil
Ginkgolides
effects of, experimental diseases, 445–446
as PAF antagonists, 433–447
ability to displace specific [^{3}H]PAF binding, and ionic conditions, 440, 444
interactions with PAF binding sites, 434–435
materials, 435–438
methods, 435–438
specificity of, 442–443
on PAF binding sites, equilibrium dissociation constant, 435
as potential therapeutic agents, 434
structure of, 434
Glomerular mesangial cells
contraction of, 553
culture, protocols for, 547–550
cultured
characterization of, 550–551
eicosanoid biochemistry in, 544–553
freezing, 551–552
immunocytochemical studies, 551
viability assays, 552–553
cyclooxygenase activity, 545
eicosanoid biosynthesis, 545–546
eicosanoid receptors, 546
functions, eicosanoids and, 547
transmembrane signaling, 546
Glomerular mesangium
function, 544–545
structure, 544–545
γ-Glutamyltransferase
comments on, 9
reaction, 9
systematic name, 9
Glutathione transferase
activity, 306–307
distinction from leukotriene C_4 synthase, 311–312
Glycerolipids
cellular, incorporation of [^{3}H]arachidonic acid into, 160–161
classes, HPLC separation of, 161–163
G protein, in regulation of agonist binding to sulfidopeptide leukotriene receptor, 432
Guinea pig. *See also* Liver, guinea pig; Lung, guinea pig
administration of PAF and PAF antagonists to, 461–464
in vivo experiments in, for PAF antagonist activity assay, 460–464
PAF-induced platelet aggregation in lung, bronchoconstriction, and blood pressure depression in, simultaneous measurement of, 461–463

H

5-HEA. *See* 5-Hydroxyeicosanoic acid
Heart. *See also* Cardiac anaphylaxis; Langendorff heart
ex vivo perfused, 617–619
guinea pig
anaphylaxis induction in, 619
perfused at constant flow or pressure, effects of leukotrienes in, 613–614
perfused at constant pressure
effects of platelet activating factor in, 614
effects of prostaglandins in, 613
isolated coronary-perfused whole-organ preparation, 610–612
animal choice and preparation for, 611
constant-flow version, 611–612
constant-pressure version, 611–612
with intact sympathetic innervation, 612–613
left ventricular contractile force measurement, 612
physiological solutions used for, 611

spontaneous heart rate determination, 612
neutrophil-perfused, 615–616
platelet-perfused, 614–615
rabbit, *ex vivo* perfusion, 617–619
Heme, degradation, 373
Heme oxygenase, 373
Hemostasis, lipoxygenase products of platelet and neutrophil in, 584, 586
HETE. *See* Hydroxyeicosatetraenoic acid
1-*O*-Hexadecyl-2-[^{3}H]-arachidonoyl-GPC
hydrolysis of [^{3}H]arachidonic acid from, 216–223
preparation of, 216–217
HHT. *See* (12*S*)-Hydroxy-(5*Z*,8*E*,10*E*)-heptadecatrienoic acid
High-density lipoproteins, PAF acetylhydrolase associated with, 348–350
High-performance liquid chromatography
of arachidonate-containing phosphoglycerides, 159–160, 164–167
of arachidonic acid metabolites in bovine corneal microsomes, 378
assay of arachidonate 5-lipoxygenase, 298–299, 313–314
assay of arachidonic acid metabolite production by neutrophils, 594–599
bonded-phase, prostaglandin H_2 purification by, 249–252
for chiral analysis of eicosanoids, 187–195
of 12,20-DiHETE production by neutrophils, 594–596
equipment
for analysis of phospholipid molecular species, 213
for phospholipid analysis, 213
of 12-HETE-1,20-dioic acid production from 12,20-DiHETE, in neutrophils, 597–599
of lipoxins, 167–175
of lipoxygenase products
advantages of, 115–116
analysis of biological samples, 110–115
apparatus, 102–103
assessments of recoveries and linearity of UV signal, 108–110
composition of mobile phase, 104
determination of recoveries, 101–102
determination of UV signal vs. mass correlation, 102
elution program, 105–108
equipment, 99–100
HPLC cartridges and columns, 100
materials, 98–99
from polymorphonucleocytes, 570–572
preparation of mobile phases, 100–101
procedures, 102–115
solvent mixtures, composition of, 101
of PAF, 133
to quantitate eicosanoids, 69–70
separation of glycerolipid classes, 161–163
separation of sulfidopeptide leukotrienes, with on-line UV detection, 90
Hormones, nephron responsiveness to, segmentation, 371–372
Hydroperoxyeicosatetraenoic acid, preparation of, 273–277
5-Hydroperoxyeicosatetraenoic acid
conversion to leukotriene A_4, 296, 338
formation, 491
preparation of, 302–305
synthesis, 296, 312, 338, 343
Hydroperoxy fatty acids, nomenclature for, 1–9
Hydroxyacetophenones, 415
(5*S*)-Hydroxy-(6*R*)-*S*-cysteinylglycine-(7*E*,9*E*,11*Z*,14*Z*)-eicosatetraenoic acid. *See* Leukotriene, D_4
5-Hydroxyeicosanoic acid
oxygen-18 internal standard of, mass spectrum of, 95
trimethylsilyl ether, pentafluorobenzyl ester of
gas chromatographic separation of, 96
mass spectrum of, 95–96
Hydroxyeicosatetraenoic acid, 187
19 and 20, formation, 385
capillary GC properties of, 16
enzymatic conversion to DiH(P)ETE, 194–195
formation, 365
hydrogenation of, prior to GC–MS analysis, 16
methyl esters and aromatic derivatives, resolution of, on DNBPG column, 191–192
NCI mass spectrum, 15

quantitative analysis of, by NCI-MS, 16–17
regioisomeric, formation, 385
$^{2}H_{8}$-12(*S*)-Hydroxyeicosatetraenoic acid, trimethylsilyl ether, pentafluorobenzyl ester of, NCI mass spectrum of, 16–17
5-Hydroxyeicosatetraenoic acid
formation, 312–313, 343
by neutrophils, from platelet precursors, 585
generated from endogenous substrate upon stimulation of blood *ex vivo*, measured using reversed-phase HPLC, 114–115
metabolism, 568
^{18}O-labeled, 93
structure, 92
reversed-phase HPLC, 101
recovery measurements, 109–110
(5*S*)-Hydroxy-(6*E*,8*Z*,11*Z*,14*Z*)-eicosatetraenoic acid. *See* 5-Hydroxyeicosatetraenoic acid
12-Hydroxyeicosatetraenoic acid
generated from endogenous substrate upon stimulation of blood *ex vivo*, measured using reversed-phase HPLC, 114–115
production
by intact islets, 186
by islet microsomes, 186
in platelets, time course of, 589
12(*R*)-Hydroxyeicosatetraenoic acid, 378
bioassay for, 378
in corneal epithelium, 373–374
effects of
on aqueous humor protein concentration, 381–382
on vascular reactivity in rabbit eye, 382–383
Na^{+},K^{+}-ATPase assay, 378–381
neovascularization assay, 382–384
ocular neovascularization response to, 382–384
12(*R*)-Hydroxy-5,8,14-eicosatetraenoic acid. *See* 12(*R*)-Hydroxyeicosatetraenoic acid
12(*S*)-Hydroxyeicosatetraenoic acid, TMS ether, PFB ester derivative of, NCI mass spectrum of, 16–17
(12*S*)-Hydroxy-(5*Z*,8*Z*,10*E*,14*Z*)-eicosatetraenoic acid. *See* 12-Hydroxyeicosatetraenoic acid
(12*S*)-Hydroxy-(5*Z*,8*Z*,10*E*,14*Z*)-eicosatetraenoic acid. *See* 12-Hydroxyeicosatetraenoic acid
15-Hydroxyeicosatetraenoic acid
ω-hydroxylation of, 264–265
by rabbit lung cytochrome *P*-450 prostaglandin ω-hydrolase, 264–265
transformation of, in lipoxin formation, 167–168
12-Hydroxyeicosatetraenoic acid-1,20-dioic acid, production from 12,20-DiHETE, by neutrophils, HPLC assay for, 597–599
Hydroxy fatty acids
D/L configuration of, assignment of, 188
nomenclature for, 1–9
(12*S*)-Hydroxy (5*Z*,8*E*,10*E*)-heptadecatrienoic acid
conjugated diene chromophore, 99
in guinea pig lung perfusates, reversed-phase HPLC of, 110–113
(12*S*)-Hydroxy-(5*Z*,8*E*,10*E*)-heptadecatrienoic acid, generated upon stimulation of blood *ex vivo*, reversed-phase HPLC of, 114–115
ω-Hydroxylases, 320
20-Hydroxyleukotriene B_4. *See* Leukotriene, B_4, 20-hydroxy
19-Hydroxyprostaglandin B_2, on-line extraction and reversed-phase HPLC of, 102
15-Hydroxyprostaglandin D dehydrogenase ($NADP^{+}$)
comments on, 7
reaction, 7
systematic name, 7
Hydroxyprostaglandin dehydrogenase (NAD)
comments on, 7
reaction, 7
systematic name, 7
15-Hydroxyprostaglandin dehydrogenase (NAD^{+})
comments on, 7
reaction, 7
systematic name, 7

15-Hydroxyprostaglandin dehydrogenase ($NADP^+$)
comments on, 7
reaction, 7
systematic name, 7
15-Hydroxyprostaglandin I dehydrogenase ($NADP^+$)
comments on, 7
reaction, 7
systematic name, 7
(5*S*)-Hydroxy-(6*R*)-*S*-cysteinyl-(7*E*,9*E*,11*Z*,14*Z*)-eicosatetraenoic acid. *See* Leukotriene, E_4
(5*S*)-Hydroxy-(6*R*)-*S*-glutathionyl-(7*E*,9*E*,11*Z*,14*Z*)-eicosatetraenoic acid. *See* Leukotriene, C_4
Hypotension, 433–434
PAF-induced, 460–464

I

ICI 198,615, 415
binding to leukotriene D_4 receptor, 430–433
radiolabeled, binding assay, for leukotriene D_4 receptor, 422–433
radiolabeled, used to determine potency and competitiveness of binding of leukotriene D_4 antagonists, 415–421
structure of, 416
tritiated, binding assay, for leukotriene D_4 receptor, 417–421
Immunoaffinity purification
antiserum for
immobilization of, 65–66
isolation of IgG fraction, 64–65
titer of, 65
columns for, preparation and use of, 63–69
of eicosanoids, 18
extraction-purification of samples, 67–69
solid phase, 67–68
gel for, binding capacity of, 67
materials, 63–64
of porcine leukocyte arachidonate 5-lipoxygenase, 338–343
procedure, 63–64
reagents, 63–64
as step in three-stage analytical process, 63–70
Immunoassays, 90
of arachidonic acid metabolites, 90
of leukotrienes C_4 and E_4, 82–89
of lipoxygenase products, 98
Immunoglobulin E antigen, effect on mast cells, 518–520
Immunoprecipitation, of E-type prostaglandins, 68–69
Indo-1, calcium solutions prepared using, 219–220
Inflammatory exudate, mouse peritoneal, lipoxygenase products in, reversed-phase HPLC chromatogram of, 113–115
Inflammatory reaction, 514
in glomerular disease, 544
lipoxygenase products of platelet and neutrophil in, 584, 586
Inositol trisphosphate, phospholipase C-catalyzed production of, by membranes from polymorphonuclear neutrophils, measurement of, 232–237
Inositol 1,4,5-trisphosphate, generation, from endogenous substrates, by membranes isolated from human polymorphonuclear neutrophils, 226
Interleukin 3, in mast cell biology and chemistry, 514

K

Kadsurenone
chemistry of, 449–450
inhibition of PAF-induced platelet aggregation, IC_{50} values for, 451–454
as PAF antagonist, 436, 446–454
in special disease models, 454
specificity, 451
pharmacology of, 449–454
structure of, 450
12-Keto-(5*Z*,8*E*,10*E*)-heptadecatrienoic acid
conjugated diene chromophore, 99
in guinea pig lung perfusates, reversed-phase HPLC of, 110–113
12-Keto-HT. *See* 12-Keto-(5*Z*,8*E*,10*E*)-heptadecatrienoic acid
6-Ketoprostaglandin $F_{1\alpha}$
antiserum for, 64
enzyme immunoassay

binding parameters, using acetylcholinesterase as label, 29
interassay variation of, at low, medium, and high concentration, 33
tetradeuterated, formation of dinor metabolite of, 46
vascular microsome production of, 407
11-Ketoreductase
characterization of, 52
identification of, 52
Keyhole limpet hemocyanin
as immunogenic carrier for leukotrienes, 116
thiolated, prepared by reaction with *S*-acetyl-mercaptosuccinic anhydride, 119

L

L-652,569, 444
L-652,731
inhibition of PAF-induced platelet aggregation, IC_{50} values for, 451–454
as PAF antagonist, 451–454
structure of, 450
L-653,150
inhibition of PAF-induced platelet aggregation, IC_{50} values for, 451–454
as PAF antagonist, 451–454
structure of, 450
L-659,989
inhibition of PAF-induced platelet aggregation, IC_{50} values for, 451–454
as PAF antagonist, 451–454
structure of, 450
L-662,025
inhibition of PAF-induced platelet aggregation, IC_{50} values for, 451–454
as PAF antagonist, 451–454
structure of, 450
Labeling techniques, nonisotopic, 24
Langendorff heart, 610–611, 614–615
ischemia–reperfusion in, 616–617
Lauric acid-hydroxylating cytochrome *P*-450, in monkey kidney cells, 267
Leukocytes
human peripheral blood, preparation, 559–560
isolation of, 314–315
5-lipoxygenase stimulatory factors from, preparation of, 316–317
in myocardial injury of ischemia–reperfusion, 617–619
porcine, preparation of cytosol fraction of, 340–341
Leukotriene
A_4
albumin-bound, leukotriene C_4 produced by mast cells after exposure to, 575–577
enzymatic hydrolysis of, 324, 486
characterization of products, 325–327
formation, 302, 312–313, 338, 343, 491
incubation conditions for cultured cells, 562
metabolism of
in cultured vascular cells, 564–565
in neutrophil–mast cell interaction, 567–577
methyl ester
hydrolysis, 562
saponification of, 557–559
release of
in neutrophil–mast cell interaction, 567–577
by neutrophils, 568–575
stability, 330
tetradeuterated methyl ester, preparation of, 70–76
semireduction step, 733–74
Wittig coupling step, 71–73
transformation products, general procedures for incubations and extractions, 325
activity, in lung, 621
antibodies against, preparation of, 116–124
B_4, 514
action of, in isolated perfused rat lung, 605–607
assays, 76
biological functions of, 319–320
biosynthesis, by erythrocyte–neutrophil interactions, 553–559
20-carboxyl
dinor derivative of, reversed-phase HPLC, 107
preparation of, 321–322
chromatographic behavior, 104

20-COOH metabolites of, reversed-phase HPLC, 105–107
coupling of
to BSA, 117, 121–122
via carbodiimidazole, 117, 123
to carrier proteins, 116
mixed anhydride, 117, 121–122
to thiolated KLH, via maleimidohexanoic acid chloride, 121
formation, 486
GC–MS analysis, 79
quantitative, 76–81
GC–NCI–MS, 76–81
generated by neutrophils, reversed-phase HPLC chromatograms of, 110–111
generated from endogenous substrate upon stimulation of blood *ex vivo*, measured using reversed-phase HPLC, 114–115
generation and release of, from cultured mast cells, 516
20-hydroxy
conversion to 20-CHO-LTB_4 and 20-COOH-LTB_4, 321
enzymatic oxidation of, 320
metabolism of, by PMN microsomes, 322–323
on-line extraction and reversed-phase HPLC of, 102
reversed-phase HPLC, 105–107
recovery measurements, 109–110
inactivation of, 319–320
isotopically labeled, 77–78
metabolism
in isolated perfused rat lung, 608–609
by PMN microsomes, 322–323
metabolites, reversed-phase HPLC, 281
origin of, verifying, 557
pentafluorobenzyl (PFB) ester, trimethylsilyl (TMS) ether derivative, 76–77
chromatography of, 81
formation, 79
NCI mass spectrum of, 15–16, 79
in pulmonary vasoregulation, 607
quantitation, 557
reversed-phase HPLC, 101
recovery measurements, 109–110
role in inflammation, 76
selected reaction monitoring assay for, using tandem mass spectrometry, 16
biosynthesis, 268, 296, 307, 324
C_4, 514
action of, in isolated perfused rat lung, 605–607
antisera, 82
cross-reactivity of, 88–89
binding sites for, 421
biosynthesis, during polymorphonuclear leukocyte–vascular cell interactions, 559–567
coupling of
to acetylcholinesterase, 85–86
to carrier protein, 116
via glutaraldehyde, 117–118
to KLH, via maleimidohexanoic acid chloride, 117–119
to modified BSA, via carbodiimide, 117–118
enzyme immunoassay, 82–89
coupling procedures for, 83–86
materials for, 83
preincubation procedure, 87
preparation of tracers for, 85–86
procedure, 83–89
production of antisera, 83–85
reagents for, 83
sensitivity, 87
optimization, 87–89
specificity, 89
time-course evolution of titer, 87
formation, 306–307, 335
by mast cells, 516, 575–577
7-*cis*-hexahydro-, coupling of, to KLH, via maleimidohexanoic acid chloride, 117–119
metabolism, 414
in isolated perfused rat lung, 608–609
from murine-transformed mast cells, quantification of, 97–98
in pulmonary vasoregulation, 607
quantification, 563
radioimmunoassay, 563
radiolabeled, binding assay, for leukotriene D_4 receptor, 422–433
receptors, 429

in regulation of glomerular hemodynamics, 544
reversed-phase HPLC, 101, 105–107
recovery measurements, 109–110
cardiac effects of, in isolated guinea pig heart, 613–614
cofactor requirements, 268
in complex biological samples, analysis of, 124
coupling of, to carrier proteins, 116
D_4
antagonists, 415
potency of, 416
antisera against, 82
binding sites for, 421
coupling of
to BSA, via eicosanoid carboxyl, 117, 120
to carrier proteins, 116
to KLH, via glutaraldehyde, 117, 120
mixed anhydride, 117, 120
to thyroglobulin, via glutaraldehyde, 117, 120
formation, 414
metabolism, 414
in isolated perfused rat lung, 608–609
pharmacological effects of, 414
radiolabeled
binding assay, for leukotriene D_4 receptor, 422–433
used to determine potency and competitiveness of binding of leukotriene D_4 antagonists, 415–421
receptor
antagonists
assessment of, 414–421
receptor binding assays, 417–421
binding to, 414
radioligand binding assay for, 421–433
association and dissociation of radiolabeled agonist and antagonist, 423–424
association and dissociation of radioligand binding to membranes, 425–427
cell culture, 422
competition of radiolabeled ligands, 425
data calculation, 425
equilibrium saturation binding of radioligand, 424–425
membrane preparation for, 422–423
methods, 422–425
radioligand binding conditions, 423–425
saturation binding of radioligands, 427–429
specificity of receptor binding sites, 430–433
in regulation of glomerular hemodynamics, 544
reversed-phase HPLC, 105–107
recovery measurements, 109–110
tritiated, binding assay, to measure competition by agonists, partial agonists, and stereoselectivity of leukotriene D_4 analogs, 417–420
double-antibody radioimmunoassays, 124
E_4
action of, in isolated perfused rat lung, 606
albumin conjugates of, purification of, by gel filtration, 84–85
antisera against, 82
binding to leukotriene D_4 receptors, 415
coupling
to acetylcholinesterase, 85–86
to carrier proteins, 116
to enzyme, using SMCC, 86
by its amino group, 84
by its carboxylic groups, 84
to amino groups of enzyme, 86
to thiolated albumin, 84–85
enzyme immunoassay, 82–89
coupling procedures for, 83–86
effect of preincubation on dose–response curve of, 87–88
materials for, 83
preincubation procedure, 87
preparation of tracers, 85–86
procedure, 83–89
production of antisera, 83–85
reagents for, 83
sensitivity, 87
optimization, 87–89

specificity, 89
time-course evolution of titer, 87
formation, 414
metabolism, in isolated rat hepatocytes, 283–286
metabolites
from isolated rat hepatocytes, 282–283
reversed-phase HPLC, 281
radiolabeled, binding assay, for leukotriene D_4 receptor, 422–433
receptors, 429
reversed-phase HPLC, 105–107
recovery measurements, 109–110
thiolated, coupling, to maleimidated enzyme, 86
function of, 268
and gastric acid secretion, 505
induction of thromboxane release, in perfused lung, 587
as mediator of cardiac anaphylaxis, 620
as mediator of inflammation, 277
metabolism, by isolated rat hepatocytes, 277–286
metabolites
generation, 280
purification of, 281–286
reagents for, 278–279
NCI mass spectrum, 15
nomenclature for, 1–9
peptido
enzyme immunoassays for, 89
in vivo production of, 89
reversed-phase HPLC, 104–105, 107–108
radioimmunoassay, 82, 123–124
receptor, sulfidopeptide, binding assay, 421–433
reversed-phase HPLC, 98
determination of UV signal vs. mass correlation, 102
sulfidopeptide
biological activity of, 90
in biological fluids, assay of, 93–97
direct analysis by mass spectrometry, 90
HPLC separation with on-line UV detection, 90
from mast cells, quantification of, 97–98
properties, 422
quantitation of, 82, 90
by gas chromatography–mass spectrometry, 90–98
by negative-ion chemical ionization-mass spectrometry, 92
radiolabeled, structure, 92
receptor binding assays, 421–433
transcellular biosynthesis, 554
ultraviolet chromophores, 99
Leukotriene A_4 hydrolase, 324
activity, 486
assay of, 289
cDNA, cloning of, 486–491
cDNA clones
antibody selection assay, 494
characterization, 494
DNA sequence analysis, 494–495
identification of, using antiserum, 486–489
distribution of, 287
from guinea pig liver cytosol, purification of, 328–330
human, cDNA, cloning, 295
human lung
molecular weights of, 290
physical properties of, 290–291
purification procedure, 289–291
stability, 291
hydration of leukotriene A_4, time course of reaction in presence of bovine serum albumin, 292
inactivator specificities of, 293–295
kinetic properties of, 291–294, 330–334
λgt10 cDNA library, screening with 48-mer, 486, 489–491
in liver tissue, subcellular distribution of, 327
methyl ester
preparation, 287
purification of, 287–288
saponification of, 288–289
N-terminal amino acid sequence, 291
occurrence in mammalian tissues, 324
plasmid for expression of, in *E. coli*, 491
properties, 334, 491
reaction catalyzed by, 7, 286–295
substrate specificity of, 293–295, 328
systematic name, 9

Leukotriene C_4 synthase, 306–312
activity, 306, 335
assay, 307–309, 336
comments on, 9
definition of units, 309
distinction from cytosolic and microsomal glutathione transferases, 311–312
from guinea pig lung microsomes, 335–337
properties of, 337
kinetic properties, 311
molecular properties, 311
from mouse mastocytoma cells, 307
preparation of, 310–311
nomenclature, 306–307
properties of, 311–312
purification, 336–337
from rat basophilic leukemia cells, 307
microsomal, 335
reaction, 9
sensitivity toward inhibitors, 311
specific activity, 309
systematic name, 9
Leukotriene–protein conjugate, immunization of rabbits with, 85, 123
Leukotriene-specific sera, properties of, 117
LiChrospher Si 100, 213–214
Lignans
as PAF antagonists, 446–454
tetrahydrofuran, 447
Lipid mediators
assays, 13–14
hydroxyl groups, conversion to trimethylsilyl (TMS) ethers, 13–14
mass spectrometry, 13
quantification of, 18–19
stable isotope analogs, 14
structure, 13
Lipids
extraction, from cell preparations, procedure, 592–593
neutral, chromatography, 198–201
tissue, extraction of, 197–198
Bligh and Dyer procedure, 197–198
Folch procedure, 197–198
Lipoxene, 2
Lipoxin
A_4, 167
biological activities, 168
7-*cis*-11-*trans*, 173–174
HPLC, 171–173
formation, biosynthetic pathway for, 168
isomers of, 168–169
properties of, 174
(6*S*)-11-*trans*, 173–174
11-*trans*, 173–174
B_4, 167
biological activities, 168
formation, biosynthetic pathway for, 168
isomers of, 168–169
transisomers, 174
HPLC, 171
from biological sources, identification of, 173–175
biosynthesis, pathways for, 167–168
C values (equivalent chain length) for, 173
extraction of, 169–170
formation, in coincubation experiments of PMN and platelets stimulated with ionophore, 574
HPLC of, 167–175
materials for, 169
procedure, 169–173
nomenclature for, 6
structure, 167, 170
ultraviolet absorbance, 170
ultraviolet chromophores, 99
Lipoxygenase
definition of enzyme units, 270
nomenclature for, 268
products
analysis of, 98
generated by neutrophils, reversed-phase HPLC chromatograms of, 110–111
in guinea pig lung perfusates, reversed-phase HPLC of, 110–113
immunoassays, 98
mass spectrometric assays, 98
in mouse peritoneal lavage fluid, reversed-phase HPLC chromatogram of, 113–115
pathophysiological role of, 98
standards of, 98–99
UV chromophore, 98–99

reaction of eicosanoids with, 195
reactions catalyzed, 268
stereofidelity of, 189–190
5-Lipoxygenase. *See* Arachidonate 5-lipoxygenase
Liver
guinea pig
homogenates, preparation of, 325
subcellular fractionation, 325
mouse
homogenates, preparation of, 325
subcellular fractionation, 325
rat
clofibrate-inducible lauric acid ω-hydroxylase, 266–267
microsomal fractions
isolation of, 386
preparation of, 218–219
Long-chain acyl-CoA synthetase, nonspecific, enzyme and protein assays, 238
Long-chain-fatty-acid-CoA ligase, nonspecific, purifications of, 237
Low-density lipoproteins
isolation of, 352
PAF acetylhydrolase associated with, 348–350
LT. *See* Leukotriene
Lung
animal, preparation, for eicosanoid research, 621
compartments, eicosanoid metabolism in, 609
edema, caused by eicosanoids, in isolated perfused rat lung preparation, 605–607
eicosanoid metabolism in, 621
ferret, eicosanoid metabolism in, 608
guinea pig
eicosanoid metabolism in, 608
membrane
inhibition of tritiated ligand binding by selective agonists and leukotriene D_4 antagonists, 417–421
kinetics of radioligand binding to, 425–427
preparation of, 416–417
saturation binding of radioligands to, 427–429
microsomes, preparation of, 336
parenchymal preparation for eicosanoid studies, 625–628
perfusate, arachidonic acid metabolites in, reversed-phase HPLC of, 110–113
preparation, for eicosanoid research, 625–628
hamster, eicosanoid metabolism in, 608
human
parenchymal preparation for eicosanoid production studies, 621–623
preparation, for eicosanoid research, 621–625
perfused, thromboxane release, leukotrienes as agonists in, 587
rabbit
cytochrome *P*-450, 266–267
eicosanoid metabolism in, 608
microsomes, detergent solubilization of, 254–255
preparation, 254–255
rat
isolated, perfused
advantages and disadvantages of, 605
in arachidonate studies, 599–610
as bioassay organ for assessment of eicosanoid action, 605–607
eicosanoid metabolism in, 608–609
localization of lipid mediator synthesis/metabolism in, 610
permeability–surface area product, 604–605
physiological parameters for, 603–605
pulmonary microvascular pressure measurement in, 603–604
stability of, 602–603
vascular permeability, assessment of, 604–605
weight, continuous measurement of, 604
isolation procedure, 600
perfusate for, choice of, 601–602
perfusion procedure, 600–601
LX. *See* Lipoxin
LY 171,883, 415
phosphodiesterase inhibitory properties of, 421

Lymphocyte–platelet mixture, metabolic profile of, 580–581
Lymphocytes. *See also* Platelet–lymphocyte interactions
arachidonic acid metabolism in, 582–583
effect of, on platelet function, 582, 584
eicosanoid generation by, 581
human, preparation of, 578–579
prostacyclin synthase activity, assay, 582
Lymphoma cells. *See* U937 cells
Lysophosphatidylcholine, chromatography, 199–201
Lysophosphatidylethanolamine, chromatography, 199–201

M

Macrophages
arachidonate in, 158
phospholipase A_2
partial purification of, 223–225
preparation of, 221–223
properties of, 224–225
specific for *sn*-2-arachidonic acid, 216–225
assay, 216–223
RAW 264.7, as source for arachidonoyl-hydrolyzing phospholipase A_2, 221–223
Mass spectrometry. *See* Gas chromatography–mass spectrometry; Negative-ion chemical ionization mass spectrometry
Mast cells
bone marrow-derived
activated, production of lipid metabolites in, 517–518
cultured, characterization of, 515
culture of, 514–520
differentiation of 5-lipoxygenase activation and exocytosis in, 517–520
exocytosis of preformed and newly generated lipid mediators, 516–517
human, culture, 515
mouse, culture, 515
phosphatidylinositol turnover, 517
cyclooxygenase product produced by, 51
metabolism of albumin-bound leukotriene A_4 into leukotriene C_4, 575–577
murine
preparation of, 160
PT-18 subline, 160
PAF derived from, analysis of, 141–142
sulfidopeptide leukotrienes, quantitation, 97–98
Medullary thick ascending limb of Henle's loop cells
arachidonic acid metabolism, 367, 369–370
hormonal stimulation of, 371–372
metabolites, 365
cultured, from neonatal rabbits, preparation of, 368
cytochrome *P*-450 activity, 372
immunodissection, 367
isolation, 366–368
by centrifugal elutriation, 367–368, 372
morphological and histochemical characteristics, 366
Na^+,K^+-ATPase activity, 365–366
Menthyl chloroformate, 193
Meriones unguiculatus. *See* Brain, gerbil
α-Methoxy-α-trifluoromethylhenacetyl chloride, 193
Methyl arachidonate, epoxidation of, 361–363
Methyl-[11,12,14,15-$^2H_4$]-(5*S*,6*S*)-*oxido*-(7*E*,9*E*,11*Z*,14*Z*)-eicosatetraenoate, 70–76
Metyrapone, effect on production of EET and DHET in rat liver microsomes, 183, 185–186
MK 571, 415
Monoclonal antibodies, anti-5-lipoxygenase
conjugation to agarose gel, 341
preparation of, 341
Monohydroxyeicosatetraenoic acids, synthesis, 373
N-Monomethylphosphatidylethanolamine, chromatography, 199–201
Mouse. *See* Liver, mouse; Murine
Murine mastocytoma cells, propagation, 309–310
Murine monoclonal anti-rabbit IgG antibodies, coating of plates with, 26
Myeloperoxidase, axial heme-binding site in, 479

N

NADPH–cytochrome *P*-450 (*c*) reductase, 372–373
  in corneal epithelial microsomes, measurement of, 376
NADPH–cytochrome *P*-450 reductase
  characterization, 390
  purification, 390
  in reconstituted cytochrome *P*-450 isoenzyme system, 391–393
  storage, 392
NCI-MS. *See* Negative-ion chemical ionization mass spectrometry
Negative-ion chemical ionization mass spectrometry, 76
Neutrophil aggregation, PAF-induced, 459–460
Neutrophils. *See also* Platelet–neutrophil interactions
  arachidonate-containing phosphoglycerides in, incorporation and subsequent remodeling of, 165–166
  arachidonate in, 158–159
  collection of, 587–588
  12,20-DiHETE production by, HPLC assay of, 594–596
  12-HETE-1,20-dioic acid production from 12,20-DiHETE, HPLC assay for, 597–599
  human, preparation of, 160, 569–570
  leukotriene B_4 production by, analysis of, 78–79
  lipoxygenase products in, reversed-phase HPLC analysis of, 110–111
  measurement of PAF in, 139–140, 149–151
  perfusion of isolated heart with, for study of leukocyte–heart interactions in eicosanoid research, 615–616
  suspensions
    diisopropyl fluorophosphate treatment of, 596–597
    for studies of eicosanoid metabolism, preparation of, 593–594
NIH-3T3 murine fibroblasts
  eicosanoid metabolism, influence of various transfected gene products or drugs on, 541–543
  PGE_2 production, 541–543
  phospholipase C activity, effect of *ras* gene product, 543
Nonsteroidal antiinflammatory drugs, blockade of PGE_2 synthesis by, 546
Nordihydroguaiaretic acid, effect on production of EET and DHET in rat liver microsomes, 182–183, 185–186

O

Oncogenes, transformation of fibroblasts with, effect on PGE_2 production, 541–543
ONO 6240, 444
ONO-6248, 446
ONO11120
  structure, 409
  TxA_2/PGH_2 receptor antagonist activity in platelets and vasculature, 408–409
ONO-RS-411, 415
Oxo fatty acids, nomenclature for, 1–8
6-Oxoprostaglandin $F_{1\alpha}$
  derivatization, 21–22
  isolation, 21–22
  MO, Tris-TMS ether, PRB ester derivative of, HCI mass spectrum of, 17–18
  quantitative analysis of, by capillary GC–NCI–MS, 20–23
15-Oxoprostaglandin Δ^{13}–reductase
  comments on, 8
  reaction, 8
  systematic name, 8

P

Palmitate, ω-hydroxylation of, by rabbit lung cytochrome *P*-450 prostaglandin ω-hydrolase, 264–265
Parietal cells
  acid production, effect of prostaglandins on, 512
  and arachidonate metabolism, 505–513
  digestion, 510
  enriched preparation, 506

applications of, 513
function, 505
PGE_2 production, effect of gastric secretagogues on, 513
preparation of, 508–512
and prostaglandins, 505–513
purification of, clumping in, 511
separation on Percoll gradients, 511–512
sieving, 510
washing, 510
PG. *See* Prostaglandin
PGI_2. *See* Prostacyclin
Phosphatidic acid, 196
Phosphatidylcholine, chromatography, 199–201
Phosphatidylethanolamine
benzoylated, molecular species separation of, 210–213
brain, intact, separation into individual molecular species, 202–204
chromatography, 198–201
Phosphatidylglycerol, chromatography, 199–201
Phosphatidyl[2-^{3}H]inositol 4,5-biphosphate, phospholipase C-catalyzed hydrolysis of, measurement of, 230–232
Phosphatidylinositol
chromatography, 198–201
phospholipase C hydrolysis of, 206
Phosphatidylserine, chromatography, 198–201
Phosphoglyceride
arachidonate-containing, 157–167
molecular species, 164
Phosphoinositides, phospholipase C-catalyzed hydrolysis of, measurement of, 230–232
Phosphoinositide-specific phospholipase C activity, 206
measurement of, 226–237
Phospholipase, 578
Phospholipase A_2, arachidonoyl-hydrolyzing, 216–225
calcium solutions for, 219–221
Phospholipase C
catalyzed formation of IP_3 by isolated membranes, measurement of, 232–237
catalyzed hydrolysis of phosphoinositides in suspension, measurement of, 230–232
from human amnion
isolation of, 227–229
purification to near homogeneity, 226
hydrolysis of phospholipids, 206
phosphoinositide-specific, 226–237
assays for, 226–237
hydrolysis of phosphatidylinositol, 206
in production of eicosanoids, 226
receptor-coupled activation of, 236–237
Phospholipid molecular species, analysis of, 195–215
comments about methods, 213–215
Phospholipids
alkenylacyl subclass, selective degradation, 205
arachidonate-containing, 157–158
sn-1, 158
chemical structure, 195
classes, 196
quantitation of, 214–215
separation of, 198–201
fractionation procedures, 196–197
handling, 197
head groups, 196
isolation of, activation of lipolytic enzymes during, 154
molecular species
analysis of
HPLC equipment for, 213
by reversed-phase chromatography, 196
silica columns for, 213–214
derivatized, separation of, 205–206
intact, separation of, 201–205
separation of, 196
oxidized, hydrolysis, by PAF acetylhydrolase, 356–357
phospholipase C hydrolysis of, 206
sample, injection onto HPLC column, 214
subclasses, 196
Pinane thromboxane A_2
activity against thromboxane synthase, 409
TxA_2/PGH_2 receptor antagonist activity in platelets and vasculature, 408–409
Pirkle columns, 191–193

Plasma
arachidonic acid metabolites in, following stimulation of whole blood *ex vivo*, 114–115
11-dehydrothromboxane B_2 in, 43, 45
eicosanoids in, 44
$9\alpha,11\beta$-PGF_2 in, purification of, 55–56
RIA measurements of 11-dehydro-TxB_2 in, 39–41
thromboxane B_2 in, 42, 45
Platelet activating factor
acether
bioassay of
by platelet aggregation, calibration plot for, 128–129
by rabbit platelet aggregation, 125–130
quantitation of, 125
activity, 152, 344, 446
in isolated perfused rat lung, 606
in mast cells, 516–517
aggregation of platelets with, 126–129
analogs, 446
antagonists, 433, 455
equilibrium dissociation constant, 449
assays, 125, 130, 134
lipid extraction before, 133, 137–138
binding sites, ginkolides and, 433–446
bioassay, 153
advantages and disadvantages, 130
problems associated with, 152
by release of [^{3}H]serotonin, 130–134, 148
dose response of, 131
equipment, 132
materials, 131–132
problems with, 133–134
procedure, 132–133
quantitation, 133
reagents, 131–132
in biological fluids or extracts of whole cells, purification before assay, 133
from biological sources, measurements of, 149–151
biosynthesis, 152
C_{16}-, 1-*O*-hexadecyl-3-acetylglycerol PFBO derivative from, NCI mass spectrum of, 29–20
cardiac effects of, in isolated guinea pig heart perfused at constant pressure, 614
chemical derivatization, 137–139
degradation of, 344
derivatives, 152–153
for positive-ion electron impact ionization–mass spectrometry, 137
detection, liquid chromatographic methods utilizing chromogenic and fluorogenic derivatives, 148
deuterium-labeled, synthesis of, 143
diglyceride derived from, electron-impact ionization mass spectrum of *tert*-butyldimethylsilyl derivative of, 140, 142
in disease, 434
enzymatic hydrolysis to 1-*O*-hexadecyl-2-acetylglycerol, 145
extraction, 153–154
efficiency of, 156
GC–NCI–MS, 19, 143–152
half-life of, determination of, 347–348
homeostasis, 152
HPLC, 133
hydrolysis of phosphate ester and isolation of diglyceride, 137–139
identification of, criteria for, 125, 129
isolation of, 154–157
from biological mixture, 137
mass spectrometric analysis, 133–134, 142–143
in mast cell biology and chemistry, 514
as mediator of cardiac anaphylaxis, 620
molecular species analysis, 149–152
vs. other platelet agonists, distinguishing characteristics, 125
pathophysiological role, 455
pentafluorobenzoyl chloride (PFB)-diglyceride
analysis of, 143–152
isolation and derivatization of, 143–144
pentafluorobenzoyl derivatives
NCI mass spectra of, 146–147
synthesis of, 145–146
from PMNs, analysis of, 149–151
precursors and metabolites, analysis of, 146
produced in response to calcium ionophore stimulus of murine mast cells, measurement of, 141–142
production of, in endothelial cells, 520–521

purification of, 133
quantitation of
by gas chromatography–mass spectrometry, 134–142
by gas chromatography–negative-ion chemical ionization mass spectrometry, 142–152
receptor, 433
antagonists, 446–454
methods for screening, *in vitro*, 447–449
binding of [^{3}H]-PAF, modulated by mono- and divalent cations, 440, 444
radioreceptor studies, 448–449
standard solutions, preparation of, 127
tert-butyldimethylsilyl derivatives of
mass spectrum, 140, 142
selective ion-monitoring assays of, 148
thin-layer chromatography, 133, 154–156
trimethylsilyl derivative of
electron-impact ionization mass spectrum of, 139
in neutrophils, 139–140
Platelet activating factor acetylhydrolase, 344–357
activity, 345
assay for, 345–347
association with lipoproteins, 348–350
catalytic behavior of, effect of lipoproteins on, 350–351
cellular sources of, 345
effect of metals and chelators on, 348
properties of, 348–352
purification of, 352–355
reaction catalyzed by, 344
substrate specificity of, 355–357
surface dilution kinetics, 350–351
Platelet activation
quantitation of platelet-activating factor based on, 125–130
syndromes associated with, thromboxane A_2 formation in, 42–43
TxA_2/PGH_2-mediated, 406–407
Platelet aggregation, 397, 433–434
assay, 447–448
bioassay of PAF-acether using, 125–130
bioassay of PAF using, advantages and disadvantages, 130–131
calibration of, 128
monitoring, using aggregometer, 126–129
PAF-induced, 458–459
inhibition of, 444
Platelet aggregation inhibitor, testing for presence of, 128
Platelet aggregometry, 407, 591–592
Platelet–lymphocyte interactions
arachidonic acid metabolism in, 582–583
direction of metabolic cooperation, 581
eicosanoid generation by, 581
general strategies for, 579–582
Platelet membranes
crude, preparation of, 399
preparation, 448–449
radioligand binding studies with TxA_2/PGH_2 receptors in, 412–413
solubilized
preparation of, 399–400
radioligand binding studies with TxA_2/PGH_2 receptors in, 412–414
Platelet–neutrophil interactions, 585
Platelets
arachidonate in, 158–159
arachidonic acid metabolism in, 582–583
procedure for studying, 589–599
aspirin treatment of, 588–589
cAMP production, measurement of, 407
collection of, 587–588
eicosanoid generation by, 581
function, effect of lymphocytes on, 584
human
intact, radioligand binding studies with TxA_2/PGH_2 receptors in, 410–411
preparation of, 578–579
labeling, with radioactive arachidonate, procedure for, 590–591
in myocardial injury of ischemia–reperfusion, 617–619
perfusion of isolated heart with, for study of platelet–vessel wall interactions in eicosanoid research, 614–615
rabbit
collection, 132
labeling, 132
stimulation by PAF, 130
washed
aggregation method, 126–127
preparation of, 126
washing, 132
radiolabeled, experimental use of, 591–592

rat, binding assays to characterize TxA_2/ PGH_2 receptors, 403–404
washed
binding of radiolabeled ligands to, 401–402
preparation of, 398–399
suspensions of, preparation of, 588
Platelet stimulation, 406
Polyclonal antisera, developed using eicosanoids, 24
Polymorphonuclear leukocytes
cultured, measurement of cellular integrity, 563–564
and endothelial cells, biochemical interactions between, 559
fMLP stimulation of, 566–567
human peripheral blood, preparation, 559–560
incorporation of label into cellular glutathione pools, 564
Polymorphonuclear leukocytes/vascular cell coincubations, 562–563
$[^{35}S]LTC_4$ production during, 567
Polymorphonuclear leukocytes/vascular cell interactions, quantitative aspects of, 565–567
Polymorphonuclear neutrophils
human, isolation of, 556
PAF and alkyl-PC derived from, amounts and major molecular species of, 150
Polymorphonucleocytes
leukotriene A_4, measurements, by alcohol trapping, 572–575
lipoxygenase products, HPLC analysis of, 570–572
preparation of, 569–570
Polyunsaturated fatty acids
dioxygenation of, 268
ultraviolet chromophores, 99
Potato, selection, for arachidonate 5-lipoxygenase purification, 297
Prostacyclin
biosynthesis, 578
cell system for study of, 535
in glomerular mesangial cells, 545–546
interaction between platelets and lymphocytes in, 578–584
by lymphocytes, use of platelet-derived prostaglandin H_2 in, 583
metabolism, in isolated perfused rat lung, 608–609
platelet receptors, radioligand-binding studies of, 407
production, in platelet-perfused heart, 615
in pulmonary vasoregulation, 607
in regulation of glomerular hemodynamics, 544
Prostacyclin synthase, 407, 578
activity, in lymphocytes, 583–584
Prostaglandin
assays, negative-ion methodology, 16–18
cardiac effects, in isolated guinea pig heart perfused at constant pressure, 613
and gastric acid secretion, 505
as mediator of cardiac anaphylaxis, 620
NCI mass spectrum, 15
in regulation of glomerular hemodynamics, 544
transcellular biosynthesis, 554
Prostaglandin D_2, 514
biological actions, 51
deuterated
conversion of, to $[^2H_7]9\alpha,11\beta$-PGF_2, 53–54
preparation of, 52–53
isomerization, in presence of albumin, 58–59
11-ketoreductase metabolism of, 58–61
metabolism, 51
metabolites, 51
pathophysiological role of, 51
platelet receptors, radioligand-binding studies, 408
in systemic mastocytosis, 51
Prostaglandin D_2-Mox, enzyme immunoassay, binding parameters, using acetylcholinesterase as label, 29
Prostaglandin D synthase
type I
comments on, 5
reaction, 5
systematic name, 5
type II
comments on, 5
reaction, 5
systematic name, 5

Prostaglandin E_1
immunoprecipitation of, 68–69
platelet receptors, radioligand-binding studies of, 407
quantitation of, 68–69
Prostaglandin E_2
enzyme immunoassay, binding parameters, using acetylcholinesterase as label, 29
metabolism, in isolated perfused rat lung, 608–609
production
in glomerular mesangial cells, 545–546
by parietal cells, 505
in regulation of glomerular hemodynamics, 544
Prostaglandin E 9-ketoreductase
comments on, 8
reaction, 8
systematic name, 8
Prostaglandin endoperoxide synthase, 578
activity, regulation of, 469
characterization of structural domains of, 479
cloning of cDNAs for, 469
materials for, 470
methods, 470–479
preparation of oligonucleotide probes for, 470–471
comments on, 5
location of heme-binding site associated with peroxidase activity, 479
mouse, primary structure of, 477–479
mouse 3T3 cell, cDNA coding for, 469
proteolytic digestion of, 479
reactions catalyzed by, 5, 469
serine residue selectively acetylated by aspirin, 479
sheep
cDNA coding for, 469
primary structure of, 477–479
systematic name, 5
Prostaglandin E synthase
comments on, 5
reaction, 5
systematic name, 5
Prostaglandin F_2
compounds formed via noncyclooxygenase oxidation of arachidonic acid, vs. isomeric PGF2 compounds arising from PGD2, 61
conversion to $9\alpha,11\beta$-PGF_2, in *in vitro* cell incubations, 55
generation of
ex vivo
from arachidonic acid in biological fluids, 54
in plasma, 60
by noncyclooxygenase oxidation of arachidonic acid, 60–61
mass spectrometric assay of, 51–62
Prostaglandin $F_{2\alpha}$
enzyme immunoassay, binding parameters, using acetylcholinesterase as label, 29
in regulation of glomerular hemodynamics, 544
synthesis, in glomerular mesangial cells, 545–546
$9\alpha,11\beta$-Prostaglandin F_2
biological actions, 51–52
in biological fluids
analysis of, 54–56
mass spectrometric assay of, 57–62
deuterium-labeled, preparation of, 52–54
differentiation from *cis*-hydroxyl-PGF_2 compounds formed by noncyclooxygenase oxidative mechanism, 62
GC–MS analysis of, 56
measurement of, using boronation reaction, 61
in plasma
from normal individual, 60
during pharmacological intervention possibly associated with release of prostaglandins *in vivo*, 59
production
after incubation of PGD_2 with human eosinophils, 58
in humans, 51
during systemic mast cell activation, 51
purification of, from biological fluids, 55–56
quantification of, to assess endogenous production of PGDs in humans, 52
RIA for, 62

in skin blister fluid, GC–NCI–MS, 57
trimethylsilyl ether derivative, 56
12-epi-9α,11β-Prostaglandin F_2, formation, 58–59
Prostaglandin F synthase
comments on, 6
reaction, 6
systematic name, 6
Prostaglandin H_2
biosynthesis, 247–248
bonded-phase rechromatography methods for, 245
chromatography, 245–246
isolated from analytical cyano-bonded column, ammonia DCI mass spectrum of, 250–252
mass spectrometry, 246
pharmacological effects, 397, 406
platelet-derived, use of, by lymphocytes, in biosynthesis of PGI_2, 583
preparation of, 245–252
purification
by bonded-phase HPLC, 249–252
semipreparative, 245–246
by silicic acid chromatography, 248–249
purified
stability of, 252
storage of, 252
purity, 252
receptors, radioligand binding assays for, 397–405
rechromatography, 252
stable analogs, 397–398
Prostaglandin H synthase
apoprotein
assay method (cyclooxygenase activity), 480–481
limited or complete proteolysis of, 480–485
preparation of, 481–482
oxidative inactivation, 480
proteolytic fragments, peptide mapping of, 484–485
trypsin-cleaved, reversed-phase HPLC of, 483–484
trypsin digestion of, 482
Prostaglandin I synthase
comments on, 5
reaction, 5
systematic name, 5
Prostanoid, activated, preparation of, 25
Psoriatic scales, PAF and alkyl-PC derived from, amounts and major molecular species of, 150
PTA-OH, radiolabeled, 397–398
affinities and specific binding for platelet and vascular receptors, 405
binding assays for thromboxane A_2 receptors, advantage of, 401
binding to washed human platelets, 401–403
in evaluation of *in vitro* role of TxA_2/PGH_2 in cellular function, 409–410
for qualitative and quantitative study of TxA_2/PGH_2 receptors, 397–405
synthesis and characterization of, 400
TxA_2/PGH_2 receptor antagonist activity in platelets and vasculature, 408–409

R

R-17525, binding to leukotriene D_4 receptor, 430–433
Rabbit. *See also* Lung, rabbit
iris and ciliary body
[^{3}H]PAF binding, 440–444
PAF binding, two-step saturation isotherm, 445
PAF binding sites, 445
PAF receptor sites in, 437–438
platelet membranes
binding of [^{3}H]PAF to, 438–439
effect of ginkgolides on [^{3}H]PAF binding in, 435–436
PAF binding, 443–444
preparation, 435–436
Radioimmunoassay, 24
of arachidonic acid metabolites, 62, 90
of 11-dehydrothromboxane B_2, 34–42, 44
of 2,3-dinorthromboxane B_2, 44
of leukotrienes, 82, 123–124, 563
nonisotopic alternative to, 24, 34
of platelet production of thromboxane B_2, 407
Rat. *See also* Liver, rat; Lung, rat; Platelets, rat

administration of PAF and PAF antagonists to, 461–464
endothelial cells
binding assays to characterize TxA_2/PGH_2 receptors, 403–404
preparation of, 400
hepatocytes
isolated, leukotriene metabolites by, 278
isolation, 279–280
purification of, 279–280
Rat basophilic leukemia cell line, PAF and alkyl-PC derived from, amounts and major molecular species of, 150
Resolving agents, for separation of racemic eicosanoids, 193
Reversed-phase HPLC
for analysis of lipoxygenase products, 98
of arachidonic acid metabolites, 370–371
in guinea pig lung perfusate, 110–113
of cytochrome *P*-450 prostaglandin ω-hydroxylase reaction products, 262–264
of 5-hydroxyeicosatetraenoic acid, 101, 109–110
of leukotriene B_4 and E_4 metabolites, 281
of leukotrienes, 98, 101–115
of lipoxygenase and cyclooxygenase products generated by zymosan stimulation of human blood, 114
of trypsin-cleaved prostaglandin H synthase, 483–484
RNA
isolation, 496
Northern blots, 496

S

S-145
partial agonist activity in platelets, 409
TxA_2/PGH_2 receptor antagonist activity in platelets and vasculature, 408–409
Seminal vesicle microsomes, ovine, preparation of, 247
Serotonin, radiolabeled, release of, bioassay of platelet activating factor based on, 125, 130–134
Sheep vesicular gland
cDNA library, screening of, 476–477
λgt10 cDNA library, preparation of, 471–476
prostaglandin endoperoxide synthase, purification of, 470–471
Silica columns, 213–214
Silyl ethers, GC-MS characteristics, 14
SKF 104,353, 415
binding to leukotriene D_4 receptor, 430–433
Slow-reacting substance of anaphylaxis, 90, 414
cardiac effects of, 613
SMCC. *See* Succinimidyl-4-(*n*-maleimidomethyl)cyclohexane 1-carboxylate
Sphingomyelin, 195
chromatography, 199–201
Spodoptera frugiperda insect cells, maintenance of, 497
SQ29,548
in evaluation of *in vitro* role of TxA_2/PGH_2 in cellular function, 409–410
radiolabeled
affinities and specific binding for platelet and vascular receptors, 405
for qualitative and quantitative study of TxA_2/PGH_2 receptors, 397–405
tritiated, 397–398
for TxA_2/PGH_2 receptor binding studies, 410–414
used to characterize TxA_2/PGH_2 receptors, 402–404
TxA_2/PGH_2 receptor antagonist activity in platelets and vasculature, 408–409
SR 2640, 415
SR-63441, 446
Stable isotope dilution mass spectrometric assay, quantification of prostaglandin F2, 51–62
Substituent groups
number of, nomenclature for, 2
position of, nomenclature for, 2
Substituent name, 2
Succinimidyl-4-(*n*-maleimidomethyl)cyclohexane 1-carboxylate
coupling of leukotriene E_4 to enzyme using, 83, 86
source, 83
Sulfidoleukotrienes, analysis of, 575

Surface dilution kinetics, 350–351
Systemic mastocytosis, 51

T

Tamm-Horsfall protein identification, for immunodissection of mTALH cells, 368–369
Thrombin
  effect on mast cells, 518–520
  platelet aggregation and release reaction in response to, effect of lymphocytes on, 582, 584, 587
Thrombosis, lipoxygenase products of platelet and neutrophil in, 584, 586
Thromboxane
  A_2
    altered metabolism vs. increased biosynthesis of, distinguishing between, 45
    metabolites, in plasma and urine, combined analysis of, 45
    pathophysiological role of, 50
    pharmacological effects, 397, 406
    as proaggregatory substance, 42
    production
      in platelet-perfused heart, 615
      RIA for 11-dehydro-TxB_2 as index of, 34, 41
    in pulmonary vasoregulation, 607
    radioligand binding assays for, 397–405
    in regulation of glomerular hemodynamics, 544
    role in human syndromes of vascular occlusion, 42
    stable analogs, 397–398
    synthesis
      cell system for study of, 535
      in glomerular mesangial cells, 545–546
    as vasoconstrictor, 42
  activated, preparation of, 25
  assays, negative-ion methodology, 16–18
  B_2
    antiserum for, 64
      titration of, 27–28
    enzyme immunoassay
      binding parameters, using acetylcholinesterase as label, 29
      precision profile of, at different stages of enzymatic reaction, 30–31
      sensitivity of
        effect of antiserum concentration on, 32
        effect of temperature on, 32
    enzyme immunoassays, interassay variation of, at low, medium, and high concentration, 33
    metabolism of
      2,3-dinor-TxB_2 and 11-dehydro-TxB_2 as indexes of two major pathways of, 43
      in isolated perfused rat lung, 608–609
    MO, Tris-TMS ether, PRB ester derivative of, HCI mass spectrum of, 17–18
    in plasma, assessment, 42
    plasma concentrations of, confounded by *ex vivo* platelet activation, 45
    platelet production of, radioimmunoassay, 407
    urinary excretion of, as index of systemic TxA2 formation, 42
  biosynthesis, biochemical assessment of, 42
  metabolites
    derivatization of, 49–50
    extraction, 46–48
    gas chromatography–mass spectrometry, internal standards, 46
    immunopurification techniques, 48
    measurement of, by gas chromatography–mass spectrometry, 42–50
    methoxime derivatization, 49
    pentafluorobenzyl (PFB) ester, formation, 49–50
    quantitation, by GC–MS, 50
    radioimmunoassays, comparison with GC–MS determinations, 44–45
    trimethylsilyl (TMS) ether, formation, 50
  NCI mass spectrum, 15
  production, monitoring, 41–42
Thromboxane A_2/prostaglandin H_2 receptors
  agonists, 397–398
  antagonists, 397–398, 406–414
    criteria for establishing specificity of, 406–408

interaction with TxA_2/PGH_2 receptor site, 408
in smooth muscle, 408
tissue heterogeneity of, 406
Thromboxane A synthase
comments on, 6
reaction, 6
systematic name, 6
Thromboxane synthase, 407
Thyroid peroxidase, axial heme-binding site in, 479
Trachea, guinea pig, preparation, for eicosanoid studies, 625–628
Transcellular biosynthesis, 553–559
Transcellular metabolism, of eicosanoids, 585–599
Tx. *See* Thromboxane

U

U46619
radiolabeled, 397–398
binding assay
in pig aortic smooth muscle membrane, 405
for thromboxane A_2 receptors, 404–405
for qualitative and quantitative study of TxA_2/PGH_2 receptors, 397–405
as TxA_2 mimetic, 407
U937 cells, thromboxane A_2 synthesis by, 535–538
Ultraviolet chromophores, of lipoxygenase (and cyclooxygenase) products, 98–99
Urine, RIA measurements of 11-dehydro-TxB_2 in, 39–41

V

Vascular cells
cultured, measurement of cellular integrity, 563–564
incorporation of label into cellular glutathione pools, 564
from rat, preparation of, 400
Vascular smooth muscle
constriction of, 397
contraction, 406
membranes, porcine, preparation of, 400
Vascular smooth muscle cells, rat
binding assays to characterize TxA_2/PGH_2 receptors, 403
preparation of, 400
Venipuncture, for collection of neutrophils and platelets, 587–588

W

WEB 2086
concentrations and doses, in PAF antagonism experiments, 457–459
inhibition of PAF-induced platelet aggregation, IC_{50} values for, 451–454
as PAF antagonist, 451–465
in isolated perfused rat lung, 606
in vitro experiments, 458–460
in vivo experiments, 460–464
physicochemical properties of, 455–457
solutions
preparation of, 456–457
stability of, 457
structure of, 456
use in disease-related experimental models, 465
WEB 2170
concentrations and doses, in PAF antagonism experiments, 457–459
as PAF antagonist, 455–465
in vitro experiments, 458–460
physicochemical properties of, 455–457
solutions
preparation of, 456–457
stability of, 457
structure of, 456
use in disease-related experimental models, 465
WI-38 cells, thromboxane A_2 synthesis by, 538–539
WY48252, binding to leukotriene D_4 receptor, 430–433

Y

YM 16638, 415

Z

Zymosan, stimulation of human blood with, reversed-phase HPLC analysis of lipoxygenase and cyclooxygenase products generated by, 114